Molecular Basis of Health and Disease

Undurti N. Das

Molecular Basis of Health and Disease

Springer

Undurti N. Das, MD, FAMS
UND Life Sciences
13800 Fairhill Road, #321
Shaker Heights, OH 44120, USA
undurti@hotmail.com

School of Biotechnology
Jawaharlal Nehru Technological University
Kakinada 533003, India

ISBN 978-94-007-0494-7 e-ISBN 978-94-007-0495-4
DOI 10.1007/978-94-007-0495-4
Springer Dordrecht Heidelberg London New Work

Library of Congress Control Number: 2011921317

Cover design: deblik, Berlin

Printed on acid-free paper

Springer is part of Springer Science+Business Media (www.springer.com)

To
My Wife Lakshmi
and
Daughter Arundhati, Son Aditya
and Son-in-Law Dr. Kalasagar Madugula

Preface

Several studies have suggested that low-grade systemic inflammation plays a significant role in the pathogenesis of obesity, insulin resistance, essential hypertension, type 2 diabetes, atherosclerosis, coronary heart disease, metabolic syndrome, dyslipidemia, lupus, rheumatoid arthritis and other autoimmune diseases, schizophrenia, depression, Alzheimer's disease and cancer. This is supported by the observation that plasma C-reactive protein (CRP), tumor necrosis factor-α (TNF-α), and interleukin-6 (IL-6), markers of inflammation, levels are elevated in these subjects.

With ageing, plasma levels of CRP, IL-6 and TNF-α tend to increase and produce insulin resistance and secondary hyperinsulinemia. Alzheimer's disease, schizophrenia, and depression are also associated with an increase in plasma and cerebrospinal fluid CRP, IL-6, TNF-α, and lipids peroxides. In all these conditions, similar, if not identical, changes in the plasma, RBC, and tissue concentrations of polyunsaturated fatty acids and anti-oxidants have been described. Similarity in the molecular events at the cellular level suggest that methods designed to suppress inappropriate inflammation and augment resolution of inflammation and tissue repair could be of therapeutic benefit in these conditions.

In this context, it is of particular significance that alterations in the metabolism of essential fatty acids and the formation of their anti-inflammatory metabolites such as lipoxins, resolvins, protectins, maresins and nitrolipids seem to be responsible for the onset of low-grade systemic inflammation in these diseases. In view of this understanding the factors and co-factors, both endogenous and exogenous, that have the ability to modulate the metabolism of essential fatty acids and the formation of their anti-inflammatory products is important. Since these anti-inflammatory lipid compounds suppress the production of pro-inflammatory eicosanoids, it appears that a disturbed balance between these pro- and anti-inflammatory products of polyunsaturated fatty acids play a significant role in the pathobiology of several adult diseases.

This is particularly relevant to the pathobiology of the metabolic syndrome that has been attributed to lack of exercise, increase in the consumption of energy-dense food and environmental changes. It is likely that insulin resistance, low-grade systemic inflammation, low-birth weight (especially in the Indian sub-continent), maternal malnutrition (both over and under-nutrition), perinatal and early childhood high carbohydrate and saturated fat diet and low polyunsaturated fatty acid intake could be responsible for this disease.

There is reasonable evidence to suggest that obesity, insulin resistance, type 2 diabetes mellitus, hypertension which are all the components of the metabolic syndrome, may occur as a result of dysfunction of specific hypothalamic nuclei and their peptide and monoaminergic neurotransmitters, an issue that needs due attention.

Human brain is rich in polyunsaturated fatty acids (PUFAs) and so they are likely to play a significant role in the pathogenesis of the metabolic syndrome, neurological conditions such as Alzheimer's disease, schizophrenia and depression.

PUFAs play a significant role in brain growth and development, modulate the actions of various neurotransmitters that have an important role in the pathobiology of the metabolic syndrome and Alzheimer's disease, schizophrenia and depression suggesting that perinatal supplementation of PUFAs could be of significant help in the prevention of these diseases since brain development occurs predominantly during the second and third trimester of pregnancy and first 5 years of life. Thus, metabolic syndrome could be a disorder of the brain. This explains why breast fed subjects have low incidence of these diseases since human breast milk is rich in PUFAs.

Vagus nerve has a regulatory role in insulin secretion, modulates inflammation, influences the levels of (BDNF) brain-derived neurotrophic factor and its stimulation increases the secretion of incretins from the gut, suggesting that vagus nerve stimulation could exploited in the treatment of insulin resistance, type 2 diabetes mellitus and metabolic syndrome and Alzheimer's disease, schizophrenia and depression; in addition to its already established role in the treatment of resistant epilepsy.

Cancer is also a low-grade systemic inflammatory condition. Some PUFAs selectively kill tumor cells without harming normal cells. Hence, it is possible to use monoclonal antibodies against growth factors that are complexed with PUFAs in the treatment of cancer. Thus, a combination of PUFAs, BDNF, vagus nerve stimulation, and other strategies could be adopted to prevent and manage several adult diseases.

I trust that several of new concepts proposed in this book would interest many scientists and encourage them to test them out.

Shaker Heights, OH Undurti N. Das

About the Author

Undurti N. Das is an M.D. in Internal Medicine from Osmania Medical College, India; a Fellow of the National Academy of Medical Sciences, India, and Shanti Swaroop Bhatnagar prize awardee. His current interests include the epidemiological aspects of diabetes mellitus, hypertension, CVD and metabolic syndrome. Dr. Das was formerly a scientist at Efamol Research Institute, Kentville, Canada; Professor of Medicine at Nizam's Institute of Medical Sciences, India and Research Professor of Surgery and Nutrition at SUNY (State University of New York) Upstate Medical University, Syracuse, USA. At present, he is the Chairman and Research Director of UND Life Sciences, USA. Dr. Das is also the Editor-in-Chief of: Lipids in Health and Disease. Dr. Das has more than 400 international publications and has been awarded 4 USA patents. Dr. Das is in receipt of Ramalingaswami Fellowship of the Department of Biotechnology of India during the tenure of writing this book. Previous books by Dr. U N Das include: *A Perinatal Strategy for Preventing Adult Disease: The Role of Long-Chain Polyunsaturated Fatty Acids*, Kluwer Academic Press, 2002; and *Metabolic Syndrome Pathophysiology: The Role of Essential Fatty Acids*, Wiley-Blackwell, 2010. **Address**: UND Life Sciences, 13800 Fairhill Road, #321, Shaker Heights, OH 44120, USA, Tel.: +1-216-231-5548, Fax: +1-928-833-0316, e-mail: Undurti@hotmail.com; School of Biotechnology, Jawaharlal Nehru Technological University, Kakinada 533003, India

Contents

Chapter 1
Introduction

World Health Organization defined health [1–3] as "Health is a state of complete physical, mental and social well-being and not merely the absence of disease or infirmity." This definition has not been amended since 1948.

However, in 1986, the WHO, in the Ottawa Charter for Health Promotion, said that health is "a resource for everyday life, not the objective of living. Health is a positive concept emphasizing social and personal resources, as well as physical capacities." Classification systems such as the WHO Family of International Classifications (WHO-FIC), which is composed of the International Classification of Functioning, Disability, and Health (ICF) and the International Classification of Diseases (ICD) also define health. Overall health is achieved through a combination of physical, mental, emotional, and social well-being, which, together is commonly referred to as the Health Triangle.

As a public relations slogan, the WHO definition appears most useful. Although this definition has been around for a more than two decades, this can be considered as incomplete. This is so since, WHO definition uses words such as "complete," "social well-being," and "disease and infirmity" whose meaning are not self-evident. These terms need explanation. Moreover, the definition does not explain what health does to organisms possessing it or how it may be measured. We need these terms interpreted to understand the definition. For instance, what techniques we you have to produce physical, mental, and social well-being in those who are free from disease or infirmity? Or for that matter what techniques we have to measure physical, mental and social well-being of an individual? Hence, in order to justify the WHO definition, we should develop or have techniques to measure these indices.

On the other hand, the open-ended nature of the WHO definition of health encourages certain activities, such as the "positive mental health" movement seeking growth, zest, and creativity of the mind. It backs popular health beliefs in the benefits of cold showers, jogging and consuming vitamin pills, as well as the more organized, physician-supported spas in Europe. It may even underlie the alleged Scottish custom of prescribing soothing draughts of milk and whisky, the milk being reduced and cut off as well-being improves. Some of these activities are beneficial, but the value of others is doubtful. Even to say that a particular activity is beneficial to the body and mind, we need to have well-defined and well-quantified measurements to measure

U. N. Das, *Molecular Basis of Health and Disease*,
DOI 10.1007/978-94-007-0495-4_1, © Springer Science+Business Media B.V. 2011

health and disease(s) that can be used across individuals and societies to compare and contrast and wherever necessary impart or take necessary remedial measures necessary to restore health and keep disease away. The definition of health as given by WHO can be considered as an *Asymptotic or open-ended concept*.

The other concept of health is called as the *Elastic concept*. In this concept importance is given to the ability of an individual or the community to resist threats of disease and pictures a positive interaction between the person or community and the environment. Thus, this definition can be redefined as "Health is the perfect adjustment of an organism to its environment." The concepts of herd immunity, attained when a certain proportion of the population is immunized, and of mental illness being a diseased state of an entire family are public health examples of this definition. Thus, this definition implies that imperfect adjustment causes ill health or disease. This concept also depends on a satisfactory picture of disease, whose presence or absence determines the absence or presence of health [4].

Measuring Health and Disease

Definition is a first step in measurement; since, it sets clear limits which will tell whether persons fall between or outside them. Ability to measure also goes further to indicate a more precise position on a scale. Hence, it is important to seek satisfactory definitions of health and disease. In this context, it is important to ask whether these concepts are truly independent or are merely different parts of the same entity. One way to clarify an ill-defined idea is to decide how it might be measured. As it is said, "If it exists, it should be measured." But, in our enthusiasm to follow this advice with enthusiasm we should not try to measure things before we are sure of their (it) existence and definition. The mere attempt at measurement clarifies what is being measured. Indeed, vague entities, such as time or intelligence are often thought of by the way they are measured. One useful principle or guideline is that the same scale rarely measures entirely different entities. Different instruments are needed to measure different entities or indices. But, it should be understood that the same thermometer measures both heat and cold since, they are merely different sections of the same scale. In a similar fashion, the big question is whether we can use the same or different instruments to measure health and disease?

In the instance of an individual, measurement of health or disease begins with questioning and examining him/her. Appropriate answers and apparently normal appearance such as rosy conjunctiva and normal appearing cheeks, glistening eyes, and an alert expression suggest good health. But, it may be difficult to judge the well-being of the mind; temper and good disposition need prolonged observation and are usually not assessed in the initial/first medical examination.

The physical examination and laboratory tests are performed to exclude disease and disability. In today's world, to an increasing extent, measurements for disease dominate the diagnostic examination of patients. No single action or test establishes more than the presence or absence of disease. For instance, at times physicians acknowledge this situation by summarizing a system as "nothing abnormal detected," (NAD) rather than by saying that the central nervous system is in excellent health.

Current efforts to assess the health of subjects who visit the physician or the hospital for assessment of their health and disease mainly entail testing for the presence of disease, with no truly independent measure of health. Many times, absence of disease is equated to health and such a person is considered to be healthy. This conclusion suggests that our long-lasting dualism about health and disease may be akin to the belief, held centuries ago, that heat and cold were separate entities.

In general, most persons associate disease with conditions of the body which shorten the expectation of life or cause unusual symptoms or signs, discomfort, disability, or death. It is also believed that each disease had a specific cause and thus, diagnosis, treatment, and prevention of disease are considered as the three basic elements of the medical system. This implies that certain conditions are diseases when they can be recognized and understood by physicians. In recent years, the multiple causation of disease became a more widely held doctrine that envisions the interplay of host, agent, and environment. This led to development of the concepts of comprehensive medical care, psychosomatic medicine, and multi-professional teams to deal with various diseases in a more comprehensive fashion. For example, to deal with the diabetic patients nowadays a team approach is made that consists of the physician, nutritionist, vascular surgeon, ophthalmologist, physiotherapist and cardiologist. In addition, it is noteworthy that even when medical interventions are made with a team of experts and yet fail to cure some conditions, the physician dominates in deciding whether disease is absent or present and the type of approach or treatment that need to be given. Thus, the concept of disease is closely related to what physicians do in society and to the degree of advancement of medical practice.

In the present day environment, a multitude of investigations and equipment are used to help the physician to determine the presence or absence or the severity of the disease(s). These include from the traditional devices, such as sphygmomanometers, electrocardiographs to more advanced CT (computerized axial tomography) and MRI (magnetic resonance imaging) scans. With few exceptions, each device tends to encourage the physician to classify as unhealthy an increasing proportion of the population. For instance, elevated blood pressure or serum cholesterol values present in a given subject who that may be entirely asymptomatic are typical examples of such results. In these instances, the subject in question may be otherwise normal but based on these investigational reports they may be termed as not entirely normal. Yet another example is the measurement of high-sensitive C-reactive protein (hs-CRP). A person who is otherwise normal may have high levels of hs-CRP that may suggest that he/she is at high-risk of developing coronary heart disease, stroke or metabolic syndrome. Based on the report of hs-CRP he/she may be classified as a patient while based on physical examination and subjective symptoms the person may be entirely normal. Another classical example is the presence of cysticercosis in CT and MRI scans of an otherwise asymptomatic individual especially in tropical/sub-tropical countries like India. When such a diagnosis is made based on the reports of the investigations performed, the subject is offered relevant treatment to prevent further progress of the disease or its possible side effects in future. Thus, as our knowledge of human body increases and our grasp of physiology and pathology of various diseases improves and as new devices are developed that enhance our ability to diagnose conditions/diseases much before they occur; the tendency on the part of the physician would be to classify as unhealthy an increasing proportion of the population. But, one unfortunate undesirable effect of these advances in medicine has been an ever increasing cost of medical care. The advances in modern medicine instead of decreasing the cost of medical care and make it affordable to the vast majority of the population, it escalated the cost of prevention, treatment and hospitalization.

The other example is the measurement of plasma glucose levels and its cut-off normal values. In 1970–1980s, the normal fasting plasma glucose value was determined as 140 mg% (7.77 mmol/dl). But as we gained more knowledge based on the long-term follow-up of subjects and correlation studies done between plasma glucose levels and the future development of coronary heart disease, it became evident the normal values of plasma glucose need to be much lower and so the normal cut-off value has been determined to be 110 mg%. Further studies led to the conclusion that even this value is high and so at present the normal fasting plasma glucose is determined to be 100 mg%. At this juncture, it is pertinent to mention that some physicians/scientists believe that even this value is not correct and the fasting plasma glucose levels should much lower than 100 mg%. This is so since, it was observed that even when the fasting plasma glucose is within the accepted normal range of 60–100 mg%, the chances of developing coronary heart disease is higher among those whose fasting plasma glucose is at the upper limit of the normal. Thus, an fasting blood glucose >85 mg/dl had a relative risk of cardiovascular death for men of 1.4 even after adjusting for age, smoking habits, serum lipids, blood pressure, and physical fitness [5]. A meta-regression analysis of published data from 20 studies of 95,783 individuals followed for 12.4 years showed the progressive relationship between glucose levels and cardiovascular risk extends below the diabetic threshold [6].

This seems to be true even to those who did not have diabetes but developed stress hyperglycemia after coronary heart disease. In a systematic review and meta analysis that assessed the risk of in-hospital mortality or congestive heart failure after myocardial infarction in patients with and without diabetes who. had stress hyperglycemia on admission, it was noted that patients without diabetes who had glucose concentrations more than or equal to range 6.1–8.0 mmol/l (4 mmol/l = 72 mg/dl or 1 mmol = 18 mg/dl of glucose; thus 6.1 mmol/l = 109.8 mg/dl) had a 3.9-fold (95% CI 2.9–5.4) higher risk of death than patients without diabetes who had lower glucose concentrations. Glucose concentrations higher than values in the range of 8.0–10.0 mmol/l on admission were associated with increased risk of congestive heart failure or cardiogenic shock in patients without diabetes. In patients with diabetes glucose concentrations more than or equal to range 10.0–11.0 mmol/l the risk of death was moderately increased (relative risk 1.7 (1.2–2.4)). Thus, stress hyperglycemia with myocardial infarction is associated with an increased risk of in-hospital mortality in patients with and without diabetes; the risk of congestive heart failure or cardiogenic shock is also increased in patients without diabetes [7].

Is There a Better Definition of Health?

Based on the preceding discussion, it is clear that either WHO definition or the *Open ended* and *Elastic concepts* of health are insufficient definitions of health implying that a new definition of health and disease that is more comprehensive is probably needed.

Since the open-ended definition of WHO was to emphasize that the absence of known disease that is not sufficient and its major defect is that it infers that health and disease are different and mutually exclusive entities, not parts of the same spectrum. This definition also ignores the possibility that presently unknown diseases could be harbored by the population that appear to be relatively healthy. The best examples are AIDS (acquired immunodeficiency syndrome), progressive multifocal leukoencephalopathy (PML), subacute sclerosing panencephalitis (SSPE), scrapie, kuru and Creutzfeldt-Jakob disease (CJD). These diseases have long incubation periods and hence, may not be evident at the time of examination but may be present in an otherwise healthy looking person. Since, health and disease are different sides of the same coin their definitions should be complementary and the better definition should also take into account the time factor. For example, both immunized and non-immunized infants may be healthy except for the fact that both are healthy for different reasons: one for having been immunized and the other because of innate immunity or healthy for the time being till gets the disease at could happen after a while. Based on these criteria Wylie [4] has adopted the definition offered by Spencer and defined health as "Health is the perfect, continuing adjustment of an organism to its environment." Conversely, disease would be an imperfect, continuing adjustment. Based on this definition, it is argued that biochemical changes, such as elevated blood glucose levels, and the clinical index such as elevated blood pressure will be considered imperfect adjustments. Thus, it is argued that the person who is free from disease will almost certainly feel well; if he has well-being however, he may or may not have disease. This is akin to the argument that one may have cancer without disease [8]. It has been argued that many people may have small or unrecognizable cancers without knowing about them or suffering from the affects of cancer. This argument led to the suggestion that there could be two stages in the natural history of cancer-one a dormant stage and the other more proliferative, angiogenic or lethal phase. The first or the dormant phase of cancer is due to mutations that convert a normal cell into a cancer cell that is seen as *in situ* tumours in many individuals at autopsy, but for these *in situ* tumors to grow and become lethal they need additional signals in the form of angiogenic factors that feed them with new blood vessels, oxygen and nutrients so that they grow to become lethal. Thus, it can be said that an otherwise normal individual may have *in situ* cancer but has the potential to die of cancer in future. Such individuals may be labeled as normal at the time of examination and by definition is healthy due to his/her ability to be able to show continuing adjustment to the environment in this instance to cancer.

Thus, health is the general condition of a person in all aspects and is also a level of functional and/or metabolic efficiency of an organism and implicitly human. Overall health is achieved through a combination of physical, mental, emotional, and social well-being.

Determinants of Health

In order to remain healthy and to avoid the risk of developing diseases, one need to follow certain guidelines. There are four general determinants of health including: human biology, environment, lifestyle, and healthcare services [9]. Thus, health is maintained and improved not only through the application of advances made in health sciences from the new knowledge gained, but also through the efforts and intelligent lifestyle choices of the individual and society. The type of environmental factor(s) may vary from place to place. The best example is water quality, especially for the health of infants and children in developing countries. It is a well known fact that in developing countries, due to poor quality of water many infants and children succumbs to infections especially, gastrointestinal infections [10]. On the other hand, in the developed countries, the lack of neighborhood recreational space leads to lower level of physical activity and higher levels obesity; therefore, lower overall wellbeing [11].

Maintaining Health

Achieving and maintaining optimal health is an ongoing process that includes the elements: observations of daily living, social activity, hygiene, stress management, health care and workplace wellness.

Observations of Daily Living

To a large extent, personal health depends on one's active, passive and assisted observations of their health and their subjective observation about their ability to do routine work. Su information, sometimes, may give valuable clues about the underlying diseases. A cardiac patient who complains that the shoes are tighter than usual may be having enhanced cardiac failure and so needs the addition of a diuretic or if already on a diuretic the dose needs to be adjusted. Thus, close subjective observations, a thorough clinical examination and performing relevant investigations are useful to make the life of a patient more comfortable.

Social Activity

Maintenance of personal health depends partially on the social structure of one's life. It has been documented that maintenance of strong social relationships can be linked to good health conditions, longevity and a positive attitude. This could be attributed to changes in the neurotransmitters that are linked to personality and intelligence traits [12]. The same can be said of people who engage in volunteer work. A volunteer while gaining plenty of social benefits also helps them to forget their own troubles and develop a positive attitude towards life.

Hygiene

Hygiene is the practice of keeping the body clean so as to prevent infection and illness. Regular bathing, brushing and flossing teeth, washing hands especially before eating, using clean utensils for food preparation, using clean plates and vessels for keeping food and eating food are some of the hygienic practices. Following the hygienic practices helps to prevent infection and illness. Regular bathing will help to clean the body and removal of dead skin cells and washing away the dead skin with germs so that their chance of entering the body is prevented.

Stress Management

In general, it is assumed that psychological stress may have a negative impact on health. Prolonged psychological stress may impair immune response, impair cognitive function, and is also believed to precipitate coronary heart disease and premature death. Hence, methods designed to reduce stress and improve psychological well being have been recommended. Some of these stress reducing techniques include learning to cope with problems better, time and task management skills and appropriate advice by concerned experts.

Health Care

Prevention, treatment and management of illness and preservation of mental, psychological and physical well being is possible through appropriate use of the medical, nursing and allied health professional services and facilities.

Workplace Wellness Programs

As increasing number of people both male and female are working, it is important that the working class remain healthy both to continue to work and maintain their livelihood but also to contribute to the welfare of the family and society. It is also important that workers remain healthy so that they can work efficiently and contribute to the productivity of the organizations in which they are working. In addition, companies need to provide facilities to their workers to improve the health and wellbeing of their employees to enhance their morale, loyalty and productivity. As a result, increasing number of companies are providing onsite fitness centers, health presentations, wellness newsletters, access to health coaching, tobacco and alcohol cessation programs and information and training related to nutrition, weight and stress management. Some organizations are also providing annual checkups, health screening, and boy mass index monitoring to encourage their safe to remain healthy, and productive.

Public Health

Public health is defined as "the science and art of preventing disease, prolonging life and promoting health through organized efforts and informed choices of society, organizations, public and private, communities and individuals" [13]. Public health is concerned with threats to the overall health of a community based on population health analysis. The population in question can be few living in a specific area, in a village, city, country or as large as all the inhabitants of several countries or continents (for instance, in the case of a pandemic). Thus, public health is concerned with endemic diseases and epidemics. Public health has many sub-fields that include: epidemiology (that studies the incidence and prevalence of diseases), biostatistics (that deals with the relevance of various factors in the causation or association with a disease that could also include the effectiveness of a particular type of intervention in the prevention and management of a disease including the use of drugs), health services (that looks after the prevention and management of various diseases including assessing the health of the community in general. Vaccination for the prevention of various diseases could come under this category), environmental, social and behavioral health and occupational health (these services look after specific areas of public health that take into account the social factors, environmental factors that could affect the health of the community such as pollution, emissions and chemical effluents form the factories, etc.). It may also be mentioned here that certain other public health hazards of modern civilization such as drug addiction and alcohol dependence are also considered as public health issues since they ultimately affect the community as a whole. For instance, if drug addiction especially for heroin, etc., becomes common a community it could threaten the peaceful atmosphere in the society since such drug addicts could resort to violence and robbery and thus, crime rate in a given community might increase. Similarly, alcohol dependence and addiction if assumes alarming proportions it not only threatens the health of the individual but also disrupts the social fabric of the family and eventually could threaten the community peace. Furthermore, such alcoholics may become dependent of the society and the government for their own living and may develop alcohol-induced diseases that may increase the burden on the health delivery system.

Thus, the focus of public health is to intervene to prevent rather than treat a disease. Public health services look at the whole community as their target and try to improve the health awareness and health of the individual at the micro level with emphasis on the community at the macro level. Public health departments do surveillance of cases and the promotion of healthy behaviors. In addition, it is vital to identify the first or the initial cases of infectious diseases and isolate them to prevent the spread of the disease in the community especially during the outbreak of infectious diseases. The best example is none other than the excellent work done by many public health departments of various countries in the identification and management of patients with AIDS (acquired immunodeficiency syndrome) and thus, its effective control throughout the world.

Role of Science in Health

Health science is the branch of science that focuses on health. This discipline does study and research of the human body and health-related issues to understand how humans (and animals) function and tried to apply the knowledge thus, gained to improve health and prevent and cure diseases. Thus, health sciences comprises of several sub-fields such as anatomy, physiology, biochemistry, genetics, epidemiology.

Applied Health Sciences

As the name indicates, this field of science tries to apply the knowledge gained by the study of the human body (including animal body) and tries to apply the knowledge thus, gained to improve health. Some of the fields that come under this branch of science include: biomedical engineering, biotechnology, nutrition, nursing, pharmacology and pharmacy, physical therapy and medicine.

References

[1] Brockington F (1967) World health, 2nd edn. J and A Churchill Ltd., London, p ix
[2] WHO Preamble to the Constitution of the World Health Organization as adopted by the International Health Conference, New York, 19–22 June, 1946; Signed on 22 July 1947 by the representatives of 61 States (Official Records of the World Health Organization, no 2, p 100); and entered into force on 7 April 1948
[3] WHO (2006) Constitution of the world health organization – basic documents, 45th edn. Suppl, October 2006
[4] Wylie CM (1970) The definition and measurement of health and disease. Public Health Rep 85:100–104
[5] Bjornholt JV, Erikssen G, Aaser E, Sandvik L, Nitter-Hauge S, Jervell J, Erikssen J, Thaulow E (1999) Fasting blood glucose: an underestimated risk factor for cardiovascular death. Diabetes Care 22:45–49
[6] Coutinho M, Gerstein HC, Wang Y, Yusuf S (1999) The relationship between glucose and incident cardiovascular events. Diabetes Care 22:233–240

[7] Capes SE, Hunt D, Malmberg K, Gerstein HC (2000) Stress hyperglycaemia and increased risk of death after myocardial infarction in patients with and without diabetes: a systematic overview. Lancet 355:773–778
[8] Folkman J, Kalluri R (2004) Cancer without disease. Nature 427:787
[9] Lalonde M (1974) A new perspective on the health of Canadians. Ministry of Supply and Services, Ottawa
[10] World Health Report (WHO) (2004) WWDR2: Water, a shared responsibility. (UNESCO-WWAP, 2006)
[11] Björk J, Albin M, Grahn P, Jacobsson H, Ardö J, Wadbro J, Östergren PO, Skärbäck E (2008) Recreational values of the natural environment in relation to neighbourhood satisfaction, physical activity, obesity and wellbeing. J Epidemiol Community Health 62:e2
[12] Way BM, Taylor SE (2010) Social influences on health: is serotonin a critical mediator? Psychosom Med 72:107–112
[13] Winslow CEA (1920) The untilled fields of public health. Science 51:23–33

Chapter 2
Health and Disease as Two Sides of the Same Coin

It is evident from the discussion in the preceding chapter that health cannot be simply stated as merely the absence of disease. Since the definition "Health is the perfect adjustment of an organism to its environment" seems to be more appropriate and implies that imperfect adjustment causes ill health or disease, it is important that appropriate clinical, biochemical, molecular and genetic measurements are developed that set clear limits which will tell whether persons fall between or outside them. On the other hand, the definition of health that suggests that "Health is the perfect, continuing adjustment of an organism to its environment" indicates that disease would be an imperfect, continuing adjustment. Thus, biochemical changes that are outside the normal range will be considered imperfect adjustments and indicates that the person is having a disease. Thus, this definition implies that if the abnormal biochemical changes are restored to normal, then that person is pronounced as normal. Even by this definition of health, it is clear that relevant clinical, biochemical, molecular and genetic measurements are developed for each disease or for a group of diseases so that such indices will form the benchmark both to define health and measure a particular disease or a group of diseases and if possible to define the severity and grade of the disease(s) and may also be used as prognostic markers.

Low-Grade Systemic Inflammation Occurs in Many Diseases

In this context, it is interesting to note that coronary heart disease (CHD), stroke, diabetes mellitus, hypertension, cancer, depression, schizophrenia, Alzheimer's disease, and collagen vascular diseases that are a severe burden on the health care system throughout the world [1] are all characterized by low-grade systemic inflammation [2–20]. This implies that prevention or suppression of inflammation reduces burden of these diseases. In other words, a person remains healthy as long as low-grade systemic inflammation does not occur and when inflammation does set in due to external and/or internal reasons, it leads to a disease. Hence, understanding the molecular basis of inflammation, factors that regulate inflammation and methods designed or strategies adopted to suppress inflammation could form the basis of restoring health.

U. N. Das, *Molecular Basis of Health and Disease*,
DOI 10.1007/978-94-007-0495-4_2,

This does not mean inflammation is always harmful. Inflammation is fundamentally a protective response whose ultimate goal is to eliminate the injury-inducing agent (that could be a microorganism, physical stimuli, chemical agent, etc.), prevent tissue damage and/or initiate the repair process. Without inflammation there is no life since, in the absence of adequate inflammation cell/tissue injury would go unchecked, the damage done to the cells/tissues/organs would never heal, and ultimately this may lead to the death of the organism itself. Thus, inflammation is both beneficial and potentially harmful.

Recent studies suggested that low-grade systemic inflammation plays a significant role in the pathogenesis of obesity, insulin resistance, essential hypertension, type 2 diabetes, atherosclerosis, coronary heart disease, collagen vascular diseases and cancer. This is supported by the observation that plasma C-reactive protein (CRP), tumor necrosis factor-α (TNF-α), and interleukin-6 (IL-6), markers of inflammation, levels are elevated in these subjects [2–20], suggesting that low-grade systemic inflammation occurs in them. Similarly, Alzheimer's disease, depression, and schizophrenia are also low-grade systemic inflammatory conditions since, elevated levels of pro-inflammatory cytokines occurs in the plasma and brains of these patients [21–23].

Chronic inflammation is a major causative factor of human malignancies. Pro-inflammatory cytokines influence tumor microenvironment and promote cell growth and survival and angiogenesis such that tumor cell growth is facilitated. Thus, immune system activation could be a double edged sword: immune surveillance may check tumor development whereas aberrant immune activation promotes malignant growth [24]. Recent studies showed that low-grade systemic inflammation plays a role in many, hitherto believed to be degenerative conditions. It is not yet clear whether low-grade systemic inflammation occurs with ageing process. If so, it will be interesting to study whether suppressing low-grade systemic inflammation can slow ageing process itself. Since inflammation is a fundamental process of all living organisms, it remains to be seen how it can influence several other cellular processes such as longevity, cancer, etc.

In view of this, it is important to understand pathophysiological mechanisms of inflammation, its mediators, and how, why and where inflammation is involved in the pathogenesis of several clinical conditions. In the chapters that follow, a detailed discussion of the inflammatory process, various mediators of inflammation and conditions in which low-grade systemic inflammation plays a role in their pathobiology are discussed.

References

[1] Lopez A, Mathers C, Ezzati M, Jamison D, Murray C (2006) Global and regional burden of disease and risk factors, 2001: systematic analysis of population health data. Lancet 367:1714–1717

[2] Das UN (2006) Hypertension as a low-grade systemic inflammatory condition that has its origins in the perinatal period. J Assoc Physicians India 54:133–142

[3] Luc G, Bard J-M, Juhan-Vague I et al (2003) C-reactive protein, interleukins-6, and fibrinogen as predictors of coronary heart disease. The PRIME study. Arterioscler Thromb Vasc Biol 23:1255–1261
[4] Das UN (2001) Is obesity an inflammatory condition? Nutrition 17:953–966
[5] Das UN (2006) Aberrant expression of perilipins and 11-β-HSD-1 as molecular signatures of metabolic syndrome X in South East Asians. J Assoc Physicians India 54:637–649
[6] Ridker PM, Burning JE, Cook NR, Rifai N (2003) C-reactive protein, the metabolic syndrome, and risk of incident cardiovascular events. Circulation 107:391–397
[7] Das UN (2007) Is metabolic syndrome X a disorder of the brain with the initiation of low-grade systemic inflammatory events during the perinatal period? J Nutr Biochem 18:701–713
[8] Das UN (2008) Folic acid and polyunsaturated fatty acids improve cognitive function and prevent depression, dementia, and Alzheimer's disease – but how and why? Prostaglandins Leukot Essent Fatty Acids 78:11–19
[9] Das UN (2007) Is depression a low-grade systemic inflammatory condition? Am J Clin Nutr 85:1665–1666
[10] Dougan M, Dranoff G (2008) Inciting inflammation: the RAGE about tumor promotion. J Exp Med 205:267–270
[11] Visser M, Bouter LM, McQuillan GM et al (1999) Elevated C-reactive protein levels in overweight and obese adults. JAMA 282:2131
[12] Hotamisligil GS (1999) The role of TNF-alpha and TNF receptors in obesity and insulin resistance. J Intern Med 245:621
[13] Pradhan AD, Manson JE, Rifai N, Buring JE, Ridker PM (2001) C-reactive protein, interleukin-6, and risk of developing type 2 diabetes mellitus. JAMA 286:327
[14] Das UN (1999) GLUT-4, tumor necrosis factor, essential fatty acids and daf-genes and their role in glucose homeostasis, insulin resistance, non-insulin dependent diabetes mellitus and longevity. J Assoc Physicians India 47:431
[15] Fichtlscherer S, Rosenberger G, Walter DH et al (2000) Elevated C-reactive protein levels and impaired endothelial vasoreactivity in patients with coronary artery disease. Circulation 102:1000
[16] Cleland SJ, Sattar N, Petrie JR et al (2000) Endothelial dysfunction as a possible link between C-reactive protein levels and cardiovascular disease. Clin Sci (Colch) 98:531
[17] Das UN (2002) A perinatal strategy for preventing adult diseases: the role of long-chain polyunsaturated fatty acids. Kluwer Academic, Boston
[18] Das UN (2010) Metabolic syndrome pathophysiology: the role of essential fatty acids fatty acids and their metabolites. Wiley-Blackwell, Ames
[19] Das UN (2009) Cross talk among leukocytes, platelets, and endothelial cells and its relevance to atherosclerosis and coronary heart disease. Curr Nutr Food Sci 5:75–93
[20] Das UN (2005) Pathophysiology of metabolic syndrome X and its links to the perinatal period. Nutrition 21:762–773
[21] Popp J, Bacher M, Kölsch H, Noelker C, Deuster O, Dodel R, Jessen F (2009) Macrophage migration inhibitory factor in mild cognitive impairment and Alzheimer's disease. J Psychiatr Res 43:749–753
[22] Patel NS, Paris D, Mathura V, Quadros AN, Crawford FC, Mullan MJ (2005) Inflammatory cytokine levels correlate with amyloid load in transgenic mouse models of Alzheimer's disease. J Neuroinflammation 2:9
[23] Sutton ET, Thomas T, Bryant MW, Landon CS, Newton CA, Rhodin JA (1999) Amyloid-beta peptide induced inflammatory reaction is mediated by the cytokines tumor necrosis factor and interleukin-1. J Submicrosc Cytol Pathol 31:313–323
[24] Rakoff-Nahoum S (2006) Why cancer and inflammation? Yale J Biol Med 79:123–130

Chapter 3
Inflammation

Introduction

Inflammation is the complex biological response of vascular tissue to harmful stimuli such as pathogens, damaged cells or irritants [1] that consists of both vascular and cellular responses. Inflammation is a protective attempt by the organism to remove the injurious stimuli and initiate healing process and to restore both structure and function. It should be understood that infection and inflammation are not synonymous: infection is caused by an exogenous pathogen while inflammation is the response of the organism to the pathogen. Inflammation may be local or systemic, and it can be acute or chronic. During the inflammatory process, the reaction of blood vessels is unique that leads to the accumulation of fluid and leukocytes in extravascular tissues. The reaction of blood vessels can be in the form of vasodilatation that is seen in the form of hyperemia at the site(s) of injury, that increases blood supply to the injured tissue/organ so that cellular elements, antibodies and nutrients can reach the site of injury in adequate amounts to eliminate the inflammation-inducing agent and/or repair process can be initiated once inflammation subsides. Thus, both injury and repair are two sides of the inflammatory process that are very closely intertwined such that it is difficult to separate these two processes. In fact, in majority of the instances, both inflammation to injury and repair occur almost simultaneously. There is now reasonable evidence to suggest that inflammation and repair are initiated, perpetuated and suppressed by different molecules though some factors seem to be common to both these phases. It is possible that inflammation process may subside once the inciting agent is removed but repair of the damaged tissues may not occur adequately. In certain other situations, the injurious agent is successfully removed but the repair process may be defective that could lead to deposition of excess of fibrous tissue that result in structural deformity of the injured tissue and as a consequence the dysfunction of the tissue/organ may occur. This results in dysfunction of the tissue or organ. For example, hepatitis B virus may be successfully eliminated by the immune system of the body but during the process of regeneration of the liver tissue that underwent necrosis, excess deposition of fibrous tissue may be deposited leading cirrhosis the liver, which causes portal hypertension, development of esophageal

U. N. Das, *Molecular Basis of Health and Disease*,
DOI 10.1007/978-94-007-0495-4_3,

varices and finally liver failure and death. On the other hand, when the hepatitis B infection is significant it can cause massive necrosis of the liver and consequent liver cell failure, hepatic encephalopathy and death. Thus, both inflammation and repair processes are both beneficial and harmful depending on the degree of inflammation and the nature and extent of repair process.

But, in general, inflammation is a protective response whose ultimate goal is to eliminate the injury-inducing agent (that could be a microorganism, physical stimuli, chemical agent, etc.), prevent tissue damage and/or initiate the repair process and restore physiological function of the tissue or organ affected by the inflammatory process. Without inflammation there is no life since, in the absence of adequate inflammation cell/tissue injury would go unchecked, the damage done to the cells/tissues/organs would not heal, and ultimately this could lead to the death of the organism. Thus, inflammation is both beneficial and potentially harmful. In order to know the significance of the existing clinical laboratory tools of inflammation and to develop newer diagnostic tools, it is important to understand pathophysiological mechanisms of inflammation.

Phases of Inflammatory Response

The inflammatory response consists of two components: a vascular response and a cellular response, both of which are integral and essential parts of the inflammatory reaction. These two phases of both acute and chronic inflammation are mediated by chemical factors that could be proteins, lipids, or lipoproteins in nature. These biologically active chemical mediators of phases of inflammation are secreted by cells that participate in the inflammatory process either directly and/or responding to the inflammatory stimulus. These chemical mediators acting singly, in combinations, or in sequence, amplify the tissues/organs response to the stimulus and influence the course of inflammation. In addition, cells or tissues that are undergoing necrosis or apoptosis during the inflammation/repair also liberate chemicals that have the ability to take part in the inflammatory process itself. As already mentioned, for the inflammatory process to subside after the initiating stimulus is withdrawn or eliminated from the site of inflammation, for the repair process to set in certain anti-inflammatory chemicals and signals need to be secreted by the local tissues or cells such as macrophages and leukocytes so that tissue damage is minimized. Thus, ultimately recovery of a tissue/organ from the inflammatory process and regaining of its function depends on the balance between pro- and anti-inflammatory chemicals. Once inflammation is terminated either by endogenous mediators/repair processes and/or by drugs that include: antibiotics, anti-inflammatory drugs, chemical and surgical measures and the offending agent is removed, all the debris at the site of inflammation is either broken down or dissipated and the tissues/organs in question revert to their natural physiological state depending on the degree of damage and repair that has occurred.

Since circulating cells and chemical mediators participate in both acute and chronic inflammation, it is possible to measure the expression of certain molecules on the surface of these circulating cells, chemicals that are released by these circulating cells or both as markers of inflammation. In instances of acute inflammation and when the inflammatory process is on the surface of the body no specific tests are necessary to measure the presence or the degree of inflammation since, the acute characteristics of inflammation such as *rubor, tumor, calor, dolor, and functiolaesa* (redness, swelling, heat, pain, and loss of function respectively) are evident. But, specific tests or special measures become necessary when chronic inflammation occurs especially, deep inside the body or in the internal organs. This is especially so since, at present, it is believed that many diseases of the modern society such as obesity, hyperlipidemia, essential hypertension, type 2 diabetes mellitus, coronary heart disease (CHD), metabolic syndrome, schizophrenia, Alzheimer's disease and depression are diseases of low-grade systemic inflammation [2]. In view of this, age-old markers of inflammation such as ESR (erythrocyte sedimentation rate), CRP (C-reactive protein measured by conventional means), body temperature, etc., may not be suitable for measuring low-grade systemic inflammation. Hence, several studies are examining the possibility of utilizing more sophisticated and sensitive markers of inflammation such as high-sensitive CRP (hs-CRP), adhesion molecules, pro-inflammatory cytokines, etc., to know the existence of low-grade systemic inflammation, measure its severity, to predict their development and to prognosticate its course.

Components of Acute Inflammation

Acute inflammation that is a rapid response to an injurious agent has mainly three components: (a) alterations in the diameter of the blood vessels generally vasodilatation whose main purpose is to increase blood flow to the site of inflammation; (b) structural changes in the microvasculature such that it permits plasma proteins and leukocytes to leak from the circulation to participate in the pathobiology of inflammation both in injury and repair processes; (c) accumulation and activation of leukocytes at the site of inflammation and release of chemical mediators of inflammation and wherever possible these leukocytes try to eliminate the offending organism or agent. Acute inflammation is triggered by bacterial, viral, fungal and parasitic infections and their respective toxins; trauma; physical and chemical agents such as burns, radiation, and environmental or synthetic chemicals; foreign bodies such as splinters, thorns, sutures; abnormal immune reactions especially hypersensitivity reactions. Although, it is known that inflammation triggered by these various agents could have some very distinct features, in general, all acute inflammatory reactions share some common basic features as discussed below and shown in Fig. 3.1 and Table 3.1.

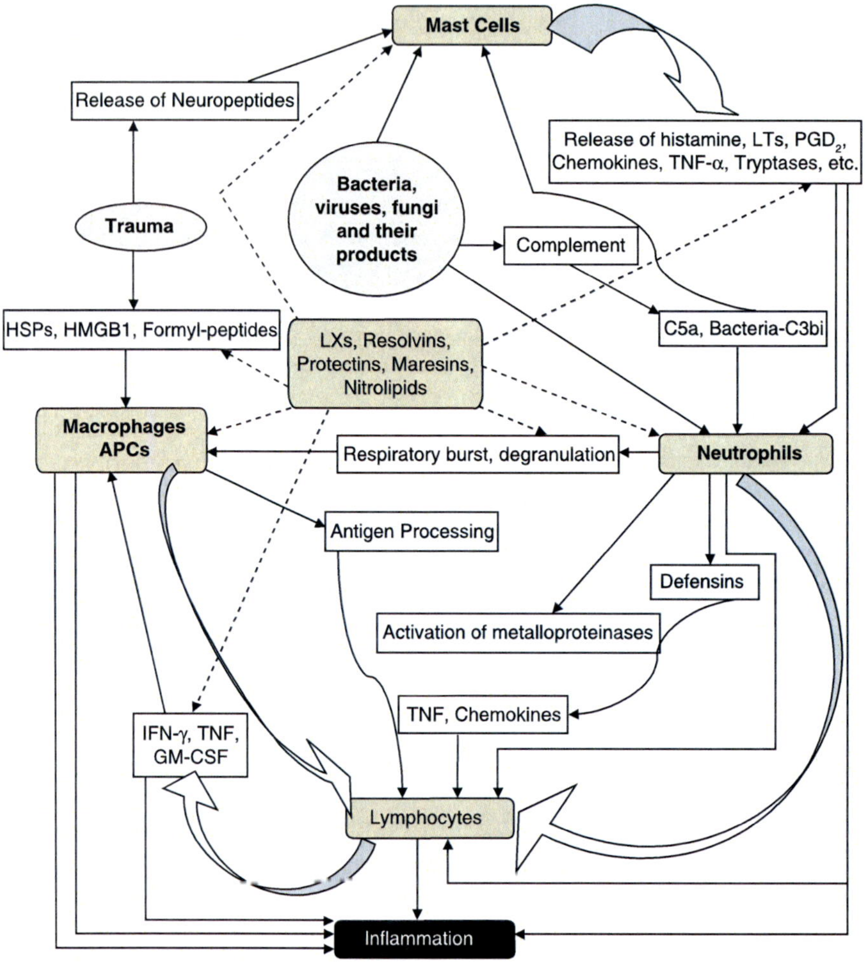

Fig. 3.1 Scheme showing the role of various cells and their products in inflammation. Scheme showing the role of various immunocytes and their products in the pathobiology of inflammation. .... indicates molecules that have a negative influence on inflammation, suppress the production of pro-inflammatory molecules or inhibit inflammation and enhance repair process. ↻ Indicates that these molecules are produced or these cells interact with each other. ⟶ Indicates that these molecules are produced by the respective cells and are involved in the recruitment and activation of various cells and inflammation

Vascular Changes

Vasodilatation: This is an essential component of inflammation and is an early manifestation of acute inflammation. Sometimes, early vasodilatation is followed by transient vasoconstriction. The purpose of vasodilatation is to increase blood flow to

Table 3.1 Components of inflammatory response: circulating cells, and proteins, cells of blood vessels, and cells and proteins of the extracellular matrix. It is clear that some proteins/molecules are common to several cells. This list is by no means exhaustive

Cellular components	Corresponding proteins/molecules
Connective tissue cells: Mast cells, fibroblasts, Macrophages	Histamine,serotonin, lysosomal enzymes
Vascular tissue cells: Smooth muscle cells, Endothelial cells	Nitric oxide, eicosanoids, reactive oxygen species, growth factors, cytokines, CRP, etc.
Circulating cells: Polymorphonuclear leukocytes, lymphocytes, platelets, monocytes, basophils, eosinophils	Platelet activating factor, growth factors, reactive oxygen species, nitric oxide, eicosanoids, cytokines, histamine, serotonin, kinins, adhesion molecules, carbon monoxide, complement system, coagulation and fibrinolysis system, etc.
Connective tissue matrix: Elastin fibres, collagen fibres, proteoglycans	Several matrix metalloproteinases, etc.

the site of inflammation to carry circulating proteins, antibodies in case of infections, nutrients, adequate oxygen and other mediators and/or molecules that are important participants in the pathobiology of inflammation. It is important to note that both during acute phase of inflammation and repair process, the site of injury or infection needs biologically active molecules such as prostaglandins, kinins, nitric oxide, etc., and nutrients. Once the inflammatory process and repair have been successfully completed, the blood vessels return to their original shape, size and diameter. Initially, the existing blood vessels undergo dilatation but at a later stage depending on the demand, necessity and the mediators that are released at the site of inflammation newer capillary beds are opened. In this context, it is noteworthy that when a dormant tumor starts growing it demands increased blood supply that is met by the generation of new blood vessels called as angiogenesis. This angiogenesis occurs due to the release of angiogenic factors by the tumor cells. Thus, tumor can be considered as a local inflammatory event. Vasodilatation that occurs during inflammation is followed by increased permeability of the microvasculature that allows outpouring of protein-rich fluid and extravasation of leukocytes to the site(s) of inflammation. Prior to the extravasation of leukocytes, as a result of leakage of protein and vasodilatation stasis of blood flow occurs that is reflected by an increase in the concentration of red blood cells in the smaller vessels resulting in increased viscosity of the blood. As a result, leukocytes, especially polymorphonuclear leukocytes (PMNs) accumulate along the vascular endothelium and slowly escape from the blood vessels into the interstitial tissue. The attachment of the leukocytes to the endothelial cells occurs due to the increased expression of various adhesion molecules both on the surface of leukocytes and endothelial cells.

Mediators of Vascular Changes

The exact mechanism(s) and the mediators involved in vasodilatation process during inflammation are still not clear. Recent studies showed that nitric oxide (NO) produced by endothelial cells and possibly the cells that are infiltrating to the site of inflammation such as leukocytes, macrophages; monocytes and lymphocytes seem to have a dominant role in inducing vasodilatation of inflammation. NO is a potent vasodilator and platelet anti-aggregator and is one of the important mediators of vasodilatation seen during inflammation. Several other mediators of vasodilatation may include carbon monoxide (CO), hydrogen sulfide (H_2S), prostaglandins (PGs) especially prostacyclin (PGI_2) and other eicosanoids, bradykinin and other kinins, and histamine. The final degree of vasodilatation at a given site of inflammation could depend on the amount of each of these possible mediators released from various cells, the balance between vasodilator and vasoconstrictor mediators released and their respective inactivators. These mediators are released by macrophages, monocytes, infiltrating leukocytes, lymphocytes, endothelial cells, and other cells present at the site of inflammation. Furthermore, there is a close interaction between these various vasoactive molecules. For instance, it was observed that myeloperoxidase (MPO) released by activated PMNs not only generates cytotoxic oxidants but also impacts deleteriously on NO-dependent signaling cascades and thus could influence vasodilatation during inflammation. MPO increased tyrosine phosphorylation and p38 mitogen-activated protein kinase activation; MPO-treated PMNs released increased amounts of free radicals, and enhanced PMN degranulation [3]. MPO, a highly abundant, PMN-derived heme protein facilitates oxidative NO consumption and impairs vascular function in animal models of acute inflammation [4]. Furthermore, myeloperoxidase (MPO) deficiency is a common inherited disorder linked to increased susceptibility to infection and malignancy that reiterates its importance in inflammation and infection [5].

It is known that MPO participates in the eradication of Mycobacterium tuberculosis, a chronic inflammatory condition that is common in the developing world. It is likely that MPO may activate cells to synthesize and release various cytokines that are essential for immunity. In a study performed in patients with active pulmonary tuberculosis (TB), it was observed that a statistically significant elevation of TNF-α (tumor necrosis factor-α) and IL-12 and MPO in the serum was present. Although in this study no statistically significant relationship between the cytokines and MPO production was noted, the increase in TNF-α and IL-12 serum concentration with simultaneous elevation of serum MPO in the group of the highest enzyme concentration suggested that some correlation between the enzyme and the cytokines might exist. The results of this study suggest possible involvement of MPO in the antituberculous, immunological response, and imply its connection with TNF-α and IL-12 activation [6]. The involvement of MPO in inflammation is further supported by the observation that inflammatory oxidants are the key contributors to Parkinson's disease (PD)- and MPTP-(1-methyl-4-phenyl-1,2,3,6-tetrahydropyridine)-induced neurodegeneration. Studies showed that MPO, a key oxidant-producing enzyme during inflammation, is upregulated in the ventral midbrain of human PD and MPTP mice. It was also observed that ventral midbrain dopaminergic neurons of mutant mice deficient in MPO are more resistant to MPTP-induced cytotoxicity than their wild-type littermates. This is further supported by the observation that in this PD model MPO-specific biomarkers 3-chlorotyrosine and hypochlorous acid-modified proteins increase in the brains of MPTP-injected mice [7]. Thus, MPO participates in the MPTP neurotoxic process and suggests that MPO serves as an important mediator of inflammation and its inhibitors could of significant benefit in the management of PD.

In a similar fashion, another gas that is produced in the body that seems to have an important role in inflammation is hydrogen sulfide (H_2S). Hydrogen sulfide is a well-known toxic gas, has been recognized as a signal molecule as well as a cytoprotectant. It is produced by three enzymes, cystathionine beta-synthase, cystathionine gamma-lyase and 3-mercaptopyruvate sulfurtransferase along with cysteine aminotransferase. H_2S is not only released immediately after its formation, it can be stored as bound sulfane sulfur, which may release H_2S in response to physiological stimuli. As a signal molecule, H_2S modulates neuronal transmission, relaxes smooth muscle, regulates release of insulin and is involved in inflammation. It is not only a toxic gas, but it also has cytoprotective functions especially in the nervous system and cardiovascular system where it protects them oxidative stress [8, 9].

Male Wistar rats that were subjected to acute endotoxemia[(induced by Escherichia coli lipopolysaccharide (LPS) 6 mg/kg) intravenously for 6 h) developed circulatory failure (hypotension and tachycardia) and an increase in serum levels of alanine aminotransferase and aspartate aminotransferase (markers for hepatic injury), lipase (indicator of pancreatic injury) and creatine kinase (indicator of neuromuscular injury). In the liver, endotoxemia induced a significant increase in the myeloperoxidase (MPO) activity, and in the expression and activity of the H_2S-synthesizing enzymes. Inhibition of H_2S synthesis either prior to or after the injection of LPS dose-dependently reduced the hepatocellular, pancreatic and neuromuscular injury caused by endotoxemia, decreased increase in MPO activity and the formation of H_2S in the liver but not the circulatory failure. These observations led to the suggestion that enhanced formation of H_2S contributes to the pathophysiology of the organ injury in endotoxemia [10]. These results are supported by other studies that showed that prophylactic, as well as therapeutic treatment with the H_2S inhibitors significantly reduced the severity of experimentally-induced pancreatitis [11]. The possible proinflammatory effect of H_2S is further strengthened by the report that mice administered sodium hydrosulfide (H_2S donor drug) resulted in marked histological signs of lung inflammation, increased lung and liver MPO activity, and raised plasma TNF-α concentration, while inhibition of H_2S inhibitor, DL-propargylglycine, showed marked anti-inflammatory activity. Significantly higher (150.5 ± 43.7 μM vs. 43.8 ± 5.1 μM, $P < 0.05$) plasma H_2S levels were noted in humans with septic shock [12]. Thus, H_2S seems to have significant proinflammatory activity. H_2S has been shown to upregulate the production of proinflammatory mediators such as TNF-α and exacerbate the systemic inflammation in sepsis through a mechanism involving NF-κB activation [13]. In contrast, nitric oxide (NO) downregulated the biosynthesis of H_2S highlighting the existence of "crosstalk" between NO and H_2S [14].

In contrast to these results, some studies suggested that H_2S has anti-inflammatory actions [15–20]. Many studies have used simple sulfide salts as the source of H_2S, which give a rapid bolus of H_2S in aqueous solutions and thus may not accurately reflect the enzymatic generation of H_2S. In view of this, Whiteman et al [21] studied the effects of sodium hydrosulfide and a slow-releasing H_2S donor (GYY4137) on the release of pro- and antiinflammatory mediators in lipopolysaccharide (LPS)-treated murine RAW264.7 macrophages. In this study it was noted that GYY4137 (slow-releasing H_2S donor) significantly and concentration-dependently inhibited LPS-induced release of proinflammatory mediators such as IL-1β, IL-6, TNF-α, nitric oxide, and PGE_2 but increased the synthesis of the antiinflammatory chemokine IL-10 through NF-κB/ATF-2/HSP-27-dependent pathways. In contrast, NaHS produced a biphasic effect on proinflammatory mediators and, at high concentrations, increased the synthesis of IL-1β, IL-6, NO, PGE_2 and TNF-α. These results suggest that the effects of H_2S on the inflammatory process are complex and dependent not only on local H_2S concentration but also on the rate of H_2S generation. It need to be emphasized here that such a dual affect may be seen even with other molecules such as NO, TNF-α and ILs.

In addition, there is a close interaction between superoxide anion ($O_2^{\cdot -}$) and myeloperoxidase produced by leukocytes during acute inflammation and NO. Superoxide anion and myeloperoxidase can inactivate NO and thus, reduce its half-life and activity [22, 23]. On the other hand, NO, when produced in adequate amounts inactivate superoxide anion and protect the cells and tissue from the cytotoxic action of superoxide anion [24]. Furthermore, NO inactivates NADPH oxidase and thus, inhibits the production and release of superoxide anion [25, 26]. Thus, there is a cross-talk among various mediators of inflammation such as NO, superoxide anion, myeloperoxidase, H_2S and cytokines.

Vascular Leakage

Leakage of circulating protein into the extravascular tissue results in edema, one of the hallmarks of inflammation. This leakage of proteinaceous fluid is due to the formation of endothelial gaps in venules, direct endothelial damage, necrosis or detachment, leukocyte-mediated endothelial injury that ultimately results in the loss of circulating protein into the extravascular tissue [27].

Although, exact details as to the chemical mediators and the sequence of their production is not clear, it is clear that cytokines such as interleukin-1 (IL-1), tumor necrosis factor-α (TNF-α), interferon-γ (IFN-γ), vascular endothelial growth factor (VEGF), histamine, substance P, free radicals, nitric oxide, myeloperoxidase, and other yet unidentified chemicals play a significant role in vasodilatation, vascular leakage, and diapedesis of leukocytes [2, 28]. PMN-induced damage to vascular endothelial cells is believed to be due to increased production of reactive oxygen species

(ROS), inducible nitric oxide (iNO) and its metabolites (such as OCl), ozone, and release of cytokines [2, 29]. The main purpose of ROS, iNO, H_2S and ozone appears to be to kill and eliminate the invading microorganisms [30–32]. In view of their ability to diffuse across cell membranes and tissues and potent actions, they produce collateral damage to the surrounding cells/tissues. In addition to their pro-inflammatory actions, ROS, iNO, IL-1, TNF, IFN, and VEGF modulate vascular reactivity, endothelial cell proliferation and function, smooth muscle cell function and proliferation, expression of adhesion molecules, leukocyte function, and extracellular matrix production. These actions ultimately influence the inflammatory process, repair of the inflamed tissues/organs, and functional integrity of the target tissues/organs. Based on these evidences, monoclonal antibodies that neutralize the actions of IL-1, TNF-α, IFN, and VEGF have been developed. For example, it is now known that age-related macular degeneration (AMD) is due to increased production of VEGF in the retinal tissue. Recent studies showed that anti-VEGF therapies are of significant benefit in AMD [33, 34]. On the other hand, monoclonal antibodies against IL-1, and TNF-α failed to show any significant benefit in acute systemic inflammatory condition such as sepsis and septic shock [35–37] suggesting that our understanding of inflammation is still inadequate to develop therapeutically meaningful approaches.

In this context, the role of free radicals in vascular reactivity during inflammation is interesting. Free radicals including hydrogen peroxide (H_2O_2), $O_2^{\cdot-}$, NO, nitrated lipids and H_2S have vasoactive actions. NO is a vasodilator, whereas $O_2^{\cdot-}$ and other free radicals have vasoconstrictor actions [38–40]. In fact, it is believed that $O_2^{\cdot-}$ could be the vasoconstrictor that produces coronary vasospasm leading acute angina [41]. The fact that NO and $O_2^{\cdot-}$ have contrasting actions on the vascular reactivity, the final diameter of the blood vessels depends on the balance between NO and $O_2^{\cdot-}$ produced at the site of inflammation. Since tissue antioxidant defenses such as superoxide dismutase (SOD), catalase, and glutathione try to neutralize, suppress, or antagonize the actions of free radicals, the tissue destructive properties and vasoconstrictor actions of free radicals are determined to a large extent on the tissue concentrations of these antioxidants. Furthermore, NO neutralizes the actions of $O_2^{\cdot-}$ and hence, the balance between these two molecules could be yet another modulator of inflammation (see Fig. 3.2).

Cellular Events

Leukocyte Extravasation and Chemotaxis

Leukocytes, monocytes and macrophages are needed at the site of injury and inflammation to eliminate the inciting agent responsible for inflammation and initiate the repair process. Leukocytes, monocytes and macrophages and at sites such as liver (Kupffer cells), skin (fibroblasts, eosinophils and basophils) and lungs (mast cells) ingest the offending agent, kill bacteria and other microbial organisms, and

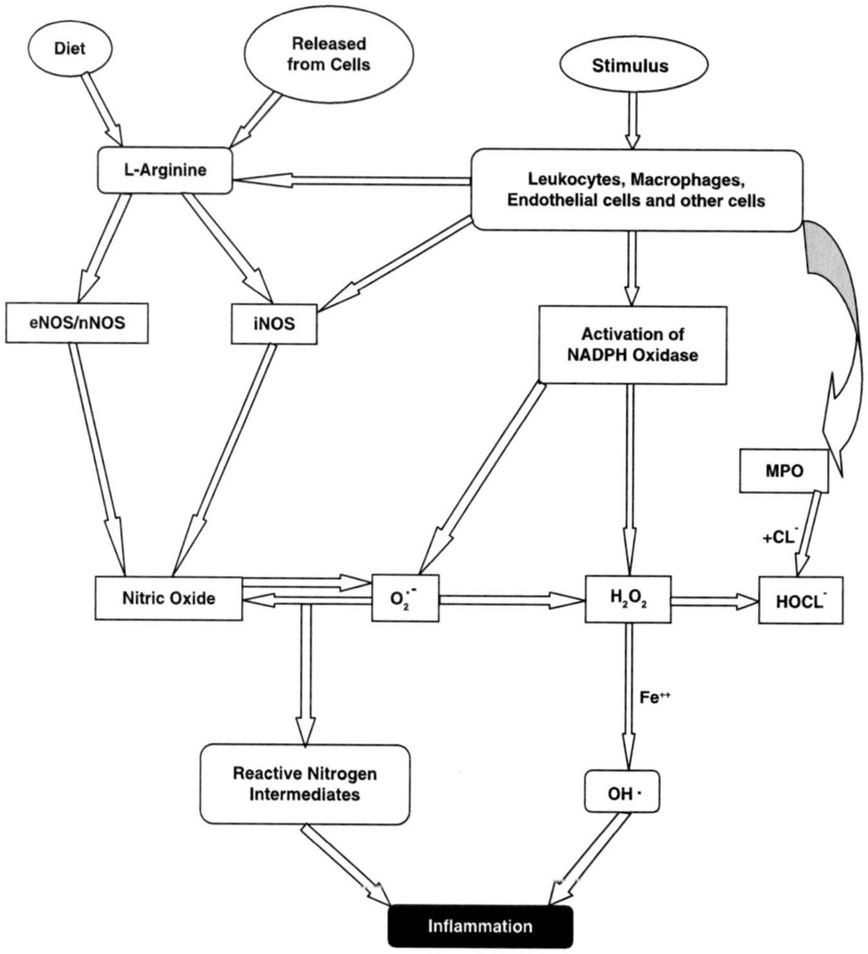

Fig. 3.2 Scheme showing generation of ROS and NO and formation of RNI (reactive nitrogen intermediates). Stimulus could be injury, foreign particles, or release of various pro-inflammatory cytokines. There is a close interaction between NADPH oxidase and MPO (see the text). Superoxide anion can inactivate NO and, in turn, NO can inactivate superoxide anion. NO and superoxide anion interact to form reactive nitrogen intermediates that are potent inflammatory substances

remove the necrotic tissue, debris and foreign material and during this process these cells liberate various biologically active molecules such as free radicals, nitric oxide, H_2S, eicosanoids, histamine, serotonin, kinins, etc. These molecules are needed both for the initiation and perpetuation of the inflammation process and could induce tissue damage. Once the offending stimulus is removed or neutralized (by the release of appropriate antibodies by the infiltrating lymphocytes), repair process has to be initiated. Some of the molecules that have been identified which seem to suppress and initiate the resolution of the inflammation and enhance repair process include metabolites of arachidonic acid (AA, 20:4 ω-6), eicosapentaenoic acid (EPA, 20:5 ω-3), and docosahexaenoic acid (DHA, 22:6 ω-3) such as lipoxins, resolvins, protectins and maresins. These are small molecular weight lipid molecules that have been shown to suppress inflammation, inhibit leukocyte activation and enhance repair process [42–48].

Leukocytes need to extravagate from inside the blood vessels in order to bring about these actions. For this purpose, leukocytes adhere to the endothelial lining of the blood vessels, transmigrate across the endothelium (a process called as diapedesis), and migrate in interstitial tissues toward the chemotactic stimulus and reach the site of inflammation or injury [49]. For this extravasation to occur and for the leukocytes to adhere and transmigrate from the blood into tissues, both leukocytes and endothelial cells express complementary adhesion molecules, whose expression, in turn, is regulated largely by cytokines. The adhesion receptors involved in this process belong to are four major molecular families, namely: selectins, immunoglobulin superfamily, integrins, and mucin-like glycoproteins. Important adhesion molecules that are expressed on endothelial cells, their complementary leukocyte receptor and their major function(s) are given in Table 3.2. The multi-step process of leukocyte migration through blood vessels involves: leukocyte rolling, activation and adhesion of leukocytes to endothelium, transmigration of leukocytes across the endothelium, piercing the basement membrane, and finally migration towards chemoattractants emanating from the site of injury or inflammation. Although almost all molecules may have a role in several of these processes, certain molecules play a more dominant role in specific processes. For instance, selectins play a major role in rolling; chemokines in activating the neutrophils to increase avidity of integrins; integrins in firm adhesion; and CD31 (PECAM-1) in transmigration [50].

Recent studies showed that neutrophil chemotaxis plays an essential role in innate immunity. Using a small-molecule functional screening, it was identified that

Table 3.2 A list of major adhesion molecules that are expressed on the surface of endothelial cells and their complementary adhesion molecules on leukocytes

Endothelial molecule	Leukocyte receptor	Major function
P-selectin	Sialyl-Lewis X, PSGL-1	Rolling neutrophils, Monocytes and lymphocytes
E-selectin	Sialyl-Lewis X	Rolling, adhesion of neutrophils, monocytes and T cells to activated endothelium
ICAM-1	CD11/CD18 (integrins) LFA-1, Mac-1	Adhesion and transmigration of leukocytes
VCAM-1	$\alpha4\beta1$ (VLA4) (integrins) $\alpha4\beta7$ (LPAM-1)	Adhesion of eosinophils, monocytes, and Lymphocytes
GlyCam-1	L-selectin	Lymphocyte homing
CD31 (PECAM)	CD31	Leukocyte migration through endothelium

ICAM-1, VCAM-1, and CD31 belong to the immunoglobulin family of proteins; *PSGL-1* = P-selectin glycoprotein ligand 1

NADPH oxidase–dependent reactive oxygen species as a key regulator of neutrophil chemotactic migration. Inhibition of neutrophil-NADPH oxidase led to the formation of more frequent multiple pseudopodia and lost their directionality as they migrated up a chemoattractant concentration gradient. Knocking down NADPH oxidase also led to defective chemotaxis. Consistent with these results, adoptively transferred CGD murine neutrophils showed impaired in vivo recruitment to sites of inflammation. These results suggest that reactive oxygen species regulate neutrophil functions including chemotaxis [51].

The induction of adhesion molecules on endothelial cells may occur by a number of mechanisms. For example, histamine, thrombin, and platelet activating factor (PAF) stimulate the redistribution of P-selectin from its intracellular stores to the cell surface; whereas macrophages, mast cells, and endothelial cells secrete pro-inflammatory cytokines such as IL-1, TNF-α, and chemokines that act on endothelial cells and induce the expression of several adhesion molecules. This results in the expression of E-selectin on the surface of endothelial cells. Simultaneously, leukocytes express carbohydrate ligands for the selectins that allow them to bind to the endothelial selectins [52]. This binding of leukocytes to endothelium is a low-affinity interaction that is easily disrupted by the flow of blood, which leads to the alternate process of binding, disruption of the binding, and binding once again of leukocytes to endothelial cells that results in rolling of leukocytes on the surface of endothelium (see Table 3.2).

On the other hand, IL-1 and TNF-α and possibly other such pro-inflammatory cytokines induce the expression of ligands for integrins such as VCAM-1 and ICAM-1. Chemokines produced at the sites of inflammation or injury act on endothelial cells such that proteoglycans (such as heparan sulfate glycosaminoglycans) are expressed at high concentrations on their surface, whereas they activate leukocytes to convert low-affinity integrins such as VLA4 and LFA-1 to high-affinity state. These events lead to firm binding of activated leukocytes to activated endothelial cells such that leukocytes stop rolling, their cytoskeleton is reorganized, and they spread out on the endothelial surface. Binding of activated leukocytes to endothelial surface induces endothelial dysfunction and damage due to ROS and iNO produced by leukocytes. These adherent leukocytes migrate through interendothelial spaces towards the site of injury or infection by binding to PECAM-1 (platelet endothelial cell adhesion molecule) that belongs to the immunoglobulin superfamily or CD31. Leukocytes pierce the basement membrane by secreting collagenases and other enzymes that can digest collagen.

One mechanism by which leukocytes emigrate towards the sites of injury or inflammation is by a process called as chemotaxis that is induced by chemotaxins. These chemoattractants can be either endogenous or exogenous molecules. The most common exogenous chemoattractants are bacterial products, some of which are peptides that contain N-formyl-methionine terminal amino acid. Some of the endogenous chemoattractants include (but not limited to): components of the complement system such as C5a, lipoxygenase pathway products such as leukotriene B_4 (LTB_4), and cytokines such as IL-8. Although the exact mechanism by which leukocytes sense and are attracted towards the chemosensory agents is not clear, studies suggested that majority of these chemoattractants bind to specific seven transmembrane G-protein-coupled receptors (GPCRs) on the surface of leukocytes [52]. GPCRs, in turn, activate phospholipase C (PLC), phosphoinositol-3-kinase (PI3K) and protein kinases. Both PLC and PI3K act on cell membrane phospholipids to generate lipid second messengers such as inositol triphosphate (IP3) that increase cytosolic calcium (Ca^{2+}) and activate small GTPases of the Rac/Rho/cdc2 family as well as numerous kinases. GTPases induce polymerization of actin that helps in the motility of the leukocytes. In this context, it is interesting to note that eNO synthase activation is critical for vascular leakage during acute inflammation [53]. It was noted that in eNO synthase-deficient ($eNOS^{-/-}$) mice the early phase (0–6 h) inflammation induced by intraplantar injection of carrageenan is eliminated, and the secondary phase (24–96 h) of the inflammatory response is markedly reduced compared to WT (wild type) mice. Zymosan-induced inflammatory cell extravasation was similar in WT and $eNOS^{-/-}$ mice, whereas extravasation of plasma protein was lower in $eNOS^{-/-}$ mice. Inhibition of phosphatidylinositol 3-kinase and hsp90 also blocked protein leakage but not leukocyte influx [53]. These and other studies clearly established the critical role of eNOS in vascular leakage during acute inflammation [54]. But, it is not yet clear as to the exact relationship between selectins, VCAM-1 and ICAM-1, GPCRs, small GTPases of the Rac/Rho/cdc2 family as well as numerous kinases, and eNOS and how the interaction between these molecules influences the inflammatory process.

There are four main factors that enhance the risk of coronary heart disease which is also associated with low-grade systemic inflammation. These are: smoking, hyperglycemia, dyslipidemia and hypertension.

Human umbilical vein endothelial cells (HUVEC) exposed to smokers' serum showed decreased nitric oxide (NO) production and endothelial nitric oxide synthase (eNOS) activity in the presence of increased eNOS expression. Similar results have been obtained with human coronary artery endothelial cells (HCAECs) also. HCAECs incubated with smokers' serum alone showed significantly lower NO production and eNOS activity but higher eNOS expression compared with nonsmokers. In smokers, addition of polyethylene glycol-superoxide dismutase (PEG-SOD, 300 U/ml), PEG-SOD+PEG-catalase (1,000 U/ml), or tetrahydrobiopterin significantly improved NO levels and eNOS activity. These results suggest that oxidative stress plays a central role in smoking-mediated dysfunction of NO biosynthesis in endothelial cells. Furthermore, these data support other studies suggesting a role for hydrogen peroxide in the upregulation of eNOS. Thus, smokers produce more free radicals that, in turn, lead to endothelial dysfunction [55].

It is known that endothelial cells exposed to constant high concentrations of glucose upregulate the expression of adhesion molecules, a phenomenon that has been related to excess generation of oxidative stress. It has also been suggested that oxidative injuries, related to high glucose, induce the activation of the enzyme poly ADP ribose polymerase (PARP), which can promote the expression of adhesion molecules and the generation of inflammation. In vivo and in vitro evidence suggests that oscillation of glucose may play an autonomous and direct role in favoring the development of cardiovascular complications in diabetes. In a study that investigated the effects of constantly high and intermittently high glucose on nitrotyrosine formation (a marker of nitrosative stress) and adhesion molecule (ICAM-1, VCAM-1 and E-selectin), as well as on interleukin (IL)-6 expression in human umbilical vein endothelial cells, revealed that oscillating glucose was more effective in triggering the generation of nitrotyrosine and inducing the expression of adhesion molecules and IL-6 than stable high glucose and this effect was found to be completely dependent on mitochondrial free radicals over-production. Pharmacological inhibition of PARP suppressed nitrotyrosine formation, adhesion molecule expression and IL-6 to the levels seen in the normal glucose conditions [56, 57]. Thus, PARP activation is essential in both promoting nitrosative stress and upregulating adhesion molecules and inflammation in endothelial cells exposed to oscillating high glucose conditions that is typical of poorly controlled hyperglycemia seen in diabetics.

Studies performed in diabetic normolipemic and egg yolk diet-induced hyperlipemic diabetic rabbits were compared with those from normoglycemic animals on similar diets after 4 weeks of hyperlipemia revealed that the frequency of aortic endothelial cells expressing VCAM-1 or E-selectin was significantly increased compared with normolipemic controls; this frequency was further increased in the aortas of hyperlipemic diabetic rabbits. VCAM-1 and E-selectin expression was more frequent in normolipemic diabetic rabbit aortas than in hyperlipemic, normoglycemic vessels. The potentiation of expression of the adhesion molecules in diabetic and hyperlipemia animals may explain the enhanced atherosclerosis associated with diabetes mellitus and hyperlipemia [58]. These results are supported by the observation that patients with hypertriglyceridemia had significantly higher levels of sVCAM-1 compared with patients with hypercholesterolemia and control subjects. Levels of sICAM-1 were significantly increased in both the hypercholesterolemic and hypertriglyceridemic groups compared with the control group. Levels of sE-selectin were significantly increased in hypercholesterolemic patients compared with control subjects. Surprisingly, comparison of soluble CAMs before and after treatment for hyperlipidemia with statins showed a significant reduction only in sE-selectin (but not for sVCAM-1 or sICAM-1. These results indicate that though severe hyperlipidemia is associated with increased levels of soluble CAMs, aggressive lipid-lowering treatment may be of only limited effects on their levels [59]. Accumulation of monocyte/macrophages and T lymphocytes in arterial intima is a hallmark of early atherogenesis that could be attributed to the increased expression of adhesion molecules by the endothelial cells. But, it was found that feeding experimental animals with 1 week of cholesterol feeding, neither macrophages nor T lymphocytes were detected, although endothelial expression of P-selectin and VCAM-1 was observed. After 3 weeks, macrophages were detectable in 75% and T lymphocytes were present in 25% of the rabbits. Expression of P-selectin and VCAM-1 was sustained until 10 weeks. Infiltration of T lymphocytes was restricted in areas in which macrophages were accumulated and did not appear to precede macrophage infiltration. E-selectin expression was not detectable before accumulation of mononuclear leukocytes; however, very few endothelial cells covering foam cell lesions expressed

E-selectin after 6 weeks. Similar results were noted in Watanabe heritable hyperlipidemic rabbits aged 1, 2, and 3 months [60]. These results suggest that localized expression of P-selectin and VCAM-1 may play a key role in the initial recruitment of macrophages and T lymphocytes in early atherogenesis and that the increased expression of adhesion molecules precedes the recruitment of macrophages and cells at sites of atherosclerosis. Since increased expression of adhesion molecules is a sign of inflammation and their expression is enhanced when animals are fed high cholesterol and in subjects with hypercholesterolemia, it is implies that cholesterol has pro-inflammatory actions. In contrast, supplementation with EPA/DHA and use of statins produced a significant reduction in the expression of adhesion molecules [61–65]. It is interesting that acute hypertriglyceridemia is a leukocyte activator by direct interaction between TRLs (triglyceride-rich lipoproteins) and leukocytes and uptake of fatty acids. TG- and cholesterol-mediated leukocyte activation (probably due to the activation of NADPH oxidase) could be the proinflammatory and proatherogenic mechanism of hyperlipidemia [66], in part, as a result of the generation of oxidative stress.

Hypertension, a risk factor for the development of coronary heart disease, is also a low-grade systemic inflammatory condition. Patients with hypertension have increased levels of pro-inflammatory cytokines such as IL-6, TNF-α, and high sensitive C-reactive protein (hs-CRP), low concentrations of anti-oxidants superoxide dismutase (SOD) [67–71]. In addition to having increased free radical generation (such as superoxide anion and H_2O_2) subjects with hypertension also showed lower concentrations of endothelial NO (eNO), a potent vasodilator and platelet anti-aggregator [67]. These biochemical abnormalities reverted to normal after the control of blood pressure by anti-hypertensive medicines. It is noteworthy that currently available anti-hypertensive medicines showed anti-oxidant actions [67]. This suggests that one of the mechanisms by which they are of benefit in hypertension could be attributed to their anti-oxidant action. In addition, it was noted that NO is a potent inhibitor of angiotensin converting enzyme (ACE) activity and thus, lowers the production of pro-inflammatory angiotensin-II, a potent vasoconstrictor molecule and pro-oxidant agent [67].

Free radicals themselves are known to modulate the tone of vascular smooth muscles directly and also indirectly by altering the half-life of prostacyclin (PGI_2) and nitric oxide (NO), enhanced free radical generation by angiotensin-II may lead to an increase in peripheral vascular resistance and hypertension [67]. It is likely that $O_2^{\cdot -}$ itself could be an endothelial-derived vasoconstrictor [72] and participate in the pathogenesis of hypertension [73, 74]. NADPH oxidase is the most important source of $O_2^{\cdot -}$ in vascular and other cells. Angiotensin II stimulates free radical generation [67] by up regulating several subunits of membrane bound NADPH oxidases [75, 76]. These results are supported by the recent reports that reduction of extracellular superoxide dismutase (SOD) in the central nervous system promoted T-cell activation and vascular inflammation, modulated sympathetic outflow and induced hypertension [77]; active oxygen species and thromboxane A_2 reduced angiotensin-II type 2 receptor-induced vasorelaxation in diabetic rats [78]; tumor necrosis factor-α (TNF-α) plays a role in activation of the PMN NADPH oxidase, thereby contributing to systemic oxidative stress, inflammation, and the development of hypertension [79]; and in healthy middle-aged and older adults, impaired endothelium-dependent dilatation is decreased by higher PMN count mediated by reduced responsiveness to NO and increased myeloperoxidase-associated reductions in tetrahydrobiopterin and NO bioavailability [80].

Thus, inflammation need not always be related only to infections, injury and surgery but seems to be at the heart of diseases such as hyperlipidemia, hypertension, diabetes mellitus, coronary heart disease and diseases associated with smoking. There is evidence to suggest that even ageing, atherosclerosis, Alzheimer's disease, depression, schizophrenia and cancer are also associated with low-grade systemic inflammation. Leukocyte activation, macrophage function and T cell responses are at the centre of inflammatory process and low-grade systemic inflammatory conditions. Naturally occurring, endogenous molecules that keep the inappropriate inflammation and inflammatory events such as leukocyte activation, macrophage function and T cell responses assume great significance in our efforts to devise methods to keep inflammation under control and prevent, reverse or halt inflammation associated diseases. One such endogenous factor(s) that has immunomodulatory influence and ability to control inflammation is the small molecular weight biologically active lipids formed from essential fatty acids (EFAs)/polyunsaturated fatty acids (PUFAs) such as lipoxins, resolvins, protectins and maresins. The actions and the importance of these lipid molecules in inflammation and various diseases is discussed in subsequent sections/chapters.

Leukocyte Activation

In order to kill microbes that produce inflammation, leukocytes generate ROS by a process that is termed as activation. Products of necrotic cells, antigen-antibody complexes, cytokines, and chemokines also induce leukocyte activation. Different classes of leukocyte cell surface receptors recognize different stimuli. For instance, chemokines, lipid mediators, and N-formyl-methionyl peptides increase integrin avidity, and produce cytoskeletal changes that aids leukocyte chemotaxis; microbial lipopolysaccharide (LPS) binds to toll-like receptors (TLRs) on leukocyte membrane leading to their activation and production of cytokines and ROS that are essential for the killing of microbes; and binding of microbial products to mannose receptor augments leukocyte phagocytic process that aids in the elimination of the invading organisms. Activation of leukocytes by various stimuli triggers several signaling pathways that result in increases in cytosolic Ca^{2+} and activation of protein kinase C (PKC) and phospholipase A_2 (PLA_2) that are ultimately seen in the form of various functional responses of leukocytes. In this context, it is interesting to note that PLA_2 activation leads to the release of lipids such as arachidonic acid (AA, 20:4 ω-6), eicosapentaenoic acid (EPA, 20:5 ω), and docosahexaenoic acid (DHA, 22:6 ω-3) from the cell membrane lipid pools. Studies showed that AA, and possibly EPA and DHA themselves could increase cytosolic Ca^{2+} and PKC concentrations in various cells [2, 81, 82]. Furthermore, AA by itself has the ability to activate leukocytes

[83]. These results suggest that dietary lipids have the ability to modulate leukocyte responses and the inflammatory process. Products of AA, EPA, and DHA such as prostaglandins (PGs), leukotrienes (LTs), lipoxins (LXs), resolvins, protectins and maresins have both positive and negative influences on leukocyte activation, chemotaxis, inflammation and its resolution [84–86]. Some of the products that are released by activated leukocytes include: AA and its metabolites, lysosomal enzymes, ROS, NO, myeloperoxidase, various cytokines, various leukocyte adhesion molecules and other surface receptors such as TLRs, GPCRs, receptors for opsonins, etc.

Phagocytosis and Killing of Microbes by ROS

In order to eliminate the invading microorganisms, leukocytes first have to phagocyte them and then release appropriate amounts of ROS, NO, and myeloperoxidase to kill them. Leukocytes use mannose receptors and scavenger receptors to bind and ingest bacteria, though they can engulf bacteria and other particles without attachment to specific receptors. Opsonins greatly enhance the efficiency of phagocytosis. Once the bacteria or other foreign particles are recognized by leukocytes, they are engulfed for killing them.

Killing and degradation of the ingested bacteria or foreign particles both by leukocytes and macrophages is accomplished by ROS, NO, myeloperoxidase and ozone. In general, phagocytosis stimulates NADPH oxidase accompanied by a burst of oxygen consumption, glycogenolysis, and increased glucose oxidation via the hexose-monophosphate shunt pathway. ROS, NO and ozone have the ability to kill bacteria. The azurophilic granules of neutrophils contain myeloperoxidase (MPO), which, in the presence of a halide such as $Cl^{\cdot -}$, converts H_2O_2 to hypochlorite (HOCL). HOCL is a potent antimicrobial agent by binding covalently to cellular constituents or by oxidation of proteins and lipids [87]. Once leukocytes have performed their function of killing the bacteria, they are rapidly withdrawn from the site of injury, infection or inflammation or undergo apoptosis and are ingested by macrophages. Lipoxins, resolvins, protectins and maresins are some of the lipid molecules that seem to have a significant role in the resolution of inflammation.

It should be noted that bacterial killing could also occur by oxygen-independent mechanisms. For instance, hitherto it is believed that neutrophils kill ingested microorganisms by releasing high concentrations of ROS and bringing about myeloperoxidase-catalyzed halogenation as described above. However, in knockout mice lacking the neutral proteases cathepsin G and elastase, these ROS do not kill microbes despite normal production of oxygen free radicals and halogenation. The passage of electrons is electrogenic and the charge generated across the wall of the phagocytic vacuole must be compensated if electron transport is to continue. This compensation is largely accomplished by the passage of Cl-, which enters the vacuole from the granules, where it is present at a concentration of about 500 mM, into the cytosol. The pH of the vacuole is regulated by a Na^+/H^+ exchanger, NHE1, which pumps Na+ out of the vacuole in exchange for cytosolic H^+ together with a flux of K^+ into the vacuole through the BKCa channel. These ion fluxes and pH changes serve to promote microbial killing and digestion by optimizing conditions for the action of the enzymes released from the cytoplasmic granules. Thus, it was shown that mice made deficient in neutrophil-granule proteases but normal in respect of ROS production and iodinating capacity are unable to resist staphylococcal and candidal infections [88–90]. It was noted that activation of neutrophils provokes the influx of high amounts of ROS into the endocytic vacuole that results in an accumulation of anionic charge that is compensated by a surge of K^+ ions. These K^+ ions cross the membrane in a pH-dependent manner inducing a steep rise in ionic strength that results in the release of cationic granule proteins, including elastase and cathepsin G. It is the release of these proteases that is primarily responsible for the destruction of bacteria. Thus, there appears to be a close relationship between ROS and the release of proteases, and bactericidal action of neutrophils. But, it looks as though; proteases are primarily responsible for bactericidal action but not ROS themselves. These observations have important clinical implications since, the relative importance of MPO and NADPH oxidase generated ROS in fight against various infections is a contentious issue. It was demonstrated that mice that have no MPO activity in their neutrophils and monocytes developed normally, were fertile, and showed normal clearance of *Staphylococcus aureus*. However, these animals showed increased susceptibility to *Candida albicans* infection [91]. Furthermore, lack of MPO significantly enhanced the dissemination of *Candida albicans* into various organs, suggesting that MPO is important for early host defense against fungal infections. In contrast, both MPO ($MPO^{-/-}$) and NADPH oxidase deficient (X-linked chronic granulomatous disease, X-CGD) mice were found to be susceptible to pulmonary infections with *Candida albicans* and *Aspergillus fumigatus* compared with normal mice, and the X-CGD mice exhibited shorter survival than $MPO^{-/-}$ mice [92]. This increased mortality in the X-CGD mice was associated with a 10- to 100-fold increased outgrowth of the fungi in their organs. These results suggest that $O_2^{\cdot-}$ produced by NADPH oxidase is more important than HOCL produced by MPO against pulmonary infection with those fungi. At the highest dose of *Candida albicans*, the mortality of $MPO^{-/-}$ mice was comparable to X-CGD mice, but was the same as normal mice at the lowest dose [93]. At the middle dose, the number of fungi disseminated into various organs of the $MPO^{-/-}$ mice was comparable to the X-CGD mice in 1 week after infection, but it was significantly lower in 2 weeks. These results suggest that both MPO and NADPH oxidase are equally important for early host defense against large inocula of *Candida albicans*. Hereditary MPO deficiency is common that has an estimated incidence of 1 in 2,000 in the United States. The results of the studies discussed above [91–93] suggest that MPO-deficient individuals could exhibit similar problems as CGD patients if exposed to a large amount of fungi/microorganisms [91–95]. It is likely that MPO deficient diabetics are more susceptible to fungal infections, if the dose of inocula is small, compared to normal.

In this context, it is important to note that TNF-α and lymphotoxin-α (LT), which are members of the TNF family, play crucial roles in the defense against infection with *Candida albicans* [96]. The TNF- and LT-deficient animals had a significantly increased mortality following *C. albicans* infection compared with control mice, and this was due to a 10- to 1,000-fold increased outgrowth of the yeast in their organs. No differences between $TNF^{-/-} - LT^{-/-}$ mice and $TNF^{+/+} + LT^{+/+}$ were observed when mice were rendered neutropenic, suggesting that activation of neutrophils mediates the beneficial effects of endogenous TNF and LT. A dramatic delay in the neutrophil recruitment at the sites of Candida infection in the $TNF^{-/-} - LT^{-/-}$ mice was noted and the neutrophils of deficient animals were less potent to phagocytize Candida blastospores than control neutrophils. In contrast, the killing of Candida and the oxygen radical production did not differ between neutrophils of $TNF^{-/-} - LT^{-/-}$ and $TNF^{+/+} + LT^{+/+}$ mice. Peak circulating IL-6 was significantly higher in $TNF^{-/-} - LT^{-/-}$ mice during infection. Peritoneal macrophages of $TNF^{-/-} - LT^{-/-}$ mice did not produce TNF, and synthesized significantly lower amounts of IL-1α, IL-1β, IL-6, and macrophage-inflammatory protein-1α than macrophages of $TNF^{+/+} + LT^{+/+}$ animals did. These results suggest that endogenous TNF and/or LT contribute to host resistance to disseminated candidiasis, and are essential for the recruitment of neutrophils and phagocytosis of *C. albicans* [97].

The rHuIL-1-α increased the release of lysozyme, beta-glucuronidase and myeloperoxidase while rHuTNF-α only increased lysozyme release [98]. Human neutrophils when exposed to recombinant human TNF alpha (rTNF-α) or rTNF-β generated HOCl (especially when incubated with FMLP) that was rapid, with 80% of total HOCl accumulation occurring within 15 min after FMLP addition. Comparison of HOCl generation with superoxide anion and myeloperoxidase release showed that the amount of HOCl generated was limited primarily by the amount of myeloperoxidase released rather than by the degree of respiratory burst activation. These results indicate that human neutrophils stimulated with FMLP after a brief incubation with rTNF-α or rTNF-β generate cytotoxic and microbicidal concentrations of chlorinated oxidants [99]. Thus, there is a close interaction and relationship between TNF-α and other cytokines and their ability to induce the generation of superoxide anion, activate myeloperoxidase and HOCL generation in leukocytes and host resistance to infections due to bacteria and fungi.

CD40 is a member of the TNF-receptor superfamily. This receptor has been found to be essential in mediating a broad variety of immune and inflammatory responses including T cell-dependent immunoglobulin class switching, memory B cell development, and germinal center formation. AT-hook transcription factor AKNA is reported to coordinately regulate the expression of this receptor and its ligand, which may be important for homotypic cell interactions. Adaptor protein TNFR2 interacts with this receptor and serves as a mediator of the signal transduction. In the macrophage, the primary signal for activation is IFN-γ from Th1 type CD4 T cells. The secondary signal is CD40L on the T cell which binds CD40 on the macrophage cell surface. As a result, the macrophage expresses more CD40 and TNF receptors on its surface which helps increase the level of activation that leads to the induction of potent microbicidal substances in the macrophage, including reactive oxygen species and nitric oxide, leading to the destruction of ingested microbe. Thus, CD40L interaction with CD40 is required for normal cellular immune responses such as T cell-mediated activation of monocytes/macrophages, proinflammatory cytokine production, and leukocyte extravasation. The $CD40L^{-/-}$ mice had a significantly increased yeast load in the kidneys compared to CD40L$^{+/+}$ mice late during infection. Similar effects were observed in CD40L$^{+/+}$ mice in which CD40 ligation was blocked by a neutralizing anti-CD40 antibody. In addition, the peak TNF-α plasma concentrations, *C. albicans*-stimulated production of NO by peritoneal macrophages were significantly lower in the CD40L$^{-/-}$ mice than in CD40L$^{+/+}$ mice. These results suggest that absence of CD40/CD40L interactions results in increased susceptibility to disseminated infection with *C. albicans* through decreased NO-dependent killing of Candida by macrophages [100].

Mediators of Inflammation

There are many chemical mediators of inflammation. Although the exact function and the source of some of the chemical mediators are not very clear, certain generalizations are possible. It is also likely that there could be some as yet unidentified chemical mediators or inhibitors of inflammation. Some of the important mediators of inflammation include: histamine, serotonin, lysosomal enzymes, eicosanoids (such as prostaglandins, leukotrienes and thromboxanes), platelet activating factors (PAFs), reactive oxygen species (ROS), NO, HOCL, myeloperoxidase, various cytokines, kinin system, coagulation/fibrinolysis system, and the complement system. Some of the general properties of the mediators of inflammation are given below.

Plasma-derived mediators such as complement proteins and kinins are present in plasma in precursor forms that must be activated by a series of proteolytic cleavages, to acquire their biologic properties. On the other hand, cell-derived mediators need to be secreted (e.g., histamine in mast cell granules) or are synthesized de novo (e.g., prostaglandins) in response to a given stimulus. The major cellular sources of these mediators are platelets, neutrophils, monocytes/macrophages, lymphocytes, and mast cells, but mesenchymal cells such as endothelium, smooth muscle, fibroblasts, and most epithelia can also be induced to elaborate some, if not all, of these mediators. The invading microorganisms trigger the production of most of these mediators or host derived products such as complement, kinins, etc., that are themselves activated by microbes or tissues under attack. These mediators, generally, bind to their specific receptors on target cells to produce their actions. In some instances, some of the mediators have direct enzymatic activity or induce the production of reactive oxygen species (ROS) or nitric oxide (NO) that, in turn, either mediate their actions or induce tissue damage. It is interesting to note that in majority of the instances, one mediator triggers the release of another mediator that acts on the target tissue. These secondary mediators either potentiate the action of the initial mediator or paradoxically abrogate its action. Thus, the ultimate degree and duration of inflammation depends on the balance between such pro- and anti-inflammatory mediators. In some instances, the anti-inflammatory chemicals or signals initiated may not only act on the target tissue but also on other tissues to suppress inflammation. Thus, both pro- and anti-inflammatory mediators may act on specific or diverse tissues. Once released or activated, most of the mediators are inactivated or decay quickly. For instance, eicosanoids have a short half-life, whereas specific or non-specific enzymes inactivate kinins. On the other hand, ROS and NO are scavenged by specific or non-specific antioxidants [2]. This suggests that under normal physiological conditions, there are both positive and negative checks and balances and when an imbalance sets in this well-balanced system pathological events occur.

Histamine, serotonin, bradykinin, complement system and coagulation cascade are well known for their involvement in infections, inflammatory process and sepsis and septic shock. A brief discussion about these molecules in inflammation and other conditions is given here.

Histamine

Histamine is a biogenic amine involved in local immune responses as well as regulating physiological function in the gut and acting as a neurotransmitter [101]. Histamine triggers the inflammatory response. As part of an immune response to foreign pathogens, histamine is produced by basophils and by mast cells found in nearby connective tissues. Histamine increases the permeability of the capillaries to leukocytes and proteins. It is found in virtually all animal body cells. Histamine has two basic centres, namely the aliphatic amino group and whichever nitrogen atom of the imidazole ring does not already have a proton. Under physiological conditions, the aliphatic amino group (having a pK_a around 9.4) will be protonated, whereas the second nitrogen of the imidazole ring ($pK_a \approx 5.8$) will not be protonated [102]. Thus, histamine is normally protonated to a singly-charged cation (see Fig. 3.3a). Histamine is derived from the decarboxylation of the amino acid histidine, a reaction

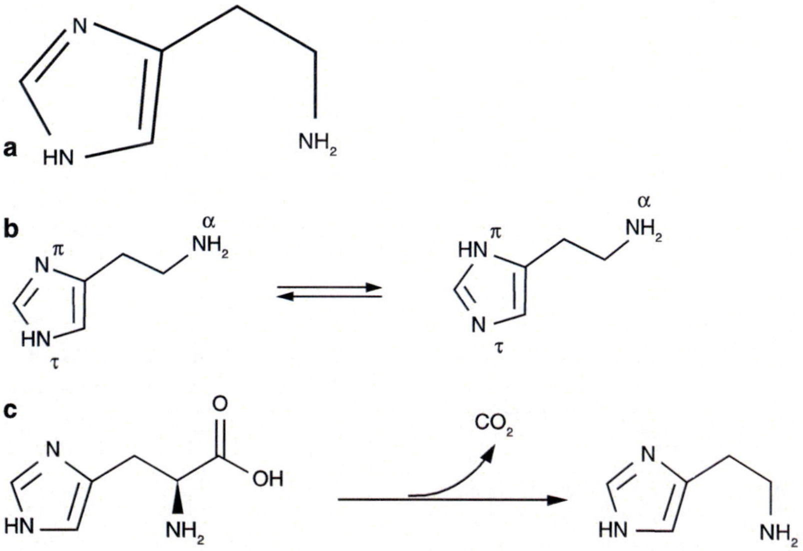

Fig. 3.3 **a** Structure of histamine. **b** Tautomers of histamine. **c** Formation of histamine from histidine by the action of histidine decarboxylase enzyme

catalyzed by the enzyme L-histidine decarboxylase (see Fig. 3.3a). It is a hydrophilic vasoactive amine. Once formed, histamine is either stored or rapidly inactivated. Histamine released into the synapses is broken down by acetaldehyde dehydrogenase. It is the deficiency of this enzyme that triggers an allergic reaction as histamines pool in the synapses. Histamine is broken down by histamine-N-methyltransferase and diamine oxidase. Some forms of foodborne disease, so-called "food poisonings," are due to conversion of histidine into histamine in spoiled food, such as fish. Fermented foods and beverages naturally contain histamine due to this same conversion.

Most histamine in the body is generated in granules in mast cells or basophils. Mast cells are especially numerous at sites of potential injury-the nose, mouth, and feet, internal body surfaces, and blood vessels. Non-mast cell histamine is found in several tissues, including the brain, where it functions as a neurotransmitter. Another important site of histamine storage and release is the enterochromaffin-like (ECL) cell of the stomach.

The most important mechanism by which histamine is released is immunologic. Cells sensitized by IgE antibodies attached to their membrane; degranulate when exposed to appropriate antigen, certain amines and alkaloids displace histamine granules and cause its release. Antibiotic like polymyxin can stimulate the release of histamine.

Histamine exerts its actions by combining with specific cellular histamine receptors. There are four histamine receptors designated as H_1 to H_4. These histamine receptors are located in specific tissues and seem to have distinct functions as well as given in Table 3.3.

Allergens bind to IgE-loaded mast cells in the nasal mucosa that leads to sneezing results from histamine-associated sensory neural stimulation; hypersecretion from

Table 3.3 Type, location and function of various histamine receptors found in the human body

Type	Location	Function
H_1 histamine receptor	Found on smooth muscle, endothelium, and central nervous system	Causes vasodilatation, bronchoconstriction, bronchial smooth muscle contraction, separation of endothelial cells (responsible for hives), and pain and itching due to insect stings; the primary receptors involved in allergic rhinitis symptoms and motion sickness; sleep regulation
H_2 histamine receptor	Located on parietal cells	Primarily stimulate gastric acid secretion
H_3 histamine receptor	Found on central nervous system and to a lesser extent peripheral nervous system	Decreased neurotransmitter release: histamine, acetylcholine, norepinephrine, serotonin
H_4 histamine receptor	Found primarily in the basophils and in the bone marrow. It is also found on thymus, small intestine, spleen, and colon	Plays a role in chemotaxis

glandular tissue; nasal mucosal congestion due to vascular engorgement associated with vasodilatation and increased capillary permeability [103].

Some of the actions of histamine, in addition to its role in inflammation, include: sleep regulation, sexual and erectile function and a possible role in schizophrenia.

Histamine is released as a neurotransmitter. The cell bodies of neurons that release histamine are found in the posterior hypothalamus, in various tuberomamillary nuclei from where the histaminergic neurons project throughout the brain, to the cortex through the medial forebrain bundle. Histaminergic action is known to modulate sleep. Classically, H_1 receptor antagonists produce sleep. Likewise, destruction of histamine releasing neurons, or inhibition of histamine synthesis leads to an inability to maintain vigilance. In contrast, H_3 receptor antagonists increase wakefulness.

It has been shown that histaminergic cells have the most wakefulness-related firing pattern of any neuronal type thus far tested. These neurons fire rapidly during waking, fire more slowly during periods of relaxation/tiredness and completely stop firing during REM and NREM (non-REM) sleep. Histaminergic cells can be recorded firing just before an animal shows signs of waking.

While histamine has stimulatory effects upon neurons, it also has suppressive actions that protect against the susceptibility to convulsion, drug sensitization, denervation supersensitivity, ischemic lesions and stress [104]. It has also been suspected that histamine may have a role in memory [105].

Histamine H_2 receptor antagonists are known to cause libido loss and erectile failure [106], while injection of histmine into the corpus cavernosum in men with psychogenic impotence produces full or partial erections in 74% of them [107]. It has been suggested that H_2 receptor antagonists may reduce testosterone uptake [106] and thus, cause sexual difficulties.

Histamine metabolites are increased in the cerebrospinal fluid of patients with schizophrenia while H_1 receptor binding sites are decreased. Furthermore, many antipsychotics increase histamine turnover [108].

Serotonin

Serotonin or 5-hydroxytryptamine (5HT) is a monoamine neurotransmitter that is primarily found in the gastrointestinal tract, platelets, and central nervous system. Approximately 80% of the human body's total serotonin is located in the enterochromaffin cells in the gut, where it is used to regulate intestinal movements [109]. The reminder is synthesized in serotonergic neurons in the CNS where it has various functions, including regulation of mood, appetite, sleep, muscle contraction, and cognitive functions including memory and learning (see Fig. 3.4a and e for the structure and metabolism of serotonin). Modulation of serotonin at synapse is thought to be a major action of several classes of pharmacological antidepressants. Serotonin secreted from the enterochromaffin cells eventually finds its way out into the blood. There, it is actively taken up by platelets, which store it. When the platelets bind to clot, they disgorge serotonin, where it serves as a vasoconstrictor and helps to regulate hemostasis, blood clotting and participate in inflammation. Serotonin also is a

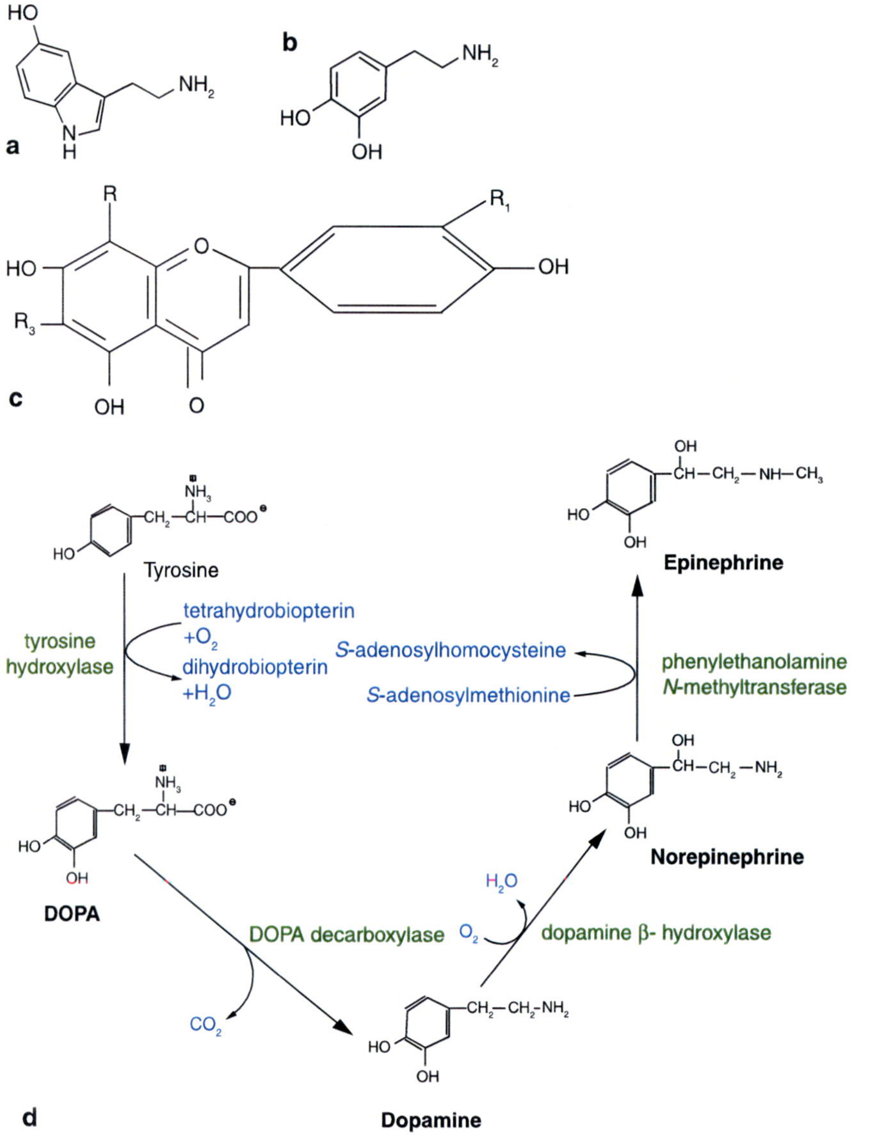

Fig. 3.4 **a** Structure of serotonin. **b** Structure of dopamine. **c** Structure of acetylcholine. **d** Structures of epinephrine and nor-epinephrine (adrenaline and nor-adrenaline respectively) and their formation from tyrosine. **e** Metabolism of serotonin

growth factor, which explains its role in wound healing. Serotonin is metabolized to 5-HIAA by the liver, and excreted by the kidneys. Serotonin is also found in fungi and plants. Serotonin's presence in insect venoms and plant spines serves to cause pain, which is a side effect of serotonin injection. Serotonin is produced by pathogenic

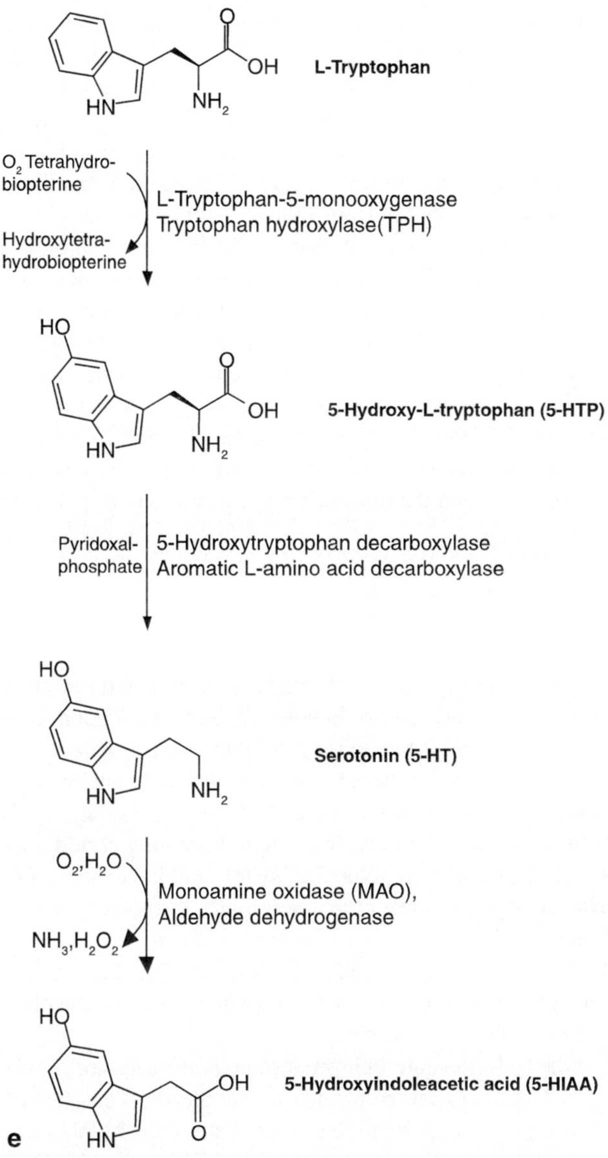

Fig. 3.4 (continued)

amoebas that could be responsible for intestinal inflammation and diarrhea seen in acute and chronic amoebiasis.

Serotonin functions as a neurotransmitter in the nervous systems of simple as well as complex animals such as *C. elegans.* serotonin is released as a signal in response to positive events, i.e., finding a new grazing ground or in male animals finding a hermaphrodite to mate. When a well-fed worm feels bacteria on its cuticle, dopamine is released, which slows it down; if it is starved, serotonin also is released, which slows the animal down further. This mechanism increases the amount of time animals spend in the presence of food [110]. The released serotonin activates the muscles used for feeding, while octopamine suppresses them [111]. Serotonin diffuses to serotonin-sensitive neurons, which control the animal's perception of nutrient availability. This system has been partially conserved during the 700 million years of evolution which separate *C. elegans* from humans. When humans smell food, dopamine is released to increase the appetite. But unlike in worms, serotonin does not increase anticipatory behaviour in humans; instead the serotonin released while consuming activates 5-HT2C receptors on dopamine-producing cells. This halts their dopamine release, and thereby serotonin decreases appetite. Drugs which block 5-HT2C receptors make the body unable to shut off appetite, and are associated with increased weight gain [112], especially in people who have a low number of receptors [113]. The expression of 5-HT2C receptors in the hippocampus follows a diurnal rhythm, just as the serotonin release in the ventromedial nucleus, which is characterized by a peak at morning when the motivation to eat is strongest [114].

Effects of Food Content

In humans, serotonin levels are affected by diet. An increase in the ratio of tryptophan to phenylalanine and leucine will increase serotonin levels. Fruits with a good ratio include dates, papaya and banana. Foods with a lower ratio inhibit the production of serotonin. These include whole wheat and rye bread. Eating a diet rich in whole grain carbohydrates and low in protein will increase serotonin by secreting insulin, which helps in amino acid competition. However, increasing insulin for a long period may trigger the onset of insulin resistance, obesity, type 2 diabetes, and lower serotonin levels. Myo-inositol, a carbocyclic polyol present in many foods, is known to play a role in serotonin modulation.

The gut is surrounded by enterochromaffin cells which release serotonin in response to food in the lumen. This makes the gut contract around the food. Platelets in the veins draining the gut collect excess serotonin.

If irritants are present in the food the enterochromaffin cells release more serotonin to make the gut move faster, i.e., to cause diarrhea so that the gut is emptied of the noxious substance. If serotonin is released in the blood faster than the platelets can absorb it, the level of free serotonin in the blood is increased. This activates 5HT3 receptors in the chemoreceptor trigger zone that stimulate vomiting. The enterochromaffin cells not only react to bad food, they are also very sensitive to irradiation and cancer chemotherapy. Drugs that block 5HT3 are very effective in controlling the nausea and vomiting produced by cancer treatment [115].

Serotonin is not only involved in the perception of food availability, but also of social rank. When injected with serotonin, the animal behaves like a dominant animal, while octopamine causes subordinate behavior [116]. The effect of 5-HT_1 receptors predominates in subordinate animals while 5-HT_2 receptors predominate in dominants [117]. In humans, levels of 5-HT_{1A} receptor activation in the brain show negative correlation with aggression [118] and a mutation in the gene that codes for the 5-HT_{2A} receptor may double the risk of suicide for those with that genotype [119]. Most of the brain serotonin is not degraded after use, but is collected by serotonergic neurons by serotonin transporters on their cell surface. Studies have revealed that nearly 10% of total variance in anxiety-related personality depends on variations in the description of where, when and how many serotonin transporters the neurons should deploy [120], and the effect of this variation was found to interact with the environment in depression [121, 122].

Serotonin is necessary for normal male mating behavior [123, 124]. The serotonergic signaling used to adapt the worm's behaviour to fast changes in the environment affects insulin-like signaling and the TGF-β signaling pathway, which control long-term adaption.

In the fruitfly where insulin both regulates blood sugar and acts as a growth factor, serotonergic neurons regulate the adult body size by affecting insulin secretion [29, 30, 125, 126]. In humans, though insulin regulates blood sugar and IGF regulates growth, serotonin controls the release of both hormones so that serotonin suppresses insulin release from the beta cells in the pancreas [127], and exposure to SSRIs reduces fetal growth. Human serotonin can also act as a growth factor directly. Liver damage increases cellular expression of 5-HT2A and 5-HT2B receptors [128]. Serotonin present in the blood then stimulates cellular growth to repair liver damage [129]. 5HT2B receptors also activate osteoblasts, which build up bone [130]. However, serotonin also activates osteoclasts, which degrade bone [131].

Serotonin in addition evokes endothelial nitric oxide synthase activation and stimulates through a 5-HT1B receptor meditated mechanism the phosphorylation of p44/p42 mitogen-activated protein kinase activation in bovine aortic endothelial cell cultures [121]. In blood, serotonin is collected from plasma, by platelets which store it. It is thus active wherever platelets bind in damaged tissue, as a vasoconstrictor to stop bleeding, and also as a fibrocyte mitotic (growth factor), to aid healing [132].

Some serotonergic agonist drugs also cause fibrosis anywhere in the body, particularly the syndrome of retroperitoneal fibrosis, as well as cardiac valve fibrosis. Three groups of serotonergic drugs have been epidemiologically linked with these syndromes. They are the serotonergic vasoconstrictive anti-migraine drugs (ergotamine and methysergide) [133], the serotonergic appetite suppressant drugs (fenfluramine, chlorphentermine, and aminorex), and certain anti-parkinsonian dopaminergic agonists, which also stimulate serotonergic 5-HT_{2B} receptors. These include pergolide and cabergoline, but not the more dopamine-specific lisuride [134].

Genetically altered C. elegans that lack serotonin have an increased reproductive lifespan, may become obese, and sometimes present with arrested development at a dormant larval state [135, 136].

Serotonin in mammals is made by two different tryptophan hydroxylases (TPHs): TPH1 produces serotonin in the pineal gland and the enterochromaffin cells, while TPH2 produces it in the raphe nuclei and in the myenteric plexus. Genetically altered mice that lack TPH1 develop progressive loss of heart strength early on. They have pale skin and breathing difficulties, are easily tired, and eventually die of heart failure [137]. Genetically altered mice that lack TPH2 are normal when they are born. However, after 3 days they appear to be smaller and weaker, and have softer skin than their siblings. In a purebred strain 50% of the mutants died during the first 4 weeks, but in a mixed strain 90% survived. Normally the mother weans the litter for 3 weeks, but the mutant animals needed 5 weeks. After that they caught up in growth and had normal mortality rates. Subtle changes in the autonomic nervous system are present, but the most obvious difference from normal mice is the increased aggressiveness and impairment in maternal care of young [138]. Despite the blood-brain barrier, the loss of serotonin production in the brain is partially compensated by intestinal serotonin. The behavioral changes become greatly enhanced if one crosses TPH1- with TPH2-lacking mice and gets animals that lack TPH entirely [139].

In humans, defective signaling of serotonin in the brain may be the root cause of sudden infant death syndrome (SIDS). Genetically modified mice that produce low levels of serotonin suffered drops in heart rate and other symptoms of SIDS, and many of the animals died at an early age. Thus, low levels of serotonin in the brainstems, which control heartbeat and breathing, may cause sudden death [128]. Thus indicates that if serotonergic neurons are abnormal in infants, there is a risk of sudden infant death syndrome (SIDS) [140, 141].

Location of Serotonergic Neurons

The neurons of the raphe nuclei are the principal source of 5-HT release in the brain [142]. The raphe nuclei are neurons grouped into about nine pairs and distributed along the entire length of the brainstem, centered around the reticular formation [143]. Axons from the neurons of the raphe nuclei form a neurotransmitter system, reaching almost every part of the central nervous system. Axons of neurons in the lower raphe nuclei terminate in the cerebellum and spinal cord while the axons of the higher nuclei spread out in the entire brain.

Serotonin is released into the space between neurons, and diffuses over a relatively wide gap (>20 μm) to activate 5-HT receptors located on the dendrites, cell bodies and presynaptic terminals of adjacent neurons.

5-HT Receptors

5-HT receptors are located on the cell membrane of nerve cells and other cell types in animals and mediate the effects of serotonin as the endogenous ligand and of a broad range of pharmaceutical and hallucinogenic drugs. With the exception of the 5-HT$_3$ receptor, a ligand gated ion channel, all other 5-HT receptors are G protein coupled seven transmembrane receptors that activate an intracellular second messenger cascade [143].

Serotonergic action is terminated primarily via uptake of 5-HT from the synapse. This is through the specific monoamine transporter for 5-HT, SERT, on the presynaptic neuron. Various agents can inhibit 5-HT reuptake including MDMA, amphetamine, cocaine, extromethorphan, tricyclic antidepressants, and selective serotonin reuptake inhibitors (SSRIs). Monoamine transporter, PMAT, has been shown to have significant 5-HT clearance capacity. The PMAT also is believed to transport dopamine and norepinephrine.

Serotonylation

Serotonin can also signal through a nonreceptor mechanism called serotonylation. In this serotonin modifies proteins [144]. This process underlies serotonin effects upon platelet-forming cells (thrombocytes) in which it links to GTPases that then trigger the release of vesicle contents by exocytosis [145]. A similar process underlies the pancreatic release of insulin. The effects of serotonin upon vascular smooth muscle "tone" (this is the biological function from which serotonin originally got its name) depend upon the serotonylation of proteins involved in the contractile apparatus of muscle cells [145].

Biosynthesis of Serotonin

Serotonin is synthesized from the amino acid L-tryptophan by two enzymes: tryptophan hydroxylase (TPH) and amino acid decarboxylase (DDC). The TPH-mediated reaction is the rate-limiting step in the pathway. TPH has been shown to exist in two forms: TPH1, found in several tissues, and TPH2, which is a brain-specific isoform [146] (see Fig. 3.4a and e).

Serotonin taken orally does not pass into the serotonergic pathways of the central nervous system because it does not cross the blood-brain barrier. However, tryptophan and its metabolite 5-hydroxytryptophan (5-HTP), from which serotonin is synthesized, can cross the blood-brain barrier and may function as effective serotonergic agents. 5-hydroxyidoleacetic acid (5-HIAA), a metabolite of serotonin, is excreted in the urine and is sometimes along with serotonin produced in excess amounts by certain tumors, and hence their levels could be used as a marker of the presence of the tumor and in assessing their prognosis.

Drugs Targeting the 5-HT System

Several classes of drugs target the 5-HT system including some antidepressants, antipsychotics, anxiolytics, antiemetics, and anti-migraine drugs as well as the psychodelic drugs and empathogens.

The most prescribed drugs in many parts of the world are drugs which alter serotonin levels especially in the management of depression, generalized anxiety disorder and social phobia. The monoamine oxidase inhibitors (MAIOs) prevent the breakdown of monoamine neurotransmitters including serotonin, and therefore increase concentrations of the neurotransmitter in the brain. These drugs can decrease bone mass and increase the risk of osteoporosis. However, it is not yet clear whether it is due to SSRI action on peripheral serotonin production and or action in the gut or in the brain.

Serotonin Modulates Inflammation and Immune Response

Serotonin is not only a neurotransmitter but also has immunomodulatory functions and thus, may have a significant role in inflammation.

Administration of 5-hydroxytryptamine (serotonin) or its precursor, 5-hydroxy-L-tryptophan (5-HTPH), produced marked depression of T cell dependent, humoral, hemolytic, primary immune response in mice. Serotonin caused a marked reduction of the thymus weight [147]. Serotonin content decreased in ventral part of the anterior hypothalamus within 20 min after immunization of rats with sheep red blood cells [148] suggesting that hypothalamic serotonin content could influence immune response. Elevation of active serotonin level results in the inhibition of immune response and the nuclei raphe serotoninergic system inhibited the immune response via the hypothalamus-hypophysis-adrenals axis in experimental animals. This inhibitory action of serotonin on immune response is attributed to its ability to attenuate suppressor cell function [149]. Serotonin in a concentration range of 10^{-7}–10^{-3}M inhibited oxidative burst of human phagocytes and exerted a dose dependent inhibition of the myeloperoxidase activity. These results suggest that serotonin could affect the oxidative burst of phagocytes and decrease in the generation of reactive oxygen species [150]. Serotonin significantly inhibited the production of TNF and IL-12, whereas IL-10, NO and PGE_2 production were increased. These immunomodulatory effects of serotonin were mimicked by 5-HT(2) receptor agonist but were not abrogated by 5-HT(2) receptor antagonist, suggesting the implication of other 5-HT receptors. Inhibitors of cyclooxygenase and antibody to PGE_2 abrogated the inhibitory and stimulatory effect of serotonin on TNF and IL-10 production, respectively, whereas NO synthase inhibitor eliminated serotonin-stimulated IL-10 increase. Furthermore, PGE_2 significantly increased alveolar macrophage IL-10 and NO production. These results suggest that serotonin can alter the cytokine network through the production of PGE_2 [151]. It is known that serotoninergic receptors (5-HTR) are expressed by a broad range of inflammatory cell types, including dendritic cells (DCs). 5-HT induced oriented migration in immature but not in LPS-matured DCs via activation of 5-HTR_1 and 5-HTR_2 receptor subtypes. 5-HT increased migration of pulmonary DCs to draining lymph nodes in vivo. Serotonin enhanced the production of pro-inflammatory cytokine IL-6. 5-HT influenced chemokine release by human monocyte-derived DCs and 5-HT induced maturation of DCs and enabled them to secrete high amounts of IL-10 from low IL-12p70 secreting phenotype. Furthermore, 5-HT favored the outcome of a Th2 immune response both in vitro and in vivo [152]. These and other results suggest that 5-HT is a potent regulator of human dendritic cell function and immune response and has pro-inflammatory actions.

The ability of serotonin to enhance inflammatory reactions in the skin, lung and gastrointestinal tract are, in part, mediated by its action on mast cells. For instance, mouse bone marrow-derived mast cells (mBMMC) and human CD34(+)-derived MC (huMC) expressed mRNA for multiple 5-HT receptors. Though serotonin did not induce degranulation of mBMMC and huMC, it did induce mBMMC and huMC adherence to fibronectin; immature and mature mBMMC and huMC migration and their chemotaxis. 5-HT did induce accumulation of MC in the dermis of 5-HT(1A)R(+/+) mice, but not in 5-HT(1A) receptor knockout mouse[5-HT(1A)R(−/−)]. These results demonstrated that both mouse and human MC respond to 5-HT through the 5-HT(1A) receptor and 5-HT promotes inflammation by increasing MC at the site of tissue injury [153]. By virtue of its actions on immunocytes and neurotransmitter functions, serotonin is expected to have a significant role in inflammation.

Dopamine

Dopamine is a neurotransmitter (see Fig. 3.4a for the structure of dopamine and Fig. 3.4d for its formation from tyrosine) has cardiovascular properties and is used in patients with systemic inflammatory response syndrome (SIRS) to maintain hemodynamic stability. Polymorphonuclear leukocytes (PMNLs) isolated from healthy volunteers and patients with SIRS and treated with varying doses of dopamine and a dopamine D-1 receptor agonist and was assessed every 6 h revealed a significant delay in PMNL apoptosis in patients with SIRS compared with controls. Treatment of isolated PMNLs from both healthy controls and patients with SIRS with dopamine induced apoptosis. PMNL ingestive and cytocidal capacity were both decreased in patients with SIRS compared with controls and treatment with dopamine significantly increased phagocytic function [154]. These data demonstrate that dopamine induces PMNL apoptosis and modulates its function both in healthy controls and in patients with SIRS.

PMN and HUVEC (human umbilical vein endothelial cells)of healthy subjects stimulated with lipopolysaccharide (LPS) and TNF-α showed a significant increase in transendothelial migration and upregulation of CD11b/CD18 and upregulation of E-selectin/ICAM-1 expression compared with normal EC (endothelial cells) respectively. Dopamine decreased PMN transmigration, attenuated PMN CD11b/CD18 and the endothelial molecules E-selectin and ICAM-1 compared with stimulated PMN/EC that were not treated dopamine. The chemoattractant effect of IL-8 was also attenuated [155], suggesting that dopamine attenuates the interaction between PMN and the endothelium, and consequently, modulates PMN exudation and thus, may function as an anti-inflammatory molecule.

Infusion of dopamine in septic mice increased splenocyte apoptosis and decreased splenocyte proliferation and IL-2 release of septic mice without any effect on sepsis-induced changes in leukocyte distribution. An inhibitory effect of dopamine infusion on splenocyte proliferation and the release of the TH1-cytokines IL-2 and IFN-γ was reported in sham operated control mice. These effects corresponded to the decreased survival of dopamine-treated septic animals [156], indicating that dopamine modulates cellular immune functions. In this context, it is interesting to note that obese subjects have decreased dopamine receptors and decreased dopamine levels in the brain [157] and hence, are thought to have "reward deficiency syndrome". Since dopamine has anti-inflammatory actions, the decrease in the dopamine receptor number or content in the brain of subjects with obesity could trigger low-grade inflammation that may affect hypothalamus and eventually lead to hypothalamic dysfunction and the development of metabolic syndrome.

Catecholamines

In general, catecholamines are considered as hormones concerned with fright and flight (see Fig. 3.4d for structure and formation of catecholines). Sympathetic nervous system activation leads to enhanced production and release of catecholamines that leads to an elevation in blood pressure, blood glucose, tachycardia, increased sweating, and catabolism. In subjects with hypertension, increase in the activity of sympathetic nervous system has been noted and this could induce peripheral vascular resistance. It is interesting to note that catecholamines have also been observed to have immunomodulatory actions and pro-inflammatory properties.

Patients with stress hyperglycemia and type 2 diabetes mellitus have increase in noradrenaline and adrenaline and decrease in serotonin and its metabolites [158–160] in the brain and increased production and release of catecholamines from the phagocytes in the peripheral circulation. This assumes importance in the light of the observation that sympathetic activation is associated with metabolic syndrome and increased risk of cardiovascular disease. In a study of 104 type 2 diabetic patients (50 female and 54 male) and the diagnosis of metabolic syndrome based on the National Cholesterol Education Program Adult Treatment Panel III criteria, it was noted that blood concentrations of hs-CRP, IL-6 and plasminogen activator inhibitor-1 were higher in diabetic patients with than in those without metabolic syndrome. Both the 24-h mean LF (low frequency, both sympathetic and parasympathetic activities) and the low frequency to high frequency (LF-to-HF ratio) were also significantly higher in diabetic patients with than in those without metabolic syndrome. The LF-to-HF ratio at 6:00 a.m. was significantly higher in diabetic patients with a high CRP concentration (>3.0 mg/l) than in those with a low (<1.0 or = or) or moderate CRP ($\leq$3.0 mg/l) concentration. These results suggest that type 2 diabetic patients with metabolic syndrome have elevated markers of inflammation and evidence of cardiac sympathetic predominance [159].

It has been reported that CSF (cerebrospinal fluid) lymphocytes when cloned and examined showed that they contain catecholamines and catecholamines affect the proliferation and differentiation of lymphocyte populations in culture [161]. In this study, it was reported that the catecholamine concentrations in $CD4^+$ T cells, CSF lymphocytes and B cells corresponded to 1.7×10^{-4} M, 1.3×10^{-5} M, and 1.7×10^{-6} M respectively. These lymphocytes contained the complete system for the synthesis of catecholamines and showed a cellular uptake mechanism for catecholamines. Incubation of lymphocytes with L-Dopa and dopamine exerted a dose-dependent inhibition of both lymphocyte proliferation and differentiation. In fact, incubation with dopamine at concentrations from 10 to 500 μM completely abolished the production of antibodies by B cells. A dose-dependent inhibition of interferon-γ synthesis by T cells was also observed when incubated with dopamine. These evidences showing the presence of catecholamines in lymphocytes and the effect of L-dopa and dopamine on lymphocyte proliferation and differentiation, indicate that catecholamines produced by lymphocytes act in an autocrine or paracrine way and are important regulatory molecules [162] and, thus, are potentially important in an ongoing immune response. These results are supported by the findings that immune system cells carry β-adrenergic and dopaminergic receptors that are a prerequisite for subsequent interaction with catecholamines [161]. It is interesting to note that both IL-1 and TNF inhibited cardiac contractile responsiveness to β-adrenergic stimulation [163]. This is an important observation since, the primary endogenous mediators of the inotropic state of the heart are catecholamines, particularly in cardiac failure and hence, this immune cytokine mediated effect may contribute to reversible impairment of cardiac function in vivo.

In a study that evaluated whether phagocytes are capable of de novo production of catecholamines, suggesting an autocrine/paracrine self-regulatory mechanism by catecholamines during inflammation, as has been described for lymphocytes above, it was observed that exposure of phagocytes (PMNLs) to lipopolysaccharide led to a release of catecholamines and an induction of catecholamine-generating and degrading enzymes, indicating the presence of the complete intracellular machinery for the generation, release and inactivation of catecholamines [164]. In an extension of this study, it was also reported that blockade of α2-adrenoreceptors or catecholamine-generating enzymes greatly suppressed lung inflammation, whereas the opposite was the case either for an α2-adrenoreceptor agonist or for inhibition of catecholamine-degrading enzymes. This study emphasizes that fact that phagocytes are a source of catecholamines, which enhance the inflammatory response. In this context, it is noteworthy that mineralocorticoid receptors exist in neutrophils that exhibited anti-inflammatory effects that are at work when neutrophils interact with endothelial cells [165]. Furthermore, aldosterone abrogated NF-κB-mediated TNF-α production in IL-8- and GM-CSF-stimulated neutrophils.

Since adrenaline and noradrenaline have pro-inflammatory actions, one of the reasons for the existence of low-grad systemic inflammation in metabolic syndrome could be due to sympathetic over activity. Since normally a balance is maintained between sympathetic and parasympathetic nervous systems, it implies that in metabolic syndrome plasma or tissue and leukocyte acetylcholine (Ach) levels will be low that has anti-inflammatory action whereas the production and release of catecholamines and consequent sympathetic over activity will be present.

In addition, these evidences suggest that both lymphocytes and PMNLs contain the complete system for the synthesis, degradation and uptake of various hormones and neurotransmitters such as catecholamines, dopamine, mineralocorticoids, serotonin, dopamine, neuropeptide Y (NPY), ghrelin, melanocortin, acetylcholine, gut peptides such as glucagon-like peptide-1 (GLP-1) and gastric inhibitory peptide (GIP), leptin and cholecystokinin. Thus, it is perfectly feasible and practical to use peripheral PMNs and lymphocytes, in addition to macrophages and monocytes, to study the role of these various hormones, neurotransmitters and peptides in several clinical conditions.

Acetylcholine

Acetylcholine (Ach) (see Fig. 3.4c for the structure of Ach), the principal vagal neurotransmitter, suppresses inflammation and is termed as the "cholinergic anti-inflammatory pathway," and these neural signals transmitted via the vagus nerve that inhibits cytokine release act through a mechanism that requires the alpha7 subunit-containing nicotinic acetylcholine receptor (alpha7nAChR). Vagus nerve regulation of peripheral functions is controlled by brain nuclei and neural networks. Studies showed that brain acetylcholinesterase activity controls systemic and organ specific TNF production during endotoxemia. Peripheral administration of the acetylcholinesterase inhibitor galantamine significantly reduced serum TNF levels through vagus nerve signaling, and protected against lethality during murine endotoxemia. Administration of a centrally-acting muscarinic receptor antagonist abolished the suppression of TNF by galantamine, indicating that suppressing acetylcholinesterase activity, coupled with central muscarinic receptors, controls peripheral cytokine responses. Administration of galantamine to alpha7nAChR knockout mice failed to suppress TNF levels, indicating that the alpha7nAChR-mediated cholinergic anti-inflammatory pathway is required for the anti-inflammatory effect of galantamine. Thus, inhibition of brain acetylcholinesterase suppresses systemic inflammation through a central muscarinic receptor-mediated and vagal- and alpha7nAChR-dependent mechanism [166–168]. Ach modulates the production and actions of other hypothalamic monoamines serotonin, dopamine, and acetylcholine and peptides: NPY, BDNF, and melanocortins and thus, participates in the regulation of energy homeostasis.

Melanocortin

The proopiomelanocortin (POMC) gene is transcribed in several tissues including the corticotrophs of the anterior pituitary, neurons of the arcuate nucleus of the hypothalamus, and cells in the dermis and the lymphoid system. In these cells, POMC propeptide is processed posttranslationally to result in a series of small peptides. Thus, pituitary corticotrophs express prohormone convertase 1 (PC1) but not PC2, resulting in the production of N-terminal peptide, joining peptide, ACTH and lipotropin. Expression of PC2 in the hypothalamus leads to the production of α-, β-, and γ-MSH (melanocyte stimulating hormone) but not ACTH. The action of these melanocortin peptides is mediated by five G protein-coupled seven transmembrane domain receptors[melanocortin receptor type 1 (MC3R to MC5R)]. Both MC3R and MC4R are highly expressed in the central nervous system and play an important role in the control of food intake and energy balance. In particular, there are two distinct subsets of neurons in the arcuate nucleus of the hypothalamus that express MC3R and MC4R that together with their downstream target sites make up the central melanocortin system. POMC neurons produce the anorectic peptide α-MSH together with cocaine- and amphetamine-related transcript (CART), whereas a separate group expresses the orexigenic neuropeptide Y (NPY) and agouti-related protein (AgRP). AgRP is a potent MC3R and MC4R antagonist. Activation of the NPY/AgRP neurons increases food intake and decreases energy expenditure, whereas activation of POMC neurons decreases food intake and increases energy expenditure. The long isoform of the leptin receptor is highly expressed on the arcuate neurons, and leptin regulates these two neuronal populations in a reciprocal manner: suppressed levels of leptin after a fast decrease POMC mRNA and increase AgRP mRNA in the hypothalamus. From the arcuate nucleus, POMC and AgRP have extensive projections to several hypothalamic regions, including the lateral hypothalamus and the paraventricular nucleus. Cell bodies within the lateral hypothalamus contain the orexigenic peptide melanin concentrating hormone, and neurons of the paraventricular nucleus express TRH (thyrotropin releasing hormone). Thus, via this second order signaling, the melanocortin peptides exert their effects. In addition melanocortins have potent anti-inflammatory effects that are mediated by direct effects on cells of the immune system as well as indirectly by affecting the function of resident non-immune cells and suppress NF-κB activation, expression of adhesion molecules and chemokine receptors, production of pro-inflammatory cytokines and other mediators. Thus α-MSH modulates inflammatory cell proliferation, activity and migration [169, 170].

Leptin

Leptin is not only involved in the pathobiology of obesity and metabolic syndrome but also has pro-inflammatory actions. In inflammatory condition such as ankylosing spondylitis (AS), leptin, IL-6 and TNF-α mRNA expression of PMBCs (peripheral blood mononuclear cells) were significantly higher than controls. Similar significances were also found in the measurements for leptin and cytokine levels of supernatants, and leptin levels correlated well with IL-6 expression in these patients. Stimulation of PBMCs by exogenous leptin significantly increased the production of IL-6 and TNF-α in PBMCs from patients with AS in a dose-dependent fashion and these increases were much exacerbated compared to controls [171] implying its pro-inflammatory effect in the pathogenesis of ankylosing spondylitis.

These results are interesting in the light of the known fact that consumption of dietary fats is amongst the most important environmental factors leading to obesity. Both in rodents and humans, the consumption of fat-rich diets blunts leptin and insulin anorexigenic signaling in the hypothalamus by a mechanism dependent on the in situ activation of inflammation. It was reported that consumption of dietary fats induces apoptosis of neurons and a reduction of synaptic inputs in the arcuate nucleus and lateral hypothalamus. This effect is dependent upon diet composition, and not on caloric intake. The presence of an intact TLR4 receptor protects cells from further apoptotic signals. In diet-induced inflammation of the hypothalamus, activation of pro-inflammatory pathways occurs that play a central role in the development of resistance to leptin and insulin [172]. The increase in the concentrations of leptin in response to high fat diet [173] may aggravate inflammation that, in turn, induces apoptosis of hypothalamic nuclei leading to the initiation and progression of the metabolic syndrome. Insulin resistance and hyperinsulinemia seen in obesity and other related conditions may, in fact, be a protective event since insulin has anti-inflammatory actions. Thus, hyperinsulinemia is beneficial though its presence implies the beginning of the metabolic syndrome. Both hyperleptinemia and hyperinsulinemia lead to reduced sympathetic activity [173] that also contributes to the pathophysiology of obesity and development of metabolic syndrome.

Neuropeptide Y

Neuropeptide Y (NPY) is a classical sympathetic co-mediator that has been shown to regulate immunological functions including T cell activation and migration of blood leukocytes. Leukocytes expressed high amounts of NPY mRNA and peptide, similar to expression levels in sympathetic ganglia. During acute allograft rejection, leukocytic NPY expression drastically dropped to approximately 1% of control levels suggesting that it modulates immune response and inflammation [174]. It was reported that NPY and NPY-related receptor specific peptides reduced granulocyte accumulation into carrageenan-induced air pouch (an experimental model of inflammation) attenuated phagocytosis attained via Y1 receptor, decreased peroxide production, albeit mediated via Y2 and Y5 receptors activation and increased nitric oxide production via Y1 receptor [175]. These results emphasize the fact that NPY has anti-inflammatory actions. It is important to note that NPY-induced modulation of the immune and inflammatory responses is regulated by tissue-specific expression of different receptor subtypes (Y1–Y6) and the activity of the enzyme dipeptidyl peptidase 4 (DP4, CD26) which terminates the action of NPY on Y1 receptor subtype. It is noteworthy that NPY suppressed paw edema in adult and aged, but not in young rats. Furthermore, plasma DP4 activity decreased, while macrophage DP4 activity, as well as macrophage CD26 expression increased with aging. Further studies showed that anti-inflammatory effect of NPY is mediated via Y1 and Y5 receptors. In contrast to the in vivo situation, NPY stimulated macrophage nitric oxide production in vitro only in young rats, and this effect was mediated via Y1 and Y2 receptors. Thus, age-dependant modulation of inflammatory reactions by NPY is determined by plasma, but not macrophage DP4 activity at different ages [176]. It is known that with age the production of TNF-α and IL-6 increase, appetite decreases and the tendency develop metabolic syndrome is increased. Thus, the decrease in DP4 activity with age could be compensatory phenomena to counteract age-associated pro-inflammatory process. But, unfortunately NPY is an orexigenic peptide and thus, may overcompensate age-associated decline in appetite and paradoxically promote the development of metabolic syndrome.

In an animal model of colitis, it was noted that there was an increase in enteric neuronal NPY and nNOS expression in WT (wild-type) mice. WT mice showed more inflammation compared to NPY$^{(-/-)}$ as indicated by higher clinical and histological scores, and myeloperoxidase (MPO) activity. WT mice had increased nitrite, decreased glutathione (GSH) levels and increased catalase activity indicating more oxidative stress. The lower histological scores, myeloperoxidase (MPO) and chemokine KC in dextran sodium sulphate (3% DSS) or streptomycin pre-treated Salmonella typhimurium-treated nNOS$^{(-/-)}$ and NPY$^{(-/-)}$/nNOS$^{(-/-)}$ mice support the contention that loss of NPY-induced nNOS attenuated inflammation. NPY-treated rat enteric neurons in vitro exhibited increased nitrite and TNF-α production [177]. These results indicate that NPY mediated increase in nNOS is a determinant of oxidative stress and subsequent inflammation. These results emphasize the close interaction between NPY, NOS and pro-inflammatory cytokine TNF-α and their modulatory influence on inflammation and metabolic syndrome.

Gastrin-releasing peptide (GRP, 10^{-10} M), NPY (10^{-10} M), somatostatin (10^{-10} M) and vasoactive intestinal peptide (VIP, 10^{-9} M) modulate the production of IL-1β, IL-6 and TNF-α by peripheral whole blood cells from healthy young and old people. GRP, NPY, somatostatin and VIP stimulated the production of IL-1β in old subjects, and NPY, somatostatin and VIP in young ones. The production of IL-6 was enhanced by GRP, NPY and VIP in young and old people. The TNF-α production was stimulated by NPY and somatostatin in young subjects, and by NPY, somatostatin and VIP in old ones, whereas GRP produced a decrease of TNF-α in young persons. GRP in old subjects and VIP in young and old subjects stimulated LPS-induced IL-6 production by whole blood cells. On the contrary, GRP and VIP inhibited LPS-induced TNF-α production in young controls [178]. Thus, neuropeptides modulate the production of pro-inflammatory cytokines by peripheral blood cells at physiological concentrations emphasizing the close relationship between appetite and food intake regulating neuropeptides and inflammation. Paradoxically, it was reported that cytokines IL-1β, IL-6, and TNF-α did not alter either basal or stimulated NPY release from the hypothalamic slices [179] suggesting that, at least, in some instances of anorexia such as cancer cachexia wherein the concentrations of these cytokines are increased, anorexia is not due to their effect on NPY levels. It is important to note that NPY is present in human adipose tissue, insulin increases NPY secretion, and adipocyte treatment with rh-NPY downregulated leptin secretion but had no effect on adiponectin and TNF-α secretion [180]. This implies that the anti-lipolytic action of NPY promotes an increase in adipocyte size in hyperinsulinemic conditions and adipocyte-derived NPY mediates reduction of leptin secretion that may have implications for central feedback of adiposity signals.

Ghrelin

Ghrelin is an orexigenic peptide produced by the gut. Response to ghrelin involves specific inhibition of fatty acid biosynthesis induced by AMP-activated protein kinase (AMPK) resulting in decreased hypothalamic levels of malonyl-CoA and increased carnitine palmitoyltransferase 1 (CPT1) activity. In addition, fasting downregulates fatty acid synthase (FAS) in a region-specific manner and that this effect is mediated by an AMPK and ghrelin-dependent mechanisms. Thus, decreasing AMPK activity in the ventromedial nucleus of the hypothalamus (VMH) is sufficient to inhibit ghrelin's effects on FAS expression and feeding. Modulation of hypothalamic fatty acid metabolism specifically in the VMH in response to ghrelin is a physiological mechanism that controls feeding [181]. In addition, ghrelin that acts as an endogenous ligand for the growth hormone secretagogue receptor (GHS-R), are expressed in human T lymphocytes and monocytes, where ghrelin acts via GHS-R to specifically inhibit the expression of proinflammatory anorectic cytokines such as IL-1β, IL-6, and TNF-α. Ghrelin inhibited while leptin upregulated GHS-R expression on human T lymphocytes suggesting a reciprocal regulatory network by which ghrelin and leptin control immune cell activation and inflammation. Ghrelin also exerts potent anti-inflammatory effects and attenuated endotoxin-induced anorexia in a murine endotoxemia model [182]. Thus, ghrelin functions as a key signal, coupling the metabolic axis to the immune system.

It was reported that pretreatment of phagocytic leukocytes with a GHS-R antagonist,[D-Lys3]-GHRP-6, abolished the stimulatory effects of trout ghrelin and des-VRQ-trout ghrelin on superoxide production. Ghrelin increased mRNA levels of superoxide dismutase and GH expressed in trout phagocytic leukocytes. Immunoneutralization of GH by addition of anti-salmon GH serum to the medium blocked the stimulatory effect of ghrelin on superoxide production [183], suggesting that ghrelin stimulates phagocytosis in fish leukocytes through a GHS-R-dependent pathway, and also that the effect of ghrelin is mediated, at least in part, by GH secreted by leukocytes. Furthermore when the serum levels of ghrelin and its relationship with disease activity and nutritional status were evaluated in patients with inflammatory bowel disease (IBD), it was noted that serum ghrelin levels were significantly higher in patients with active ulcerative colitis and Crohn's disease than in those in remission (108 ± 11 pg/ml vs. 71 ± 13 pg/ml for ulcerative colitis patients, $P < 0.001$; 110 ± 10 pg/ml vs. 75 ± 15 pg/ml for Crohn's disease patients, $P < 0.001$). Circulating ghrelin levels in patients with these two diseases were positively correlated with sedimentation, fibrinogen and CRP and were negatively correlated with IGF-1, BMI, fat mass (%), and fat free mass (%). These results indicate that ghrelin level may be important in determination of the activity in IBD patients and evaluation of nutritional status [184].

Ghrelin and GH secretagogue receptor 1b were expressed in PBMCs (peripheral blood mononuclear cells) of subjects with metabolic syndrome. Ghrelin gene expression correlated positively with the expressions of TNF-α ($P < 0.001$), IL-1β ($P < 0.001$) and IL-6 ($P = 0.026$), but was not associated with the plasma ghrelin concentration. At baseline, the plasma ghrelin levels were associated with fasting serum insulin concentrations, insulin sensitivity index and high-density lipoprotein cholesterol. Weight, BMI or waist circumference and acute insulin response in intravenous glucose tolerance test were found to best strong predictors of the ghrelin concentration. These results indicate an autocrine role for ghrelin within an immune microenvironment in view of its expression in PBMCs [185]. Thus, ghrelin expression in PBMCs could be used as a marker of low-grade systemic inflammation seen in metabolic syndrome along with plasma TNF-α and IL-6.

It is noteworthy that TNF-α is a glycoprotein hormone with important functions in inflammation and apoptosis, serves as a pro-inflammatory cytokine in the defense against viral, bacterial and parasitic infections and autoimmune disorders, and influences energy homeostasis and has an anorexigenic effect on the hypothalamus. TNF-α is also involved in the pathogenesis of psychiatric disorders such as depression or narcolepsy. On the other hand, ghrelin is a peptide hormone which primarily regulates eating behavior through modulation of expression of orexigenic peptides in the hypothalamus. Ghrelin administration increases food intake and body weight, while weight loss in turn increases ghrelin levels. Ghrelin possesses anti-inflammatory properties and has antidepressant and anxiolytic properties. Therefore, it is suggested that TNF-α and ghrelin seem to have opposite effects regarding the hypothalamic regulation of eating behavior, modulation of the immune response and the state of mental health. In a similar fashion, hypothalamic monoamines serotonin, dopamine, and acetylcholine and peptides such as NPY, BDNF, and melanocortins not only modulate eating behavior but also participate in the regulation of immune response and inflammation.

Gut Peptides

Incretins are a group of gastrointestinal hormones that cause an increase in insulin release from pancreatic β cells after eating, even before blood glucose levels become elevated. They also slow the rate of absorption of nutrients into the blood stream by reducing gastric emptying and may directly reduce food intake. Incretins inhibit glucagon release from the α cells of the Islets of Langerhans. The two main candidate molecules that fulfill criteria for an incretin are glucagon-like peptide-1 (GLP-1) and gastric inhibitory peptide (GIP). Both GLP-1 and GIP are rapidly inactivated by the enzyme dipeptidyl peptidase-4 (DPP-4).

GLP-1 is derived from the transcription product of the proglucagon gene and its major source is the intestinal L cell. GLP-1 has a half life of less than 2 min, due to rapid degradation by the enzyme dipeptidyl pepidase-4. The known physiological functions of GLP-1 include:

- increases insulin secretion from the pancreas in a glucose-dependent manner
- decreases glucagon secretion from the pancreas
- increases β cell mass and insulin gene expression
- inhibits acid secretion and gastric emptying in the stomach
- decreases food intake by increasing satiety
- promote insulin sensitivity

In addition to its role in the regulation of insulin secretion, GLP-1 also has immunomodulatory and anti-inflammatory actions.

GLP-1 binding and GLP-1 receptor mRNA expression is detected in both astrocytes and microglia. GLP-1 treatment induced morphological changes in microglia from the ramified type to the amoeboid type, suggesting an increase in the production and release of endogenous GLP-1. GLP-1 prevented the LPS-induced IL-1β mRNA expression, increased cAMP concentration and cAMP response element-binding protein phosphorylation in astrocytes indicating that it is a modulator of inflammation in the central nervous system [186].

Pro-inflammatory cytokines IL-1β, IFN-γ, and TNF-α inhibited the proliferation of pancreatic β cells in vitro through the extracellular signal-regulated kinase 1/2 (ERK1/2) activation, the signaling pathway involved in β cell replication. GLP-1 completely reversed the cytokine-induced inhibition of ERK phosphorylation and increased β cell proliferation threefold in cytokine-treated cultures. While pro-inflammatory cytokines reduced islet cell ERK1/2 activation and β cell proliferation in pancreatic islet culture, GLP-1 was capable of reversing this effect [187], suggesting that GLP-1 not only has anti-inflammatory actions but is also capable of preventing the loss of pancreatic β cells and may, in fact, enhance their proliferation and thus, preserve insulin secreting ability of β cells.

In addition, inhibition of DPP-4 that increases the circulating levels of incretins GLP-1 and GIP has been shown to preserve islet mass in rodent models of type 1 diabetes. DPP-4 inhibitor, sitagliptin, treatment of NOD mice before and after islet transplantation resulted in prolongation of islet graft survival by decreasing insulitis and reducing migration of isolated splenic CD4+ T-cells, possibly, by the activation of protein kinase A and Rac1. These results indicate that both GLP-1 and GIP enhance graft survival through a pathway involving cAMP/PKA/Rac1 activation [188] and thus shown immunosuppressive and anti-inflammatory properties.

Cholecystokinin

The autonomic nervous system plays an important role in sensing luminal contents in the gut by way of hard-wired connections and chemical messengers, such as cholecystokinin (CCK). Ingestion of dietary fat stimulates CCK receptors, and leads to attenuation of the inflammatory response by way of the efferent vagus nerve and nicotinic receptors. Vagotomy and administration of antagonists for CCK and nicotinic receptors significantly blunted the inhibitory effect of high-fat enteral nutrition on hemorrhagic shock-induced TNF-α and IL-6 release. Furthermore, the protective effect of high-fat enteral nutrition on inflammation-induced intestinal permeability was abrogated by vagotomy and administration of antagonists for CCK and nicotinic receptors, suggesting that there exists a neuroimmunologic pathway, controlled by nutrition [189]. This anti-inflammatory action of CCK could be a self-defense protective pathway developed in order to prevent inflammation that occurs due to the consumption of high fat diet.

Thus there appears to be impressive pro- and anti-inflammatory actions exhibited by various hypothalamic monoaminergic and peptide molecules and those produced

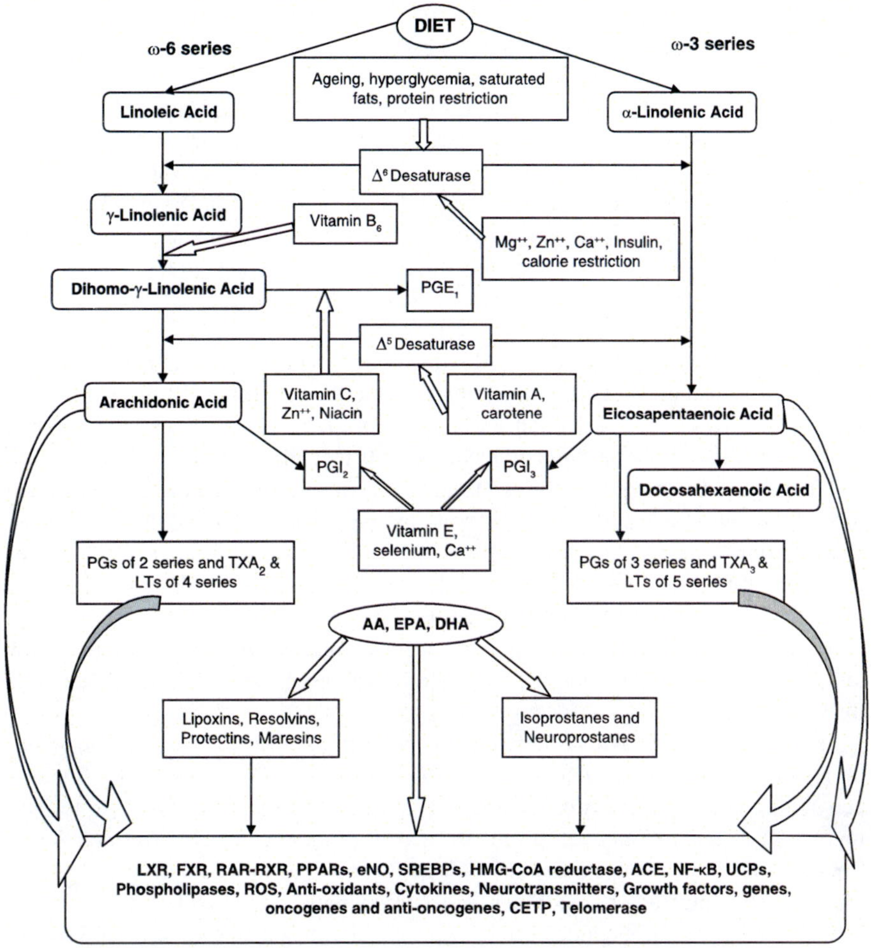

Fig. 3.5 Scheme showing the metabolism of essential fatty acids and their actions and factors that influence the formation of their products. Resolvins are formed from AA, EPA, and DHA that have anti-inflammatory action and inhibit leukocyte migration. LXs and resolvins reduce inflammation. PGE_2, PGE_3, $PGF_{2\alpha}$, $PGF_{3\alpha}$, LTA_4, and LTA_5 are proinflammatory in nature. PGE_1 appears to have anti-inflammatory action. TXA_2 and TXA_3 are platelet aggregators and vasoconstrictors, whereas PGI_2 and PGI_3 are potent platelet anti-aggregators and vasodilators. For details see the text

by the gut that not only participate in the regulation of appetite, satiety and food intake but also in the modulation of immune response (see Fig. 3.5). Based on these findings, it is no wonder that obesity, insulin resistance, hypertension, dyslipidemia and metabolic syndrome are low-grade systemic inflammatory conditions.

It is evident from the preceding discussion that neurotransmitters, hypothalamic peptides, and gut hormones have immunomodulatory and ability to influence the pathobiology of inflammation. These results may explain the involvement of immunocytes in the pathogenesis of neurological conditions such asby the gut that depression, schizophrenia, Alzheimer's disease, and other similar conditions. Several other low-grade systemic inflammatory conditions such as obesity, atherosclerosis, hypertension, type 2 diabetes mellitus, insulin resistance, dyslipidemia and metabolic syndrome and cancer are also associated with alterations in the function of leukocytes, macrophages, lymphocytes, gut hormones, neurotransmitters and hypothalamic peptides. These evidences that are discussed in the subsequent chapters emphasize the relationship between inflammation, hypothalamus, gut, environment and genes. The fact that alterations in immune response and inflammatory events (mainly in the form of low-grade systemic inflammation as evidenced by an increase in the plasma or tissue levels of high sensitive-C-reactive protein, hs-CRP; IL-6 and TNF-α) are common in several diseases such as obesity, insulin resistance, type 2 diabetes mellitus, hypertension, osteoporosis, atherosclerosis, metabolic syndrome, dyslipidemia, cancer, schizophrenia, Alzheimer's disease, and depression and cancer, it is reasonable to state that there could be some molecular events that are common to these diseases and may show overlapping features. If this is true, it suggests that methods designed to suppress low-grade systemic inflammation and restore to normal the imbalanced immune functions may be useful in the management of these diseases. This implies that some drugs that are useful in some of these diseases that are already in use may be of benefit in other conditions as well. For instance, statins that are useful in the management of dyslipidemia and prevent, arrest or regress atherosclerosis also show anti-inflammatory actions and hence, could be of significant benefit in Alzheimer's disease, osteoporosis, and cancer.

In this context, reviewing briefly the involvement of other molecules in the pathobiology of inflammation such as essential fatty acids and their metabolites, platelet activating factor (PAF), cytokines and chemokines, nitric oxide, leukocyte lysosomal enzymes, reactive oxygen species and neuropeptides may prove to be interesting. This is especially so since essential fatty acids and their metabolites, platelets and platelet activating factor, cytokines and chemokines, nitric oxide, reactive oxygen species and neuropeptides seem to play a significant role in several diseases enumerated above.

Kinins

Kinins are structurally related polypeptides, such as bradykinin and kallikrein. They are members of the autacoid family. They act locally to induce vasodilation and contraction of smooth muscle. Kinin is a component of the kinin-kallikrein system. The precursors of kinins are kininogen. Aspirin inhibits the activation of kallenogen by interfering with the formation of kallikrien enzyme which is essential in the process of activation. Kinins are generally pro-inflammatory in nature and vasoactive peptides that are released during tissue damage and may contribute to neuronal degeneration, inflammation, and edema formation after brain injury by acting on discrete bradykinin receptors, B1R and B2R that are G-protein-coupled receptors. Kinins play an important role in regulation of pain and hyperalgesia after tissue injury and inflammation. It is generally accepted that the B2 receptor is constitutively expressed, whereas the B1 receptor is induced in response to inflammation. Up-regulation of kinin receptors seems to be involved in the development of the early phase of inflammation and inflammatory pain. The up-regulation of B1 receptors may contribute to acute inflammatory pain [190].

The kinin-kallikrein system or simply kinin system plays a role not only in inflammation but also in blood pressure control and coagulation. Its important mediators bradykinin and kallidin that are vasodilators and act on many cell types.

High-molecular weight kininogen (HMWK) and low-molecular weight kininogen (LMWK) are precursors of the polypeptides. They have no activity by themselves. HMWK is produced by the liver together with prekallikrein that acts mainly as a cofactor on coagulation and inflammation, and has no intrinsic catalytic activity. LMWK is produced locally by numerous tissues, and secreted together with tissue kallikrein.

Bradykinin (BK), which acts on the B2 receptor and slightly on B1, is produced when kallikrein releases it from HMWK. It is a nonapeptide with the amino acid sequence Arg-Pro-Pro-Gly-Phe-Ser-Pro-Phe-Arg. Recent studies showed that induction of a focal cryolesion in the brain of mice produced blood-brain barrier (BBB) disruption, and inflammatory processes and significantly induced B1R and B2R gene transcripts in the lesioned hemispheres of wild-type mice. The volume of the cortical lesions and neuronal damage at 24 h after injury in $B1R^{-/-}$ mice were significantly smaller than in wild-type controls. Treatment with the B1R antagonist after lesion induction likewise reduced lesion volume in wild-type mice that was accompanied by a remarkable reduction of BBB disruption and tissue inflammation. In contrast, genetic deletion or pharmacological inhibition of B2R had no significant impact on lesion formation or the development of brain edema. These results suggest that B1R is involved in inflammation that occurs due to acute brain injuries [191].

Kallidin (KD) is released from LMWK by tissue kallikrein. It is a decapeptide. KD has the same amino acid sequence as bradykinin with the addition of a Lysine at the N-Terminus, thus is sometimes referred to as Lys-Bradykinin. HMWK and LMWK are formed by alternative splicing of the same gene.

Kallikreins (tissue and plasma kallikrein) are serine proteases that liberate kinins (BK and KD) from the kininogens, which are plasma proteins that are converted into vasoactive peptides. Prekallikrein is the precursor of plasma kallikrein. It can only activate kinins after being activated itself by factor XIIa or other stimuli.

Carboxypeptidases are present in two forms: N circulates and M is membrane-bound. They remove arginine residues at the carboxy-terminus of BK and KD.

Angiotensin converting enzyme (ACE), also termed *kininase II*, inactivates a number of peptide mediators, including bradykinin. It is better known for activating angiotensin.

Neutral endopeptidase also deactivates kinins and other mediators.

Inhibition of ACE with ACE inhibitors leads to decreased conversion of angiotensin I to angiotensin II (a vasoconstrictor) but also to an increase in bradykinin due to decreased degradation. This explains why some patients on ACE inhibitors for the management of hypertension develop a dry cough, and some react with angioedema.

It is thought that some, if not all, beneficial actions of ACE-inhibitors are due to their action on the kinin-kallikrein system. This includes their effects in arterial hypertension, in ventricular remodeling (after myocardial infarction) and possibly diabetic nephropathy.

Defects of the kinin-kallikrein system in diseases are not generally well recognized. They are involved in the pathogenesis of inflammation and regulation of blood pressure. It is known that kinins are inflammatory mediators that cause dilation of blood vessels and increased vascular permeability. Kinins are small peptides produced from kininogen by kallikrein and are broken down by kininases. They act on phospholipase and increase arachidonic acid release and thus, enhance the production of prostaglandin (PGE_2) that also plays a significant role in inflammation. It is possible that the involvement of kinin-kallikrein system in inflammation is, in part, due to its stimulatory action on the prostaglandin synthesis.

C1-inhibitor is a serine protease inhibitor (serpin) protein. C1-INH is the most important physiological inhibitor of plasma kallikrein, fXIa and fXIIa. C1-INH also inhibits proteinases of the fibrinolytic, clotting, and kinin pathways. Deficiency of C1-INH permits plasma kallikrein activation, which leads to the production of the vasoactive peptide bradykinin.

Tissue kallikrein (KLK1) processes low-molecular weight kininogen to produce vasoactive kinins, which exert biological functions via kinin receptor signaling. Tissue kallikrein acts through kinin B2 receptor signaling and exhibits a wide spectrum of beneficial effects by reducing cardiac and renal injuries, restenosis and ischemic stroke, and by promoting angiogenesis and skin wound healing, independent of blood pressure reduction. Protection by tissue kallikrein in oxidative organ damage is attributed to the inhibition of apoptosis, inflammation, hypertrophy and fibrosis. Tissue kallikrein also enhances neovascularization in ischemic heart and limb. Moreover, tissue kallikrein/kinin infusion not only prevents but also reverses kidney injury, inflammation and fibrosis in salt-induced hypertensive rats. Furthermore, delayed kallikrein infusion for 24 h after cerebral ischemia in rats is effective in reducing neurological deficits, infarct size, apoptosis and inflammation. Human tissue kallikrein has been found to be effective in the treatment of patients with acute brain infarction when injected within 48 h after stroke onset. Kallikrein promotes skin wound healing and keratinocyte migration by direct activation of protease-activated receptor 1 [192]. These results suggest that kallikrein system has both adverse and beneficial actions.

Essential Fatty Acids and Their Products

There is now evidence available to suggest that essential fatty acids (EFAs) and their metabolites that include eicosanoids, lipoxins, resolvins, protectins, maresins and nitrolipids play a significant role in the pathobiology of inflammation. Products of EFAs possess both pro- and anti-inflammatory actions suggesting that the balance between these pro- and anti-inflammatory products could determine either the resolution or persistence of inflammation. The metabolism and actions pertinent to inflammation are mentioned briefly here and more detailed discussion is presented in the chapter on essential fatty acids.

Cis-Linoleic acid (LA, 18:2 ω-6) and α-linolenic acid (ALA, 18:3 ω-3), are the dietary essential fatty acids (EFAs). LA is converted to gamma-linolenic acid (GLA, 18:3, ω-6) by the action of the enzyme Δ^6 desaturase, and GLA is elongated (by the action of the enzyme elongase) to form di-homo-GLA (DGLA, 20:3, ω-6), the precursor of the 1 series of prostaglandins. DGLA can also be converted to arachidonic acid (AA, 20:4, ω-6) by the action of the enzyme Δ^5 desaturase. AA forms the precursor of 2 series of prostaglandins, thromboxanes (TXs) and the 4 series leukotrienes (LTs). ALA is converted to eicosapentaenoic acid (EPA, 20:5, ω-3) by Δ^6 and Δ^5 desaturases. EPA forms the precursor of the 3 series of prostaglandins and the 5 series of leukotrienes. EPA can be elongated to form docosahexaenoic acid (DHA, 22:6, ω-3) (see Fig. 3.5 for metabolism of EFAs). Several of these PGs, LTs and TXs have pro-inflammatory actions.

AA, EPA and DHA also form precursors to anti-inflammatory compounds lipoxins (LXs), resolvins (RSVs), protectins, maresins and nitrolipids [193, 194] (See Fig. 3.5 for metabolism of essential fatty acids). Eicosanoids mediate virtually every step of inflammation, are found in inflammatory exudates, and their synthesis is increased at sites of inflammation. Non-steroidal anti-inflammatory drugs (NSAIDs) such as aspirin not only inhibit cyclo-oxygenase (COX) activity that has been held responsible for their anti-inflammatory action but also enhance the production of lipoxins that have anti-inflammatory action. Based on the role of eicosanoids in inflammation, COX-2 inhibitors have been developed that are expected to reduce inflammation in vivo without gastric side effects but, were found to enhance the risk of coronary heart disease [195]. In the presence of aspirin, AA, EPA, and DHA are converted to form epi-lipoxins, lipoxins, and resolvins that, in turn, enhance the formation of eNO [193, 194, 196–200]. Lipoxins possess potent anti-inflammatory actions (reviewed in 193, 194). In addition, NO not only blocks the interaction between leukocytes and the vascular endothelium during inflammation but also stimulates the formation of PGI_2, a potent vasodilator and platelet anti-aggregator, from AA [201, 202], while PGI_2 augments the production of NO [203]. It should be noted here that AA, EPA and DHA can augment the production of NO form various tissues especially endothelial cells (reviewed in [193, 194]).

Cyclo-oxygenase (COX), Lipoxygenase (LO) Pathways and Generation of Lipoxins, Resolvins, Protectins and Maresins

There are two cyclo-oxygenase enzymes, the constitutively expressed COX-1 and the inducible enzyme COX-2 that leads to the generation of prostaglandins (PGs). Different types of PGs are formed depending on the substrate fatty acid from which

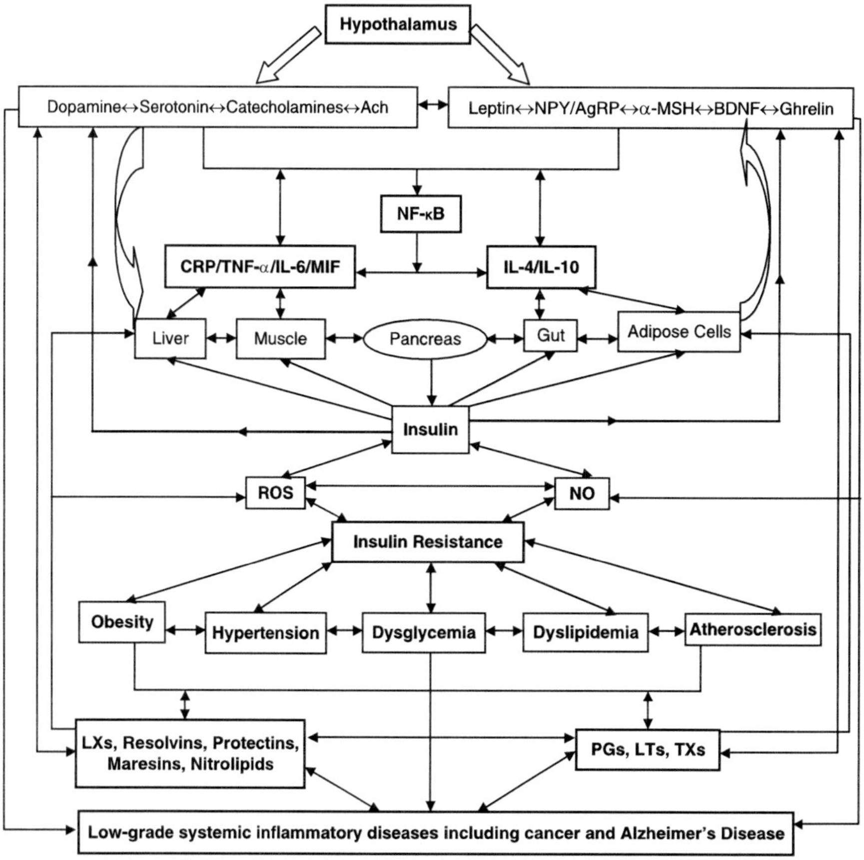

Fig. 3.6 Scheme showing the relationship among neurotransmitters, hypothalamic peptides, pro- and anti-inflammatory cytokines and lipid molecules and their involvement in various diseases. For details see the text

they are derived. It is noteworthy that different types of PGs have different actions and sometimes diametrically opposite actions. For example, PGE_2, $PGF_{2\alpha}$, thromboxane A_2 (TXA_2), and leukotrienes (LTs) have pro-inflammatory actions whereas PGE_1 and prostacyclin (PGI_2) may show anti-inflammatory actions; while TXA_2 and $PGF_{2\alpha}$ induce platelet aggregation and vasoconstriction; PGE_2 is a platelet aggregator but causes vasodilatation. Furthermore, the distributions of COX-1 and COX-2 enzymes have restricted tissue distribution. For instance, platelets contain thromboxane synthetase, and hence TXA_2, a platelet-aggregator and vasoconstrictor, is the major product in these cells. On the other hand, vascular endothelial cells lack thromboxane synthetase but possess PGI_2 synthetase that leads to the formation of

PGI_2 that is a potent platelet anti-aggregator and vasodilator. The production of PGI_2 by endothelial cells is also essential since it helps to prevent platelet aggregation and endothelial-leukocyte interaction and thus, abrogates thrombosis and atherosclerosis. PGI_2 potentiates the permeability-increasing and chemotactic effects of other mediators and thus, may participate in inflammation. The balance between TXA_2 and PGI_2 plays a significant role in thrombus formation in coronary and cerebral blood vessels. PGs have a role in the pathogenesis of pain and fever of inflammation. PGE_2 is hyperalgesic, causes a marked increase in pain produced by intradermal injection of suboptimal concentrations of histamine and bradykinin, and is involved in cytokine-induced fever during infections. PGD_2, PGE_2 and $PGF_{2\alpha}$, major metabolites of the COX pathway in mast cells cause vasodilatation and increase the permeability of postcapillary venules, thus potentiating edema formation.

COX-2 enzyme is absent in most tissues under normal "resting" conditions and is expressed in response to various pro-inflammatory stimuli, whereas COX-1 is constitutively expressed in most tissues. This suggests that PGs produced by COX-1 serve a homeostatic function (such as fluid and electrolyte balance in the kidneys, cytoprotection in the gastrointestinal tract) and are also involved in inflammation, whereas COX-2 stimulates the production of the PGs that are involved in inflammatory reactions.

There are three types of lipoxygenases and are present in only a few types of cells (for more detailed discussion of EFA metabolism see the next chapter). 5-lipoxygenase (5-LO) is the predominant enzyme in neutrophils. The main product, 5-HETE, which is chemotactic for neutrophils, is converted into leukotrienes (LTs). LTB_4 is a potent chemotactic agent and activator of neutrophils and induces aggregation and adhesion of leukocytes to vascular endothelium, generation of ROS, and release of lysosomal enzymes. The cysteinyl-containing leukotrienes C_4, D_4, and E_4 (LTC_4, LTD_4, and LTE_4) induce vasoconstriction, bronchospasm, and vascular permeability. The vascular leakage, as with histamine, is restricted to venules. LTs are more potent than histamine in increasing vascular permeability and causing bronchospasm. LTs mediate their actions by binding to cysteiny leukotreine 1 (CysLT1) and CysLT2 receptors.

Lipoxins (LXs) are generated from AA, EPA and DHA by transcellular biosynthetic mechanisms (involving two cell populations). Leukocytes, particularly neutrophils, produce intermediates in LX synthesis, and these are converted to LXs by platelets interacting with leukocytes. LXA_4 and LXB_4 are generated by the action of platelet 12-lipoxygenase on neutrophil-derived LTA_4. Cell-cell contact enhances transcellular metabolism, and blocking adhesion inhibits LX production. LXs inhibit leukocyte recruitment and the cellular components of inflammation. They inhibit neutrophil chemotaxis and adhesion to endothelium [193, 194, 197]. LXs serve as endogenous negative regulators of LT synthesis and action and thus, play a role in the resolution of inflammation. The inverse relationship that exists between the amounts of LXs and LTs formed suggests that the balance between these two molecules is crucial in the determination of degree of inflammation and in its final resolution.

Aspirin-triggered 15 Epimer LXs (ATLs) and Resolvins and Formation of Protectins and Maresins

It is interesting to note that aspirin-triggered 15-epimer LXs (ATLs) are potent counter regulators of polymorphonuclear neutrophils (PMNs)-mediated injury and acute inflammation [204, 205]. Acetylation of COX-2 by aspirin prevents the formation of prostanoids, but the acetylated enzyme remains active in situ to generate 15R-hydroxyeicosatetraenoic acid (15R-HETE) from AA that is released and converted by activated inflammatory cells such as PMNs to the 15-epimeric LXs. These LXs possess potent anti-inflammatory properties [204–207]. This cross-talk between endothelial cells and PMNs leading to the formation of 15R-HETE and its subsequent conversion to 15-epimeric LXs by aspirin-acetylated COX-2 is a protective mechanism to prevent local inflammation on the vessel wall by regulating the motility of PMNs, eosinophils, and monocytes [207]. Furthermore, endothelial cells oxidize AA via P450 enzyme system to form 11,12-epoxy-eicosatetraenoic acid(s) that blocks endothelial cell activation [206], while non-enzymatic oxidation products of EPA inhibit phagocyte-endothelium interaction and suppress the expression of adhesion molecules [208]. This suggests that when specific COX-2 inhibitors are used, the beneficial LXs may not be formed and so PMNs are able to interact with endothelial cells, release reactive oxygen species that, in turn, inhibit the formation of NO, PGI_2 and lipoxins by the endothelial cells that leads to endothelial damage, thrombus formation, atherosclerosis and vascular diseases including coronary artery disease.

Akin to the formation of 15R-HETE and 15-epimeric LXs from AA, similar compounds are also formed from EPA and DHA. Human endothelial cells, when treated with EPA and aspirin, converted EPA to 18R-HEPE, 18-HEPE, and 15R-HEPE. Similar to the ability of PMNs to convert aspirin triggered, COX-2 derived 15R-HETE to 15-epi-LXA_4 and EPA to 5-series LXs, activated human PMNs converted 18R-HEPE to 5,12,18R-triHEPE and 15R-HEPE to 15-epi-LXA_5 by their 5-lipoxygenase. Both 18R-HEPE and 5,12,18R-triHEPE inhibited LTB_4-stimulated PMN transendothelial migration similar to 15-epiLXA_4. 5,12,18R-triHEPE effectively competed with LTB_4 for its receptors and inhibited PMN infiltration suggesting that it can suppress LT-mediated responses if present in adequate amounts at the sites of inflammation. Similar to aspirin, other NSAIDs such as acetaminophen and indomethacin also induced the formation of 18R-HEPE and 15R-HEPE when tested with recombinant COX-2 and EPA, suggesting that NSAIDs permit oxygenation of AA and EPA by activated endothelial cells at sites of inflammation to form the novel anti-inflammatory compounds [209].

In a similar fashion, murine brain cells expressing COX-2 and treated with aspirin transformed enzymatically DHA to 17R series of hydroxy DHAs (HDHAs) that, in turn, is converted enzymatically by PMNs to di- and tri-hydroxy containing docosanoids [210] called as protectins or neuroprotectins. It is interesting to note that similar small molecular weight compounds are generated from AA, EPA, and DHA: 15R-hydroxy containing compounds from AA, 18R series from EPA, and 17R-hydroxy series from DHA and all these compounds have potent anti-inflammatory actions and are involved in resolution of the inflammatory process and hence have been termed as "resolvins". Resolvins have the ability to inhibit cytokine generation, leukocyte recruitment, leukocyte diapedesis, and exudate formation and their endogenous function could be to suppress inflammation. This is supported by the observation that resolvins inhibit brain ischemia-reperfusion injury [196, 210]. Hence, it is likely that some of the cardiovascular protective and anti-inflammatory actions of EPA and DHA can be related to their conversion to resolvins, lipoxins and protectins (neuroprotectins). In view of this, any defect in their synthesis or their inappropriate degradation may lead to continuation of the inflammatory process and/or continuation of acute inflammation to chronic phase. In addition, DHA has been shown to form precursor to another group of novel anti-inflammatory compounds called as maresins that may also serve as endogenous anti-inflammatory and neuro- and cyto-protective compounds [211–213].

Platelet Activating Factor (PAF)

PAF is another bioactive phospholipid-derived mediator of inflammation. Chemically, PAF is acetyl-glyceryl-ether-phosphorylcholine (AGEPC), a phospholipid with a glycerol backbone, a long-chain fatty acid in the A position, an unusually short chain substituent in the B location, and a phosphatidylcholine moiety (see Fig. 3.7). PAF mediates its effects via a single G-protein-coupled receptor, and a family of inactivating PAF acetylhydrolases regulates its effects. Platelets, basophils, mast cells, neutrophils, monocytes/macrophages, and endothelial cells can elaborate PAF. PAF not only causes platelet activation but also causes vasoconstriction and bronchoconstriction, and at extremely low concentrations induces vasodilatation and increased venular permeability with potency many times greater than that of histamine. PAF also causes leukocyte adhesion to endothelium by enhancing integrin-mediated leukocyte binding, chemotaxis, degranulation, and the oxidative burst. PAF boosts the synthesis of eicosanoids by leukocytes and other cells. Thus, PAF can elicit all the cardinal features of inflammation [214]. PAF receptor antagonists inhibit inflammation in some experimental models.

PAF is synthesised continuously by cells but at low levels, controlled by the activity of PAF acetyl hydrolases (see Fig. 3.7b for the metabolism of PAF). However, it is produced in much greater quantities by inflammatory cells when required in response to cell-specific stimuli. Studies with the purified acetyltransferase have shown that with cells in the resting state, the enzyme can utilize arachidonoyl-CoA to produce the membrane-bound PAF precursor 1-alkyl-2-arachidonoylglycero-phosphocholine with even greater facility than the generation of PAF *per se*. Only when the cells are subjected to acute inflammatory stimulation does the activated enzyme produce PAF in appreciable amounts, probably after phosphorylation by a protein kinase, while simultaneously arachidonate is released for eicosanoid production. However, a second lyso-PAF acetyltransferase (LPCAT1) operates under non-inflammatory conditions. This is a constitutively expressed enzyme, while LPCAT2 is inducible.

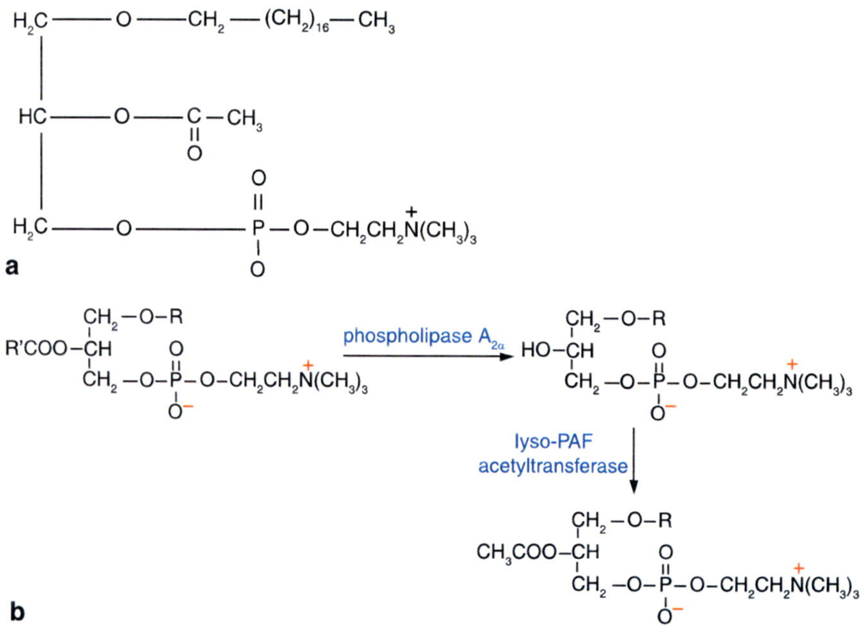

Fig. 3.7 **a** Structure of platelet activating factor. **b** Formation of PAF by the action of a membrane bound acetyltransferase that catalyzes the transfer an acetyl residue from acetyl-CoA to 1-alkyl-sn-glycero-3-phosphocholine (lyso-PAF), generated by the action of phospholipase A_2 on phosphatidylcholine

PAF can also be produced by acetylation of 1-alkyl-*sn*-glycero-3-phosphate (lysophosphatidic acid), which is subsequently converted to 1-alkyl-2-acetylglycerol and thence to PAF, i.e., by a mechanism analogous to the biosynthesis of phosphatidylcholine. This "*de novo*" pathway is also believed to be non-inflammatory.

PAF-like molecules with some biological activity can be produced in tissues by non-enzymatic oxidation of polyunsaturated fatty acids in phosphatidylcholine. Such compounds are formed in lipoproteins and are present in human atherosclerotic lesions. They bring about platelet aggregation at nanomolar concentrations and are probably involved in thrombosis and acute coronary events. Such oxidatively truncated phospholipids (and PAF) are pro-apoptotic by a mechanism that is independent of the PAF receptor, and they have a substantial influence on regulated cell death.

PAF induces aggregation of platelets at concentrations as low as 10^{-11} M, and it has vasoactive properties [216]. It is now recognised that PAF binds to its specific receptor, and activates the cytoplasmic phospholipase A_2 and phospholipase C. The result of the latter is an increase in intracellular Ca^{2+} downstream of the cell and activation of protein kinase C that has many different types of non-inflammatory biological events and functions, including glycogen degradation, reproduction, brain function and blood circulation.

PAF as a mediator of inflammation, and in the mechanism of the immune response and has been implicated in the pathogenesis of allergic reactions, asthma, stroke, myocardial infarction, colitis and multiple sclerosis. PAF can act directly as a chemotactic factor and stimulate the release of other inflammatory agents such as prostaglandins and leukotrienes that have a significant role in asthma [216–219]. In addition, PAF enhances TNF production by macrophages [220]. PAF enhances calcium mobilization, interacts with GM-CSF (granulocyte macrophage-colony stimulating factor) to enhance the availability of AA, PLA_2 activation and leukotrienes synthesis [221, 222]. PAF also has the ability to enhance IL-6 production that could account for its pro-inflammatory action [223].

PAF receptor signalling can be either pro- or anti-apoptotic, depending upon the nature of the *sn*-1 alkyl moiety, probably because of differential binding of each isomer to the receptor [224].

The plasma PAF-acetylhydrolase is associated with both LDL and HDL particles and functions on the lipid-aqueous interface, where it is termed the "lipoprotein-associated phospholipase A_2" (group VII family, LP-PLA_2), which is secreted by macrophages and is a 45 KDa protein that circulates in plasma in its active form. This enzyme hydrolyses phospholipids containing hydro-peroxyoctadecadienoyl and F_2-isoprostane residues. The effect is to remove from lipoproteins and from atherosclerotic plaques any oxidized phospholipids that might otherwise contribute to their inflammatory properties. LP-PLA_2 is closely associated with coronary heart disease and is considered as a marker to predict the occurrence of heart disease even in those who do not have other risk factors and is also a marker of atherosclerosis [225–228], conditions that are associated with low-grade systemic inflammation.

Cytokines in Inflammation

Cytokines are proteins produced by many cell types including activated lymphocytes and macrophages, endothelial cells, epithelial cells, and connective tissue cells and have are capable of modulating the functions of various other cells. Cytokines not only have a regulatory role in cellular immune responses but also participate in both acute and chronic inflammation. TNF, IL-1, IL-6, MIF (macrophage migration inhibitory factor) are the major cytokines that are involved in inflammation and have pro-inflammatory actions. On the other hand, IL-4 and IL-10 have anti-inflammatory actions, restrict inflammation and thus, they antagonize the actions of IL-1, IL-6 TNF-α and MIF. Activated macrophages and T cells produce these pro-inflammatory cytokines. But recent studies showed that a variety of other cells and tissues are also capable of producing these cytokines. For instance, endothelial cells, adipose tissue, Kupffer cells, and glial cells are capable of producing them. Endotoxin and other microbial products, immune complexes, physical injury, and a variety of inflammatory stimuli stimulate the secretion of TNF and IL-1. They activate endothelial cells, stimulate leukocytes, and fibroblasts, and induce systemic acute-phase reactions. Activation of endothelial cells by TNF, IL-6, IL-1 and MIF induces a spectrum of changes-mostly regulated at the level of gene transcription, and induce the synthesis of endothelial adhesion molecules and chemical mediators of inflammation such as other cytokines, chemokines, growth factors, eicosanoids, and nitric oxide (NO) [229–232]. These events increase the thrombotic tendency on the surface of the endothelium. TNF primes neutrophils, leading to augmented responses of these cells to other mediators, and stimulates neutrophils to produce ROS [233]. IL-1, IL-6, TNF-α and MIF induce the systemic acute-phase responses associated with infection or injury such as fever, loss of appetite, slow-wave sleep, the release of neutrophils into the circulation, the release of corticotropin and corticosteroids. Excess production of these cytokines may produce hemodynamic effects of septic shock such as hypotension, decreased vascular resistance, increased heart rate, and decreased blood pH that may ultimately cause death [234–236]. Sustained and increased production of TNF-α as it occurs during chronic intracellular infections such as tuberculosis and neoplastic diseases, lipid and protein mobilization occurs leading to the development of cachexia in these patients. IL-1, IL-6, and TNF-α suppress appetite and this contributes to cachexia [237]. Increased production of IL-1, IL-6, TNF-α and MIF (macrophage migration inhibitory factor) is also seen in rheumatoid arthritis and systemic lupus erythematosus (SLE), and other collagen vascular diseases [238]. This discovery led to the development anti-TNF-α antibodies and TNF-α receptor blockers that are useful in these conditions.

Several studies showed that plasma levels of IL-6, TNF-α and MIF are increased in patients with obesity, type 2 diabetes mellitus, hypertension, hyperlipidemia, insulin resistance, Alzheimer's disease, depression, schizophrenia that lends support to the concept that these diseases are low-grade systemic inflammatory conditions [239–247]. Both insulin and metformin appear to have the ability to suppress the production of MIF (and also, IL-6 and TNF-α) and thus, are anti-inflammatory in nature [242, 248, 249].

Chemokines in Inflammation

Chemokines are a family of small (8–10 kD) proteins that act primarily as chemoattractants for specific types of leukocytes [250–252]. In all, about 40 different chemokines and 20 different receptors for chemokines have been identified that have been classified into four major groups, according to the arrangement of the conserved cysteine (C) residues in the mature proteins. Chemokines mediate their action by binding to seven transmembrane G-protein-coupled receptors that usually exhibit overlapping ligand specificities, and leukocytes generally express more than one receptor type. Certain chemokine receptors (e.g., CXCR-4, CCR-5) act as co-receptors for a viral envelope glycoprotein and are thus, involved in binding and entry of the viruses into cells. Chemokines stimulate leukocyte recruitment in inflammation and control the normal migration of cells through various tissues [253]. Some chemokines are produced transiently in response to inflammatory stimuli and promote the recruitment of leukocytes to the sites of inflammation [254–256], whereas others are produced constitutively in tissues and participate in organogenesis [257–260]. In both situations, chemokines are displayed at high concentrations attached to proteoglycans on the surface of endothelial cells and in the extracellular matrix.

Nitric Oxide (NO)

NO was originally discovered as a factor that is released from endothelial cells that caused vasodilatation and hence was called as endothelium-derived relaxing factor [261]. NO is a soluble gas produced not only by endothelial cells, but also by a variety of cells such as macrophages and neurons in the brain. It is now realized that NO is produced by many cells (if not all) and that it participates in inflammation. NO activates guanylate cyclase that induces smooth muscle relaxation by: (i) increasing intracellular cGMP that inhibits calcium entry into the cell, and decreases intracellular calcium concentrations; (ii) activates K^+ channels, which leads to hyperpolarization and relaxation; (iii) stimulates a cGMP-dependent protein kinase that activates myosin light chain phosphatase, the enzyme that dephosphorylates myosin light chains, which leads to smooth muscle relaxation.

NO acts in a paracrine manner on target cells through induction of cyclic guanosine monophosphate (cGMP) that, in turn, initiates a series of intracellular events leading to the desired response such as relaxation of vascular smooth muscle cells, neurotransmission, tumoricidal, cytotoxic, and bactericidal actions. The half-life of NO is only few seconds and hence, it has to be produced in close proximity to where its action is needed.

L-arginine forms the precursor of NO and is synthesized by the action of nitric oxide synthase (NOS) enzyme [262–264] (see Fig. 3.8). There are three different types of NOS-endothelial (eNOS), neuronal (nNOS), and inducible (iNOS). NOS exhibit two patterns of expression: eNOS and nNOS are constitutively expressed at low levels and can be activated rapidly by an increase in cytoplasmic calcium ions. Influx of calcium into cells leads to a rapid production of NO. In contrast, iNOS is induced in macrophages and other cells when are activated by cytokines such as TNF-α and IFN-γ. It is paradoxical that NO has both beneficial and harmful actions. Endothelial NO is essential to keep blood vessels patent and its deficiency may predispose to the development of cardiovascular diseases including coronary heart disease. On the other hand, iNO produced by activated macrophages could play a role in sepsis and septic shock and cause hypotension and participate in inflammation.

NO is a potent vasodilator and prevents platelet aggregation, inhibits vascular smooth muscle cell proliferation, reduces platelet adhesion and inhibits several features of mast cell-induced inflammation, and serves as an endogenous regulator of leukocyte recruitment. Inhibition of endogenous NO production promotes leukocyte rolling and adhesion in postcapillary venules. On the other hand, delivery of exogenous NO reduces leukocyte recruitment. Thus, under normal physiological

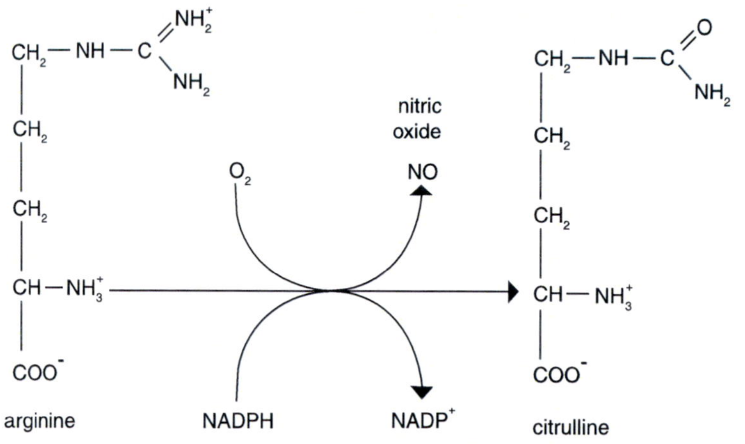

Fig. 3.8 Scheme showing the formation of nitric oxide from its precursor L-arginine. Nitric oxide synthase produces NO by catalyzing a five-electron oxidation of guanidino nitrogen of L-arginine (L-Arg). Oxidation of L-Arg to L-citrulline occurs via two successive monooxygenation reactions producing N^{ω}-hydroxy-L-arginine (NOHLA) as an intermediate. 2 mol of O_2 and 1.5 mol of NADPH are consumed per mole of NO formed.
L-Arg + NADPH + H^+ + O_2 → NOHLA + $NADP^+$ + H_2O
NOHLA + ½NADPH + ½H^+ + O_2 → L-citrulline + ½$NADP^+$ + NO + H_2O

conditions NO is an inhibitor of inflammatory response and possibly, increased production of NO in inflammatory conditions could be a compensatory mechanism to block inflammatory responses [265]. Increased production of NO seen in response to various inflammatory stimuli might itself perpetuate inflammation due to the conversion of NO to peroxynitrite radical that has potent pro-inflammatory actions (see Fig. 3.2). Recently, NOS activity has been demonstrated in several bacterial species, including pathogens as Bacillus anthraces and Staphylococcus aureus [266, 267]. Bacterial NOS (bNOS) has been shown to protect bacteria against oxidative stress, diverse antibiotics, and host immune response [267, 268] (see Table 3.4 for the types of NOS, their location and actions).

Table 3.4 Different forms of NO synthase and their location and actions

Name	Gene(s)	Location	Function
Neuronal NOS (nNOS or NOS1)	NOS1	Nervous tissue skeletal muscle type II	Cell communication
Inducible NOS (iNOS or NOS2)	NOS2A, *NOS2B*, *NOS2C*	Immune system cardiovascular system	Immune defense against pathogens
Endothelial NOS (eNOS or NOS3 or cNOS)	NOS3	Endothelium	Vasodilation
Bacterial NOS (bNOS)	Multiple	Various Gram (+) species	Defense against oxidative stress, antibiotics, immune attack

Decreased production of eNO has been documented in obesity, insulin resistance, hyperlipidemia, atherosclerosis, coronary heart disease, diabetes mellitus, and hypertension [269–275].

NO is an Endogenous Anti-infective Molecule

NO and its derivatives have microbicidal actions and thus, NO functions as an endogenous mediator of host defense against infections [276–280]. Enhanced production of NO by macrophages and other immune cells has been shown to inhibit the growth of several bacteria, viruses, fungi, and other organisms. NO can kill bacteria, *Mycobacterium tuberculosis*, and viruses. NO interacts with reactive oxygen species to generate nitrogen intermediates that kill the invading pathogens. This is supported by the observation that: (a) reactive nitrogen intermediates derived from NO possess antimicrobial activity; (b) NO interacts with ROS to form multiple antimicrobial metabolites; (c) in response to infections the production of NO is increased by macrophages and other immune cells; and (d) inactivation of iNOS enhances the incidence of infections and augments the multiplication of microbial organisms in experimental animals. Several studies suggested that NO has an important role in the pathobiology of malaria. TNF induced more reactive nitrogen intermediates (NO and its derivatives) in malaria-infected mice than in normal mice, and appreciably more was in the form of nitrate than was in the form of nitrite. NG-methyl-L-arginine (a specific inhibitor of NO generation) inhibited the in vivo generation of reactive nitrogen intermediates by TNF in a dose-dependent manner, implying that these molecules were arginine derived, suggesting that TNF, lymphotoxin, and interleukin-1 contribute to host pathology and parasite suppression through generation of NO [281–283]. Furthermore, resistance to malarial infection seems to depend on the ability of immunocytes to produce NO. Studies have also suggested that high plasma levels of NO could be involved in the pathogenesis of cerebral malaria [284], though some studies did not support this contention [285]. In fact, some studies indicated that low NO bioavailability contributes to the genesis of cerebral malaria. Mice deficient in vascular NO synthase showed parasitemia and mortality similar to that observed in control mice. Exogenous NO provided marked protection against cerebral malaria and mice treated with exogenous NO were clinically indistinguishable from uninfected mice at a stage when control infected mice were moribund. Administration of exogenous NO restored NO-mediated signaling in the brain, decreased proinflammatory biomarkers in the blood, and markedly reduced vascular leak and petechial hemorrhage into the brain. These results led to the conclusion that low rather than high NO bioavailability contributes to the genesis of cerebral malaria [286]. Similar results were noted in patients with cerebral malaria [287]. There is also evidence to suggest that TNF, MIF and oxidative stress also have a role in the pathogenesis of cerebral malaria [288–290]. It is noteworthy that NO has tumoricidal actions [291–292], though in an occasional instance it has also been reported to have antiapoptotic actions [294–296]. In general, NO seems to prevent apoptosis of normal cells such as hepatocytes induced by TNF, while inducing apoptosis of tumor cells. The ability of NO to induce apoptosis or prevent apoptosis of tumor cells may depend on the dose of NO, the balance between TNF and NO and superoxide anion and NO and other factors. These results suggest that NO has both apoptotic and anti-apoptotic actions depending on the local conditions. In a similar fashion, NO has both pro- and anti-inflammatory actions-eNO protects endothelial cells from inflammatory stimuli while NO released by macrophages shows potent pro-inflammatory actions. This may, in part, be due to the amount of NO released: high amounts of NO have pro-inflammatory actions while constitutional NO has anti-inflammatory and cytoprotective actions.

NO and Cellular Senescence

NO is associated with endothelial senescence. NO activates telomerase and delays endothelial senescence [297], while endothelial replicative aging results in decreased endothelial expression of eNOS [298]. On the other hand, stable expression of hTERT (human telomerase reverse transcriptase) resulted in endothelial cells with a younger phenotype with greater amount of eNOS and NO activity. Furthermore, senescent endothelial cells exhibited increased ICAM-1 expression and decreased eNOS activity, while introduction of telomerase catalytic component significantly extended the life span and inhibited the functional alterations associated with endothelial senescence, which contributes to atherosclerosis [299]. These results indicate that a decrease in NO generation and enhanced telomerase activity in endothelial cells promotes inflammation as evidenced by increased ICAM-1 expression.

Asymmetrical dimethylarginine (ADMA), an endogenous inhibitor of nitric oxide synthase (NOS), accumulation that is associated with cardiovascular disease significantly accelerated senescence, accelerated the shortening of telomere length and reduced the telomerase activity. ADMA accumulation in endothelial cells was associated with an increase of oxidative stress in the form of enhanced intracellular reactive oxygen species (ROS) generation with a simultaneous decrease in NO generation and ADMA-increased oxidative stress was accompanied by a decrease in the activity of dimethylarginine dimethylaminohydrolase (DDAH), the enzyme that degrades ADMA. Exogenous ADMA also stimulated secretion of MCP-1 and interleukin-8. These data suggest that ADMA accelerates senescence, via increased oxygen radical formation and by inhibiting nitric oxide elaboration and enhancing telomere shortening [300]. These results are supported by the observation that aspirin inhibited endothelial senescence, increased telomerase activity, decreased reactive oxygen species and increased nitric oxide (NO) and cGMP levels. Furthermore, aspirin reduced the elaboration of asymmetric dimethylarginine (ADMA) and up-regulated the activity of dimethylarginine dimethylaminohydrolase, the enzyme that degrades

ADMA. NO synthase inhibitor l-NAME completely inhibited aspirin-delayed senescence. These findings demonstrate that aspirin delays endothelial senescence by enhancing NO formation/generation [301]. Even the beneficial actions of exercise in atherosclerosis and hypertension could be attributed to increase in NO generation and stabilization of telomere length [302].

NO and Brain-derived Neurotrophic Factor (BDNF)

In this context, it is interesting to note that eNO enhances both VEGF and BDNF generation. ENOS knock-out ($eNOS^{-/-}$) mice showed decreased angiogenesis and exhibited a reduced response to vascular endothelial growth factor (VEGF)-induced angiogenesis in a corneal assay, showed decreased brain-derived neurotrophic factor (BDNF) expression. In addition, cultured subventricular zone (SVZ) progenitor cell proliferation and migration, neurosphere formation, proliferation, telomerase activity, and neurite outgrowth were significantly reduced in $eNOS^{-/-}$ mice compared with wild-type mice. Interestingly, BDNF treatment of SVZ cells derived from $eNOS^{-/-}$ mice restored the decreased neurosphere formation, proliferation, neurite outgrowth, and telomerase activity in cultured $eNOS^{-/-}$ SVZ neurospheres, indicating that eNOS regulates BDNF expression and may serve as a downstream mediator for VEGF and angiogenesis [303]. These results are interesting since BDNF is now believed to play a significant role in the pathogenesis of type 2 diabetes mellitus [304–309].

NO has many useful actions as well. It is a potent platelet anti-aggregator and vasodilator and prevents atherosclerosis. Production of appropriate amounts of eNO is possible only when endothelial cells are healthy. Hence, plasma concentrations or endothelial production of NO can be used as a marker of endothelial cell integrity and health. In obesity, hypertension, type 2 diabetes mellitus, insulin resistance, hyperlipidemias, and CHD, the plasma concentrations of NO are low that suggests that these conditions are due to endothelial dysfunction. NO levels can be made to revert to normal by reduction in body weight that can be achieved by diet restriction and exercise, control of hypertension, normalization of plasma glucose levels in type 2 DM, and reduction of plasma lipid levels. Thus, measurement of plasma levels of NO could be used as a marker not only of endothelial function but also to judge adequacy of treatment given to patients in these conditions. Since many factors could influence the synthesis and half-life of NO, it is important to keep a note of them. For instance, decreased production of NO could be due to a deficiency of its precursor, L-arginine, and/or lack or deficiency of co-factors such as tetrahydrobiopterin (BH_4). Hence, at times simple lack or deficiency of these co-factors may lead to low plasma levels of NO. Hence, before a judgment as to the cause of decreased NO levels is made, one has to take these factors into consideration.

Although NO is unstable, its concentrations in the plasma and in cultured cells in vitro could be measured using various colorimetric techniques and specific NO probes. NO is measured as its stable metabolites nitrite and nitrate in the plasma that gives an indication as to the concentrations of NO that is released by endothelial cells. Highly sensitive NO probes are commercially available to measure intracellular concentrations of NO and NO that is released by cells in vitro. These techniques enable one to study the effect of various chemicals, drugs, and factors that influence the generation of NO to evaluate its role in various conditions.

Leukocyte Lysosomal Enzymes

Lysosomal granules are present in neutrophils and monocytes. There are of two types of lysosomal granules: smaller specific (secondary) granules and larger azurophil (primary) granules. The smaller specific secondary granules contain lysozyme, collagenase, gelatinase, lactoferrin, plasminogen activator, histaminase, and alkaline phosphatase. On the other hand, the large azurophil primary granules contain myeloperoxidase, lysozyme, defensins, acid hydrolases, and a variety of neutral proteases such as elastase, cathepsin G, proteinase 3, and nonspecific collagenases [310]. Both types of granules release their contents into phagocytic vacuoles that form around engulfed material to bring about their actions. These granule contents can also be released into the extracellular space. The release of the contents of lysosomal granules contributes to inflammation. It is important to note that different granule enzymes show different functions. For instance, acid proteases degrade bacteria and debris within the phagolysosomes under acidic pH conditions, whereas neutral proteases degrade various extracellular components. Neutral proteases attack and degrade collagen, basement membrane, fibrin, elastin, and cartilage that ultimately result in tissue destruction that are typically seen in acute and chronic inflammatory processes. Neutral proteases cleave C3 and C5 directly resulting in the release of anaphylatoxins, and kinin-like peptide from kininogen. Neutrophil elastase degrades virulence factors of bacteria and thus helps in the control of bacterial infections [311]. Both monocytes and macrophages contain acid hydrolases, collagenase, elastase, phospholipase, and plasminogen activator by virtue of which they participate in chronic inflammatory reactions. In view of the destructive nature of lysosomal enzymes, it is important to control leukocytes infiltration at the site of injury and infection. If the leukocyte infiltration remains unchecked, it can lead to further increase in vascular permeability and tissue destruction. In order to control the harmful effects of these proteases, a number of antiproteases are present in the serum and tissue fluids. One of the best examples is α_1-antitrypsin that inhibits neutrophil elastase. A deficiency of α_1-antitrypsin leads to uncontrolled action of leukocyte elastase that causes pulmonary damage resulting in emphysema. α_2-macro-globulin is another antiprotease found in serum and various secretions.

Reactive Oxygen Species (ROS)

ROS or oxygen-derived free radicals are released by leukocytes, macrophages, Kupffer cells and other similar cells present in various organs into the extracellular compartment on exposure to various noxious agents such as microbes, foreign objects, in response to chemokines, immune complexes, or following a phagocytic challenge [312–315]. The production of ROS is due to the activation of the NADPH oxidative system. Known ROS species are mainly: superoxide anion ($O_2^{\cdot -}$), hydrogen peroxide (H_2O_2), and hydroxyl radical (OH). ROS are produced mainly within the cell, and are capable of reacting with NO to form reactive nitrogen intermediates that are cytotoxic to various organelles of cells [316]. Since ROS and reactive nitrogen intermediates are highly toxic, their release into the extracellular space even in low concentrations may prove to be harmful and even at low concentrations are capable of increasing the expression of chemokines (e.g., IL-8), cytokines, and endothelial leukocyte adhesion molecules, events that amplify the inflammatory cascade [317–319]. Both ROS and reactive nitrogen species destroy bacteria, viruses, fungi, and cancer cells. At the other end of the spectrum, increased production of ROS and reactive nitrogen intermediates are potentially harmful and cause acute and chronic inflammation, sepsis, and other pathological conditions. ROS and reactive nitrogen intermediates (RNI) can cause endothelial cell damage resulting in increased vascular permeability, insulin resistance, and thrombosis. Activated adherent neutrophils not only produce ROS and RNI but also stimulate xanthine oxidase in endothelial cells that, in turn, further accentuates generation of superoxide anion. ROS and RNI inactivate antiproteases such as α_1-antitrypsin that leads to unopposed protease activity, which could result in increased destruction of extracellular matrix. ROS by themselves damage many cells and tissues including but not limited to parenchymal cells. It is now believed that several clinical conditions are due to excess production of ROS. For instance, ROS and RNI are considered to be responsible for diseases such as rheumatoid arthritis, lupus, and ulcerative colitis, ischemia-reperfusion injury to myocardium following coronary bypass surgery and cerebral cortical damage after ischemic stroke; and several low-grade systemic inflammatory conditions such as insulin resistance, metabolic syndrome, atherosclerosis, schizophrenia, Alzheimer's disease, depression and schizophrenia. In view of this, efforts are being made to develop anti-oxidants and free radical quenchers that might mitigate these diseases and processes. For instance, excess production of ROS in endothelial cells (or close to endothelial cells) produces damage to these cells that results in endothelial dysfunction. Obesity, hypertension, type 2 diabetes mellitus, hyperlipidemias, and CHD, which are components of metabolic syndrome, are all characterized by endothelial dysfunction. This proposal is supported by the fact that increased generation of ROS is seen in obesity, hypertension, type 2 diabetes mellitus, hyperlipidemias, and insulin resistance. But, it is not yet clear why and how increased generation of ROS occurs. It is important to know as to when this increase in the generation of ROS starts so that appropriate timing of preventive or therapeutic measures can be taken.

Several antioxidants are present in the serum, various tissue fluids and cells to abrogate the harmful actions of ROS. These antioxidants include: (1) the copper-containing serum protein ceruloplasmin; (2) the iron-free fraction of serum, transferrin; (3) the enzyme superoxide dismutase (SOD), which is found or can be activated in a variety of cell types; (4) the enzyme catalase, which detoxifies H_2O_2; and (5) glutathione peroxidase, another powerful H_2O_2 detoxifier. Thus, the influence of ROS in inflammatory conditions depends on the balance between the production and the inactivation of these metabolites by cells and tissues.

NO also has an important role in the pathogenesis of both acute and chronic inflammation. Excess production of NO especially, by macrophages is harmful to several tissues. Activation of iNOS that occurs in response to various stimuli by itself sometimes is sufficient to initiate and perpetuate the inflammatory process. But, more often than not, excess production of both ROS and NO occurs in majority of the inflammatory conditions.

Neuropeptides in Inflammation

Neuropeptides are known to play a significant role in the initiation and propagation of inflammation. Substance P and neurokinin A that are produced both in the central and peripheral nervous systems have the ability to influence transmission of pain signals, regulation of blood pressure, stimulation of secretion by endocrine cells, and increasing vascular permeability [320–322]. The involvement of these neuropeptides in the inflammatory process explains the neurogenic component of inflammation. Sensory neurons produce certain pro-inflammatory molecules that link the sensing of dangerous stimuli to the development of protective host responses that form the basis of neurogenic inflammation [320].

Obesity, type 2 diabetes, hypertension, hyperlipidemia, insulin resistance, Alzheimer's disease, depression, schizophrenia and cancer are low-grade systemic inflammatory conditions

Recent studies suggested that low-grade systemic inflammation plays a significant role in the pathogenesis of obesity, type 2 diabetes, hypertension, hyperlipidemia, insulin resistance, Alzheimer's disease, depression, schizophrenia and cancer. This is based on the observation that the plasma concentrations of CRP, TNF-α, IL-6 and MIF (resistin in obesity and type2 diabetes mellitus), which are markers of inflammation are elevated whereas the concentrations of adiponectin that shows anti-inflammatory actions are reduced in obesity, type 2 diabetes mellitus and metabolic syndrome [42, 43, 323–331].

The elevated plasma concentrations of CRP (C-reactive protein), TNF-α, IL-6 and MIF may produce their harmful effects in type 2 diabetes mellitus, hypertension, and obesity by inducing endothelial dysfunction. TNF-α and IL-6 damage endothelial cells, cause apoptosis of endothelial cells, and trigger procoagulant activity and fibrin deposition [323–326]. It was shown that forearm blood flow responses to acetylcholine (ACh) were inversely correlated with CRP serum levels indicative of endothelial dysfunction [327]. High CRP concentrations were associated with decreased endothelial nitric oxide (eNO) generation [328]. Previously, I and others showed that NO levels were low in patients with diabetes mellitus [271, 272]. These results suggest that elevated CRP, IL-6 and TNF-α concentrations may lead to decrease in eNO production and consequently endothelial dysfunction. Since NO is a potent vasodilator and platelet anti-aggregator, low eNO may, in turn, lead to increase in peripheral vascular resistance and higher incidence of thrombosis and atherosclerosis.

However, it is still debated whether inflammation is the primary event or it is secondary to the development of type 2 diabetes. For instance, CRP levels do not correlate with the extent of atherosclerosis. This suggests that CRP levels reflect the body's response to inflammation elsewhere. On the other hand, CRP functions as a chemoattractant, increases the expression of adhesion molecules, and activates complement proteins, which are important mediators of inflammation. Furthermore, CRP binds to LDL cholesterol and increases the uptake of LDL by macrophages. Studies in animals revealed that CRP enhances the size of myocardial infarction, stroke and methods designed to neutralize CRP levels or action minimize damage to the heart and brain [332–334]. These results suggest that inflammation plays a role in the pathobiology of type 2 diabetes and metabolic syndrome.

Both IL-6 and TNF-α increase neutrophil superoxide anion generation [83, 335]. Superoxide anion ($O_2^{\cdot-}$) inactivates NO and prostacyclin (PGI_2) and thus causes endothelial dysfunction, enhances thrombosis and atherosclerosis [336, 337], which are common in type 2 diabetes. On the other hand, optimal production of NO inactivates $O_2^{\cdot-}$ and thus prevents/arrests thrombosis and atherosclerosis [337, 338]. This indicates that an increase in oxidative stress could be a factor that contributes to the development of type 2 diabetes, hypertension, and other components of metabolic syndrome [42, 43, 339].

Adipose tissue produces several biologically active molecules that have important actions on immune response and inflammation. Three of these molecules are adiponectin, resistin, and corticosterone. Adiponectin has anti-inflammatory actions and its plasma concentrations are inversely related to insulin resistance and the severity of typ2 diabetes whereas resistin induces insulin resistance and has pro-inflammatory actions [42, 43, 340, 341]. Transgenic mice over expressing 11β hydroxysteroid dehydrogenase types 1 (11βHSD-1) selectively in adipose tissue developed abdominal obesity and exhibited insulin-resistant diabetes (type 2 diabetes), hyperlipidemia, and hyperphagia [342], suggesting that type 2 diabetes behaves like localized Cushing's syndrome.

Several other studies also revealed that elevated plasma concentrations of CRP and possibly, IL-6 and TNF-α predict the future development of type 2 diabetes mellitus, hypertension, and coronary heart disease [343–345]. A reduction in the levels of CRP, IL-6 and TNF-α achieved by diet control, exercises, and statin therapy improved prognosis in these patients, indicating that measurement of these inflammatory markers could be useful to predict both the development of metabolic syndrome and its response to therapy.

Diagnosis of Low-grade Systemic Inflammation

It is evident from the preceding discussion that many biological molecules are involved in the pathobiology of inflammation. At bedside, it is relatively simple to diagnose acute inflammation that is characterized by *rubor, tumor, calor, dolor, and functiolaesa* (redness, swelling, heat, pain, and loss of function respectively). Since these acute inflammatory events are easily visible, perhaps, no specific laboratory tests are necessary to measure the presence or absence of inflammation. But, when the inflammatory process is low-grade and localized to the internal organs it is difficult, if not impossible, to detect and confirm the presence of inflammation. This is especially true when there is low-grade systemic inflammation. Examples of diseases in which low-grade systemic inflammation is common include: obesity, insulin resistance, type 2 diabetes mellitus, hypertension, coronary heart disease (CHD), hyperlipidemia, Alzheimer's disease, depression, schizophrenia, atherosclerosis and cancer. In these conditions, enhanced plasma levels of high-sensitive CRP, IL-6, and TNF-α is seen. These patients also have low circulating NO levels and simultaneously increased generation of reactive oxygen species (ROS) and MPO. Increased ROS decreases anti-oxidant content of the cells/tissues due to their utilization. Hence, they may show decreased vitamin E, superoxide dismutase, and glutathione levels. This suggests the delicate balance between the pro- and anti-oxidants is tilted more in favor of the pro-oxidants leading tissue damage and the onset and progression of disease. A brief description of the markers that can be used to assess the presence of low-grade systemic inflammation in various conditions is given below.

Hs-CRP

CRP is a 135,000-dalton non-immunoglobulin protein and the most reliable marker of inflammation that raises several 100-fold in response to acute injury, infection, or other inflammatory stimuli and is more reliable than erythrocyte sedimentation rate (ESR). One of the most attractive features of CRP is its pre-analytical stability in serum or plasma, at room temperature or frozen, and for long periods. A commonly assigned cutoff value for CRP is ~10 mg/l, with concentrations of 10–40 mg/l associated with mild inflammation and concentrations of 40–200 mg/l associated with acute inflammation and bacterial infections. The commonly used procedures usually detect values ~3 mg/l. The concentrations of CRP in low-grade systemic inflammatory conditions are much lower than those measured in acute inflammation. The high-sensitivity CRP (hs-CRP) assays detect concentrations accurately and reproducibly down to 0.3 mg/l.

The cutoff points recommended for the risk assessment of majority of the low-grade systemic inflammatory conditions are ≥1.0 mg/l (low risk), 1.0–3.0 mg/l (average risk), and ≥3.0 mg/l (high risk). These cut off points can be applied irrespective of sex and race. The units of measure recommended group are milligrams per liter, and hs-CRP results should be expressed to 1 decimal point. The plasma levels of hs-CRP can be used in the assessment of the risk of development of the low-grade systemic inflammatory conditions, and also in the evaluation of their prognosis and response to treatment.

Cytokines and Chemokines

A number of studies showed that other inflammatory markers could be used to predict the development of various cardiovascular diseases and other low-grade systemic inflammatory conditions including atherosclerosis and to predict their prognosis. Interleukin-1 (IL-1), IL-6, IL-8, IL-10, tumor necrosis factor-α (TNF-α), and monocyte chemoattractant protein (MCP-1) are some such factors that have been studied. Both adhesion molecules such as intracellular adhesion molecule-1 (ICAM-1) and soluble vascular adhesion molecule-1 and proinflammatory cytokines IL-1, IL-6, IL-8, IL-10, and TNF-α have been associated with a risk of new coronary events in ischemic heart diseases and with clinical recurrence of symptoms and other low-grade systemic inflammatory conditions [346–362]. But, these markers are less advantageous compared to hs-CRP because these markers are relatively unstable in serum; serum and plasma samples need to be rapidly separated from the cellular constituents of blood, and assayed rapidly or the samples need to be frozen to prevent degradation of the cytokines and adhesion molecules. Typically, these assays are performed using ELISA technique. If more automated assay methods become available and development of an automated microplate system, chemiluminescent assays may make their measurements more attractive. Multiplex assays for several cytokines are also an attractive option to use in the clinical setting.

Recent studies showed that fibrinogen was consistently associated with long-term risk of CHD [363], although its association differs among studies. This in part could be due to the differences in the analytical methods employed. Serum amyloid A is another inflammatory marker that can be use in CHD and other low-grade systemic inflammatory conditions [364–368], although some of these results have been inconsistent. In some studies, serum amyloid A but not hs-CRP was found to be associated with the extension of CHD, suggesting that both markers have a similar association with events but may possess different roles in the pathogenesis of atherosclerosis but not in the prediction of future events.

IL-18, originally described as interferon-inducing factor, is present atherosclerotic plaques [369]. IL-18 has been shown to be associated with future cardiovascular death in a 3.9-year-long follow-up of patients with stable angina and unstable angina pectoris. The predictive value of IL-18 was similar to that of hs-CRP, suggesting that it does not add significant value in predicting future CHD compared to hs-CRP [370]. Similar increases in the plasma IL-18 was reported in other low-grade systemic inflammatory conditions such as obesity, type 2 diabetes, hypertension, schizophrenia, Alzheimer's disease and depression [371–378].

Myeloperoxidase is a pro-inflammatory leukocyte enzyme that is present in abundant amounts in the ruptured plaque. Recent studies showed that myeloperoxidase could be associated with the recurrence of CHD and other cardiovascular events even in those who were negative for troponins [379]. It is interesting to note that the predictive value of myeloperoxidase was found to be independent of both troponin and hs-CRP levels [380]. It remains to be seen whether myeloperoxidase could be used routinely to predict prognosis of patients with CHD. Similar increase in MPO can also be found in other inflammatory conditions such as depression and schizophrenia [42, 43, 381].

Conventional Markers of Inflammation

Leukocytosis is known to be an excellent marker of inflammation. Recent studies revealed that higher leukocyte count could be associated with a greater cardiovascular risk. Since there are many extraneous factors that can influence leukocyte count, one need to be careful in using leukocyte count as a marker of predicting or prognosticating cardiovascular risk and other low-grade systemic inflammatory conditions.

Elevated fibrinogen levels have been shown to be a major independent risk factor for cardiovascular diseases and stroke outcomes [382, 383]. Higher fibrinogen levels enhanced the CHD risk of patients with hypertension, cigarette smokers, and people with diabetes and are also found in other low-grade systemic inflammatory conditions such as depression, schizophrenia, Alzheimer's disease, hypertension, type 2 diabetes mellitus, and atherosclerosis [384–387].

Role of Pro-inflammatory Markers in the Pathophysiology of the Low-grade Systemic Inflammatory Conditions

It is evident from the preceding discussion that hs-CRP and other proinflammatory indices could be used as an independent risk factor for cardiovascular diseases, atherothrombosis, Alzheimer's disease, depression, schizophrenia, hypertension, and type 2 diabetes mellitus. Collagen vascular diseases such as lupus and rheumatoid arthritis are known inflammatory conditions and even when these diseases appear to be relatively silent low-grade inflammatory process may be present. High levels of hs-CRP, IL-6, IL-18, TNF-α, amyloid A, MPO, fibrinogen, and leukocytosis seem to predict future cardiovascular risk, schizophrenia and other low-grade systemic inflammatory conditions in otherwise apparently healthy men and women. The mechanism(s) involved in their ability to serve such a powerful tool to detect patients at risk for cardiovascular diseases and other inflammatory conditions needs to be clarified. It is likely that these markers, especially CRP and MPO, are closely linked to the underlying pathophysiology namely, low-grade systemic inflammation. High concentrations of hs-CRP may simply be a reflection of the underlying inflammatory process that is ultimately responsible for the initiation and progression of the disease at a later date. Since low-grade systemic inflammation occurs as a result of failure of the anti-thrombotic properties of endothelium, it is possible that increase in hs-CRP and inflammatory markers is an indication that endothelial cells are no longer able to perform their anti-thrombotic actions adequately. In other words, increase in the concentrations of hs-CRP and other pro-inflammatory markers is an indication that endothelial cells have failed to produce anti-thrombotic molecules such as NO and PGI2 and that inflammation is responsible for this failure.

This is supported by the observation that CRP induced matrix metalloproteinase-1 (MMP-1) expression through the Fc gamma RII and extracellular signal-related kinase pathway, upregulated IL-8 in human aortic endothelial cells via NF-κB, promoted monocyte chemoattractant protein-1-mediated chemotaxis by upregulating CC-chemokine receptor 2 expression in monocytes, and attenuated endothelial progenitor cell survival, differentiation, and function via inhibiting NO generation, events that initiate and perpetuate inflammation [42, 43, 388]. Studies using human CRP transgenic animal models showed that CRP promoted atherothrombosis and increased plasminogen activator inihibitor-1. There is evidence to suggest that CRP is not only produced by liver but also by endothelial cells indicating that local increased production of CRP could be responsible for endothelial dysfunction. CRP binds to Fc gamma receptors on leukocytes. CRP significantly upregulated surface expression of Fc gamma receptors, CD32, as well as CD64 on human aortic endothelial cells. CRP is co-localized with CD32 and CD64. The increase in IL-8, intercellular adhesion molecule 1, and vascular adhesion molecule-1, and the decrease in eNO and PGI_2 induced by CRP was abrogated by specific antibodies to CD32 and CD64. These results suggest that the biological effects of CRP are mediated via binding and internalization through Fc gamma receptors, CD32 and CD64 [389]. CRP selectively enhanced intracellular generation of ROS in monocytes and neutrophils [390], decreased PGI_2 release from human aortic endothelial cells by inactivating PGIS (prostacyclin synthase) via nitration [391], and also directly inhibited NO generation by cytokine-stimulated vascular smooth muscle cells [392], and most importantly, induced apoptosis in human coronary vascular smooth muscle cells [393]. All these actions of CRP ultimately lead to the development and progression of atherosclerosis and CHD (see Fig. 3.9). It is likely that other pro-inflammatory markers such as IL-6, IL-18, TNF, IL-1, IL-2, fibrinogen, MPO, ROS and chemokines may have similar actions at their sites of increased formation and lead to the initiation and progression of various low-grade systemic inflammatory conditions. These pro-inflammatory markers may also suppress the production and action of anti-inflammatory molecules such as NO, lipoxins, resolvins, protectins

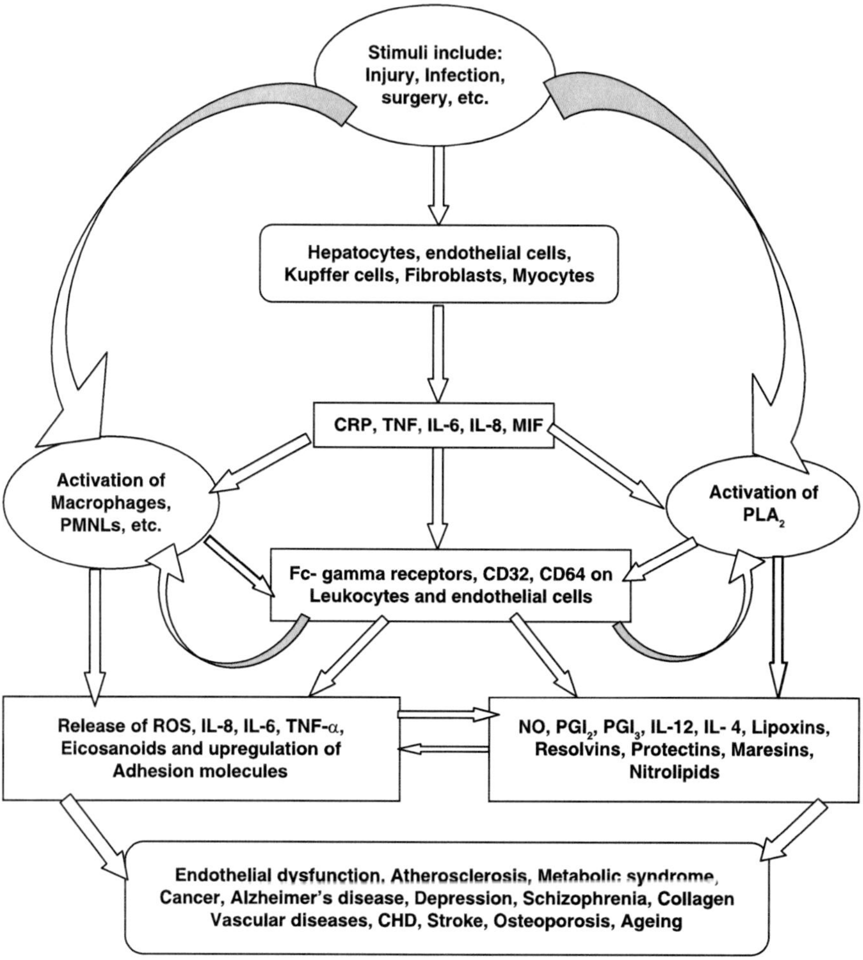

Fig. 3.9 Actions of pro-inflammatory molecules and their relevance to low-grade systemic inflammatory conditions including ageing

and maresins and anti-oxidant enzymes and thus, tilt the balance more in favor of inflammation.

Despite all these evidences, it is not clear what triggers the initiation of the low-grade systemic inflammatory process. It is likely that some endogenous anti-inflammatory molecule(s) are not produced in adequate amounts at a given point of time and at the right time and thus, the inflammatory process could be triggered. I propose that such endogenous anti-inflammatory molecule(s) that have the ability to regulate inflammatory process and suppress these low-grade systemic inflammatory conditions are unsaturated fatty acids and their anti-inflammatory metabolites such as lipoxins, resolvins, protectins and maresins.

References

[1] Ferrero-Miliani L, Nielsen OH, Andersen PS, Girardin SE (2007) Chronic inflammation: importance of NOD2 and NALP3 in interleukin-1beta generation. Clin Exp Immunol 147:227–235

[2] Das UN (2006) Clinical laboratory tools to diagnose inflammation. Adv Clin Chem 41:189–229

[3] Lau D, Mollanau H, Eiserich JP et al (2005) Myeloperoxidase mediates neutrophil activation by association with CD11/CD18 integrins. Proc Natl Acad Sci U S A 102:431–436

[4] Baldus S, Heitzer T, Eiserich JP et al (2004) Myeloperoxidase enhances nitric oxide catabolism during myocardial ischemia and reperfusion. Free Radic Biol Med 37:902–911

[5] DeLeo FR, Goedken M, McCormick SJ, Nauseef WM (1998) A novel form of hereditary myeloperoxidase deficiency linked to endoplasmic reticulum/proteasome degradation. J Clin Invest 101:2900–2909

[6] Koziol-Montewka M, Kolodziejek A, Oles J (2004) Study on myeloperoxidase role in antituberculous defense in the context of cytokine activation. Inflammation 28:53–58

[7] Choi DK, Pennathur S, Perier C, Tieu K, Teismann P, Wu DC, Jackson-Lewis V, Vila M, Vonsattel JP, Heinecke JW, Przedborski S (2005) Ablation of the inflammatory enzyme myeloperoxidase mitigates features of Parkinson's disease in mice. J Neurosci 25:6594–6600

[8] Kimura H (2010) Hydrogen sulfide: its production, release and functions. Amino Acids (in press)

[9] Kimura H (2010) Hydrogen sulfide: from brain to gut. Antioxid Redox Signal 12:1111–1123

[10] Collin M, Anuar FB, Murch O, Bhatia M, Moore PK, Thiemermann C (2005) Inhibition of endogenous hydrogen sulphide formation reduces the organ injury caused by endotoxemia. Br J Pharmacol 146:498–505

[11] Bhatia M, Wong FL, Fu D, Lau HY, Moochhala SM, Moore PK (2005) Role of hydrogen sulfide in acute pancreatitis and associated lung injury. FASEB J 19:623–625

[12] Li L, Bhatia M, Zhu YZ, Zhu YC, Ramnath RD, Wang ZJ, Anuar FB, Whiteman M, Salto-Tellez M, Moore PK (2005) Hydrogen sulfide is a novel mediator of lipopolysaccharide-induced inflammation in the mouse. FASEB J 19:1196–1198

[13] Zhang H, Zhi L, Moochhala S, Moore PK, Bhatia M (2007) Hydrogen sulfide acts as an inflammatory mediator in cecal ligation and puncture-induced sepsis in mice by upregulating the production of cytokines and chemokines via NF-kappaB. Am J Physiol Lung Cell Mol Physiol 292:L960–L971

[14] Anuar F, Whiteman M, Siau JL, Kwong SE, Bhatia M, Moore PK (2006) Nitric oxide-releasing flurbiprofen reduces formation of proinflammatory hydrogen sulfide in lipopolysaccharide- treated rat. Br J Pharmacol 147:966–974

[15] Sodha NR, Clements RT, Feng J, Liu Y, Bianchi C, Horvath EM, Szabo C, Stahl GL, Sellke FW (2009) Hydrogen sulfide therapy attenuates the inflammatory response in a porcine model of myocardial ischemia/reperfusion injury. J Thorac Cardiovasc Surg 138:977–984

[16] Wallace JL, Vong L, McKnight W, Dicay M, Martin GR (2009) Endogenous and exogenous hydrogen sulfide promotes resolution of colitis in rats. Gastroenterology 137:569–578

[17] Chen YH, Wu R, Geng B, Qi YF, Wang PP, Yao WZ, Tang CS (2009) Endogenous hydrogen sulfide reduces airway inflammation and remodeling in a rat model of asthma. Cytokine 45:117–123

[18] Jain SK, Bull R, Rains JL, Bass PF, Levine SN, Reddy S, McVie R, Bocchini JA (2010) Low levels of hydrogen sulfide in the blood of diabetes patients and streptozotocin-treated rats causes vascular inflammation? Antioxid Redox Signal (in press)

[19] Ekundi-Valentim E, Santos KT, Camargo EA, Denadai-Souza A, Teixeira SA, Zanoni CI, Grant AD, Wallace JL, Muscarà MN, Costa SK (2010) Differing effects of exogenous and endogenous hydrogen sulphide in carrageenan-induced knee joint synovitis in the rat. Br J Pharmacol (in press)

[20] Spiller F, Orrico MI, Nascimento DC, Czaikoski PG, Souto FO, Alves-Filho JC, Freitas A, Carlos D, Montenegro MF, Neto AF, Ferreira SH, Rossi MA, Hothersall JS, Assreuy J,

Cunha FQ (2010) Hydrogen sulfide improves neutrophil migration and survival in sepsis via K+ATP channel activation. Am J Respir Crit Care Med (in press)
[21] Whiteman M, Li L, Rose P, Tan CH, Parkinson DB, Moore PK (2010) The effect of hydrogen sulfide donors on lipopolysaccharide-induced formation of inflammatory mediators in macrophages. Antioxid Redox Signal (in press)
[22] Baldus S, Heitzer T, Eiserich JP et al (2004) Myeloperoxidase enhances nitric oxide catabolism during myocardial ischemia and reperfusion. Free Radic Biol Med 37:902–911
[23] Wang HD, Pagano PJ, Du Y, Cayatte AJ, Quinn MT, Brecher P, Cohen RA (1998) Superoxide anion from the adventitia of the rat thoracic aorta inactivates nitric oxide. Circ Res 82:810–818
[24] Rubanyi GM, Ho EH, Cantor EH, Lumma WC, Botelho LH (1991) Cytoprotective function of nitric oxide: inactivation of superoxide radicals produced by human leukocytes. Biochem Biophys Res Commun 181:1392–1397
[25] Fujii H, Ichimori K, Hoshiai K, Nakazawa H (1997) Nitric oxide inactivates NADPH oxidase in pig neutrophils by inhibiting its assembling process. J Biol Chem 272:32773–32778
[26] Selemidis S (2008) Suppressing NADPH oxidase-dependent oxidative stress in the vasculature with nitric oxide donors. Clin Exp Pharmacol Physiol 35:1395–1401
[27] McDonald DM, Thurston G, Baluk P (1999) Endothelial gaps as sites for plasma leakage in inflammation. Microcirculation 6:7–22
[28] Ferrara N (2002) Role of vascular endothelial growth factor in physiologic and pathologic angiogenesis: therapeutic implications. Semin Oncol 29:10–14
[29] Zeck-Kapp G, Kapp A, Busse R, Riede UN (1990) Interaction of granulocytes and endothelial cells upon stimulation with tumor necrosis factor-alpha: an ultrastructural study. Immunobiology 181:267–275
[30] Rosen H, Klebanoff SJ (1985) Oxidation of microbial iron-sulfur centers by the myeloperoxidase-H_2O_2-halide antimicrobial system. Infect Immun 47:613–618
[31] Turco J, Liu H, Gottlieb SF, Winkler HH (1998) Nitric oxide- mediated inhibition of the ability of Rickettsia prowazekii to infect mouse fibroblasts and mouse macrophage like cells. Infect Immun 66:558–566
[32] Krinsky NI (1974) Singlet excited oxygen as a mediator of the antibacterial action of leukocytes. Science 186:363–365
[33] Gragoudas ES, Adamis AP, Cunningham Jr ET, Feinsod M, Guyer DR, for the VEGF Inhibition Study in Ocular Neovascularization Clinical trial Group (2004) Pegaptanib for neovascular age related macular degeneration. N Engl J Med 351:2805–2816
[34] Bressler NM (2009) Antiangiogenic approaches to age-related macular degeneration today. Ophthalmology 116(10 Suppl):S15–S23
[35] Das UN (2000) Recent advances in the pathobiology of septicemia and septic shock. J Assoc Physicians India 48:1181–1184
[36] Cohen J, Carlet J (1996) INTERSEPT: an international, multicenter, placebo-controlled trial of monoclonal antibody to human tumor necrosis factor-alpha in patients with sepsis. International Sepsis Trial Study Group. Crit Care Med 24:1431–1440
[37] Abraham E, Anzueto A, Gutierrez G, Tessler S, San Pedro G, Wunderink R, Dal Nogare A, Nasraway S, Berman S, Cooney R, Levy H, Baughman R, Rumbak M, Light RB, Poole L, Allred R, Constant J, Pennington J, Porter S (1998) Double-blind randomised controlled trial of monoclonal antibody to human tumour necrosis factor in treatment of septic shock. NORASEPT II Study Group. Lancet 351:929–933
[38] Katusic ZS, Vanhoutte PM (1989) Superoxide anion is an endothelium derived contracting factor. Am J Physiol 257:433–437
[39] Kumar KV, Das UN (1993) Are free radicals involved in the pathobiology of human essential hypertension? Free Rad Res Commun 19:59–66
[40] Nakazono L, Watanabe N, Matsuno K, Sasaki J, Sato T, Inoue M (1991) Does superoxide underlie the pathogenesis of hypertension? Proc Natl Acad Sci U S A 88:10045–10048
[41] Das UN (1992) Can free radicals induce coronary vasospasm and acute myocardial infarction? Med Hypotheses 39:90–94

[42] Das UN (2002) A Perinatal strategy for preventing adult disease: The role of long-chain polyunsaturated fatty acids. Kluwer Academic Publishers, Boston, MA
[43] Das UN (2010) Metabolic syndrome pathophysiology: the role of essential fatty acids. Wily-Blackwell, Ames, IA
[44] Spite M, Norling LV, Summers L, Yang R, Cooper D, Petasis NA, Flower RJ, Perretti M, Serhan CN (2009) Resolvin D2 is a potent regulator of leukocytes and controls microbial sepsis. Nature 461:1287–1291
[45] Serhan CN, Yacoubian S, Yang R (2008) Anti-inflammatory and pro-resolving lipid mediators. In: Abbas AK, Galli SJ, Howley PM (eds) Annu Rev Pathol Mech Dis 3:279–312
[46] Schwab JM, Chiang N, Arita M, Serhan CN (2007) Resolvin E1 and protectin D1 activate inflammation-resolution programmes. Nature 447:869–874
[47] Serhan CN, Chiang N, Van Dyke TE (2008) Resolving inflammation: dual anti-inflammatory and pro-resolution lipid mediators. Nat Rev Immunol 8:249–261
[48] Serhan CN, Yang R, Martinod K, Kasuga K, Pillai PS, Porter TF, Oh SF, Spite M (2009) Maresins: novel macrophage mediators with potent anti-inflammatory and pro-resolving actions. J Exp Med 206:15–23
[49] Muller WA (2002) Leukocyte-endothelial cell interactions in the inflammatory response. Lab Invest 82:521
[50] Kumar V, Abbas AK, Fausto N (2005) Acute and chronic inflammation. In: Robbins and cotran pathologic basis of disease, 7th edn. Elsevier Saunders, St. Louis, MO, pp 47–86
[51] Hattori H, Subramanian KK, Sakai J, Jia Y, Li Y, Porter TF, Loison F, Sarraj B, Kasorn A, Jo H, Blanchard C, Zirkle D, McDonald D, Pai SY, Serhan CN, and Luo HR (2010) Small-molecule screen identifies reactive oxygen species as key regulators of neutrophil chemotaxis. Proc Natl Acad Sci U S A 107:3546–3551
[52] Cicchetti G, Allen PG, Glogauer M (2002) Chemotactic signaling pathways in neutrophils: from receptor to actin assembly. Crit Rev Oral Biol Med 13:220–228
[53] Bucci M, Roviezzo F, Posadas I et al (2005) Endothelial nitric oxide synthase activation is critical for vascular leakage during acute inflammation in vivo. Proc Natl Acad Sci U S A 102:904–908
[54] Connelly L, Madhani M, Hobbs AJ (2005) Resistance to endotoxic shock in endothelial nitric-oxide synthase (eNOS) knock-out mice: a pro-inflammatory role for eNOS-derived no in vivo. J Biol Chem 280:10040–10046
[55] Barua RS, Ambrose JA, Srivastava S, DeVoe MC, Eales-Reynolds LJ (2003) Reactive oxygen species are involved in smoking-induced dysfunction of nitric oxide biosynthesis and up-regulation of endothelial nitric oxide synthase: an in vitro demonstration in human coronary artery endothelial cells. Circulation 107:2342–2347
[56] Piconi L, Quagliaro L, Da Ros R, Assaloni R, Giugliano D, Esposito K, Szabó C, Ceriello A (2004) Intermittent high glucose enhances ICAM-1, VCAM-1, E-selectin and interleukin-6 expression in human umbilical endothelial cells in culture: the role of poly(ADP-ribose) polymerase. J Thromb Haemost 2:1453–1459
[57] Quagliaro L, Piconi L, Assaloni R, Da Ros R, Maier A, Zuodar G, Ceriello A (2005) Intermittent high glucose enhances ICAM-1, VCAM-1 and E-selectin expression in human umbilical vein endothelial cells in culture: the distinct role of protein kinase C and mitochondrial superoxide production. Atherosclerosis 183:259–267
[58] Richardson M, Hadcock SJ, DeReske M, Cybulsky MI (1994) Increased expression in vivo of VCAM-1 and E-selectin by the aortic endothelium of normolipemic and hyperlipemic diabetic rabbits. Arterioscler Thromb 14:760–769
[59] Hackman A, Abe Y, Insull W Jr, Pownall H, Smith L, Dunn K, Gotto AM Jr, Ballantyne CM (1996) Levels of soluble cell adhesion molecules in patients with dyslipidemia. Circulation 93:1334–1338
[60] Sakai A, Kume N, Nishi E, Tanoue K, Miyasaka M, Kita T (1997) P-selectin and vascular cell adhesion molecule-1 are focally expressed in aortas of hypercholesterolemic rabbits before intimal accumulation of macrophages and T lymphocytes. Arterioscler Thromb Vasc Biol 17:310–316

[61] Nomura S, Kanazawa S, Fukuhara S (2003) Effects of eicosapentaenoic acid on platelet activation markers and cell adhesion molecules in hyperlipidemic patients with Type 2 diabetes mellitus. J Diabetes Complications 17:153–159
[62] Rezaie-Majd A, Prager GW, Bucek RA, Schernthaner GH, Maca T, Kress HG, Valent P, Binder BR, Minar E, Baghestanian M (2003) Simvastatin reduces the expression of adhesion molecules in circulating monocytes from hypercholesterolemic patients. Arterioscler Thromb Vasc Biol 23:397–403
[63] Stulc T, Vrablík M, Kasalovà Z, Marinov I, Svobodovà H, Ceska R (2008) Leukocyte and endothelial adhesion molecules in patients with hypercholesterolemia: the effect of atorvastatin treatment. Physiol Res 57:184–194
[64] Bolewski A, Lipiecki J, Plewa R, Burchardt P, Siminiak T (2008) The effect of atorvastatin treatment on lipid profile and adhesion molecule levels in hypercholesterolemic patients: relation to low-density lipoprotein receptor gene polymorphism. Cardiology 111:140–146
[65] Benítez MB, Cuniberti L, Fornari MC, Gómez Rosso L, Berardi V, Elikir G, Stutzbach P, Schreier L, Wikinski R, Brites F (2008) Endothelial and leukocyte adhesion molecules in primary hypertriglyceridemia. Atherosclerosis 197:679–687
[66] Alipour A, van Oostrom AJ, Izraeljan A, Verseyden C, Collins JM, Frayn KN, Plokker TW, Elte JW, Castro Cabezas M (2008) Leukocyte activation by triglyceride-rich lipoproteins. Arterioscler Thromb Vasc Biol 28:792–797
[67] Kumar KV, Das UN (1993) Are free radicals involved in the pathobiology of human essential hypertension? Free Radic Res Commun 19:59–66
[68] Pedrinelli R, Dell'Omo G, Di Bello V, Pellegrini G, Pucci L, Del Prato S, Penno G (2004) Low-grade inflammation and microalbuminuria in hypertension. Arterioscler Thromb Vasc Biol 24:2414–2419
[69] Syrenicz A, Garanty-Bogacka B, Syrenicz M, Gebala A, Dawid G, Walczak M (2006) Relation of low-grade inflammation and endothelial activation to blood pressure in obese children and adolescents. Neuro Endocrinol Lett 27:459–464
[70] Das UN (2006) Hypertension as a low-grade systemic inflammatory condition that has its origins in the perinatal period. J Assoc Physicians India 54:133–142
[71] Kuklinska AM, Mroczko B, Musial WJ, Sawicki R, Kozieradska A, Waszkiewicz E, Szmitkowski M (2009) High-sensitivity C-reactive protein and total antioxidant status in patients with essential arterial hypertension and dyslipidemia. Adv Med Sci 54:225–232
[72] Katusic ZS, Vanhoutte PM (1989) Superoxide anion is an endothelium derived contracting factor. Am J Physiol 257:433–437
[73] Nakazono L, Watanabe N, Matsuno K, Sasaki J, Sato T, Inoue M (1991) Does superoxide underlie the pathogenesis of hypertension? Proc Natl Acad Sci U S A 88:10045–10048
[74] Jun T, Ke-yan F, Catalano M (1996) Increased superoxide anion production in humans: a possible mechanism for the pathogenesis of hypertension. J Hum Hypertens 10:305–309
[75] Wolf G (2000) Free radical production and angiotensin. Curr Hypertens Rep 2:167–173
[76] Zhang H, Schmeisser A, Garlichs CD, Plotze K, Damme U, Mugge A, Daniel WG (1999) Angiotensin II-induced superoxide anion generation in human vascular endothelial cells: role of membrane-bound NADH-NAD(P)H-oxidases. Cardiovasc Res 44:215–222
[77] Lob HE, Marvar PJ, Guzik TJ, Sharma S, McCann LA, Weyand C, Gordon FJ, Harrison DG (2010) Induction of hypertension and peripheral inflammation by reduction of extracellular superoxide dismutase in the central nervous system. Hypertension 55:277–283
[78] Retailleau K, Belin de Chantemèle EJ, Chanoine S, Guihot AL, Vessières E, Toutain B, Faure S, Bagi Z, Loufrani L, Henrion D (2010) Reactive oxygen species and cyclooxygenase 2-derived thromboxane A2 reduce angiotensin II type 2 receptor vasorelaxation in diabetic rat resistance arteries. Hypertension 55:339–344
[79] Mazor R, Itzhaki O, Sela S, Yagil Y, Cohen-Mazor M, Yagil C, Kristal B (2010) Tumor necrosis factor-alpha: a possible priming agent for the polymorphonuclear leukocyte-reduced nicotinamide-adenine dinucleotide phosphate oxidase in hypertension. Hypertension 55:353–362

[80] Walker AE, Seibert SM, Donato AJ, Pierce GL, Seals DR (2010) Vascular endothelial function is related to white blood cell count and myeloperoxidase among healthy middle-aged and older adults. Hypertension 55:363–369
[81] Padma M, Das UN (1996) Effect of cis-unsaturated fatty acids on cellular oxidant stress in macrophage tumor (AK-5) cells in vitro. Cancer Lett 109:63–75
[82] Das UN (1991) Arachidonic acid as a mediator of some of the actions of phorbolmyristate acetate, a tumor promotor and inducer of differentiation. Prostaglandins Leukot Essent Fatty Acids 42:241–244
[83] Das UN, Padma M, Sangeetha P et al (1990) Stimulation of free radical generation in human leukocytes by various stimulants including tumor necrosis factor is a calmodulin dependent process. Biochem Biophys Res Commun 167:1030–1036
[84] Serhan CN, Hong S, Gronert K, Colgan SP, Devchand PR, Mirick G, Moussignac R-L (2002) Resolvins: a family of bioactive products of omega-3 fatty acid transformation circuits initiated by aspirin treatment that counter proinflammatory signals. J Exp Med 196:1025–1037
[85] Claria J, Serhan CN (1995) Aspirin triggers previously undescribed bioactive eicosanoids by human endothelial cell-leukocyte interactions. Proc Natl Acad Sci U S A 92:9475–9479
[86] Serhan CN (2009) Systems approach to inflammation resolution: identification of novel anti-inflammatory and pro-resolving mediators. J Thromb Haemost 7(Suppl 1):44–48
[87] Hampton MB, Kettle AJ, Winterbourn CC (1998) Inside the neutrophil phagosome: oxidants, myeloperoxidase, and bacterial killing. Blood 92:3007
[88] Reeves EP, Lu H, Jacobs HL, Messina CG, Bolsover S, Gabella G, Potma EO, Warley A, Roes J, Segal AW (2002) Killing activity of neutrophils is mediated through activation of proteases by K+ flux. Nature 416:291–297
[89] Segal AW (2006) How superoxide production by neutrophil leukocytes kills microbes. Novartis Found Symp 279:92–98 discussion 98–100, 216–219
[90] Ahluwalia J, Tinker A, Clapp LH, Duchen MR, Abramov AY, Pope S, Nobles M, Segal AW (2004) The large-conductance Ca2+-activated K+ channel is essential for innate immunity. Nature 427:853–858
[91] Aratani Y, Koyama H, Nyui S, Suzuki K, Kura F, Maeda N (1999) Severe impairment in early host defense against Candida albicans in mice deficient in myeloperoxidase. Infect Immun 67:1828–1836
[92] Aratani Y, Kura F, Watanabe H et al (2002) Relative contributions of myeloperoxidase and NADPH-oxidase to the early host defense against pulmonary infections with Candida albicans and Aspergillus fumigatus. Med Mycol 40:557–563
[93] Aratani Y, Kura F, Watanabe H et al (2004) In vivo role of myeloperoxidase for the host defense. Jpn J Infect Dis 57:S15
[94] Aratani Y, Kura F, Watanabe H, Akagawa H, Takano Y, Suzuki K, Dinauer MC, Maeda N, Koyama H (2002) Critical role of myeloperoxidase and nicotinamide adenine dinucleotide phosphate-oxidase in high-burden systemic infection of mice with Candida albicans. J Infect Dis 185:1833–1837
[95] Lehrer RI, Cline MJ (1969) Leukocyte myeloperoxidase deficiency and disseminated candidiasis: the role of myeloperoxidase in resistance to Candida infection. J Clin Invest 48:1478–1488
[96] Netea MG, van Tits LJ, Curfs JH, Amiot F, Meis JF, van der Meer JW, Kullberg BJ (1999) Increased susceptibility of TNF-alpha lymphotoxin-alpha double knockout mice to systemic candidiasis through impaired recruitment of neutrophils and phagocytosis of Candida albicans. J Immunol 163:1498–1505
[97] Vonk AG, Netea MG, van Krieken JH, van der Meer JW, Kullberg BJ (2002) Delayed clearance of intraabdominal abscesses caused by Candida albicans in tumor necrosis factor-alpha- and lymphotoxin-alpha-deficient mice. J Infect Dis 186:1815–1822
[98] Ferrante A, Nandoskar M, Walz A, Goh DH, Kowanko IC (1988) Effects of tumour necrosis factor alpha and interleukin-1 alpha and beta on human neutrophil migration, respiratory burst and degranulation. Int Arch Allergy Appl Immunol 86:82–91

[99] Test ST (1991) Effect of tumor necrosis factor on the generation of chlorinated oxidants by adherent human neutrophils. J Leukoc Biol 50:131–139
[100] Netea MG, Meer JW, Verschueren I, Kullberg BJ (2002) CD40/CD40 ligand interactions in the host defense against disseminated Candida albicans infection: the role of macrophage-derived nitric oxide. Eur J Immunol 32:1455–1463
[101] Marieb E (2001) Human anatomy & physiology. Benjamin Cummings, San Francisco, CA, p 414
[102] Paiva TB, Tominaga M, Paiva ACM (1970) Ionization of histamine, N-acetylhistamine, and their iodinated derivatives. J Med Chem 13:689–692
[103] Monroe E, Daly A, Shalhoub R (1997) Appraisal of the validity of histmine-induced wheal and flare to predict the clinical efficacy of antihistamines. J Allergy Clin Immunol 99:S789–S806
[104] Yanai K, Tashiro M (2007) The physiological and pathophysiological roles of neuronal histamine: an insight from human positron emission tomography studies. Pharmacol Ther 113:1–15
[105] Alvarez EO (2009) The role of histamine on cognition. Behav Brain Res 199:183–189
[106] White JM, Rumbold GR (1988) Behavioural effects of histamine and its antagonists: a review. Psychopharmacology 95:1–14
[107] Carà AM, Lopes-Martins RA, Antunes E, Nahoum CR, De Nucci G (1995) The role of histamine in human penile erection. Br J Urol 75:220–224
[108] Ito C (2004) The role of the central histaminergic system on schizophrenia. Drug News Perspect 17:383–387
[109] Berger M, Gray JA, Roth BL (2009) The expanded biology of serotonin. Annu Rev Med 60:355–66
[110] Niacaris T, Avery L (2003) Serotonin regulates repolarization of the C. elegans pharyngeal muscle. J Exp Biol 206(Pt 2):223–231
[111] Stahl SM, Mignon L, Meyer JM (2009) Which comes first: atypical antipsychotic treatment or cardiometabolic risk? Acta Psychiatr Scand 119:171–179
[112] Buckland PR, Hoogendoorn B, Guy CA, Smith SK, Coleman SL, O'Donovan MC (2005) Low gene expression conferred by association of an allele of the 5-HT2C receptor gene with antipsychotic-induced weight gain. Am J Psychiatry 162:613–615
[113] Holmes MC, French KL, Seckl JR (1997) Dysregulation of diurnal rhythms of serotonin 5-HT2C and corticosteroid receptor gene expression in the hippocampus with food restriction and glucocorticoids. J Neurosci 17:4056–4065
[114] Leibowitz SF (1990) The role of serotonin in eating disorders. Drugs 39(Suppl 3):33–48
[115] de Wit R, Aapro M, Blower PR (2005) Is there a pharmacological basis for differences in 5-HT3-receptor antagonist efficacy in refractory patients? Cancer Chemother Pharmacol 56:231–238
[116] Kravitz EA (1988) Hormonal control of behavior: amines and the biasing of behavioral output in lobsters. Science 241:1775–1781
[117] Yeh SR, Fricke RA, Edwards DH (1996) The effect of social experience on serotonergic modulation of the escape circuit of crayfish. Science 271:366–369
[118] Caspi N, Modai I, Barak P, Waisbourd A, Zbarsky H, Hirschmann S, Ritsner M (2001) Pindolol augmentation in aggressive schizophrenic patients: a double-blind crossover randomized study. Int Clin Psychopharmacol 16:111–115
[119] Greg B (2000) Suicide linked to serotonin gene. CMAJ 162:1343
[120] Lesch KP, Bengel D, Heils A, Sabol SZ, Greenberg BD, Petri S, Benjamin J, Müller CR, Hamer DH, Murphy DL (1996) Association of anxiety-related traits with a polymorphism in the serotonin transporter gene regulatory region. Science 274:1527–1531
[121] Caspi A, Sugden K, Moffitt TE, Taylor A, Craig IW, Harrington H, McClay J, Mill J, Martin J, Braithwaite A, Poulton R (2003) Influence of life stress on depression: moderation by a polymorphism in the 5-HTT gene. Science 301:386–389
[122] Levinson DF (2006) The genetics of depression: a review. Biol Psychiatry 60:84–92

[123] Srinivasan S, Sadegh L, Elle IC, Christensen AG, Faergeman NJ, Ashrafi K (2008) Serotonin regulates C. elegans fat and feeding through independent molecular mechanisms. Cell Metab 7:533–544
[124] Loer CM, Kenyon CJ (1993) Serotonin-deficient mutants and male mating behavior in the nematode Caenorhabditis elegans. J Neurosci 13:5407–5417
[125] Lipton J, Kleemann G, Ghosh R, Lints R, Emmons SW (2004) Mate searching in Caenorhabditis elegans: a genetic model for sex drive in a simple invertebrate. J Neurosci 24:7427–7434
[126] Kaplan DD, Zimmermann G, Suyama K, Meyer T, Scott MP (2008) A nucleostemin family GTPase, NS3, acts in serotonergic neurons to regulate insulin signaling and control body size. Genes Dev 22:1877–1893
[127] Ruad AF, Thummel CS (2008) Serotonin and insulin signaling team up to control growth in Drosophila. Genes Dev 22:1851–1855
[128] Lesurtel M, Graf R, Aleil B, Walther DJ, Tian Y, Jochum W, Gachet C, Bader M, Clavien PA (2006) Platelet-derived serotonin mediates liver regeneration. Science 312:104–107
[129] Matondo RB, Punt C, Homberg J, Toussaint MJ, Kisjes R, Korporaal SJ, Akkerman JW, Cuppen E, de Bruin A (2009) Deletion of the serotonin transporter in rats disturbs serotonin homeostasis without impairing liver regeneration. Am J Physiol Gastrointest Liver Physiol 296:G963–G968
[130] Collet C, Schiltz C, Geoffroy V, Maroteaux L, Launay JM, deVernejoul MC (2008) The serotonin 5-HT2B receptor controls bone mass via osteoblast recruitment and proliferation. FASEB J 22:418–427
[131] Yadav VK, Ryu JH, Suda N, Tanaka KF, Gingrich JA, Schütz G, Glorieux FH, Chiang CY, Zajac JD, Insogna KL, Mann JJ, Hen R, Ducy P, Karsenty G (2008) Lrp5 controls bone formation by inhibiting serotonin synthesis in the duodenum. Cell 135:825–837
[132] Marieb EN (2009) Essentials of human anatomy & physiology, 8th edn. Pearson/Benjamin Cummings, San Francisco, CA, p 336
[133] Baskin SI (1991) Principles of cardiac toxicology. CRC Press, Boca Raton, FL
[134] Schurad B, Horowski R, Jähnichen S, Görnemann T, Tack J, Pertz HH (2006) Proterguride, a highly potent dopamine receptor agonist promising for transdermal administration in Parkinson's disease: interactions with alpha(1)-, 5-HT(2)- and H(1)-receptors. Life Sci 78:2358–2364
[135] Ben Arous J, Laffont S, Chatenay D (2009) Molecular and sensory basis of a food related two-state behavior in C. elegans. PLoS ONE 4:e7584
[136] Sze JY, Victor M, Loer C, Shi Y, Ruvkun G (2000) Food and metabolic signaling defects in a Caenorhabditis elegans serotonin-synthesis mutant. Nature 403:560–564
[137] Côté F, Thévenot E, Fligny C et al (2003) Disruption of the nonneuronaltph1 gene demonstrates the importance of peripheral serotonin in cardiac function. Proc Natl Acad Sci U S A 100:13525–13530
[138] Alenina N, Kikic D, Todiras M, Mosienko V, Qadri F, Plehm R, Boyé P, Vilianovitch L, Sohr R, Tenner K, Hörtnagl H, Bader M (2009) Growth retardation and altered autonomic control in mice lacking brain serotonin. Proc Natl Acad Sci U S A 106:10332–10337
[139] Savelieva KV, Zhao S, Pogorelov VM et al (2008) Genetic disruption of both tryptophan hydroxylase genes dramatically reduces serotonin and affects behavior in models sensitive to antidepressants. PLoS One 3:e3301
[140] Enrica A, Elisabetta C, Boris M, Tiziana R, Antonio C, Mumna AB, Renato C, Cornelius G (2008) Sporadic autonomic dysregulation and death associated with excessive serotonin autoinhibition. Science 321:130–133
[141] Paterson DS, Trachtenberg FL, Thompson EG, Belliveau RA, Beggs AH, Darnall R, Chadwick AE, Krous HF, Kinney HC (2006) Multiple serotonergic brainstem abnormalities in sudden infant death syndrome. JAMA 296:2124–2132
[142] Frazer A, Hensler JG (1999) Understanding the neuroanatomical organization of serotonergic cells in the bran provides insight into the functions of this neurotransmitter. In: Bernard WA,

Fisher SK, Wayne AR, Uhler MD (Hrsg) Basic neurochemistry, 6th edn. Lippincott Williams and Wilkins, Philadelphia, PA
[143] Hannon J, Hoyer D (2008) Molecular biology of 5-HT receptors. Behav Brain Res 195:198–213
[144] Walther DJ, Peter JU, Winter S, Höltje M, Paulmann N, Grohmann M, Vowinckel J, Alamo-Bethencourt V, Wilhelm CS, Ahnert-Hilger G, Bader M (2003) Serotonylation of small GTPases is a signal transduction pathway that triggers platelet alpha-granule release. Cell 115:851–862
[145] Watts SW, Priestley JR, Thompson JM (2009) Serotonylation of vascular proteins important to contraction. PLoS One 4:e5682
[146] Walther DJ, Peter JU, Bashammakh S, Hörtnagl H, Voits M, Fink H, Bader M (2003) Synthesis of serotonin by a second tryptophan hydroxylase isoform. Science 299:76
[147] Bliznakov EG (1980) Serotonin and its precursors as modulators of the immunological responsiveness in mice. J Med 11:81–105
[148] Mashek K, Devoino LV, Kadletsova O, Idova GV, Morozova NB (1985) Changes in the level of serotonin in the brain and immunocompetent organs during the formation of the immune response. Fiziol Zh SSSR Im I M Sechenova 71:992–997
[149] Devoino L, Morozova N, Cheido M (1988) Participation of serotoninergic system in neuroimmunomodulation: intraimmune mechanisms and the pathways providing an inhibitory effect. Int J Neurosci 40:111–128
[150] Ciz M, Komrskova D, Pracharova L, Okenkova K, Cizova H, Moravcova A, Jancinova V, Petrikova M, Lojek A, Nosal R (2007) Serotonin modulates the oxidative burst of human phagocytes via various mechanisms. Platelets 18:583–590
[151] Menard G, Turmel V, Bissonnette EY (2007) Serotonin modulates the cytokine network in the lung: involvement of prostaglandin E2. Clin Exp Immunol 150:340–348
[152] Muller T, Durk T, Blumental B, Grimm M, Cicko S, Panther E, Sorichter S, Herouy Y, Di Virgilio F, Ferrari D, Norgauer J, Idzko M (2009) 5-hydroxytryptamine modulates migration, cytokine and chemokine release and T-cell priming capacity of dendritic cells in vitro and in vivo. PLoS One 4:e6453
[153] Kushnir-Sukhov NM, Gilfillan AM, Coleman JW, Brown JM, Bruening S, Toth M, Metcalfe DD (2006) 5-hydroxytryptamine induces mast cell adhesion and migration. J Immunol 177:6422–6432
[154] Sookhai S, Wang JH, McCourt M, O'Connell D, Redmond HP (1999) Dopamine induces neutrophil apoptosis through a dopamine D-1 receptor-independent mechanism. Surgery 126:314–322
[155] Sookhai S, Wang JH, Winter D, Power C, Kirwan W, Redmond HP (2000) Dopamine attenuates the chemoattractant effect of interleukin-8: a novel role in the systemic inflammatory response syndrome. Shock 14:295–299
[156] Oberbeck R, Schmitz D, Wilsenack K, Schuler M, Husain B, Schedlowski M, Exton MS (2006) Dopamine affects cellular immune functions during polymicrobial sepsis. Intensive Care Med 32:731–739
[157] Wang G-J, Volkow ND, Logan J, Pappas NR, Wong CT, Zhu W, Netusil N, Fowler JS (2001) Brain dopamine and obesity. Lancet 357:354–357
[158] Esposito K, Nappo F, Marfella R et al (2002) Inflammatory cytokine concentrations are acutely increased by hyperglycemia in humans. Role of oxidative stress. Circulation 106:2067–2072
[159] Aso Y, Wakabayashi S, Nakano T, Yamamoto R, Takebayashi K, Inukai T (2006) High serum high-sensitivity C-reactive protein concentrations are associated with relative cardiac sympathetic overactivity during the early morning period in type 2 diabetic patients with metabolic syndrome. Metabolism 55:1014–1021
[160] Dungan KM, Braithwaite SS, Preiser J-C (2009) Stress hyperglycemia. Lancet 373:1798–1807
[161] Bergquist J, Tarkowski A, Ekman R, Ewing A (1994) Discovery of endogenous catecholamines in lymphocytes and evidence for catecholamine regulation of lymphocyte function via an autocrinc loop. Proc Natl Acad Sci U S A 91:12912–12916

[162] Josefsson E, Bergquist J, Ekman R, Tarkowski A (1996) Catecholamines are synthesized by mouse lymphocytes and regulate function of these cells by induction of apoptosis. Immunology 88:140–146

[163] Gulik T, Chung MK, Pieper SJ, Lange LG, Schreiner GF (1989) Interleukin 1 and tumor necrosis factor inhibit cardiac myocyte β-adrenergic responsiveness. Proc Natl Acad Sci U S A 86:6763–6767

[164] Flierl MA, Rittirsch D, Nadeau BA, Chen AJ, Sarma JV, Zetoune FS, McGuire SR, List RP, Day DE, Hoesel LM, Gao H, Rooijen NV, Huber-Lag MS, Neubig RR, Ward PA (2007) Phagocyte- derived catecholamines enhance acute inflammatory injury. Nature 449:721–726

[165] Bergmann A, Eulenberg C, Wellner M, Rolle S, Luft F, Kettritz R (2010) Aldosterone abrogates nuclear factor κB-mediated tumor necrosis factor α production in human neutrophils via the mineralocorticoid receptor. Hypertension 55:370–379

[166] Pavlov VA, Parrish WR, Rosas-Ballina M, Ochani M, Puerta M, Ochani K, Chavan S, Al-Abed Y, Tracey KJ (2009) Brain acetylcholinesterase activity controls systemic cytokine levels through the cholinergic anti-inflammatory pathway. Brain Behav Immun 23:41–45

[167] Das UN (2007) Acetylcholinesterase and butyrylcholinesterase as possible markers of low-grade systemic inflammation. Med Sci Monit 13:RA214–RA221

[168] Van Maanen MA, Vervoordeldonk MJ, Tak PP (2009) The cholinergic anti-inflammatory pathway: towards innovative treatment of rheumatoid arthritis. Nat Rev Rheumatol 5:229–232

[169] Luger TA, Brzoska T (2007) alpha-MSH related peptides: a new class of anti-inflammatory and immunomodulating drugs. Ann Rheum Dis 66(Suppl 3):52–55

[170] Taylor AW, Kitaichi N (2008) The diminishment of experimental autoimmune encephalomyelitis (EAE) by neuropeptide alpha-melanocyte stimulating hormone (alpha-MSH) therapy. Brain Behav Immun 22:639–646

[171] Park MC, Chung SJ, Park YB, Lee SK (2009) Pro-inflammatory effect of leptin on peripheral blood mononuclear cells of patients with ankylosing spondylitis. Joint Bone Spine 76:170–175

[172] Moraes JC, Coope A, Morari J, Cintra DE, Roman EA, Pauli JR, Romanatto T, Carvalheir JB, Oliveira AL, Saad MJ, Velloso LA (2009) High-fat diet induces apoptosis of hypothalamic neurons. PLoS One 4:e5045

[173] Matsumoto T, Miyatsuji A, Miyawaki T, Yanagimoto Y, Moritani T (2003) Potential association between endogenous leptin and sympatho-vagal activities in young obese Japanese women. Am J Hum Biol 15:8–15

[174] Holler J, Zakrzewicz A, Kaufmann A, Wilhelm J, Fuchs-Moll G, Dietrich H, Padberg W, Kuncova J, Kummer W, Grau V (2008) Neuropeptide Y is expressed by rat mononuclear blood leukocytes and strongly down-regulated during inflammation. J Immunol 181:6906–6912

[175] Dimitrijevic M, Stanojevic S, Mitic S, Vujic V, Kovacevic-Jovanovic V, Mitic K, von Horsten S, Kosec D (2006) Neuropeptide Y (NPY) modulates oxidative burst and nitric oxide production in carrageenan-elicited granulocytes from rat air pouch. Peptides 27:3208–3215

[176] Dimitrijevic M, Stanojevic S, Mitic S, Kustriovic N, Vujic V, Miletic T, Kovacevic-Jovanovic V (2008) The anti-inflammatory effect of neuropeptide Y (NPY) in rats is dependent on dipeptidyl peptidase 4 (DP4) activity and age. Peptides 29:2179–2187

[177] Chandrasekharan B, Bala V, Kolachala VL, Vijaykumar M, Jones D, Gewirtz AT, Sitaraman SV, Srinivasan S (2008) Targeted deletion of neuropeptide Y (NPY) modulates experimental colitis. PLoS One 3:e3304

[178] Hernanz A, Tato E, De la Fuente M, de Miguel E, Arnalich F (1996) Differential effects of gastrin-releasing peptide, neuropeptide Y, somatostatin and vasoactive intestinal peptide on interleukin-1 beta, interleukin-6 and tumor necrosis factor-alpha production by whole blood cells from healthy young and old subjects. J Neuroimmunol 71:25–30

[179] King PJ, Widdowson PS, Doods H, Williams G (2000) Effect of cytokines on hypothalamic neuropeptide Y release in vitro. Peptides 21:143–146

[180] Kos K, Harte AL, James S, Snead DR, O'Hare JP, McTernan PG, Kumar S (2007) Secretion of neuropeptide Y in human adipose tissue and its role in maintenance of adipose tissue mass. Am J Physiol Endocrinol Metab 293:E1335–E1340

[181] Lopez M, Lage R, Saha AK et al (2008) Hypothalamic fatty acid metabolism mediates the orexigenic action of ghrelin. Cell Metab 7:389–399
[182] Dixit VD, Schaffer EM, Pyle RS, Collins GD, Sakthivel SK, Palaniappan R, Lillard JW Jr, Taub DD (2004) Ghrelin inhibits leptin- and activation-induced proinflammatory cytokine expression by human monocytes and T cells. J Clin Invest 114:57–66
[183] Yada T, Mutoh K, Azuma T, Hyodo S, Kangawa K (2006) Ghrelin stimulates phagocytosis and superoxide production in fish leukocytes. J Endocrinol 189:57–65
[184] Ates Y, Degertekin B, Erdil A, Yaman H, Dagalp K (2008) Serum ghrelin levels in inflammatory bowel disease with relation to disease activity and nutritional status. Dig Dis Sci 53:2215–2221
[185] Mager U, Kolehmainen M, d Mello VD, Schwab U, Laaksonen DE, Rauramaa R, Gylling H, Atalay M, Pulkkinen L, Uusitupa M (2008) Expression of ghrelin gene in peripheral blood mononuclear cells and plasma ghrelin concentrations in patients with metabolic syndrome. Eur J Endocrinol 158:499–510
[186] Iwai T, Ito S, Tanimitsu K, Udagawa S, Oka J (2006) Glucagon-like peptide-1 inhibits LPS-induced IL-1beta production in cultured rat astrocytes. Neurosci Res 55:352–360
[187] Blandino-osano M, Perez-Arana G, Mellado-Gil JM, Segundo C, Aguilar-Diosdado M (2008) Anti-proliferative effect of pro-inflammatory cytokines in cultured beta cells is associated with extracellular signal-regulated kinase 1/2 pathway inhibition: protective role of glucagon-like peptide-1. J Mol Endocrinol 41:35–44
[188] Kim SJ, Nian C, Doudet DJ, McIntosh CH (2009) Dipeptidyl peptidase IV inhibition with MK0431 improves islet graft survival in diabetic NOD mice partially via T-cell modulation. Diabetes 58:641–651
[189] Luyer MD, Greve JW, Hadfoune M, Jacobs JA, Dejong CH, Buurman WA (2005) Nutritional stimulation of cholecystokinin receptors inhibits inflammation via the vagus nerve. J Exp Med 202:1023–1029
[190] Hamza M, Wang XM, Adam A, Brahim JS, Rowan JS, Carmona GN, Dionne RA (2010) Kinin B1 receptors contributes to acute pain following minor surgery in humans. Mol Pain 6:12
[191] Raslan F, Schwarz T, Meuth SG, Austinat M, Bader M, Renne T, Roosen K, Stoll G, Sirén AL, Kleinschnitz C (2010) Inhibition of bradykinin receptor B1 protects mice from focal brain injury by reducing blood-brain barrier leakage and inflammation. J Cereb Blood Flow Metab (in press)
[192] Chao J, Shen B, Gao L, Xia CF, Bledsoe G, Chao L (2010) Tissue kallikrein in cardiovascular, cerebrovascular and renal diseases and skin wound healing. Biol Chem 391:345–355
[193] Das UN (2006) Essential fatty acids – a review. Curr Pharm Biotechnol 7:467–482
[194] Das UN (2006) Essential fatty acids: biochemistry, physiology, and pathology. Biotechnology J 1:420–439
[195] Ritter JM, Harding I, Warren JB (2009) Precaution, cyclooxygenase inhibition, and cardiovascular risk. Trends Pharmacol Sci 30:503–508
[196] Serhan CN, Hong S, Gronert K, Colgan SP, Devchand PR, Mirick G, Moussignac R-L (2002) Resolvins: a family of bioactive products of omega-3 fatty acid transformation circuits initiated by aspirin treatment that counter proinflammatory signals. J Exp Med 196:1025–1037
[197] Claria J, Serhan CN (1995) Aspirin triggers previously undescribed bioactive eicosanoids by human endothelial cell-leukocyte interactions. Proc Natl Acad Sci U S A 92:9475–9479
[198] Chiang N, Gronert K, Clish CB, O'Brien JA, Freeman MW, Serhan CN (1999) Leukotriene B_4 receptor transgenic mice reveal novel protective roles for lipoxins and aspirin-triggered lipoxins in reperfusion. J Clin Invest 104:309–316
[199] Gilroy DW (2005) New insights into the anti-inflammatory actions of aspirin-induction of nitric oxide through the generation of epi-lipoxins. Mem Inst Oswaldo Cruz 100(Suppl 1):49–54
[200] Paul-Clark MJ, Van Cao T, Moradi-Bidhendi N, Cooper D, Gilroy DW (2004) 15-epi-lipoxin A4-mediated induction of nitric oxide explains how aspirin inhibits acute inflammation. J Exp Med 200:69–78

[201] Wen F, Watanabe K, Yoshida M (2000) Nitric oxide enhances PGI(2)production by human pulmonary artery smooth muscle cells. Prostaglandins Leukot Essent Fatty Acids 62:369–378
[202] Swierkosz TA, Mitchell JA, Warner TD, Botting RM, Vane JR (1995) Co-induction of nitric oxide synthase and cyclo-oxygenase: interactions between nitric oxide and prostanoids. Br J Pharmacol 114:1335–1342
[203] Plum J, Huang C, Grabensee B, Schrör K, Meyer-Kirchrath J (2002) Prostacyclin enhances the expression of LPS/INF-gamma-induced nitric oxide synthase in human monocytes. Nephron 91:391–398
[204] Chiang N, Gronert K, Clish CB, O'Brien JA, Freeman MW, Serhan CN (1999) Leukotriene B_4 receptor transgenic mice reveal novel protective roles for lipoxins and aspirin-triggered lipoxins in reperfusion. J Clin Invest 104:309–316
[205] Levy BD, Clish CB, Schmidt B, Gronert K, Serhan CN (2001) Lipid mediator class switching during acute inflammation signals in resolution. Nat Immunol 2:612–619
[206] Bandeira-Melo C, Serra MF, Diaz BI, Cordeiro RSB, Silva PMR, Lenzi HL, Bakhle YS, Serhan CN, Martins MA (2000) Cyclooxygenase-2-derived prostaglandin E_2 and lipoxin A_4 accelerate resolution of allergic edema in Angiostronglus costaricensis-infected rats: relationship with concurrent eosinophilia. J Immunol 164:1029–1036
[207] Serhan CN, Maddox JF, Petasis NA, Akritopoulou-Zanze I, Papayianni A, Brady HR, Colgan SP, Madara JL (1995) Design of lipoxin A_4 stable analogs that block transmigration and adhesion of human neutrophils. Biochemistry 34:14609–14615
[208] Sethi S, Eastman AY, Eaton JW (1996) Inhibition of phagocyte-endothelium interactions by oxidized fatty acids: a natural anti-inflammatory mechanism? J Lab Clin Med 128:27–38
[209] Serhan CN, Clish CB, Brannon J, Colgan SP, Chiang N, Gronert K (2000) Novel functional sets of lipid-derived mediators with antiinflammatory actions generated from omega-3 fatty acids via cyclooxygenase 2-nonsteroidal antiinflammatory drugs and transcellular processing. J Exp Med 192:1197–1204
[210] Marcheselli VL, Hong S, Lukiw WJ, Tian XH, Gronet K, Musto A, Hardy M, Gimenez JM, Chiang N, Serhan CN, Bazan NG (2003) Novel docosanoids inhibit brain ischemia-reperfusion-mediated leukocyte infiltration and pro-inflammatory gene expression. J Biol Chem 278:43807–43817
[211] Serhan CN, Yang R, Martinod K, Kasuga K, Pillai PS, Porter TF, Oh SF, Spite M (2009) Maresins: novel macrophage mediators with potent antiinflammatory and proresolving actions. J Exp Med 206:15–23
[212] Serhan CN (2009) Systems approach to inflammation resolution: identification of novel anti-inflammatory and pro-resolving mediators. J Thromb Haemost 7(Suppl 1):44–48
[213] Das UN, Puskàs LG (2009) Transgenic fat-1 mouse as a model to study the pathophysiology of cardiovascular, neurological and psychiatric disorders. Lipids Health Dis 8:61
[214] Prescott SM, Zimmerman GA, Stafforini DM, McIntyre TM (2000) Platelet-activating factor and related lipid mediators. Annu Rev Biochem 69:419–445
[215] Hamasaki Y, Mojarad M, Saga T, Tai HH, Said SI (1984) Platelet-activating factor raises airway and vascular pressures and induces edema in lungs perfused with platelet-free solution. Am Rev Respir Dis 129:742–746
[216] Vemulapalli S, Chiu PJ, Barnett A (1984) Cardiovascular and renal action of platelet-activating factor in anesthetized dogs. Hypertension 6:489–493
[217] Kawaguchi H, Yasuda H (1986) Platelet-activating factor stimulates prostaglandin synthesis in cultured cells. Hypertension 8:192–197
[218] Bruynzeel PL, Koenderman L, Kok PT, Hameling ML, Verhagen J (1986) Platelet-activating factor (PAF-acether) induced leukotriene C4 formation and luminol dependent chemiluminescence by human eosinophils. Pharmacol Res Commun 18(Suppl):61–69
[219] Bruijnzeel PL, Kok PT, Hamelink ML, Kijne AM, Verhagen J (1987) Platelet-activating factor induces leukotriene C4 synthesis by purified human eosinophils. Prostaglandins 34:205–214
[220] Dubois C, Bissonnette E, Rola-Pleszczynski M (1989) Platelet-activating factor (PAF) enhances tumor necrosis factor production by alveolar macrophages. Prevention by PAF receptor antagonists and lipoxygenase inhibitors. J Immunol 143:964–970

[221] McColl SR, Krump E, Naccache PH, Poubelle PE, Braquet P, Braquet M, Borgeat P (1991) Granulocyte-macrophage colony-stimulating factor increases the synthesis of leukotriene B4 by human neutrophils in response to platelet-activating factor. Enhancement of both arachidonic acid availability and 5-lipoxygenase activation. J Immunol 146:1204–1211
[222] Wirthmueller U, de Weck AL, Dahinden CA (1990) Studies on the mechanism of platelet-activating factor production in GM-CSF primed neutrophils: involvement of protein synthesis and phospholipase A2 activation. Biochem Biophys Res Commun 170:556–562
[223] Shindo K, Koide K, Fukumura M (1997) Enhancement of leukotriene B4 release in stimulated asthmatic neutrophils by platelet activating factor. Thorax 52:1024–1029
[224] Thivierge M, Rola-Pleszczynski M (1992) Platelet-activating factor enhances interleukin-6 production by alveolar macrophages. J Allergy Clin Immunol 90:796–802
[225] Ha SW, Lee CS (2010) Differential effect of platelet activating factor on 1-methyl-4-phenylpyridinium-induced cell death through regulation of apoptosis-related protein activation. Neurochem Int (in press)
[226] Anuurad E, Ozturk Z, Enkhmaa B, Pearson TA, Berglund L (2010) Association of lipoprotein-associated phospholipase A2 with coronary artery disease in African-Americans and Caucasians. J Clin Endocrinol Metab (in press)
[227] Khakpour H, Frishman WH (2009) Lipoprotein-associated phospholipase A2: an independent predictor of cardiovascular risk and a novel target for immunomodulation therapy. Cardiol Rev 17:222–229
[228] Wilensky RL, Macphee CH (2009) Lipoprotein-associated phospholipase A(2) and atherosclerosis. Curr Opin Lipidol 20:415–420
[229] Mantovani A, Sozzani S, Introna M (1997) Endothelial activation by cytokines. Ann NY Acad Sci 832:93–116
[230] Das UN (1998) Oxidants, anti-oxidants, essential fatty acids, eicosanoids, cytokines, gene/oncogene expression and apoptosis in systemic lupus erythematosus. J Assoc Physicians India 46:630–634
[231] Binder C, Schulz M, Hiddemann W, Oellerich M (1999) Induction of inducible nitric oxide synthase is an essential part of tumor necrosis factor-alpha-induced apoptosis in MCF-7 and other epithelial tumor cells. Lab Invest 79:1703–1712
[232] Koblish HK, Hunter CA, Wysocka M, Trinchieri G, Lee WM (1998) Immune suppression by recombinant interleukin (rIL)-12 involves interferon gamma induction of nitric oxide synthase 2 (iNOS) activity: inhibitors of NO generation reveal the extent of rIL-12 vaccine adjuvant effect. J Exp Med 188:1603–1610
[233] Das UN, Padma M, Sagar PS et al (1990) Stimulation of free radical generation in human leukocytes by various agents including tumor necrosis factor is a calmodulin dependent process. Biochem Biophys Res Commun 167:1030–1036
[234] Natanson C, Eichenholz PW, Danner RL, Eichacker PQ, Hoffman WD, Kuo GC, Banks SM, MacVittie TJ, Parrillo JE (1989) Endotoxin and tumor necrosis factor challenges in dogs simulate the cardiovascular profile of human septic shock. J Exp Med 169:823–832
[235] Das UN (2007) Insulin in the critically ill with focus on cytokines, reactive oxygen species, HLA-DR expression. J Assoc Physicians India 55(Suppl):56–65
[236] Zanotti S, Kumar A, Kumar A (2002) Cytokine modulation in sepsis and septic shock. Expert Opin Investig Drugs 11:1061–1075
[237] Argiles JM, Lopez-Soriano J, Busquets S, Lopez-Soriano FJ (1997) Journey from cachexia to obesity by TNF. FASEB J 11:743–751
[238] Ayoub S, Hickey MJ, Morand EF (2008) Mechanisms of disease: macrophage migration inhibitory factor in SLE, RA and atherosclerosis. Nat Clin Pract Rheumatol 4:98–105
[239] Verschuren L, Kooistra T, Bernhagen J, Voshol PJ, Ouwens DM, van Erk M, de Vries-van der Weij J, Leng L, van Bockel JH, van Dijk KW, Fingerle-Rowson G, Bucala R, Kleemann R (2009) MIF deficiency reduces chronic inflammation in white adipose tissue and impairs the development of insulin resistance, glucose intolerance, and associated atherosclerotic disease. Circ Res 105:99–107

[240] Toso C, Emamaullee JA, Merani S, Shapiro AM (2008) The role of macrophage migration inhibitory factor on glucose metabolism and diabetes. Diabetologia 51:1937–1946
[241] Herder C, Kolb H, Koenig W, Haastert B, Müller-Scholze S, Rathmann W, Holle R, Thorand B, Wichmann HE (2006) Association of systemic concentrations of macrophage migration inhibitory factor with impaired glucose tolerance and type 2 diabetes: results from the co-operative health research in the region of Augsburg, Survey 4 (KORA S4). Diabetes Care 29:368–371
[242] Dandona P, Aljada A, Ghanim H, Mohanty P, Tripathy C, Hofmeyer D, Chaudhuri A (2004) Increased plasma concentration of macrophage migration inhibitory factor (MIF) and MIF mRNA in mononuclear cells in the obese and the suppressive action of metformin. J Clin Endocrinol Metab 89:5043–5047
[243] Vozarova B, Stefan N, Hanson R, Lindsay RS, Bogardus C, Tataranni PA, Metz C, Bucala R (2002) Plasma concentrations of macrophage migration inhibitory factor are elevated in Pima Indians compared to Caucasians and are associated with insulin resistance. Diabetologia 45:1739–1741
[244] Bacher M, Deuster O, Aljabari B, Egensperger R, Neff F, Jessen F, Popp J, Noelker C, Reese JP, Al-Abed Y, Dodel R (2010) The role of macrophage migration inhibitory factor in Alzheimer's disease. Mol Med 16:116–121
[245] Edwards KM, Bosch JA, Engeland CG, Cacioppo JT, Marucha PT (2010) Elevated Macrophage Migration Inhibitory Factor (MIF) is associated with depressive symptoms, blunted cortisol reactivity to acute stress, and lowered morning cortisol. Brain Behav Immun 24:1202–1208.
[246] Conboy L, Varea E, Castro JE, Sakouhi-Ouertatani H, Calandra T, Lashuel HA, Sandi C (in press) Macrophage migration inhibitory factor is critically involved in basal and fluoxetine-stimulated adult hippocampal cell proliferation and in anxiety, depression, and memory-related behaviors. Mol Psychiatry
[247] de la Fontaine L, Schwarz MJ, Riedel M, Dehning S, Douhet A, Spellmann I, Kleindienst N, Zill P, Plischke H, Gruber R, Müller N (2006) Investigating disease susceptibility and the negative correlation of schizophrenia and rheumatoid arthritis focusing on MIF and CD14 gene polymorphisms. Psychiatry Res 144:39–47
[248] Das UN (2001) Is insulin and anti-inflammatory molecule? Nutrition 17:409–413
[249] Das UN (2003) Is insulin an endogenous cardioprotector? Curr Opin Crit Care 9:375–383
[250] Rossi D, Zlotnik A (2000) The biology of chemokines and their receptors. Annu Rev Immunol 18:217–242
[251] Zlotnik A, Yoshie O (2000) Chemokines: a new classification system and their role in immunity. Immunity 12:121–127
[252] Yoshie O, Imai T, Nomiyama H (2001) Chemokines in immunity. Adv Immunol 78:57–110
[253] Johnston B, Butcher EC (2002) Chemokines in rapid leukocyte adhesion triggering and migration. Semin Immunol 14:83–92
[254] Bacon KB, Schall TJ (1996) Chemokines as mediators of allergic inflammation. Int Arch Allergy Immunol 109:97–109
[255] Struyf S, Gouwy M, Dillen C, Proost P, Opdenakker G, Van Dame J (2005) Chemokines synergize in the recruitment of circulating neutrophils into inflamed tissue. Eur J Immunol 35:1583–1591
[256] Nanki T, Takada K, Komano Y, Morio T, Kanegane H, Nakajima A, Lipsky PE, Miyasaka N (2009) Chemokine receptor expression and functional effects of chemokines on B cells: implication in the pathogenesis of rheumatoid arthritis. Arthritis Res Ther 11:R149
[257] McGrath KE, Koniski AD, Maltby KM, McGann JK, Palis J (1999) Embryonic expression and function of the chemokine SDF-1 and its receptor, CXCR4. Dev Biol 213:442–456
[258] Zou YR, Kottmann AH, Kuroda M, Taniuchi I, Littman DR (1998) Function of the chemokine receptor CXCR4 in haematopoiesis and in cerebellar development. Nature 393:595–599
[259] Rossi D, Zlotnik A (2000) The biology of chemokines and their receptors. Ann Rev Immunol 18:217–242

[260] Ohl L, Henning G, Krautwald S, Lipp M, Hardtke S, Bernhardt G, Pabst O, Förster R (2003) Cooperating mechanisms of CXCR5 and CCR7 in development and organization of secondary lymphoid organs. J Exp Med 197:1199–1204
[261] Furchgott RF, Zawadzki JV (1980) The obligatory role of endothelial cells in the relaxation of arterial smooth muscle by acetylcholine. Nature 288:373–376
[262] Bredt DS, Snyder SH (1990) Isolation of nitric oxide synthetase, a calmodulin-requiring enzyme. Proc Natl Acad Sci U S A 87:682–685
[263] Hiki K, Hattori R, Kawai C, Yui Y (1992) Purification of insoluble nitric oxide synthase from rat cerebellum. J Biochem 111:556–558
[264] Evans T, Carpenter A, Cohen J (1992) Purification of a distinctive form of endotoxin-induced nitric oxide synthase from rat liver. Proc Natl Acad Sci U S A 89:5361–5365
[265] Laroux FS, Pavlick KP, Hines IN, Kawachi S, Harada H, Bharwani S, Hoffman JM, Grisham MB (2001) Role of nitric oxide in inflammation. Acta Physiol Scand 173:113–118
[266] Gusarov I, Starodubtseva M, Wang ZQ, McQuade L, Lippard SJ, Stuehr DJ, Nudler E (2008) Bacterial nitric-oxide synthases operate without a dedicated redox partner. J Biol Chem 283:13140–13147
[267] Shatalin K, Gusarov I, Avetissova E, Shatalina Y, McQuade LE, Lippard SJ, Nudler E (2008) Bacillus anthracis-derived nitric oxide is essential for pathogen virulence and survival in macrophages. Proc Natl Acad Sci U S A 105:1009–1013
[268] Gusarov I, Shatalin K, Starodubtseva M, Nudler E (2009) Endogenous nitric oxide protects bacteria against a wide spectrum of antibiotics. Science 325:1380–1384
[269] Das UN (2010) Essential fatty acids and their metabolites in the context of hypertension. Hypertension Res 33:782–785
[270] Das UN, Krishna Mohan I, Vijay Kumar K, Sravan Kumar G, Chandra Sekhar C (1993) Beneficial effect of L-arginine in non-insulin dependent diabetes mellitus: a potential role for nitric oxide. Med Sci Res 21:669–670
[271] Mohan KI, Das UN (1997) Oxidant stress, anti-oxidants and nitric oxide in non-insulin dependent diabetes mellitus. Med Sci Res 25:55–57
[272] Makino N, Maeda T, Sugano M, Satoh S, Watanabe R, Abe N (2005) High serum TNF-alpha level in Type 2 diabetic patients with microangiopathy is associated with eNOS down-regulation and apoptosis in endothelial cells. J Diabetes Complications 19:347–355
[273] Ikeda K, Nara Y, Tagami M, Yamori Y (1997) Nitric oxide deficiency induces myocardial infarction in hypercholesterolaemic stroke-prone spontaneously hypertensive rats. Clin Exp Pharmacol Physiol 24:344–348
[274] Gauthier TW, Scalia R, Murohara T, Guo JP, Lefer AM (1995) Nitric oxide protects against leukocyte-endothelium interactions in the early stages of hypercholesterolemia. Arterioscler Thromb Vasc Biol 15:1652–1659
[275] Feron O, Dessy C, Moniotte S, Desager JP, Balligand JL (1999) Hypercholesterolemia decreases nitric oxide production by promoting the interaction of caveolin and endothelial nitric oxide synthase. J Clin Invest 103:897–905
[276] Turco J, Liu H, Gottlieb SF, Winkler HH (1998) Nitric oxide-mediated inhibition of the ability of Rickettsia prowazekii to infect mouse fibroblasts and mouse macrophage like cells. Infect Immun 66:558–566
[277] Saura M, Zaragoza C, McMillan A, Quick RA, Hohenadl C, Lowenstein JM, Lowenstein CJ (1999) An antiviral mechanism of nitric oxide: inhibition of a viral protease. Immunity 10:21–28
[278] Darrah PA, Hondalus MK, Chen Q, Ischiropoulos H, Mosser DM (2000) Cooperation between reactive oxygen and nitrogen intermediates in killing of Rhodococcus equi by activated macrophages. Infect Immun 68:3587–3593
[279] Bose M, Farnia P, Sharma S, Chattopadhya D, Saha K (1999) Nitric oxide dependent killing of mycobacterium tuberculosis by human mononuclear phagocytes from patients with active tuberculosis. Int J Immunopathol Pharmacol 12:69–79
[280] Barak Y, Schreiber F, Thorne SH, Contag CH, Debeer D, Matin A (2010) Role of nitric oxide in Salmonella typhimurium-mediated cancer cell killing. BMC Cancer 10:146

[281] Rockett KA, Awburn MM, Aggarwal BB, Cowden WB, Clark IA (1992) In vivo induction of nitrite and nitrate by tumor necrosis factor, lymphotoxin, and interleukin-1: possible roles in malaria. Infect Immun 60:3725–3730

[282] Seguin MC, Klotz FW, Schneider I, Weir JP, Goodbary M, Slayter M, Raney JJ, Aniagolu JU, Green SJ (1994) Induction of nitric oxide synthase protects against malaria in mice exposed to irradiated Plasmodium berghei infected mosquitoes: involvement of interferon gamma and CD8+ T cells. J Exp Med 180:353–358

[283] Naotunne TS, Karunaweera ND, Mendis KN, Carter R (1993) Cytokine-mediated inactivation of malarial gametocytes is dependent on the presence of white blood cells and involves reactive nitrogen intermediates. Immunology 78:555–562

[284] Gouado I, Pankoui MJB, Fotso KH, Zambou O, Nguélé S, Combes V, Grau GE, Amvam ZPH (2009) Physiopathologic factors resulting in poor outcome in childhood severe malaria in Cameroon. Pediatr Infect Dis J 28:1081–1084

[285] Dhangadamajhi G, Mohapatra BN, Kar SK, Ranjit M (2009) Genetic variation in neuronal nitric oxide synthase (nNOS) gene and susceptibility to cerebral malaria in Indian adults. Infect Genet Evol 9:908–911

[286] Gramaglia I, Sobolewski P, Meays D, Contreras R, Nolan JP, Frangos JA, Intaglietta M, van der Heyde HC (2006) Low nitric oxide bioavailability contributes to the genesis of experimental cerebral malaria. Nat Med 12:1417–1422

[287] Lopansri BK, Anstey NM, Weinberg JB, Stoddard GJ, Hobbs MR, Levesque MC, Mwaikambo ED, Granger DL (2003) Low plasma arginine concentrations in children with cerebral malaria and decreased nitric oxide production. Lancet 361:676–678

[288] Gimenez F, Barraud de Lagerie S, Fernandez C, Pino P, Mazier D (2003) Tumor necrosis factor alpha in the pathogenesis of cerebral malaria. Cell Mol Life Sci 60:1623–1635

[289] Clark IA, Awburn MM, Whitten RO, Harper CG, Liomba NG, Molyneux ME, Taylor TE (2003) Tissue distribution of migration inhibitory factor and inducible nitric oxide synthase in falciparum malaria and sepsis in African children. Malaria J 2:6

[290] Clark I, Awburn M (2002) Migration inhibitory factor in the cerebral and systemic endothelium in sepsis and malaria. Crit Care Med 30(5 Suppl):S263–S267

[291] Sumitani K, Kamijo R, Nagumo M (1997) Cytotoxic effect of sodium nitroprusside on cancer cells: involvement of apoptosis and suppression of c-myc and c-myb proto-oncogene expression. Anticancer Res 17(2A):865–871

[292] Lee MH, Jang MH, Kim EK, Han SW, Cho SY, Kim CJ (2005) Nitric oxide induces apoptosis in mouse C2C12 myoblast cells. J Pharmacol Sci 97:369–376

[293] Lee YJ, Lee KH, Kim HR, Jessup JM, Seol DW, Kim TH, Billiar TR, Song YK (2001) Sodium nitroprusside enhances TRAIL-induced apoptosis via a mitochondria-dependent pathway in human colorectal carcinoma CX-1 cells. Oncogene 20:1476–1485

[294] Salvucci O, Carsana M, Bersani I, Tragni G, Anichini A (2001) Antiapoptotic role of endogenous nitric oxide in human melanoma cells. Cancer Res 61:318–326

[295] Li J, Bombeck CA, Yang S, Kim YM, Billiar TR (1999) Nitric oxide suppresses apoptosis via interrupting caspase activation and mitochondrial dysfunction in cultured hepatocytes. J Biol Chem 274:17325–17333

[296] Wenzel U, Kuntz S, De Sousa UJ, Daniel H (2003) Nitric oxide suppresses apoptosis in human colon cancer cells by scavenging mitochondrial superoxide anions. Int J Cancer 106:666–675

[297] Vasa M, Breitschopf K, Zeiher AM, Dimmeler S (2000) Nitric oxide activates telomerase and delays endothelial cell senescence. Circ Res 87:540–542

[298] Matsushita H, Chang E, Glassford AJ, Cooke JP, Chiu CP, Tsao PS (2001) eNOS activity is reduced in senescent human endothelial cells: preservation by hTERT immortalization. Circ Res 89:793–798

[299] Minamino T, Miyauchi H, Yoshida T, Ishida Y, Yoshida H, Komuro I (2002) Endothelial cell senescence in human atherosclerosis: role of telomere in endothelial dysfunction. Circulation 105:1541–1544

[300] Scalera F, Borlak J, Beckmann B, Martens-Lobenhoffer J, Thum T, Täger M, Bode-Böger SM (2004) Endogenous nitric oxide synthesis inhibitor asymmetric dimethyl L-arginine accelerates endothelial cell senescence. Arterioscler Thromb Vasc Biol 24:1816–1822
[301] Bode-Böger SM, Martens-Lobenhoffer J, Täger M, Schröder H, Scalera F (2005) Aspirin reduces endothelial cell senescence. Biochem Biophys Res Commun 334:1226–1232
[302] Werner C, Fürster T, Widmann T, Pöss J, Roggia C, Hanhoun M, Scharhag J, Büchner N, Meyer T, Kindermann W, Haendeler J, Böhm M, Laufs U (2009) Physical exercise prevents cellular senescence in circulating leukocytes and in the vessel wall. Circulation 120:2438–2447
[303] Chen J, Zacharek A, Zhang C, Jiang H, Li Y, Roberts C, Lu M, Kapke A, Chopp M (2005) Endothelial nitric oxide synthase regulates brain-derived neurotrophic factor expression and neurogenesis after stroke in mice. J Neurosci 25:2366–2375
[304] Das UN (2010) Obesity: genes, brain, gut, and environment. Nutrition 26:459–473
[305] Rao AA, Sridhar GR, Srinivas B, Das UN (2008) Bioinformatics analysis of functional protein sequences reveals a role for brain-derived neurotrophic factor in obesity and type 2 diabetes mellitus. Med Hypotheses 70:424–429
[306] Stranahan AM, Lee K, Martin B, Maudsley S, Golden E, Cutler RG, Mattson MP (2009) Voluntary exercise and caloric restriction enhance hippocampal dendritic spine density and BDNF levels in diabetic mice. Hippocampus 19:951–961
[307] Sornelli F, Fiore M, Chaldakov GN, Aloe L (2009) Adipose tissue-derived nerve growth factor and brain-derived neurotrophic factor: results from experimental stress and diabetes. Gen Physiol Biophys 28:179–183
[308] Yamanaka M, Itakura Y, Tsuchida A, Nakagawa T, Taiji M (2008) Brain-derived neurotrophic factor (BDNF) prevents the development of diabetes in prediabetic mice. Biomed Res 29:147–153
[309] Tonra JR, Ono M, Liu X, Garcia K, Jackson C, Yancopoulos GD, Wiegand SJ, Wong V (1999) Brain-derived neurotrophic factor improves blood glucose control and alleviates fasting hyperglycemia in C57BLKS-Lepr(db)/lepr(db) mice. Diabetes 48:588–594
[310] Borregaard N, Cowland JB (1997) Granules of the human neutrophilic polymorphonuclear leukocyte. Blood 89:3503–3521
[311] Belaaouaj A (2002) Neutrophil elastase-mediated killing of bacteria: lessons from targeted mutagenesis. Microbes Infect 4:1259–1264
[312] Root RK, Metcalf J, Oshino N, Chance B (1975) H_2O_2 release from human granulocytes during phagocytosis. I. Documentation, quantitation, and some regulating factors. J Clin Invest 55:945–955
[313] Tauber AI, Babior BM (1977) Evidence for hydroxyl radical production by human neutrophils. J Clin Invest 60:374–379
[314] Goldstein IM, Cerqueira M, Lind S, Kaplan HB (1977) Evidence that the superoxide-generating system of human leukocytes is associated with the cell surface. J Clin Invest 59:249–254
[315] Goldstein IM, Kaplan HB, Radin A, Frosch M (1976) Independent effects of IgG and complement upon human polymorphonuclear leukocyte function. J Immunol 117:1282–1287
[316] Pacher P, Beckman JS, Liaudet L (2007) Nitric oxide and peroxynitrite in health and disease. Physiol Rev 87:315–424
[317] Schmidt KN, Amstad P, Cerutti P, Baeuerle PA (1995) The roles of hydrogen peroxide and superoxide as messengers in the activation of transcription factor NF-kappa B. Chem Biol 2:13–22
[318] Yamaguchi Y, Matsumura F, Liang J, Okabe K, Ohshiro H, Ishihara K, Matsuda T, Mori K, Ogawa M (1999) Neutrophil elastase and oxygen radicals enhance monocyte chemoattractant protein-expression after ischemia/reperfusion in rat liver. Transplantation 68:1459–468
[319] Lorne E, Zmijewski JW, Zhao X, Liu G, Tsuruta Y, Park YJ, Dupont H, Abraham E (2008) Role of extracellular superoxide in neutrophil activation: interactions between xanthine oxidase and TLR4 induce proinflammatory cytokine production. Am J Physiol Cell Physiol 294:C985–C993

[320] Severini C, Improta G, Falconieri-Erspamer G, Salvadori S, Erspamer V (2002) The tachykinin peptide family. Pharmacol Rev 54:285–322
[321] Harrison S, Geppetti P. Substance P (2001) Int J Biochem Cell Biol 33:555–576
[322] Richardson JD, Vasko MR (2002) Cellular mechanisms of neurogenic inflammation. J Pharmacol Exp Ther 302:839–845
[323] Visser M, Bouter LM, McQuillan GM et al (1999) Elevated C-reactive protein levels in overweight and obese adults. JAMA 282:2131
[324] Hotamisligil GS (1999) The role of TNF-alpha and TNF receptors in obesity and insulin resistance. J Intern Med 245:621
[325] Pradhan AD, Manson JE, Rifai N, Buring JE, Ridker PM (2001) C-reactive protein, interleukin-6, and risk of developing type 2 diabetes mellitus. JAMA 286:327–334
[326] Das UN (1999) GLUT-4, tumor necrosis factor, essential fatty acids and daf-genes and their role in glucose homeostasis, insulin resistance, non-insulin dependent diabetes mellitus and longevity. J Assoc Physicians India 47:431–435
[327] Fichtlscherer S, Rosenberger G, Walter DH et al (2000) Elevated C-reactive protein levels and impaired endothelial vasoreactivity in patients with coronary artery disease. Circulation 102:1000–1006
[328] Cleland SJ, Sattar N, Petrie JR et al (2000) Endothelial dysfunction as a possible link between C-reactive protein levels and cardiovascular disease. Clin Sci (Colch) 98:531–535
[329] Das UN (2007) Vagus nerve stimulation, depression, and inflammation. Neuropsychopharmacology 32:2053–2054
[330] Das UN (2007) Is depression a low-grade systemic inflammatory condition? Am J Clin Nutr 85:1665–1666
[331] Pitsavos C, Panagiotakos DB, Tzima N, Lentzas Y, Chrysohoou C, Das UN, Stefanadis C (2007) Diet, exercise, and C-reactive protein levels in people with abdominal obesity: the ATTICA epidemiological study. Angiology 58:225–233
[332] Taibes G (2002) Does inflammation cut to the heart of the matter? Science 296:242–245
[333] Pepys MB, Hirschfield GM, Tennent GA, Gallimore JR, Kahan MC, Bellotti V, Hawkins PN, Myers RM, Smith MD, Polara A, Cobb AJ, Ley SV, Aquilina JA, Robinson CV, Sharif I, Gray GA, Sabin CA, Jenvey MC, Kolstoe SE, Thompson D, Wood SP (2006) Targeting C-reactive protein for the treatment of cardiovascular disease. Nature 440:1217–1221
[334] Gill R, Kemp JA, Sabin C, Pepys MB (2004) Human C-reactive protein increases cerebral infarct size after middle cerebral artery occlusion in adult rats. J Cereb Blood Flow Metab 24:1214–1218
[335] Mullen PG, Windsor AC, Walsh CJ et al (1995) Tumor necrosis factor-alpha and interleukin-6 selectively regulate neutrophil function in vitro. J Surg Res 58:124–130
[336] Gryglewski RJ, Palmer RMJ, Moncada S (1986) Superoxide anion is involved in the breakdown of endothelium derived vascular relaxing factor. Nature 320:454–456
[337] Kumar KV, Das UN (1997) Effect of cis-unsaturated fatty acids, prostaglandins and free radicals on angiotensin-converting enzyme activity in vitro. Proc Soc Exp Biol Med 214:374–379
[338] Das UN (2000) Beneficial effect(s) of n-3 fatty acids in cardiovascular diseases: but, why and how? Prostaglandins Leukot Essen Fatty Acids 63:351–362
[339] Das UN (2005) Is angiotensin-II an endogenous pro-inflammatory molecule? Med Sci Monit 11:RA155–RA162
[340] Pellme F, Smith U, Funahashi T, Matsuzawa Y, Brekke H, Wiklund O, Taskinen MR, Jansson PA (2003) Circulating adiponectin levels are reduced in non-obese but insulin resistant first-degree relatives of type 2 diabetic patients. Diabetes 52:1182–1186
[341] Krakoff J, Funahashi T, Stehouwer CDA, Schalkwijk CG, Tanaka S, Matsuzawa Y, Kobes S, Tataranni PA, Hanson RL, Knowler WC, Lindsay RS (2003) Inflammatory markers, adiponectin, and risk of type 2 diabetes in the Pima Indian. Diabetes Care 26:1745–1751
[342] Masuzaki H, Paterson J, Shinyama H, Morton NM, Mullins JJ, Seckl JR, Flier JS (2001) A transgenic model of visceral obesity and the metabolic syndrome. Science 294:2166–2170

[343] Tracy RP (2003) Inflammation, the metabolic syndrome and cardiovascular risk. Int J Clin Pract Suppl 134:10–17

[344] Sattar N, Scherbakova O, Ford I, O'Reilly DS, Stanley A, Forrest E, Macfarlane PW, Packard CJ, Cobbe SM, Shepherd J west of Scotland coronary prevention study (2004) Elevated alanine aminotransferase predicts new-onset type 2 diabetes independently of classical risk factors, metabolic syndrome, and C-reactive protein in the west of Scotland coronary prevention study. Diabetes 53:2855–2860

[345] Matsumoto K, Sera Y, Abe Y, Ueki Y, Tominaga T, Miyake S (2003) Inflammation and insulin resistance are independently related to all-cause of death and cardiovascular events in Japanese patients with type 2 diabetes mellitus. Atherosclerosis 169:317–321

[346] Wallen NH, Held C, Rehnqvist N, Hjemdahl P (1999) Elevated serum intercellular adhesion molecule-1 and vascular adhesion molecule-1 among patients with stable angina pectoris who suffer cardiovascular death or non-fatal myocardial infarction. Eur Heart J 20:1039–1043

[347] Mulvihill NT, Foley JB, Murphy RT, Curtin R, Crean PA, Walsh M (2001) Risk stratification in unstable angina and non-Q wave myocardial infarction using soluble cell adhesion molecules. Heart 85:623–627

[348] Ridker PM, Rifai N, Pfeiffer M, Sacks F, Lepage S, Braunwald E (2000) Elevation of tumor necrosis factor-alpha and increased risk of recurrent coronary events after myocardial infarction. Circulation 101:2149–2153

[349] Patti G, Di Sciascio G, D'Ambrosio A, Dicuonzo G, Abbate A, Dobrina A (2002) Prognostic value of interleukin-1 receptor antagonist in patients undergoing percutaneous coronary intervention. Am J Cardiol 89:372–376

[350] Biasucci LM, Vitelli A, Liuzzo G, Altamura S, Caligiuri G, Monaco C, Rebuzzi AG, Ciliberto G, Maseri A (1996) Elevated levels of interleukin-6 in unstable angina. Circulation 94:874–877

[351] Biasucci LM, Liuzzo G, Fantuzzi G, Caligiuri G, Rebuzzi AG, Ginnetti F, Dinarello CA, Maseri A (1999) Increasing levels of interleukin (IL)-1Ra and IL-6 during the first 2 days of hospitalization in unstable angina are associated with increased risk of in-hospital coronary events. Circulation 99:2079–2084

[352] Lee KS, Chung JH, Choi TK, Suh SY, Oh BH, Hong CH (2009) Peripheral cytokines and chemokines in Alzheimer's disease. Dement Geriatr Cogn Disord 28:281–287

[353] Pellicanò M, Bulati M, Buffa S, Barbagallo M, Di Prima A, Misiano G, Picone P, Di Carlo M, Nuzzo D, Candore G, Vasto S, Lio D, Caruso C, Romano GC (2010) Systemic immune responses in alzheimer's disease: in vitro mononuclear cell activation and cytokine production. J Alzheimers Dis (in press)

[354] Reale M, Iarlori C, Feliciani C, Gambi D (2008) Peripheral chemokine receptors, their ligands, cytokines and Alzheimer's disease. J Alzheimers Dis 14:147–159

[355] Mattace-Raso FU, Verwoert GC, Hofman A, Witteman JC (2010) Inflammation and incident-isolated systolic hypertension in older adults: the Rotterdam study. J Hypertens 28:892–895

[356] Gacka M, Dobosz T, Szymaniec S, Bednarska-Chabowska D, Adamiec R, Sadakierska-Chudy A (2010) Proinflammatory and atherogenic activity of monocytes in type 2 diabetes. J Diabetes Complications 24:1–8

[357] Antonelli A, Fallahi P, Rotondi M, Ferrari SM, Romagnani P, Ghiadoni L, Serio M, Taddei S, Ferrannini E (2008) High serum levels of CXC chemokine ligand 10 in untreated essential hypertension. J Hum Hypertens 22:579–581

[358] Touyz RM, Savoia C, He Y, Endemann D, Pu Q, Ko EA, Deciuceis C, Montezano A, Schiffrin EL (2007) Increased inflammatory biomarkers in hypertensive type 2 diabetic patients: improvement after angiotensin II type 1 receptor blockade. J Am Soc Hypertens 1:189–199

[359] Domenici E, Willé DR, Tozzi F, Prokopenko I, Miller S, McKeown A, Brittain C, Rujescu D, Giegling I, Turck CW, Holsboer F, Bullmore ET, Middleton L, Merlo-Pich E, Alexander RC, Muglia P (2010) Plasma protein biomarkers for depression and schizophrenia by multi analyte profiling of case-control collections. PLoS One 5:e9166

[360] O'Brien SM, Scully P, Dinan TG (2008) Increased tumor necrosis factor-alpha concentrations with interleukin-4 concentrations in exacerbations of schizophrenia. Psychiatry Res 160:256–262

[361] Drexhage RC, Padmos RC, de Wit H, Versnel MA, Hooijkaas H, van der Lely AJ, van Beveren N, deRijk RH, Cohen D (2008) Patients with schizophrenia show raised serum levels of the pro-inflammatory chemokine CCL2: association with the metabolic syndrome in patients? Schizophr Res 102:352–355

[362] Zhang XY, Zhou DF, Cao LY, Zhang PY, Wu GY, Shen YC (2004) Changes in serum interleukin-2, -6, and -8 levels before and during treatment with risperidone and haloperidol: relationship to outcome in schizophrenia. J Clin Psychiatry 65:940–947

[363] Thompson SG, Kienast J, Pyke SD, Haverkate F, van de Loo JC (1995) Hemostatic factors and the risk of myocardial infarction or sudden death in patients with angina pectoris. European Concerted Action on Thrombosis and Disabilities Angina Pectoris Study Group. N Engl J Med 332:635–641

[364] Morrow DA, Rifai N, Antman EM, Weiner DL, McCabe CH, Cannon CP, Braunwald E (2000) Serum amyloid A predicts early mortality in acute coronary syndromes: a TIMI 11A substudy. J Am Coll Cardiol 35:358–362

[365] Johnson BD, Kip KE, Marroquin OC, Ridker PM, Kelsey SF, Shaw LJ, Pepine CJ, Sharaf B, Bairey Merz CN, Sopko G et al (2004) Serum amyloid A as a predictor of coronary artery disease and cardiovascular outcome in women: the National Heart, Lung, and Blood Institute-Sponsored Women's Ischemia Syndrome Evaluation (WISE). Circulation 109:726–732

[366] Higaki J, Murphy GM Jr, Cordell B (1997) Inhibition of beta-amyloid formation by haloperidol: a possible mechanism for reduced frequency of Alzheimer's disease pathology in schizophrenia. J Neurochem 68:333–336

[367] Du JL, Liu JF, Men LL, Yao JJ, Sun LP, Sun GH, Song GR, Yang Y, Bai R, Xing Q, Li CC, Sun CK (2009) Effects of five-year intensive multifactorial intervention on the serum amyloid A and macroangiopathy in patients with short-duration type 2 diabetes mellitus. Chin Med J (Engl) 122:2560–2566

[368] Stettler C, Witt N, Tapp RJ, Thom S, Allemann S, Tillin T, Stanton A, O'Brien E, Poulter N, Gallimore JR, Hughes AD, Chaturvedi N (2009) Serum amyloid A, C-reactive protein, and retinal microvascular changes in hypertensive diabetic and nondiabetic individuals: an Anglo-Scandinavian Cardiac Outcomes Trial (ASCOT) substudy. Diabetes Care 32:1098–1100

[369] Mallat Z, Corbaz A, Scoazec A, Besnard S, Leseche G, Chvatchko Y, Tedgui A (2001) Expression of interleukin-18 in human atherosclerotic plaques and relation to plaque instability. Circulation 104:1598–1603

[370] Mallat Z, Henry P, Fressonnet R, Alouani S, Scoazec A, Beaufils P, Chvatchko Y, Tedgui A (2002) Increased plasma concentrations of interleukin-18 in acute coronary syndromes. Heart 88:467–469

[371] Hernesniemi JA, Karhunen PJ, Oksala N, Kähönen M, Levula M, Rontu R, Ilveskoski E, Kajander O, Goebeler S, Viiri LE, Hurme M, Lehtimäki T (2009) Interleukin 18 gene promoter polymorphism: a link between hypertension and pre-hospital sudden cardiac death: the Helsinki Sudden Death Study. Eur Heart J 30:2939–2946

[372] Rabkin SW (2009) The role of interleukin 18 in the pathogenesis of hypertension-induced vascular disease. Nat Clin Pract Cardiovasc Med 6:192–199

[373] Trøseid M, Seljeflot I, Arnesen H (2010) The role of interleukin-18 in the metabolic syndrome. Cardiovasc Diabetol 9:11

[374] Hivert MF, Sun Q, Shrader P, Mantzoros CS, Meigs JB, Hu FB (2009) Circulating IL-18 and the risk of type 2 diabetes in women. Diabetologia 52:2101–2108

[375] Osborn O, Gram H, Zorrilla EP, Conti B, Bartfai T (2008) Insights into the roles of the inflammatory mediators IL-1, IL-18 and PGE2 in obesity and insulin resistance. Swiss Med Wkly 138:665–673

[376] Yu JT, Tan L, Song JH, Sun YP, Chen W, Miao D, Tian Y (2009) Interleukin-18 promoter polymorphisms and risk of late onset Alzheimer's disease. Brain Res 1253:169–175
[377] Merendino RA, Di Rosa AE, Di Pasquale G, Minciullo PL, Mangraviti C, Costantino A, Ruello A, Gangemi S (2002) Interleukin-18 and CD30 serum levels in patients with moderate-severe depression. Mediators Inflamm 11:265–267
[378] Tanaka KF, Shintani F, Fujii Y, Yagi G, Asai M (2000) Serum interleukin-18 levels are elevated in schizophrenia. Psychiatry Res 96:75–80
[379] Baldus S, Heeschen C, Meinertz T, Zeiher AM, Eiserich JP, Munzel T, Simoons ML, Hamm CW CAPTURE Investigators (2003) Myeloperoxidase serum levels predict risk in patients with acute coronary syndromes. Circulation 108:1440–1445
[380] Brennan ML, Penn MS, Lente FV et al (2003) Prognostic value of myeloperoxidase in patients with chest pain. N Engl J Med 349:1595–1604
[381] Lahdelma L, Jee KJ, Joffe G, Tchoukhine E, Oksanen J, Kaur S, Knuutila S, Andersson LC (2006) Altered expression of myeloperoxidase precursor, myeloid cell nuclear differentiation antigen, Fms-related tyrosine kinase 3 ligand, and antigen CD11A genes in leukocytes of clozapine-treated schizophrenic patients. J Clin Psychopharmacol 26:335–358
[382] Danesh J, Collins R, Appleby P, Peto R (1998) Association of fibrinogen, C-reactive protein, albumin, or leukocyte count with coronary heart disease: meta-analyses of prospective studies. JAMA 279:1477–1482
[383] Ernst E, Resch KL (1993) Fibrinogen as a cardiovascular risk factor: a meta analysis and review of the literature. Ann Intern Med 118:956–963
[384] Folsom AR, Wu KK, Rasmussen M, Chambless LE, Aleksic N, Nieto FJ (2000) Determinants of population changes in fibrinogen and factor VII over 6 years: the Atherosclerosis Risk in Communities (ARIC) Study. Arterioscler Thromb Vasc Biol 20:601–606
[385] Lee AJ, Fowkes FG, Lowe GD, Connor JM, Rumley A (1999) Fibrinogen, factor VII and PAI-1 genotypes and the risk of coronary and peripheral atherosclerosis: Edinburgh Artery Study. Thromb Haemost 81:553–560
[386] Maes M, Delange J, Ranjan R, Meltzer HY, Desnyder R, Cooremans W, Scharpé S (1997) Acute phase proteins in schizophrenia, mania and major depression: modulation by psychotropic drugs. Psychiatry Res 66:1–11
[387] Vallianou NG, Evangelopoulos AA, Panagiotakos DB, Georgiou AT, Zacharias GA, Vogiatzakis ED, Avgerinos PC (2010) Associations of acute-phase reactants with metabolic syndrome in middle-aged overweight or obese people. Med Sci Monit 16:CR56–CR60
[388] Venugopal SK, Devaraj S, Jialal I (2005) Effect of C-reactive protein on vascular cells: evidence for a proinflammatory, proatherogenic role. Curr Opin Nephrol Hypertens 14:33–37
[389] Devaraj S, Du Clos TW, Jialal I (2005) Binding and internalization of C-reactive protein by Fcgamma receptors on human aortic endothelial cells mediates biological effects. Arterioscler Throm Vasc Biol 25:1359–1363
[390] Zeller JM, Sullivan BL (1992) C-reactive protein selectively enhances the intracellular generation of reactive oxygen products by IgG-stimulated monocytes and neutrophils. J Leukoc Biol 52:449–455
[391] Venugopal SK, Devaraj S, Jialal I (2003) C-reactive protein decreases prostacyclin release from human aortic endothelial cells. Circulation 108:1676–1678
[392] Ikeda U, Takahashi M, Shimada K (2003) C-reactive protein directly inhibits nitric oxide production by cytokine-stimulated vascular smooth muscle cells. J Cardiovasc Pharmacol 42:607–611
[393] Blaschke F, Bruemmer D, Yin F et al (2004) C-reactive protein induces apoptosis in human coronary vascular smooth muscle cells. Circulation 110:579–587

Chapter 4
Essential Fatty Acids—Biochemistry, Physiology and Clinical Significance

Introduction

The lipids of all higher organisms contain appreciable quantities of polyunsaturated fatty acids (PUFAs) with methylene-interrupted, i.e., with two or more double bonds of the *cis*-configuration separated by a single methylene group. The term "homo-allylic" is occasionally used to describe this molecular feature.

In higher plants, the number of double bonds in fatty acids only rarely exceeds three, but in algae and animals there can be up to six. Two principal families of PUFAs occur in nature that are derived biosynthetically from linoleic (9-*cis*,12-*cis*-octadecadienoic, 18:2 ω-6) and α-linolenic (9-*cis*,12-*cis*,15-*cis*-octadecatrienoic, 18:3 ω-3) acids. Both of the parent fatty acids can be synthesized in plants, but not in animal tissues, and they are therefore essential dietary components and hence, called as essential fatty acids (EFAs). PUFAs are important as constituents of the phospholipids, where they confer distinctive properties to the membranes, in particular by decreasing their rigidity and thus rendering the membrane more fluid.

Essential fatty acids (EFAs) are important constituents of all cell membranes and alter membrane fluidity and thus, determine and influence the behaviour of membrane-bound enzymes and receptors. EFAs are essential for survival of humans and as are not synthesized in the body; have to be obtained in our diet [1–3]. EFAs are of two types as they occur in the body, the ω-6 series derived from *cis*-linoleic acid (LA, 18:2) and the ω-3 series derived from α-linolenic acid (ALA, 18:3). There is another sequence of fatty acids derived from oleic acid (OA, 18:1 ω-9), and OA is not an EFA. The ω-9, ω-6, and ω-3 (also referred to as *n*-9, *n*-6 and *n*-3) series of fatty acids are metabolized by the same set of enzymes to their respective long-chain metabolites. Further discussion is focused on *n*-6 LA and *n*-3 A fatty acids and their metabolites, since they are the EFAs. It is important to note that while some of the actions and functions of EFAs require their conversion to eicosanoids and other products, in some instances the fatty acids themselves are active.

U. N. Das, *Molecular Basis of Health and Disease*,

DOI 10.1007/978-94-007-0495-4_4,

Metabolism of EFAs

EFAs are polyunsaturated fatty acids (PUFAs) since they contain two or more double bonds. There are at least four independent families of PUFAs. They include:

- The "ω-3" series derived from α-linolenic acid (ALA, 18:3, ω-3).
- The "ω-6" series derived from *cis*-linoleic acid (LA, 18:2, ω-6).
- The "ω-9" series derived from oleic acid (OA, 18:1, ω-9).
- The "ω-7" series derived from palmitoleic acid (PA, 16:1, ω-7).

The n-6 Polyunsaturated Fatty Acids

Linoleic acid is a ubiquitous component of plant lipids, and of all the seed oils of commercial importance. For instance, corn, sunflower and soybean oils usually contain over 50% of linoleate, and safflower oil contains up to 75%. Although all the linoleate in animal tissues must be acquired from the diet, it is usually the most abundant di- or polyenoic fatty acid in mammals (and in most lipid classes) typically at levels of 15–25%, although it can amount to as much as 75% of the total fatty acids of heart cardiolipin. It is also a significant component of fish oils, although fatty acids of the (*n*-3) family tend to predominate in this instance. Analogues of linoleic acid with *trans*-double bonds are occasionally found in seed oils. For example 9*c*,12*t*-18:2 is reported from *Dimorphotheca* and *Crepis* species, and 9*t*,12*t*-18:2 is found in *Chilopsis linearis*. The remaining members of the (*n*-6) family of fatty acids are synthesised from linoleate in animal and plant tissues by a sequence of elongation and desaturation reactions.

LA is converted to γ-linolenic acid (GLA, 18:3, *n*-6) by the enzyme Δ^6 desaturase (d-6-d). γ-linolenic acid (GLA or 6-*cis*,9-*cis*,12-*cis*-octadecatrienoic acid or 18:3, *n*-6) is usually a minor component of animal tissues in quantitative terms (<1%), as it is rapidly converted to higher metabolites. It is found in the seed oils of evening primrose, borage and blackcurrant. Evening primrose oil contains about 10% GLA.

11-*cis*,14-*cis*-Eicosadienoic acid (20:2, *n*-6) is a common minor component of animal tissues. 8-*cis*,11-*cis*,14-*cis*-Eicosatrienoic acid (dihomo-γ-linolenic acid or 20:3(*n*-6)) is the immediate precursor of arachidonic acid, and of a family of eicosanoids (prostaglandins of 1 series). However, it does not accumulate to a significant extent in animal tissue lipids, and is typically about 1–2% of the phospholipid fatty acids.

DGLA can be converted to arachidonic acid (AA, 20:4, n-6) by the enzyme Δ^5 desaturase (d-5-d). Arachidonic acid (5-*cis*,8-*cis*,11-*cis*,14-*cis*-eicosatetraenoic acid) is the most important metabolite of linoleic acid in animal tissues, both in quantitative and biological terms. It is often the most abundant polyunsaturated component of the phospholipids, and can comprise as much as 40% of the fatty acids of phosphatidylinositol. As such, it has an obvious role in regulating the physical properties of membranes, but the free acid also is involved in the mechanism by

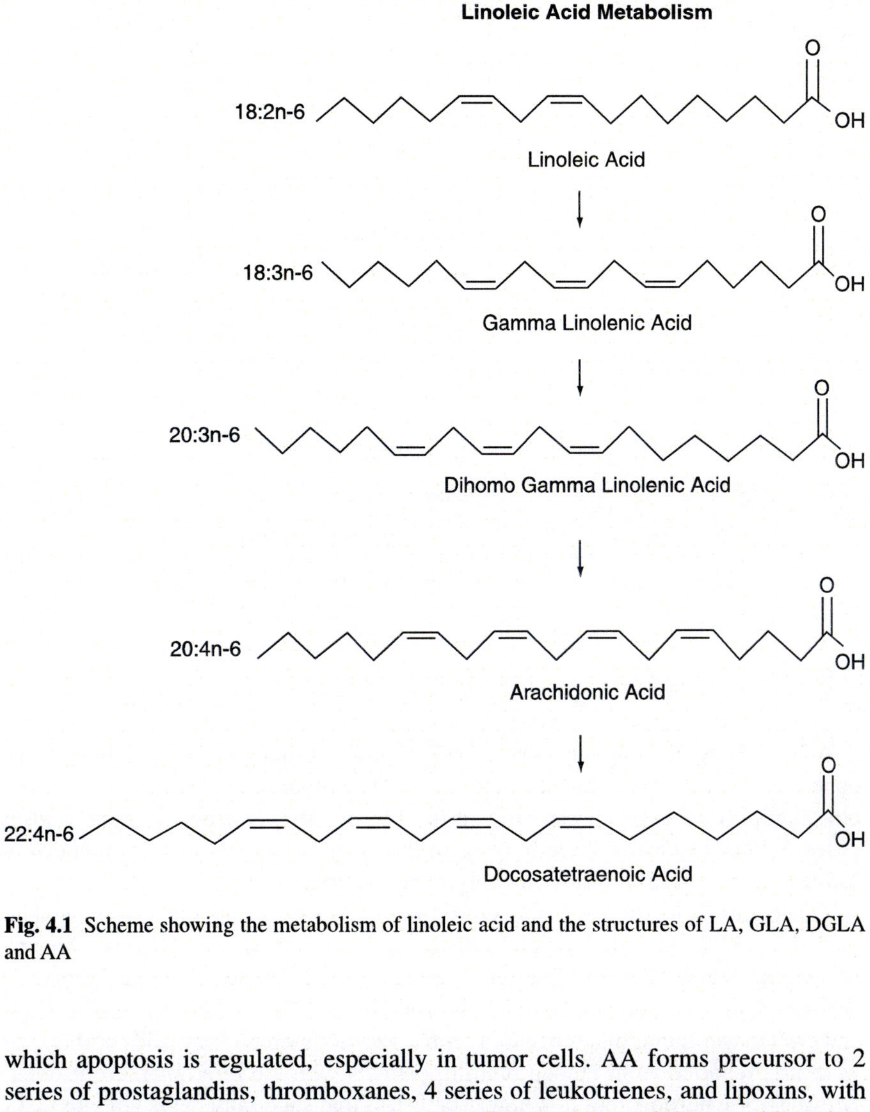

Fig. 4.1 Scheme showing the metabolism of linoleic acid and the structures of LA, GLA, DGLA and AA

which apoptosis is regulated, especially in tumor cells. AA forms precursor to 2 series of prostaglandins, thromboxanes, 4 series of leukotrienes, and lipoxins, with phosphatidylinositol being the primary source. These compounds have a wide variety of essential biological functions (see Figs. 4.1 and 4.2 for the metabolism of LA).

In addition, 2-arachidonoylglycerol and anandamide (*N*-arachidonoylethanolamine) have important biological properties as endocannabinoids, although they are minor lipids in quantitative terms. While arachidonate is found in all fish oils, polyunsaturated fatty acids of the (*n*-3) families tend to be present in much larger amounts. Arachidonic acid is frequently found as a constituent of mosses, liverworts and ferns, and in some plants such as *Agathis robusta*. The fungus *Mortierella alpina* is a commercial source or arachidonate via a fermentation process.

Fig. 4.2 Scheme showing the formation of various metabolites from dietary LA

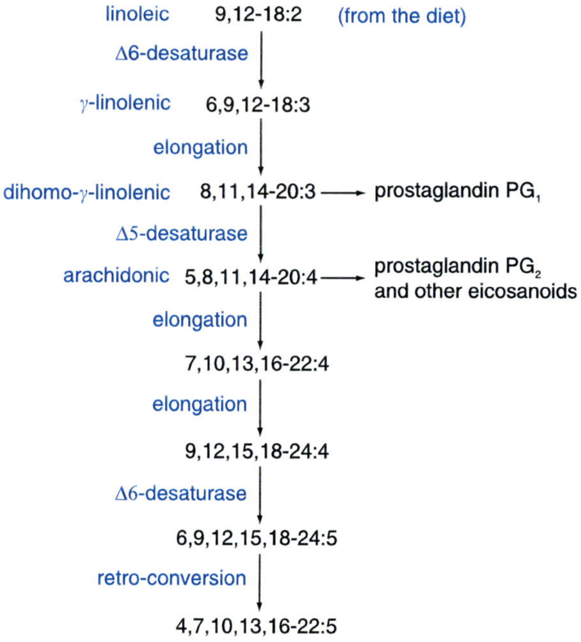

4,7,10,13,16-Docosapentaenoic acid (22:5, *n*-6) is usually a relatively minor component of animal lipids, but it is the main C22 polyunsaturated fatty acid in the phospholipids of testes. It can amount to 70% of the lysobisphosphatidic acid in testes. In this instance, C22 fatty acids of the (*n*-3) family are present at relatively low levels, in contrast to most other reproductive tissues.

Other fatty acids of the (*n*-6) family that are found in animal tissues include 22:3(*n*-6) and 22:4(*n*-6). 7,10,13,16-docosatetraenoic or adrenic acid, is a significant component of the phospholipids of the adrenal glands and of testes. Tetra- and pentaenoic fatty acids of the (*n*-6) family from C24 to C28 have been found in testes, and even longer homologues occur in retina. Very-long-chain fatty acids of this type were first reported from human brain in patients with the rare inherited disorder, Zellweger's syndrome, but it is now established that such fatty acids with up to 38 carbon atoms and with from 3 to 6 methylene-interrupted double bonds are present at low levels in the brain of normal young humans, with 34:4(*n*-6) and 34:5(*n*-6) tending to predominate. The function of these is not known.

The most highly unsaturated fatty acid of the (*n*-6) family to have been characterized are 28:7(*n*-6) (4,7,10,13,16,19,22-octacosaheptaenoate), which has been found in the lipids of marine dinoflagellates and herring muscle, and 4,7,10,13,16,19,22,25,28-tetratriacontanona-enoic acid (34:9, *n*-6) from the freshwater crustacean species *Bathynella natans*.

The n-3 Polyunsaturated Fatty Acids

α-Linolenic acid (9-*cis*,12-*cis*,15-*cis*-octadecatrienoic acid or 18:3 *n*-3 abbreviated as ALA) is a major component of the leaves and especially of the photosynthetic apparatus of algae and higher plants, where most of it is synthesised. It can amount to 65% of the total fatty acids of linseed oil, where its relative susceptibility to oxidation has practical commercial value in paints and related products. In contrast, soybean and rapeseed oils have up to 7% of linolenate, and this reduces the value of these oils for cooking purposes. ALA is the biosynthetic precursor of jasmonates in plants, which appear to have functions that parallel those of the eicosanoids in animals. In animal tissue lipids, ALA tends to be a minor component (<1%), the exception being grass-eating non-ruminants such as the horse or goose, where it can amount to as much as 10% of the adipose tissue lipids. As with linoleate, the remaining members of the (*n*-3) family of fatty acids are synthesised from ALA in animal and plant tissues by a sequence of elongation and desaturation reactions, while shorter-chain components may also be produced by *alpha* or *beta*oxidation.

11,14,17-Eicosatrienoic acid (20:3(*n*-3)) can usually be detected in the phospholipids of animal tissues but rarely at above 1% of the total. Somewhat higher concentrations may be found in fish oils.

Stearidonic acid (6,9,12,15-octadecatetraenoic or 18:4 *n*-3) is occasionally found in seed oils as a minor component, and it occurs in algae and fish oils. 3,6,9,12,15-Octadecapentaenoic acid or 18:5(*n*-3) is a significant component of the lipids of dinoflagellates, and it can enter the marine food chain from this source.

8,11,14,17-Eicosatetraenoic acid (20:4 *n*-3) is found in most fish oils and as a minor component of animal phospholipids. It is frequently encountered in algae and mosses, but rarely in higher plants.

5,8,11,14,17-Eicosapentaenoic acid (EPA or 20:5 *n*-3) is one of the most important fatty acids of the (*n*-3) family. It occurs widely in algae and in fish oils, which are major commercial sources, but there are few definitive reports of its occurrence in higher plants. It is an important constituent of the phospholipids in animal tissues, especially in brain, and it is the precursor of the 3 series of prostaglandins, thromboxanes and 5 series of leukotrienes and anti-inflammatory compounds called as resolvins. These anti-inflammatory compounds resolvins are being investigated for their role in several inflammatory conditions such as rheumatoid arthritis, lupus, inflammatory bowel diseases such as ulcerative colitis, and neurological conditions such as schizophrenia, Alzheimer's disease, and vascular conditions such as diabetic retinopathy and age related macular degeneration.

7,10,13,16,19-Docosapentaenoic acid (22:5 *n*-3) is an important constituent of fish oils, and it is usually present in animal phospholipids at a level of 2–5%.

4,7,10,13,16,19-Docosahexaenoic acid (DHA or 22:6 *n*-3) is usually the end point of ALA metabolism in animal tissues (see Figs. 4.3 and 4.4 for the metabolism of ALA). It is a major component of fish oils, especially from tuna eyeballs, and of animal phospholipids, those of brain synapses and retina containing particularly high proportions. Indeed, there is evidence that increased levels of this fatty acid are

Fig. 4.3 Scheme showing the metabolism of ALA to DHA

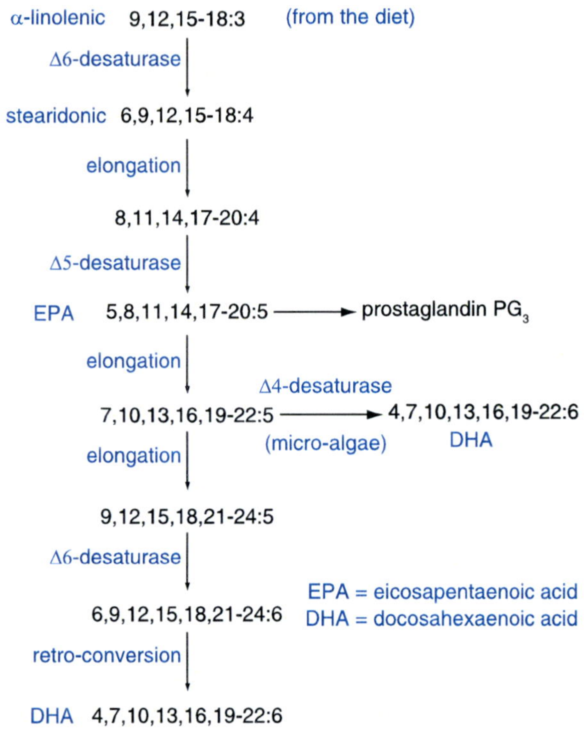

correlated with improved cognitive and behavioral function in the development of the human infant. Dietary supplements may also benefit the elderly. While DHA is found in high concentrations in many species of algae, especially those of marine origin, it is not present in higher plants. DHA is not a substrate for the prostaglandin synthase/cyclooxygenase enzymes, and indeed it inhibits them. However, via the action of lipoxygenases, it is the precursor of the docosanoids, termed resolvins or protectins, which have potent antiinflammatory and immuno-regulatory actions.

The concentration of DHA in tissues has been correlated with a number of human disease states, and it is essential to many neurological functions. DHA is particularly important for the function of retina where it is a major structural component of the photoreceptor outer segment membranes. DHA binds strongly to specific sites on rhodopsin, the primary light receptor in the eye, modifying its stability and activity. It affects signalling mechanisms involved in phototransduction, enhancing activation of membrane-bound retinal proteins, and it may be involved in rhodopsin regeneration. In some cases, sight defects have been ameliorated with DHA supplementation. It is intimately involved with phosphatidylserine metabolism in neuronal tissue. DHA has specific effects on gene transcription that regulate a number of proteins involved in fatty acid synthesis and desaturation. It has been demonstrated to have beneficial effects upon inflammatory disorders of the intestine and in reducing the risk of colon cancer, which may be mediated through associations with specific

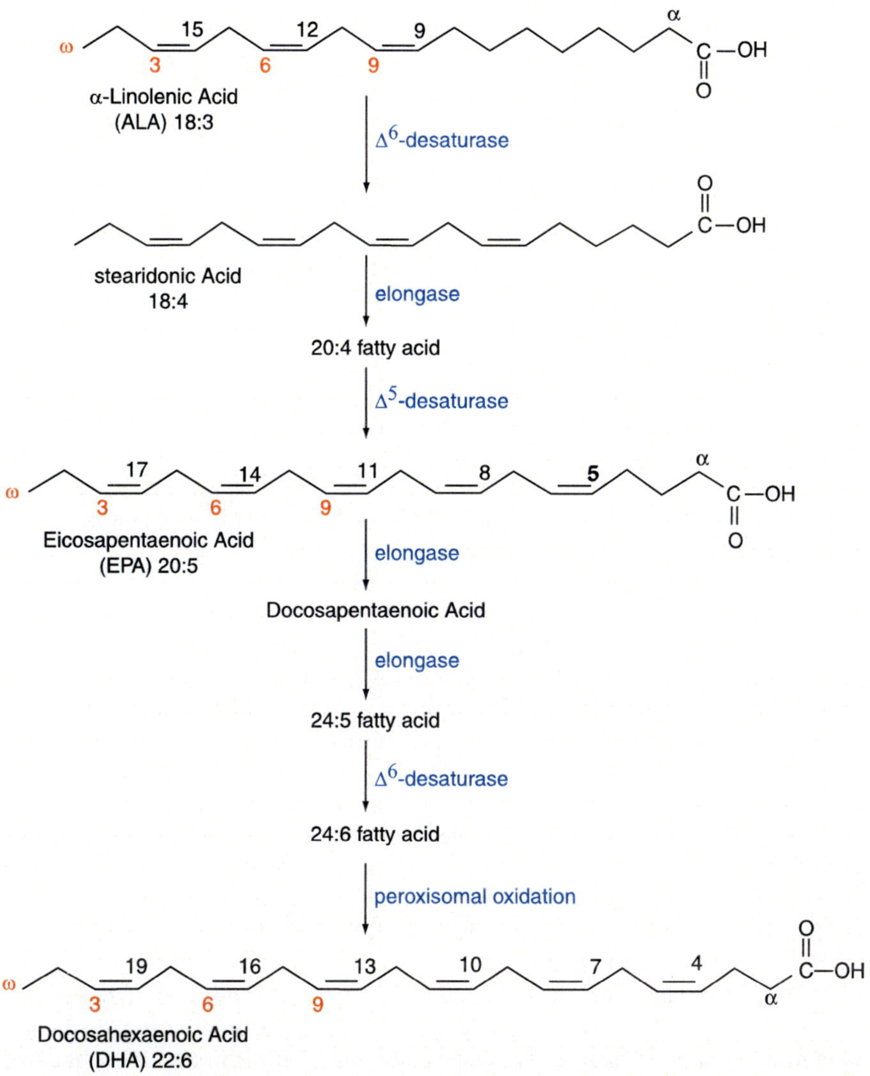

Fig. 4.4 Scheme showing the metabolism of ALA to DHA and their structures

signalling proteins in membranes. As a phospholipid constituent, it has profound effects on the properties of membranes, modulating their structure and function. DHA is believed to be more compact than more saturated chains with an average length of 8.2 Å at 41 °C compared to 14.2 Å for oleic chains. This is the result of adoption of a conformation with pronounced twists of the chain, which reduce the distance between the ends. The methyl group with its extra bulk is located in the interior region. In mixed-chain phospholipids, a further consequence is a marked increase in the conformational disorder of the saturated chain. There appears to be an incompatibility between the rigid structure of cholesterol and the highly flexible chains of DHA, promoting the lateral segregation of membranes into PUFA-rich/cholesterol-poor and PUFA-poor/cholesterol-rich regions. The latter may ultimately become the membrane microdomains known as rafts. PUFA-rich/cholesterol-poor membrane microdomains are technically less easy to study than rafts, but they may also contain distinctive proteins and have important biological functions. It is likely that as a result of the changes in the conformation of signalling proteins when they move between these very different domains may have the potential to modulate cell function in a manner that may explain some of the health benefits of dietary consumption of DHA. Other fatty acids of the (*n*-3) family that are found in nature include 22:3(*n*-3) from animal tissues and 16:3(*n*-3), which is a common constituent of leaf lipids. Other *n*-3 fatty acids such as 16:4 *n*-3, 16:4 *n*-3, 21:5 *n*-3, 24:5 *n*-3 and 24:6 *n*-3 are occasionally present in marine organisms, including fish. Heptaenoic fatty acids of the (*n*-3) family (38:7 *n*–3) and (40:7 *n*–3) have been reported from brains of patients with a defined genetic defect, but the most highly unsaturated fatty acid of the n-3 family yet found is 4,7,10,13,16,19,22,25-octacosaoctaenoate (28:8 *n*-3) from marine dinoflagellates.

LA, GLA, DGLA, AA, ALA, EPA and DHA are all PUFAs, but only LA and ALA are EFAs (see Figs. 4.1, 4.2 and 4.3 for metabolism of EFAs). AA and EPA also are converted to their respective LTs. PGs, TXs, and LTs are biologically active and involved in diseases such as atherosclerosis, bronchial asthma, inflammatory bowel disease, collagen vascular disease, and several other inflammatory conditions. In the present discussion, the term "PUFAs" is used to refer to all unsaturated fatty acids: LA, GLA, DGLA, AA, ALA, EPA, and DHA; and the term EFAs refers to LA and ALA. Although many scientists use the terms EFAs and PUFAs rather interchangeably for the sake of convenience, it should be understood that all EFAs are PUFAs but all PUFAs are not EFAs. Many actions of EFAs are also brought about by PUFAs and so EFA-deficiency can be corrected to a large extent by PUFAs, and hence, PUFAs are termed as "functional EFAs". In view of this, the terms EFAs and PUFAs are used interchangeably.

EFAs/PUFAs play a significant role in collagen vascular diseases, hypertension, diabetes mellitus, metabolic syndrome, psoriasis, eczema, atopic dermatitis, coronary heart disease, atherosclerosis, and cancer [4–8]. This is in addition to the role of PGs, TXs and LTs in these conditions. The molecular mechanism(s) by which various stimuli preferentially induce the release of AA, EPA and/or DHA and convert them to their respective products is not clear. AA, EPA and DHA also give rise to anti-inflammatory molecules such as lipoxins (LXs), resolvins (RSVs), protectins and maresins that have potent anti-inflammatory, anti-fibrotic and immunomodulatory actions. Thus, PUFAs form precursors to both pro- and anti-inflammatory molecules and the balance between these mutually antagonistic compounds could determine the degree, extent and final outcome of the inflammatory process and the underlying disease process(es).

In addition, biologically active compounds formed due to the nitration of unsaturated fatty acids called as nitrolipids have also been identified. Nitration of linoleate by nitric oxide-derived reactive species forms novel derivatives including nitrolinoleate that stimulates smooth muscle relaxation, blocks platelet activation, inhibits human neutrophil functions and suppresses inflammation. Similar nitration of GLA, DGLA, AA, ALA, EPA and DHA and other long-chain polyunsaturated fatty acids could occur that may have similar actions on platelets, vascular smooth muscle, vascular endothelial cells, leukocytes and thus, modulate inflammation and immune response. Thus, it is likely that PUFAs have many important and critical actions not only by themselves but also by giving raise to various biologically active compounds.

Dietary Sources of EFAs

The main dietary sources of EFAs/PUFAs are:

LA: Cereals, eggs, poultry, most vegetable oils, whole-grain breads, baked goods, and margarine. Sunflower, saffola, and corn oils are rich in LA [1].

ALA: Canola oil, flaxseed oil, linseed and rapeseed oils, walnuts, and leafy green vegetables. Human milk is rich in EFAs and GLA, DGLA, AA, EPA, and DHA. Olive oil is rich in OA, whereas palm and coconut oils contain virtually none. The average daily intake of EFAs, in general, is around 7–15 g/day in Europe and USA.

GLA: Human milk contains 0.3–1.0% of its fat as GLA. Thus, breast fed babies get significant amounts of GLA. Evening primrose oil (EPO), borage oil, black currant oil, and hemp seed oil contain substantial amounts of GLA. GLA is present in EPO at concentrations of 7–14% of total fatty acids; in borage seed oil it is 20–27%; and in black currant seed oil at 15–20%. GLA is also found in some fungal sources.

DGLA: Moderate amounts are found in human milk, liver, testes, adrenals, and kidneys.

AA: Human milk contains modest amounts and cow's milk small amounts. Meat, egg yolks, some seaweeds, and some shrimps contain substantial amounts. Average daily intake of AA is estimated to be in the region of 100–200 mg/day, more than enough to account for the total daily production of various PGs, which is estimated to be about 1 mg/day.

Adrenic acid (22:44 ω-6): The main sources of adrenic acid are adrenals, kidneys, testes, and brain.

EPA and DHA: The major source of these two fatty acids is marine fish. These fatty acids may be denatured and converted into trans fats that are harmful to the body during processing [6, 7]. Substantial fall in the intake of ω-3 fatty acids: EPA and DHA could be one of the major changes in Western nutrition in the last 50 years that contributed to the increasing incidence of atherosclerosis, CHD, hypertension, metabolic syndrome X, obesity, collagen vascular diseases and cancer.

The Activities of Δ^6 and Δ^5 Desaturases Are Low in Humans

Dietary LA and ALA are metabolized by the enzymes Δ^6 and Δ^5 desaturases to their respective metabolites as shown in Figs. 4.1, 4.2 and 4.3. LA, ALA, and OA are metabolized by the same set of Δ^6 and Δ^5 desaturases and elongases. As a result, these 3 series compete with one another for the same set of enzymes, though the enzymes seem to prefer ω-3 to ω-6 and ω-6 over ω-9. Hence, under normal physiological conditions the metabolites of ω-9 are formed only in trivial amounts in the cells. Thus, presence of significant amounts of 20:3 ω-9 suggests that there is deficiency of ω-3 and ω-6 and is used as an indicator of EFA deficiency. The activities of Δ^6 and Δ^5 desaturases are slow in humans ($\Delta^5 > \Delta^6$). As a result, the conversion of LA and ALA to their respective metabolites may be inadequate under certain circumstances. In such an instance, it is necessary to supplement GLA and DGLA (to bypass Δ^6 desaturase) and AA (though this can also be obtained from diet and if it becomes necessary to bypass Δ^5 desaturase)and EPA and DHA (to bypass Δ^6 and Δ^5 desaturases). Generally, supplementation of AA is not necessary since; it can be obtained from the diet. But, patients with coronary heart disease, hypertension and diabetes mellitus have been found to be deficient in AA (in addition to other PUFAs) suggesting that the activity of Δ^5 desaturase may be defective and that their dietary intake of AA is low). It is also possible that in these patients the absorption of AA in the intestines could be defective. But, this is very unlikely since AA is a low-molecular weight lipid and so its absorption may not be an issue. Western diet is rich in ω-6 fatty acids compared to ω-3 fatty acids (ω-6 to ω-3 ratio is 10:1), whereas the recommended ratio is ~1:1 [1–3]. It has been suggested that this imbalance in the intake of ω-6 and ω-3 fatty acids could be one of the factors contributing to the increased incidence of cardiovascular diseases, metabolic syndrome and cancer.

It is also important to note that in human tissues, the rates of conversion of ALA to longer-chain metabolites is very low, suggesting that a high proportion of the latter must come from the diet (meat, eggs and fish) in normal circumstances. In particular, the rate of DHA synthesis is so low that it has been argued that dietary supplementation is essential to maintain sufficient levels in brain and retina, although vegetarians whose intake of EPA and DHA is low do not appear to suffer any deficiency symptoms. During the natural processes of turnover and renewal of cell membranes in retinal cells, there appears to be some mechanisms exist to ensure that DHA is conserved and used for the normal structure and function of retinal cells and brain cells.

It is particularly interesting to know that the nematode *Caenorhabditis elegans* possesses all of the enzymes required for the synthesis of AA and EPA fatty acids *de novo*. Similarly, the fungus, *Mortierella alpina*, and some mosses and red algae also have all the enzymes for the synthesis of AA and EPA. Several studies have been performed using *Caenorhabditis elegans* as the model organism to understand the essentiality and functions of various PUFAs by developing organisms in which Δ^6 and Δ^5 enzymes have been knocked down.

Modulators of EFAs/PUFAs Metabolism

Several factors are known to influence the metabolism of EFAs/PUFAs. Saturated fats, cholesterol, trans-fatty acids, alcohol, adrenaline, and glucocorticoids inhibit Δ^6 and Δ^5 desaturases [1–3, 7–10]. Pyridoxine, zinc, nicotinic acid, and magnesium are co-factors for normal Δ^6 desaturase activity.

Protein and Insulin Augment Δ^6 Desaturase Activity

A 96-h fast or diabetes caused a decrease in the conversion of LA to GLA i.e., a decrease in the activity of Δ^6 desaturase that was restored to normalcy by protein feeding both in the fasted and diabetic animals. Insulin administration or casein feeding restored the LA conversion to GLA to normal levels in diabetic rats. However, insulin dosage sufficient to produce hypoglycemia to normal animals prevented the enhancing effect of this protein. Furthermore, the administration of insulin to diabetic rats fed a protein diet did not increase LA desaturation. Nevertheless, the simultaneous administration of casein and a nonhypoglycemic dosage of insulin to diabetic rats produced an additive effect. These findings suggest that glucose metabolism decreases, whereas dietary proteins increase, LA desaturation. They also suggest that the enhancement of LA desaturation by insulin is probably mediated via a glycolysis-independent mechanism [11]. In an extension of these studies, it was reported that 96-h fasted animals showed a slight decrease in the conversion of LA to GLA with a slight decrease in plasma insulin levels. When these fasted animals were fed glucose for 50 h enhanced the conversion of LA to GLA that was transient. In contrast, feeding a lipid-free diet did not modify the conversion of LA to GLA. On the other hand, feeding a carbohydrate-free diet for 96 h resulted in increased linoleic acid desaturation but decreased glucokinase and pyruvate kinase activity, thus apparently eliminating a putative correlation between the fatty acid desaturating activity and glycolytic activity or blood insulin levels, suggesting that dietary proteins play an important role in determining the activity of Δ^6 desaturase [12].

In this context, it is also important to note that different desaturases behave differently to the dietary components. For example, the conversion of stearic acid to oleic acid by Δ^9 desaturase is increased by a high carbohydrate diet, while the activity of Δ^6 desaturase (the converts LA to GLA) is increased by a high protein diet. The amount of protein needed to enhance the activity of Δ^6 desaturase must be of the order of 35% or higher [13]. Since this activation of the Δ^6 desaturase enzyme can be blocked by protein synthesis inhibitors, it is reasonable to assume that protein-rich diet activates the enzyme directly by the induction of the enzyme. It is interesting that the induction of Δ^6 desaturase in old animals by the protein-rich diet is much higher compared to the young. This suggests that to retain the activity of the Δ^6 desaturase in the older age group, it is essential that they consume protein-rich diet ($\geq$35%). Based on these results, it can be said that insulin and high-protein diet activates Δ^6 desaturase whereas diabetics have reduced Δ^6 desaturase activity.

Ageing and Season Influence Δ^6 Desaturase Activity

The activity of Δ^6 desaturase seems to vary with age and season [13]. In rat testicular tissue Δ^6 desaturase activity was found to be very low after 6 months of age. In liver tissue the activity of Δ^6 desaturase was not significantly reduced at least the first 2 months of life. But, it was reported that the activity of Δ^6 desaturase was significantly reduced in liver of 1-year-old animals. These results suggest that the activity of Δ^6 desaturase and, possibly, Δ^5 desaturase falls with age. In contrast, the activity of Δ^9 desaturase that synthesizes oleic acid from stearic acid was found to be enhanced with age. Although, the mechanisms involved in such differential changes in the activities of Δ^6 and Δ^9 with age is not clear, this has some important clinical implications in the pathogenesis of some diseases. These results imply that with advancing age, plasma and tissue levels of GLA, DGLA, AA and EPA and DHA might fall while those of oleic acid may increase. Oleic acid has been shown to enhance the proliferation and prevent apoptosis of breast cancer cells in vitro [14], while others reported that the anti-cancer action of olive oil is due to its high content of oleic acid [15–17]. In contrast, EPA and DHA have consistently been reported to have potent anti-cancer actions against a variety of tumor cells both in vitro and in vivo [18–28]. It is also possible that fall in the activities of Δ^6 and Δ^5 desaturases and consequent decrease in the levels of GLA, DGLA, AA, EPA and DHA with age may explain the high incidence of coronary heart disease and cancer in the older age group since these long-chain fatty acids are believed to be of significant benefit in these conditions [29–33].

In addition, it was reported that the activating effect of dietary protein on the Δ^6 desaturase is blunted to a significant degree during the summer period [13]. The exact reason for this change in the activity of the enzyme during summer is not clear. But, in corollary to this, we observed that the tumoricidal action of GLA and other PUFAs is much less during summer and is maximum or optimum during the winter period (Das UN, unpublished data). These results suggest that for some unexplained reason both the metabolism and actions of PUFAs are much less during summer and are optimum to highest during winter.

Oncogenic Viruses, Radiation, SREBP and PPARs Influence EFA Metabolism

Oncogenic viruses and radiation inhibit Δ^6 desaturase activity. Fat- free diet and partial caloric restriction enhance Δ^6 desaturase activity. Activity of Δ^6 and Δ^5 desaturases are regulated by sterol regulatory element binding protein-1 (SREBP-1) and peroxisome proliferator-activated receptor-α (PPAR-α), two reciprocal transcription factors for fatty acid metabolism, and some of their (SREBP-1 and PPAR-α) lipogenic functions are brought about by their action on PUFAs [34, 35].

Sterol regulatory element-binding proteins (SREBPs) are membrane-bound transcription factors that increase the synthesis of fatty acids as well as cholesterol in animal cells. All three SREBP isoforms (SREBP-1a, -1c, and -2) are subject to feedback regulation by cholesterol, which blocks their proteolytic release from membranes. SREBPs are also negatively regulated by unsaturated fatty acids. Unsaturated fatty acids decreased the nuclear content of SREBP-1, but not SREBP-2, in cultured human embryonic kidney (HEK)-293 cells. The potency of unsaturated fatty acids increased with increasing chain length and degree of unsaturation. Oleate, linoleate, and arachidonate were all effective, but the saturated fatty acids palmitate and stearate were not effective. The mRNAs encoding SREBP-1a and SREBP-1c were markedly reduced, and the proteolytic processing of these SREBPs was inhibited by unsaturated fatty acids. When administered together, sterols and unsaturated fatty acids potentiated each other in reducing nuclear SREBP-1. In the absence of fatty acids, sterols did not cause a sustained reduction of nuclear SREBP-1, but they did reduce nuclear SREBP-2. Unsaturated fatty acids (PUFAs) down-regulated nuclear SREBPs with their greatest inhibitory effects on SREBP-1a and SREBP-1c, whereas sterols have their greatest inhibitory effects on SREBP-2 [36]. LA, AA, EPA and DHA were the most effective fatty acids that decreased the amount of mature SREBP-1 and mRNA levels of SREBP-1c, SREBP-1a, FAS (fatty acid synthase) and acetyl-CoA carboxylase, while SREBP-2 gene or mature protein expression was not altered. Furthermore, these fatty acids decreased the rate of fatty acid synthesis. Based on these evidences and other reports it is clear that PUFAs decrease gene and protein expression of SREBP-1 and FAS mRNA, through interference with LXR activity [37].

Statins Enhance EFA Metabolism

In a recent study, it was reported that statins that inhibit cholesterol synthesis also increased the conversion of LA to its derivatives such as AA, both in vivo and in vitro by markedly enhancing Δ^5 desaturase activity. Both PPAR-α and PPAR-γ agonists did increase the Δ^5 desaturase mRNA levels, while PPAR-α agonist showed a synergistic effect with simvastatin with a concomitantly increase in PPAR-α expression and β-oxidation. Simvastatin increased SREBP-1 levels with respect to controls but did not influence PPAR-α and LA β-oxidation. These results suggest that SREBP-1 is also involved in the regulation of Δ^5 desaturase gene by simvastatin [38].

Trans-fats, Saturated Fats and Cholesterol Inhibit Δ^6 Desaturase

Activities of Δ^6 and Δ^5 desaturases are decreased in diabetes mellitus, hypertension, hyperlipidemia, and metabolic syndrome. Trans-fats interfere with the metabolism of EFAs and promote inflammation, atherosclerosis and coronary heart disease [1–3,

39–44]. The pro-inflammatory action of trans-fats could be attributed to their ability to interfere with EFA metabolism. Several PUFAs, especially EPA and DHA inhibit the production of pro-inflammatory cytokines: interleukin-6 (IL-6), tumor necrosis factor-α (TNF-α), IL-1, and IL-2 [1–3, 45]. Saturated fatty acids and cholesterol interfere with the metabolism of EFAs and thus, promote the production of pro-inflammatory cytokines, which explains their ability to cause atherosclerosis and coronary heart disease (CHD). Thus, trans-fats, saturated fats, and cholesterol have pro-inflammatory actions whereas PUFAs possess anti-inflammatory properties. Interference with the metabolism of EFAs by saturated fats, cholesterol and trans-fats reduces the formation of GLA, DGLA, AA, EPA, and DHA that are essential for the formation of biologically beneficial prostacyclin (PGI_2), PGI_3, lipoxins, resolvins, protectins and maresins. Deficiency and/or absence of PGI_2, PGI_3, lipoxins, resolvins, protectins and maresins could initiate and accelerate the progression of atherosclerosis, persistence of inflammation, CHD and failure of the healing process [1–3].

Zinc Modifies EFA and PG Metabolism

It is known that other dietary factors such as zinc (Zn^{2+}), magnesium (Mg^{2+}), niacin, calcium, vitamin C, selenium and vitamin E have also been shown to influence EFA metabolism [1–3]. Acrodermatitis enteropathica, a rare autosomal recessive disorder of zinc deficiency, is due to the genetic defect that has been mapped to 8q24 and the defective gene identified as SLC39A4, encodes the zinc transporter Zip4. In this disorder, severe Zn^{2+} deficiency develops causing dermatitis that responds to oral Zn^{2+} supplementation [46, 47]. Dermatitis similar to that is seen in acrodermatitis enteropathica is also seen in subjects with essential fatty acid deficiency. It was reported that patients with acrodermatitis enteropathica have defective metabolism of unsaturated fatty acids [48–50] in the form of extremely reduced LA and its metabolites in triglycerides and sterol-esters. In contrast, n-3 fatty acids were increased in sterol-esters and phospholipids. Zinc supplementation led to quick clinical improvement, and linoleic and arachidonic acids increased rapidly in triglycerides and sterol-esters to the values of healthy infants. These findings suggest that in these patients there is impaired enteral absorption of LA and that Zn^{2+} is essential for LA metabolism. Many of the features of zinc deficiency and of essential fatty acid (EFA) deficiency are similar in both animals and humans. EFAs are important in zinc absorption, while Zn^{2+} seems to be necessary for the conversion of LA to GLA, and the mobilisation of DGLA for the synthesis of 1 series PGs. Zinc may also be important in the conversion of DGLA to AA and AA to 2 series PG formation [51–58]. These interactions between Zn^{2+} and metabolism of EFAs may explain the beneficial actions of Zn^{2+} and EFAs in dermatitis and acrodermatitis enteropathica.

Magnesium is an Essential Co-factor for Normal Δ^6 Desaturase

Magnesium (Mg^{2+}) deficiency decreases AA content of plasma lipoproteins, although the tissue AA content remains unchanged. It appears as though the loss of extracellular Mg^{2+} is responsible for this change in AA content. This is so since, in the isolated rabbit heart studies, in vitro perfusion conditions which produce loss of intracellular Mg^{2+} also resulted in disturbances of AA that resulted in increased generation of PGs from AA without any change in the Km or Vmax of cyclo-oxygenase. In addition, the incorporation of AA into tissue phospholipids is significantly reduced, but not other fatty acids such as OA, SA (stearic acid) and LA, suggesting that the activity of the enzymes (CoA synthetases and acyl transferases) which mediate AA incorporation is reduced during Mg^{2+} depletion. Since protein kinase C (PKC)-mediated phosphorylation of both CoA synthetase and acyl transferase reduces their activity, and as PKC has an Mg^{2+} binding site, it is possible that loss of intracellular Mg^{2+} could lead to the activation of PKC, with the consequent reduction of AA reacylation enzyme activity [59, 60]. These results are significant since, Mg^{2+}-induced vascular relaxation in the presence of nitric oxide (NO) agonists is enhanced suggesting an increase in NO production. In contrast, removal of Mg^{2+} from the organ bath was associated with the reduction in the intensity of vessel relaxation confirming the proposal that Mg^{2+} ion favorably influences the NO production by the vascular endothelium [61]. It is also interesting to note that Mg^{2+} ion concentration influences vascular smooth muscle (VSM) contractility, endothelium-dependent relaxing factor (EDRF) release and prostaglandin production. For example, isolated rabbit aorta in the presence of physiologic salt solution (PSS) containing 1.2 mM Mg^{2+}, endothelium-dependent relaxation and endothelium-independent contraction to acetylcholine (Ach) were not affected by incubation in elevated glucose (44 mM), indomethacin (10^{-5} M) treatment, or both. On the other hand, in solution containing 0.6 mM magnesium, the endothelium-dependent relaxation to ACh was impaired in elevated glucose. Indomethacin treatment did not affect endothelium-dependent relaxation in the control solution (containing 1.2 mM of Mg^{2+}) but partially restored the response to ACh in elevated glucose. Under basal conditions and in the presence of NO synthase inhibition, ACh induced a contraction. In low Mg^{2+}-containing medium, this contraction was potentiated by the presence of endothelial cells in control and even more in elevated glucose concentration (44 mM). The endothelium-dependent contractions were abolished by pretreatment with indomethacin. However, in control Mg^2 conditions (1.2 mM); an endothelium-dependent component to the ACh contraction was observed only in elevated glucose concentration. Responses to sodium nitroprusside (SNP), KCI, and serotonin, and the concentration to ACh (in the absence of endothelium) were not influenced by elevated glucose, indomethacin treatment, or both [62]. These results suggest that when the Mg^{2+} is present in optimum amounts, the formation of NO and possibly, cyclo-oxygenase-derived vasodilators such as PGI_2 and PGI_3 are formed in adequate amounts to produce vasodilatation, whereas in Mg^{2+} deficiency states, the formation of both NO and PGI_2 and PGI_3 will be defective leading to vasoconstriction. Thus, Mg^{2+} seems to have the ability to enhance both NO and PGI_2 and PGI_3 generation from vascular endothelial cells. These results have important therapeutic implications. For instance, in patients with pre-eclampsia $MgSO_4$ is an effective treatment option [63, 64] to reduce blood pressure and relax the uterus. This beneficial action could be related to the ability of Mg^{2+} to enhance NO and PGI_2 and PGI_3 generation and thus, reduce hypertension in these patients. These proposals are supported by the observation that Mg^{2+} influences intracellular Ca^{2+} mobilization by inhibiting both Ca^{2+} influx and intracellular Ca^{2+} release and enhances PGI_2 production in vascular smooth muscle cells and human umbilical vascular endothelial cells [65] and also increased NO generation [66–70].

Experimental Mg^{2+} deficiency produced an increase in plasma total cholesterol and triglyceride levels while HDL-cholesterol was decreased, lipid peroxidation was increased, increased cytosolic Ca^{2+}, enhanced hydro and endoperoxide levels as a consequence of the increase of AA release and eicosanoid synthesis, inhibited mitochondrial respiratory activity and activated Ca^{2+}-dependent proteases which may activate the conversion of xanthine dehydrogenase to xanthine oxidase which generates active O_2 species (such as superoxide anion), and decreased the conversion of LA tovs AA which was consistent with the decrease of Δ^6 desaturase activity in rat liver microsomes [71]. The inhibitory effect of Mg^+ deficiency on Δ^6 desaturase activity is somewhat similar to the inhibitory action of saturated fats, trans-fats and cholesterol on Δ^6 desaturase [72]. Thus, Mg^{2+} appears to be an important co-factor for the normal activity of Δ^6 desaturase activity.

Calcium Enhances PGI_2 Synthesis and Interacts with PUFAs

Similar to the actions of Mg^{2+} on PUFA metabolism and formation of PGI_2 and NO, even Ca^{2+} may have similar, if not identical, action. When the effect of AA on intracellular Ca^{2+} concentration in human osteoblasts MG63 was studied, it was noted that AA caused a concentration-dependent increase in Ca^{2+}, mainly due to inward Ca^{2+} transport from extracellular environment and also triggered Ca^{2+} release from intracellular stores. It is interesting that the Ca^{2+} response to AA was inhibited by the cyclooxygenase (COX) inhibition, but both PGE_1 and PGE_2 caused an increase in intracellular Ca^{2+} that was far lower than that obtained with AA. The Ca^{2+} response to AA was not inhibited by calcium antagonist, nifedipine, suggesting that AA did not activate a voltage-dependent Ca^{2+} channel [73]. Thus, AA has the ability to mobilize Ca^{2+} in human osteoblasts, and possibly, other cells [74] and thus, could augment PGI_2 and NO formation. Furthermore, AA may influence Ca^{2+} transport across both plasma and endoplasmic membranes. Furthermore, they suggest that osteoblast activity may be modulated by AA. The Ca^{2+} mobilization induced by AA may be involved in the initiation of superoxide production by human neutrophils [75]. Rat aortic tissue slices when exposed to increasing the Ca^{2+} concentration of the medium between 0–5 mM, the release of PGI_2 was augmented and the release of

TXA_2 was diminished. Potassium free medium caused a very large increase in PGI_2 release of the tissue slices [76].

Vitamin C and Ethanol Enhance the Formation of PGE_1

Vitamin C has been shown to augment Δ^6 desaturase activity and enhance the conversion of DGLA to PGE_1 and AA to PGI_2 [1–3, 77–80]. On the other hand, ethanol has been shown to have, at least, two actions on EFA and PG metabolism. Ethanol enhances the conversion of DGLA to PGE_1 but it blocks the activity of the Δ^6 desaturase, an enzyme that is essential for replenishment of DGLA stores from dietary precursors. The acute effect of ethanol is therefore an increased production of PGE_1 but chronic consumption will lead to depletion ofs DGLA and PGE_1, while withdrawal from ethanol will lead to a precipitous fall in PGE_1 [1–3, 81, 82].

Both vitamin E and selenium appear to inhibit the activity of Δ^5 desaturase activity but enhance the conversion of AA and EPA to PGI_2 and PGI_3 respectively [1–3, 83–85].

Vitamin B_6 is another co-factor that is needed for the normal activity of Δ^6 and Δ^5 desaturases [81, 86–90]. Hence, in the presence of low concentrations of vitamin B_6 the formation of GLA and DGLA from LA and the product of DGLA namely PGE_1 will be low [86, 87].

Nicotinic acid has been shown to stimulate PGE_2, TXA_2 and LTE_4 (leukotriene E_4) synthesis [91].

Actions of EFAs/PUFAs and Their Metabolites

Cell Membrane Fluidity

Cell membrane fluidity is determined by its lipid composition: increasing its content of saturated fatty acids and cholesterol renders the membrane more rigid, whereas increasing unsaturated fatty acids makes it more fluid. This is an important function of lipids since the number of receptors and their affinity to their respective hormones/growth factors/proteins depends on the fluidity of the cell membrane. For instance, a rigid cell membrane shows reduced number of insulin receptors and their affinity to insulin that, in turn, causes insulin resistance. In contrast, increase in cell membrane fluidity due to its high content of unsaturated fatty acids and reduced cholesterol, increases the number of insulin receptors on the membrane and their affinity to insulin and thus, decreases insulin resistance [1–3, 92–107]. The changes in cell membrane fluidity could result in alterations in phagocytosis in leukocytes, ability of the cells to produce various PGs, LTs, TXs; their sensitivity or resistance to viral, bacterial and other infections, uptake of glucose by cells due to changes in the expression and affinity of GLUT receptors, binding to their specific receptors and actions of various neurotransmitters, sensitivity of tumor cells to the actions of cytotoxic agents and radiation, ability of viruses to replicate in the cells, actions of various enzymes and membrane bound enzymes such as Na^+-K^+-ATPase [92–107]. Thus, the incorporation of various PUFAs, trans-fats and cholesterol in the cell membranes alter the properties and actions of various cells not only by influencing the cell membrane fluidity but also the actual properties of the cell organelles depending the type of cell that is under study.

Availability of appropriate amounts of ω-3 and ω-6 fatty acids and various growth factors is essential for the growth of brain during the perinatal period and adolescence [1–3, 108, 109]. Deficiency of ω-3 EPA and DHA and ω-6 AA during the critical period impairs brain growth and the development of appropriate synaptic connections that, in turn, could lead to developmental disorders of the brain and neuropsychological conditions: dementia, depression, schizophrenia, Alzheimer's disease, and neurodegenerative diseases: Huntington's disease, Parkinson's disease, spinocerebellar degeneration, etc., and may impair memory formation and consolidation.

Furthermore, brain is rich in PUFAs such as AA, EPA and DHA which constitutes as much as 30–50% of the total fatty acids in the brain, where they are predominantly associated with membrane phospholipids. For proper neuronal development and increase in cell membrane surface area, growth of neurite processes from the cell body is critical. Nerve growth cones are highly enriched with AA-releasing phospholipases, which have been implicated in neurite outgrowth. Syntaxin 3, a plasma membrane protein that has an important role in the growth of neurites, is activated by AA, DHA and other PUFAs. Furthermore, AA stimulated syntaxin 3 to form the ternary SNARE complex (soluble N-ethylmaleimide-sensitive factor attachment protein receptor), which is needed for the fusion of plasmalemmal precursor vesicles into the cell surface membrane that leads to membrane fusion [108–113]. These results imply that when the concentrations of PUFAs are inadequate, during the critical period of brain growth development and maturation, it could lead to inappropriate synaptic connections in the brain that could predispose to the development of neuropsychological conditions such as dementia, depression,indexDepression schizophrenia, Alzheimer's disease, and neurodegenerative diseases: Huntington's disease, Parkinson's disease, spinocerebellar degeneration, etc.

Recent studies suggest that syntaxin may also have a role in insulin resistance. It is known that the accumulation of cytosolic lipid droplets in muscle and liver cells are linked to the development of insulin resistance and type 2 diabetes mellitus. Such droplets are formed as small structures that increase in size through fusion, a process that is dependent on intact microtubules and the motor protein dynein. Approximately 15% of all droplets are involved in fusion processes at a given time. These lipid droplets are associated with proteins involved in fusion processes in the cell: NSF (N-ethylmaleimide-sensitive-factor), alpha-SNAP (soluble NSF attachment protein) and the SNAREs (SNAP receptors), SNAP23 (synaptosomal-associated protein of 23 kDa), syntaxin-5 and VAMP4 (vesicle-associated membrane protein 4). Knockdown of the genes for SNAP23, syntaxin-5 or VAMP4, or microinjection of a dominant-negative mutant of alpha-SNAP, decreased the rate of fusion and the size of the lipid droplets. Thus, the SNARE system has an important role in lipid droplet fusion. Oleic acid treatment decreased the insulin sensitivity of heart muscle cells, and this sensitivity is completely restored by transfection with SNAP23. Thus, SNAP23 could be a link between insulin sensitivity and the inflow of fatty acids to the cell [114]. It is possible that unlike oleic acid, which is a monounsaturated fatty acid (18:1), PUFAs such as GLA, DGLA, AA, EPA and DHA may enhance insulin sensitivity by acting on SNAP23, syntaxin-5 or VAMP4. This is so since, AA stimulated syntaxin 3 to form the ternary SNARE complex (soluble N-ethylmaleimide-sensitive factor attachment protein receptor), which is needed for the fusion of plasmalemmal precursor vesicles into the cell surface membrane that leads to membrane fusion.

EFAs/PUFAs Have Second Messenger Actions

Growth factors and hormones activate phospholipase A_2 (PLA_2) leading to the release of DGLA, AA, EPA, and DHA from the cell membrane lipid pool. Fatty acids thus released are utilized for the formation of eicosanoids to bring about some of their actions. For example, the tumoricidal action of TNF-α is dependent on its ability to induce PLA_2, and inhibitors of PLA_2 completely inhibited this action. I observed that TNF-α-resistant tumor cells can be rendered sensitive to TNF-α by the addition of various PUFAs especially, GLA. PUFAs enhance the activity of protein kinase C (PKC), a well-known second messenger; activate macrophages, polymorphonuclear leukocytes (PMNs), modulate T_H1 and T_H2 balance, and increase free radical generation by these cells [1–3, 7, 8].

The interaction between growth factors and PUFAs is interesting. Studies showed that growth factors could modulate the uptake and action of PUFAs by various cells and thus, regulate their utilization and metabolism including the formation of biologically active metabolites derived from PUFAs such as eicosanoids, lipoxins and resolvins. For example, n-3 and n-6 PUFAs at concentration >10 μM inhibited the proliferation of 21HKE, the human kidney epithelial cells, which has retained phenotypic characteristics of normal kidney epithelial cells. In contrast, the proliferation was stimulated by n-3 and n-6 PUFAs at concentrations <10 μM. The stimulatory effect of n-3 and n-6 PUFAs was found to be enhanced in the presence of EGF. Specific tyrosine kinase inhibitors totally abrogated the growth stimulatory effects of PUFAs, both in the presence of EGF or absence of EGF suggesting interaction with tyrosine kinase signal transduction pathways especially involving the EGF-R [115]. This is supported by the observation that an EGF-R blocking antibody caused suppression of LA-stimulated proliferation of a human prostate cancer cell line in serum-free medium [116]. These results suggest that EGF and PUFAs interact with each other that may have a role in the regulation of normal and tumor cell growth.

Studies revealed that PUFAs can trigger tyrosine phosphorylation of EGFR in endothelial cells, while saturated fatty acids were inactive. This activation of EGFR by PUFAs was found to be independent of any autocrine secretion of EGF or other related mediators. PUFA-induced EGFR autophosphorylation triggered EGFR signaling pathway activation and subsequent p42/p44 mitogen-activated protein kinase, suggesting that EGFR is a primary target of fatty acids. Insulin resistance, hypertension, and obesity are associated with elevated levels of nonesterified fatty acids (NEFAs). These data suggest that EGFR is activated by PUFAs and could, possibly, act as a sensor for unsaturated fatty acids [117].

The vascular endothelial vascular endothelial growth factors (VEGFs), comprising VEGF-A, placental growth factor (PLGF) and VEGF-B, signal via the endothelial receptors VEGF receptor 1 (VEGFR1) and VEGFR2, and are major regulators of blood vessel physiology. VEGF-A is required for vasculogenesis and angiogenesis, and is induced by hypoxia and nutrient deprivation. PLGF is involved in pathological angiogenesis, whereas VEGF-B is poorly angiogenic in most tissues except heart, an event that is surprising as both PLGF and VEGF-B signal through the same receptors, VEGFR1 and neuropilin 1 (NRP1). Mice lacking VEGF-B (Vegfb2/2 mice) are healthy and fertile, whereas cardiac overexpression of VEGF-B causes cardiomyocyte hypertrophy and ceramide accumulation. VEGF-B is highly expressed in heart, oxidative skeletal muscle and brown adipose tissue, all of which mainly use fatty acids as energy source. Subsequent studies revealed that VEGF-B specifically controlled endothelial uptake of fatty acids via transcriptional regulation of vascular fatty acid transport proteins. As a consequence, Vegfb$^{-/-}$ mice showed less uptake and accumulation of lipids in muscle, heart and brown adipose tissue, and instead shunted lipids to white adipose tissue. This regulation was mediated by VEGF receptor 1 and neuropilin 1 expressed by the endothelium [118]. The involvement of VEGF-B in PUFAs uptake certainly indicates that there is a close interaction between growth factors such as VEGFs, angiogenesis (both physiological and pathological) and PUFAs and their metabolites. In this context, it is noteworthy that n-3 PUFAs inhibit VEGF expression in colon cancer cells by their ability to negatively regulate COX-2/PGE_2 pathway [119]. Furthermore, EPA suppresses choroidal neovascularization by inhibiting the expression of related inflammatory molecules [120]. These evidences suggest that growth factors and PUFAs play a significant role in the pathobiology of atherosclerosis, diabetic retinopathy, obesity and cardiovascular diseases. In this context, it will be interesting to study how antibodies to various growth factors could modify the uptake and metabolism of various PUFAs and PUFAs, in turn, influence the binding, affinity and actions of the anti-growth factor antibodies. It is also important to know how when a combination of growth factors and PUFAs and/or anti-growth factors + PUFAs are given together to normal and tumor cells in vitro and in vivo will modify the growth characteristics of the treated cells.

PUFAs Behave as Endogenous Anti-infective Molecules

Several studies showed that various EFAs/PUFAs and their metabolites have anti-bacterial, anti-viral, anti-fungal and anti-parasitic actions both in vitro and in vivo. For example, ALA rapidly killed cultures of *Staphylococcus aureus*, and hydrolyzed linseed oil (which contains both LA and ALA) inactivated methicillin-resistant *S. aureus*. ALA promotes adhesion of *Lactobacillus casei* to mucosal surfaces and thus, augments their growth. *Lactobacilli*, in turn, suppress the growth of pathogenic bacteria like *Helicobacter pylori*, *Shigella flexneri*, *Salmonella typhimurium*, *Pseudomonas aeroginosa, Clostridium difficile*, and *Escherichia coli*. Kodicek [121] showed that both LA and ALA have bacteriostatic effect on both gram-positive and gram-negative bacteria. Lacey and Lord [122] observed that cultures of *Staphylococcus aureus* seeded on to human skin were rapidly killed after the skin has been covered with ALA and suggested that it has all the attributes of an ideal anti-bacterial agent. A variety of bacteria were found to be sensitive to the growth inhibitory actions of LA and ALA *in vitro* [123]. Hydrolyzed linseed oil, which contains 52% ALA and pure ALA were found to be capable of killing methicillin-resistant *Staphylococcus aureus* [124]. Both LA and AA can inactivate animal herpes, influenza, Sendai, and Sindbis virus [125].vs LA administered orally as safflower oil (which ~70% LA) produced remission of mycosis fungoides, a rare skin disease of viral etiology, in dogs that correlated with an increase in the plasma levels of LA and AA [126]. AA, EPA, and DHA induce death of *Plasmodium* both *in vitro* and *in vivo* [127]. An analog of myristic acid (14:0) showed selective toxicity to African Trypanosomes [128]. It is possible that PUFAs may possess anti-Trypanosomal action. In addition, PGE_1 and PGA, derived from DGLA, AA, and EPA, inhibit viral replication and behave as anti-viral compounds [129, 130], suggesting that EFAs /PUFAs function as endogenous anti-viral compounds and could be of benefit in AIDS (acquired immunodeficiency syndrome) [131–133]. Thus, PUFAs and their products have anti-bacterial, anti-fungal, anti-viral, and anti-parasitic actions. Lymphocytes and macrophages contain significant amounts of PUFAs and release PUFAs on appropriate stimulation. PUFAs stimulate NADPH-dependent superoxide production by macrophages, neutrophils and lymphocytes that are capable of killing the invading microorganisms [134] and this could be yet another mechanism by which EFAs/PUFAs bring about their anti-bacterial, anti-viral, anti-fungal and anti-parasitic actions.

It is not known whether local application or intravenous infusion of PUFAs could be of help in the treatment of various bacterial, viral and fungal infections. Since, neutrophils, T cells and macrophages release PUFAs on stimulation; it is possible that this could be one of the defense mechanisms adopted by the body to fight infections [135–137]. Recent studies showed that AA, EPA, and DHA could give rise to anti-inflammatory compounds such as lipoxins (LXs) and resolvins that are essential to limit and resolve inflammation [1–3]. These studies imply that a deficiency of LXs, resolvins, protectins and maresins could lead to the perpetuation of inflammation and tissue damage. Hence, it will be interesting to study whether a sub-clinical deficiency of PUFAs, decreased formation of LXs, resolvins, protectins and maresins occurs in subjects who develop various types of infections and their complications such as sepsis. Based on the evidence that PUFAs inactivate enveloped viruses [125, 131–133], it will be interesting to study the effect of these fatty acids on flu viruses and to evaluate whether increased intake of these fatty acids reduce the risk of flu. These and other evidences led to the proposal that PUFAs could be useful in the prevention and treatment of Trypanosomiasis [138].

PUFAs Inhibit ACE Activity and Enhance Endothelial Nitric Oxide Generation

One of the areas where EFAs /PUFs could play significant role is hypertension and coronary heart disease (CHD). In these two diseases, angiotensin converting enzyme (ACE) has a dominant role. It should be noted here that ACE enzyme is present in several tissues including brain.

Angiotensin I-converting enzyme, an exopeptidase, is a circulating enzyme that participates in the renin-angiotensin system (RAS), which mediates extracellular volume and arterial vasoconstriction. It is secreted by pulmonary and renal endothelial cells and catalyzes the conversion of decapeptide angiotensin I to octapeptide angiotensin II. It has two primary functions: (a) ACE catalyses the conversion of angiotensin I to angiotensin II, a potent vasoconstrictor in a substrate concentration dependent manner; and (b) ACE degrades bradykinin, a potent vasodilator, and other vasoactive peptides. These two actions make ACE inhibition a goal in the treatment of hypertension, heart failure, diabetic nephropathy, and type 2 diabetes mellitus. Inhibition of ACE results in the decreased formation of angiotensin II and decreased metabolism of bradykinin, leading to systematic dilation of the arteries and veins and a decrease in arterial blood pressure. In addition, inhibiting angiotension II formation diminishes angiotensin II-mediated aldosterone secretion from the adrenal cortex, leading to a decrease in water and sodium reabsorption and a reduction in extracellular volume.

The ACE gene, ACE, encodes 2 isozymes: (a) the somatic isozyme is expressed in many cells including: the lung, vascular endothelial cells, epithelial kidney cells, and testicular Leydig cells, whereas (b) the germinal is expressed only in sperm. Brain has ACE enzyme which takes part in local RAAS and converts plaquogenic (Aβ42) to more soluble and removal forms of β-amyloid (Aβ40); latter is predominantly a function of N domain portion on the ACE enzyme. Inhibition of ACE with ACE Inhibitors, especially those that cross the blood brain barrier (BBB) and with preferentially select N terminal activity would cause accumulation of Aβ42 (amyloid β42) which is plaquogenic causing progression of dementia; preferential C domain active BBB crossing ACE would likely have less of this latter effect. Aβ42 displays enhanced neurotoxicity relative to Aβ40.

Blood-borne ANG II, produced principally in the lungs by the action of ACE on blood-borne angiotensin I (ANG I), acts up on angiotensin type 1 (AT_1) receptors on neurons in the circumventricular organs of the brain to stimulate sodium appetite and to increase sympathetic nerve activity. In addition, discrete regions of the brain are capable of producing ANG II locally. For example, very high concentrations of ACE are present in the circumventricular organs, particularly the subfornical organ and the area postrema, and lower but still significant concentrations are found in cardiovascular-related regions protected by the blood-brain barrier, such as the hypothalamus and the brain stem. Brain ACE activity supports baseline renal sympathetic nerve activity (RSNA) in normal rats. It was also reported that ACE activity, along with AT_1 receptor binding, increases in the brains of rats with heart failure after myocardial infarction and that transgenic rats deficient in brain angiotensinogen have less left ventricular remodeling and less impairment of sympathetic regulation than control Sprague-Dawley rats 8 weeks after a myocardial infarction [139]. These observations strongly suggest that the brain renin-angiotensin system plays a key role in the augmented neurohumoral drive in heart failure.

PUFAs inhibited leukocyte ACE activity [140, 141] suggesting that they could function as endogenous regulators of ACE activity, and thus, regulate the formation of (angiotensin-II) Ang-II. PUFAs enhance nitric oxide generation [142]. This implies that whenever tissue/cell concentrations of PUFAs are low the formation of Ang-II will be high whereas that of endothelial nitric oxide (eNO) will be low. Plasma concentrations of PUFA and eNO are low in hypertension, diabetes mellitus, renal diseases, rheumatoid arthritis, lupus; psoriasis, eczema, atopic and non-atopic dermatitis; atherosclerosis, insulin resistance, obesity; dementia, schizophrenia, bipolar disorders, Huntington's disease, Alzheimer's disease; peptic ulcer disease; and cancer [1–3, 7, 8, 143, 144]. Furthermore, a 25-nucelotide ACE deletion polymorphism increases ACE activity and such individuals showed a higher risk of developing stroke, obesity, emphysema, bipolar affective disorders, and cancers [143, 144]. This suggests that an altered ACE activity and EFA/PUFA metabolism could play a significant role in many diseases.

Transgenic rats overexpressing both human renin and angiotensinogen genes (dTGR) develop hypertension, inflammation, and renal failure and showed decreased formation epoxy-eicosatrienoic acids (5,6-, 8,9-, 11,12- and 14,15-EETs) and hydroxyeicosa-tetraenoic acids (19- and 20-HETEs) from AA. These EETs and HETEs inhibited IL-6 and TNF-α-induced activation of NF-κB and prevented vascular inflammation [145] suggesting that AA and other PUFAs not only regulate ACE activity and Ang-II levels but also possess anti-inflammatory properties. This is supported by the recent reports that AA forms precursor to anti-inflammatory bioactive lipid metabolites: lipoxins. Lipoxins are potent anti-inflammatory molecules that suppress leukocyte activation, inhibit free radical generation, augment NO release and are able to block the synthesis of pro-inflammatory cytokines: IL-6 and TNF-β.

Furthermore, AA can be metabolized by endothelial 15-lipoxygenase (15-LO) to several vasodilatory eicosanoids such as 11,12,15-trihydroxyeicosatrienoic acid (11,12,15-THETA) and its unstable precursor 15-hydroxy-11,12-epoxyeicosatrienoic acid (15-H-11,12-EETA). Rabbit aorta, mesenteric arteries, and the combination of 15-LO and cytochrome P450 2J2 converted AA to two distinct HEETA metabolites that were resistant to acidic hydrolysis but were hydrolyzed by recombinant sEH (soluble eicosahydroxylase) to polar metabolite 13,14,15-THETA. Erythro- and threo-diastereomers of 13-H-trans-14,15-EETA induced vascular relaxation via K^+ channel activation to cause SMC hyperpolarization, suggesting that 13-H-14,15-EETA is an endothelium derived vascular relaxing factor (EDHF) [146]. In addition, AA forms precursor to vasodilator and platelet anti-aggregator PGI_2. Thus, AA seems to gives rise to a variety of vasodilator and anti-hypertensive molecules that also have renoprotective actions. Since, AA can also form precursor to pro-inflammatory PGs, TXs, and LTs, it is reasonable to assume that the balance between the pro- and anti-inflammatory products formed froms AA will ultimately dictate the role of AA in a disease process. It is still not clear as to the factors that regulate the formation of these pro- and anti-inflammatory molecules from AA. A better understanding of these factors and devising methods to manipulate AA metabolism such that anti-inflammatory molecules are preferentially formed in various inflammatory conditions would be a significant advance.

EPA and AA stimulate eNO synthesis [1–3, 142]. NO has potent anti-atherosclerotic and anti-inflammatory actions. Aspirin enhances the formation of eNO through the generation of epi-lipoxins that may explain its anti-inflammatory action [147]. Epi-lipoxins that have potent anti-inflammatory actions enhance the generation of NO that, in turn, prevents the interaction between leukocytes and the vascular endothelium. NO stimulates the formation of PGI_2 from AA [148] and lipoxins are derived from AA, EPA, and DHA. Aspirin inhibits TXA_2 formation, a potent platelet aggregator and vasoconstrictor, and enhances PGI_2 formation, a platelet anti-aggregator and vasodilator, and thus brings about its anti-atherosclerotic actions. These results emphasize the close interaction between PUFAs, NO synthase, and COX enzymes [149] (see Fig. 4.5). In view of this, efforts are being made to develop NO donating aspirin so that it would preserve PGI_2 synthesis by the endothelial cells and at the same time release NO in adequate amounts to inhibit atherosclerosis, enhance wound healing and suppress inflammation, prevent cancer and suppress tumor growth, inhibit angiogenesis. Such a NO donating aspirin is also expected to be of significant benefit in preventing CHD and stroke [150–154].

PUFAs and Cytokines

ALA, DGLA, EPA, and DHA; LXs, resolvins and protectins suppress pro-inflammatory cytokine IL-1, IL-2, IL-6, macrophage migration inhibitory factor (MIF), HMGB1 (high mobility group box 1) and TNF-α production by T cells and other cells [1–3, 155–157], and thus could function as endogenous anti-inflammatory molecules. PGE_2, $PGF_{2\alpha}$, TXA_2 and LTs derived from AA also modulate IL-6 and TNF-α production. These results imply that levels of IL-6 and TNF-α at the sites of inflammation and injury may depend on the local levels of various PUFAs and eicosanoids formed from them. In particular, the suppressive action of DHA on IL-1β and TNF-α production by stimulated human retinal vascular endothelial cells [158] is interesting since this suggests that it (DHA) and possibly, other PUFAs play an important role in the prevention of atherosclerosis, macular degeneration, and diabetic retinopathy [159, 160]. The ability of EPA and DHA to suppress the production of pro-inflammatory cytokines and induce their anti-inflammatory actions are mediated by their ability to increase PPAR-γ mRNA and protein activity [161].

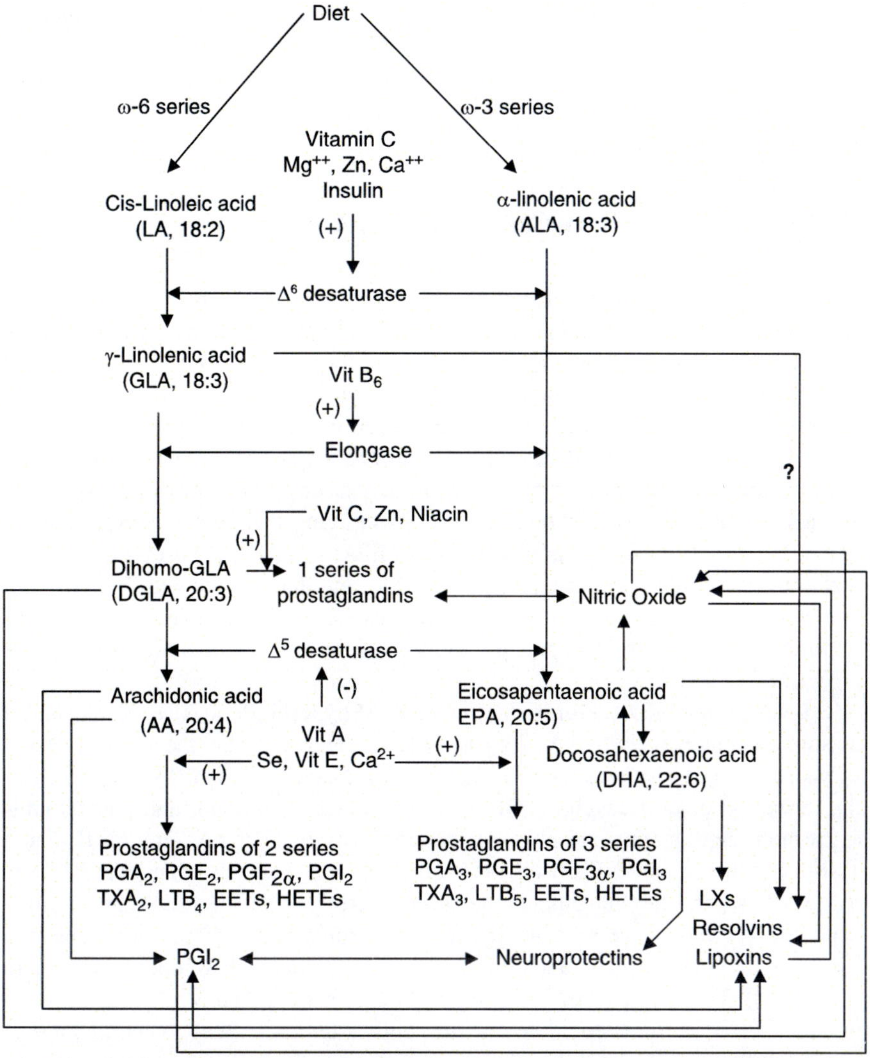

Fig. 4.5 Scheme showing the metabolism of essential fatty acids and co-factors that enhance the activity of Δ^6 and Δ^5 desaturases and elongases and formation of PGs. Ethanol blocks both Δ^6 and Δ^5 desaturases. (+) Indicates enhancement of the activity of the enzyme or increase in the formation of the product. (−) Indicates either in the inhibition of the activity of the enzyme or decrease in the formation of the product

IL-1, IL-6, MIF (macrophage migration inhibitory factor) and TNF-α induce insulin resistance, have cytotoxic actions, are neurotoxic, and produce cachexia seen in tuberculosis, cancer, and AIDS. EPA and other PUFAs ameliorate cachexia induced by TNF-α in animal tumor models [162]. Lipodystrophy and insulin resistance seen with the use of retroviral agents is due to increased levels of TNF-α and decreased concentrations of adiponectin [163]. PUFAs decrease TNF-α and enhance adiponectin levels and thus, could be of benefit to prevent/reverse insulin resistance [1–3, 164], and side effects of retroviral drugs.

PUFAs Decrease HMG-CoA Reductase Activity

The two sterol regulatory element-binding proteins (SREBPs): SREBP-1 and SREBP-2, each ~1,150 amino acids in length, control the transcription of the genes for the low-density lipoprotein (LDL) receptor and 3-hdroxy-3-methylglutaryl coenzyme A (HMG-CoA) synthase. The proteolytic processing of both SREBPs is blocked by sterol overloading and enhanced when sterols are depleted by statins, the HMG-CoA synthesis inhibitors [165]. Cholesterol depletion that occurs due to the use of statins leads to proteolytic activation of transcription factors of the SREBPs and also induces PPAR-γ expression [166], implying that PPAR-γ expression is controlled by SREBPs. Similar to statins, AA, EPA, and DHA are useful in the treatment of hyperlipidemias, have anti-proliferative action on tumor cells both *in vitro* and *in vivo*, bind to DNA and regulate the expression of genes and oncogenes. More importantly, PUFAs are also potent inhibitors of the HMG-CoA reductase enzyme [167, 168].

Statins are not only useful in the treatment of hyperlipidemias but also enhance plasma arachidonic acid (AA, 20:4 ω-6) levels and decrease the ratio of eicosapentaenoic acid (EPA; 20:5 ω-3) to AA significantly [169–171], and enhance the formation of prostacyclin (PGI_2) [172], a potent vasodilator and platelet anti-aggregator. Both statins and polyunsaturated fatty acids (PUFAs) inhibit IL-6 and TNF-α production and NF-κB activation, increase the synthesis of endothelial nitric oxide (eNO); and are anti-inflammatory in nature [1–3, 7, 8, 173–176] that may explain why they are useful in atherosclerosis, coronary heart disease, osteoporosis, stroke, Alzheimer's disease, and inflammatory conditions such as lupus (reviewed in [1–3, 7, 8]). These evidences indicate that PUFAs mediate many, if not all, actions of statins [168] and this could be one mechanism by which they lower cholesterol levels. Furthermore, when a combination of statins and PUFAs are given together a synergistic beneficial effect was noted in patients with combined hyperlipemia [177–179].

Recent studies suggested that statins augment the concentrations of lipoxins (LXs), the potent anti-inflammatory products derived from AA, in the heart [180, 181] lending support to the proposal that the beneficial actions of statins could be attributed to their action on PUFA metabolism. It is not yet known but highly probable that statins may also enhance the formation of other anti-inflammatory compounds from EPA and DHA such as resolvins, protectins and maresins.

PUFAs have inhibitory effects on SREBP-1a and SREBP-1c. In CaCo-2 cells, PUFAs decreased gene and protein expression of SREBP-1 and FAS mRNA by interfering with LXR activity, and in rats PUFAs enhanced cholesterol losses via bile acid synthesis [182, 183]. In the intestine, dietary PUFAs suppress SREBP-1c mRNA without altering expression of its target genes, fatty acid synthase, acetyl-CoA carboxylase, or ATP citrate lyase and decreased intestinal fatty acid synthesis by a posttranscriptional mechanism independent of the SREBP pathway [184]. Feeding mice on fish oil diet for 2 weeks decreased serum cholesterol and triacylglycerol levels, by 50% and 60% respectively, hepatic FPP (farnesyl diphosphate synthase, a SREBP target enzyme that is subject to negative-feedback regulation by sterols in co-ordination with HMG-CoA reductase) synthase and HMG-CoA reductase mRNAs were decreased by 70% and 40% respectively. PUFAs down regulate hepatic cholesterol synthesis by impairing the SREBP pathway [185]. PUFAs reduce SREBP-mediated gene transcription by increasing intracellular cholesterol content through the hydrolysis of cellular sphingomyelin, and the lipid second messenger ceramide, a product of sphingomyelin hydrolysis, decreased SRE-mediated gene transcription of SREBP-1 and SREBP-2 [186].

HMG-CoA reductase catalyzes the synthesis of mevalonate, which is the rate-limiting step in the mevalonate pathway. Mevalonate is the precursor of cholesterol and a variety of isoprenoid containing compounds. These isoprenoid precursors are necessary for the posttranslational lipid modification (prenylation) and hence, the function of *Ras* and other small GTPases. Hence, inhibition of mevalonate pathway has the potential to disrupt the function of oncogenic forms of *Ras*. This explains the ability of both statins and PUFAs to suppress *Ras* activity, anti-proliferative action and induce apoptosis of tumor cells. In addition, small GTPases, which are prenylated products of the mevalonate pathway, have negative control on the expression of BMPs (bone morphogenetic proteins). In view of this, inhibition of the mevalonate pathway by PUFAs will prevent the function of small GTPases and enhance the expression of various BMPs. Various BMPs are known to be essential for neuronal growth, proliferation, and differentiation. Thus, PUFAs modulate brain growth and development, and neuronal differentiation. This action is in addition to their (PUFAs) ability to form an important constituent of neuronal cell membranes and involvement in memory formation and consolidation [187–189], explaining the beneficial action of PUFAs in the prevention and treatment of dementia and Alzheimer's disease [190–196]. The beneficial actions of PUFAs in Alzheimer's disease, schizophrenia and dementia could e attributed to the formation of anti-inflammatory compounds such as lipoxins, resolvins, protectins and maresins whose formation is discussed below.

Lipoxins, Resolvins, Protectins and Maresins

PUFAs form precursors to LXs (lipoxins), resolvins, protectins and maresins that are potent anti-inflammatory, anti-fibrotic and wound healing enhancing active lipid molecules [1–3, 7, 8, 197–204]. Aspirin converts AA, EPA and DHA to form aspirin-triggered 15 epimer LXs (ATLs) that are potent inhibitors of acute inflammation [1–3, 7, 8, 197, 198]. Acetylation of COX-2 by aspirin prevents the formation of prostanoids, but the acetylated enzyme remains active *in situ* to generate 15*R*-hydroxy-eicosatetraenoic acid (15R-HETE) from AA that is released and converted by activated PMNs to the 15-epimeric LXs [197, 198]. This interaction between endothelial cells and PMNs leading to the formation of 15R-HETE and its subsequent conversion to 15-epimeric LXs by aspirin-acetylated COX-2 is a protective mechanism to prevent local inflammation on the vessel wall by regulating the motility of PMNs, eosinophils, and monocytes [1–3, 7, 8, 197, 198]. Endothelial cells oxidize AA (and possibly EPA and DHA) via P450 enzyme system to form 11,12-epoxy-eicosatetraenoic acid(s) that blocks endothelial cell activation, suggesting that COX-2 enzyme is essential for the formation of LXs. Deficiency or absence of LXs leads to interaction between PMN and endothelial cells as a result of which endothelial damage occurs that results in the initiation and progression of atherosclerosis, thrombus formation and coronary artery disease, and persistence of inflammation.

Compounds similar to 15R-HETE and 15-epimeric LXs are also formed from EPA ands DHA. These include conversion of EPA to 18R-HEPE (18R-hydroxy-eicosapentaenoic acid), 18-HEPE, and 15R-HEPE. Activated human PMNs, in turn, converted 18R-HEPE to 5,12,18R-triHEPE and 15R-HEPE to 15-epi-LXA_5 by 5-lipoxygenase. Both 18R-HEPE and 5,12,18R-triHEPE inhibited LTB_4-stimulated PMN transendothelial migration similar to 15-epiLXA_4. 5,12,18R-triHEPE competed with LTB_4 for its receptors and inhibited PMN infiltration, and thus, 5,12,18R-triHEPE suppresses LT-mediated responses when present at the sites of inflammation [199].

Murine brain cells transformed enzymatically DHA to 17R series of hydroxy DHAs (HDHAs) that, in turn, is converted enzymatically by PMNs to di- and tri-hydroxy containing docosanoids [200]. Similar small molecular weight compounds (similar to HDHAs) are generated from AA and EPA. Thus, 15R-hydroxy containing compounds are formed from AA, 18R series from EPA, and 17R-hydroxy series from DHA that have potent anti-inflammatory actions and induce resolution of the inflammatory process and hence are called "resolvins" (see Figs. 4.6, 4.7, 4.8, 4.9, 4.10 and 4.11 for the structures and formation of lipoxins, resolvins and protectins). Resolvins inhibited cytokine generation, leukocyte recruitment, leukocyte diapedesis, and exudate formation. AA, EPA, and DHA-derived resolvins from acetylated COX-2 are formed due to communication between endothelial cells and PMNs. Resolvins inhibit brain ischemia-reperfusion injury [201]. Thus, lipoxins and resolvins formed from AA, EPA, and DHA have cardio-protective, neuroprotective, and other cytoprotective actions.

Of the several 17-hydroxy-containing bioactive mediators derived from DHA that were termed docosatrienes and 17S series resolvins, 10,17S-dihydroxydocosatriene termed as neuroprotectin D1 (NPD1) that reduced infiltration of PMNs, showed anti-inflammatory and neuroprotective properties [201–204]. NPD1 inhibited oxidative stress-induced apoptosis of human retinal pigment epithelial cells [203]. LXs, resolvins, protectins and maresins NPD1) have the ability to enhance wound healing [204], and promote brain cell survival via the induction of antiapoptotic and neuroprotective gene-expression programs [202, 203, 205, 206].

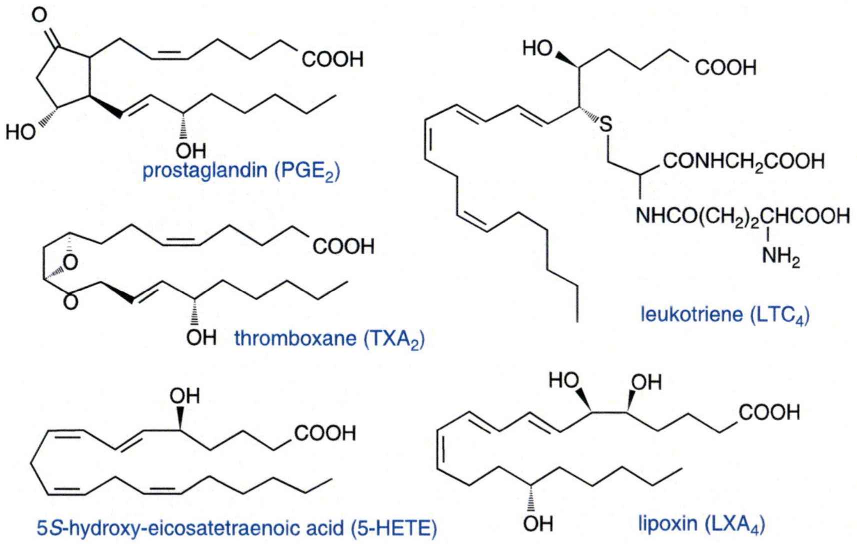

Fig. 4.6 Structures of PGE_2, LTB_4, 5-HETE and LXA_4

Based on the preceding discussion, it is clear that under physiological conditions COX-1 and COX-2 enzymes could selectively induce the formation of beneficial eicosanoids PGE_1, PGI_2, and LXs, resolvins, protectins (such as neuroprotectin D1) to prevent inflammation, protect various cells and tissues from the harmful actions of pro-inflammatory cytokines, superoxide anion and invading organisms, to successfully resolve inflammation, enhance wound healing and prevent or minimize scar tissue formation so that structural integrity and functional capacity of the various cells/tissues and organs is preserved. Failure to produce adequate amounts of LXs, resolvins, protectins and maresins or interference with their action and/or a simultaneous increase in the production of pro-inflammatory eicosanoids and cytokines could lead to initiation and persistence of inflammation, tissue damage scar formation, and abnormalities in the structural integrity and functional capacity of cells/tissues and organs that could result in various diseases [1–3, 7, 8]. Thus, the balance between pro-inflammatory cytokines and eicosanoids and anti-inflammatory cytokines and lipids appears to be crucial in determining the severity and duration of inflammation, and the final recovery from injury, infection and inflammation process.

NO Reacts with PUFAs to Yield Nitrolipids

Recent studies revealed that nitro fatty acids are present in the membrane phospholipids of human tissues both *in vitro* and *in vivo*, and at concentrations that had the potential to exert biological effects. It should be noted here that the formation of nitro

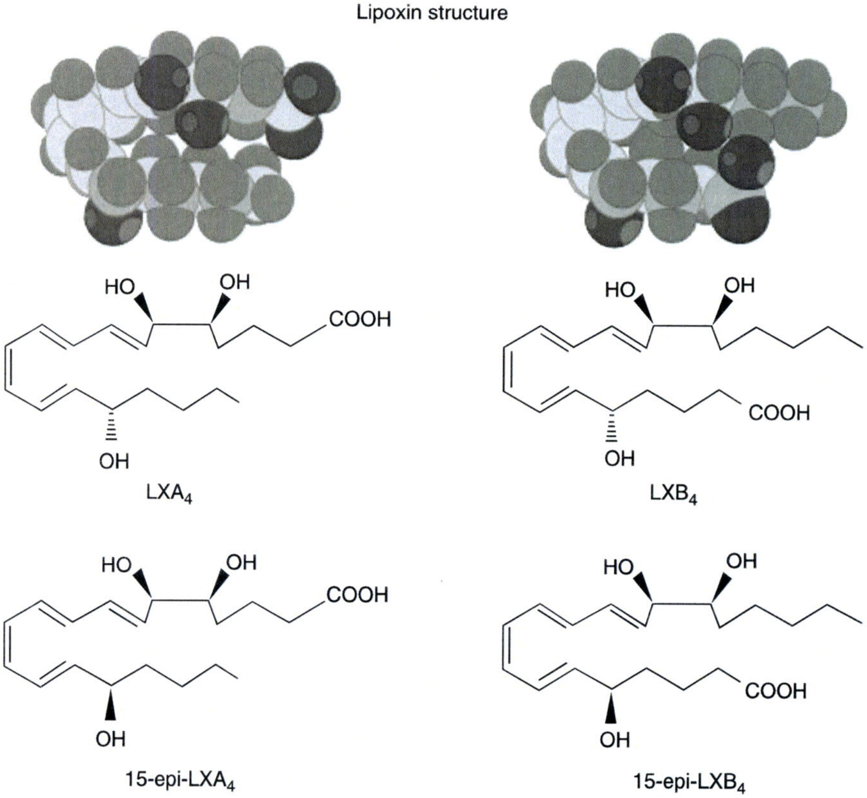

Fig. 4.7 Structures of various lipoxins

fatty acids *in vitro* is well known wherein their presence was demonstrated earlier in studies of lipid oxidation products induced by air pollutants.

It is now known that nitrated derivatives of palmitoleic, oleic, linoleic, linolenic, arachidonic and eicosapentaenoic acids together with their nitrohydroxy derivatives are present in human plasma and urine. Of all, the two most abundant species are derived from oleic acid, i.e., 9- and 10-nitro-9-*cis*-octadecenoic acids (see Fig. 4.12). In the plasma, they occur in the free form; most bound reversibly to thiol-containing proteins and glutathione, and as cholesterol esters and the basal levels in plasma of healthy humans is closer to 1 nM.

Analogous compounds derived from linoleate have been detected at significant concentrations. All the possible nitro-linoleate isomers have been found in tissues, but 10-nitro- and 12-nitro-9-*cis*,12-*cis*-octadecadienoic acids are the main ones found; it appears that the 9-isomer is relatively unstable and is rapidly degraded.

In addition, both nitro and nitro-hydroxy derivatives of oleate, linoleate and linolenate have been characterized. The structures of the nitrohydroxy derivatives of oleate

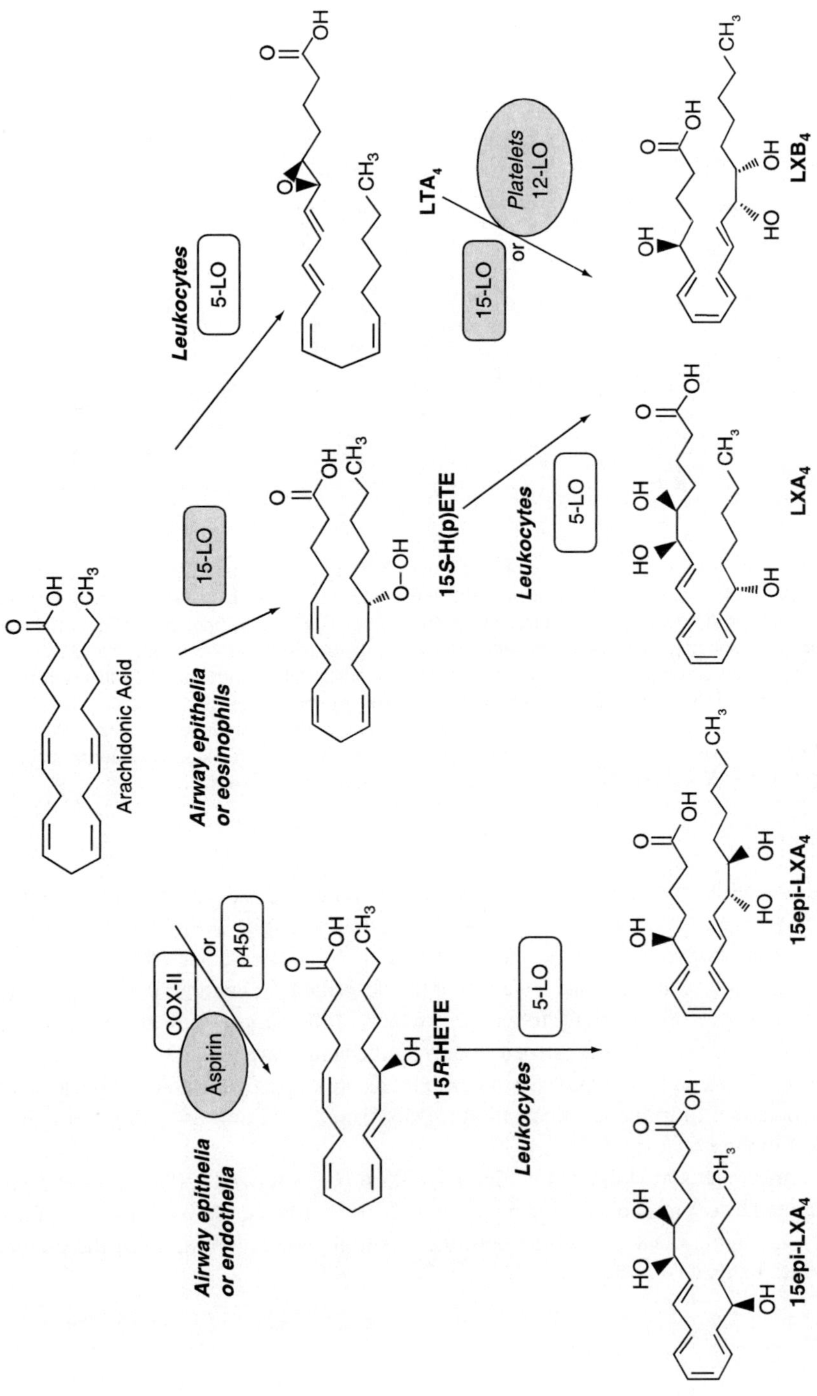

Fig. 4.8 Scheme showing the formation of lipoxins from arachidonic acid

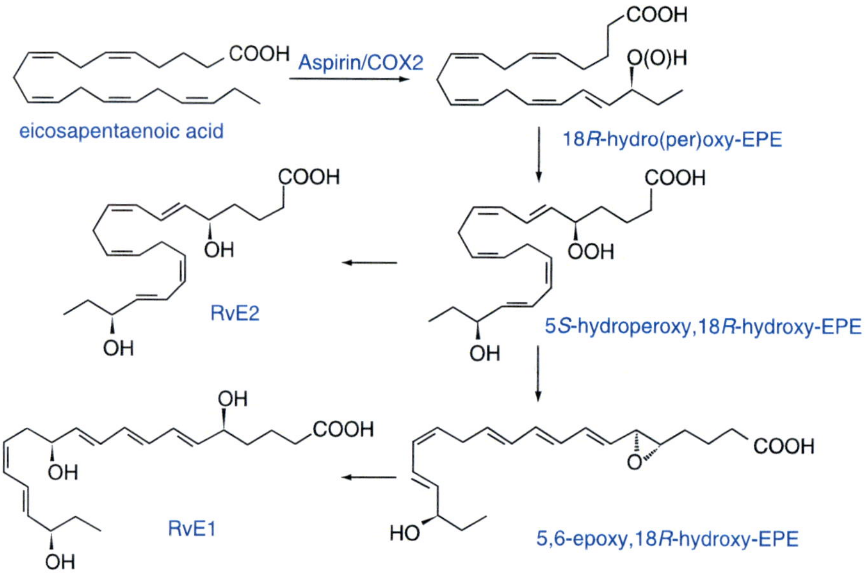

Fig. 4.9 Scheme showing the formation of resolvin E (RvE) derived from EPA. In the endothelial cells, the COX-2 enzyme that has been acetylated introduces an 18*R* hydroperoxy-group into the EPA molecule (c.f. the role of aspirin in the biosynthesis of the epi-lipoxins). This is reduced to the corresponding hydroxy compound before a 5*S*-hydroperoxy group is introduced into the molecule by the action of 5-lipoxygenase as in the biosynthesis of leukotrienes. A further reduction step produces 15S,18R-dihydroxy-EPE or resolvin E2. Alternatively, the 5*S*-hydrpperoxy, 18*R*-hydroxy-EPE intermediates is converted to a 5,6-epoxy fatty acids in polymorphonuclear leukocytes I humans and eventually to 5*S*,12*R*,18*R*-trihydroxy-6*Z*,8*E*,10*E*,14*Z*,16*E*-eiocsapentaenoic acid or resolvin E1 by process similar to the formation of leukotrienes in leukocytes

and linoleate are given in Figs. 4.12 and 4.13. In essence, these are formed by addition of reactive nitrogen species across one of the double bonds. Nitroeicosatetraenoic, α,β-nitrohydroxyeicosatrienoic and *trans*-arachidonic acids, derived from arachidonic acid via such reactions have also been described. In general, there appears to be some degree of selectivity in terms of which of the various isomers are detected in tissues. For example, the nitroeicosatetraenoic acids have the NO_2 groups in positions 9, 12, 14, and 15 mainly. Such compounds are of particular interest because of their potential to influence eicosanoid metabolism in addition to having biological effects in their own right.

Further related metabolites, which have been characterized and are presumed to be formed by comparable mechanisms, include nitro-allyl derivatives of various fatty acids, including oleate, in which both the position and configuration of the double bond is changed (see Fig. 4.14).

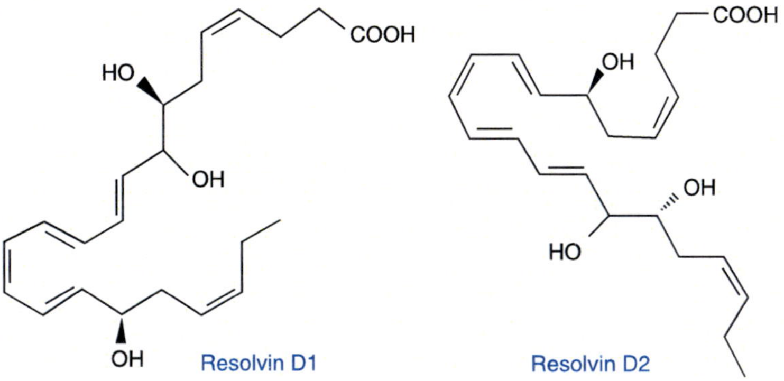

Fig. 4.10 Structures of Resolvin D1 and D2. DHA is converted to 17R-resolvins by a similar aspirin-triggered mechanism similar to the scheme shown in Fig. 2a. In the absence of aspirin, COX-2 of endothelial cells converts DHA to 13S-hydroxy-DHA. In the presence of aspirin, the initial product is 17R-hydroxy-DHA, which is converted to 7S-hydroperoxy, 17R-hydroxy-DHA by the action of a lipoxygenase, and thence via an epoxy intermediate to epimeric resolvins D1 and D2. An alternative lipoxygenase-generated intermediate, 4S-hydroperoxy, 17R-hydroxy-DHA, is transformed via an epoxide to epimeric resolvins D3 and D4. 17S Resolvins of the D series are produced in cells in the absence of aspirin by a reaction catalyzed in the first step by a lipoxygenase

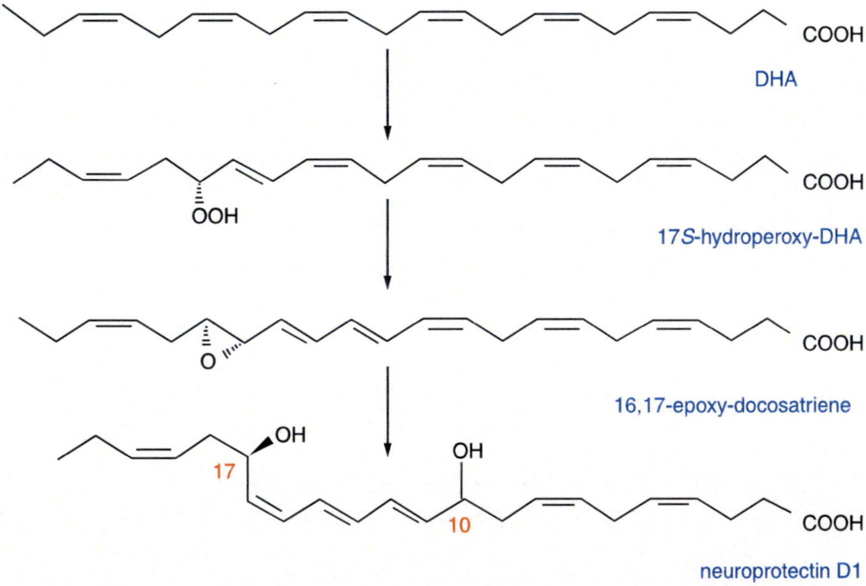

Fig. 4.11 Scheme showing the synthesis of neuroprotectin D1. Resolvins are generated in brain tissue in response to aspirin treatment, and in addition docosatrienes termed neuroprotectins are also produced. The lipoxygenase product 17*S*-hydroperoxy-DHA is converted first to a 16(17)-epoxide and then to the 10, 17-dihydroxy docosatriene denoted as 10, 17 *S*-DT or NPD1. As with the leukotrienes, there are three double bonds in conjugation, hence the term "triene", though there are six double bonds in total

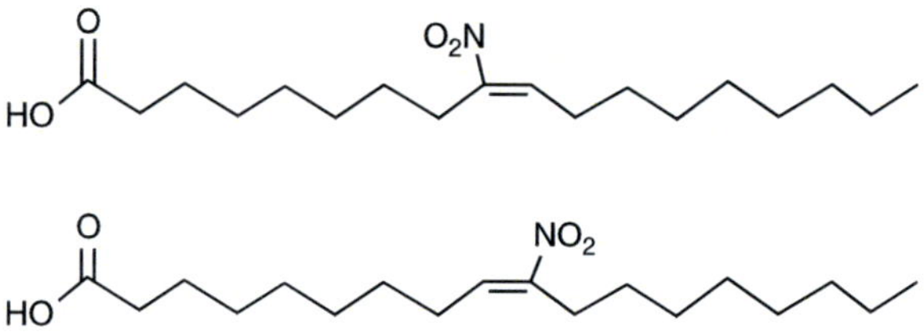

Fig. 4.12 Two regioisomers of OA-NO_2: 9- and 10-nitro-9-*cis*-octadecenoic acids

Formation of Nitro Fatty Acids (Nitrolipids) in Tissues

Formation of nitro fatty acids occurs in tissues through the non-enzymatic reactions of free radicals such as nitric oxide ($NO^{\bullet}$), and $NO^{\bullet}$-derived oxides of nitrogen (e.g., nitrogen dioxide ($NO_2^{\bullet}$)) and peroxynitrite ($ONOO^{\bullet}$). These operate in conjunction with superoxide ($O_2^{\bullet}$), hydrogen peroxide (H_2O_2) and lipid peroxyl radicals ($LOO^{\bullet}$) that are formed during inflammatory process. Many different mechanisms are involved in the production of the secondary radicals and in their subsequent reactions. These are controlled by such factors as the concentration of the $NO^{\bullet}$ radicals,

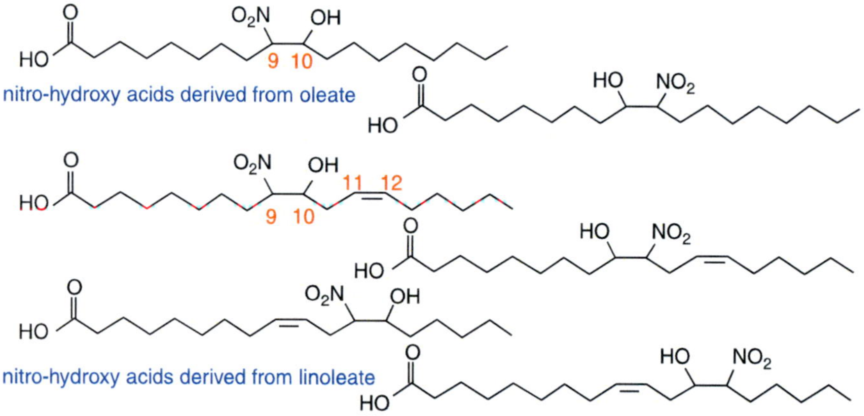

Fig. 4.13 Structures of nitroalkenes derived from oleic and linoleic acids

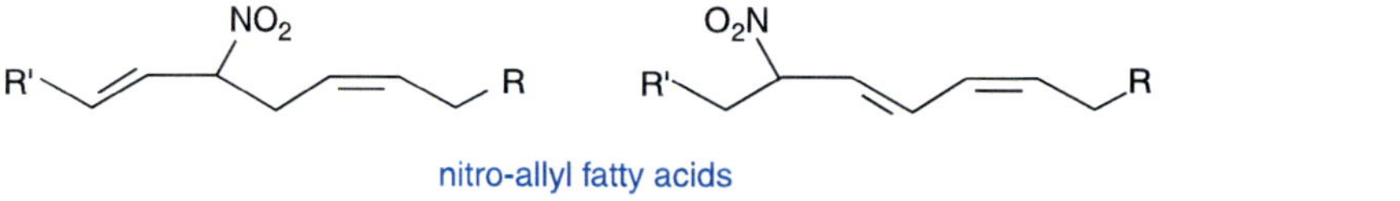

Fig. 4.14 Nitro-allyl derivatives of various fatty acids, including oleate, in which both the position and configuration of the double bond is changed as given above. Compare this with Fig. 4.12

the site of their production, oxygen tension, and the concentrations and membrane environment of the target molecules and of any catalysts and antioxidants. These reactions are somewhat similar to the formation of isoprostanes, which are also non-enzymatic and the reaction is with intact lipids rather than the free acids. In addition, nitrolipids could be formed in foods and in such an event they could reach tissues via the digestive system.

The $NO_2^•$ radical can arise from various endogenous and exogenous sources in humans. For example, immune responses to inflammatory stimuli induce nitric oxide synthase in certain cells that form $NO^•$, which is then oxidized to $NO_2^•$. $NO_2^•$ is a common air pollutant and can be absorbed via the lungs. Meat and other foods may contain appreciable quantities of nitrite (added as a preservative), and nitrate can be reduced to nitrite by aerobic bacteria in the mouth. In the stomach, nitrite decomposes rapidly in the acidic environment to form $NO^•$ and $NO_2^•$ and other bioactive nitrogen oxides, and these are absorbed from the intestines and thence enter into the circulation. The various mechanisms by which $NO_2^•$. forms as described above may be relevant to the development of cancer in these organs. For example, increased formation of nitrites in preserved food may be responsible for high incidence of gastrointestinal cancer in some regions of the world.

Although the detailed mechanisms by which nitro fatty acid formation in human and other animal tissues occur is not clear, the biosynthetic mechanisms proposed are largely extrapolated from chemical studies *in vitro* (see Figs. 4.15, 4.16 and 4.17). The $NO_2^•$ radicals can react with unsaturated lipids and lipid radicals to form all the types of products found in tissues. Thus at low oxygen tensions, homolytic attack to the double bond yields nitroalkyl radicals, which combine with other $NO_2^•$ radicals to form nitro-nitrite intermediates. Loss of nitrous acid (HNO_2) from these intermediates results in the formation of nitroalkenes, while hydrolysis leads to the production of nitro-alcohols. In an alternative reaction, abstraction of a hydrogen atom from the nitroalkyl radicals leads to the formation of nitro-allyl derivatives (see Fig. 4.15).

As an $NO_2^•$ radical can also initiate lipid oxidation reactions, yields of nitration versus oxidation will depend on the concentration of oxygen. For example at elevated oxygen levels, the $NO_2^•$ radical can interact with an unsaturated fatty acid to form a carbon-centered radical, which can interact with oxygen to form a lipid hydroperoxide. Unstable alkyl peroxynitrite intermediates can also be formed through the reactions of lipid peroxyl radical ($LOO^•$) and $NO^•$, of peroxynitrile radicals, and of a lipid hydroperoxide reaction with N_2O_4 or with HNO_2, the last leading to the production of nitro-epoxy fatty acids.

However, nitro fatty acid radicals can also be produced, which may lose HNO_2 to re-generate the unsaturated fatty acid but with one of the double bonds isomerized from the *cis* to the *trans* configuration (see Fig. 4.16).

A further mechanism for nitroalkene formation is addition of a nitronium ion (NO_2^+), which can be formed by reaction of a transition metal with peroxynitrite, by electrophilic substitution at the double bond (see Fig. 4.17).

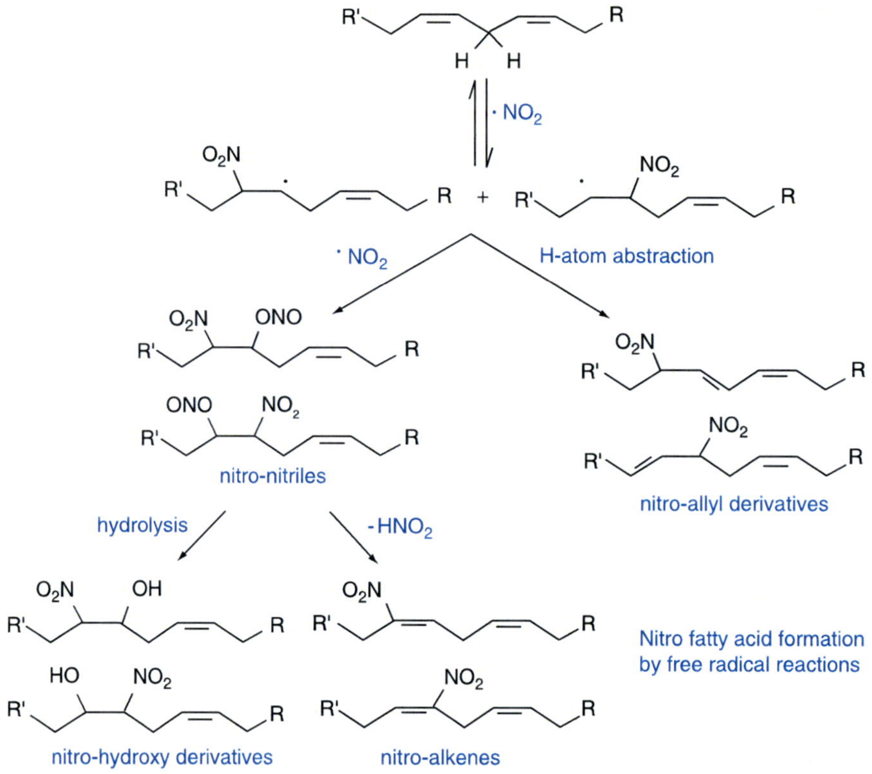

Fig. 4.15 At low oxygen tensions, homolytic attack to the double bond yields nitroalkyl radicals, which combine with other $NO_2^{\bullet}$ radicals to form nitro-nitrite intermediates. Loss of nitrous acid (HNO_2) from these intermediates results in the formation of nitroalkenes, while hydrolysis leads to the production of nitro-alcohols as given above. In an alternative reaction, abstraction of a hydrogen atom from the nitroalkyl radicals leads to the formation of nitro-allyl derivatives (see Fig. 4.13)

Actions of Nitro Fatty Acids (Nitrolipids)

It is known for quite some time that that NO is involved in many biological processes, but the potential role of nitro fatty acids in mediating these reactions has only recently recognized.

In plasma, nitro fatty acids are stabilized by incorporation into lipoproteins, while in erythrocytes and other cells the membrane environment is similarly protective and may provide a reservoir of these compounds. However, nitroalkenoic fatty acids decay rapidly in phosphate buffers, and presumably in the cytoplasm of cells, due to solvation reactions with release of nitric oxide radicals. Thus, it is possible that nitrated unsaturated fatty acids are powerful electrophiles that mediate reversible nitroalkylation reactions with thiol groups of glutathione and of thio-amino acid residues of proteins, thereby regulating the structure and function of the latter. Indeed,

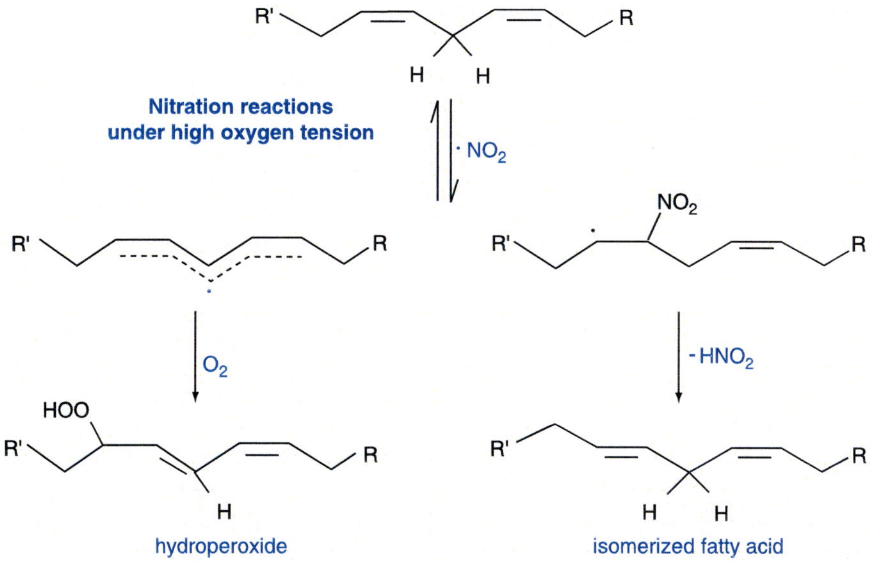

Fig. 4.16 Nitro fatty acid radicals can also be produced, which may lose HNO_2 to re-generate the unsaturated fatty acid but with one of the double bonds isomerized from the *cis* to the *trans* configuration as shown above

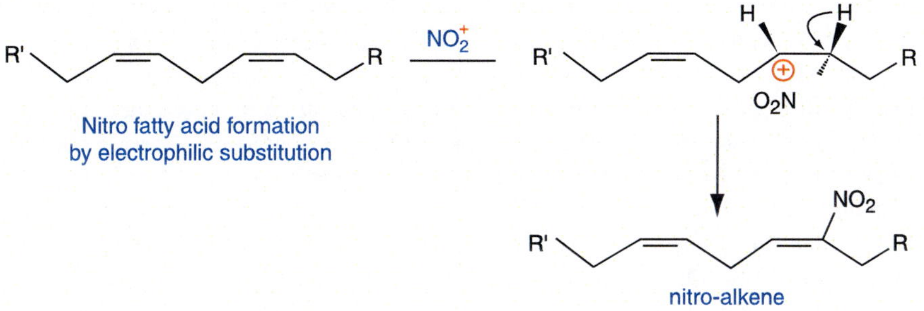

Fig. 4.17 A further mechanism for nitroalkene formation is addition of a nitronium ion (NO_2^+), which can be formed by reaction of a transition metal with peroxynitrite, by electrophilic substitution at the double bond as shown above

nitro-linoleate isomers in red cells and plasma constitute the single largest pool of bioactive oxides of nitrogen in the vasculature and are potent vasodilators that suggest that they may play a significant role in hypertension and other cardiovascular diseases.

In addition, intact nitro-linoleate isomers function as signalling mediators via receptor-dependent pathways as high-affinity endogenous ligands for peroxisome proliferator-activated receptors (PPARγ), and they activate receptor-dependent gene expression at physiological concentrations. 12-Nitrolinoleate is a much more potent activator of PPARγ than any other regioisomer. In neutrophils and platelets, nitro fatty acids activate cAMP-dependent protein kinase signalling pathways and by such means have an anti-inflammatory role in cells. Similarly, both nitro-oleate and nitro-linoleate have been shown to be endogenous anti-inflammatory signalling mediators in a number of biological processes including the inhibition of the lipopolysaccharide-induced secretion of pro-inflammatory cytokines in macrophages, actions that are independent of nitric oxide formation or of activation of PPARs. Nitro-oleic acid is an irreversible inhibitor of the enzyme xanthine oxidoreductase, which generates proinflammatory oxidants and secondary nitrating species. In this instance, it has been established that the carboxyl group, nitration at the nine or ten olefinic carbons, and the double bond are all required for the inhibitory action. Therefore, nitro lipids antagonize the pro-inflammatory cell-signalling pathways that involve oxidized lipids by a variety of mechanisms.

Nitrated derivative of AA have also been shown to have anti-inflammatory properties via effects upon gene transcription. It is possible that these nitrated derivatives of AA could influence the formation and action of various eicosanoids. Similarly, the *trans*-arachidonate isomers formed as by-products of nitration reactions are emerging as biomarkers that target various biological systems. It is likely that the peroxynitrite *per se* has profound effects on the enzymes of prostanoid biosynthesis.

Thus, these nitroalkene derivatives produce vascular relaxation, inhibit neutrophil degranulation and superoxide formation, and inhibit platelet activation, possess endogenous PPAR-γ ligand activity and release NO [1–3, 7, 8, 207–211]. These actions of nitrolipids prevent platelet aggregation, thrombus formation and atherosclerosis. Since nitrolipids or nitroalkenes also possess anti-inflammatory actions, they may have a significant role in low-grade systemic inflammatory conditions such as type 2 diabetes, hypertension, hyperlipidemias, insulin resistance, and metabolic syndrome. Since, nitrolipids are present both in the plasma and urine in substantial amounts; it will be interesting to measure their levels in these conditions. These evidences suggest that PUFAs not only form precursors to various eicosanoids, resolvins, LXs, protectins, and maresins but also react with various other molecules and form novel compounds that have biological activity. It is not yet clear whether nitrolipids interact with eicosanoids, lipoxins, resolvins, protectins and maresins.

Interaction(s) Among n-3, n-6 Fatty Acids, NO and Nitrolipids

There is substantial interaction(s) among n-3, n-6 PUFAs, NO and nitrolipids that is relevant to the role of PUFAs, NO, nitrolipids and eicosanoids in the pathobiology of various diseases. In perfused vascular tissue, DGLA increases the conversion of EPA to PGI_3, a potent vasodilator and platelet anti-aggregator [212]. AA augmented the conversion of EPA to PGI_3 in the tissues [213–215]. In contrast, EPA inhibits the activity of the enzyme Δ^5 desaturase that results in an increase in the concentrations of DGLA in the tissues (especially in the endothelial cells). This increase in tissue levels of DGLA leads to the formation of increased amounts of PGE_1, a vasodilator and platelet anti-aggregator (Fig. 4.18). Thus, EPA indirectly enhances the formation

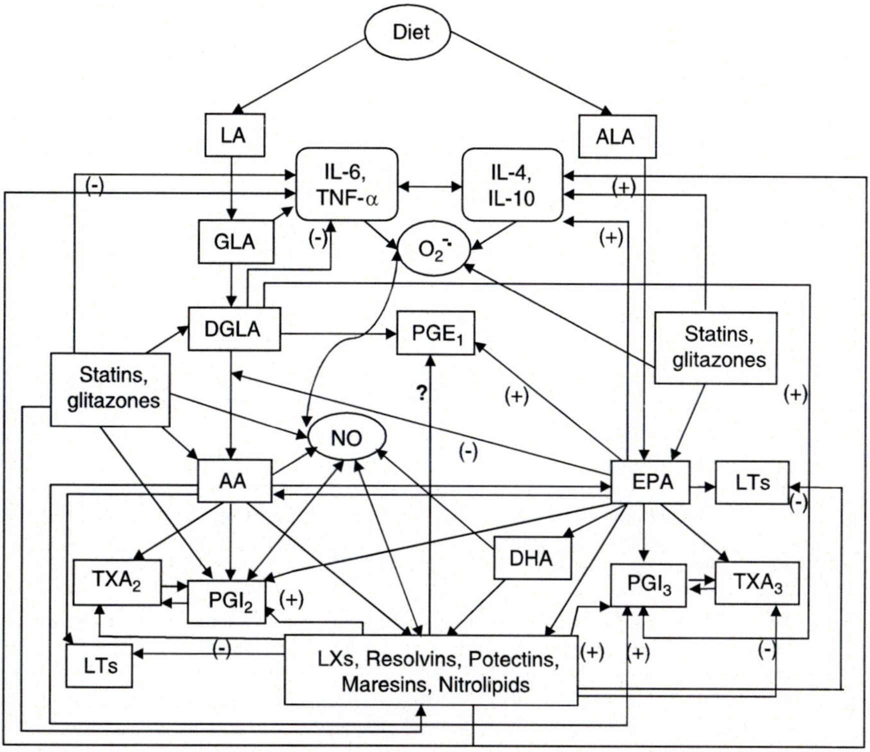

Fig. 4.18 Scheme showing interaction(s) among n-3 and n-6 fatty acids and their effect on the formation of PGI_2, PGI_3, PGE_1, and LXs, resolvins, protectins, maresins and nitrolipids. (−) Indicates inhibition or block in the synthesis, formation or release. (+) Indicates enhancement in the formation or release. It is likely that LXs, resolvins, protectins,indexProtectins maresins and nitrolipids enhance the formation and/or action of PGI_2, PGI_3 and suppress that of TXA_2 and TXA_3. LXs, resolvins, protectins, maresins and nitrolipids suppress the formation of LTs

of PGE_1. In contrast, trans-fats interfere with the formation of DGLA, AA, EPA, and DHA from their respective dietary precursors by blocking the activity of Δ^6 and Δ^5 desaturases and thus, prevent the formation of PGE_1, PGI_2, PGI_3, LXs, resolvins, protectins and maresins and at the same time may augment the formation and/or action of proinflammatory PGs, TXs, and LTs. Thus, trans-fats could enhance the susceptibility of an individual to atheroma and CHD. The beneficial actions of statins (HMG-CoA reductase inhibitors) and glitazones (PPARs agonists) are mediated to some extent by EFAs /PUFAs [215] and their metabolites LXs, resolvins, protectin and maresins are anti-inflammatory molecules [1–3, 7, 8]. Cholesterol and saturated fatty acids block the activities of Δ^6 and Δ^5 desaturases similar to trans-fats [1–3, 7, 8] and thus, inhibit the conversion of dietary LA and ALA to their respective long-chain metabolites including lipoxins, resolvins, protectins and maresins. Thus, it can be said that even cholesterol and saturated fats also possess pro-inflammatory actions partly, by inhibiting the formation and actions of PUFAs and their products LXs, resolvins, protectins, maresins and nitrolipids and by interfering with the beneficial actions of statins and glitazones. This may explain why enhanced consumption of cholesterol, saturate fats, and trans-fats not only render the cell membrane rigid but also initiate and augment the atherosclerotic process and other low-grade systemic inflammatory conditions. The significance of these interactions among n-3, n-3 PUFAs, NO, nitrolipids, lipoxins, resolvins, protectins maresins and the formation and actions of various cytokines is discussed in the following chapters.

References

[1] Das UN (2006) Essential fatty acids: biochemistry, physiology, and pathology. Biotechnol J 1:420–439
[2] Das UN (2006) Essential fatty acids—a review. Curr Pharm Biotechnol 7:467–482
[3] Das UN (2006) Biological significance of essential fatty acids. J Assoc Physicians India 54:309–319
[4] Das UN, Mohan IK, Raju TR (2001) Effect of corticosteroids and eicosapentaenoic acid/docosahexaenoic acid on pro-oxidant and anti-oxidant status and metabolism of essential fatty acids in patients with glomerular disorders. Prostaglandins Leukot Essent Fatty Acids 65:197–203
[5] Das UN (2004) Long-chain polyunsaturated fatty acids interact with nitric oxide, superoxide anion, and transforming growth factor-β to prevent human essential hypertension. Eur J Clin Nutr 58:195–203
[6] Das UN (2003) Can perinatal supplementation of long-chain polyunsaturated fatty acids prevent diabetes mellitus? Eur J Clin Nutr 57:218–226
[7] Das UN (2002) A perinatal strategy for preventing adult diseases: the role of long-chain polyunsaturated fatty acids. Kluwer Academic, Boston
[8] Das UN (2010) Metabolic syndrome pathophysiology: the role of essential fatty acids. Wiley-Blackwell, Ames
[9] Mozaffarian D, Pischon T, Hankinson SE, Rifai N, Joshipura K, Willett WC, Rimm EB (2004) Dietary intake of trans fatty acids and systemic inflammation in women. Am J Clin Nutr 79:606–612
[10] Brenner RR (1982) Nutritional and hormonal factors influencing desaturation of essential fatty acids. Prog Lipid Res 20:41–48
[11] Peluffo RO, Dumm NTD, De Alaniz MJT, Brenner RR (1971) Effect of protein and insulin on linoleic acid desaturation of normal and diabetic Rats. J Nutr 101:1075–1084
[12] DeGomez Dumm INT, De Alaniz MJT, Brenner RR (1970) Effect of diet on linoleic acid desaturation and on some enzymes of carbohydrate metabolism. J Lipid Res 11:96–101
[13] Peluffo RO, Brenner RR (1974) Influence of dietary protein on 6- and 9-desaturation of fatty acids in rats of different ages and in different seasons. J Nutr 104:894–900
[14] Hardy S, Langelier Y, Prentki M (2000) Oleate activates phosphatidylinositol 3-kinase and promotes proliferation and reduces apoptosis of MDA-MB-231 breast cancer cells, whereas palmitate has opposite effects. Cancer Res 60:6353–6358
[15] Menendez JA, Lupu R (2006) Mediterranean dietary traditions for the molecular treatment of human cancer: anti- oncogenic actions of the main olive oil's monounsaturated fatty acid oleic acid (18;1n-9). Curr Pharm Biotechnol 7:495–502
[16] Menendez JA, Papadimitropoulou A, Vellon L, Lupu R (2006) A genomic explanation connecting "Mediterranean diet", olive oil and cancer: oleic acid, the main monounsaturated fatty acid of olive oil, induces formation of inhibitory "PEA3 transcription factor- PEA3 DNA binding site" complexes at the Her-2/neu (erbB-2) oncogene promoter in breast, ovarian and stomach cancer cells. Eur J Cancer 42:2425–2432

[17] Menendez JA, Vellon L, Colomer R, Lupu R (2005) Oleic acid, the main monounsaturated fatty acid of olive oil, suppresses Her-2/neu (erbB-2) expression and synergistically enhances the growth inhibitory effects of trastuzumab (Herceptin) in breast cancer cells with Her-2/neu oncogene amplification. Ann Oncol 16:359–371

[18] Gabor H, Abraham S (1986) Effect of dietary menhaden oil on tumor cell loss and the accumulation of mass of a transplantable mammary adenocarcinoma in BALB/c mice. J Natl Cancer Inst 76:1223–1229

[19] Kinoshita K, Noguchi M, Earashi M, Tanaka M, Sasaki T (1994) Inhibitory effects of purified eicosapentaenoic acid and docosahexaenoic acid on growth and metastasis of murine transplantable mammary tumor. In Vivo 8:371–374

[20] Sagar PS, Das UN, Koratkar R, Ramesh G, Padma M, Kumar GS (1992) Cytotoxic action of cis-unsaturated fatty acids on human cervical carcinoma (HeLa) cells: relationship to free radicals and lipid peroxidation and its modulation by calmodulin antagonists. Cancer Lett 63:189–198

[21] Sravan Kumar G, Das UN (1997) Cytotoxic action of alpha-linolenic and eicosapentaenoic acids on myeloma cells in vitro. Prostaglandins Leukot Essent Fatty Acids 56:285–293

[22] Iigo M, Nakagawa T, Ishikawa C, Iwahori Y, Asamoto M, Yazawa K, Araki E, Tsuda H (1997) Inhibitory effects of docosahexaenoic acid on colon carcinoma 26 metastasis to the lung. Br J Cancer 75:650–655

[23] Sagar PS, Das UN (1995) Cytotoxic action of cis-unsaturated fatty acids on human cervical carcinoma (HeLa) cells in vitro. Prostaglandins Leukot Essent Fatty Acids 53:287–299

[24] Calviello G, Palozza P, Piccioni E, Maggiano N, Frattucci A, Franceschelli P, Bartoli GM (1998) Dietary supplementation with eicosapentaenoic and docosahexaenoic acid inhibits growth of Morris hepatocarcinoma 3924A in rats: effects on proliferation and apoptosis. Int J Cancer 75:699–705

[25] Yuri T, Danbara N, Tsujita-Kyutoku M, Fukunaga K, Takada H, Inoue Y, Hada T, Tsubura A (2003) Dietary docosahexaenoic acid suppresses N-methyl-N-nitrosourea-induced mammary carcinogenesis in rats more effectively than eicosapentaenoic acid. Nutr Cancer 45:211–217

[26] Wu M, Harvey KA, Ruzmetov N, Welch ZR, Sech L, Jackson K, Stillwell W, Zaloga GP, Siddiqui RA (2005) Omega-3 polyunsaturated fatty acids attenuate breast cancer growth through activation of a neutral sphingomyelinase-mediated pathway. Int J Cancer 117:340–348

[27] Kato T, Kolenic N, Pardini RS (2007) Docosahexaenoic acid (DHA), a primary tumor suppressive omega-3 fatty acid, inhibits growth of colorectal cancer independent of p53 mutational status. Nutr Cancer 58:178–187

[28] Lim K, Han C, Dai Y, Shen M, Wu T (2009) Omega-3 polyunsaturated fatty acids inhibit hepatocellular carcinoma cell growth through blocking beta-catenin and cyclooxygenase-2. Mol Cancer Ther 8:3046–3055

[29] Das UN, Puskàs LG (2009) Transgenic fat-1 mouse as a model to study the pathophysiology of cardiovascular, neurological and psychiatric disorders. Lipids Health Dis 8:61

[30] Das UN (2008) Essential fatty acids and their metabolites could function as endogenous HMG-CoA reductase and ACE enzyme inhibitors, anti-arrhythmic, anti-hypertensive, anti-atherosclerotic, anti-inflammatory, cytoprotective, and cardioprotective molecules. Lipids Health Dis 7:37

[31] Das UN (2008) Can essential fatty acids reduce the burden of disease(s)? Lipids Health Dis 7:9

[32] Das UN (2007) A defect in the activity of Delta6 and Delta5 desaturases may be a factor in the initiation and progression of atherosclerosis. Prostaglandins Leukot Essent Fatty Acids 76:251–268

[33] Das UN (2005) A defect in the activity of Delta6 and Delta5 desaturases may be a factor predisposing to the development of insulin resistance syndrome. Prostaglandins Leukot Essent Fatty Acids 72:343–350

[34] Matsuzaka T, Shimano H, Yahagi N, Amemiya-Kudo M, Yoshikawa T, Hasty AH, Tamura Y, Osuga J, Okazaki H, Iizuka Y, Takahashi A, Sone H, Gotoda T, Ishibashi S, Yamada N (2002) Dual regulation of mouse Delta(5)- and Delta(6)-desaturase gene expression by SREBP-1 and PPARalpha. J Lipid Res 43:107–114
[35] Shimomura I, Shimano H, Korn BS, Bashmakov Y, Horton JD (1998) Nuclear sterol regulatory element-binding proteins activate genes responsible for the entire program of unsaturated fatty acid biosynthesis in transgenic mouse liver. J Biol Chem 273:35299–35306
[36] Hannah VC, Ou J, Luong A, Goldstein JL, Brown MS (2001) Unsaturated fatty acids down-regulate srebp isoforms 1a and 1c by two mechanisms in HEK-293 cells. J Biol Chem 276:4365–4372
[37] Field FJ, Born E, Murthy S, Mathur SN (2002) Polyunsaturated fatty acids decrease the expression of sterol regulatory element-binding protein-1 in CaCo-2 cells: effect on fatty acid synthesis and triacylglycerol transport. Biochem J 368(Pt 3):855–864
[38] Risé P, Ghezzi S, Carissimi R, Mastromauro F, Petroni A, Galli C (2007) Delta5 desaturase mRNA levels are increased by simvastatin via SREBP-1 at early stages, not via PPARalpha, in THP-1 cells. Eur J Pharmacol 571:97–105
[39] Cook HW (1981) The influence of trans acids on desaturation and elongation of fatty acids. Lipids 16:920–926
[40] Cantwell MM, Flynn MA, Cronin D, O'Neill JP, Gibney MJ (2005) Contribution of foods to trans unsaturated fatty acid intake in a group of Irish adults. J Hum Nutr Diet 18:377–385
[41] Lopez-Garcia E, Schultze MB, Meigs JB, Manson JE, Rifai N, Stampfer MJ, Willett WC, Hu FB (2005) Consumption of trans fatty acids is related to plasma biomarkers of inflammation and endothelial dysfunction. J Nutr 135:562–566
[42] Mozaffarian D, Pischon T, Hankinson SE, Rifai N, Joshipura K, Willett WC, Rimm EB (2004) Dietary intake of trans fatty acids and systemic inflammation in women. Am J Clin Nutr 79:606–612
[43] Mahfouz M (1981) Effect of dietary trans fatty acids on the delta 5, delta 6 and delta 9 desaturases of rat liver microsomes in vivo. Acta Biol Med Ger 40:1699–1705
[44] Mahfouz MM, Johnson S, Holman RT (1980) The effect of isomeric trans-18:1 acids on the desaturation of palmitic, linoleic and eicosa-8,11,14-trienoic acids by rat liver microsomes. Lipids 15:100–107
[45] Das UN (2006) Clinical laboratory tools to diagnose inflammation. Adv Clin Chem 41:189–229
[46] Maverakis E, Fung MA, Lynch PJ, Draznin M, Michael DJ, Ruben B, Fazel N (2007) Acrodermatitis enteropathica and an overview of zinc metabolism. J Am Acad Dermatol 56:116–124
[47] Prasad AS (1985) Clinical, endocrinological and biochemical effects of zinc deficiency. Clin Endocrinol Metab 14:567–589
[48] Robertson AF, Putz J (1967) Serum and erythrocyte fatty acids in a case of acrodermatitis enteropathica. J Pediatr 70:279–281
[49] Cash R, Berger CK (1969) Acrodermatitis enteropathica: defective metabolism of unsaturated fatty acids. J Pediatr 74:717–729
[50] Koletzko B, Bretschneider A, Bremer HJ (1985) Fatty acid composition of plasma lipids in acrodermatitis enteropathica before and after zinc supplementation. Eur J Pediatr 143:310–314
[51] Horrobin DF, Cunnane SC (1980) Interactions between zinc, essential fatty acids and prostaglandins: relevance to acrodermatitis enteropathica, total parenteral nutrition, the glucagonoma syndrome, diabetes, anorexia nervosa and sickle cell anaemia. Med Hypotheses 6:277–296
[52] Hamilton RM, Gillespie CT, Cook HW (1981) Relationships between levels of essential fatty acids and zinc in plasma of cystic fibrosis patients. Lipids 16:374–376
[53] Cunnane SC, Horrobin DF (1980) Parenteral Linoleic and gamma-linolenic acids ameliorate the gross effects of zinc deficiency. Proc Soc Exp Biol Med 164:583–588

[54] Bettger WJ, Reeves PG, Moscatelli EA, Reynolds G, O'Dell BL (1979) Interaction of zinc and essential fatty acids in the rat. J Nutr 109:480–488
[55] Cunnane SC, Horrobin DF, Manku MS (1984) Essential fatty acids in tissue phospholipids and triglycerides of the zinc-deficient rat. Proc Soc Exp Biol Med 177:441–446
[56] Ayala S, Brenner RR (1983) Essential fatty acid status in zinc deficiency. Effect on lipid and fatty acid composition, desaturation activity and structure of microsomal membranes of rat liver and testes. Acta Physiol Lat Am 33:193–204
[57] Clejan S, Castro-Magana M, Collipp PJ, Jonas E, Maddaiah VT (1982) Effects of zinc deficiency and castration on fatty acid composition and desaturation in rats. Lipids 17:129–135
[58] Eder K, Kirchgessner M (1995) Activities of liver microsomal fatty acid desaturases in zinc-deficient rats force-fed diets with a coconut oil/safflower oil mixture of linseed oil. Biol Trace Elem Res 48:215–229
[59] Weis MT, Saunders C (1993) Magnesium and arachidonic acid metabolism. Magnes Res 6:179–190
[60] Weis MT (1993) Magnesium influences incorporation of[3H]arachidonic acid in perfused rabbits hearts. Am J Physiol 265(1 Pt 2):H83–H90
[61] Volpe MA, Carneiro JJ, Magna LA, Viaro F, Origuela EA, Evora PR (2003) The role of magnesium in the endothelial dysfunction caused by global ischemia followed by reperfusion: in vitro study of canine coronary arteries. Scand Cardiovasc J 37:288–296
[62] Félétou M, Rasetti C, Duhault J (1994) Magnesium modulates endothelial dysfunction produced by elevated glucose incubation. J Cardiovasc Pharmacol 24:470–478
[63] Magpie Trial Follow-Up Study Collaborative Group (2007) The Magpie trial: a randomised trial comparing magnesium sulphate with placebo for pre-eclampsia. Outcome for women at 2 years. BJOG 114:300–309
[64] Altman D, Carroli G, Duley L, Farrell B, Moodley J, Neilson J, Smith D; Magpie Trial Collaboration Group (2002) Do women with pre-eclampsia, and their babies, benefit from magnesium sulphate? The Magpie trial: a randomised placebo-controlled trial. Lancet 359:1877–1890
[65] Satake K, Lee JD, Shimizu H, Uzui H, Mitsuke Y, Yue H, Ueda T (2004) Effects of magnesium on prostacyclin synthesis and intracellular free calcium concentration in vascular cells. Magnes Res 17:20–27
[66] Yuan J, Zhou J, Chen BC, Zhang X, Zhou HM, Du DF, Chang S, Chen ZK (2005) Magnesium supplementation prevents chronic cyclosporine nephrotoxicity via adjusting nitric oxide synthase activity. Transplant Proc 37:1892–1895
[67] Volpe MA, Carneiro JJ, Magna LA, Viaro F, Origuela EA, Evora PR (2003) The role of magnesium in the endothelial dysfunction caused by global ischemia followed by reperfusion: in vitro study of canine coronary arteries. Scand Cardiovasc J 37:288–296
[68] Ikari A, Kano T, Suketa Y (2002) Magnesium influx enhanced by nitric oxide in hypertensive rat proximal tubule cells. Biochem Biophys Res Commun 294:710–713
[69] Karmin O, Cheung F, Sung FL, Zhu DY, Siow YL (2000) Effect of magnesium tanshinoate B on the production of nitric oxide in endothelial cells. Mol Cell Biochem 207:35–39
[70] Pearson PJ, Evora PR, Seccombe JF, Schaff HV (1998) Hypomagnesemia inhibits nitric oxide release from coronary endothelium: protective role of magnesium infusion after cardiac operations. Ann Thorac Surg 65:967–972
[71] Mahfouz MM, Kummerow FA (1989) Effect of magnesium deficiency on delta 6 desaturase activity and fatty acid composition of rat liver microsomes. Lipids 24:727–732
[72] Garg ML, Sebokova E, Thomson AB, Clandinin MT (1988) Delta 6-desaturase activity in liver microsomes of rats fed diets enriched with cholesterol and/or omega 3 fatty acids. Biochem J 249:351–356
[73] Soldati L, Terranegra A, Baggio B, Biasion R, Arcidiacono T, Priante G, Cusi D, Vezzoli G (2006) Arachidonic acid influences intracellular calcium handling in human osteoblasts. Prostaglandins Leukot Essent Fatty Acids 75:91–96

[74] Soldati L, Lombardi C, Adamo D, Terranegra A, Bianchin C, Bianchi G, Vezzoli G (2002) Arachidonic acid increases intracellular calcium in erythrocytes. Biochem Biophys Res Commun 293:974–978
[75] Hardy SJ, Robinson BS, Ferrante A, Hii CS, Johnson DW, Poulos A, Murray AW (1995) Polyenoic very-long-chain fatty acids mobilize intracellular calcium from a thapsigargin-insensitive pool in human neutrophils. The relationship between Ca2+ mobilization and superoxide production induced by long- and very-long-chain fatty acids. Biochem J 311(Pt 2):689–697
[76] Nàdasy GL, Székàcs B, Juhàsz I, Fehér J, Kovàch AG, Monos E (1991) Role of endothelium, oxygen and ionic milieu in the prostacyclin and thromboxane production of rat aortic tissue slices. Acta Physiol Hung 78:77–87
[77] Gardiner NS, Duncan JR (1988) Enhanced prostaglandin synthesis as a mechanism for inhibition of melanoma cell growth by ascorbic acid. Prostaglandins Leukot Essent Fatty Acids 34:119–126
[78] Srivastava KC (1985) Ascorbic acid enhances the formation of prostaglandin E1 in washed human platelets and prostacyclin in rat aortic rings. Prostaglandins Leukot Med 18:227–233
[79] Horrobin DF, Oka M, Manku MS (1979) The regulation of prostaglandin E1 formation: a candidate for one of the fundamental mechanisms involved in the actions of vitamin C. Med Hypotheses 5:849–858
[80] Manku MS, Oka M, Horrobin DF (1979) Differential regulation of the formation of prostaglandins and related substances from arachidonic acid and from dihomogammalinolenic acid. II. Effects of vitamin C. Prostaglandins Med 3:129–137
[81] Horrobin DF (1980) A biochemical basis for alcoholism and alcohol-induced damage including the fetal alcohol syndrome and cirrhosis: interference with essential fatty acid and prostaglandin metabolism. Med Hypotheses 6:929–942
[82] Manku MS, Oka M, Horrobin DF (1979) Differential regulation of the formation of prostaglandins and related substances from arachidonic acid and from dihomogammalinolenic acid. I. Effects of ethanol. Prostaglandins Med 3:119–128
[83] Masukawa T, Goto J, Iwata H (1983) Impaired metabolism of arachidonate in selenium deficient animals. Experientia 39:405–406
[84] Schiavon R, Freeman GE, Guidi GC, Perona G, Zatti M, Kakkar VV (1984) Selenium enhances prostacyclin production by cultured endothelial cells: possible explanation for increased bleeding times in volunteers taking selenium as a dietary supplement. Thromb Res 34:389–396
[85] Kunisaki M, Umeda F, Inoguchi T, Nawata H (1992) Vitamin E binds to specific binding sites and enhances prostacyclin production by cultured aortic endothelial cells. Thromb Haemost 68:744–751
[86] Das UN (1981) Pyridoxine, thrombosis and prostaglandins. Lancet ii:638
[87] Das UN (1982) Haemostatic vitamins and prostaglandins. Med J Aust 2:316–317
[88] Cabrini L, Bochicchio D, Bordoni A, Sassi S, Marchetti M, Maranesi M (2005) Correlation between dietary polyunsaturated fatty acids and plasma homocysteine concentration in vitamin B6-deficient rats. Nutr Metab Cardiovasc Dis 15:94–99
[89] Bertrandt J, Klos A, Debski B (2004) Content of polyunsaturated fatty acids (PUFAs) in serum and liver of rats fed restricted diets supplemented with vitamins B2, B6 and folic acid. Biofactors 22:189–192
[90] Bergami R, Maranesi M, Marchetti M, Sangiorgi Z, Tolomelli B (1999) Influence of dietary n-3 polyunsaturated fatty acids on plasma lipemic effect of vitamin B6 deficiency. Int J Vitam Nutr Res 69:315–321
[91] Saareks V, Mucha I, Sievi E, Riutta A (1999) Nicotinic acid and pyridoxine modulate arachidonic acid metabolism in vitro and ex vivo in man. Pharmacol Toxicol 84:274–280
[92] Bashford CL, Harrison SJ, Radda GK (1975) The relation between lipid mobility and the specific hormone binding of thyroid membranes. Biochem J 146:473–479
[93] Doi O, Doi F, Schroeder F, Alberts AW, Vagelos PR (1978) Manipulation of fatty acid composition of membrane phospholipid and its effects on cell growth in mouse LM cells. Biochim Biophys Acta 509:239–250

[94] Tsao YK, Lands WE (1980) Cell growth with trans fatty acids is affected by adenosine 3′,5′-monophosphate and membrane fluidity. Science 207:777–779
[95] McMurchie EJ, Raison JK (1979) Membrane lipid fluidity and its effect on the activation energy of membrane-associated enzymes. Biochim Biophys Acta 554:364–374
[96] Hashimoto M, Hossain S, Yamasaki H, Yazawa K, Masumura S (1999) Effects of eicosapentaenoic acid and docosahexaenoic acid on plasma membrane fluidity of aortic endothelial cells. Lipids 34:1297–1304
[97] Breton M, Wolf C, Colard O (1983) Linoleate incorporation into rat liver membranes phospholipids: effect on plasma membrane ATPase activities and physical properties. Biochem Biophys Res Commun 117:809–816
[98] Shih JC, Ohsawa R (1983) Differential effect of cholesterol on two types of 5-hydroxytryptamine binding sites. Neurochem Res 8:701–710
[99] Tomonaga A, Hirota M, Snyderman R (1983) Effect of membrane fluidizers on the number and affinity of chemotactic factor receptors on human polymorphonuclear leukocytes. Microbiol Immunol 27:961–972
[100] Bruneau C, Staedel-Flaig C, Crémel G, Leray C, Beck JP, Hubert P (1987) Influence of lipid environment on insulin binding in cultured hepatoma cells. Biochim Biophys Acta 928:287–296
[101] Bruneau C, Hubert P, Waksman A, Beck JP, Staedel-Flaig C (1987) Modifications of cellular lipids induce insulin resistance in cultured hepatoma cells. Biochim Biophys Acta 928:297–304
[102] Ginsberg BH, Chatterjee P, Yorek MA (1991) Insulin sensitivity is increased in Friend erythroleukemia cells enriched in polyunsaturated fatty acid. Receptor 1:155–166
[103] Guinea R, Carrasco L (1991) Effects of fatty acids on lipid synthesis and viral RNA replication in poliovirus-infected cells. Virology 185:473–476
[104] Lu L, Okada N, Nakatani S, Yoshikawa K (1995) Eicosapentaenoic acid-induced changes in membrane fluidity and cell adhesion molecules in cultured human keratinocytes. Br J Dermatol 133:217–222
[105] Valentine RC, Valentine DL (2004) Omega-3 fatty acids in cellular membranes: a unified concept. Prog Lipid Res 43:383–402
[106] Sipka S, Dey I, Buda C, Csongor J, Szegedi G, Farkas T (1996) The mechanism of inhibitory effect of eicosapentaenoic acid on phagocytic activity and chemotaxis of human neutrophil granulocytes. Clin Immunol Immunopathol 79:224–228
[107] Nugent C, Prins JB, Whitehead JP, Wentworth JM, Chatterjee VK, O'Rahilly S (2001) Arachidonic acid stimulates glucose uptake in 3T3-L1 adipocytes by increasing GLUT1 and GLUT4 levels at the plasma membrane. Evidence for involvement of lipoxygenase metabolites and peroxisome proliferator-activated receptor gamma. J Biol Chem 276:9149–9157
[108] Das UN (2003) Long-chain polyunsaturated fatty acids in the growth and development of the brain and memory. Nutrition 19:62–65
[109] Calderon F, Kim HY (2004) Docosahexaenoic acid promotes neurite growth in hippocampal neurons. J Neurochem 90:979–988
[110] Rickman C, Davletov B (2005) Arachidonic acid allows SNARE complex formation in the presence of Munc18. Chem Biol 12:545–553
[111] Darios F, Davletov B (2006) Omega-3 and omega-6 fatty acids stimulate cell membrane expansion by acting on syntaxin 3. Nature 440:813–817
[112] Connell E, Darios F, Broersen K, Gatsby N, Peak-Chew SY, Rickman C, Davletov B (2007) Mechanism of arachidonic acid action on syntaxin-Munc18. EMBO Rep 8:414–419
[113] Pongrac JL, Slack PJ, Innis SM (2007) Dietary polyunsaturated fat that is low in (n-3) and high in (n-6) fatty acids alters the SNARE protein complex and nitrosylation in rat hippocampus. J Nutr 137:1852–1856
[114] Boström P, Andersson L, Rutberg M, Perman J, Lidberg U, Johansson BR, Fernandez-Rodriguez J, Ericson J, Nilsson T, Borén J, Olofsson SO (2007) SNARE proteins mediate

fusion between cytosolic lipid droplets and are implicated in insulin sensitivity. Nat Cell Biol 9:1286–1293

[115] Mollerup S, Haugen A (1996) Differential effect of polyunsaturated fatty acids on cell proliferation during human epithelial in vitro carcinogenesis: involvement of epidermal growth factor receptor tyrosine kinase. Br J Cancer 74:613–618

[116] Connolly JM, Rose DP (1992) Interaction between epidermal growth factor-mediated autocrine regulation and linoleic acid-stimulated growth of a human prostate cancer cell line. Prostate 20:151–158

[117] Vacaresse N, Lajoie-Mazenc I, Augé N, Suc I, Frisach MF, Salvayre R, Nègre-Salvayre A (1999) Activation of epithelial growth factor receptor pathway by unsaturated fatty acids. Circ Res 85:892–899

[118] Hagberg C, Falkevall A, Wang X et al (2010) Vascular endothelial growth factor B controls endothelial fatty acid uptake. Nature 464:917–924

[119] Calviello G, Di Nicuolo F, Gragnoli S, Piccioni E, Serini S, Maggiano N, Tringali G, Navarra P, Ranelletti FO, Palozza P (2004) n-3 PUFAs reduce VEGF expression in human colon cancer cells modulating the COX-2/PGE2 induced ERK-1 and -2 and HIF-1alpha induction pathway. Carcinogenesis 25:2303–2310

[120] Koto T, Nagai N, Mochimaru H, Kurihara T, Izumi-Nagai K, Satofuka S, Shinoda H, Noda K, Ozawa Y, Inoue M, Tsubota K, Oike Y, Ishida S (2007) Eicosapentaenoic acid is anti-inflammatory in preventing choroidal neovascularization in mice. Invest Ophthalmol Vis Sci 48:4328–4334

[121] Kodicek E (1949) The effect of unsaturated fatty acids on gram-positive bacteria. Symp Soc Exp Biol 3:217–232

[122] Lacey RW, Lord VL (1981) Sensitivity of staphylococci to fatty acids: novel inactivation of linolenic acid by serum. J Med Microbiol 14:41–49

[123] Galbraith H, Miller TB, Paton AM, Thompson JK (1971) Antibacterial activity of long chain fatty acids and the reversal with calcium, magnesium, ergocalciferol and cholesterol. J Appl Bacteriol 34:803–813

[124] McDonald MI, Graham I, Harvey KJ, Sinclair A (1981) Antibacterial activity of hydrolysed linseed oil ad linolenic acid against methicillin-resistant Staphylococcus aureus. Lancet 2:1056

[125] Kohn A, Gitelman J, Inbar M (1980) Unsaturated free fatty acids inactivate animal enveloped viruses. Arch Virol 66:301–307

[126] Iwamoto KS, Bennett LR, Norman A, Villalobos AE, Hutson CA (1992) Linoleate produces remission in canine mycosis fungoides. Cancer Lett 64:17–22

[127] Krugloak M, Deharo E, Shalmiev G, Sauvain M, Moretti C, Ginsburg H (1995) Antimalarial effects of C18 fatty acids on Plasmodium falciparum in culture and on Plasmodium vinckei petteri and Plamodium yoelii nigeriensis in vivo. Exp Parasitol 81:97–105

[128] Doering TL, Raper J, Buxbaum LU, Adams SP, Gordon JI, Hart GW, Englund PT (1991) An analog of myristic acid with selective toxicity for African Trypanosomes. Science 252:1851–1854

[129] Giron DJ (1982) Inhibition of viral replication in cell cultures treated with prostaglandins E1. Proc Soc Exp Biol Med 170:25–28

[130] Santoro MG, Benedetto A, Carruba G, Garaci E, Jaffe BM (1980) Prostaglandin A compounds as antiviral agents. Science 209:1032–1034

[131] Das UN (1985) Can essential fatty acid deficiency predispose to AIDS? Can Med Assoc J 132:900

[132] Das UN (1985) Anti-biotic-like action of essential fatty acids. Can Med Assoc J 132:1350

[133] Das UN (2005) Essential fatty acids and acquired immunodeficiency syndrome. Med Sci Monit 11:RA206–RA211

[134] Bromberg Y, Pick E (1984) Unsaturated fatty acids stimulate NADPH-dependent superoxide production by cell-free system derived from macrophages. Cell Immunol 88:213–221

[135] Sun CQ, O'Connor CJ, Robertson A (2003) Antibacterial actions of fatty acids and monoglycerides against Helicobacter pylori. FEMS Immunol Med Microbiol 36:9–17

[136] Das UN (2006) Do unsaturated fatty acids function as endogenous anti-bacterial and anti-viral molecules? Am J Clin Nutr 83:390–391
[137] Giamarellos-Bourboulis EJ, Mouktaroudi M, Adamis T, Koussoulas V, Baziaka F, Perrea D, Karavannacos PE, Giamarellou H (2004) n-6 polyunsaturated fatty acids enhance the activities of ceftazidime and amikacin in experimental sepsis caused by multidrug-resistant Pseudomonas aeroginosa. Antimicrob Agents Chemother 48:4713–4717
[138] Das UN (2010) Do long-chain unsaturated fatty acids function as endogenous anti-Trypanosomal molecules? Med Hypotheses 74:676–678
[139] Wang H, Huang BS, Ganten D, Leenen FH (2004) Prevention of sympathetic and cardiac dysfunction after myocardial infarction in transgenic rats deficient in brain angiotensinogen. Circ Res 94:843–849
[140] Das UN (2005) Is angiotensin II an endogenous pro-inflammatory molecule? Med Sci Monit 11:RA155–RA162
[141] Kumar KV, Das UN (1997) Effect of cis-unsaturated fatty acids, prostaglandins, and free radicals on angiotensin-converting enzyme activity in vitro. Proc Soc Exp Biol Med 214:374–379
[142] Okuda Y, Kawashima K, Sawada T, Tsurumaru K, Asano M, Suzuki S, Soma M, Nakajima T, Yamashita K (1997) Eicosapentaenoic acid enhances nitric oxide production by cultured human endothelial cells. Biochem Biophys Res Commun 232:487–491
[143] Bunk S (2002) ACEs wild. Scientist 16:22–24
[144] Moskowitz DW (2002) Is angiotensin I-converting enzyme a 'master' disease gene? Diabetes Technol Ther 4:683–711
[145] Kaergel E, Muller DN, Honeck H, Theuer J, Shagdarsuren E, Mullally A, Luft FC, Schunck W-H (2002) P450-dependent arachidonic acid metabolism and angiotensin-II-induced renal damage. Hypertension 40:273–279
[146] Chawengsub Y, Gauthier KM, Nithipatikom K, Hammock BD, Falck JR, Narsimhaswamy D, Campbell WB (2009) Identification of 13-hydroxy-14,15-epoxyeicosatrienoic acid as an acid-stable endothelium-derived hyperpolarizing factor in rabbit arteries. J Biol Chem 284:31280–31290
[147] Gilroy DW (2005) New insights into the anti-inflammatory actions of aspirin- induction of nitric oxide through the generation of epi-lipoxins. Mem Inst Oswaldo Cruz 100(Suppl 1):49–54
[148] Wang W, Diamond SL (1997) Does elevated nitric oxide production enhance the release of prostacyclin from shear stressed aortic endothelial cells? Biochem Biophys Res Commun 233:748–751
[149] Das UN (2005) COX-2 inhibitors and metabolism of essential fatty acids. Med Sci Monit 11:R233–R237
[150] Pieper GM, Siebeneich W, Olds CL, Felix CC, Del Soldato P (2002) Vascular protective actions of a nitric oxide aspirin analog in both in vitro and in vivo models of diabetes mellitus. Free Radic Biol Med 32:1143–1156
[151] Nath N, Kashfi K, Chen J, Rigas B (2003) Nitric oxide-donating aspirin inhibits beta-catenin/T cell factor (TCF) signaling in SW480 colon cancer cells by disrupting the nuclear beta-catenin-TCF association. Proc Natl Acad Sci U S A 100:12584–12589
[152] Gao J, Liu X, Rigas B (2005) Nitric oxide-donating aspirin induces apoptosis in human colon cancer cells through induction of oxidative stress. Proc Natl Acad Sci U S A 102:17207–17212
[153] Ouyang N, Williams JL, Rigas B (2008) NO-donating aspirin inhibits angiogenesis by suppressing VEGF expression in HT-29 human colon cancer mouse xenografts. Carcinogenesis 29:1794–1798
[154] Gresele P, Marzotti S, Guglielmini G, Momi S, Giannini S, Minuz P, Lucidi P, Bolli GB (2010) Hyperglycemia-induced platelet activation in type 2 diabetes mellitus is resistant to aspirin but not to a nitric oxide-donating agent. Diabetes Care 33:1262–1268
[155] Kumar GS, Das UN (1994) Effect of prostaglandins and their precursors on the proliferation of human lymphocytes and their secretion of tumor necrosis factor and various interleukins. Prostaglandins Leukot Essent Fatty Acids 50:331–334

[156] Arita M, Bianchini F, Aliberti J, Sher A, Chiang N, Hong S, Yang R, Petasis NA, Serhan CN (2005) Stereochemical assignment, antiinflammatory properties, and receptor for the omega-3 lipid mediator resolvin E1. J Exp Med 201:713–722
[157] Dooper MM, van Riel B, Graus YM, M'Rabet L (2003) Dihomo-gamma-linolenic acid inhibits tumour necrosis factor-alpha production by human leucocytes independently of cyclooxygenase activity. Immunology 110:348–357
[158] Chen W, Esselman WJ, Jump DB, Busik JV (2005) Anti-inflammatory effect of docosahexaenoic acid on cytokine-induced adhesion molecule expression in human retinal vascular endothelial cells. Invest Ophthalmol Vis Sci 46:4342–4347
[159] Das UN (2008) Pathological retinal angiogenesis and polyunsaturated fatty acids. Agro Food Ind Hi Tech 19:44–49
[160] Das UN (2009) Polyunsaturated fatty acids in pathological retinal angiogenesis. Curr Nutr Food Sci 5:94–111
[161] Li H, Ruan XZ, Powis SH, Fernando R, Mon WY, Wheeler DC, Moorhead JF, Varghese Z (2005) EPA and DHA reduce LPS-induced inflammation responses in HK-2 cells: evidence for a PPAR-gamma-dependent mechanism. Kidney Int 67:867–874
[162] Beck SA, Smith KL, Tisdale MJ (1991) Anti-cachectic and anti-tumor effect of eicosapentaenoic acid and its effect on protein turnover. Cancer Res 51:6089–6093
[163] Lihn AS, Richelsen B, Pedersen SB, Hangaard SB, Rathje GS Madsbad S, Andersen O (2003) Increased expression of TNF-alpha, IL-6, and IL-8 in HALS implications for reduced adiponectin expression and plasma levels. Am J Physiol Endocrinol Metab 285:E1072–E1080
[164] Miller J, Carr A, Emery S, Law M, Mallal S, Baker D, Smith D, Kaldor J, Cooper DA (2003) HIV lipodystrophy: prevalence, severity and correlates of risk in Australia. HIV Med 4:293–301
[165] Sheng Z, Otani H, Brown MS, Goldstein JL (1995) Independent regulation of sterol regulatory element-binding proteins 1 and 2 in hamster liver. Proc Natl Acad Sci U S A 92:935–938
[166] Fajas L, Schoonjans K, Gelman L, Kim JB, Najib J, Martin G, Fruchart JC, Briggs M, Spiegelman BM, Auwerx J (1999) Regulation of peroxisome proliferator-activated receptor gamma expression by adipocyte differentiation and determination factor 1/sterol regulatory element binding protein 1: implications for adipocyte differentiation and metabolism. Mol Cell Biol 19:5495–5503
[167] El-Sohemy A, Archer MC (1999) Regulation of mevalonate synthesis in low density lipoprotein receptor knockout mice fed n-3 or n-6 polyunsaturated fatty acids. Lipids 34:1037–1043
[168] Das UN (2000) Essential fatty acids and osteoporosis. Nutrition 16:286–290
[169] Nakamura N, Hamazaki T, Jokaji H, Minami S, Kobayashi M (1998) Effect of HMG-CoA reductase inhibitors on plasma polyunsaturated fatty acid concentration in patients with hyperlipidemia. Int J Clin Lab Res 28:192–195
[170] Hannah VC, Ou J, Luong A, Goldstein JL, Brown MS (2001) Unsaturated fatty acids down-regulate srebp isoforms 1a and 1c by two mechanisms in HEK-293 cells. J Biol Chem 276:4365–4372
[171] Duncan RE, El-Sohemy A, Archer MC (2005) Regulation of HMG-CoA reductase in MCF-7 cells by genistein, EPA, and DHA, alone and in combination with mevastatin. Cancer Lett 224:221–228
[172] Levine L (2003) Statins stimulate arachidonic acid release and prostaglandin I2 production in rat liver cells. Lipids Health Dis 2:1
[173] Dobrucki LW, Kalinowski L, Dobrucki IT, Malinski T (2001) Statin-stimulated nitric oxide release from endothelium. Med Sci Monit 7:622–627
[174] McGown CC, Brookes ZL (2007) Beneficial effects of statins on the microcirculation during sepsis: the role of nitric oxide. Br J Anaesth 98:163–175
[175] Yasuda H, Yuen PS, Hu X, Zhou H, Star RA (2006) Simvastatin improves sepsis-induced mortality and acute kidney injury via renal vascular effects. Kidney Int 69:1535–1542

[176] Nordoy A, Bonaa KH, Sandset PM, Hansen JB, Nilsen H (2000) Effect of omega-3 fatty acids and simvastatin on hemostatic risk factors and postprandial hyperlipemia in patients with combined hyperlipemia. Arterioscler Thromb Vasc Biol 20:259–265

[177] Nordoy A, Hansen JB, Brox J, Svensson B (2001) Effects of atorvastatin and omega-3 fatty acids on LDL subfractions and postprandial hyperlipemia in patients with combined hyperlipemia. Nutr Metab Cardiovasc Dis 11:7–16

[178] Nordoy A, Svensson B, Hansen JB (2003) Atorvastatin and omega-3 fatty acids protect against activation of the coagulation system in patients with combined hyperlipemia. J Thromb Haemost 1:690–697

[179] Nordoy A (2002) Statins and omega-3 fatty acids in the treatment of dyslipidemia and coronary heart disease. Minerva Med 93:357–363

[180] Levy BD (2006) Myocardial 15-epi-lipoxin A4 generation provides a new mechanism for the immunomodulatory effects of statins and thiazolidinediones. Circulation 114:873–875

[181] Birnbaum Y, Ye Y, Lin Y et al (2006) Augmentation of myocardial production of 15-epi-lipoxin A4 by pioglitazone and atorvastatin in the rat. Circulation 114:929–935

[182] Field FJ, Born E, Murthy S, Mathur SN (2002) Polyunsaturated fatty acids decrease the expression of sterol regulatory element-binding protein-1 in CaCo-2 cells: effect on fatty acid synthesis and triacylglycerol transport. Biochem J 368(Pt 3):855–864

[183] Xu J, Cho H, O'Malley S, Park JH, Clarke SD (2002) Dietary polyunsaturated fats regulate rat liver sterol regulatory element binding proteins-1 and -2 in three distinct stages and by different mechanisms. J Nutr 132:3333–3339

[184] Field FJ, Born E, Mathur SN (2003) Fatty acid flux suppresses fatty acid synthesis in hamster intestine independently of SREBP-1 expression. J Lipid Res 44:1199–1208

[185] Le Jossic-Corcos C, Gonthier C, Zaghini I, Logette E, Shechter I, Bournot P (2005) Hepatic farnesyl diphosphate synthase expression is suppressed by polyunsaturated fatty acids. Biochem J 385(Pt 3):787–794

[186] Worgall TS, Johnson RA, Seo T, Gierens H, Deckelbaum RJ (2002) Unsaturated fatty acid-mediated decreases in sterol regulatory element-mediated gene transcription are linked to cellular sphingolipid metabolism. J Biol Chem 277:3878–3885

[187] Das UN (2003) Long-chain polyunsaturated fatty acids in the growth and development of the brain and memory. Nutrition 19:62–65

[188] Das UN (2003) Can memory be improved? A discussion on the role of ras, GABA, acetylcholine, NO, insulin, TNF-α, and long-chain polyunsaturated fatty acids in memory formation and consolidation. Brain Dev 25:251–261

[189] Das UN (2003) Long-chain polyunsaturated fatty acids in memory formation and consolidation: further evidence and discussion. Nutrition 19:988–993

[190] Calon F, Lim GP, Yang F, Morihara T, Teter B, Ubeda O, Rostaing P, Triller A, Salem Jr. N, Ashe KH, Frautschy SA, Cole GM (2004) Docosahexaenoic acid protects from dendritic pathology in an Alzheimer's disease mouse model. Neuron 43:633–645

[191] Lukiw WJ, Cui J-G, Marcheselli VL, Bodker M, Botkjaer A, Gotlinger K, Serhan CN, Bazan NG (2005) A role for docosahexaenoic acid-derived neuroprotectin D1 in neural cell survival and Alzheimer disease. J Clin Invest 115:2774–2783

[192] Calon F, Lim GP, Morihara T, Yang F, Ubeda O, Salem N Jr, Frautschy SA, Cole GM (2005) Dietary n-3 polyunsaturated fatty acid depletion activates caspases and decreases NMDA receptors in the brain of a transgenic mouse model of Alzheimer's. Eur J Neurosci 22:617–626

[193] Akbar M, Calderon F, Wen Z, Kim HY (2005) Docosahexaenoic acid: a positive modulator of Akt signaling in neuronal survival. Proc Natl Acad Sci U S A 102:10858–10863

[194] Hashimoto M, Tanabe Y, Fujii Y Kikuta T, Shibata H, Shido O (2005) Chronic administration of docosahexaenoic acid ameliorates the impairment of spatial cognition learning ability in amyloid beta-infused rats. J Nutr 135:549–555

[195] Aravindakshan M, Ghate M, Ranjekar PK, Evans DR, Mahadik SP (2003) Supplementation with a combination of omega-3 fatty acids and antioxidants (vitamins E and C) improves the outcome of schizophrenia. Schizophr Res 62:195–204

[196] Das UN (2004) Can perinatal supplementation of long-chain polyunsaturated fatty acids prevents schizophrenia in adult life? Med Sci Monit 10:HY33–HY37
[197] Claria J, Serhan CN (1995) Aspirin triggers previously undescribed bioactive eicosanoids by human endothelial cell-leukocyte interactions. Proc Natl Acad Sci U S A 92:9475–9479
[198] Chiang N, Gronert K, Clish CB, O'Brien JA, Freeman MW, Serhan CN (1999) Leukotriene B4 receptor transgenic mice reveal novel protective roles for lipoxins and aspirin-triggered lipoxins in reperfusion. J Clin Invest 104:309–316
[199] Serhan CN, Clish CB, Brannon J, Colgan SP, Chiang N, Gronert K (2000) Novel functional sets of lipid-derived mediators with antiinflammatory actions generated from omega-3 fatty acids via cyclooxygenase 2-nonsteroidal antiinflammatory drugs and transcellular processing. J Exp Med 192:1197–1204
[200] Serhan CN, Hong S, Gronert K, Colgan SP, Devchand PR, Mirick G, Moussignac R-L (2002) Resolvins: a family of bioactive products of omega-3 fatty acid transformation circuits initiated by aspirin treatment that counter proinflammation signals. J Exp Med 196:1025–1037
[201] Marcheselli VL, Hong S, Lukiw WJ, Tian XH, Gronet K, Musto A, Hardy M, Gimenez JM, Chiang N, Serhan CN, Bazan NG (2003) Novel docosanoids inhibit brain ischemia-reperfusion-mediated leukocyte infiltration and pro-inflammatory gene expression. J Biol Chem 278:43807–43817
[202] Hong S, Gronert K, Devchand PR, Moussignac RL, Serhan CN (2003) Novel docosatrienes and 17S-resolvins generated from docosahexaenoic acid in murine brain, human blood, and glial cells. Autacoids in anti-inflammation. J Biol Chem 278:14677–14687
[203] Mukherjee PK, Marcheselli VL, Serhan CN, Bazan NG (2004) Neuroprotectin D1: a docosahexaenoic acid-derived docosatriene protects human retinal pigment epithelial cells from oxidative stress. Proc Natl Acad Sci U S A 101:8491–8496
[204] Gronert K, Maheshwari N, Khan N, Hasan IR, Dunn M, Schwartzman ML (2005) A role for the mouse 12/15-lipoxygenase pathway in promoting epithelial wound healing and host defense. J Biol Chem 280:15267–15278
[205] Serhan CN, Yang R, Martinod K, Kasuga K, Pillai PS, Porter TF, Oh SF, Spite M (2009) Maresins: novel macrophage mediators with potent antiinflammatory and proresolving actions. J Exp Med 206:15–23
[206] Serhan CN (2009) Systems approach to inflammation resolution: identification of novel anti-inflammatory and pro-resolving mediators. J Thromb Haemost 7(Suppl 1):44–48
[207] Baker PRS, Lin Y, Schopfer FJ, Woodcock SR, Groeger AL, Batthyany C, Swooney S, Long MH, Iles KE, Baker LMS, Branchaud BP, Chen Y, Freeman BA (2005) Fatty acid transduction of nitric oxide signaling: multiple nitrated unsaturated fatty acid derivatives exist in human blood and urine and serve as endogenous peroxisome proliferator-activated receptor ligands. J Biol Chem 280:42464–42475
[208] Coles B, Bloodsworth A, Clark SR, Lewis MJ, Cross AR, Freeman BA, O'Donnell VB (2002) Nitrolinoleate inhibits superoxide generation, degranulation, and integrin expression by human neutrophils. Circ Res 91:375–381
[209] Lima ES, Bonim MG, Augusto O, Barbeiro HV, Souza HP, Abdalla DSP (2005) Nitrated lipids decompose to nitric oxide and lipid radicals and cause vasorelaxation. Free Radic Biol Med 39:532–539
[210] Wright MM, Schopfer FJ, Baker PRS, Vidyasagar V, Powell P, Chumley P, Iles KE, Freeman BA, Agarwal A (2006) Fatty acid transduction of nitric oxide signaling: nitrolinoleic acid potently activates endothelial heme oxygenase 1 expression. Proc Natl Acad Sci U S A 103:4299–4304
[211] Juan H, Sametz W (1985) Dihomo-gamma-linolenic acid increases the metabolism of eicosapentaenoic acid in perfused vascular tissue. Prostaglandins Leukot Med 19:79–86
[212] Das UN (1985) Minerals, trace elements, and vitamins interact with essential fatty acids and prostaglandins to prevent hypertension, thrombosis, hypercholesterolemia, and atherosclerosis and their attendant complications. Med Sci Res 13:684–687

[213] Bordet JC, Guichardant M, Lagarde M (1988) Hydroperoxides produced by n-6 lipoxygenation of arachidonic and linoleic acids potentiate synthesis of prostacyclin related compounds. Biochim Biophys Acta 958:460–468
[214] Bordet JC, Guichardant M, Lagarde M (1986) Arachidonic acid strongly stimulates prostaglandin I3 (PGI3) production from eicosapentaenoic acid in human endothelial cells. Biochem Biophys Res Commun 135:403–410
[215] Das UN (2001) Essential fatty acids as possible mediators of the actions of statins. Prostaglandins Leukot Essent Fatty Acids 65:37–40

Chapter 5
Cell Membrane Organization

Introduction

The cell membrane also called as the plasma membrane or plasmalemma is one biological membrane that separates the interior of the cell from the outside environment. The cell membrane surrounds all cells and is selectively permeable, controlling the movement of substances in and out of cells. One of the main functions of the cell membrane is also to take messages from outside the cell (environment) and convey the same to the internal structures of the cell such as nucleus (DNA), mitochondria, etc., so that appropriate responses can be elicited from the cell to these outside stimuli. The cell contains a variety of biological molecules that include proteins, lipids and a variety of enzyme systems that are involved in various cellular processes such as adhesion, ion channel conductance and cell signaling. In certain cells, the molecules that are present inside the cells control their mobility and ability to produce certain biologically active molecules in response to a variety of external and sometimes, internal stimuli such as leukocyte movement, macrophage synthesis and secretion of cytokines, T cell response to antigens, etc. The plasma membrane also serves as the attachment point for the intracellular cytoskeleton and if present and necessary, the extracellular cell wall [1].

Fluid Mosaic Model of the Membrane

To achieve the desired functions of the cell membrane it need to be a fluid bilayer. Thus, cell membrane is a fluid bilayer of phospholipids and globular proteins. Globular proteins that lie side by side and extend through the outer plasma membrane serve as facilitated transport channels for substances like glucose and amino acids. Other globular proteins will serve as active (energy expending) transport gated channels for ions like sodium (Na^+) and potassium (K^+). Many of the plasma membrane proteins have polysaccharides, glycolipids and protein chains projecting from its surface. Some serve as "cellular cement" for adhering to adjacent cells in a tissue layer. Other proteins allow for the recognition of cell type that is important for the immune system to recognize "self." A receptor is an element of a transport channel.

U. N. Das, *Molecular Basis of Health and Disease*,
DOI 10.1007/978-94-007-0495-4_5, © Springer Science+Business Media B.V. 2011

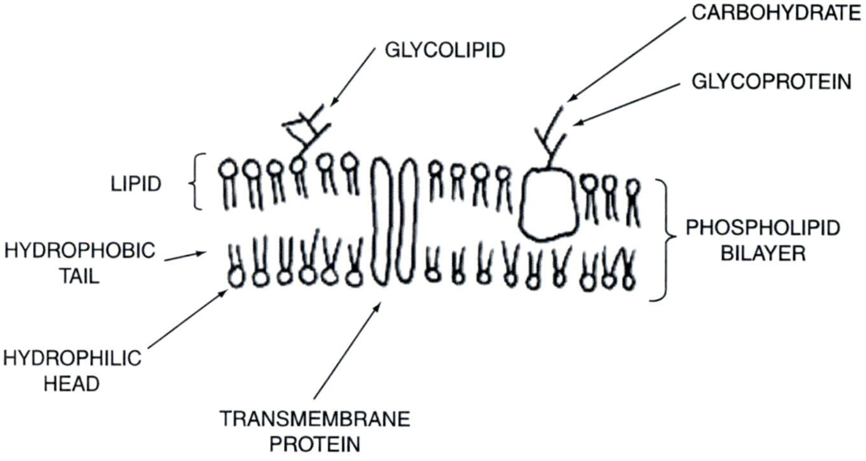

Fig. 5.1 Scheme showing the structure of the cell membrane

Many surface proteins serve as external membrane receptors for peptide hormones such as insulin, which enhances the cellular uptake of glucose from the extracellular fluid (ECF) by causing facilitated glucose transport via its receptors called as GLUT (glucose transporter) receptors or channels that appear in the target cell's plasma membrane. The hormone insulin and its receptor have a lock and key relationship (insulin is the key), as do viruses and their receptor, and enzymes and substrates. The G proteins that are present in and adjacent to the membrane act as second messengers. Mutated genes can cause receptors to be absent, receptors may lose function, or the receptors may over-function that could lead to various disease depending on the type of receptor(s) affected. Receptors also bind to neurotransmitters and lipoproteins.

Within the plasma membrane bilayer of phospholipids, cholesterol is embedded among the fatty acids tails of the cell membrane (see Fig. 5.1). It renders the membrane rigid by limiting the movement of the fatty acid tails of the phospholipids. Membrane not only covers the cell but also covers many of the internal organelles such as the nucleus, lysosomes and mitochondria. Both the fatty acid tails and cholesterol, are nonpolar and hydrophobic, but mix with each other. Water cannot stay in the interior as the membrane is polar. However, steroids hormones such as estrogen and testosterone are nonpolar; they will dissolve in and pass through the membrane [2].

The Phospholipid (PL) Bilayer—Its Structure, Properties and Functions

The plasma membranes of animal cells contain four major phospholipids (phosphatidylcholine, phosphatidylethanolamine, phosphatidylserine, and sphingomyelin) that together account for more than half of the lipid in most membranes. These phospholipids (PLs) are asymmetrically distributed between the two halves of the membrane bilayer. The outer leaflet of the plasma membrane consists mainly of phosphatidylcholine (PC) and sphingomyelin, whereas phosphatidylethanolamine (PE) and phosphatidylserine (PS) are the predominant PLs of the inner leaflet. A fifth phospholipid (PL), phosphatidylinositol (PI), is also localized to the inner half of the plasma membrane (see Figs. 5.2, 5.3, 5.4, 5.5, 5.6, 5.7, 5.8 and 5.9 for the

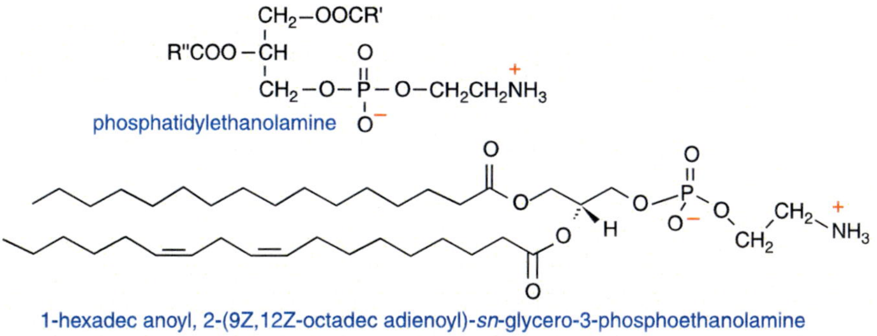

Fig. 5.2 Structure of phosphatidylethanolamine with one specific molecular species illustrated as an example

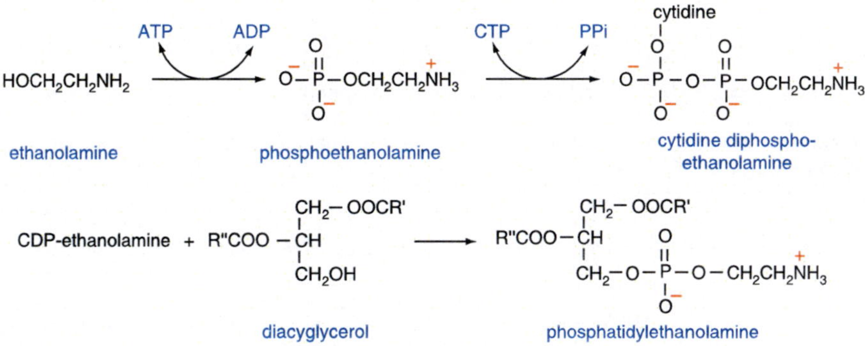

Fig. 5.3 Major pathway of biosynthesis of phosphatidylethanolamine

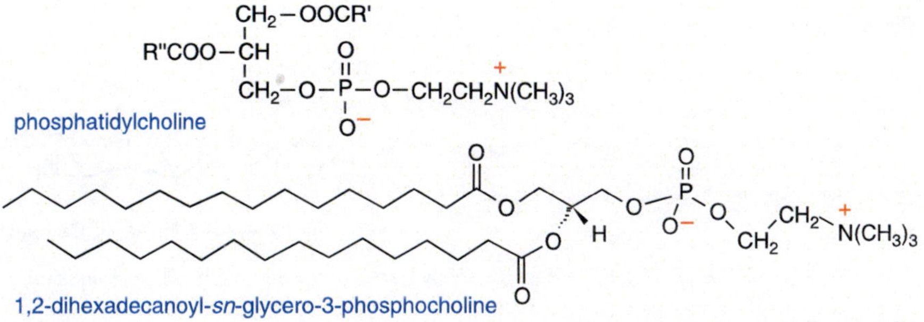

Fig. 5.4 Structure of phosphatidylcholine and that of a specific molecular species illustrated as an example

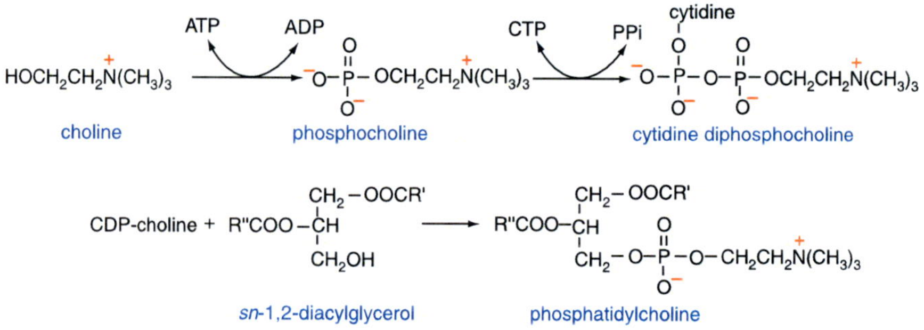

Fig. 5.5 Major pathway of synthesis of phosphatidylcholine

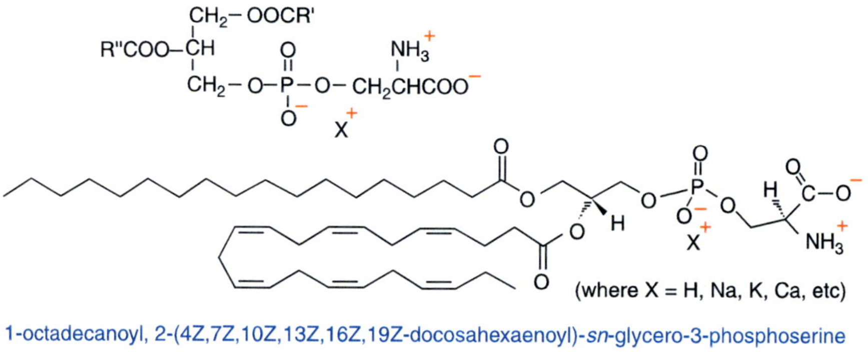

Fig. 5.6 Structure of phosphatidylserine and that of a specific molecular species illustrated as an example

phosphatidylcholine + L-serine —PS synthase I→ phosphatidylserine + choline

phosphatidylserine —PS decarboxylase→ phosphatidylethanolamine + CO_2

phosphatidylethanolamine + serine —PS synthase II→ phosphatidylserine + ethanolamine

Fig. 5.7 Biosynthetic pathway of phosphatidylserine

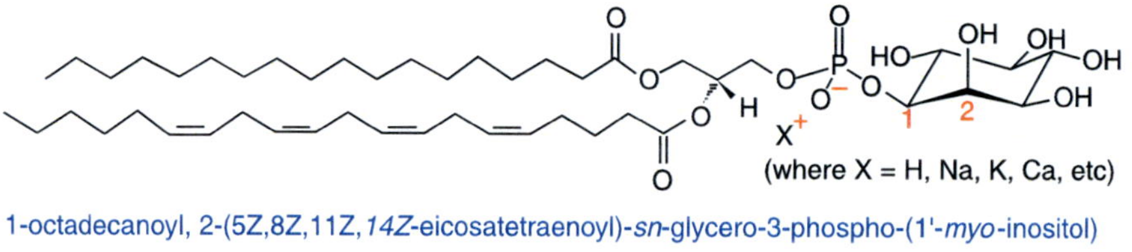

Fig. 5.8 The structure of 1-stearoyl,2-arachidoyl molecular species that is of considerable importance in the brain

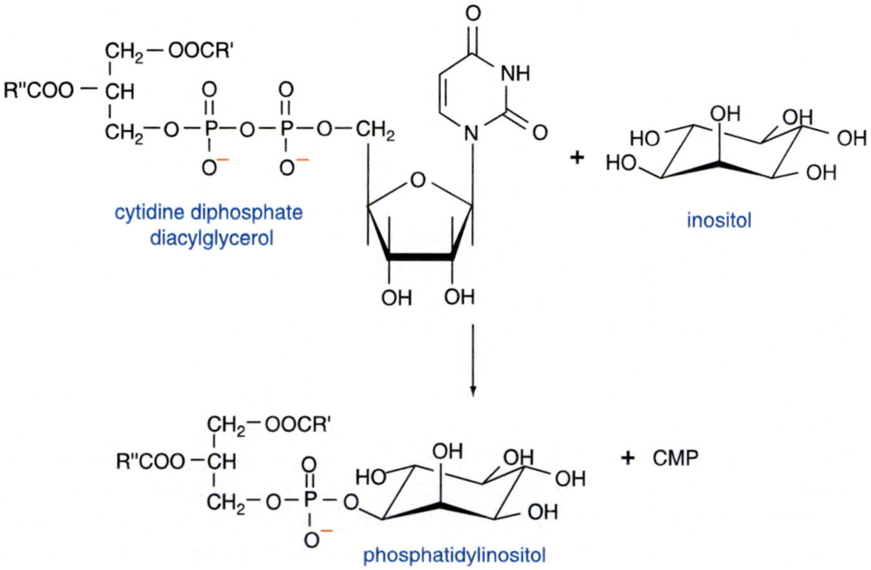

Fig. 5.9 Phosphatidylinositol is formed from the precursor cytidine diphosphate diacylglycerol by the reaction with inositol and catalyzed by the enzyme CDP-diacylglycerol inositol phosphatidyltransferase (CMP)

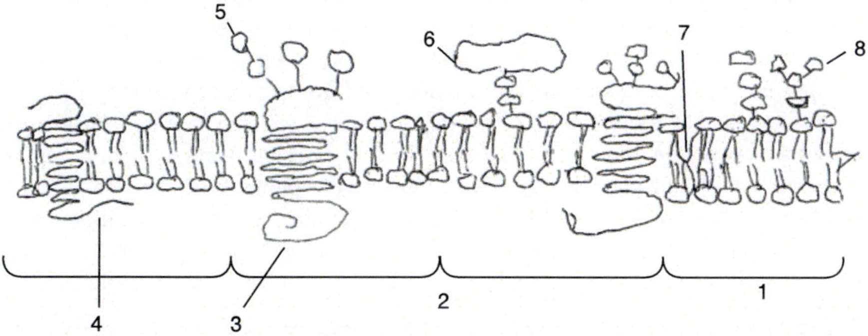

Fig. 5.10 Structure of lipid raft-caveolae organization. *1*. Non-raft membrane *2*. Lipid raft *3*. Lipid raft associated transmembrane protein *4*. Non-raft membrane protein *5*. Glycosylation modification (on glycoproteins and glycolipids) *6*. GPI-anchored protein *7*. Cholesterol *8*. Glycolipid

structure and biosynthesis of PC, PE, PS, PI). PI is a quantitatively minor membrane component but, plays an important role in cell signaling. The head groups of both PS and PI are negatively charged and hence, their predominance in the inner leaflet results in a net negative charge on the cytosolic face of the plasma membrane (Fig. 5.10).

In addition to the PL, the plasma membranes of animal cells contain glycolipids and cholesterol. The glycolipids are found exclusively in the outer leaflet of the plasma membrane, with their carbohydrate portions exposed on the cell surface. They are relatively minor membrane components, constituting only about 2% of the lipids of most plasma membranes. Cholesterol, on the other hand, is a major membrane constituent of animal cells, being present in about the same molar amounts as the PLs.

Two general features of phospholipid bilayers that are critical to membrane function are: (a) the structure of PL is responsible for the barrier function of membranes between two aqueous compartments since, the interior of the PL bilayer is occupied by hydrophobic fatty acid chains that renders the membrane impermeable to water-soluble molecules; and (b) the bilayers of the naturally occurring PLs are viscous fluids, not solids. The fatty acids of most natural phospholipids have one or more double bonds, which introduce kinks into the hydrocarbon chains and make them difficult to pack together. The long hydrocarbon chains of the fatty acids therefore move freely in the interior of the membrane, so the membrane itself is soft and flexible. In addition, both PLs and proteins are free to diffuse laterally within the membrane—a property that is critical for many membrane functions.

The rigid ring structure of cholesterol renders it to play a distinct role in membrane structure. Cholesterol will not form a membrane by itself, but inserts into a bilayer of PLs with its polar hydroxyl group close to the PL head groups. Depending on the temperature, cholesterol has distinct effects on membrane fluidity. At high temperatures, cholesterol interferes with the movement of the PL fatty acid chains, making the outer part of the membrane less fluid and reducing its permeability to small molecules. At low temperatures, however, cholesterol has the opposite effect: by interfering with interactions between fatty acid chains, cholesterol prevents membranes from freezing and maintains membrane fluidity. Plant cells lack cholesterol, but they contain related compounds (sterols) that fulfill a similar function.

It is known that not all lipids diffuse freely in the plasma membrane. Instead, discrete membrane domains appear to be enriched in cholesterol and the sphingolipids (sphingomyelin and glycolipids). These clusters of sphingolipids and cholesterol are thought to form "rafts" that move laterally within the plasma membrane and may associate with specific membrane proteins. Lipid rafts may play an important role in processes such as cell signaling and the uptake of extracellular molecules by endocytosis.

Cell Membrane Properties

The cell membrane amphipathic phospholipids which spontaneously arrange so that the hydrophobic "tail" regions are shielded from the surrounding polar fluid, causing the more hydrophilic "head" regions to associate with the cytosolic and extracellular faces of the resulting bilayer leads to the formation of a continuous, spherical lipid bilayer. The purpose of this arrangement of hydrophilic heads and hydrophobic tails of the lipid bilayer is to prevent polar solutes (e.g., amino acids, nucleic acids, carbohydrates, proteins, and ions) from diffusing across the membrane, but generally allows for the passive diffusion of hydrophobic molecules. This affords the cell the ability to control the movement of these substances via transmembrane protein complexes such as pores and gates.

Membranes serve diverse functions in eukaryotic and prokaryotic cells. One important role is to regulate the movement of materials into and out of cells. The PL bilayer structure (fluid mosaic model) with specific membrane proteins accounts for the selective permeability of the membrane and passive and active transport mechanisms. In addition, membranes in prokaryotes and in the mitochondria and chloroplasts of eukaryotes facilitate the synthesis of ATP through chemiosmosis.

The apical membrane of a polarized cell is the surface of the plasma membrane that faces the lumen. This is particularly evident in epithelial and endothelial cells, but also describes other polarized cells, such as neurons.

The basolateral membrane of a polarized cell is the surface of the plasma membrane that forms its basal and lateral surfaces. It faces towards the interstitium, and away from the lumen.

"Basolateral membrane" is a compound phrase referring to the terms *basal (base) membrane* and *lateral (side) membrane*, which, especially in epithelial cells, are identical in composition and activity. Proteins (such as ion channels and pumps) are free to move from the basal to the lateral surface of the cell or *vice versa* in accordance with the fluid mosaic model. Tight junctions that join epithelial cells near their apical surface prevent the migration of proteins from the basolateral membrane to the apical membrane. The basal and lateral surfaces thus remain roughly equivalent to one another, yet distinct from the apical surface.

Integral Membrane Proteins

The cell membrane contains many integral membrane proteins, which pepper the entire surface. These structures, which can be visualized by electron microscopy or fluorescence microscopy, can be found on the inside of the membrane, the outside, or membrane spanning. These may include integrins, cadherins, desmosomes, clathrin-coated pits, caveolaes, and different structures involved in cell adhesion.

Cell Membrane-Cytoskeleton Integration

The cytoskeleton is found underlying the cell membrane in the cytoplasm and provides scaffolding for membrane proteins to anchor to, as well as forming organelles that extend from the cell. Indeed, cytoskeletal elements interact extensively and intimately with the cell membrane. Anchoring proteins restricts them to a particular cell surface—for example, the *apical surface* of epithelial cells that line the vertebrate gut—and limits how far they may diffuse within the bilayer. The cytoskeleton is able to form appendage-like organelles, such as cilia, which are microtubule-based extensions covered by the cell membrane, and filopodia, which are actin-based extensions. These extensions are ensheathed in membrane and project from the surface of the cell in order to sense the external environment and/or make contact with the substrate or other cells. The apical surfaces of epithelial cells are dense with actin-based finger-like projections known as microvilli, which increase cell surface area and thereby increase the absorption rate of nutrients. The localized decoupling of the cytoskeleton and cell membrane results in formation of a bleb.

Cell membranes contain a variety of biological molecules, notably lipids and proteins. Material is incorporated into the membrane, or deleted from it, by a variety of mechanisms such as:

1. Fusion of intracellular vesicles with the membrane (exocytosis) not only excretes the contents of the vesicle but also incorporates the vesicle membrane's components into the cell membrane. The membrane may form blebs around extracellular material that pinch off to become vesicles (endocytosis).
2. If a membrane is continuous with a tubular structure made of membrane material, then material from the tube can be drawn into the membrane continuously.
3. Although the concentration of membrane components in the aqueous phase is low (stable membrane components have low solubility in water), there is an exchange of molecules between the lipid and aqueous phases.

Cell Membrane Lipids

The cell membrane consists of three classes of amphipathic lipids: phospholipids, glycolipids, and steroids. The amount of each depends upon the type of cell, but in the majority of cases phospholipids are the most abundant as already discussed above. For example, in RBC 30% of the plasma membrane is lipid.

The fatty chains in phospholipids and glycolipids usually contain an even number of carbon atoms, typically between 16 and 20. The 16- and 18-carbon fatty acids are the most common. Fatty acids may be saturated or unsaturated, with the configuration of the double bonds nearly always *cis*. The length and the degree of unsaturation of fatty acid chains have a profound effect on membrane fluidity as unsaturated lipids create a kink, preventing the fatty acids from packing together as tightly, thus decreasing the melting temperature (increasing the fluidity) of the membrane. The ability of some organisms to regulate the fluidity of their cell membranes by altering lipid composition is called homeoviscous adaptation.

The entire membrane is held together via non-covalent interaction of hydrophobic tails; however the structure is quite fluid and not fixed rigidly in place. Under physiological conditions phospholipid molecules in the cell membrane are in the liquid crystalline state. It means the lipid molecules are free to diffuse and exhibit rapid lateral diffusion along the layer in which they are present. However, the exchange of phospholipid molecules between intracellular and extracellular leaflets of the bilayer is a very slow process. Lipid rafts and caveolae are examples of cholesterol-enriched microdomains in the cell membrane.

In animal cells cholesterol is normally found dispersed in varying degrees throughout cell membranes, in the irregular spaces between the hydrophobic tails of the membrane lipids, where it confers stiffening and strengthening effect on the membrane. Thus, the ratio between cholesterol and unsaturated fatty acids and also the amount of saturated fatty acids present in the membrane plays a significant role in determining the properties of the cell membrane. The ratio and the amount of cholesterol, saturated fatty acids and unsaturated fatty acids present in the cell membrane determines not only the fluidity of the membrane but also the number of receptors of a given protein/growth factor/hormone but also the affinity of the receptor to its specific growth factor/hormone. The changes in the fluidity of the cell membrane may also have its impact on the expression of gene(s) and their function. For example, it is known that unsaturated fatty acids increase the fluidity of the cell membrane and as a consequence increase the number of receptors for insulin and the affinity of the insulin receptor to insulin and at the same time could also change the expression of certain oncogenes such as *ras* and *myc*. Sometimes, the expression of receptors and their affinity to its growth factor/hormone may or may not be related to the changes in the expression of genes/oncogenes.

Plasma Membrane Carbohydrates

Plasma membranes also contain carbohydrates, predominantly glycoproteins, but with some glycolipids (cerebrosides and gangliosides). For the most part, no glycosylation occurs on membranes within the cell; rather generally glycosylation occurs on the extracellular surface of the plasma membrane.

The glycocalyx is an important feature in all cells, especially epithelia with microvilli. Recent data suggest the glycocalyx participates in cell adhesion, lymphocyte homing, and many others.

The penultimate sugar is galactose and the terminal sugar is sialic acid, as the sugar backbone is modified in the Golgi. Sialic acid carries a negative charge, providing an external barrier to charged particles.

Plasma Membrane Proteins

The cell membrane plays host to a large amount of protein that is responsible for its various activities. The amount of protein differs between species and according to function, however the typical amount in a cell membrane is 50%. These proteins are important to a cell. Approximately a third of the genes in yeast code specifically for them, and this number is even higher in multicellular organisms.

The cell membrane plays an important role in communicating between the outside environment and the cellular constituents and as such participates in cell-cell communication. For this purpose, a large variety of protein receptors and identification proteins, such as antigens, are present on the surface of the membrane. Functions of membrane proteins can also include cell-cell contact, surface recognition, cytoskeleton contact, signaling, enzymatic activity, or transporting substances across the membrane.

Most membrane proteins are inserted in some way into the membrane. For this to occur, an N-terminus "signal sequence" of amino acids directs proteins to the endoplasmic reticulum, which inserts the proteins into a lipid bilayer. Once inserted, the proteins are then transported to their final destination in vesicles, where the vesicle fuses with the target membrane.

Cell Membrane Permeability

The permeability of a membrane depends mainly on the electric charge of the molecule and to a lesser extent the molar mass of the molecule. Electrically neutral and small molecules pass the membrane easier than charged, large ones. The inability of charged molecules to pass through the cell membrane results in pH parturition of substances throughout the fluid compartments of the body [3].

There are two types of cell membrane structures that are important to know. They are cell membrane lipid rafts and caveolae. A brief description of these structures is given here.

Lipid Raft

The plasma membrane of cells is made of a combination of glycosphingolipids and protein receptors organized in glycolipoprotein microdomains termed lipid rafts [4–6]. These specialized membrane microdomains compartmentalize cellular processes by serving as organizing centers for the assembly of signaling molecules, influencing membrane fluidity and membrane protein trafficking, and regulating neurotransmission and receptor trafficking [6, 7]. Lipid rafts are more ordered and tightly packed than the surrounding bilayer, but float freely in the membrane bilayer [7].

Properties of Lipid Rafts

One key difference between lipid rafts and the plasma membranes from which they are derived is lipid composition. Lipid rafts contain twice the amount of cholesterol found in the surrounding bilayer and are also enriched in sphingolipids such as sphingomyelin, which is typically elevated by 50% compared to the plasma membrane. To offset the elevated sphingolipid levels, phosphatidylcholine levels are decreased which results in similar choline-containing lipid levels between the rafts and the surrounding plasma membrane. Cholesterol interacts preferentially, although not exclusively, with sphingolipids due to their structure and the saturation of the hydrocarbon chains. Although not all of the phospholipids within the raft are fully saturated, the hydrophobic chains of the lipids contained in the rafts are more saturated and tightly packed than the surrounding bilayer [7]. Cholesterol is the dynamic "glue" that holds the raft together [6]. Due to the rigid nature of the sterol group, cholesterol partitions preferentially into the lipid rafts where acyl chains of the lipids tend to be more rigid and in a less fluid state [7]. One important property of membrane lipids is their amphipathic character. Amphipathic lipids have a polar, hydrophilic head group and a non-polar, hydrophobic region [7, 8]. It should be noted that cholesterol has the ability to pack in between the lipids in rafts, serving as a molecular spacer and filling any voids between associated sphingolipids [9]. http://en.wikipedia.org/wiki/Lipid_raft-cite_note-6

Lipid rafts can be related to the immiscibility of ordered (Lo phase) in model membranes and disordered (Ld or Lα phase) liquid phases [10]. The cause of this immiscibility is thought to minimize the free energy between the two phases. There is a difference in thickness of the lipid rafts and the surrounding membrane which results in hydrophobic mismatch at the boundary between the two phases. This phase height mismatch increases line tension which may lead to the formation of larger and more circular raft platforms to minimize the energetic cost of maintaining the rafts as a separate phase. Other spontaneous events, such as curvature of the membrane and fusing of small rafts into larger rafts, can also minimize line tension [7]. Lipid rafts can be extracted from a plasma membrane based on the resistance of lipid rafts to non-ionic detergents at low temperatures. When such a detergent is added to cells, the fluid membrane will dissolve while the lipid rafts may remain intact and could be extracted.

Because of their composition and detergent resistance, lipid rafts are also called detergent-insoluble glycolipid-enriched complexes (GEMs) or DIGs [11] or Detergent Resistant Membranes (DRMs). However the validity of the detergent resistance methodology of membranes has been questioned due to ambiguities in the lipids and proteins recovered and the observation that they can also cause solid areas to form where there were none previously [12].

Types of Lipid Rafts

Two types of lipid rafts have been described: planar lipid rafts (also referred to as non-caveolar, or glycolipid, rafts) and caveolae. Planar rafts are in continuation with the plane of the plasma membrane (not invaginated) and by their lack of distinguishing morphological features. Caveolae, on the other hand, are flask shaped invaginations of the plasma membrane that contain caveolin proteins and are the most readily-observed structures in lipid rafts. Caveolins are widely expressed in the brain micro-vessels of the nervous system, endothelial cells, astrocytes, oligodendrocytes, Schwann cells, dorsal root ganglia and hippocampal neurons. Planar rafts contain flotillin proteins and are found in neurons where caveolae is absent. Both types of lipid rafts are enriched in cholesterol and sphingolipids. Flotillin and caveolins have the ability to recruit signaling molecules into lipid rafts, thus playing an important role in neurotransmitter signal transductions [13]. It is likely that these microdomains spatially organize signaling molecules to promote kinetically favorable interactions which are necessary for signal transduction and it is also possible that these microdomains separate signaling molecules and thus, inhibit unwarranted interactions among them and dampen signaling responses [14].

Lipid Rafts and Signal Transduction

The movement of signal or stimulus can be simple, like that associated with receptor molecules. More complex signal transduction involves the coupling of ligand-receptor interactions to many intracellular events. These events include phosphorylations by tyrosine kinases and/or serine/threonine kinases in which lipid rafts seem to have an active participation [15]. The specificity and fidelity of signal transduction are essential for cells to respond efficiently to changes in their environment. This is achieved in part by the differential localization of proteins that participate in signalling pathways. In the plasma membrane, one approach of compartmentalization utilizes lipid rafts [16]. It is possible that small lipid rafts (microdomains) can form concentrating platforms after ligand binding activation for individual receptors [17, 18]. If receptor activation takes place in a lipid raft, the signalling complex is protected from non-raft enzymes such as membrane phosphatases. Overall, raft binding recruits proteins to a new micro-environment so that the phosphorylation state can be modified by local kinases and phosphatases to give downstream signalling [19]. Lipid rafts are involved in many signal transduction processes, such as Immunoglobulin E (IgE), T cell antigen receptor, B cell antigen receptor, EGF receptor, and insulin receptor signalling [20–27].

Immunoglobulin E Signalling

Immunoglobulin E (IgE) signaling is the first convincingly demonstrated lipid rafts involving signaling process [28–30]. It was reported that IgE first binds to Fc-epsilon receptors (FcϵR) residing in the plasma membrane of mast cells and basophils through its Fc segment. FcϵR is a tetramer consist of one α, one β and two γ chains [30]. It is monomeric and binds one IgE molecule. The α chain binds IgE and the other three chains contain immune receptor tyrosine-based activation motifs (ITAM). The oligomeric antigens that bind to receptor-bound IgE crosslink two or more of these receptors that, in turn, recruits doubly acylated non-receptor Src-like tyrosine kinase Lyn to phosphorylate ITAMs. The Syk family tyrosine kinases bind these phosphotyrosine residues of ITAMs to initiate the signaling cascade [29, 31]. Syk can, in turn, activate other proteins such as LAT. In addition, through crosslinking LAT can recruit other proteins into the raft and further amplify the signal [32].

T-cell Antigen Receptor Signaling and Lipid Rafts

T cell antigen receptor (TCR) is a molecule found on the surface of T lymphocytes (T cells). It is composed of $\alpha\beta$-heterodimers, CD3 ($\gamma\delta\epsilon$) complex and ξ-homodimer. The α- and β- subunits contains extracellular binding sites for peptides that are presented by the major histocompatibility complex (MHC) class I and class II proteins on the surface of antigen presenting cells (APCs). The CD3 and ξ-subunits contain cytoplasmic ITAM motifs. During the signaling process, MHCs binding to TCRs brings two or more receptors together. This crosslinking, similar to IgE signaling, then recruit doubly acylated non-receptor Src-like tyrosine kinases to phosphorylate ITAM tyrosine residues. In addition to recruiting Lyn, TCR signaling also recruits Fyn [33, 34]. Following this procedure, ZAP-70 (which is also different with IgE signalling) binds to phosphorylated ITAMs, which leads to its own activation and LAT activation. LAT activation is the source of signal amplification. Another difference between IgE and T cell antigen receptor signalling is that Lck activation by TCR could results in more severe raft clustering [35, 36] thus more signal amplification. To down-regulating the signal, one possible way is the binding of cytosolic kinase Csk to the raft associated protein CBP. Csk may then suppress the Src-family kinases through phosphorylation [37].

Thus, during the interaction of T cell and antigen presenting cell (APC), a highly organized structure is formed at the interface of the two cells, where cholesterol and sphingolipids are enriched, and form a liquid ordered phase that facilitates the signaling proteins on and off. Lipid rafts are also involved in virus entry and assembly. The lipid raft of the plasma membrane seem to ensure the correct intracellular traffic of proteins and lipids, such as protein-protein interactions by concentrating certain proteins in these microdomains, while excluding others. In view of this, it is logical to propose that disruption of lipid rafts is related to different diseases and aging, and could also be exploited as pharmaceutical targets for anti-virus and anti-inflammation [38]. For example, human immunodeficiency virus (HIV) envelope (Env) mediated bystander apoptosis that causes the progressive, and irreversible loss of CD4+ T cells in HIV-1 infected patients is gp41 dependent and related to the membrane hemifusion between envelope expressing cells and target cells. Caveolin-1 (Cav-1), the scaffold protein of a specific membrane lipid raft caveolae, interacts with gp41. Cav-1 modulated Env-induced bystander apoptosis through the interaction with gp41 in SupT1 cells and CD4+ T lymphocytes. Cav-1 significantly suppressed Env-induced membrane hemifusion, caspase-3 activation and augmented Hsp70 upregulation. Furthermore, a peptide containing the Cav-1 scaffold domain sequence markedly inhibited bystander apoptosis and apoptotic signal pathways, suggesting the potential role of Cav-1 in limiting HIV pathogenesis [39] and the development of a novel therapeutic strategy in treating HIV-1 infected patients. Thus, a better understanding of the role of lipid rafts in T cell signalling and T cell responses in various diseases could form a reasonable approach to develop newer methods of treatment for various immunological diseases.

B-cell Antigen Receptor Signaling and Lipid Rafts

B cell antigen receptor (BCR) is a complex between a membrane bound Ig (mIg) molecule and a disulfide-linked Igα- Igβheterodimer of two polypetides [40, 41]. Igα and Igβ each contain an amino acid motif, called ITAM. The process of B cell antigen receptor signalling is similar to Immunoglobulin E signalling and T-cell antigen receptor signalling. Lipid rafts play an important role in B cell activation that include signaling by BCR, modulation of that signaling by co-receptors, signaling by CD40, endocytosis of antigen bound to the BCR and its routing to late endosomes to facilitate loading of antigen-derived peptides onto class II MHC molecules, routing of those peptide/MHC-II complexes to the cell surface, and their manipulation of cholesterol is one of the most widely used techniques for studying lipid rafts. Sequestration (using filipin, nystatin or amphotericin), depletion and removal (using methyl-B-cyclodextrin) and inhibition of cholesterol synthesis (using HMG-CoA reductase inhibitors, statins) are the methods employed to manipulate cholesterol content in lipid rafts.

Using fluorescence resonance energy transfer between the same probes (homo-FRET or fluorescence anisotropy), it was reported that a fraction (20–40%) of GPI-anchored proteins are organized into high density clusters of 4–5 nm radius, each consisting of a few molecules and different GPI-anchored proteins [42].

Caveolae

Caveolae are a special type of lipid raft that are small (50–100 nm) invaginations of the plasma membrane mainly in endothelial cells and adipocytes. Some cell types, like neurons, may completely lack caveolae. These flask-shaped structures are rich in proteins as well as lipids such as cholesterol and sphingolipids and have several functions in signal transduction [43]. They play a role in endocytosis, oncogenesis, and the uptake of pathogenic bacteria and certain viruses [44–46]. Caveolae are one source of clathrin-independent endocytosis involved in turnover of adhesive complexes. Formation and maintenance of caveolae is primarily due to the protein caveolin, a 21 kD protein. This protein has both a cytoplasmic C-terminus and a cytoplasmic N-terminus, linked together by a hydrophobic hairpin that is inserted into the membrane. The presence of caveolin leads to the local change in morphology of the membrane. Because of their specific lipid content, caveolae are sometimes considered as a caveolin-positive subset of lipid rafts. Some known inhibitors of the caveolae pathway are filipin III, genistein and nystatin.

Lipid Rafts, Caveolae and Polyunsaturated Fatty Acids (PUFAs)

In the previous chapter, I discussed the role of PUFAs in the regulation of immune response and inflammation. PUFAs, especially n-3 fatty acids and their metabolites such as lipoxins, resolvins, protectins and maresins are known to suppress leukocyte function and thus modulate inflammatory and immune responses. In addition to the ability of PUFAs and their products to modulate the activity of intracellular signalling pathways, binding to TLRs (Toll-like receptors), control of gene expression, activation of transcription factors, induction of cell death and production of reactive oxygen and nitrogen species, they can also modulate the activity of lipid-raft-associated proteins.

When gramicidin (gA) analogues of different lengths together with bilayers of different thicknesses were used to assess whether docosahexaenoic acid (DHA) could exert its effects through a bilayer-mediated mechanism, it was noted that DHA increases gA (gramicidin) channel appearance rates and lifetimes and decreased the free energy of channel formation. The appearance rate and lifetime changes increased with increasing channel-bilayer hydrophobic mismatch and were not related to differing DHA bilayer absorption coefficients, suggesting that DHA alters bilayer elastic properties, not just lipid intrinsic curvature. This indicates that elasticity changes are important for DHA's bilayer-modifying actions. On the other hand, oleic acid (OA), which has little effect on membrane protein function, exerted no such effects despite OA's adsorption coefficient being an order of magnitude greater than DHA's. These results suggest that DHA (and other PUFAs) may modulate membrane protein function by bilayer-mediated mechanisms that do not involve specific protein binding but rather changes in bilayer material properties [47]. Studies performed in fat-1 transgenic mice (that are enriched in n-3 PUFAs) showed that membrane raft accumulation in $CD4^+$ cells was enhanced compared to the wild-type control However, the localization of protein kinase C theta, phospholipase C gamma-1, and F-actin into the immunological synapse (IS) was suppressed. On the other hand, both the phosphorylation status of phospholipase C gamma-1 at the IS and cell proliferation were suppressed in fat-1 cells [48], suggesting that n-3 PUFAs alter lipid rafts and thus, suppress inflammation.

It is known that perturbations in caveolae lipid composition could displace proteins from lipid microdomains, thereby altering their functionality and subsequent downstream signaling. This is supported by the observation that colonic caveolae in mice fed n-6 or n-3 PUFAs enriched diets significantly altered colonic caveolae microenvironment by increasing phospholipid n-3 fatty acyl content and reducing both cholesterol (by 46%) and caveolin-1 (by 53%), without altering total cellular levels. Concomitantly, localization of caveolae-resident signaling proteins H-Ras and eNOS in colonic caveolae was decreased by n-3 PUFA, by 45% and 56%, respectively, whereas the distribution of non-caveolae proteins K-Ras and clathrin was unaffected. Furthermore, EGF-stimulated H-Ras, but not K-Ras activation was significantly suppressed following n-3 PUFA feeding, in parallel with the selective alterations in their microlocalization. These findings clearly showed that caveolae lipid composition could be altered by diets enriched in PUFAs in vivo and thereby alter caveolae protein localization and functionality [49]. Thus, the ability of dietary PUFAs to alter the composition of caveolae could be one important mechanisms by which these fatty acids are able to bring about their beneficial actions. Similar results with regard to the alteration in the composition of lipid rafts of mouse T cells that were fed n-3 PUFAs were reported. Mice fed diets containing either 5 g/100 g corn oil (control) or 4 g/100 g fish oil[contains (n-3) PUFA] + 1 g/100 g corn oil for 14 days revealed that splenic T-cell lipid raft sphingomyelin content (mol/100 mol) was decreased ($P < 0.05$) in T cells isolated from (n-3) PUFA-fed mice. Dietary (n-3) PUFA were selectively incorporated into T-cell raft and soluble membrane phospholipids. Phosphatidylserine and glycerophosphoethanolamine, which are highly localized to the inner cytoplasmic leaflet, were enriched to a greater extent with unsaturated fatty acids compared with sphingomyelin, phosphatidylinositol and glycerophosphocholine, suggesting that dietary (n-3) PUFA differentially modulate T-cell raft and soluble membrane phospholipid and fatty acyl composition and thus, alter their function [48, 50] such as IL-2 production [51], down-regulation of the cyclin D1 promoter activity and inhibition of smooth muscle cell proliferation through the mitogen-activated protein kinase pathway due to increased concentration of EPA and DHA of caveolin-1 and caveolin-3 in caveolae [52].

Dietary n-3 PUFAs enhances endothelial NO. EPA treatment profoundly altered lipid composition and fatty acyl substitutions of phospholipids in caveolae. Caveolin-1 that was solely located in caveolae fractions in control cells, while EPA treatment displaced caveolin-1 from caveolae. Endothelial NOS (nitric oxide synthase) that was detected in the caveolin-enriched fractions and noncaveolae fractions in control cells was found to be translocated from caveolae fractions to soluble fraction following EPA treatment. Furthermore, eNOS activity in human umbilical vascular endothelial cells (HUVEC) was increased after EPA treatment, suggesting that eNOS translocation was paralleled by a stimulated capacity for NO production in the cells [53]. These results indicated that n-3 PUFAs altered caveolae microenvironment, thereby modifying location and function of proteins in caveolae. In a similar fashion, even DHA could alter the lipid composition of caveolae/lipid rafts and thus, regulate cytokine signaling [54], produced selective displacement of caveolin-1 and eNOS from caveolae and thus, enhance eNOS activation [55], modify TNF-α-induced endothelial cell activation [56].

Thus, PUFAs when given orally or by infusion in sufficient amounts are able to get incorporated into the cell membrane, lipid rafts and caveolae, and possibly, into other membranes such as mitochondrial membrane alter their properties such that are able to suppress the production of pro-inflammatory cytokines such as IL-6, TNF-α, enhance the production of NO, and are also able to bind to several nuclear receptors such as PPARs, RARs, RXR, HNF-4α, LXR and thus, prevent atherosclerosis, the underlying cause for cardiovascular diseases and stroke, produce their immunomodulatory actions, suppress inflammation and bring about their beneficial actions [57–67]. A summary of the actions of PUFAs is given in Tables 5.1 and 5.2 for easy reference.

It may be mentioned here that there is evidence to suggest that peroxidized products of PUFAs and eicosanoids bind to DNA and regulate gene expression [68–86]. This suggests that possibly, PUFAs, their oxidized products such as lipid peroxides, and PGs, lipoxins, resolvins, protectins and maresins and nitrolipids could bind to specific regions of DNA that code for specific genes and regulate their expression and synthesis of specific proteins and thus, bring about their various actions. In

Table 5.1 Summary of effects of PUFAs on nuclear receptors involved in the regulation of lipogenesis

Nuclear receptor	Effects on gene regulation	Expected changes		
		TG	HDL	LDL
PPAR-α	↑	↓↓	↑	↓
LXR	↓	↓↓	↓	↓
FXR	↑	↓↓	↑	↑
HNF-4α	↓	↓↓	↓	↔
Net effects		↓↓↓↓	↔	↔

FXR Farnesol X receptor, *HDL* High-density lipopoprotein, *HNF-4α* Hepatocyte nuclear factor-4α, *LDL* Low-density lipoprotein, *LXR* Liver X receptor, *PPAR-α* Peroxisome proliferator-activated receptor, ↑ Increase, ↓ Decrease, ↔ Neutral effect

Table 5.2 Summary of the actions of PUFAs (especially of ω-3 fatty acids) that are responsible for their beneficial action in the prevention of cardiovascular diseases and low-grade systemic inflammatory conditions

Action on	Effect
Plasma triglyceride concentration-fasting and post-prandial	↓↓
Plasma cholesterol	↓↔
HDL cholesterol	↑↔
LDL cholesterol	↓↔
Blood pressure	↓
Endothelial production of NO	↑
ACE activity	↓
HMG-CoA activity	↓
Thrombosis	↓
Platelet aggregation	↓
Leukocyte activation	↓
Cell-surface expression of adhesion molecules	↓
Production of chemoattractants	↓
Cardiac arrhythmias	↓
Heart rate variability	↑
Atheromatous plaque stability	↑
Production of lipoxins and resolvins	↑
Production of free radicals and formation of lipid peroxides	↓
Production of PGI_2, PGI_3, PGE_1	↑
Production of TXA_2, LTs	↓
Synthesis of pro-inflammatory cytokines such as TNF-α and MIF	↓
Production of anti-inflammatory cytokines such as IL-10	↑
Production of growth factors	↓
Insulin sensitivity	↑
Endothelial integrity	↑
Telomere length	↑

addition, the metabolites of PUFAs such as eicosanoids bind to their specific receptors on the cell membrane and convey their messages to the cytoplasmic structures. Thus, PUFAs and their various products seem to have a multitude of actions on various cells/tissues and organs that explains their actions in various physiological and pathological processes such as cell cycle regulation and mitosis, inflammation, tissue repair, cardiovascular responses, atherosclerosis, cancer, metabolic syndrome, cardiovascular diseases, osteoporosis, stem cell biology, immune response regulation, differentiation of certain cells and neurotransmission and neurological conditions [57–87].

References

[1] Alberts B, Johnson A, Lewis J, Raff M, Roberts K, Walter P (2002) Molecular biology of the cell, 4th edn. Garland, New York
[2] Lodish H, Berk A, Matsudaira P, Kaiser CA, Krieger M, Scott MP, Zipurksy SL, Darnell J (2004) Molecular cell biology, 5th edn. WH Freeman, New York
[3] Cooper GM (2000) The cell: a molecular approach, 2nd edn. ASM Press, Washington
[4] Thomas S, Pais AP, Casares S, Brumeanu TD (2004) Analysis of lipid rafts in T cells. Mol Immunol 41:399–409
[5] Thomas S, Kumar RS, Brumeanu TD (2004) Role of lipid rafts in T cells. AITE 52:215–224
[6] Korade Z (2008) Lipid rafts, cholesterol, and the brain. Neuropharmacology 55:1265–1273
[7] Pike LJ (2009) The challenge of lipid rafts. J Lipid Res 50(Suppl):S323–S328
[8] Simons K, Ehehalt R (2002) Cholesterol, lipid rafts, and disease. J Clin Invest 110:597–603
[9] Fantini J, Garmy N, Mahfoud R, Yahi N (2002) Lipid rafts: structure, function and role in HIV, Alzheimer's and prion diseases. Expert Rev Mol Med 4:1–22
[10] Rietveld A, Simons K (1998) The differential miscibility of lipids as the basis for the formation of functional membrane rafts. Biochim Biophys Acta 1376:467–479
[11] Fivaz M, Abrami L, van der Goot FG (1999) Landing on lipid rafts. Trends Cell Biol 9:212–213
[12] Heerklotz H (2002) Triton promotes domain formation in lipid raft mixtures. Biophys J 83:2693–2701
[13] Simons K, Ikonen E (1997) Functional rafts in cell membranes. Nature 387:569–572
[14] Allen JA (2007) Lipid raft microdomains and neurotransmitter signalling. Nature 8:128–140
[15] Kurzchalia TV, Parton RG (1999) Membrane microdomains and caveolae. Curr Opin Cell Biol 11:424–431
[16] Janes PW, Ley SC, Magee AI, Kabouridis PS (2000) The role of lipid rafts in T cell antigen receptor (TCR) signalling. Semin Immunol 12:23–24
[17] Schmitz G, Grandl M (2008) Update on lipid membrane microdomains. Curr Opin Clin Nutr Metab Care 11:106–112
[18] Ilangumaran S, Borisch B, Hoessli DC (1999) Signal transduction via CD44: role of plasma membrane microdomains. Leuk Lymphoma 35:455–469
[19] Kail S, Derek T (2000) Lipid rafts and signal transduction. Nat Rev Mol Cell Biol 1:31–39
[20] Brown DA, London E (2000) Structure and function of sphingolipid- and cholesterol-rich membrane rafts. J Biol Chem 275:17221–17224
[21] Brown DA, London E (1998) Functions of lipid rafts in biological membranes. Annu Rev Cell Dev Biol 14:111–136
[22] Brown D (2002) Structure and function of membrane rafts. Int J Med Microbiol 291:433–437
[23] Lingwood D, Simons K (2010) Lipid rafts as a membrane-organizing principle. Science 327:46–50
[24] Taghibiglou C, Bradley CA, Gaertner T, Li Y, Wang Y, Wang YT (2009) Mechanisms involved in cholesterol-induced neuronal insulin resistance. Neuropharmacology 57:268–276
[25] Rivera M, Muto A, Feigel A, Kondo Y, Dardik A (2009) Venous and arterial identity: a role for caveolae? Vascular 17(Suppl 1):S10–S14
[26] Lajoie P, Goetz JG, Dennis JW, Nabi IR (2009) Lattices, rafts, and scaffolds: domain regulation of receptor signaling at the plasma membrane. J Cell Biol 185:381–385

[27] Kinoshita MO, Furuya S, Ito S, Shinoda Y, Yamazaki Y, Greimel P, Ito Y, Hashikawa T, Machida T, Nagatsuka Y, Hirabayashi Y (2009) Lipid rafts enriched in phosphatidylglucoside direct astroglial differentiation by regulating tyrosine kinase activity of epidermal growth factor receptors. Biochem J 419:565–575
[28] Field KA, Holowka D, Baird B (1995) FceRI-mediated recruitment of p53/56^{lyn} to detergent-resistant membrane domains accompanies cellular signaling. Proc Natl Acad Sci U S A 92:9201–9205
[29] Sheets ED, Holowka D, Baird B (1999) Membrane organization in immunoglobulin E receptor signaling. Curr Opin Chem Biol 3:95–99
[30] Baird B, Sheets ED, Holowka D (1999) How does the plasma membrane participate in cellular signaling by receptors for immunoglobulin E? Biophys Chem 82:109–119
[31] Field KA, Holowka D, Baird B (1995) FceRI-mediated recruitment of p53/56^{lyn} to detergent-resistant membrane domains accompanies cellular signaling. Proc Natl Acad Sci U S A 92:9201–9205
[32] Goitsuka R, Kanazashi H, Sasanuma H, Fujimura Y, Hidaka Y, Tatsuno A, Ra C, Hayashi K, Kitamura D (2000) A BASH/SLP-76-related adaptor protein MIST/Clnk involved in IgE receptor-mediated mast cell degranulation. Int Immunol 12:573–580
[33] Janes PW, Ley SC, Magee AI, Kabouridis PS (2000) The role of lipid rafts in T cell antigen receptor (TCR) signalling. Semin Immunol 12:23–24
[34] Langlet C, Bernard A-M, Drevot P, He H-T (2000) Membrane rafts and signaling by the multichain immune recognition receptors. Curr Opin Immunol 12:250–255
[35] Zhang W, Trible RP, Samelson LE (1998) LAT palmitoylation: its essential role in membrane microdomain targeting and tyrosine phosphorylation during T cell activation. Immunity 9:239–246
[36] Brdi kab T, Jan erný, Ho ej ía V (1998) T cell receptor signalling results in rapid tyrosine phosphorylation of the linker protein LAT present in detergent-resistant mMembrane microdomains. Biochem Biophys Res Commun 248:356–360
[37] Carl LA, Cooper JA (2000) Signal transduction: molecular switches in lipid rafts. Nature 404:945–947
[38] Luo C, Wang K, Liu de Q, Li Y, Zhao QS (2008) The functional roles of lipid rafts in T cell activation, immune diseases and HIV infection and prevention. Cell Mol Immunol 5:1–7
[39] Wang XM, Nadeau PE, Lo YT, Mergia A (2010) Caveolin-1 modulates HIV-1 envelope induced bystander apoptosis through gp41. J Virol 84:6515–6526
[40] Gupta N, DeFranco AL (2007) Lipid rafts and B cell signaling. Semin Cell Dev Biol 18:616–626
[41] Gupta N, DeFranco AL (2003) Visualizing lipid raft dynamics and early signaling events during antigen receptor-mediated B-lymphocyte activation. Mol Biol Cell 14:432–444
[42] Sharma P, Varma R, Sarasij RC, Ira, Gousset K, Krishnamoorthy G, Rao M, Mayor S (2004) Nanoscale organization of multiple GPI-anchored proteins in living cell membranes. Cell 116:577–589
[43] Anderson RG (1998) The caveolae membrane system. Annu Rev Biochem 67:199–225
[44] Frank P, Lisanti M (2004) Caveolin-1 and caveolae in atherosclerosis: differential roles in fatty streak formation and neointimal hyperplasia. Curr Opin Lipidol 15:523–529
[45] Li X, Everson W, Smart E (2005) Caveolae, lipid rafts, and vascular disease. Trends Cardiovasc Med 15:92–96
[46] Pelkmans L (2005) Secrets of caveolae- and lipid raft-mediated endocytosis revealed by mammalian viruses. Biochim Biophys Acta 1746:295–304
[47] Bruno MJ, Koeppe RE 2nd, Andersen OS (2007) Docosahexaenoic acid alters bilayer elastic properties. Proc Natl Acad Sci U S A 104:9638–9643
[48] Kim W, Fan YY, Barhoumi R, Smith R, McMurray DN, Chapkin RS (2008) n-3 polyunsaturated fatty acids suppress the localization and activation of signaling proteins at the immunological synapse in murine CD4+ T cells by affecting lipid raft formation. J Immunol 181:6236–6243

[49] Ma DW, Seo J, Davidson LA, Callaway ES, Fan YY, Lupton JR, Chapkin RS (2004) n-3 PUFA alter caveolae lipid composition and resident protein localization in mouse colon. FASEB J 18:1040–1042
[50] Fan YY, McMurray DN, Ly LH, Chapkin RS (2003) Dietary (n-3) polyunsaturated fatty acids remodel mouse T-cell lipid rafts. J Nutr 133:1913–1920
[51] Fan YY, Ly LH, Barhoumi R, McMurray DN, Chapkin RS (2004) Dietary docosahexaenoic acid suppresses T cell protein kinase C theta lipid raft recruitment and IL-2 production. J Immunol 173:6151–6160
[52] Bousserouel S, Raymondjean M, Brouillet A, Béréziat G, Andréani M (2004) Modulation of cyclin D1 and early growth response factor-1 gene expression in interleukin-1beta-treated rat smooth muscle cells by n-6 and n-3 polyunsaturated fatty acids. Eur J Biochem 271:4462–4473
[53] Li Q, Zhang Q, Wang M, Zhao S, Ma J, Luo N, Li N, Li Y, Xu G, Li J (2007) Eicosapentaenoic acid modifies lipid composition in caveolae and induces translocation of endothelial nitric oxide synthase. Biochimie 89:169–177
[54] Chen W, Jump DB, Esselman WJ, Busik JV (2007) Inhibition of cytokine signaling in human retinal endothelial cells through modification of caveolae/lipid rafts by docosahexaenoic acid. Invest Ophthalmol Vis Sci 48:18–26
[55] Li Q, Zhang Q, Wang M, Liu F, Zhao S, Ma J, Luo N, Li N, Li Y, Xu G, Li J (2007) Docosahexaenoic acid affects endothelial nitric oxide synthase in caveolae. Arch Biochem Biophys 466:250–259
[56] Wang L, Lim EJ, Toborek M, Hennig B (2008) The role of fatty acids and caveolin-1 in tumor necrosis factor alpha-induced endothelial cell activation. Metabolism 57:1328–1339
[57] Das UN (2002) A perinatal strategy for preventing adult diseases: the role of long-chain polyunsaturated fatty acids. Kluwer Academic, Boston
[58] Das UN (2010) Metabolic syndrome pathophysiology: the role of essential fatty acids. Wiley-Blackwell, Ames
[59] Das UN (2006) Essential fatty acids: biochemistry, physiology, and pathology. Biotechnol J 1:420–439
[60] Das UN (2006) Essential fatty acids—a review. Curr Pharm Biotechnol 7:467–482
[61] Das UN (2006) Biological significance of essential fatty acids. J Assoc Physicians India 54:309–319
[62] Das UN (2004) Long-chain polyunsaturated fatty acids interact with nitric oxide, superoxide anion, and transforming growth factor β to prevent human essential hypertension. Eur J Clin Nutr 58:195–203
[63] Das UN, Puskàs LG (2009) Transgenic fat-1 mouse as a model to study the pathophysiology of cardiovascular, neurological and psychiatric disorders. Lipids Health Dis 8:61
[64] Das UN (2008) Essential fatty acids and their metabolites could function as endogenous HMG-CoA reductase and ACE enzyme inhibitors, anti-arrhythmic, anti-hypertensive, anti-atherosclerotic, anti-inflammatory, cytoprotective, and cardioprotective molecules. Lipids Health Dis 7:37
[65] Das UN (2008) Can endogenous lipid molecules serve as predictors and prognostic markers of coronary heart disease? Lipids Health Dis 7:19
[66] Das UN (2008) Can essential fatty acids reduce the burden of disease(s)? Lipids Health Dis 7:9
[67] Das UN (2007) A defect in the activity of Delta6 and Delta5 desaturases may be a factor in the initiation and progression of atherosclerosis. Prostaglandins Leukot Essent Fatty Acids 76:251–268
[68] Das UN (1999) Essential fatty acids, lipid peroxidation and apoptosis. Prostaglandins Leukot Essent Fatty Acids 61:157–163
[69] Das UN (1983) Prostaglandins and gene action. Med Hypotheses 11:185–194
[70] Benavente J, Esteban M, Jaffe BM, Santoro MG (1984) Selective inhibition of viral gene expression as the mechanism of the antiviral action of PGA1 in vaccinia virus-infected cells. J Gen Virol 65(Pt 3):599–608

[71] Ishioka C, Kanamaru R, Sato T, Dei T, Konishi Y, Asamura M, Wakui A (1988) Inhibitory effects of prostaglandin A2 on c-myc expression and cell cycle progression in human leukemia cell line HL-60. Cancer Res 48:2813–2818
[72] Marui N, Sakai T, Hosokawa N, Yoshida M, Aoike A, Kawai K, Nishino H, Fukushima M (1990) N-myc suppression and cell cycle arrest at G1 phase by prostaglandins. FEBS Lett 270:15–18
[73] Acarregui MJ, Snyder JM, Mitchell MD, Mendelson CR (1990) Prostaglandins regulate surfactant protein A (SP-A) gene expression in human fetal lung in vitro. Endocrinology 127:1105–1113
[74] Khan I, Hossain A, Whitman GF, Sarkar NH, McDonough PG (1993) Differential induction of c-jun expression by PGF2-alpha in rat ovary, uterus and adrenal. Prostaglandins 46:139–144
[75] Anastassiou ED, Paliogianni F, Balow JP, Yamada H, Boumpas DT (1992) Prostaglandin E2 and other cyclic AMP-elevating agents modulate IL-2 and IL-2R alpha gene expression at multiple levels. J Immunol 148:2845–2852
[76] Desanctis JB, Varesio L, Radzioch D (1994) Prostaglandins inhibit lipoprotein lipase gene expression in macrophages. Immunology 81:605–610
[77] Bui T, Kuo C, Rotwein P, Straus DS (1997) Prostaglandin A2 specifically represses insulin-like growth factor-I gene expression in C6 rat glioma cells. Endocrinology 138:985–993
[78] Walton SL, Burne TH, Gilbert CL (2002) Prostaglandin F2alpha-induced nest-building behaviour is associated with increased hypothalamic c-fos and c-jun mRNA expression. J Neuroendocrinol 14:711–723
[79] Tang CH, Yang RS, Fu WM (2005) Prostaglandin E2 stimulates fibronectin expression through EP1 receptor, phospholipase C, protein kinase C alpha, and c-Src pathway in primary cultured rat osteoblasts. J Biol Chem 280:22907–22916
[80] Kim CH, Park YG, Noh SH, Kim YK (2005) PGE2 induces the gene expression of bone matrix metalloproteinase-1 in mouse osteoblasts by cAMP-PKA signaling pathway. Int J Biochem Cell Biol 37:375–385
[81] Huang JC, Wun WS, Goldsby JS, Egan K, FitzGerald GA, Wu KK (2007) Prostacyclin receptor signaling and early embryo development in the mouse. Hum Reprod 22:2851–2856
[82] Fang IM, Yang CH, Yang CM, Chen MS (2007) Linoleic acid-induced expression of inducible nitric oxide synthase and cyclooxygenase II via p42/44 mitogen-activated protein kinase and nuclear factor-kappaB pathway in retinal pigment epithelial cells. Exp Eye Res 85:667–677
[83] Renedo M, Gayarre J, García-Domínguez CA, Pérez-Rodríguez A, Prieto A, Caada FJ, Rojas JM, Pérez-Sala D (2007) Modification and activation of Ras proteins by electrophilic prostanoids with different structure are site-selective. Biochemistry 46:6607–6616
[84] Stanley DW, Goodman C, An S, McIntosh A, Song Q (2008) Prostaglandins A1 and E1 influence gene expression in an established insect cell line (BCIRL-HzAM1 cells). Insect Biochem Mol Biol 38:275–284
[85] Yamada T (2009) Regulation of the expression of inducible nitric oxide synthase by prostanoids. Yakugaku Zasshi 129:1211–1214
[86] Marei WF, Wathes DC, Fouladi-Nashta AA (2009) The effect of linolenic Acid on bovine oocyte maturation and development. Biol Reprod 81:1064–1072
[87] Das UN (2011) Influence of polyunsaturated fatty acids and their metabolites on stem cell biology. Nutrition 27:21–25

Chapter 6
Low-grade Systemic Inflammation is Present in Common Diseases/Disorders

Introduction

Obesity, coronary heart disease (CHD), stroke, type 2 diabetes mellitus, hypertension, cancer, depression, schizophrenia, Alzheimer's disease, and collagen vascular diseases are a severe burden on the health care system throughout the world [1]. Though the exact cause of these diseases is not clear, it is known that low-grade systemic inflammation is common in all of them [2–4].

It was estimated that a combination of a multidrug regimen that lowers blood pressure, induces dieresis and prevents platelet aggregation comprising of a statin, aspirin, and two blood-pressure lowering medicines reduces about 17.9 million deaths from cardiovascular diseases [4]. This assumption is based on the concept that a formulation that consists of a statin, three blood pressure lowering drugs (such as a thiazide, a β blocker, and an angiotensin converting enzyme inhibitor), each at half standard dose; folic acid (0.8 mg); and aspirin (75 mg)-called as "polypill" reduces ischemic heart disease (IHD) events by 88% and stroke by 80% [5]. It has been proposed that one third of people taking this pill from age 55 would benefit, gaining on average about 11 years of life free from an IHD event or stroke. Summing the adverse effects of the components observed in randomised trials shows that the Polypill would cause symptoms in 8–15% of people (depending on the precise formulation) [5]. It was projected that the benefit of a secondary prevention "poly-portfolio" strategy, including pharmacologic and lifestyle approaches for those with CHD or stroke using combinations of a high-dose statin, low to standard doses of antihypertensive therapy, aspirin, omega-3 fish oil, cardiac rehabilitation, and diet estimated that patients with CHD, post-myocardial infarction (MI), or stroke was projected to reduce by 84%, 91%, and 77% reductions, respectively, in CHD events from a pharmacologic approach [6]. Numbers of those needed to treat (NNT) for 5 years were 9 to 11 to prevent 1 CHD event, and 21 to prevent 1 stroke. Post-MI patients were projected to experience a 93% reduction in the risk of CHD death (NNT 16) from a pharmacologic approach and a 97% reduction in the risk of CHD death (NNT 15) with the addition of lifestyle changes. These calculations led to the proposal that secondary

U. N. Das, *Molecular Basis of Health and Disease*,
DOI 10.1007/978-94-007-0495-4_6, © Springer Science+Business Media B.V. 2011

prevention polyportfolio holds promise for reducing the burden of cardiovascular disease in the highest risk patients.

Despite the popularity of the concept of polypill, reservations have been expressed in its implementation. It was felt that the implementation of the intriguing concept of the polypill to prevent cardiovascular events remains doubtful due to a lack of evidence, expected adherence problems, the inevitable overtreatment and under treatment of individuals, and potential side effects [7]. This led to the suggestion that lifestyle changes and individual interventions are more preferable strategies.

In contrast to this pharmacological polypill, Franco et al. [8] proposed the concept of "polymeal" for which the ingredients were taken from the literature, whose recipe included wine, fish, dark chocolate, fruits, vegetables, garlic, and almonds. Data from the Framingham heart study and the Framingham offspring study were used to build life tables to model the benefits of the "polymeal" in the general population from age 50, assuming multiplicative correlations. It was suggested that combining the ingredients of the "polymeal" would reduce cardiovascular disease events by 76%. It was calculated that for men, taking the polymeal daily represented an increase in total life expectancy of 6.6 years, an increase in life expectancy free from cardiovascular disease of 9.0 years, and a decrease in life expectancy with cardiovascular disease of 2.4 years. The corresponding differences for women were 4.8, 8.1, and 3.3 years. These calculations led to the conclusion that the "polymeal" could be an effective, non-pharmacological, safe, cheap, and tasty alternative to reduce cardiovascular morbidity and increase life expectancy in the general population [8].

It is interesting to note that the contents of the "polypill" have anti-inflammatory actions. For example, aspirin is a well known anti-inflammatory drug. Though the amount of aspirin recommended is 50–125 mg, even at this dose aspirin might show anti-inflammatory action. More important, aspirin at 50–125 mg may inhibit thromboxane A_2 (TXA_2) generation but does not prevent the formation of prostacyclin, PGI_2, a potent vasodilator and platelet anti-aggregator [9–12]. Aspirin when used at 50–125 mg ay actually enhance the formation of lipoxins from AA and EPA (see Chap. 4) that have anti-inflammatory actions [13, 14]. Furthermore, aspirin has also been shown to enhance nitric oxide synthesis and thus, bring about some of its useful actions [15, 16].

Statins are known to possess an anti-inflammatory action [17–21] that is in addition to their ability to inhibit HMG-CoA reductase enzyme and lower plasma and tissue cholesterol levels.

Previously, we showed that hypertension could be an inflammatory condition and that many commonly used anti-hypertensive drugs have anti-inflammatory actions by virtue of their ability to inhibit the production of free radicals, enhance the synthesis of endothelial nitric oxide and block the conversion of angiotensin I to angiotensin II [22, 23]. Angiotensin II is known to augment free radical generation and thus, may serve as a pro-inflammatory molecule while angiotensin converting enzyme inhibitors possess anti-inflammatory properties [24].

But, any strategy that prevents stroke, cancer and other chronic diseases in addition to cardiovascular diseases is expected to reduce the burden of chronic diseases substantially. Low-grade systemic inflammation is one of the characteristic features of CHD, stroke, diabetes mellitus, hypertension, cancer, depression schizophrenia, Alzheimer's disease, and collagen vascular diseases implying that prevention or suppression of inflammation reduces burden of these diseases [9].

Low-grade Systemic Inflammation is Present in Chronic Diseases

Plasma C-reactive protein (CRP), tumor necrosis factor-α (TNF-α), and interleukin-6 (IL-6), markers of inflammation, levels are elevated in subjects with obesity, insulin resistance, essential hypertension, type 2 diabetes, CHD, cancer, Alzheimer's disease, depression, schizophrenia and cancer [25–33], suggesting that low-grade systemic inflammation occurs in all these conditions (see Fig. 6.1 in which a the role

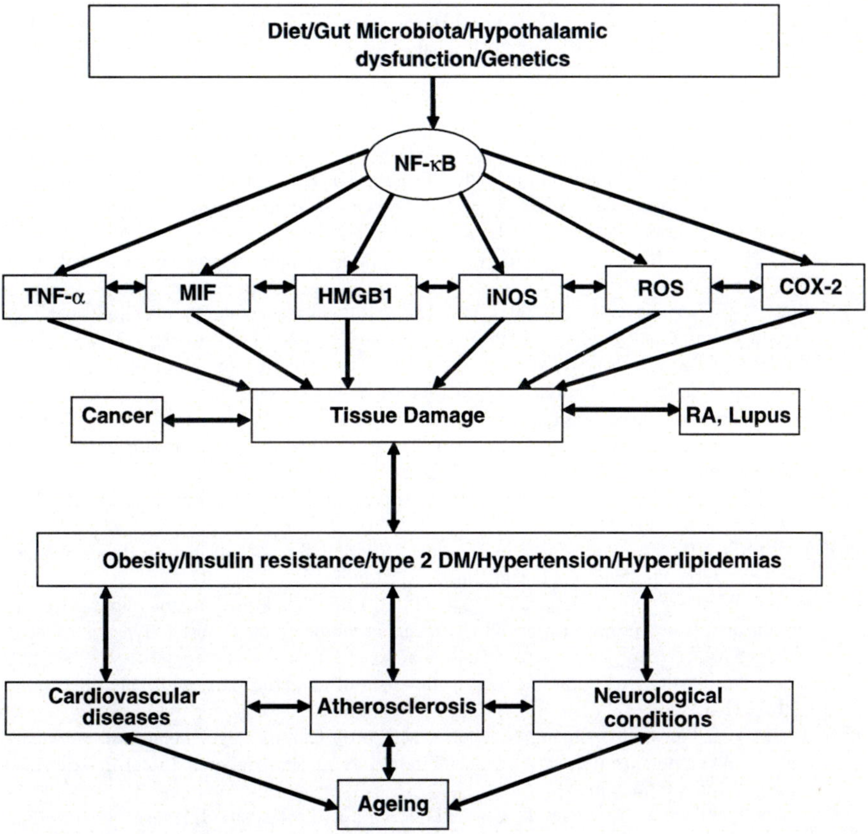

Fig. 6.1 Scheme showing the role of various inflammatory mediators in some common cardiovascular, neurological and collagen vascular diseases

of various inflammatory molecules in various common diseases is depicted). In order to understand the role of inflammation in these disorders/diseases, it is essential to have a brief introduction to them and then delve in depth as to the evidence(s) that support(s) the concept that low-grade systemic inflammation is the common underlying pathobiology in all these diseases. In the subsequent chapters, a brief introduction to obesity, hypertension, dyslipidemia, type 2 diabetes mellitus, metabolic syndrome, atherosclerosis, coronary heart disease (CHD), stroke, cancer, depression, schizophrenia, Alzheimer's disease,collagen vascular diseases, atherosclerosis, and ageing followed by a detailed discussion as to the evidences that are available at present to suggest that low-grade systemic inflammation exists in them is presented. It may be mentioned here that in all these conditions insulin resistance is also present.

References

[1] http://www.who.int/entity/healthinfo/statistics/bodgbddeathdalyestimates.xls
[2] Lopez A, Mathers C, Ezzati M, Jamison D, Murray C (2006) Global and regional burden of disease and risk factors, 2001: systematic analysis of population health data. Lancet 367:1714–1717
[3] Ezzati M, Vander Hoom S, Lawes C et al (2005) Rethinking the "Diseases of Affluence" paradigm: global patterns of nutritional risks in relation to economic development. PLoS Med 2:e133
[4] Lim SS, Gaziana TA, Gakidou E, Reddy KS, Farzadfar F, Lozana R, Rodgers A (2007) Prevention of cardiovascular disease in high-risk individuals in low-income and middle-income countries: health effects and costs. Lancet 370:2054–2625
[5] Wald NJ, Law MR (2003) A strategy to reduce cardiovascular disease by more than 80%. BMJ 326:1419–1424
[6] Robinson JG, Maheshwari N (2005) A "poly-portfolio" for secondary prevention: a strategy to reduce subsequent events by up to 97% over five years. Am J Cardiol 95:373–378
[7] Westerweel PE, van Wijk JP, Verhaar MC (2005) The polypill: not an effective strategy for reduction of cardiovascular disease. Ned Tijdschr Geneeskd 149:1741
[8] Franco OH, Bonneux L, de Laet C, Peeters A, Steyerberg EW, Mackenbach JP (2004) The Polymeal: a more natural, safer, and probably tastier (than the polypill) strategy to reduce cardiovascular disease by more than 75%. BMJ 329:1447–1450
[9] Das UN (2008) Do polyunsaturated fatty acids behave like an endogenous "polypill"? Med Hypotheses 70:430–434
[10] Walsh SW (1989) Low-dose aspirin: treatment for the imbalance of increased thromboxane and decreased prostacyclin in preeclampsia. Am J Perinatol 6:124–132
[11] Walsh SW, Wang Y, Kay HH, McCoy MC (1992) Low-dose aspirin inhibits lipid peroxides and thromboxane but not prostacyclin in pregnant women. Am J Obstet Gynecol 167(4 Pt 1):926–930
[12] Das UN (2005) COX-2 inhibitors and metabolism of essential fatty acids. Med Sci Monit 11:RA233–RA237
[13] Serhan CN, Fierro IM, Chiang N, Pouliot M (2001) Cutting edge: nociceptin stimulates neutrophil chemotaxis and recruitment: inhibition by aspirin-triggered-15-epi-lipoxin A4. J Immunol 166:v3650–v3654
[14] Mitchell S, Thomas G, Harvey K, Cottell D, Reville K, Berlasconi G, Petasis NA, Erwig L, Rees AJ, Savill J, Brady HR, Godson C (2002) Lipoxins, aspirin-triggered epi-lipoxins, lipoxin stable analogues, and the resolution of inflammation: stimulation of macrophage phagocytosis of apoptotic neutrophils in vivo. J Am Soc Nephrol 13:2497–2507

[15] Schröder H (2008) Nitric oxide and aspirin: a new mediator for an old drug. Am J Ther (in press)
[16] López-Farré A, Riesco A, Digiuni E, Mosquera JR, Caramelo C, de Miguel LS, Millàs I, de Frutos T, Cernadas MR, Montón M, Alonso J, Casado S (1996) Aspirin-stimulated nitric oxide production by neutrophils after acute myocardial ischemia in rabbits. Circulation 94:838–837
[17] Ando H, Takamura T, Ota T, Nagai Y, Kobayashi K (2000) Cerivastatin improves survival of mice with lipopolysaccharide-induced sepsis. J Pharmacol Exp Ther 294:1043–1046
[18] Grip O, Janciauskiene S, Lindgren S (2000) Pravastatin down-regulates inflammatory mediators in human monocytes in vitro. Eur J Pharmacol 410:83–92
[19] Omi H, Okayama N, Shimizu M, Fukutomi T, Imaeda K, Okouchi M, Itoh M (2003) Statins inhibit high glucose-mediated neutrophil-endothelial cell adhesion through decreasing surface expression of endothelial adhesion molecules by stimulating production of endothelial nitric oxide. Microvasc Res 65:118–124
[20] Danesh FR, Anel RL, Zeng L, Lomasney J, Sahai A, Kanwar YS (2003) Immunomodulatory effects of HMG-CoA reductase inhibitors. Arch Immunol Ther Exp (Warsz) 51:139–148
[21] Stüve O, Youssef S, Dunn S, Slavin AJ, Steinman L, Zamvil SS (2003) The potential therapeutic role of statins in central nervous system autoimmune disorders. Cell Mol Life Sci 60:2483–2491
[22] Das UN (2006) Hypertension as a low-grade systemic inflammatory condition that has its origins in the perinatal period. J Assoc Physicians India 54:133–142
[23] Kumar KV, Das UN (1993) Are free radicals involved in the pathobiology of human essential hypertension? Free Radic Res Commun 19:59–66
[24] Das UN (2005) Is angiotensin II an endogenous pro-inflammatory molecule? Med Sci Monit 11:RA155–RA162
[25] Luc G, Bard J-M, Juhan-Vague I et al (2003) C-reactive protein, interleukins-6, and fibrinogen as predictors of coronary heart disease. The PRIME study. Arterioscler Thromb Vasc Biol 23:1255–1261
[26] Das UN (2001) Is obesity an inflammatory condition? Nutrition 17:953–966
[27] Das UN (2006) Aberrant expression of perilipins and 11-β-HSD-1 as molecular signatures of metabolic syndrome X in South East Asians. J Assoc Physicians India 54:637–649
[28] Ridker PM, Buring JE, Cook NR, Rifai N (2003) C-reactive protein, the metabolic syndrome, and risk of incident cardiovascular events. Circulation 107:391–397
[29] Das UN (2007) Is metabolic syndrome X a disorder of the brain with the initiation of low-grade systemic inflammatory events during the perinatal period? J Nutr Biochem 18:701–713
[30] Das UN (2008) Folic acid and polyunsaturated fatty acids improve cognitive function and prevent depression, dementia, and Alzheimer's disease-but how and why? Prostaglandins Leukot Essent Fatty Acids 78:11–19
[31] Das UN (2007) Is depression a low-grade systemic inflammatory condition? Am J Clin Nutr 85:1665–1666
[32] Dougan M, Dranoff G (2008) Inciting inflammation: the RAGE about tumor promotion. J Exp Med 205:267–270
[33] Lawrence T, Hagemann T, Balkwill F (2007) Sex, cytokines, and cancer. Science 317:51–52

Chapter 7
Obesity

Obesity is common. In general, subjects with hypertension, type 2 diabetes, hyperlipidemias, CHD, and stroke often have obesity, though not all. Hence, a deeper understanding of the factors that cause obesity is important. Such an understanding may also shed light on the pathophysiology of the metabolic syndrome.

It is important to note that the incidence of obesity has assumed epidemic proportions both in the developed and developing countries that cannot be attributed to genetic factors since the human genes have not changed recently. This dramatic increase in the incidence of obesity is also contributing to the high prevalence and incidence of the metabolic syndrome that is being noticed recently. Lack of exercise, increased consumption of calorie-dense food, enhanced intake of saturated fats, carbonated drinks, and increase in total calorie intake are responsible for the epidemic of obesity.

The energy balance is very tightly controlled by hypothalamic factors. Hence, the gut-brain axis and the cross-talk between gut hormones, liver, adipose and muscle tissues, pancreas and hypothalamic factors plays an important role in the regulation of food intake, energy balance and development of obesity. Hence, digestive process and assimilation from the small intestine, gluconeogenesis ability of liver, various soluble factors secreted by the adipose tissue (called as adipokines) and energy utilization by the muscle tissue all play a significant role in the development of obesity. Fat deposition and so development of obesity depends on the balance between the diet consumed and energy expenditure. If the amount and type of food taken is substantially more than the energy expenditure that, in turn, depends on the type, frequency and duration of exercise a person does will lead to the onset of obesity. Thus, factors that modulate the digestive process and assimilation could impact human body weight. Since a major portion of digestion and assimilation of digested food occurs in the small intestine, it is quite but natural that bacteria that are present in this portion of the gut could also impact energy balance and obesity. Furthermore, some individuals may be genetically programmed or more susceptible to develop obesity partly due to the environmental factors, familial tendency and hypothalamic dysfunction. Thus, a better understanding of the

1. Genetics of an individual;
2. Hypothalamic factors that control energy homeostasis;

U. N. Das, *Molecular Basis of Health and Disease*,
DOI 10.1007/978-94-007-0495-4_7, © Springer Science+Business Media B.V. 2011

3. Gut factors that control digestion and assimilation of food including the various digestive enzymes and the structure and function of epithelial cells and factors that control their function;
4. Bacteria that reside in the human gut that seems to have the ability to digest polysaccharides and thus provide energy, is necessary to unravel the mechanism(s) involved in the pathobiology of obesity.

Definition of Obesity

Obesity is an excess of body fat. Obesity results when the size or number of fat cells in a person's body increases. A normal-sized person has between 30 and 35 billion fat cells. When a person gains weight, these fat cells first increase in size and later in number. When a person starts losing weight, the cells decrease in size, but the number of fat cells generally stays the same [1]. This is part of the reason as to why once a person becomes obese; it is difficult to lose the excess weight or fat.

For adults, overweight and obesity ranges are determined by using weight and height to calculate a number called the "body mass index" (BMI). BMI correlates with the amount of body fat

$$\text{BMI (kg/m}^2) = \frac{\text{weight in kilograms}}{\text{height in meters}^2}$$

- An adult who has a BMI between 25 and 29.9 is considered overweight.
- An adult who has a BMI of 30 or higher is considered obese.

For children and teens, BMI ranges above a normal weight have different labels (at risk of overweight and overweight). Additionally, BMI ranges for children and teens are defined so that they take into account normal differences in body fat between boys and girls and differences in body fat at various ages.

BMI is an indicator of potential health risks associated with being overweight or obese. For assessing someone's likelihood of developing overweight- or obesity-related diseases, the National Heart, Lung, and Blood Institute guidelines recommend looking at two other predictors:

1. The individual's waist circumference (because abdominal fat is a predictor of risk for obesity-related diseases). Thus, measuring wait to hip ratio [2] seems to be a more dependable risk factor for coronary heart disease (CHD).
2. Other risk factors the individual has for diseases and conditions associated with obesity (e.g. high blood pressure or physical inactivity).

Incidence and Prevalence of Obesity

It is estimated that globally, there are more than 1 billion overweight adults, at least 300 million of them obese—and is a major contributor to the global burden of chronic disease and disability. Often coexisting in developing countries with under-nutrition, obesity is a complex condition, with serious social and psychological dimensions, affecting virtually all ages and socioeconomic groups.

Increased consumption of more energy-dense, nutrient-poor foods with high levels of sugar and saturated fats, combined with reduced physical activity, have led to obesity rates that have risen threefold or more since 1980 in some areas of North America, the United Kingdom, Eastern Europe, the Middle East, the Pacific Islands, Australasia and China [3–5]. The obesity epidemic is not restricted to industrialized societies; this increase is often faster in developing countries than in the developed world.

Obesity and overweight pose a major risk for serious diet-related chronic diseases, including type 2 diabetes mellitus, cardiovascular disease, hypertension and stroke, and certain forms of cancer. Of special concern is the increasing incidence of child obesity.

Obesity May Be Familial

Children residing in homes with poor dietary habits and a couch potato lifestyle are much more likely to be overweight or obese when they are adolescents. Children were also more likely to be overweight if they had strong social bonds with their overweight or obese grandparents and when eating habits included factors such as no parental control over the child's diet and skipping breakfast. Children are also more likely to become overweight adolescents if their parents are obese [6, 7]. Children of parents with higher education levels were less likely to be overweight or obese, as were children with higher levels of self-esteem.

Fast Food Industry and Obesity

Increase in the incidence of obesity can be related to the growth of fast-food industry. For example, fast food consumption has increased greatly in the USA during the past three decades. A close association of frequency of fast-food restaurant visits (fast-food frequency) at baseline and follow-up with 15-year changes in bodyweight and the homoeostasis model (HOMA) for insulin resistance revealed that baseline fast-food frequency was directly associated with changes in bodyweight in both black ($p = 0.0050$) and white people ($p = 0.0013$). Changes were also directly associated with insulin resistance in both ethnic groups ($p = 0.0015$ in black people, $p < 0.0001$ in white people). These results strongly support the contention that fast-food consumption has strong positive associations with weight gain and insulin resistance, suggesting that fast food increases the risk of obesity and type 2 diabetes [8–10]. Relationship between the growth of fast food industry and obesity is given in Fig. 7.1.

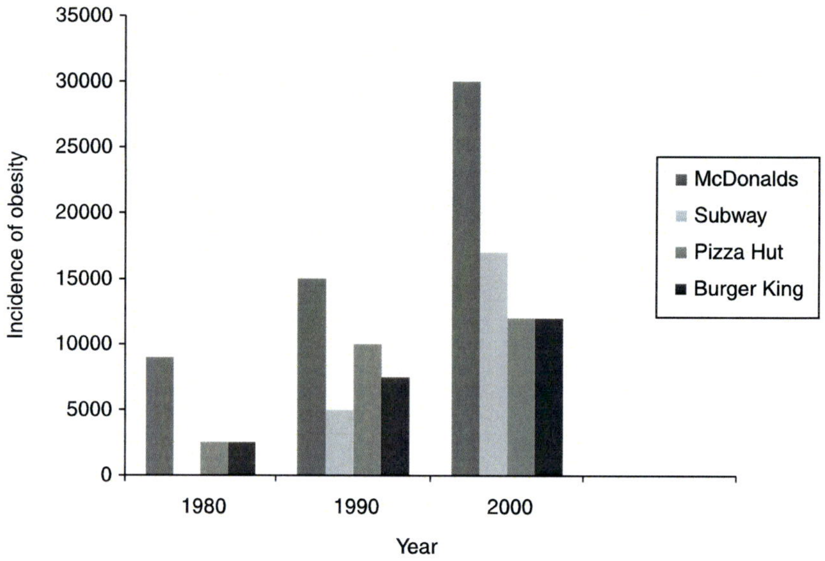

Fig. 7.1 Data showing the strong relationship between the growth of fast food industry and obesity in the USA population

Obesity Is Harmful

Obesity is a chronic disease and is the second leading cause of preventable death, exceeded only by cigarette smoking [11]. Obesity is a major risk factor for hypertension, cardiovascular disease, type 2 diabetes mellitus and some cancers in both men and women, sleep apnea, osteoarthritis, infertility, idiopathic intracranial hypertension, lower extremity venous stasis disease, gastro-esophageal reflux and urinary stress incontinence. The relationship between obesity (body mass index) and relative risk of death due to diseases associated with obesity is given in Fig. 7.2.

The number of annual deaths attributable to obesity among US adults is approximately 280,000 based on relative hazard ratio from all subjects and 325,000 based on hazard ratio from only non-smokers and never-smokers [12]. One-third of all cases of high blood pressure are associated with obesity, and obese individuals are 50% more likely to have elevated blood cholesterol levels [13]. Type 2 diabetes mellitus accounts for nearly 90% of all cases of diabetes. About 88–97% of type 2 diabetes cases diagnosed in overweight people are a direct result of obesity. Overweight and obesity also increases the risk of coronary heart disease [14, 15]. Thus, excess weight is an established risk factor for high blood pressure, type 2 diabetes mellitus, high blood cholesterol level, coronary heart disease and gallbladder disease [16].

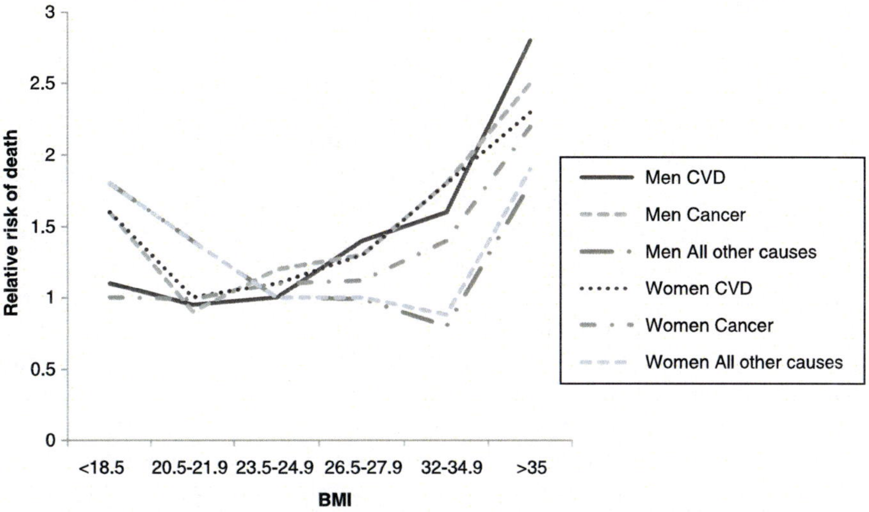

Fig. 7.2 Obesity and its relationship to mortality due to digestive and pulmonary, cardiovascular, gall bladder and type 2 diabetes mellitus-diseases that are common in these subjects

Genetic and Non-genetic Factors Contributing to Obesity

Both genetic and non-genetic factors play a role in the development of obesity. Some of them include:

- Resting metabolic rate
- Thermic response to food
- Nutrient partitioning
- Energy expenditure associated with physical activity
- Gene knockout and transgenic animals (animal models for experimental studies)

It is likely that there could be individual variations in these factors that either predispose he/she to develop or resistant to obesity. In a study wherein measurements of total energy expenditure by the doubly labeled water method was used to determine the range of variation and significant determinants of energy expenditure in healthy adults, it was noted that there was a significant difference with respect to total energy expenditure (TEE), TEE/BMR (basal metabolic rate), and TEE-BMR divided by weight and TEE-BMR between normal athletes, Pima Indians, people in developing countries and others. Multiple regression analysis showed that fat-free mass and age are the significant variables that can explain 65% of the variation in TEE, suggesting that TEE varies dramatically among healthy, free living adults [17]. It was also observed that a low rate of non-basal energy expenditure is a permissible factor for obesity. Studies revealed that the exon 8 ins/del polymorphism of UCP2 (uncoupling protein 2) and UCP2/UCP3 genetic locus are associated with childhood-onset obesity in African American, white, and Asian children [18, 19], suggesting that there is a close association between certain genetic markers and energy expenditure and their susceptibility to develop obesity.

FOXC2 is a winged helix gene that has been shown to counteract obesity, hypertriglyceridemia, and diet-induced insulin resistance in rodents. Hence, it is likely that FOXC2 could be a candidate gene for susceptibility to obesity and type 2 diabetes mellitus. Four variants were identified by sequencing the coding region, as well as 638 bp of the 5′ region and 300 bp of the 3′ region of the gene. Two single nucleotide polymorphisms (SNPs) were found in the putative promoter region, a C-512T transition and a G-350T and two SNPs were found in the 3′ region, a C1548T and a C1702T. In Pima Indians the C-512T variant was associated with BMI ($P = 0.03$) and percentage of body fat ($P = 0.02$) in male and female subjects, as well as with basal glucose turnover and fasting plasma triglycerides in women, suggesting that that variation in FOXC2 may have a role in body weight control and in the regulation of basal glucose turnover and plasma triglyceride levels in women [20].

Adiponectin is an important adipokine that is known to enhance insulin sensitivity. In a cross-sectional design study, it was noted that resting metabolic rate (RMR) was the most important predictor of adiponectin (−0.31; 29%), followed successively by insulin resistance (−0.16; 31%; model containing RMR and insulin resistance), fat mass (0.20; 34%), age (0.34; 35%), visceral fat (−0.34; 40%), and fasting triacylglycerol (−0.12, 41%). The fact that low resting metabolism (RMR) is associated with high serum adiponectin indicates that subjects with low RMR, who are at greater risk of obesity-related disorders, are especially protected by adiponectin [21].

When possible association between fat mass and obesity associated gene (FTO) and phenotypic variation in their energy expenditure (basal metabolic rate (BMR) and maximal oxygen consumption VO(2)max) and energy intake was studied no significant association between the FTO genotype and BMR or VO(2)max was noted [22]. Pima Indians heterozygous for R165Q or NT100 in MC4R (melanocortin 4 receptor) had higher BMIs and lower energy expenditure (by approximately 140 kcal/day), indicating that lower energy expenditure was a component of the increased adiposity [23]. These results suggest that obesity and type 2 diabetes mellitus are associated with variations in the expression and genotype (including single nucleotide polymorphism) UCPs, FOXC2, adiponectin, FTO, MC4R and other related genes.

Gene Expression Profile in Obesity

It was reported that that many genes could be either upregulated or down regulated in obesity [24]. Some of the upregulated genes include: vascular endothelial growth factor, fibroblast growth factor, low density lipoprotein receptor, adrenergic beta receptor kinase, glycogen synthase kinase 3 alpha, neuropeptide Y receptor Y1 and Y5 and mitogen activated protein kinases. Some of the functions of these genes include: increasing vascular supply to the growing adipose tissue, mitogen activity and regulation of appetite (neuropeptide Y), events that could contribute to increase in energy consumption and growth of adipose tissue. At the same time, genes that are down

Table 7.1 Summary of the genes that are either upregulated or down regulated in obese subjects. (From ref. [24])

Biological process	Fold	Gene symbol	Name
Up-regulated			
Cell proliferation	3.5	VEGFB	Vascular endothelial growth factor B
	2.9	FGF1	Fibroblast growth factor 1 (acidic)
Immune response	7.4	FCGR3B	Fc fragment of IgG, low-affinity IIIb, receptor for CD16
Metabolism	2.5	LRP5	Low density lipoprotein receptor-related protein 5
	2.4	ADRBK2	Adrenergic, beta, receptor kinase 2
	2.3	GSK3A	Glycogen synthase kinase 3 alpha
	2.1	PGK1	Phosphoglycerate kinase 1
Signal transduction	5.6	MAPK3	Mitogen-activated protein kinase 3
	3.1	NPY1R	Neuropeptide Y receptor Y1
	2.8	MAPK3K4	Mitogen-activated protein kinase kinase kinase 4
	2.4	MAPK9	Mitogen-activated protein kinase 9
	2.2	MAP2K6	Mitogen-activated protein kinase kinase 6
	2.1	NPY5R	Neuropeptide Y receptor Y5
Down-regulated			
Cell proliferation	5.9	FGF4	Fibroblast growth factor 4
	4.7	FGF2	Fibroblast growth factor 2 (basic)
	4.1	IGF1	Insulin-like growth factor 1
	3.2	FGF7	Fibroblast growth factor 7 (keratinocyte growth factor)
	3.0	FIGF	c-fos-induced growth factor (VEGF D)
	3.0	LDLR	Low-density lipoprotein receptor
	2.3	AR	Androgen receptor
Signal transduction	2.0	PTGER3	Prostaglandin E receptor 3 (subtype EP3)
	3.3	IRS4	Insulin receptor substrate 4
	2.2	ADRB2	Adrenergic, beta-2, receptor, surface

regulated in obese subjects include: c-fos-induced growth factor, prostaglandin E receptor, insulin receptor substrate 4, natriuretic peptide receptor 4, and adrenergic beta-2 receptor, genes that are involved in the regulation cell growth (c-fos), inflammation (prostaglandin E) and regulation sympathetic nervous system (adrenergic receptor) (see Table 7.1). Thus, there seems to be a concerted upregulation and down regulation of genes in such a way that it paves the way to the development of obesity by conserving energy.

All Adipose Cells are Not the Same

In addition to the changes in the expression of several genes noted in obesity, it should be understood that not all adipose cells in the body are the same. Depending on the location, the functions of adipose cells seem to be different. For instance, adipose cells present in the abdominal cavity are different from those present in the gluteal region. Similarly intramyocellular lipid is different from the lipid present in other cells and elsewhere. This is especially so since both abdominal obesity and increase in intramyocellular lipid content is associated with insulin resistance, one of the markers of the metabolic syndrome.

Biochemical and Functional Differences Between Adipose Cells of Different Regions

Abdominal obesity or increased visceral fat is a marker of the presence of insulin resistance and hyperinsulinemia, which are risk factors for the presence or development of hypertension, type 2 diabetes, hyperlipidemias, and CHD. Adipose tissue distribution is dependent on genetic, environmental, and hormonal factors, and is an important predictor of obesity-associated morbidity and mortality [25]. Females have more subcutaneous and gluteal-femoral region adipose tissue compared to males. On the other hand, males have higher adipose tissue localized intra-abdominally. The gluteal-femoral fat cells are enlarged in females that have a higher lipoprotein lipase (LPL) activity [26, 27]. Females do not accumulate fat in visceral depots up to a certain degree of obesity whereas males deposit excess fat in this region parallel with other depots. Gluteal region fat cells from females had higher insulin receptor binding and higher rates of non-insulin-stimulated and maximally insulin-stimulated rates of glucose transport and glucose metabolism [28]. These differences in the distribution and properties of fat between males and females could be attributed to female sex steroid hormones and their interaction with cortisol. Omental adipose tissue contains more number of glucocorticoid receptors (GR) compared to subcutaneous adipose tissue with similar Kd values whereas LPL activity in subcutaneous adipose tissue is lower compared to omental adipose tissue. A positive correlation between LPL activity and glucocorticoid binding was reported. Human adipose tissue glucocorticoid binding was higher in omental than in subcutaneous adipose tissue, whereas LPL activity was higher in omental than in subcutaneous adipose tissue [29]. Leptin mRNA expression is higher in abdominal subcutaneous adipocytes compared with omental adipocytes. A significant inverse correlation exists between adipocytes, PPAR-γ expression and body mass index (BMI) [30–33]. Cellular inhibitor of apoptosis protein-2 (cIAP-2) that regulates tumor necrosis factor-α (TNF-α) signaling was expressed at higher levels in omental than subcutaneous adipocytes [32]. This raises the possibility that depot-specific differences exist in the regulation of adipocyte apoptosis. Subcutaneous adipose tissue produces less interleukin-6 (IL-6) and corticosterone and more TNF-α in comparison to mesenteric adipose tissue [34]. PPAR-γ is involved in adipocyte development and insulin sensitivity and exerts a negative control on TNF-α synthesis, suggesting that a complex but local network of events regulate adipocytes accumulation, metabolism and function. This also emphasizes the fact that different depots of fat display distinct characteristics that are specific to each region of the body. In this context, it is interesting to note that increase in intramyocellular lipid content can influence insulin resistance [35–39]. What is the physiological significance of intramyocellular lipid droplets?

Intramyocellular Lipid (IMCL) Droplets and Perilipins

These lipid droplets encased in a thin phospholipid membrane, contain three proteins: perilipin, adipose differentiation related protein (ADRP or adipophilin) and TIP47. These three proteins together are called as PAT (perilipin/ADRP/TIP47). Because perilipin is found primarily in the adipose cells, it led to the suggestion that it could play a role in lipid deposition and/or lipolysis. Perilipin A increased the triacylglycerol content of cells by forming a barrier that reduced lipolysis, suggesting that perilipin A regulates triacylglycerol storage and lipolysis [40]. Perilipins (A, B, and C) are a family of phosphorylated proteins encoded by a single gene and detected in almost all cells that store excess cholesterol and triacylglycerol as cholesterol and triacylglycerol esters in lipid storage droplets. Adipocytes express predominantly perilipin A, with smaller amounts of perilipin B; whereas Y-1 adrenal cortical cells express primarily perilipin A, with smaller amounts of the isoform perilipin C. Under basal conditions, hormone-sensitive lipase (HSL) resides in the cytosol, and unphosphorylated perilipin upon the lipid droplet. Young rats have high rates of lipolysis and showed translocation of HSL to the lipid droplet, and demonstrated no movement of perilipin from the droplet to the cytosol, though phosphorylation of perilipin also occurred. In contrast, mature rats, upon lipolytic stimulation, showed no HSL translocation but perilipin phosphorylation and movement of perilipin away from the lipid droplet was evident. These results suggest that high rates of lipolysis requires translocation of HSL to the lipid droplet whereas low rates of lipolysis is due to movement of phosphorylated perilipin, and translocation of HSL and perilipin occur independent of each other. Since adipocytes from younger rats have markedly greater rates of lipolysis compared to those from the older rats, and translocation of HSL is needed for high rates of lipolysis, it is evident that a loss of the ability to translocate HSL to the lipid droplet is responsible for the diminished lipolysis seen with advancing age [41]. It is likely that with age the activity of perilipin increases whereas that of HSL decreases that ultimately leads to increase in lipid storage since, perilipins increases triacylglycerol storage by decreasing the rate of triacylglycerol hydrolysis [40].

Perilipins and Inflammation

Obesity is associated with increase in IMCL, low-grade systemic inflammation, insulin resistance, and perilipin expression. But, it is not clear whether perilipin has pro-inflammatory actions or not. Human mast cells, neutrophils, eosinophils, monocytes, and murine fibroblasts showed the presence of prostaglandin hydroperoxide (PGH) synthase on lipid bodies [42]. It is known that the number and size of lipid droplets increase in cells associated with inflammation, especially in monocytes and macrophages, suggesting that lipid droplets may have a role in inflammation. Recent studies showed co-localization of cytosolic phospholipase A_2 ($cPLA_2$) and its activating protein kinases, including extracellular signal-regulated kinase 1 and 2 (ERK1 and ERK2) and p85 and p38 MAPKs, on lipid droplets in monocytic U937 cells [43]. These data suggest that lipid droplets could be active sites for arachidonic acid release and eicosanoid formation [44, 45]. Furthermore, macrophages and monocytes when stimulated to make lipid droplets by feeding them with free fatty acids also made eicosanoids such as leukotrienes (LTs) and prostaglandins (PGs) that occurred on the lipid droplet's surface. On the other hand, aspirin, a COX inhibitor, prevented lipid droplet formation independent of its ability to inhibit COX enzyme [46]. These results suggest that lipid droplets play an active role in the formation of PGs and LTs that have pro-inflammatory actions. Since IMCL was dispersed into smaller droplets after caloric restriction and exercise and the decrement in droplet size correlated highly with improved insulin sensitivity [47] and exercise is anti-inflammatory in nature [48–51], it is likely that the bigger the size and higher the number of lipid droplets more amounts of pro-inflammatory eicosanoids are formed and when the droplet size and number is decreased the formation of eicosanoids falls.

Low Grade Systemic Inflammation Occurs in Obesity

It is evident from the preceding discussion that obesity could be a low-grade systemic inflammatory condition. Obesity is frequently associated with insulin resistance, hyperinsulinemia, hypertension, hyperlipidemia, and CHD, which form core components of metabolic syndrome. Perilipin, whose concentrations are increased in obesity, also have pro-inflammatory action. Furthermore, increase in IMCL is associated with enhanced levels of inflammatory markers [39], and it decreases with diet control and exercise [47] that is anti-inflammatory in nature [48–51]. Thus, obesity is associated with low-grade systemic inflammation.

Plasma levels of C-reactive protein (CRP), TNF-α, and IL-6, which are markers of inflammation, are elevated in subjects with obesity, insulin resistance, essential hypertension, type 2 diabetes, and CHD both before and after the onset of these diseases [52–59]. Overweight children and adults showed an increase in CRP concentration compared with normal weight children (reviewed in [52]). In these subjects, a direct correlation between the degree of adiposity and plasma CRP levels was noted. Elevated CRP concentrations were associated with an increased risk of CHD, ischemic stroke, peripheral arterial disease, and ischemic heart disease mortality in healthy men and women. A strong relation between elevated CRP levels and cardiovascular risk factors: fibrinogen, and HDL cholesterol was also reported.

Increased expression of IL-6 in adipose tissue and its release into the circulation is responsible for elevated CRP concentrations. This is due to the stimulatory influence of IL-6 on the production of CRP in the liver. Experiments done with transgenic mice showed that IL-6 is absolutely essential for the production of CRP [52, 58]. Overweight and obese subjects have significantly higher serum levels of TNF-α levels compared to lean subjects. Weight reduction and/or exercise decrease serum concentrations of TNF-α. The negative correlation observed between plasma TNF-α and HDL cholesterol, glycosylated hemoglobin, and serum insulin concentrations explain why CHD is more frequent in obese compared to healthy or lean subjects [52].

Subjects with elevated CRP levels were two times more likely to develop diabetes at 3–4 years of follow-up period [60]. CRP levels greater than 3.0 mg/l was significantly associated with increased incidence of myocardial infarction, stroke, coronary revascularization, or cardiovascular death [61]. Dietary glycemic load is significantly and positively associated with plasma CRP in healthy middle-aged women [62] suggesting that hyperglycemia induces inflammation. CRP binds to ligands exposed in damage tissue and activates complement [63] and this leads to increases in the size of myocardial and cerebral infarcts in rats subjected to coronary and cerebral artery ligation, respectively [64, 65]. Human CRP activates complement hence; neutralization or inhibitors of CRP could be of significant therapeutic value. 1, 6-bis (phosphocholine)-hexane is a specific small molecule inhibitor of CRP that abrogated the increase in infarct size and cardiac dysfunction produced by injection of human CRP in rats [66]. This suggests that inhibition of CRP produces cardioprotection and possibly, neuroprotection in stroke. It remains to be seen whether such inhibition of CRP will prevent or postpone the development of metabolic syndrome in high-risk subjects.

In general, the current trend is to measure plasma CRP levels as a marker of low-grade systemic inflammation in obesity, type 2 diabetes mellitus and metabolic syndrome. But, it is important to note that the plasma levels of CRP need to be interpreted with caution. The exact role of CRP under physiological conditions is not yet clear. No deficiency or polymorphism in human CRP has been reported. Several studies showed that in vitro pro-inflammatory actions of CRP could be due to bacterial endotoxin and other contaminants rather than CRP itself (reviewed in [67]). Pure human CRP does not seem to possess any pro-inflammatory actions when injected into normal healthy animals [66, 68]. CRP may contribute to innate immunity, can be anti-inflammatory, and exacerbates pre-existing tissue damage in a complement-dependent fashion [63–66]. Nevertheless, CRP may enhance post-reproductive-age diseases such as atherothrombosis, autoimmune diseases, CHD, stroke, and other conditions. Hence, it may be worthwhile to inhibit or neutralize the actions of CRP as shown recently [66].

Despite the fact that inflammatory molecules such as CRP, IL-6, TNF-α and MIF (macrophage migration inhibitory factor) are closely associated with obesity, how and why inflammation occurs in obesity is not clear. Recent evidences suggest that there is a cross-talk between adipose cells and inflammatory cells such as monocytes, macrophages and T cells as discussed below.

In a cross-sectional study of Korean adults, serum concentrations of CRP, TNF-α and IL-6 significantly correlated with weight, BMI (body mass index), waist circumference, hip circumference, and waist-hip ratio. In obese subjects, CRP and IL-6 correlated with BMI, waist circumference and visceral adipose tissue. Multiple regression analysis showed that CRP was significantly associated with BMI, whereas IL-6 was significantly related with visceral adiposity in obese subjects. The positive associations of obesity and visceral adiposity with elevated cytokine levels suggest that low-grade systemic inflammation occurs in these conditions [69]. In fact, it was reported that the peripheral blood mononuclear cells (MNC) from obese subjects are in a proinflammatory state as evidence by elevated binding of nuclear factor kappaB (NF-kB) binding to DNA and significantly lower levels of inhibitor of NF-kB-β (IkB-β) in the obese. In these patients, the mRNA expression and plasma levels of migration inhibitor factor (MIF), IL-6, TNF-α, and matrix metalloproteinase-9 (MMP-9) were elevated and the inflammatory mediators were significantly related to BMI and the degree of insulin resistance [70].

Weight Loss Ameliorates Inflammation

The relationship between obesity and inflammation is further supported by the observation that weight loss achieved by diet control exercise or surgical intervention leads to a reduction in the levels of inflammatory markers. In a study that evaluated the cross-sectional and longitudinal relation of CRP, IL-6, and TNF-α in morbidly obese patients with different stages of glucose tolerance, it was noted that weight loss after gastric surgery induced a significant shift from diabetes (37 vs. 3%) to impaired glucose tolerance (40 vs. 33%) and normal glucose tolerance (23 vs. 64%), concentrations of CRP and IL-6 decreased after weight loss whereas serum levels of TNF-α remained unchanged [71]. Multiple regression analysis revealed that the decrease in insulin resistance remained independently and significantly correlated with the decrease in IL-6 concentrations ($P<0.01$) and the decrease in body mass index with the decrease in CRP ($P<0.05$), respectively, suggesting that weight loss in morbidly obese patients induces a significant decrease of CRP and IL-6 concentrations and an improvement of the insulin resistance. Even liposuction, a common elective surgical procedure in obesity, was found to reduce serum concentrations of CRP, IL-6, IL-18 and TNF-α, patients were less insulin resistant ($p<0.05$), had increased serum levels of adiponectin ($p<0.02$) and HDL-cholesterol ($p<0.05$). A significant correlation was noted between the amount of fat aspirated and changes in insulin resistance ($r=0.28$, $p<0.05$), TNF-α ($r=0.31$, $p<0.02$), and adiponectin ($r=-0.34$, $p<0.02$), as well as between the decrease in TNF-α and the increase in adiponectin after the surgical procedure ($r=-0.45$, $p<0.01$) [72].

Similar improvements in CRP, IL-6, soluble TNF receptor (sTNFR)-1 concentration was noted in a group of obese sedentary obese women who lost weight following a 6 month program of diet control and exercise. Weight loss resulted in significant reductions in body weight, fat mass, visceral adipose tissue (VAT), and fasting glucose and insulin levels ($P<0.05$). Both glucose utilization and insulin sensitivity increased by 16% ($P<0.05$), but concentrations of TNF-α, sTNFR-2, and soluble IL-6 receptor (IL-6sR) did not change. Stepwise regression analysis revealed that changes in VAT and sTNF-R1 independently predicted changes in glucose utilization ($r=-0.49$ and cumulative $r=-0.64$, $P<0.01$), while changes in VAT and IL-6 were both independent predictors of changes in insulin sensitivity ($r=-0.57$ and cumulative $r=-0.68$, $P<0.01$). These results suggest that improvements in glucose metabolism with weight loss programs are independently associated with decreases in cytokine concentrations, suggesting that a reduction in inflammation is a potential mechanism that mediates improvements in insulin sensitivity [73].

In this context, it is interesting to note that contracting skeletal muscle is a major source of circulating IL-6 in response to acute exercise, but with a markedly lower response in trained subjects. In obese subjects, physical inactivity was associated with elevated C-peptide ($P = 0.036$), IL-6 ($P = 0.014$), and CRP ($P = 0.007$) independent of obesity, age, gender, and smoking. Furthermore, the amount of leisure-time physical activity score was inversely associated with IL-6 ($P = 0.017$) and CRP ($P = 0.005$), indicating that low levels of IL-6 and CRP reflect regular physical activity [74].

Adipose Tissue Macrophages (ATMs) and Inflammation

Despite the fact that obesity leads to an increase in inflammatory marker expression, the exact mechanism by which this increase occurs is not clear. Recent studies revealed that adipose tissue macrophages (ATMs), which make up a large proportion of the nonadipose cells in adipose tissue, infiltrate adipose tissue at later stages of obesity and could play a significant role in triggering low-grade systemic inflammation. ATMs infiltrate fat and can cause insulin resistance. Recently, it was reported that T cells are also actively regulated in adipose tissue and contribute to obesity-induced inflammation. These studies provided compelling evidence that specific rearrangements in the T cell receptor (TCR) are selected for in adipose tissue T cells, suggesting that antigens in fat may communicate with the adaptive immune system.

It is known that depending on the immune challenge, T helper cells regulate the activity of other immune cells to generate T-helper type (TH_1) responses through phagocyte activation that produce pro-inflammatory action or humoral TH_2 responses through stimulation of B cell activity. It was noted that in lean mice, resident ATMs have low inflammatory activity restrained by TH_2 cytokines. On the other hand, in obesity, new macrophages are recruited to fat, and, stimulated by TH_1 signals, these macrophages secrete proinflammatory cytokines that impair insulin signaling in adipocytes leading to the development of type 2 diabetes. This is due to increased lipolysis secondary to insulin resistance in adipocytes that leads to the release of free fatty acids into the circulation. These fatty acids render the liver and skeletal muscle insulin resistant that contributes to the development of diabetic state.

For example, it was reported that an increase in the ratio of CD8+ to CD4+ adipose tissue T cells occurs weeks before ATMs typically infiltrate fat [75–77]. It was found that large numbers of CD8+ effector T cells infiltrated obese epididymal adipose tissue in mice fed a high-fat diet, whereas the numbers of CD4+ helper and regulatory T cells were diminished. The infiltration by CD8+ T cells preceded the accumulation of macrophages, and immunological and genetic depletion of CD8+ T cells lowered macrophage infiltration and adipose tissue inflammation and ameliorated systemic insulin resistance. In addition, adoptive transfer of CD8+ T cells to CD8-deficient mice aggravated adipose inflammation. There seem to occur an interactions between CD8+ T cells, macrophages and adipose tissue. Obese adipose tissue activates CD8+

T cells, which, in turn, promote the recruitment and activation of macrophages in this tissue. These results support the proposal that CD8+ T cells have an essential role in the initiation and propagation of adipose inflammation [75]. In a study whose results support this observation [75], Winer et al. [76] showed that CD4+ T lymphocytes, resident in visceral adipose tissue (VAT), control insulin resistance in mice with diet-induced obesity (DIO). DIO VAT-associated T cells showed antigen-specific expansion. CD4+ T lymphocyte control of glucose homeostasis was compromised in DIO progression with VAT accumulation of T helper type 1 (TH_1) cells.

CD4+ (but not CD8+) T cell transfer into lymphocyte-free Rag1-null DIO mice reversed weight gain and insulin resistance, predominantly through TH_2 cells. In obese wild-type and ob/ob (leptin-deficient) mice, treatment with CD3-specific antibody or its F(ab′)2 fragment, reduced the predominance of TH_1 cells.

Foxp3+ cells reversed insulin resistance despite continuation of a high-fat diet, supporting the concept that the progression of obesity-associated metabolic abnormalities is regulated by CD4+ T cells that can be reversed by immunotherapy. These results are supported by the studies of Feuerer et al. [77] who reported that CD4+ Foxp3+ T regulatory (Treg) cells, which are one of the body's most crucial defenses against inappropriate immune responses, operating in contexts of autoimmunity, allergy, inflammation, infection and tumorigenesis [78, 79], were found in large numbers in the abdominal fat of normal mice, but their numbers were strikingly and specifically reduced in the obese mice that were insulin-resistant. Treg cells influenced the inflammatory state of adipose tissue and, thus, insulin resistance. Cytokines differentially synthesized by fat-resident regulatory and conventional T cells directly affected the synthesis of inflammatory mediators and glucose uptake by cultured adipocytes, suggesting that Treg cells could be employed to suppress low-grade systemic inflammation seen in obesity and the metabolic syndrome. It is rather intriguing that adipose tissue Treg cell numbers decrease with obesity and that boosting their numbers in obese mice can improve insulin sensitivity. This protective action of Treg cells could be linked to the production of the cytokine interleukin-10 (IL-10) in ATMs to restrain proinflammatory macrophage activity, which stimulates insulin sensitivity. IL-10 protected adipocytes from the negative effects on insulin signaling induced by TNF-α, and IL-10 can also block the production of inflammatory mediators made by adipocytes in response to TNF-α. Furthermore, the ablation of Treg cells in lean mice worsened glucose tolerance, thus supporting the unique concept that self tolerance and nutrient metabolism are linked [80].

Macrophage Differentiation Is Dependent on Fatty Acid Synthesis

In this context, it is noteworthy that macrophage colony-stimulating factor (M-CSF)-dependent differentiation of primary human monocytes from healthy volunteers induces transcription of SREBP-1c target genes required for fatty acid (FA) biosynthesis and impairs transcription of SREBP-2 target genes required for cholesterol synthesis. This transcriptional regulation leads to a dramatically increased fatty acid synthesis as driving force for enhanced phospholipid synthesis. During cell differentiation the major lipid class switches from cholesterol in monocytes to phosphatidylcholine in macrophages. This transcriptional and metabolic regulation is essential for development of macrophage filopodia and cellular organelles including primary lysosomes, endoplasmic reticulum, and Golgi network, whereas suppression of fatty acid synthesis prevented phagocytosis, implying that induction of fatty acid synthesis is a key requirement for phagocyte development and function [81]. At day 4 of M-CSF-mediated differentiation, a striking shift of fatty acid composition from saturated and polyunsaturated to monounsaturated fatty acids was detected. The C16 and C18 monounsaturated fatty acid content increased from 15% in monocytes to 38% in macrophages as a result of induction of desaturation during differentiation. There was little or no change in cholesterol synthesis or concentration during the differentiation of monocytes to macrophages.

Fatty Acid Metabolism Enhances T Cell Memory

CD8 T cells, which have a crucial role in immunity to infection and cancer, are maintained in constant numbers, but on antigen stimulation undergo expansion and then contraction of antigen-specific effector (TE) populations, followed by the persistence of long-lived memory (TM) cells. TNF receptor-associated factor 6 (TRAF6), an adaptor protein in the TNF-receptor and interleukin-1R/Toll-like receptor superfamily, regulates CD8 TM-cell development after infection by modulating fatty acid metabolism. Mice with a T-cell-specific deletion of TRAF6 mount robust CD8 TE-cell responses, but have a profound defect in their ability to generate TM cells. TRAF6-deficient CD8 T cells exhibit altered regulation of fatty acid metabolism.

Activated CD8 T cells lacking TRAF6 displayed defective AMP-activated kinase activation and mitochondrial fatty acid oxidation (FAO) in response to growth factor withdrawal. Anti-diabetic drug metformin restored FAO and CD8 TM-cell generation in the absence of TRAF6 increased CD8 TM cells in wild-type mice, and improved the efficacy of an experimental anti-cancer vaccine [82].

In an independent study, results of which lend further support to this concept, Araki et al. showed mice treated with rapamycin during the first 8 days after viral infection markedly increased the number of memory T cells 5 weeks later due to an enhanced commitment of effector T cells to become memory precursor cells. When rapamycin was given during the contraction phase of the T-cell response (days 8–35 after infection), the number of memory T cells did not increase, but there was a speeding up of the conversion of effector T cells to long-lived memory T cells with superior recall ability. Rapamycin inhibits mTOR (mammalian target of rapamycin), a protein-kinase enzyme that has mTORC1, which is rapamycin sensitive, and mTORC2, which is resistant to inhibition by rapamycin. Rapamycin seems to act on the mTORC1 complex and regulates memory-cell differentiation [83]. Thus, both rapamycin and metformin enhance T-cell memory formation. Metformin activates AMPK, an enzyme that inhibits mTOR activity. Both AMPK and mTOR sense and control the energy status of a cell (ATP:AMP ratio) and regulate key aspects of cell growth and, as part of this, glucose metabolism [84].

How the results of these studies [26–35] can be integrated to explain the role of diet, inflammation and obesity?

It is possible that following intake of diet rich in saturated fats some of it gets stored in the adipose cells. This leads to an increase in the size of the adipose cells leading to a stretch of its cell membrane. Obviously, excess dietary fat, especially cholesterol and saturated fat, is incorporated into the cell membrane of the adipose cells that leads to a change in the fluidity of the membrane that alters the expression of cell membrane receptors especially the adhesion molecules and chemokines [85–87]. This attracts circulating monocytes and T cells and renders adipose cells more antigenic and hence, circulating and/or resident macrophages and T cells recognize them as foreign and mount an immune attack by elaborating pro-inflammatory cytokines such as IL-6, TNF-α and MIF (macrophage migration inhibitory factor). These cytokines, in turn, enhance the expression of chemokines and adhesion molecules that attract and activate macrophages and T cells leading to persistence and perpetuation of inflammation. As a result, the resident and circulating monocytes may be triggered to mature into macrophages, a process during which the expression of genes that regulate fatty acid metabolism is enhanced leading to the generation of long-lived memory (TM) cells. Since, the C16 and C18 monounsaturated fatty acid content increased in macrophages as a result of induction of desaturation during this process of differentiation with little or no change in cholesterol synthesis (it is suggested that non-reduction in the concentrations of cholesterol itself could serve as a pro-inflammatory stimulus), it is predicted that reduction in cholesterol synthesis or levels could be of benefit in suppressing the activation of macrophages and inhibition of inflammation [88–90], possibly by disrupting microdomain structure, decrease in cholesterol/ganglioside ratio and caveolin expression resulting in reduced pro-inflammatory signals. Hence, it is likely that a combination of HMG-CoA reductase inhibitors and metformin may be more beneficial especially in diabetics so that TM cells are generated in adequate amounts and at the same time inflammation is under control. To verify this postulation, it is necessary to measure plasma concentrations of pro- and anti-inflammatory cytokines periodically in those who are on metformin and statins.

In contrast, it is predicted that increased intake of polyunsaturated fatty acids (PUFAs), especially n-3 eicosapentaenoic acid and docosahexaenoic acid, will render the cell membrane more fluid and decrease the expression of chemokines and adhesion molecules [91–97], either by themselves or by leading to the formation of their anti-inflammatory products such as lipoxins, resolvins, protectins, maresins and nitrolipids, that leads to decreased infiltration of adipose tissue with macrophages, monocytes and T cells. It is also predicted that increased intake of PUFAs will enhance IL-10 production and decrease the synthesis and release of IL-6, TNF-α and MIF [98–106]. As result the balance between TH_1 and TH_2 is tilted more towards the TH_2 and thus, inflammation is dampened. Such an increased incorporation of

PUFAs into the adipose cell membrane will also dampen the differentiation of monocytes to macrophages. Furthermore, PUFAs are also known to inhibit the activity of HMG-CoA reductase [107–111] and thus, lower plasma and cell/tissue levels of cholesterol.

Recently, it was reported that PUFAs form precursors to anti-inflammatory molecules such as lipoxins, resolvins, protectins and maresins [112–129]. Thus, when the incorporation of PUFAs is enhanced, the formation of these anti-inflammatory molecules is augmented that, in turn, suppresses inflammation, conversion of monocytes to macrophages is blocked and production of IL-6 TNF-α and MIF is decreased (see Fig. 7.3).

These evidences suggest that there is a close interaction among diet, TH_1 and TH_2 cells, obesity, differentiation of monocytes to macrophages, fatty acid metabolism and inflammation.

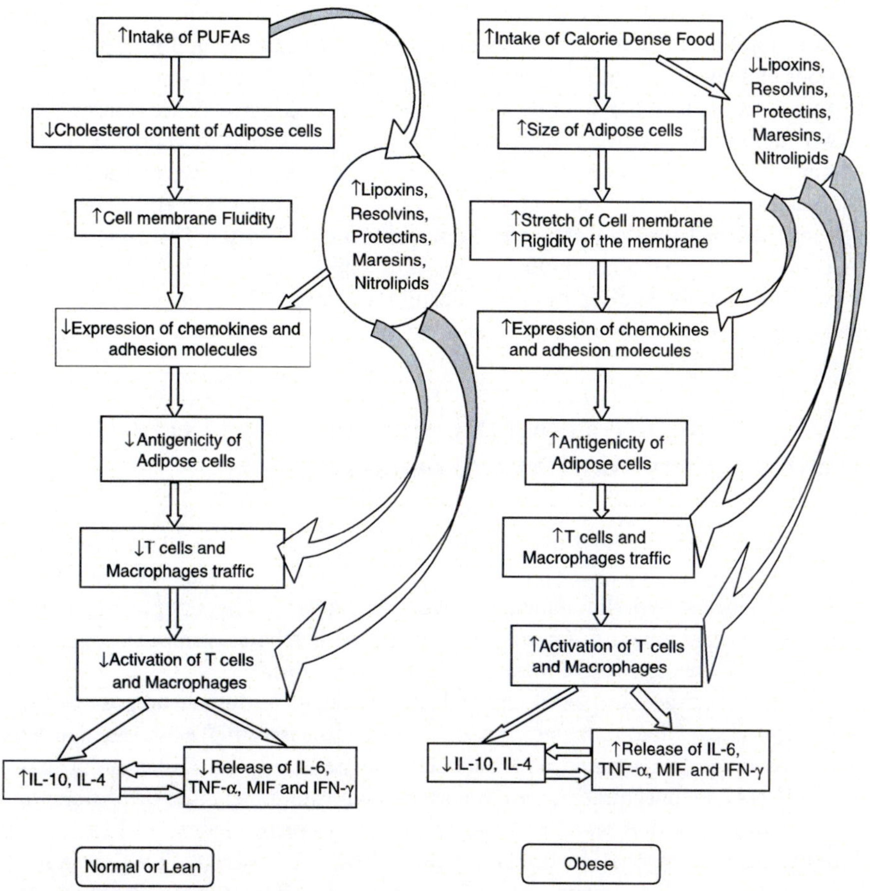

Fig. 7.3 Scheme showing the relationship between diet and BMI (body mass index) and the role of inflammatory cells in obesity

What Causes Abdominal Obesity—How and Why?

Abdominal obesity is the most common and dominant component of the metabolic syndrome and is often associated with insulin resistance, hypertension, and dyslipidemia. Brochu et al. [130] examined the metabolic characteristics of obese, sedentary postmenopausal women who were metabolically normal but obese (MNO) or as metabolically abnormal obese (MAO) based on insulin sensitivity (measured by the hyperinsulinemic/euglycemic clamp technique). MNO subjects displayed high insulin sensitivity (11.2 ± 2.6 mg/min kg lean body mass) whereas MAO showed lower insulin sensitivity (5.7 ± 1.1 mg/min kg lean body mass). Despite comparable total body fatness between these two groups (45.2 ± 5.3 vs. $44.8 \pm 6.6\%$; P = NS), MNO individuals had 49% less visceral adipose tissue than MAO subjects (141 ± 53 vs. $211 \pm 85\ cm^2$; $P < 0.01$), whereas no difference was noted between groups for abdominal subcutaneous adipose tissue (453 ± 126 vs. $442 \pm 144\ cm^2$; P = NS), total fat mass (38.1 ± 10.6 vs. 40.0 ± 11.8 kg), and physical activity energy expenditure. MNO subjects had significantly lower fasting plasma glucose and insulin concentrations and lower insulin levels during the oral glucose tolerance test, lower plasma triglycerides and higher high-density lipoprotein cholesterol concentrations than MAO individuals. Stepwise regression analysis showed that visceral adipose tissue and the age-related onset of obesity explained 22% and 13%, respectively, of the variance observed in insulin sensitivity, suggesting that visceral adipose tissue may account for the differences between MNO and MAO. This indicates that visceral adipose tissue accumulation could be one of the main culprits in the development of metabolic syndrome and insulin resistance. Hence, understanding the pathophysiology of abdominal obesity is essential.

Excess 11β-hydroxysteroid Dehydrogenase Type 1 (11β-HSD-1) Enzyme Activity May Cause Abdominal Obesity

Mice over expressing 11β-hydroxysteroid dehydrogenase type 1 (11β-HSD-1) enzyme selectively in adipose tissue develop abdominal obesity and exhibit insulin-resistant diabetes, hyperlipidemia, and hyperphagia despite hyperleptinemia [131], features that are similar to those seen in subjects with metabolic syndrome, suggesting that abdominal obesity is like localized Cushing's syndrome.

In primary cultures of paired omental and subcutaneous human adipose stromal cells, 11β-HSD-1 oxo-reductase activity was significantly higher in omental adipose stromal cells compared with subcutaneous cells despite similar endogenous NADPH/NADP concentrations. Both cortisol and insulin increased the differentiation of adipose stromal cells to adipocytes, but only cortisol increased 11β-HSD-1 activity and messenger RNA levels in a dose-dependent fashion. Cortisone was as effective as cortisol in inducing adipose stromal cells differentiation. The local conversion of cortisone to active cortisol through expression of 11β-HSD-1 was found to be higher in omental human adipose stromal cells compared with subcutaneous cells. These results imply that glucocorticoids have a differential action on different adipose tissue depots, and indicates that increased local metabolism of glucocorticoid may be responsible for abdominal obesity [132]. 11β-HSD-1 mRNA levels were higher in omental compared with subcutaneous preadipocytes in obese women [133]. In Pima Indians, when single nucleotide polymorphisms in the 11β-HSD-1 gene were genotyped, two representative SNPs (SNP1, and SNP5) were associated with Type 2 diabetes mellitus, although neither SNP was associated with obesity. SNP1 and SNP5 were associated with insulin-mediated glucose uptake rates, and SNP1 was further associated with fasting, 30-min, and 2-h plasma insulin concentrations, whereas adipocyte 11β-HSD-1 mRNA concentrations correlated positively with adiposity and insulinemia, and additionally negatively correlated with insulin-mediated glucose uptake rates. In contrast, muscle 11β-HSD-1 mRNA concentrations did not correlate with any anthropometric or metabolic variables. These results confirm that adipocyte 11β-HSD-1 mRNA concentrations are associated with adiposity, and suggest that genetic variations in the 11β-HSD-1 gene are associated with Type 2 diabetes mellitus, plasma insulin concentrations and insulin action, independent of obesity implying that 11β-HSD-1 gene is under tissue-specific regulation, and has tissue-specific consequences [134]. It was reported that though obese men had no difference in their whole-body rate of regenerating cortisol, they had a more rapid conversion of ^{3}H cortisone to ^{3}H cortisol in abdominal subcutaneous adipose tissue. Insulin infusion produced a marked decrease in adipose 11β-HSD-1 activity in lean but not in obese men. These results suggest that in vivo cortisol generation is increased selectively within adipose tissue in obesity, and this increase in 11β-HSD-1 activity is resistant to insulin-mediated down regulation [135, 136]. These studies indicate that specific and effective inhibitors of 11β-HSD-1 in adipose tissue are needed to increase insulin sensitivity and treat abdominal obesity. The observation that 11β-HSD-1 deficiency protects against the development of high-fat diet induced abdominal obesity and remains insulin sensitive are in supportive of this assertion.

11β-HSD-1$^{(-/-)}$ mice expressed lower resistin and TNF-α, but higher PPAR-γ, adiponectin, and uncoupling protein-2 (UCP-2) mRNA levels in adipose tissue, and isolated 11β-HSD-1$^{(-/-)}$ adipocytes exhibited higher basal and insulin-stimulated glucose uptake. 11β-HSD-1$^{(-/-)}$ mice also showed reduced visceral fat accumulation upon high-fat feeding [137]. These data strongly support the proposal that adipose 11β-HSD-1 deficiency prevents the development of abdominal obesity and possibly, other features of metabolic syndrome and indicates that increase in 11β-HSD-1 activity may suppress adiponectin, PPAR-γ, and UCP-2 activities (see Fig. 7.4).

Interaction Among 11β-HSD-1, TNF-α and Insulin

In this context, the close interaction between 11β-HSD-1, TNF-α, and insulin is worth noting since, obesity is associated with insulin-resistance and abnormal glucose homeostasis. TNF-α has a role in mediating the insulin-resistance of obesity through its overexpression in adipose tissue. Adipose tissue cells obtained from breast

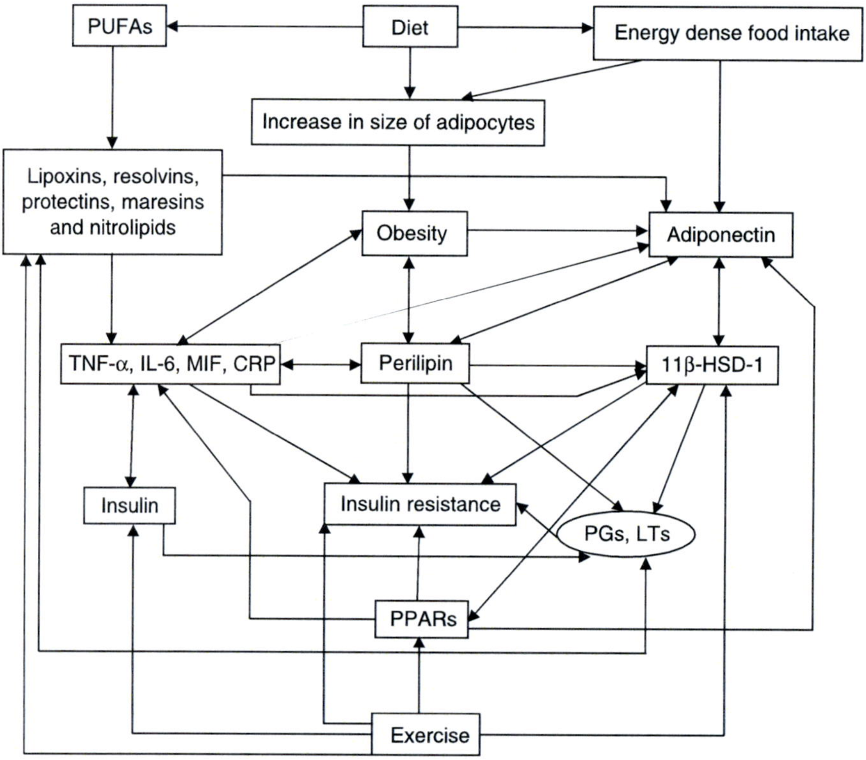

Fig. 7.4 A simplified scheme showing the relationship between obesity, perilipin, cytokines, adiponectin, PPARs, adipocyte size, exercise and insulin resistance. For details see text. Energy dense foods cause obesity by increasing the number and size of adipocytes that, in turn, increases the expression of perilipins on the lipid droplets of adipocytes. Obesity is associated with increased levels of TNF-α, IL-6, MIF and CRP, and decrease in the levels of adiponectin. High concentrations of TNF-α, IL-6, and perilipins and decrease in the levels of adiponectin cause insulin resistance. Perilipin deficient experimental animals show insulin resistance but also show near normal blood glucose levels due to decreased basal hepatic glucose production. Insulin resistance decreases the expression of PPARs. Exercise and diet control leads to weight loss, decrease in the size of adipocytes and lipid droplets, TNF-α, IL-6, MIF and CRP, an increase in adiponectin levels and decrease in insulin resistance and increased utilization of PUFAs and lead to the formation of lipoxins, resolvins, protectins, maresins and nitrolipids. A decrease in adipocyte size reduces perilipin production. TNF-α decreases perilipin production and thus, enhances lipolysis, whereas overexpression of perilipin resists TNF-α-induced lipolysis. PPAR-γ decreases TNF-α production, increases perilipin expression and adiponectin levels, reduces insulin resistance, and decreases the size of adipocytes and so a decrease in the size of lipid droplets is expected. The effects of exercise are similar to that of PPAR-γ: decreases TNF-α, IL-6, and CRP levels, increases adiponectin levels, decreases the size of adipocytes and lipid droplets, decreases insulin resistance and perilipin levels (due to decrease in the size of lipid droplets) and increases the expression of PPAR-γ. In obese subjects, plasma and tissue concentrations of PUFAs are low. During exercise, consumption of PUFAs is increased and production of beneficial PGI_2, lipoxins, resolvins, maresins, protectins and nitrolipids may be increased. PUFAs and lipoxins, resolvins, protectins, maresins and nitrolipids decrease TNF-α, IL-6 and enhance adiponectin production. PUFAs are endogenous ligands for PPARs and thus, reduce insulin resistance. Hence, PUFAs are expected to decrease perilipin

on treatment with TNF-α increased 11β-HSD-1 activity in a dose dependent fashion. In contrast, insulin had no effect under basal conditions, but inhibited the stimulatory effects of TNF-α on 11β-HSD-1 activity. These alterations in the activity of 11β-HSD-1 were seen in the level of 11β-HSD-1 mRNA suggesting that both TNF-α and insulin are mediating their actions at the level of gene transcription [138].

When primary cultures of human hepatocytes and subcutaneous and omental adipose stromal cells (ASC) were treated with TNF-α, a dose-dependent increase in 11β-HSD-1 activity was noted only in the subcutaneous and omental adipose cells, but had no effect on 11β-HSD-1 activity in hepatocytes. Insulin-like growth factor I (IGF-I), similar to insulin, caused a dose-dependent inhibition of 11β-HSD-1 activity in subcutaneous and omental stromal cells, but not in human hepatocytes. Both TNF-α and IL-1β enhanced the expression of 11β-HSD-1 activity both in subcutaneous and omental stromal cells in a time- and dose-dependent manner. PPAR-γ ligands significantly increased 11β-HSD-1 activity in omental and subcutaneous adipose cells [139]. These results suggest that tissue-specific regulation of 11β-HSD-1 occurs and the response of omental adipose cells differs from that seen in subcutaneous adipocytes. Glucocorticoids, which induce abdominal obesity, insulin resistance and possess anti-inflammatory actions, inhibit TNF-α synthesis [140] but enhance 11β-HSD-1 activity suggests that the ability of glucocorticoids to induce abdominal obesity could be related to its action on 11β-HSD-1. Subcutaneous adipocytes of lean subjects treated with TNF-α showed inhibited adiponectin release but had no effect on adiponectin release from subcutaneous or omental adipocytes from obese subjects. On the other hand, dexamethasone significantly inhibited adiponectin release [141]. Thus, there is a close positive and negative feedback regulation between glucocorticoids, TNF-α, 11β-HSD-1 activity, adiponectin secretion, insulin, and PPARs that is relevant to their role in obesity, insulin resistance, and metabolic syndrome (see Fig. 7.4). It is also interesting to note that glucocorticoids enhance the expression of perilipins [142, 143], which are associated with low-grade systemic inflammation as already discussed.

Glucocorticoids and Perilipins

Glucocorticoids produce abdominal obesity, cause insulin resistance, and possess anti-inflammatory actions, but inhibit TNF-α synthesis [141] and adiponectin release; whereas TNF-α increased 11β-HSD-1 activity in a dose dependent fashion [144]. In contrast, insulin had no effect under basal conditions, but inhibited the stimulatory effects of TNF-α on 11β-HSD-1 mRNA [138] and suppressed the production

---

production. Breast fed children have decreased tendency to develop obesity since human breast milk is rich in PUFAs and breast fed children show reduced insulin resistance. Abdominal obesity may be due to increased activity of 11β-HSD-1. Enhanced expression of 11β-HSD-1 is associated with insulin resistance, increased production of perilipins, decreased plasma adiponectin levels, and decreased PPARs expression.

of TNF-α, IL-6, IL-1, IL-2, and macrophage migration inhibitory factor (MIF), and enhanced the production of anti-inflammatory cytokines: IL-4 and IL-10 [145, 146]. This supports the original proposal that insulin has anti-inflammatory actions [145]. Glucocorticoids and TNF-α have inhibitory action on adiponectin production that enhances the action of insulin and shows anti-inflammatory action; and glucocorticoids suppress TNF-α synthesis while glucocorticoids and TNF-α have opposite actions on inflammation. But, surprisingly both glucocorticoids and TNF-α induce peripheral insulin resistance. TNF-α down regulates [147], whereas glucocorticoids enhance perilipin expression. Excess production of TNF-α causes cachexia (as seen in patients with cancer), whereas glucocorticoids produce abdominal obesity suggesting that some of their downstream events could be different and their actions on adiposity could be, in part, due to their opposite actions on perilipin expression. These results also emphasize the complexity of the pathobiology of obesity, inflammation and the interactions among various molecules involved in these processes.

Glucocorticoids, TNF-α, and Inflammation

Glucocorticoids bring about their anti-inflammatory actions by (i) the induction and activation of annexin 1 (also called as lipocortin-1) [148], (ii) the induction of mitogen-activated protein kinase (MAPK) phosphatase 1 [149], and (iii) the inhibition of cyclo-oxygenase-2 (COX-2) [150]. Annexin 1 or Lipocortin-1 physically interacts with and inhibits cytosolic phospholipase $A_2\alpha$ ($cPLA_2\alpha$) so that arachidonic acid (AA) is not released that is needed for the formation of various pro-inflammatory eicosanoids. Increased expression of $cPLA_2$ is also necessary to give rise to anti-inflammatory molecules such as prostaglandin D_2 (PGD_2) and 15deoxy$\Delta^{12-14}$$PGJ_2$, and lipoxins (LXs). Thus, the timing of expression (perhaps a pulsatile expression) of $cPLA_2$ and the local concentrations of glucocorticoids could be an important factor that determines the progression and/or resolution of inflammation. The selective inhibition of COX-2 and inducible nitric oxide synthase (iNOS) expression by glucocorticoids could explain their potent anti-inflammatory actions [150, 151]. Glucocorticoids also inhibit the production of pro-inflammatory cytokines such as IL-1, IL-6, TNF-α, and MIF [152–154]. Glucocorticoids mediate their inhibitory action on iNOS and COX enzymes through lipocortin-1 (annexin1) [148]. On the other hand, eNO activates constitutive (COX-1) resulting in optimal release of PGE_2, whereas iNO activates COX-2 resulting in markedly increased release of PGE_2 that results in inflammation [155]. This indicates that constitutive production of NO and PGE_2 are anti-inflammatory in nature whereas inducible production of NO and PGE_2 are pro-inflammatory, simply because the quantities of NO and PGE_2 are extremely high in the later instance. Low concentrations of glucocorticoids enhance MIF synthesis that, in turn, overrides glucocorticoid-mediated inhibition of secretion of other pro-inflammatory cytokines. MIF induces the production of TNF-α and vice versa. But, it is not yet clear whether corticosteroids can enhance the production of lipoxins, resolvins, protectins, maresins and nitrolipids that have anti-inflammatory actions, possibly, inhibit their production.

It is interesting to note that eicosanoid metabolism by human peripheral blood monocytes (PBM) is altered based on corticosteroids use. Isolated PBM when stimulated with calcium ionophore A23187, with or without exogenous 15(S)-hydroxyeicosatetraenoic acid (15(S)-HETE), were found to produce LTB_4 without 15(S)-HETE $\sim 40 \pm 12$ ng and 59 ± 11 ng by untreated and steroid-dependent asthmatics, respectively. In the presence of 15(S)-HETE, PBM produced sixfold smaller amounts of LTB_4, whereas low amounts of lipoxins (LXs) were produced by PBM from asthmatics only (2.7 ± 0.7 ng and 4.6 ± 2.8 ng for untreated and steroid-dependent asthmatics, respectively) [156, 157]. This suggests that PBM can metabolize 15(S)-HETE to lipoxins and corticosteroids seem to have the ability to enhance the formation of lipoxins relative to LTs. This ability of steroids to enhance lipoxins, a potent anti-inflammatory molecule, may explain, in part, the anti-inflammatory actions of steroids.

In addition, glucocorticoids accelerate the catabolism of LTC_4 (leukotriene C_4), a pro-inflammatory molecule [158]. 15-HPETE, an anti-inflammatory eicosanoid formed via lipoxygenase pathway, causes a significant increase in the rate of TNF degradation [159] an action that may also be seen with LXs. LXA_4 inhibited not only the secretion of TNF-α [160], but also prevented TNF-α-induced production of IL-1β, IL-6, cyclin E expression, and NF-κB activation [161]. Thus, glucocorticoids and lipoxins have similar actions on inflammation, both are anti-inflammatory, but their mechanisms of action appear to be different.

In this context, it is noteworthy that both TNF-α and glucocorticoids have opposite actions on PLA_2: the former stimulates [162] while the later inhibits [153]. There is evidence to suggest that activation of $cPLA_2$ is crucial to the actions of TNF-α [163]. This indicates that $cPLA_2$ and other PLA_2s play a central role in the pathobiology of inflammation and its resolution that could be attributed to the fact that PUFAs such as arachidonic acid (AA), eicosapentaenoic acid (EPA), and docosahexaenoic acid (DHA) released by PLA_2 form precursors to several pro- and anti-inflammatory compounds. On the other hand, glucocorticoids inhibit the production of TNF-α and thus, bring about some of its anti-inflammatory actions [153], whereas TNF-α increased 11β-HSD-1 activity leading to the formation of increased amounts of cortisol that, in turn, inhibits TNF-α formation and enhances, at least, partially lipoxin formation and restores normalcy. Thus, there is close positive and negative interaction between TNF-α, glucocorticoids, PUFA metabolism, lipoxin formation and the inflammatory process (see Fig. 7.4).

Diet, Genetics, Inflammation and Obesity

It is evident from the preceding discussion that increased expression of perilipins and 11β-HSD-1 in adipose cells, especially in the omental and mesenteric adipose tissue, could lead to insulin resistance, and low-grade systemic inflammation. Increased expression of perilipins in the mesenteric/omental adipose cells leads to insulin resistance and increased production of pro-inflammatory eicosanoids due to the activation of PLA_2. This ushers in low-grade systemic inflammation seen in obesity.

It appears as though consumption of even normal food; high calorie diet rich in fats (saturated and trans-fats) or protein and glucose challenge enhances generation of reactive oxygen species by leukocytes and decreases vitamin E levels [164–166]. Oxidative stress and pro-inflammatory process induces insulin resistance [167]. This increase in reactive oxygen species could be as a result of increased production of IL-6, TNF-α, IL-18, MIF and CRP. IL-6 and TNF-α activate NADPH oxidase and enhance the generation of reactive oxygen species [168]. Thus, consumption of energy dense diets induce oxidative stress that could be toxic to pancreatic β cells and also produces long-term complications seen in obesity, diabetes, hypertension, and metabolic syndrome. Continued consumption of energy dense diet from childhood effectively abrogates the anti-oxidant defenses of various cells and tissues and leads to the development of obesity, hypertension, type 2 diabetes mellitus, and the metabolic syndrome.

Previously, I proposed that physiological response to even normal food intake (containing carbohydrates, proteins, and fats and mixed meals) includes an increase in the production of TNF-α and IL-6 and consequent increase in plasma CRP and decrease of anti-inflammatory cytokines IL-4 and IL-10, and adiponectin. TNF-α and IL-6 induce oxidative stress and activate NF-κB, which induces insulin resistance and consequent hyperinsulinemia. Insulin secreted in response to food intake is not only necessary to normalize plasma glucose, lipid and amino acid concentrations but also to suppress TNF-α and IL-6 and enhance IL-4 and IL-10 synthesis. Insulin stimulates the synthesis of PUFAs that, in turn, enhance insulin action [145, 164, 169]. Increased production of TNF-α and IL-6 following food intake activates phospholipase A_2 (PLA_2) [170–172] that, in turn, releases PUFAs from the membrane lipid pool. PUFAs thus released, if adequate, suppress the synthesis and release of TNF-α and IL-6 resulting in the restoration of balance between pro- and anti-inflammatory cytokines and suppression of oxidative stress. This action could be produced by PUFAs by themselves or may occur as a result of increased formation of lipoxins, resolvins, protectins, maresins and nitrolipids. Continued consumption of energy rich diet and/or saturated and trans-fatty acids and/or sub-optimal intake of PUFAs and/or as a result of defects in the formation of anti-inflammatory lipoxins, resolvins, protectins, maresins and nitrolipids due to genetic polymorphism may cause a state of low-grade systemic inflammation and chronic oxidative stress. In contrast, dietary restriction, exercise, and weight loss suppress free radical generation and oxidative stress [173], decrease the production of TNF-α and IL-6 and enhance IL-4 and IL-10, and adiponectin synthesis, and disperse IMCL into smaller droplets, enhance the formation of PUFAs from dietary LA and ALA and augment the synthesis and release of lipoxins, resolvins, protectins, maresins and nitrolipids that leads to improved insulin sensitivity. Saturated and trans-fats and hyperglycemia interfere with the synthesis of PUFAs, and their conversion to lipoxins, resolvins, protectins, maresins and nitrolipids hence, normal inhibitory control exerted by these beneficial lipids on TNF-α, IL-6, MIF and other pro-inflammatory molecules will be defective or sub-optimal. This proposal derives support from the observation that adequate intake of EPA and DHA but not ALA was inversely associated with plasma levels of sTNF-R1 and sTNF-R2 (soluble tumor necrosis factor receptors 1 and 2) and CRP whereas ω-6 fatty acids did not inhibit the anti-inflammatory effects of ω-3 fatty acids [174]. A combined intake of ω-3 and ω-6 fatty acids produced lowest levels of inflammation.

Based on these evidences, I propose that certain individuals are genetically programmed to have increased expression of perilipins and 11β-HSD-1 (especially in the mesenteric/omental adipose cells) that predisposes them to develop abdominal obesity and the metabolic syndrome. This genetic predisposition coupled with lack of adequate exercise and consumption of energy rich diets renders them highly susceptible to develop obesity and other features of metabolic syndrome. This explains how the interaction between genetic predisposition (in the form of constitutionally increased expression of perilipins and 11β-HSD-1) interact with environmental factors (in the form of lack of exercise and consumption of energy rich diets) could lead to an explosion in the incidence of obesity and its consequences.

In addition, there is now evidence to suggest that gut, gut hormones and gut bacteria and hypothalamic factors and the interaction(s) between gut and hypothalamus play a significant role in the pathobiology of obesity.

Gut and Obesity

In general, humans are more suited to resist famine than overabundance of food (called as the thrifty gene hypothesis) and hence, it has been argued that easy and relatively inexpensive availability of energy dense food is responsible for the current obesity epidemic. This coupled with lack of exercise, enhanced intake of saturated fats, carbonated drinks, and increase in total calorie intake seems to be driving the increase in the incidence of obesity. The food that is ingested needs to be digested, assimilated and this, in turn, contributes to the total amount of calories that are available to the human body. The energy balance is very tightly controlled by hypothalamic factors. Hence, the gut-brain axis and the cross talk between gut hormones and hypothalamic factors are important in the regulation of food intake, energy balance and development of obesity. Thus, factors that modulate the digestive process and assimilation could impact human body weight. Recent studies revealed that bacteria present in the colon could impact energy balance and obesity. Furthermore, as already discussed above, some individuals may be genetically programmed or more susceptible to develop obesity partly due to the environmental factors, familial tendency and hypothalamic dysfunction. One of the environmental factors that could render an individual more susceptible to develop obesity could be perinatal nutritional environment.

Perinatal Nutritional Environment Influences Development of Obesity

Fetal nutritional environment influences the risk of developing obesity in adult life [175–177] by influencing the developing neuroendocrine hypothalamus that controls food intake, hunger and satiety. Neuropeptide Y, agouti-related peptide, proopiomelanocortin (POMC), cocaine- and amphetamine-regulated transcript (CART), and insulin receptor mRNAs and leptin receptor mRNA, the key central components of adult energy balance regulation, were already present in early gestation [178]. Hence, perinatal and early childhood nutrition is likely to influence the hypothalamic neurotransmitters and thus, determine the development of obesity in adulthood. This is supported by the observation that both obesity and type 2 diabetes mellitus could be disorders of hypothalamic dysfunction and low birth weight is associated with high prevalence of obesity, type 2 diabetes mellitus and metabolic syndrome in later life [179, 180]. Though some studies disputed these findings and suggested that postnatal nutrition and growth are more important [181], this suggests that early nutrition has a bearing on the development of obesity, type 2 diabetes mellitus and metabolic syndrome in later life.

Obesity and Type 2 Diabetes Mellitus as Disorders of the Brain

In experimental animals, ventromedial hypothalamic (VMH) lesion induces hyperphagia and excessive weight gain, fasting hyperglycemia, hyperinsulinemia, hypertriglyceridemia, and impaired glucose tolerance. Intraventricular administration of antibodies to neuropeptide Y (NPY) abolished hyperphagia in these animals. Streptozotocin-induced diabetic animals showed increase in NPY concentrations in paraventricular, VMH and lateral hypothalamic areas. VMH-lesioned rats showed selectively decreased concentrations of norepinephrine and dopamine in the hypothalamus; whereas long-term infusion of norepinephrine and serotonin into the VMH impaired pancreatic islet cell function. These changes in the hypothalamic neurotransmitters reverted to normal after insulin therapy, suggesting that dysfunction of VMH impairs pancreatic β-cell function and induces metabolic abnormalities seen in obesity and type 2 diabetes mellitus (reviewed in [179, 182]). TNF-α decrease the firing rate of the VMH neurons and is neurotoxic [183–185]. In VMH-lesioned rats, the abundance of (obese) ob mRNA increased after the gain of body weight and marked expression was observed following VMH lesion [186], suggesting that ob gene is up-regulated with fat accumulation even in non-genetically obese animals.

The tone of the parasympathetic nervous system increases after VMH lesion, whereas the sympathetic tone decreases [187, 188], as a result lipolysis is inhibited that leads to obesity. Acetylcholinesterase (AchE) activity in liver, pancreas, and stomach of VMH-lesioned obese rats was significantly increased [189], whereas radical vagotomies blocked the development of obesity in VMH-lesioned animals. This suggests that vagus is the neural pathway from the hypothalamus to the visceral fat and the pancreatic β cells to communicate the messages from VMH to produce disturbances in metabolism that leads to obesity seen in the VMH-lesioned animals [189].

Cross-Talk Between the Liver, Adipose Tissue and the Brain Through Vagus

Vagus nerve serves as the neuronal pathway in the cross talk between the liver and adipose tissue. In mouse, adenovirus-mediated expression of peroxisome proliferative-activated receptor (PPAR)-$\gamma 2$ in the liver induced acute hepatic steatosis while markedly decreasing peripheral adiposity that is accompanied by increased energy expenditure and improved systemic insulin sensitivity. These animals not only showed increased hepatic PPAR-$\gamma 2$ expression but also had decreased fasting plasma glucose, insulin, leptin and TNF-α levels indicating markedly improved insulin sensitivity and showed decreased glucose output from the liver. These animals had high tonus of the sympathetic nervous system. Resection of the hepatic branch of the vagus nerve completely blocked the decreases in peripheral adiposity and other indices indicating that the afferent vagus mediates the effects of hepatic PPAR$\gamma 2$ expression [190]. Thus, that afferent vagal nerve activation originating in the liver mediates the remote effects of hepatic PPAR$\gamma 2$ expression on peripheral tissues. Dissection of the hepatic branch of the vagus before thizolidinedione (TZD) administration reversed the increases in resting oxygen consumption as well as UCP-1 expression in the adipose tissue (both in the white and brown adipose tissue) indicating that the neuronal pathway originating in the liver is also involved in the acute systemic effects of TZD in the obese subjects in whom the hepatic PPAR$\gamma 2$ expression is upregulated. Thus, the afferent vagus from the liver and efferent sympathetic nerves to adipose tissues are involved in the regulation of energy expenditure, systemic insulin sensitivity, glucose metabolism and fat distribution between the liver and the peripheral tissues. Liver conveys information regarding energy balance to the hypothalamus, especially to the VMH neurons via the afferent vagus whereas leptin could be the humoral signal to the brain from the adipocytes. Brain integrates all the information received both from humoral and neural pathways from various sources to produce appropriate responses-either sympathetic nervous system activation and/or parasympathetic modulation to maintain energy homeostasis [182].

Cross-Talk Between the Liver and Pancreatic β Cells is Mediated by the Vagus

Obesity is associated with insulin resistance that promotes pancreatic β cell proliferation as a compensatory response. This, in turn, leads to hyperinsulinemia that is seen in early stages of type 2 diabetes mellitus and metabolic syndrome. Efferent vagal signals to the pancreas modulate insulin secretion and pancreatic β cell mass [191–193]. Mice lacking the M3 muscarinic acetylcholine receptor in pancreatic β cells showed impaired glucose tolerance and reduced insulin release. In contrast, transgenic mice selectively overexpressing M3 receptors in pancreatic β cells showed enhanced insulin release and increase in glucose tolerance and were resistant to diet-induced glucose intolerance and hyperglycemia suggesting that β cell M3 muscarinic receptors ensure proper insulin release and glucose homeostasis [191]. VMH-lesioned animals not only showed obesity and features of type 2 diabetes mellitus but also had increase in pancreatic weight, DNA content, and DNA synthesis due to proliferation of islet β and acinar cells that was completely inhibited by vagotomy. This suggests that vagal hyperactivity in the form of increased tone of parasympathetic activity produced by VMH lesions stimulated cell proliferation of rat pancreatic β and acinar cells primarily through a cholinergic receptor mechanism [192, 193]. Vagal nerve-mediated insulin hypersecretion and pancreatic β cell proliferation is due to hepatic activation of extracellular regulated kinase (ERK) signaling. Afferent splanchnic and efferent pancreatic vagal nerves play a major role in pancreatic β cell expansion during diet-induced obesity development, in *ob/ob* and streptozotocin-induced diabetic mice [194]. Thus, hepatic ERK activation transmits signals from the liver to the brain that activates the efferent vagus to the pancreas that triggers the pancreatic β cell proliferation. These results indicate that hepatic ERK activation could be useful to trigger pancreatic β cells mass both in type 1 and type 2 diabetes mellitus to regulate plasma glucose levels.

The Gut-Brain-Liver Axis Circuit is Activated by Long-Chain Fatty Acids

The gastrointestinal tract initiates a series of homeostatic mechanisms to regulate plasma glucose levels at near normal levels both during fasting and postprandial periods. Ingested nutrients stimulate the secretion of incretins from the gut that enhance insulin secretion, and initiate a gut-brain-liver axis by responding to small amounts of triglycerides in the duodenum to rapidly increase insulin secretion. Oleic acid (OA, 18:1 ω-9); linoleic acid (LA, 18:2 ω-6) ; α-linolenic acid (ALA, 18:3 ω-3) ; arachidonic acid (AA, 20:4 ω-6); eicosapentaenoic acid (EPA, 20:5 ω-3) ; and docosahexaenoic acid (DHA, 22:6 ω-3) that are cleaved from triglycerides by the gastrointestinal enzymes when given at calorically insignificant amounts markedly and rapidly increased insulin sensitivity [195, 196]. Long-chain fatty acid metabolite called LCFA-CoA is sensed by the intestine, probably by specific receptors that are yet to be indentified; and this lipid sensing in the gut is relayed to the liver such that homeostatic mechanisms are activated to maintain blood glucose homeostasis by enhancing secretion of insulin from the pancreatic β cells by the release of incretins and by the inhibition of gluconeogenesis in the liver. In this scheme sequence of events, brain plays a role through the parasympathetic nervous system, principally by vagus. The LCFA-CoA sensed by the gut signals the brain through the vagus nerve, and then back down the vagal efferent pathway that terminates in the liver (see Fig. 7.5). Though the exact mechanism by which the communication occurs

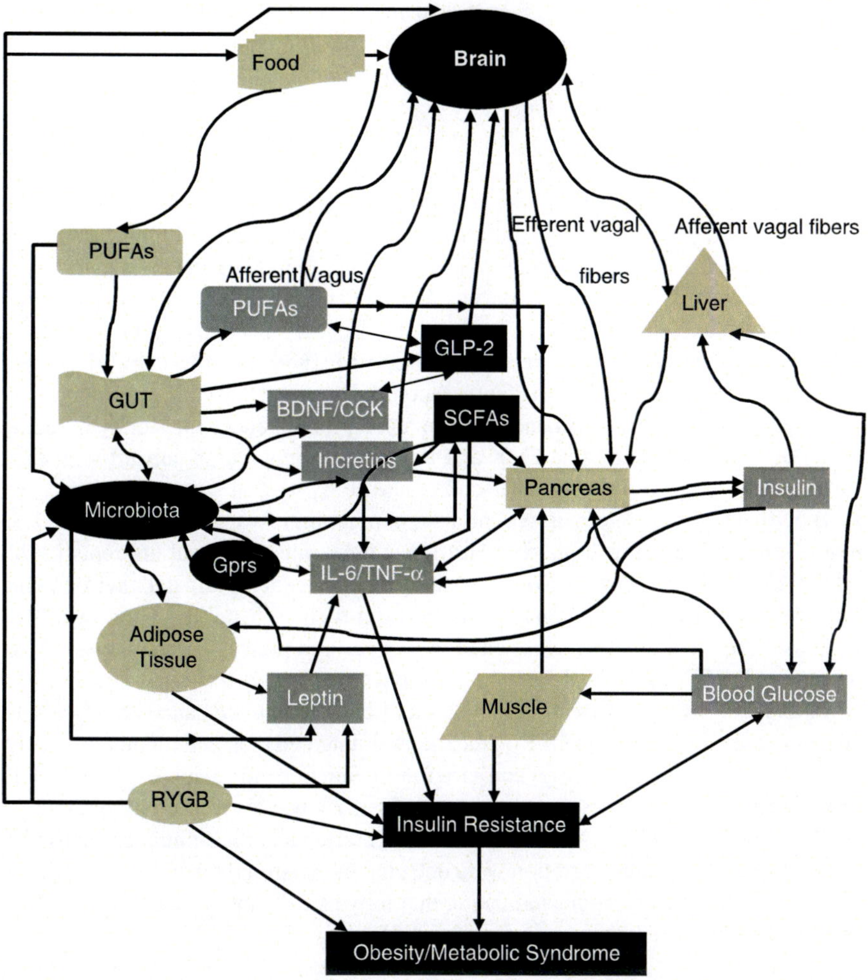

Fig. 7.5 Scheme showing interaction(s) among food, gut, gut hormones and brain and their role in glucose homeostasis/obesity and the metabolic syndrome. PUFAs = Polyunsaturated fatty acids, SFAs = Short-chain fatty acids, Gprs = G-protein coupled receptors. Human intestine contains 10 to 100 trillion bacteria. Fermentation of the dietary fiber is accomplished by members of the *Bacteroidetes* and the *Firmicutes* that generates short-chain fatty acids (SCFAs) such as acetate, propionate and butyrate. The microbial fermentation of the polysaccharides to SCFAs accounts for up to 10% of our daily caloric intake. These SCFAs serve as ligands for Gpr41, a G protein-coupled receptor expressed by a subset of enteroendocrine cells in the gut epithelium. These SCFAs are used as substrates for lipogenesis in the liver that ultimately leads to obesity. SCFAs can activate leukocytes and thus, modify inflammation. Binding of SCFAs to Gpr41 stimulates leptin expression in adipose cells. Pancreatic β-cells express Gpr40. Fatty acids butyrate, oleic acid, α-linolenic acid, γ-linolenic acid, arachidonic acid and docosahexaenoic acid bind to the Gpr40 and stimulate insulin secretion from pancreatic β-cells. RYGB surgery changes the microbiota of the gut and produces changes in the expression of genes in the hypothalamus such that satiety is induced; weight loss occurs, reduces insulin resistance and in some patients cures the metabolic syndrome. It is not known but possible that leptin, TNF-α and IL-6; CCK, PUFAs and SFAs; incretins, BDNF, insulin and glucose may influence the growth of the microbiota in the gut either directly or indirectly. For further details see the text. (This figure is modified from Ref. [182])

between the gut and the vagus is not clear there could exist a role for incretins in this process or other gut hormones/peptides such as cholecystokinin, leptin or brain-derived neurotrophic factor (BDNF) [197]. Intraduodenal perfusion of long chain fatty acids but not medium chain fatty acids reduced calorie intake that could be abolished by inhibition of fat hydrolysis. LCFA perfusion not only resulted in a reduction in calorie intake and food consumption but also a concomitant increase in plasma cholecystokinin (CCK) concentrations. The use of potent and selective CCK-A receptor antagonist completely abolished the satiation effect of LCFAs indicating that the presence of LCFAs in the duodenum would stimulate the release of CCK; CCK then acts on CCK-A receptors that are present on the abdominal vagus. Another possibility is that leptin may have a role in this process since leptin gene expression and immunoreactivity has been reported in the gastric fundus [198] and food ingestion causes rapid stimulation of gastric leptin secretion, an effect that can be reproduced by CCK administration. In experimental animals, leptin enhances the satiety inducing effect of CCK suggesting that CCK and leptin could function in concert with each other to induce satiety and regulate food intake [199].

BDNF, which regulates survival of a subpopulation of vagal sensory neurons, is expressed in developing stomach wall tissues innervated by vagal afferents [200]. BDNF interacts with leptin [201] suggesting that abnormal perinatal environments alter development of vagal sensory innervation of the GI tract by altering BDNF expression that could affect satiety and influence food intake. Thus, LCFAs, CCK, leptin and BDNF influence development of obesity.

In this context, it is interesting to note that LCFAs stimulate pancreatic β cells to secrete insulin by binding to GPR40 receptors situated on their cell membrane [202]. GPR40 functions as a specific receptor for long-chain free fatty acids, especially oleic acid, linoleic acid and docosahexaenoic acid (DHA). It was also reported that OA, LA, ALA, GLA, AA and DHA stimulated insulin secretion and the stimulatory activities of OA and LA on insulin secretion were detected more strongly in high-glucose than in low-glucose concentrations indicating that fatty acids amplify glucose-stimulated insulin secretion from pancreatic β cells [202].

In addition, gut polypeptides secreted in response to food intake, such as glucagon-like peptide-1 (GLP-1) are potent incretin hormones that enhance the glucose-dependent secretion of insulin from pancreatic β cells. The G-protein-coupled receptor, GPR120, which is abundantly expressed in intestine, serves as a receptor for unsaturated long-chain FFAs (free fatty acids; PUFAs). It was noted that stimulation of GPR120 by PUFAs promoted the secretion of GLP-1 *in vitro* and *in vivo*, and increased circulating insulin [203]. These results suggest that there is a cross-talk between GPRs (GPR40 and GPR120), GLP-1 and insulin secretion and imply that GPR120-mediated GLP-1 secretion induced by dietary PUFAs play a significant role in regulation of insulin secretion and control of plasma glucose levels.

Since intraduodenal perfusion of long chain fatty acids (PUFAs) reduced calorie intake, increased plasma cholecystokinin (CCK) concentrations and food ingestion caused leptin secretion from the gastric fundus, it can be deduced that the cross-talk between gut and brain involves CCK, PUFAs, leptin and GLP-1.

Furthermore, as they do in the intestine, LCFA-CoA molecule in the hypothalamus activates neural pathways that increase insulin sensitivity in the liver that, in turn, reduces food intake [204–206]. LCFAs and their metabolite LCFA-CoA function as a signal of nutrient intake and triggers counter-regulatory responses that originate in the hypothalamus and the gut to regulate plasma glucose concentrations. This regulatory system quickly fades in the face of continued ingestion of fat-rich diet [196, 205, 206]. Thus, fat-rich (especially saturated fat and trans-fat-rich) and energy-dense foods promote obesity and diabetes by impairing nutrient-sensing systems that are designed to limit food intake and enhance insulin sensitivity. Diet rich in polyunsaturated fatty acids (PUFAs) [196–206] trigger the gut-brain-liver circuit to limit increases in plasma glucose concentrations and restrict the development of obesity and diabetes whereas the modern diets that are rich in saturated fats not only impair the gut-brain-liver circuit described but also do not function efficiently to restrict food intake and may interfere with the metabolism of dietary essential fatty acids and prevent the formation of their long-chain metabolites and anti-inflammatory molecules such as lipoxins, resolvins, protectins, maresins and nitrolipids. This explains as to why PUFAs (LCFAs) are more beneficial compared to saturated fats.

Previously, I showed that PUFAs protect pancreatic β cells from chemical-induced apoptosis and thus, prevent the development of diabetes mellitus [207–210]. PUFAs (LCFAs) also form precursors to various endocannabinoids that play a role in the pathobiology of obesity and diabetes mellitus [211, 212]. Furthermore, PUFAs (especially ω-3 PUFAs) are anti-inflammatory whereas saturated fats and trans-fats are pro-inflammatory in nature that accounts for low-grade systemic inflammation and insulin resistance seen in obesity, type 2 diabetes mellitus and metabolic syndrome. Insulin resistance seen, thus, may override the acute-insulin sensitizing effects of intestinal LCFAs (PUFAs) reported.

The gut-brain-liver circuit described may also play a role in the improvement in insulin sensitivity, amelioration of diabetes, and decrease in food intake and weight loss reported after bariatric surgery since these beneficial effects are seen much before the weight loss is seen. We showed that there are distinct changes in the hypothalamic neurotransmitters and peptides that could account for some, if not, all of the beneficial actions seen after bariatric surgery [213, 214]. Since there are both anorexigenic and orexigenic molecules secreted by the gut, hypothalamus and adipose tissue, the final response in the form of satiety or hunger and food consumption depends on the balance between these regulatory and counter-regulatory stimuli.

BDNF and Obesity

Hypothalamic neurons play a critical role in energy homeostasis. Brain-derived neurotrophic factor (BDNF) is one factor produced by neuronal cells of the brain that regulates functions of the gut and pancreatic β cells in response to plasma levels of glucose, protein, fatty acids, insulin and leptin. BDNF, present in the hippocampus, cortex, basal forebrain, many nuclei in the brain stem and catecholamine neurons, including dopamine neurons in the substantia nigra, regulates food intake and body weight both in experimental animals and humans. Systemic administration of BDNF decreased nonfasted blood glucose in obese, non-insulin-dependent diabetic C57BLKS-Lepr(db)/lepr(db) (db/db) mice, with a concomitant decrease in body weight. The effects of BDNF on non-fasted blood glucose levels are not caused by decreased food intake but reflect a significant improvement in blood glucose control, an effect that persisted for weeks after cessation of BDNF treatment. BDNF reduced the hepatomegaly present in db/db mice, in association with reduced liver glycogen and reduced liver enzyme activity in serum, supporting the involvement of liver tissue in the mechanism of action for BDNF [215]. Administration of BDNF once or twice per week (70 mg/kg/week) to db/db mice for 3 weeks, significantly reduced blood glucose concentrations and hemoglobin A_{1c}, (HbA_{1c}) suggesting that BDNF not only reduced blood glucose concentrations but also restored systemic glucose balance even with treatment as infrequently as once per week [216]. The therapeutic efficacy of BDNF by gene transfer in mouse models of obesity and type 2 diabetes mellitus is further strengthened by the marked weight loss and alleviation of obesity-associated insulin resistance seen in the study by Cao et al. [217].

Furthermore, involvement of BDNF in type 2 diabetes mellitus in human has also been shown in several studies [218–221]. BDNF is an anorexigenic factor that is expressed in ventromedial hypothalamic (VMH) nuclei. Its concentrations in the brain are regulated by feeding status. Stress hormone corticosterone decreased the expression of BDNF in rats, and led to the atrophy of the hippocampus, suggesting that BDNF has a critical role in obesity and type 2 DM [183, 221–222].

Interaction(s) Among Insulin, Melanocortin, and BDNF

Insulin is an adiposity signal to the brain [101] by its action on the arcuate nucleus (ARC) of hypothalamus that, in turn, controls energy homeostasis [223, 224]. Insulin stimulates the synthesis of proopiomelanocortin (POMC) that acts on melanocortin receptors MC3R and MC4R in hypothalamic nuclei [225]. MC4R has a critical role in regulating energy balance, and mutations in the MC4R gene result in obesity in mice and humans. BDNF is expressed at high levels in the (VMH) where its expression is regulated by nutritional state and by MC4R signaling. Similar to MC4R mutants, mouse mutants that express the BDNF receptor TrkB at a quarter of the normal amount showed hyperphagia and excessive weight gain on higher-fat diets. BDNF infusion into the brain suppressed the hyperphagia and excessive weight gain observed on higher-fat diets in mice with deficient MC4R signaling [221]. These results suggest that MC4R signaling controls BDNF expression in the VMH and support the hypothesis that BDNF is an important effector through which MC4R signaling controls energy balance.

Ghrelin, Leptin, and BDNF

Ghrelin is a gut hormone that increases food intake. It is produced in the epithelial cells lining the fundus of the stomach, with smaller amounts produced in the placenta, kidney, pituitary and hypothalamus. Ghrelin stimulates growth hormone secretion and regulates energy balance by acting on the arcuate nucleus of hypothalamus [226]. Ghrelin increases hunger through its action on hypothalamic feeding centers. Blood concentrations of ghrelin are lowest shortly after consumption of a meal, and then rise during the fast just prior to the next meal. ICV injections of ghrelin increased glucose utilization rate of white and brown adipose tissue and strongly stimulated feeding in rats and increased body weight gain [227]. Factors that regulate ghrelin secretion and action include: plasma glucose, insulin, acetylcholine levels in the brain, leptin, BDNF, and various other neurotransmitters and peptides [227–229].

Leptin, an adiposity hormone produced by the white adipose tissue, stomach, mammary gland, placenta, and skeletal muscle, shows actions similar to that of insulin. It reflects total fat mass especially, subcutaneous fat of the body. Leptin prevents obesity by inhibiting appetite since rodents and patients lacking leptin or functional leptin receptors developed hyperphagia and obesity [230]. Leptin acts on the hypothalamus and other areas in the brain through the neuronal circuits, stimulates the enzymes involved in lipid metabolism, reduces feeding and increases energy expenditure by directly suppressing NPY (neuropeptide Y) and increasing POMC. Arcuate neurons expressing these peptides project to the paraventricular nucleus and lateral hypothalamic area, resulting in increases in corticotrophin-releasing hormone (CRH) and thyrotropin-releasing hormone (TRH) and reductions in MCH and orexins [222, 230]. Leptin acts centrally to increase insulin action in liver. Congenital leptin deficiency decreases brain weight, impairs myelination, and reduces several neuronal and glial proteins [231]. These deficits are partially reversible in adult *Lepob/ob* mice by leptin [231]. Furthermore, there is a close interaction between leptin and BDNF [201].

These evidence suggest that BDNF plays a crucial role in the regulation of appetite, obesity and development of type 2 DM both by its actions on the hypothalamic neurons and modulating the secretion and actions of leptin, ghrelin, insulin, NPY, melanocortin, serotonin, dopamine and other neuropeptides, neurotransmitters, and gut hormones. Hence, selective delivery BDNF to hypothalamus is useful in the management of obesity, type 2 diabetes mellitus and metabolic syndrome as shown recently [217].

Obesity and Type 2 Diabetes Mellitus Are Inflammatory Conditions

It is evident from the preceding discussion that obesity is a low-grade systemic inflammatory condition [34, 48, 49, 77, 101, 164, 177, 179, 182, 183, 222] and is frequently associated with insulin resistance, hyperinsulinemia, hypertension, hyperlipidemia, and coronary heart disease (CHD). Perilipins, whose concentrations are increased in obesity [48, 69], also have pro-inflammatory actions. Increase in intramyocellular lipid (IMCL), common in obesity, is associated with enhanced levels of inflammatory markers [48], and its decrease with diet control and exercise reduces the levels of inflammatory indices [232].

Plasma levels of C-reactive protein (CRP), TNF-α, and IL-6, markers of inflammation, are elevated in subjects with obesity, insulin resistance, essential hypertension, type 2 diabetes, and CHD both before and after the onset of these diseases [232–237]. Overweight children and adults showed a direct correlation between the degree of adiposity and plasma CRP levels. A strong relation between elevated CRP levels and cardiovascular risk factors: fibrinogen, and HDL cholesterol was also reported.

Increased expression of IL-6 in adipose tissue and its release into the circulation is responsible for elevated CRP concentrations since IL-6 enhances the production of CRP in the liver. Overweight and obese subjects have significantly higher serum levels of TNF-α levels compared to lean subjects. Weight reduction and/or exercise decrease serum concentrations of TNF-α. The negative correlation observed between plasma TNF-α and HDL cholesterol, glycosylated hemoglobin, and serum insulin concentrations explain why CHD is more frequent in obese compared to healthy or lean subjects [232].

BDNF and Inflammation

Since low-grade systemic inflammation occurs in obesity and type 2 diabetes mellitus and BDNF is involved in their pathobiology, it is anticipated that BDNF may modulate inflammation. Peripheral inflammation induced an increased expression of BDNF mRNA which was mediated by nerve growth factor (NGF) in the dorsal root ganglion (DRG). Significant increases in the percentage of BDNF-immunoreactive (IR) neuron profiles in the L5 dorsal root ganglion and marked elevation in the expression of BDNF-IR terminals in the spinal dorsal horn were observed following peripheral tissue inflammation produced by an intraplantar injection of Freund's adjuvant into the rat paws suggesting that peripheral tissue inflammation induces an increased BDNF synthesis in the dorsa root ganglion and an elevated anterograde transport of BDNF to the spinal dorsal horn [87]. Similar to nerve growth factor (NGF) even BDNF might have a role in inflammation and hyperalgesia as supported by the observation that after 2 h of induction of bladder inflammation there were significant increases in levels of NGF, BDNF and neurotrophin-3 mRNAs. The rapid elevation of NGF and BDNF and neurotrophin-3 corresponding to the sensory and reflex changes during bladder inflammation [238] suggests that these neurotrophic factors have a role in the inflammatory response.

In the bronchoalveolar lavage (BAL) fluid from patients with asthma after segmental allergen provocation, a significant increase in the neurotrophins NGF, BDNF, and neurotrophin-3 was noted suggesting that neurotrophins could play a role in inflammation and airway hyperresponsiveness in allergic bronchial asthma [239]. BDNF has potent effects on neuronal survival and plasticity during development and after injury. Activated human T cells, B cells, monocytes, and, in particular, T helper: T_{H1} and T_{H2}-type $CD4^+$ T cell lines that are specific for myelin autoantigens such as myelin basic protein or myelin oligodendrocyte glycoprotein secrete bioactive BDNF upon antigen stimulation. BDNF immunoreactivity is demonstrable in inflammatory infiltrates in the brain of patients with acute disseminated encephalitis and multiple sclerosis, indicating that in the nervous system, inflammatory infiltrates may have a neuroprotective effect [240]. Thus, BDNF and other neurotrophins have two functions: to protect the brain neurons from inflammatory events [241, 242] whereas in the respiratory tract, peripheral nerves and urinary bladder may function as pro-inflammatory molecules [243–245]. It is noteworthy that BDNF is not only present in brain neurons but also in several other tissues such as salivary glands, stomach, duodenum, ileum, colon, lung, heart, liver, pancreas, kidney, oviduct, uterus, bladder, and monocytes and eosinophils [246–248]. BDNF is involved in other inflammatory diseases such as rheumatological conditions [249–251], myocardial injury in the ageing heart [252], inflammatory bowel disease [253, 254], atopic dermatitis [255] and other conditions. Since, BDNF is present in many tissues and in some tissues/organs BDNF appears to induce inflammation, caution need to be exercised when the use of BDNF in the clinic is contemplated.

Another factor that seems to play a significant role in the pathobiology of obesity is gut bacteria. Recent studies revealed that the type of gut bacteria is different in the lean and obese subjects. It has been suggested that certain gut bacteria digest complex carbohydrates present in the diet and thus, increase the availability of energy to the individual that, in turn, contributes to the development of obesity simply because more energy is being extracted from the diet. A brief description of the gut bacteria and their role in obesity is discussed below. An attempt is also made to integrate the possible role of gut bacteria in low-grade systemic inflammation seen in obesity.

Gut Bacteria and Obesity

The food that is ingested needs to be digested, assimilated and this, in turn contributes to the total amount of calories that are available to the human body. This indicates that factors that modulate the digestive process and assimilation could impact human body weight. Hence, it is no surprise that human gut bacteria play a role in obesity. Trillions of bacteria collectively termed as the microbiota reside in the human gastrointestinal tract and have been shown to play a role in the pathobiology of obesity.

Gut Flora

The microbiota of the human gut is dominated by the *Firmicutes* and *Bacteroidetes.* Both these phyla of bacteria are benign, although a few are pathogenic. The *Firmicutes* is the largest bacterial phylum containing more than 250 genera. Some of the genera in the *Firmicutes* phyla include: *Lactobacillus, Mycoplasma, Bacillus, and Clostridium*. There are variations in the phylum. For instance, the *Clostridium* species are obligate anaerobes, whereas members of the *Bacillus* form spores and many of them are obligate aerobes. *Streptococcus pyogenes* that causes infections in the humans is also a member of the *Firmicutes* phylum. In contrast to the *Firmicutes*, the *Bacteroidetes* contain about 20 genera and *Bacteroidetes thetaiotaomicron* is the most abundant organism in this group. *Bacteroidetes* are obligate anaerobes and are benign inhabitants of the human gut. These *Bacteroidetes* are opportunistic pathogens and cause disease especially following intestinal surgery or perforation of the gut [256, 257]. It is likely that there could be many more unidentified gut bacteria that may have a role in human obesity.

Gut Bacteria Are Different in the Lean and Obese

In obese humans, the predominant gut bacteria are the *Firmicutes*. When obese individuals lost weight, the proportion of *Firmicutes* became more like that of lean individuals [256, 257]. The *Firmicutes* are rich in enzymes that break down hard to digest dietary polysaccharides leading to their digestion and absorption and so the host could become obese. When microbiota from the obese animals was transferred to the lean, mice given the microbiota from obese mice extracted more calories from their food and gained weight, suggesting that gut microflora play a role in the development of obesity [258, 259].

Gut microbial-community composition was found to be inherited from mothers, and compared with lean mice and regardless of kinship, *ob/ob* animals showed a 50% reduction in the abundance of *Bacteroidetes* and a proportional increase in *Firmicutes* [260], confirming the previous observations [256–259] that leanness and obesity are associated with specific gut microbiota.

Germ free (GF) mice do not develop obesity induced by western-style, high- fat, and sugar-rich diet. When adult GF mice were conventionalized (i.e. the cecal content of 8-week old conventionally-raised mouse that contain their microbiota were given to 7–10 week old GF mouse) showed 60% increase in body fat, insulin resistance and hyperleptinemia within 14 days of conventionalization, suggesting that gut microbiota influence the development of obesity [261]. The lean phenotype seen in germ-free mice has been attributed to increased skeletal muscle and liver levels of phosphorylated AMP-activated protein kinase (AMPK) and its downstream targets involved in fatty acid oxidation and elevated levels of PGC-1α (peroxisomal proliferator-activated receptor coactivator) that increase fatty acid metabolism. In contrast, GF knockout mice lacking fasting-induced adipose factor (Fiaf), a circulating lipoprotein lipase inhibitor whose expression is normally selectively suppressed in the gut epithelium by the gut microbiota and hence, are not protected from diet-induced obesity. The GF *Fiat*$^{-/-}$ animals exhibited similar levels of phosphorylated AMPK as their wild-type littermates in liver and gastrocnemius muscle, but showed reduced expression of PGC-1α and enzymes involved in fatty acid oxidation that accounted for their propensity to develop diet-induced obesity [262].

Bacterial populations from gut of genetically lean and obese pigs fed a low- or high-fiber diet (0% or 50% alfalfa meal respectively) revealed that the total bacterial culture counts in rectal samples declined 56% and 63% in lean and obese animals respectively after feeding the high- fiber diet. The number of cellulolytic bacteria in rectal samples of lean-genotype pigs fed the high-fiber diet increased; however these increases were not seen in the obese pigs [263]. These data confirm that high-fiber diet (that helps in reducing obesity) is beneficial, in part, since it is able to enhance cellulolytic bacterial content in the gut, especially in the lean animals. It is likely that the high-fiber diet fed animals showed an increase in *Bacteroidetes* and a proportional decrease in *Firmicutes* but this needs to be confirmed.

Gut Bacteria and GPR41

Gut bacteria may influence the development of obesity, in part, by altering the expression of Gpr41, a G protein-coupled receptor expressed by a subset of enteroendocrine cells in the gut epithelium. Gpr41 plays a key role in microbial-host communication circuit. Short chain fatty acids and their products formed as a result of microbial fermentation of dietary polysaccharides interact with Gpr41 leading to an increase in the production of enteroendocrine cell-derived hormones such as PYY. PYY increases absorption of short chain fatty acids which are used as substrates for lipogenesis in the liver that ultimately leads to obesity [264]. Thus, gut bacteria could (a) enhance digestion of complex polysaccharides, (b) increase the formation of short chain fatty acids that interact with Gpr41, (c) increase the production of PYY that enhances the absorption of fatty acids, and thus, (d) augment lipogenesis in the liver. These events ultimately cause obesity. These chain of events also suggest the complex nature of interaction among diet, dietary fiber, gut microbiota, gut hormones, absorption of digested food, and obesity [222].

The short chain fatty acids (SCFAs), including acetate, propionate, and butyrate, that are produced at high concentration by gut bacteria are absorbed into the bloodstream that is facilitated by PYY. These short chain fatty acids have the ability to activate leukocytes, particularly neutrophils. The orphan G protein-coupled receptors, GPR41 and GPR43, are the receptors for these short chain fatty acids. Propionate was the most potent agonist for both GPR41 and GPR43. Acetate was more selective for GPR43, whereas butyrate and isobutyrate were more active on GPR41. Both GPR41 and GPR43 were found to be coupled to inositol 1,4,5-trisphosphate formation, intracellular Ca^{2+} release, ERK1/2 activation, and inhibition of cAMP accumulation. The expression profile of GPR41 is found in a number of tissues, the expression of GPR43 is highly selective in leukocytes, particularly polymorphonuclear cells. This implies that whenever = short chain fatty acids are formed in adequate amounts by the gut bacteria, it could recruit polymorphonuclear leukocytes and activate them by binding to their GPR43 receptors [265]. Such an activation of leukocytes by the short-chain fatty acids produced by the gut bacteria may initiate an inflammatory process that could explain the relationship between diet, gut bacteria, and low-grade systemic inflammation seen in obesity.

It was also reported that human GPR43 is highly expressed not only in human neutrophils but also in human monocytes. Short-chain fatty acids induced robust calcium flux in human neutrophils, but not in human monocytes. These short-chain fatty acids induced human monocytes to release PGE_2, an effect that was enhanced in the presence of lipopolysaccharide (LPS). Furthermore, short-chain fatty acids specifically inhibited constitutive monocyte chemotactic protein-1 (MCP-1) production and LPS-induced interleukin-10 (IL-10) production in human monocytes and polymorphonuclear leukocytes without affecting the secretion of other cytokines and chemokines. In addition, short-chain fatty acids inhibited LPS-induced production of TNF-α and IFN-γ in human leukocytes. In an in vivo study, it was also noticed that short-chain fatty acids and LPS induced PGE_2 production by intraplantar injection into rat paws [266]. These results suggest that short-chain fatty acids may show pro- or anti-inflammatory actions depending on the presence or absence of LPS. For example, high fat diet is known to enhance LPS absorption by altering the permeability of the gut [267]. Further support to the relationship between gut microbiota and inflammation is derived from the observation that in human ulcerative colitis and other colitic diseases there is a change in "healthy" microbiota such as Bifidobacterium and Bacteriodes, and a concurrent reduction in short-chain fatty acids. It was reported that short-chain fatty acids-GPR43 interactions profoundly affect inflammatory responses. Stimulation of GPR43 by short-chain fatty acids leads to resolution of the inflammatory response that could be due to increased production of lipoxins, resolvins, protectins, maresins and nitrolipids. For example, GPR43-deficient mice showed exacerbated or unresolving inflammation in models of colitis, arthritis and asthma that seemed to be due to increased production of inflammatory mediators by Gpr432/2 immune cells, and increased immune cell recruitment. Germ-free mice, which are devoid of bacteria and express little or no short-chain fatty acids, showed a dysregulation of inflammatory responses in the form of enhanced myeloperoxidase production [268]. Furthermore, short-chain fatty acids stimulated leptin expression in both a mouse adipocyte cell line and mouse adipose tissue in primary culture. Acute oral administration of short-chain fatty acids increased circulating leptin levels in mice [269]. These evidences suggest that short-chain fatty acids produced by the gut microbiota, which are ligands of GPR41 influence leptin levels in the plasma and hyperleptinemia is a feature of obesity and metabolic syndrome. This also suggests that intestinal bacteria could ultimately influence obesity development through the Gpr pathway. In addition, Gpr40 are also expressed by neuronal cells in the brain, which indicates that dietary content of fatty acids and those produced by the gut bacteria may have actions on hypothalamic neurons and thus, participate in the regulation of food intake, satiety and glucose homeostasis through central actions.

It is clear from the preceding discussion that obesity is associated with enhanced intestinal permeability and metabolic endotoxaemia. It has been shown that a selective increase of Bifidobacterium spp. reduces the impact of high-fat diet-induced metabolic endotoxaemia and inflammatory disorders. It is interesting to note that *ob/ob* mice treated with prebiotics exhibited lower plasma LPS and cytokines IL-1α, TNF-α, IL-6 and IL-8 and a decreased hepatic expression of inflammatory and oxidative stress markers; showed a lower intestinal permeability and improved tight-junction integrity compared to controls. Furthermore, prebiotic treatment increased the endogenous intestinotrophic proglucagon-derived peptide (GLP-2) production whereas the GLP-2 antagonist abolished these effects. On the other hand, pharmacological GLP-2 treatment decreased gut permeability, systemic and hepatic inflammatory phenotype associated with obesity [270]. These results suggest that a selective gut microbiota change controls and increases endogenous GLP-2 production, and consequently improves gut barrier functions by a GLP-2-dependent mechanism, resulting in improvement of gut barrier functions during obesity and diabetes, implying that GLP-2 could form a useful target to develop newer therapeutic strategies in obesity.

Diet, Low-Grade Systemic Inflammation, and Obesity

A positive correlation was observed between plasma LPS concentration and fat and energy intake in a study of 1015 subjects. In a multivariate analysis, endotoxemia was independently associated with energy intake. Mice fed a high-energy diet showed an increase in plasma lipopolysaccharide (LPS) and the increase in LPS was more evident in mice fed high-fat diet compared to those that received a high-carbohydrate diet. Fat is a more efficient transporter of bacterial LPS from the gut lumen into the bloodstream [271] that, in turn, could stimulate macrophages and lymphocytes to secrete pro-inflammatory cytokines TNF-α and IL-6. Thus, high-fat diet enhances the proliferation of *Firmicutes;* augments the production of PYY, increases the absorption of LPS and this, in turn, induces low-grade systemic inflammation. It is likely that high-fat diet-induced proliferation of *Firmicutes* may also stimulate gut-associated lymphocytes (GAL) that could release enhanced amounts of TNF-α and IL-6, but this remains to be confirmed.

Gastric Bypass Surgery for Obesity Alters Gut Bacteria and Hypothalamic Factors

The relationship between gut and hypothalamus and obesity is supported by the fact that gastric bypass surgery performed for extreme obesity not only produces significant weight loss and amelioration from type 2 diabetes mellitus and insulin resistance but also changes gut microbiota and the concentrations of hypothalamic neurotransmitters. Following gastric bypass, a large shift in the bacterial population of the gut was noted. *Firmicutes* were dominant in normal-weight and obese individuals but significantly decreased in post-gastric-bypass individuals [272]. Open RYGB surgery produced greater inhibition of innate immunity [273]. This inhibition was not accounted for by phenotypic changes in lymphocytes as assessed by flow cytometry. Microarray analysis of the preoperative and day 2 specimens identified a 20-gene signature that correlated with the surgical approach. These data thus, established further relationship among gut, inflammation and obesity.

Previously, we observed significant decrease in body weight in RYGB rats after operation that was accompanied with a decrease in NPY in arcuate nucleus of hypothalamus, paraventricular nucleus and an increase in α-MSH (melanocyte stimulating hormone) in arcuate and paraventricular nuclei and a concomitant increase in serotonin receptor (5-HT1B receptor) in paraventricular nucleus [274–276]. These results emphasize the interaction among genes, brain, gut and gut bacteria and hormones, and immunocytes in the pathobiology of obesity [179, 182, 222].

Insulin Acts Not only on Peripheral Tissues but Also in the Brain

Insulin signaling has a role in the regulation of food intake, neuronal growth, and differentiation by regulating neurotransmitter release and synaptic plasticity in the central nervous system (CNS). Neuron-specific disruption of the insulin-receptor gene (NIRKO) in mice induces obesity, insulin resistance, hyperinsulinemia, and type 2 diabetes without interfering with brain development [179, 182, 222, 277]. This indicates that a decrease in the number of insulin receptors, defects in the function of insulin receptors, and insulin lack or resistance in the brain leads to the development of obesity and type 2 diabetes mellitus even when pancreatic β-cells are normal. Intraventricular injection of insulin inhibits food intake and the site of insulin action is on the hypothalamic NPY network. Insulin enhances the formation of PUFAs (polyunsaturated fatty acids or long-chain fatty acids: LCFAs), whereas PUFAs augment the action of insulin and the number of insulin receptors. Further, both insulin and PUFAs augment the formation of eNO (endothelial nitric oxide), a potent neurotransmitter that seems to transmit the messages (probably via RBCs that are known to carry NO) from VMH neurons to the pancreatic β-cells and *vice versa* to control insulin secretion. This suggests that maintaining adequate amounts of insulin and insulin receptors in the brain is necessary to control appetite, obesity (BMI), maintain normoglycemia, and control inflammation [179, 182, 222].

These results imply that factors that regulate insulin action in the brain are important in the control of obesity and type 2 diabetes mellitus, this is especially so since hypothalamus is rich in insulin receptors and drugs that specifically bind to insulin receptors in the brain decrease appetite, reduce obesity and plasma glucose levels.

Interaction Between PUFAs and BDNF and its Relationship to Obesity

The role of unsaturated fatty acids in obesity and type 2 diabetes mellitus is supported by the observation that infusion of oleic acid in the third ventricle resulted in marked decline in the plasma insulin concentration and a modest decrease in the plasma glucose concentration [204]. Oleic acid did not alter glucose uptake but suppressed the rate of glucose production and enhanced hepatic insulin action via the activation of K_{ATP} channels in the hypothalamus. Oleic acid also decreased the hypothalamic expression of NPY suggesting that UFAs (unsaturated fatty acids) control food intake via their action on hypothalamic centers. PUFAs have the ability to enhance BDNF production [278, 279] and are capable of modulating the production and actions of neurotransmitters such as serotonin, dopamine, NPY, and MSH that have regulate appetite, satiety and food intake [280, 281]. Thus, dietary PUFAs (or LCFAs) could form complexes with BDNF (since fatty acids bind rather tightly with proteins and peptides) derived from gut and reach the brain to regulate food intake, glucose and insulin production and energy homeostasis. Since PUFAs are present in several tissues including liver, muscle and pancreas, it is possible that local concentrations of PUFAs may regulate the production and action of BDNF. Thus, PUFAs and BDNF could participate in the gut-brain-liver axis (see Fig. 7.5).

Diet, Gut Peptides and Hypothalamic Neurotransmitters in Obesity

It is evident from the preceding discussion that muscle, adipose cells, pancreas and liver and hypothalamic neurons communicate with each other to maintain energy homeostasis both by neural and humoral pathways. Gut peptides: ghrelin, cholecystokinin (CCK), and incretins interact with hypothalamic neurons and signal hunger and satiety sensations via vagal afferent neurons. BDNF present in the duodenum, ileum, colon, liver and pancreas [246] interacts with PUFAs to influence insulin secretion, production of pro-inflammatory cytokines, and glucose homeostasis through vagus. Vagal afferent neurons express both leptin and CCK-1 to influence food intake by reducing meal size and enhancing satiation [282]. It is known that ghrelin and leptin interact with each other to regulate energy homeostasis and metabolism [283]. Ghrelin significantly increased NPY and AGRP mRNA expression in hypothalamus [284], suggesting that ghrelin and NPY interact with each other. Ghrelin facilitates both cholinergic and tachykininergic excitatory pathways through the vagus nerve [285]. Thus, sympathetic and parasympathetic (especially vagus) nerves carry messages from the peripheral tissues and pancreatic β cells to the hypothalamus and *vice versa* to regulate overall energy balance.

Afferent vagus from the liver and efferent sympathetic nerves to adipose tissues regulates energy expenditure, systemic insulin sensitivity, glucose metabolism, and fat distribution between the liver and the periphery, as already discussed above. [190]. Pro-inflammatory cytokine production is regulated by the efferent vagal "cholinergic anti-inflammatory pathway" mediated by acetylcholine (ACh) [286–288], which is both a neurotransmitter and regulator of release and actions of serotonin, dopamine and other neuropeptides [289]; whereas PUFAs (LCFAs) influence ACh release [290, 291], and insulin sensitivity [179, 182, 222, 292–296], suggesting that an interaction(s) exists among these molecules in the regulation of energy homeostasis. Brain insulin resistance exists in peripheral insulin resistance, especially in regions subserving appetite and reward [296]; and exercise enhanced the sensitivity of hypothalamus to the actions of leptin and insulin and the appetite-suppressive actions of exercise are mediated by the hypothalamus [297]. Exercise also increases brain BDNF levels [298].

Unsaturated fatty acids, fatty acid synthase inhibitors, leptin and insulin decrease plasma insulin and glucose concentrations and suppress hypothalamic NPY and the rate of glucose production by activating K_{ATP} channels in the hypothalamus [300–304]. Fatty acid synthase inhibitors induced increase in malonyl-CoA mediates nutrient-stimulated insulin secretion in the pancreatic β cell. Concentrations of malonyl-CoA also serve as a fuel status signal in the hypothalamic neurons. Hypothalamic neuronal PUFA content modulates the expression of NPY [305] and thus, regulates food intake. Hence, regulation of ATP-sensitive K^+ channels could be a common pathway by which nutrients modulate neuronal sensing of fuels. Exercise prevents and helps in the management of obesity and type 2 diabetes mellitus by (a) enhancing energy expenditure, (b) increasing brain BDNF levels [298], (c) decreasing plasma and pancreatic β cell content of IL-6 and TNF-α [306, 307], (d) increasing parasympathetic tone [308], (e) increasing the utilization of PUFAs, and (f) serving as an anti-inflammatory vehicle. In summary, obesity is associated with low-grade systemic inflammation and needs to be managed by adopting several measures that should include: diet control, consumption of increased amounts of PUFAs (especially n-3) and dietary fiber and moderate exercise. In future, perhaps, drugs based on BDNF, brain insulin receptor binding small molecules, and fatty acids that selectively target hypothalamic neurons.

References

[1] Spalding KL, Arner E, Westermark PO, Bernard S, Buchholz BA, Bergmann O, Blomqvist L, Hoffstedt J, Naslund E, Britton T, Concha H, Hassan M, Ryden M, Frisen J, Arner P (2008) Dynamics of fat cell turnover in humans. Nature 453:783–787

[2] Folsom AR, Stevens J, Schreiner PJ, McGovern PG (1998) Body mass index, waist/hip ratio, and coronary heart disease incidence in African Americans and whites. Atherosclerosis risk in communities study investigators. Am J Epidemiol 148:1187–1194

[3] Hill JO, Peters JC (1998) Environmental contributions to the obesity epidemic. Science 280:1371–1374

[4] Ko TC, Chan J, Chan A, Wong P, Hui S, Chow F, Tong S, Chan C (2007) Doubling over ten years of central obesity in Hong Kong Chinese working men. Chin Med J (Engl) 120:1151–1154

[5] Misra A, Misra R, Wijesuriya M, Banerjee D (2007) The metabolic syndrome in South Asians: continuing escalation & possible solutions. Indian J Med Res 125:345–354
[6] Burke V, Beilin LJ, Dunbar D (2001) Family lifestyle and parental body mass index as predictors of body mass index in Australian children: a longitudinal study. Int J Obes Relat Metab Disord 25:147–157
[7] Davis MM, McGonagle K, Schoeni RF, Stafford F (2008) Grandparental and parental obesity influences on childhood overweight: implications for primary care practice. J Am Board Fam Med 21:549–554
[8] Pereira M, Kartashov AI, Ebbeling CB, Van Horn L, Slattery ML, Jacobs DR Jr, Ludwig DS (2005) Fast-food habits, weight gain, and insulin resistance (the CARDIA study): 15-year prospective analysis. Lancet 365:36–42
[9] Bes-Rastrollo M, Sanchez-Villegas A, Gomez-Gracia E, Martinez JA, Pajares RM, Martinez-Gonzalez MA (2006) Predictors of weight gain in a Mediterranean cohort: the Seguimiento Universidad de Navarra Study. Am J Clin Nutr 83:362–370
[10] Stender S, Dyerberg J, Astrup A (2007) Fast food: unfriendly and unhealthy. Int J Obes (Lond) 31:887–890
[11] Eckel RH (1997) Obesity and heart disease. Circulation 96:3248–3250
[12] Mokdad AH, Serdula MK, Dietz WH, Bowman BA, Marks JS, Koplan JP (2000) The continuing epidemic of obesity in the United States. JAMA 284:1650–1651
[13] Carek PJ, Sherer JT, Carson DS (1997) Management of obesity: medical treatment options. Am Fam Physician 55:551–558
[14] Schulte H, Cullen P, Assmann G (1999) Obesity, mortality and cardiovascular disease in the Münster heart study (PROCAM). Atherosclerosis 144:199–209
[15] Hillier TA, Fosse S, Balkau B, Simon D, Eschwege E, Fagot-Campagna A (2006) Weight, the metabolic syndrome, and coronary heart disease in type 2 diabetes: associations among a national French sample of adults with diabetes-the ENTRED study. J Cardiometab Syndr 1:318–325
[16] Must A, Spadano J, Coakley EH, Field AE, Colditz G, Dietz WH (1999) The disease burden associated with overweight and obesity. JAMA 282:1523–1529
[17] Schultz LO, Schoeller DA (1994) A compilation of total daily energy expenditures and body weights in healthy adults. Am J Clin Nutr 60:676–681
[18] Kimm SY, Glynn NW, Aston CE, Damcott CM, Poehlman ET, Daniels SR, Ferrell RE (2002) Racial differences in the relation between uncoupling protein genes and resting energy expenditure. Am J Clin Nutr 75:714–719
[19] Yanovski JA, Diament AL, Sovik KN, Nguyen TT, Li H, Sebring NG, Warden CH (2000) Associations between uncoupling protein 2, body composition, and resting energy expenditure in lean and obese African American, white, and Asian children. Am J Clin Nutr 71:1405–1420
[20] Kovacs P, Lehn-Stefan A, Stumvoll M, Bogardus C, Baier LJ (2003) Genetic variation in the human winged helix/forkhead transcription factor gene FOXC2 in Pima Indians. Diabetes 52:1292–1295
[21] Ruige JB, Ballaux DP, Funahashi T, Mertens IL, Matsuzawa Y, Van Gaal LF (2005) Resting metabolic rate is an important predictor of serum adiponectin concentrations: potential implications for obesity-related disorders. Am J Clin Nutr 82:21–25
[22] Speakman JR, Rance KA, Johnstone AM (2008) Polymorphisms of the FTO gene are associated with variation in energy intake, but not energy expenditure. Obesity (Silver Spring) 16:1961–1965
[23] Krakoff J, Ma L, Kobes S, Knowler WC, Hanson RL, Bogardus C, Baier LJ (2008) Lower metabolic rate in individuals heterozygous for either a frameshift or a functional missense MC4R variant. Diabetes 57:3267–3272
[24] Gomez-Ambrosi J, Catalan V, Diez-Caballero A, Martinez-Cruz LA, Gil MJ, Garcia-Foncillas J, Cienfuegos JA, Salvador J, Mato JM, Frohbeck G (2004) Gene expression profile of omental adipose tissue in human obesity. FASEB J 18:215–217

[25] Flegal KM, Graubard BI, Williamson DF, Gail MH (2005) Excess deaths associated with underweight, overweight, and obesity. JAMA 293:1861–1867
[26] Bjorntorp P (1991) Adipose tissue distribution and function. Int J Obes 15(Suppl 2):67–81
[27] Krotkiewski M, Bjorntorp P, Sjostrom L, Smith U (1983) Impact of obesity on metabolism in men and women. Importance of regional adipose tissue distribution. J Clin Invest 72:1150–1162
[28] Pederson O, Hjollund E, Lindskov HO (1982) Insulin binding and action on fat cells from young healthy females and males. Am J Physiol 243:E158–E167
[29] Rebuffe-Scrive M, Enk L, Crona N, Lonnroth P, Abrahamsson L, Smith U, Bjorntorp P (1985) Fat cell metabolism in different regions in women. Effect of menstrual cycle, pregnancy, and lactation. J Clin Invest 75:1973–1976
[30] Orel M, Lichnovska R, Gwozdziewiczova S, Zlamalova N, Klementa I, Merkunova A, Hrebicek J (2004) Gender differences in tumor necrosis factor alpha and leptin secretion from subcutaneous and visceral fat tissue. Physiol Res 53:501–505
[31] Das UN (2003) Sex differences in the number of adipose cells. XX vs. XY 1:132–133
[32] Montague CI, Prins JB, Sanders L, Zhang J, Sewter CP, Digby J, Byrne CD, O'Rahilly S (1998) Depot-related gene expression in human subcutaneous and omental adipocytes. Diabetes 47:1384–1391
[33] Vidal-Puig AJ, Considine RV, Jimenez-Linan M, Werman A, Pories WJ, Caro JF, Flier JS (1997) Peroxisome proliferator-activated receptor gene expression in human tissues. Effects of obesity, weight loss, and regulation by insulin and glucocorticoids. J Clin Invest 99:2416–2422
[34] Ramos EJB, Xu Y, Romanova I, Middleton F, Chen C, Quinn R, Inui A, Das UN, Meguid MM (2003) Is obesity an inflammatory disease? Surgery 134:329–335
[35] Krssak M, Falk Petersen K, Dresner A, DiPietro L, Vogel SM, Rothman DL, Roden M, Shulman GI (1999) Intramyocellular lipid concentrations are correlated with insulin sensitivity in humans: a 1H NMR spectroscopy study. Diabetologia 42:113–116
[36] Jacob S, Machann J, Rett K, Brechtel K, Volk A, Renn W, Maerker E, Matthaei S, Schick F, Claussen CD, Haring HU (1999) Association of increased intramyocellular lipid content with insulin resistance in lean nondiabetic offspring of type 2 diabetic subjects. Diabetes 48:1113–1119
[37] Perseghin G, Scifo P, DeCobelli F, Pagliato E, Battezzati A, Arcelloni C, Vanzulli A, Testolin G, Pozza G, Del Maschio A, Luzi L (1999) Intramyocellular triglyceride content is a determinant of in vivo insulin resistance in humans: a 1H-13C nuclear magnetic resonance spectroscopy assessment in offspring of type 2 diabetic parents. Diabetes 48:1600–1606
[38] Forouchi NG, Jenkinson G, Thomas EL, Mullick S, Mierisova S, Bhonsle U, McKeigue PM, Bell JD (1999) Relation of triglyceride stores in skeletal muscle cells to central obesity and insulin sensitivity in European and south Asian men. Diabetologia 42:932–935
[39] Sinha S, Rathi M, Misra A, Kumar V, Kumar M, Jagannathan NR, Pandey RM, Dwivedi M, Luthra K (2005) Subclinical inflammation and soleus muscle intramyocellular lipids in healthy Asian Indian males. Clin Endocrinol (Oxf) 63:350–355
[40] Brasaemle DL, Rubin B, Harten IA, Gruia-Gray J, Kimmel AR, Londos C (2000) Perilipin A increases triacylglycerol storage by decreasing the rate of triacylglycerol hydrolysis. J Biol Chem 275:38486–38493
[41] Clifford GM, Londos C, Kraemer FB, Vernon RG, Yeaman SJ (2000) Translocation of hormone-sensitive lipiase and perilipin upon lipolytic stimulation of rat adipocytes. J Biol Chem 275:5011–5015
[42] Dvorak AM, Morgan E, Schleimer RP, Ryeom SW, Lichetenstein LM, Weller PF (1992) Ultrastructural immunogold localization of prostaglandin endoperoxide synthase (cyclooxygenase) to non-membrane-bound cytoplasmic lipid bodies in human lung mast cells, alveolar macrophages, type II pneumocytes, and neutrophils. J Histochem Cytochem 40:759–769
[43] Yu W, Bozza PT, Tzizik DM, Gray JP, Cassara J, Dvoak AM, Weller PF (1998) Co-compartmentalization of MAP kinases and cytosolic phospholipiase A2 at cytoplasmic arachidonate-rich lipid bodies. Am J Pathol 152:759–769

[44] Weller PF, Dvorak AM (1994) Lipid bodies: intracellular sites for eicosanoid formation. J Allergy Clin Immunol 94:1151–1156
[45] Meadows JW, Pitzer B, Brockman DE, Myatt L (2005) Expression and localizatin of adipophilin and perilipin in human fetal membranes: association with lipid bodies and ezymes involved in prostaglandin synthesis. J Endocrinol Metab 90:2344–2350
[46] Beckman M (2006) Great balls of fat. Science 311:1232–1234
[47] He J, Goodpaster BH, Kelley DE (2004) Effects of weight loss and physical activity on muscle lipid content and droplet size. Obes Res 12:761–769
[48] Das UN (2002) Obesity, metabolic syndrome X, and inflammation. Nutrition 18:430–432
[49] Das UN (2003) Metabolic syndrome X is common in Indians: but, why and how? J Assoc Physicians India 51:987–998
[50] Das UN (2004) Antiinflammatory nature of exercise. Nutrition 20:323–326
[51] Das UN (2006) Exercise and inflammation. Eur Heart J 27:1385–1386
[52] Das UN (2002) A perinatal strategy to prevent coronary heart disease. Nutrition 19:1022–1027
[53] Albert MA, Glynn RJ, Ridker PM (2003) Plasma concentration of C-reactive protein and the calculated Framingham coronary heart disease risk score. Circulation 108:161–165
[54] van der Meer IM, de Maat MPM, Hak AE et al (2002) C-reactive protein predicts progression of atherosclerosis measured as various sites in the arterial tree. The Rotterdam study. Stroke 33:2750–2755
[55] Luc G, Bard J-M, Juhan-Vague I et al (2003) C-reactive protein, interleukins-6, and fibrinogen as predictors of coronary heart disease. The PRIME study. Arterioscler Thromb Vasc Biol 23:1255–1261
[56] Engstrom G, Hedblad B, Stavenow L, Lind P, Janzon L, Lindgarde F (2003) Inflammation-sensitive plasma proteins are associated with future weight gain. Diabetes 52:2097–2101
[57] Mosca L (2002) C-reactive protein-to screen or not to screen. N Engl J Med 347:1615–1617
[58] Castell JV, Gomez-Lechon MJ, David M, Horano T, Kishimoto T, Heinrich PC (1988) Recombinant human interleukins-6 (IL-6/BSF-2/HSF) regulates the synthesis of acute phase proteins in human hepatocytes. FEBS Lett 232:347–350
[59] Barzilay JI, Abraham L, Heckbert SR, Cushman M, Kuller LH, Resnick HE, Tracy RP (2001) The relation of markers of inflammation to the development of glucose disorders in the elderly. Diabetes 50:2384–2389
[60] Kim MJ, Yoo KH, Park HS, Chung SM, Jin CJ, Lee Y, Shin YG, Chung CH (2005) Plasma adiponectin and insulin resistance in Korean type 2 diabetes mellitus. Yonsei Med J 46:42–50
[61] Ridker PM, Buring JE, Cook NR, Rifai N (2003) C-reactive protein, the metabolic syndrome, and risk of incident cardiovascular events. Circulation 107:391–397
[62] Liu S, Manson JE, Buring JE, Stampfer MJ, Willett WC, Ridker PM (2002) Relation between a diet with a high glycemic load and plasma concentrations of high-sensitivity C-reactive protein in middle-aged women. Am J Clin Nutr 75:492–498
[63] Pepys MB, Hirshfiled GM (2003) C-reactive protein: a crticial update. J Clin Invest 111:1805–1812
[64] Griselli M, Herbert J, Hutchinson WL, Taylor KM, Sohail M, Krausz T, Pepys MB (1999) C-reactive protein and complement are important mediators of tissue damage in acute myocardial infarction. J Exp Med 190:1733–1739
[65] Gill R, Kemp JA, Sabin C, Pepys MB (2004) Human C-reactive protein increases cerebral infarct size after middle cerebral artery occlusion in adult rats. J Cereb Blood Flow Metab 24:1214–1218
[66] Pepys MB, Hirschfield GM, Tennent GA, Gallimore JR, Kahan MC, Bellotti V et al (2006) Targeting C-reactive protein for the treatment of cardiovascular disease. Nature 440:1217–1221
[67] Das UN (2006) Clinical laboratory tools to diagnose inflammation. Adv Clin Chem 41:189–229
[68] Hirschfield GM, Kahan MC, Hutchinson WL, Sabin CA, Benson GM, Dhillon AP, Tennent GA, Pepys MB (2005) Transgenic human C-reactive protein is not proatherogenic in apolipoprotein E-deficient mice. Proc Natl Acad Sci U S A 102:8309–8314

[69] Park HS, Park JY, Yu R (2005) Relationship of obesity and visceral adiposity with serum concentrations of CRP, TNF-alpha and IL-6. Diabetes Res Clin Pract 69:29–35
[70] Ghanim H, Aljada A, Hofmeyer D, Syed T, Mohanty P, Dandona P (2004) Circulating mononuclear cells in the obese are in a proinflammatory state. Circulation 110:1564–1571
[71] Kopp HP, Kopp CW, Festa A, Krzyzanowska K, Kriwanek S, Minar E, Roka R, Schernthaner G (2003) Impact of weight loss on inflammatory proteins and their association with the insulin resistance syndrome in morbidly obese patients. Arterioscler Thromb Vasc Biol 23:1042–1047
[72] Giugliano G, Nicoletti G, Grella E, Giugliano F, Esposito K, Scuderi N, D'Andrea F (2004) Effect of liposuction on insulin resistance and vascular inflammatory markers in obese women. Br J Plast Surg 57:190–194
[73] Ryan AS, Nicklas BJ (2004) Reductions in plasma cytokine levels with weight loss improve insulin sensitivity in overweight and obese postmenopausal women. Diabetes Care 27:1699–1705
[74] Fischer CP, Berntsen A, Perstrup LB, Eskildsen P, Pedersen BK (2007) Plasma levels of interleukin-6 and C-reactive protein are associated with physical inactivity independent of obesity. Scand J Med Sci Sports 17:580–587
[75] Nishimura S, Manabe I, Nagasaki M, Eto K, Yamashita H, Ohsugi M, Otsu M, Hara K, Ueki K, Sugiura S, Yoshimura K, Kadowaki T, Nagai R (2009) CD8+ effector T cells contribute to macrophage recruitment and adipose tissue inflammation in obesity. Nat Med 15:914–920
[76] Winer S, Chan Y, Paltser G, Truong D, Tsui H, Bahrami J, Dorfman R, Wang Y, Zielenski J, Mastronardi F, Maezawai Y, Drucker DJ, Engelman E, Winer D, Dosch H-M (2009) Normalization of obesity-associated insulin resistance through immunotherapy. Nat Med 15:921–930
[77] Feuerer M, Herrero L, Cipolletta D, Naaz A, Wong J, Nayer A, Lee J, Goldfine AB, Benoist C, Shoelson S, Mathis D (2009) Lean, but not obese, fat is enriched for a unique population of regulatory T cells that affect metabolic parameters. Nat Med 15:930–939
[78] Zheng Y, Rudensky AY (2007) Foxp3 in control of the regulatory T cell lineage. Nat Immunol 8:457–462
[79] Sakaguchi S, Yamaguchi T, Nomura T, Ono M (2008) Regulatory T cells and immune tolerance. Cell 133:775–787
[80] Lumeng CN, Maillard I, Saltiel AR (2009) T-ing up inflammation in fat. Nat Med 15:846–847
[81] Ecker J, Liebisch G, Englmaier M, Grandl M, Robenek H, Schmitz G (2010) Induction of fatty acid synthesis is a key requirement for phagocytic differentiation of human monocytes. Proc Natl Acad Sci U S A 107:7817–7822
[82] Pearce EL, Walsh MC, Cejas PJ, Harms GM, Shen H, Wang L-S, Jones RG, Choi Y (2009) Enhancing CD8 T-cell memory by modulating fatty acid metabolism. Nature 460:103–108
[83] Araki K, Turner AP, Shaffer VO, Gangappa S, Keller SA, Bachmann MF, Larsen CP, Ahmed R (2009) mTOR regulates memory CD8 T-cell differentiation. Nature 460:109–113
[84] Prlic M, Bevan MJ (2009) A metabolic switch to memory. Nature 460:41–42
[85] Rydberg EK, Salomonsson L, Hultén LM, Norén K, Bondjers G, Wiklund O, Björnheden T, Ohlsson BG (2003) Hypoxia increases 25-hydroxycholesterol-induced interleukin-8 protein secretion in human macrophages. Atherosclerosis 170:245–252
[86] Wang N, Tabas I, Winchester R, Ravalli S, Rabbani LE, Tall A (1996) Interleukin 8 is induced by cholesterol loading of macrophages and expressed by macrophage foam cells in human atheroma. J Biol Chem 271:8837–8842
[87] Persson J, Nilsson J, Lindholm MW (2006) Cytokine response to lipoprotein lipid loading in human monocyte-derived macrophages. Lipids Health Dis 5:17
[88] Park EJ, Suh M, Thomson B, Thomson AB, Ramanujam KS, Clandinin MT (2005) Dietary ganglioside decreases cholesterol content, caveolin expression and inflammatory mediators in rat intestinal microdomains. Glycobiology 15:935–942
[89] Martín-Ventura JL, Blanco-Colio LM, Gómez-Hernàndez A, Muoz-García B, Vega M, Serrano J, Ortega L, Hernàndez G, Tuón J, Egido J (2005) Intensive treatment with atorvastatin

reduces inflammation in mononuclear cells and human atherosclerotic lesions in one month. Stroke 36:1796–1780

[90] Bulcão C, Ribeiro-Filho FF, Saudo A, Roberta Ferreira SG (2007) Effects of simvastatin and metformin on inflammation and insulin resistance in individuals with mild metabolic syndrome. Am J Cardiovasc Drugs 7:219–224

[91] De Caterina R, Cybulsky MA, Clinton SK, Gimbrone MA Jr, Libby P (1995) Omega-3 fatty acids and endothelial leukocyte adhesion molecules. Prostaglandins Leukot Essent Fatty Acids 52:191–195

[92] De Caterina R, Cybulsky MI, Clinton SK, Gimbrone MA Jr, Libby P (1994) The omega-3 fatty acid docosahexaenoate reduces cytokine-induced expression of proatherogenic and proinflammatory proteins in human endothelial cells. Arterioscler Thromb 14:1829–1836

[93] De Caterina R, Libby P (1996) Control of endothelial leukocyte adhesion molecules by fatty acids. Lipids 31(Suppl):S57–S63

[94] Weber C, Erl W, Pietsch A, Danesch U, Weber PC (1995) Docosahexaenoic acid selectively attenuates induction of vascular cell adhesion molecule-1 and subsequent monocytic cell adhesion to human endothelial cells stimulated by tumor necrosis factor-alpha. Arterioscler Thromb Vasc Biol 15:622–628

[95] Baumann KH, Hessel F, Larass I, Müller T, Angerer P, Kiefl R, von Schacky C (1999) Dietary omega-3, omega-6, and omega-9 unsaturated fatty acids and growth factor and cytokine gene expression in unstimulated and stimulated monocytes. A randomized volunteer study. Arterioscler Thromb Vasc Biol 19:59–66

[96] Shahbakhti H, Watson RE, Azurdia RM, Ferreira CZ, Garmyn M, Rhodes LE (2004) Influence of eicosapentaenoic acid, an omega-3 fatty acid, on ultraviolet-B generation of prostaglandin-E2 and proinflammatory cytokines interleukin-1 beta, tumor necrosis factor-alpha, interleukin-6 and interleukin-8 in human skin in vivo. Photochem Photobiol 80:231–235

[97] Rudolph TK, Rudolph V, Edreira MM, Cole MP, Bonacci G, Schopfer FJ, Woodcock SR, Franek A, Pekarova M, Khoo NK, Hasty AH, Baldus S, Freeman BA (2010) Nitro-fatty acids reduce atherosclerosis in apolipoprotein E-deficient mice. Arterioscler Thromb Vasc Biol 30:938–945

[98] Das UN (2000) Beneficial effect(s) of n-3 fatty acids in cardiovascular diseases: but, why and how? Prostaglandins Leukot Essent Fatty Acids 63:351–362

[99] Das UN (2005) A defect in the activity of Delta6 and Delta5 desaturases may be a factor predisposing to the development of insulin resistance syndrome. Prostaglandins Leukot Essent Fatty Acids 72:343–350

[100] Das UN (2004) Long-chain polyunsaturated fatty acids interact with nitric oxide, superoxide anion, and transforming growth factor-beta to prevent human essential hypertension. Eur J Clin Nutr 58:195–203

[101] Das UN (2001) Is obesity an inflammatory condition? Nutrition 17:953–966

[102] Tappia PS, Man WJ, Grimble RF (1995) Influence of unsaturated fatty acids on the production of tumour necrosis factor and interleukin-6 by rat peritoneal macrophages. Mol Cell Biochem 143:89–98

[103] Hayashi N, Tashiro T, Yamamori H, Takagi K, Morishima Y, Otsubo Y, Sugiura T, Furukawa K, Nitta H, Nakajima N, Suzuki N, Ito I (1998) Effects of intravenous omega-3 and omega-6 fat emulsion on cytokine production and delayed type hypersensitivity in burned rats receiving total parenteral nutrition. JPEN J Parenter Enteral Nutr 22:363–367

[104] Dooper MM, van Riel B, Graus YM, M'Rabet L (2003) Dihomo-gamma-linolenic acid inhibits tumour necrosis factor-alpha production by human leucocytes independently of cyclooxygenase activity. Immunology 110:348–357

[105] Wallace FA, Miles EA, Calder PC (2003) Comparison of the effects of linseed oil and different doses of fish oil on mononuclear cell function in healthy human subjects. Br J Nutr 89:679–689

[106] Bhattacharya A, Sun D, Rahman M, Fernandes G (2007) Different ratios of eicosapentaenoic and docosahexaenoic omega-3 fatty acids in commercial fish oils differentially alter

pro-inflammatory cytokines in peritoneal macrophages from C57BL/6 female mice. J Nutr Biochem 18:23–30
[107] Kuroda M, Endo A (1976) Inhibition of in vitro cholesterol synthesis by fatty acids. Biochim Biophys Acta 486:70–81
[108] El-Sohemy A, Archer MC (1997) Regulation of mevalonate synthesis in rat mammary glands by dietary n-3 and n-6 polyunsaturated fatty acids. Cancer Res 57:3685–3687
[109] Das UN (2001) Essential fatty acids as possible mediators of the actions of statins. Prostaglandins Leukot Essent Fatty Acids 65:37–40
[110] Das UN (2007) A defect in the activity of Delta6 and Delta5 desaturases may be a factor in the initiation and progression of atherosclerosis. Prostaglandins Leukot Essent Fatty Acids 76:251–268
[111] Das UN (2008) Essential fatty acids and their metabolites could function as endogenous HMG-CoA reductase and ACE enzyme inhibitors, anti-arrhythmic, anti-hypertensive, anti-atherosclerotic, anti-inflammatory, cytoprotective, and cardioprotective molecules. Lipids Health Dis 7:37
[112] Puustinen T, Webber SE, Nicolaou KC, Haeggström J, Serhan CN, Samuelsson B (1986) Evidence for a 5(6)-epoxytetraene intermediate in the biosynthesis of lipoxins in human leukocytes. Conversion into lipoxin A by cytosolic epoxide hydrolase. FEBS Lett 207:127–132
[113] Sok DE, Phi TS, Jung CH, Chung YS, Kang JB (1988) Soybean lipoxygenase-catalyzed formation of lipoxin A and lipoxin B isomers from arachidonic acid via 5,15-dihydroperoxyeicosatetraenoic acid. Biochem Biophys Res Commun 153:840–847
[114] Thomas E, Leroux JL, Blotman F, Chavis C (1995) Conversion of endogenous arachidonic acid to 5,15-diHETE and lipoxins by polymorphonuclear cells from patients with rheumatoid arthritis. Inflamm Res 44:121–124
[115] O'Meara YM, Brady HR (1997) Lipoxins, leukocyte recruitment and the resolution phase of acute glomerulonephritis. Kidney Int Suppl 58:S56–S61
[116] Fierro IM, Colgan SP, Bernasconi G, Petasis NA, Clish CB, Arita M, Serhan CN (2003) Lipoxin A4 and aspirin-triggered 15-epi-lipoxin A4 inhibit human neutrophil migration: comparisons between synthetic 15 epimers in chemotaxis and transmigration with microvessel endothelial cells and epithelial cells. J Immunol 170:2688–2694
[117] Carlo T, Levy BD (2008) Chemical mediators and the resolution of airway inflammation. Allergol Int 57:299–305
[118] Kure I, Nishiumi S, Nishitani Y, Tanoue T, Ishida T, Mizuno M, Fujita T, Kutsumi H, Arita M, Azuma T, Yoshida M (2010) Lipoxin A(4) reduces lipopolysaccharide-induced inflammation in macrophages and intestinal epithelial cells through inhibition of nuclear factor-kappaB activation. J Pharmacol Exp Ther 332:541–548
[119] Ye XH, Wu Y, Guo PP, Wang J, Yuan SY, Shang Y, Yao SL (2010) Lipoxin A(4) analogue protects brain and reduces inflammation in a rat model of focal cerebral ischemia reperfusion. Brain Res 1323:174–183
[120] Ryan A, Godson C (2010) Lipoxins: regulators of resolution. Curr Opin Pharmacol 10:166–172
[121] Seki H, Tani Y, Arita M (2009) Omega-3 PUFA derived anti-inflammatory lipid mediator resolvin E1. Prostaglandins Other Lipid Mediat 89:126–130
[122] Levy BD (2010) Resolvins and protectins : natural pharmacophores for resolution biology. Prostaglandins Leukot Essent Fatty Acids 82:327–332
[123] Xu ZZ, Zhang L, Liu T, Park JY, Berta T, Yang R, Serhan CN, Ji RR (2010) Resolvins RvE1 and RvD1 attenuate inflammatory pain via central and peripheral actions. Nat Med 16:592–597
[124] Seki H, Sasaki T, Ueda T, Arita M (2010) Resolvins as regulators of the immune system. ScientificWorldJournal 10:818–831
[125] Serhan CN (2008) Systems approach with inflammatory exudates uncovers novel anti-inflammatory and pro-resolving mediators. Prostaglandins Leukot Essent Fatty Acids 79:157–163

[126] Levy BD, Kohli P, Gotlinger K, Haworth O, Hong S, Kazani S, Israel E, Haley KJ, Serhan CN (2007) Protectin D1 is generated in asthma and dampens airway inflammation and hyperresponsiveness. J Immunol 178:496–502

[127] Duffield JS, Hong S, Vaidya VS, Lu Y, Fredman G, Serhan CN, Bonventre JV (2006) Resolvin D series and protectin D1 mitigate acute kidney injury. J Immunol 177:5902–5911

[128] Serhan CN (2006) Novel chemical mediators in the resolution of inflammation: resolvins and protectins. Anesthesiol Clin 24:341–364

[129] Serhan CN, Yang R, Martinod K, Kasuga K, Pillai PS, Porter TF, Oh SF, Spite M (2009) Maresins: novel macrophage mediators with potent antiinflammatory and proresolving actions. J Exp Med 206:15–23

[130] Brochu M, Tchernof A, Dionne IJ, Sites CK, Eltabbakh GH, Sims EA, Poehlman ET (2001) What are the physical characteristics associated with a normal metabolic profile despite a high level of obesity in postmenopausal women? J Clin Endocrinol Metab 86:1020–1025

[131] Masuzaki H, Paterson J, Shinyama H et al (2001) A Transgenic model of visceral obesity and the metabolic syndrome. Science 294:2166–2170

[132] Bujalska IJ, Kumar S, Hewison M, Stewart PM (1999) Differentiation of adipose stromal cells: the roles of glucocorticoids and 11beta-hydroxysteroid dehydrogenase. Endocrinology 140:3188–3196

[133] Tomlinson JW, Sinha B, Bujalska I, Hewison M, Stewart PM (2002) Expression of 11beta-hydroxysteroid Dehydrogenase type 1 in adipose tissue is not increased in human obesity. J Clin Endocrinol Metab 87:5630–5635

[134] Nair S, Lee YH, Lindsay RS, Walker BR, Tataranni PA, Bogardus C, Baier LJ, Permana PA (2004) 11beta-Hydroxysteroid dehydrogenase type 1: genetic polymorphisms are associated with type 2 diabetes in Pima Indians independently of obesity and expression in adipocyte and muscle. Diabetologia 47:1088–1095

[135] Robitaille J, Brouillette C, Houde A, Despres JP, Tchernof A, Vohl MC (2004) Molecular screening of the 11beta-HSD1 gene in men characterized by the metabolic syndrome. Obes Res 12:1570–1575

[136] Sandeep TC, Andrew R, Homer N, Andrews RC, Smith K, Walker BR (2005) Increased in vivo regeneration of cortisol in adipose tissue in human obesity and effects of the 11beta-hydroxysteroid dehydrogenase type 1 inhibitor carbenoxolone. Diabetes 54:872–879

[137] Morton NM, Paterson JM, Masuzaki H, Holmes MC, Staels B, Fievet C, Walker BR, Flier JS, Mullins JJ, Seckl JR (2004) Novel adipose tissue-mediated resistance to diet-induced visceral obesity in 11beta-hydroxysteroid dehydrogenase type 1-deficient mice. Diabetes 53:931–938

[138] Handoko K, Yang K, Strutt B, Khalil W, Killinger D (2000) Insulin attenuates the stimulatory effects of tumor necrosis factor alpha on 11beta-hydroxysteroid dehydrogenase 1 in human adipose stromal cells. J Steroid Biochem Mol Biol 72:163–168

[139] Tomlinson JW, Moore J, Cooper MS, Bujalska I, Shahmanesh M, Burt C, Strain A, Hewison M, Stewart PM (2001) Regulation of expression of 11beta-hydroxysteroid dehydrogenase type 1 in adipose tissue: tissue-specific induction by cytokines. Endocrinology 142:1982–1989

[140] Han J, Thompson P, Beutler B (1990) Dexamethasone and pentoxifylline inhibit endotoxin-induced cachectin/tumor necrosis factor synthesis at separate points in the signaling pathway. J Exp Med 172:391–394

[141] Prusty D, Park BH, Davis KE, Farmer SR (2002) Activation of MEK/ERK signaling promotes adipogenesis by enhancing peroxisome proliferator-activated receptor gamma (PPARgamma) and C/EBPalpha gene expression during the differentiation of 3T3-L1 preadipocytes. J Biol Chem 277:46226–46232

[142] Davis KE, Moldes M, Farmer SR (2004) The forkhead transcription factor FoxC2 inhibits white adipocyte differentiation. J Biol Chem 279:42453–42461

[143] Han J, Thompson P, Beutler B (1990) Dexamethasone and pentoxifylline inhibit endotoxin-induced cachectin/tumor necrosis factor synthesis at separate points in the signaling pathway. J Exp Med 172:391–394

[144] Degawa-Yamauchi M, Moss KA, Bovenkerk JE, Shankar SS, Morrison CL, Lelliott CJ, Vidal-Puig A, Jones R, Considine RV (2005) Regulation of adiponectin expression in human adipocytes: effects of adiposity, glucocorticoids, and tumor necrosis factor alpha. Obes Res 13:662–669

[145] Das UN (2001) Is insulin an anti-inflammatory molecule? Nutrition 17:409–413

[146] Dandona P, Aljada A, Mohanty P (2002) The anti-inflammatory and potential anti-atherogenic effect of insulin: a new paradigm. Diabetologia 45:924–930

[147] Wang Y, Sullivan S, Trujillo M, Lee M-J, Schneider SH, Brolin RE, Kang YH, Werber Y, Greenberg AS, Fried SK (2003) Perilipin expression in human adipose tissues: effects of severe obesity, gender, and depot. Obes Res 11:930–936

[148] Wu C-C, Croxtall JD, Perretti M, Bryant CE, Thiemermann C, Flower RJ, Vane JR (1995) Lipocortin 1 mediates the inhibition by dexamethasone of the induction by endotoxin of nitric oxide synthase in the rat. Proc Natl Acad Sci U S A 92:3473–3477

[149] Lasa M, Abraham SM, Boucheron C, Saklatvala J, Clark AR (2002) Dexamethasone causes sustained expression of mitogen-activated protein kinase (MAPK) phosphatase 1 and phosphatase-mediated inhibition of MAPK p38. Mol Cell Biol 22:7802–7811

[150] Crafford LJ, Wilder RL, Ristimaki AP, Sano M, Remmers EF, Epps HR, Hla T (1994) Cyclooxygenase-1 and -2 expression in rheumatoid arthritis synovial tissues. J Clin Invest 93:109–1101

[151] Radomski MW, Palmer RMJ, Moncada S (1990) Glucocorticoids inhibit the expression of an inducible, but not the constitutive, nitric oxide synthase in vascular endothelial cells. Proc Natl Acad Sci U S A 87:10043–10047

[152] Calandra T, Bernhagen J, Metz CN, Spiegel LA, Bacher M, Donnelly T, Cerami A, Bucala R (1995) MIF as a glucocorticoid-induced modulator of cytokine production. Nature 377:68–71

[153] Hoeck WG, Ramesha CS, Chang DJ, Fan N, Heller RA (1993) Cytoplasmic phospholipase A2 activity and gene expression are stimulated by tumor necrosis factor: dexamethasone blocks the inducible synthesis. Proc Natl Acad Sci U S A 90:4475–4479

[154] Kunicka JE, Talle MA, Denhardt GH, Brown M, Prince LA, Goldstein G (1993) Immunosuppression by glucocorticoids: inhibition of production of multiple lymphokines by in vivo administration of dexamethasone. Cell Immunol 149:39–49

[155] Salvemini D, Seibert K, Masferrer JL, Misko TP, Currie MG, Needleman P (1994) Endogenous nitric oxide enhances prostaglandin production in a model of renal inflammation. J Clin Invest 93:1940–1947

[156] Chavis C, Vachier I, Bousquet J, Godard P, Chanez P (1998) Generation of eicosanoids from 15(S)-hydroxyeicosatetraenoic acid in blood monocytes from steroid-dependent asthmatic patients. Biochem Pharmacol 56:535–541

[157] Hashimoto A, Murakami Y, Kitasato H, Hayashi I, Endo H (2007) Glucocorticoids co-interact with lipoxin A4 via lipoxin A4 receptor (ALX) up-regulation. Biomed Pharmacother 61:81–85

[158] Zaitsu M, Hamasaki Y, Tsuji K, Matsuo M, Fujita I, Aoki Y, Ishii E, Kohashi O (2003) Dexamethasone accelerates catabolism of leukotriene C4 in bronchial epithelial cells. Eur Respir J 22:35–42

[159] Dworski R, Fitzgerald GA, Oates JA, Sheller JR (1994) Effect of oral prednisone on airway inflammatory mediators in atopic asthma. Am J Respir Crit Care Med 149(4 Pt 1):953–959

[160] Ferrante JV, Ferrante A (2005) Novel role of lipoxygenases in the inflammatory response: promotion of TNF mRNA decay by 15-hydroperoxyeicosatetraenoic acid in a monocytic cell line. J Immunol 174:3169–3172

[161] Ariel A, Chiang N, Arita M, Petasis NA, Serhan CN (2003) Aspirin-triggered lipoxin A4 and B4 analogs block extracellular signal-regulated kinase-dependent TNF-alpha secretion from human T cells. J Immunol 170:6266–6272

[162] Wu SH, Lu C, Dong L, Zhou GP, He ZG, Chen ZQ (2005) Lipoxin A4 inhibits TNF-alpha-induced production of interleukins and proliferation of rat mesangial cells. Kidney Int 68:35–46

[163] Hayakawa M, Ishida N, Takeuchi K, Shibamoto S, Hori T, Oku N, Ito F, Tsujimoto M (1993) Arachidonic acid-selective cytosolic phospholipase A2 is crucial in the cytotoxic action of tumor necrosis factor. J Biol Chem 268:11290–11295
[164] Das UN (2005) Pathophysiology of metabolic syndrome X and its links to the perinatal period. Nutrition 21:762–773
[165] Mohanty P, Hamouda W, Garg R, Aljada A, Ghanim H, Dandona P (2000) Glucose challenge stimulates reactive oxygen species (ROS) generation by leucocytes. J Clin Endocrinol Metab 85:2970–2973
[166] Mohanty P, Ghanim H, Hamouda W, Aljada A, Garg R, Dandona P (2002) Both lipid and protein intakes stimulate increased generation of reactive oxygen species by polymorphonuclear leukocytes and mononuclear cells. Am J Clin Nutr 75:767–772
[167] Houstis N, Rosen ED, Lander ES (2006) Reactive oxygen species have a causal role in multiple forms of insulin resistance. Nature 440:944–948
[168] Fan J, Frey RS, Rahman A, Malik AB (2002) Role of neutrophil NADPH oxidase in the mechanism of tumor necrosis factor-α-induced NF-kB activation and intercellular adhesion molecule-1 expression in endothelial cells. J Biol Chem 277:3404–3411
[169] Das UN (1994) Insulin resistance and hyperinsulinemia: are they secondary to an alteration in the metabolism of essential fatty acids. Med Sci Res 22:243–245
[170] Seeds MC, Jones DF, Chilton FH, Bass DA (1998) Secretory and cytosolic phospholipase A2 are activated during TNF priming of human neutrophils. Biochim Biophys Acta 1389:273–284
[171] Wu T, Ikezono T, Angus CW, Shelhamer JH (1996) Tumor necrosis factor-alpha induces the 85-kDa cytosolic phospholipase A2 gene expression in human bronchial epithelial cells. Biochim Biophys Acta 1310:175–184
[172] Crowl RM, Stoller TJ, Conroy RR, Stoner CR (1991) Induction of phospholipase A2 gene expression in human hepatoma cells by mediators of the acute phase response. J Biol Chem 266:2647–2651
[173] Dandona P, Mohanty P, Ghanim H et al (2001) The suppressive effect of dietary restriction and weight loss in the obese on the generation of reactive oxygen species by leukocytes, lipid peroxidation, and protein carbonylation. J Clin Endocrinol Metab 86:355–362
[174] Pischon T, Hankinson SE, Hotamisligil GS et al (2003) Habitual dietary intake of n-3 and n-6 fatty acids in relation to inflammatory markers among US men and women. Circulation 108:155–160
[175] Das UN (2008) Perinatal nutriiton and obesity. Br J Nutr 99:1391–1392
[176] Das UN (2007) Metabolic syndrome X is a low-grade systemic inflammatory condition with its origins in the perinatal period. Curr Nutr Food Sci 3:277–295
[177] Das UN (2007) Is metabolic syndrome X a disorder of the brain with the initiation of low-grade systemic inflammatory events during the perinatal period? J Nutr Biochem 18:701–713
[178] Adam CL, Findlay PA, Chanet A, Aitken RP, Milne JS, Wallace JM (2008) Expression of energy balance regulatory genes in the developing ovine fetal hypothalamus at midgestation and the influence of hyperglycemia. Am J Physiol Regul Integr Comp Physiol 294:R1895–R1900
[179] Das UN (2002) A perinatal strategy for preventing adult disease: the role of long-chain polyunsaturated fatty acids. Kluwer Academic Publishers, Boston
[180] Barker DJ, Hales CN, Fall CH et al (1993) Type 2 (non-insulin dependent) diabetes mellitus, hypertension, and hyperlipidemia (syndrome X): relation to reduced fetal growth. Diabetologia 36:62–67
[181] Lucas A, Fewtrell MS, Cole TJ (1999) Fetal origins of adult disease-the hypothesis revisited. BMJ 319:245–249
[182] Das UN (2010) Metabolic syndrome Pathophysiology: the role of essential fatty acids. Wiley-Blackwell, Ames
[183] Das UN (2002) Obesity, metabolic syndrome X, and inflammation. Nutrition 18:430–432
[184] Gold RM, Quackenbush PM, Kapatos G (1972) Obesity following combination of rostrolateral to VMH cut and contralateral mammillary area lesion. J Comp Physiol Psychol 79:210–218

[185] King BM, Smith RL, Frohman LA (1984) Hyperinsulinemia in rats with ventromedial hypothalamic lesions: role of hyperphagia. Behav Neurosci 98:152–155
[186] Funahashi T, Shimomura I, Hiraoka H, Arai T, Takahashi M, Nakamura T, Nozaki S, Yamashita S, Takemura K, Tokunaga K et al (1995) Enhanced expression of rat obese (ob) gene in adipose tissues of ventromedial hypothalamus (VMH)-lesioned rats. Biochem Biophys Res Commun 211:469–475
[187] Paes AM, Carniatto SR, Francisco FA, Brito NA, Mathias PC (2006) Acetylcholinesterase activity changes on visceral organs of VMH lesion-induced obese rats. Int J Neurosci 116:1295–1302
[188] Sakaguchi T, Bray GA, Eddlestone G (1988) Sympathetic activity following paraventricular or ventromedial hypothalamic lesions in rats. Brain Res Bull 20:461–465
[189] Cox JE, Powley TL (1981) Prior vagotomy blocks VMH obesity in pair-fed rats. Am J Physiol 240:E573–E583
[190] Uno K, Katagiri H, Yamada T, Ishigaki Y, Ogihara T, Imai J, Hasegawa Y, Gao J, Kaneko K, Iwasaki H, Ishihara H, Sasano H, Inukai K, Mizuguchi H, Asano T, Shiota M, Nakazato M, Oka Y (2006) Neuronal pathway from the liver modulates energy expenditure and systemic insulin sensitivity. Science 312:1656–1659
[191] Gautam D, Han SJ, Duttaroy A, Mears D, Hamdan FF, Li JH, Cui Y, Jeon J, Wess J (2007) Role of the M3 muscarinic acetylcholine receptor in beta-cell function and glucose homeostasis. Diabetes Obes Metab 9(Suppl 2):158–169
[192] Edvell A, Lindstrom P (1998) Vagotomy in young obese hyperglycemic mice: effects on syndrome development and islet proliferation. Am J Physiol 274(6 Pt 1):E1034–E1039
[193] Kiba T, Tanaka K, Hoshino M, Misugi K, Inoue S (1996) Ventromedial hypothalamic lesion-induced vagal hyperactivity stimulates rat pancreatic cell proliferation. Gastroenterology 110:885–893
[194] Imai J, Katagiri H, Yamada T, Ishigaki Y, Suzuki T, Kudo H, Uno K, Hasegawa Y, Gao J, Kaneko K, Ishihara H, Niijima A, Nakazato M, Asano T, Minokoshi Y, Oka Y (2008) Regulation of pancreatic β cell mass by neuronal signals from the liver. Science 322:1250–1254
[195] Thaler JP, Cummings DE (2008) Food alert. Nature 452:941–942
[196] Wang PYT, Caspi L, Lam CKL, Chari M, Li X, Light PE, Gutierrez-Juarez R, Ang M, Schwartz GJ, Lam TKT (2008) Upper intestinal lipids trigger a gut-brain-liver axis to regulate glucose production. Nature 452:1012–1016
[197] Matzinger D, Degen L, Drewe J, Meuli J, Duebendorfer R, Ruckstuhl N, D'Amato M, Rowati L, Beglinger C (2000) The role of long chain fatty acids in regulating food intake and cholecystokinin release in humans. Gut 46:688–693
[198] Bado A, Levasseur S, Attoub S, Kermorgant S, Laigneau JP, Bortoluzzi MN, Moizo L, Lehy T, Guerre-Millo M, LeMarchand-Brustel Y, Lewin MJ (1998) The stomach is a source of leptin. Nature 394:790–793
[199] Barrachina MD, Martinez V, Wang L, Wei JT, Tache Y (1997) Synergistic interaction between leptin and cholecystokinin to reduce short-term food intake in mice. Proc Natl Acad Sci U S A 94:10455–10460
[200] Fox EA, Murphy MC (2008) Factors regulating vagal sensory development: potential role in obesities of developmental origin. Physiol Rev 94:90–104
[201] Komori T, Morikawa Y, Nanjo K, Senba E (2006) Induction of brain-derived neurotrophic factor by leptin in the ventromedial hypothalamus. Neuroscience 139:1107–1115
[202] Itoh Y, Kawamata Y, Harada M et al (2003) Free fatty acids regulate insulin secretion from pancreatic beta cells through GPR40. Nature 422:173–176
[203] Hirasawa A, Tsumaya K, Awaji T et al (2005) Free fatty acids regulate gut incretin glucagon-like peptide-1 secretion through GPR120. Nat Med 11:90–94
[204] Obici S, Feng Z, Morgan K, Stein D, Karkanias G, Rossetti L (2002) Central administration of oleic acid inhibits glucose production and food intake. Diabetes 51:271–275
[205] Obici S, Feng Z, Arduini A, Conti R, Rossetti L (2003) Inhibition of hypothalamic carnitine palmitoyltransferase-1 decreases food intake and glucose production. Nat Med 9:756–761

[206] Lam TK, Pocai A, Gutierrez-Juarez R, Obici S, Bryan J, Aquilar-Bryan L, Schwartz GJ, Rossetti L (2005) Hypothalamic sensing of circulating fatty acids is required for glucose homeostasis. Nat Med 11:320–327
[207] Suresh Y, Das UN (2001) Protective action of arachidonic acid against alloxan-induced cytotoxicity and diabetes mellitus. Prostaglandins Leukot Essent Fatty Acids 64:37–52
[208] Suresh Y, Das UN (2003) Long-chain polyunsaturated fatty acids and chemically-induced diabetes mellitus: effect of ω-6 fatty acids. Nutrition 19:93–114
[209] Suresh Y, Das UN (2003) Long-chain polyunsaturated fatty acids and chemically-induced diabetes mellitus: effect of ω-3 fatty acids. Nutrition 19:213–228
[210] Suresh Y, Das UN (2006) Differential effect of saturated, monounsaturated, and polyunsaturated fatty acids on alloxan-induced diabetes mellitus. Prostaglandins Leukot Essent Fatty Acids 74:199–213
[211] Richard D, Guesdon B, Timofeeva E (2009) The brain endocannabinoid system in the regulation of energy balance. Best Pract Res Clin Endocrinol Metab 23:17–32
[212] Di Marzo V (2008) The endocannabinoid system in obesity and type 2 diabetes. Diabetologia 51:1356–1367
[213] Romanova IV, Ramos EJ, Xu Y, Quinn R, Chen C, George ZM, Inui A, Das UN, Meguid MM (2004) Neurobiologic changes in the hypothalamus associated with weight loss after gastric bypass. J Am Coll Surg 199:887–895
[214] Xu Y, Ramos EJ, Middleton F, Romanova I, Quinn R, Chen C, Das U, Inui A, Meguid MM (2004) Gene expression profiles post Roux-en-Y gastric bypass. Surgery 136:246–252
[215] Tonra JR, Ono M, Liu X, Garcia K, Jackson C, Yancoupoulos GD, Wiegand SJ, Wong V (1999) Brain-derived neurotrophic factor improves blood glucose control and alleviates fasting hyperglycemia in C57BLKS-Lepr(db)/lepr(db) mice. Diabetes 48:588–594
[216] Ono M, Itakura Y, Nonomura T, Nakagawa T, Nakayama C, Taiji M, Noguchi H (2000) Intermittent administration of brain-derived neurotrophic factor ameliorates glucose metabolism in obese diabetic mice. Metabolism 49:129–133
[217] Cao L, Lin E-J D, Cahill MC, Wang C, Liu X, During MJ (2009) Molecular therapy of obesity and diabetes by a physiological autoregulatory approach. Nat Med 15:447–454
[218] Nonomura T, Tsuchida A, Ono-Kishino M, Nakagawa T, Taiji M, Noguchi H (2001) Brain-derived neurotrophic factor regulates energy expenditure through the central nervous system in obese diabetic mice. Int J Exp Diabetes Res 2:201–209
[219] Suwa M, Kishimoto H, Nofuji Y, Nakano H, Sasaki H, Radak Z, Kumagai S (2006) Serum brain-derived neurotrophic factor level is increased and associated with obesity in newly diagnosed female patients with type 2 diabetes mellitus. Metabolism 55:852–857
[220] Krabbe KS, Nielsen AR, Krogh-Madsen R, Plomgaard P, Rasmussen P, Erikstrup C, Fischer CP, Lindegaard B, Petersen AM, Taudorf S, Secher NH, Pilegaard H, Bruunsgaard H, Pedersen BK (2007) Brain-derived neurotrophic factor (BDNF) and type 2 diabetes. Diabetologia 50:431–438
[221] Xu B, Goulding EH, Zang K, Cepoi D, Cone RD, Jones KR, Tecott LH, Reichardt LF (2003) Brain-derived neurotrophic factor regulates energy balance downstream of melanocortin-4 receptor. Nat Neurosci 6:736–742
[222] Das UN (2010) Obesity: genes, gut and environment. Nutrition 26:459–473
[223] Tran PV, Akana SF, Malkovska I, Dallman MF, Parada LF, Ingraham HA (2006) Diminished hypothalamic bdnf expression and impaired VMH function are associated with reduced SF-1 gene dosage. J Comp Neurol 498:637–648
[224] Obici S, Feng Z, Tan J, Liu L, Karkanias G, Rossetti L (2001) Central melanocortin receptors regulate insulin action. J Clin Invest 108:1079–1085
[225] Tamura H, Kamegai J, Shimizu T, Ishii S, Sugihara H, Oikawa S (2002) Ghrelin stimulates GH but not food intake in arcuate nucleus ablated rats. Endocrinology 143:3268–3275
[226] Kamegai Tamura H, Shimizu T, Ishii S, Sugihara H, Wakabayashi I (2001) Chronic central infusion of ghrelin increases hypothalamic neuropeptide Y and Agouti-related protein mRNA levels and body weight in rats. Diabetes 50:2438–2443

[227] Saad MF, Bernaba B, Hwu CM, Jinagouda S, Fahmi S, Kogosov E, Boyadjian R (2002) Insulin regulates plasma ghrelin concentration. J Clin Endocrinol Metab 87:3997–4000
[228] Broglio F, Gottero C, Van Koetsveld P, Prodam F, Destefanis S, Benso A, Gauna C, Hofland L, Arvat E, Van der Lely AJ, Ghigo E (2004) Acetylcholine regulates ghrelin secretion in humans. J Clin Endocrinol Metab 89:2429–2433
[229] Dardennes RM, Zizzari P, Tolle V, Foulon C, Kipman A, Romo L, Iancu-Gontard D, Boni C, Sinet PM, Therese Bluet M, Estour B, Mouren MC, Guelfi JD, Rouillon F, Gorwood P, Epelbaum J (2007) Family trios analysis of common polymorphisms in the obestatin/ghrelin, BDNF and AGRP genes in patients with Anorexia nervosa: association with subtype, body-mass index, severity and age of onset. Psychoneuroendocrinology 32:106–113
[230] Zhang Y, Proenca R, Maffei M, Barone M, Leopold L, Friedman JM (1994) Positional cloning of the mouse obese gene and its human homologue. Nature 372:425–432
[231] Huang Q, Viale A, Picard F, Nahon J, Richard D (1999) Effects of leptin on melanin-concentrating hormone expression in the brain of lean and obese Lep(ob)/Lep(ob) mice. Neuroendocrinology 69:145–153
[232] Das UN (2006) Aberrant expression of perilipins and 11-β-HSD-1 as molecular signatures of metabolic syndrome X in South East Asians. J Assoc Physicians India 54:637–649
[233] Sinha S, Rathi M, Misra A, Kumar V, Kumar M, Jagannathan NR, Pandey RM, Dwivedi M, Luthra K (2005) Subclinical inflammation and soleus muscle intramyocellular lipids in healthy Asian Indian males. Clin Endocrinol (Oxf) 63:350–355
[234] Albert MA, Glynn RJ, Ridker PM (2003) Plasma concentration of C-reactive protein and the calculated Framingham coronary heart disease risk score. Circulation 108:161–165
[235] van der Meer IM, de Maat MPM, Hak AE et al (2002) C-reactive protein predicts progression of atherosclerosis measured as various sites in the arterial tree. The Rotterdam study. Stroke 33:2750–2755
[236] Engstrom G, Hedblad B, Stavenow L, Lind P, Janzon L, Lindgarde F (2003) Inflammation-sensitive plasma proteins are associated with future weight gain. Diabetes 52:2097–2101
[237] Barzilay JI, Abraham L, Heckbert SR, Cushman M, Kuller LH, Resnick HE, Tracy RP (2001) The relation of markers of inflammation to the development of glucose disorders in the elderly. Diabetes 50:2384–2389
[238] Oddiah D, Anand P, McMahon SB, Rattray M (1998) Rapid increase of NGF, BDNF and NT-3 mRNAs in inflamed bladder. Neuroreport 9:1455–1458
[239] Virchow JC, Julius P, Lommatzsch M, Luttmann W, Renz H, Braun A (1998) Neurotrophins are increased in bronchoalveolar lavage fluid after segmental allergen provocation. Am J Respir Crit Care Med 158:2002–2005
[240] Kerschensteiner M, Gallmeier E, Behrens L, Leal VV, Misgeld T, Klinkert WE, Kolbeck R, Hoppe E, Oropeza-Wekerle RL, Bartke I, Stadelmann C, Lassmann H, Wekerle H, Hohlfeld R (1999) Activated human T cells, B cells, and monocytes produce brain-derived neurotrophic factor in vitro and in inflammatory brain lesions: a neuroprotective role of inflammation? J Exp Med 189:865–870
[241] Tabakman R, Lecht S, Sephanova S, Arien-Zakay H, Lazarovici P (2004) Interactions between the cells of the immune and nervous system: neurotrophins as neuroprotection mediators in CNS injury. Prog Brain Res 146:387–401
[242] Makar TK, Trisler D, Sura KT, Sultana S, Patel N, Bever CT (2008) Brain derived neurotrophic factor treatment reduces inflammation and apoptosis in experimental allergic encephalomyelitis. J Neurol Sci 270:70–76
[243] Ricci A, Mariotta S, Saltini C, Falasca C, Giovagnoli MR, Mannino F, Graziano P, Sciacchitano S, Amenta F (2005) Neurotrophin system activation in bronchoalveolar lavage fluid immune cells in pulmonary sarcoidosis. Sarcoidosis Vasc Diffuse Lung Dis 22:186–194
[244] Hahn C, Islamian AP, Renz H, Nockher WA (2006) Airway epithelial cells produce neurotrophins and promote the survival of eosinophils during allergic airway inflammation. J Allergy Clin Immunol 117:787–794
[245] Bennedich Kahn L, Gustafsson LE, Olgart Hoglund C (2008) Brain-derived neurotrophic factor enhances histamine-induced airway responses and changes levels of exhaled nitric oxide in guinea pigs in vivo. Eur J Pharmacol 595:78–83

[246] Lommatzsch M, Braun A, Mannsfeldt A, Botchkarev VA, Botchkarev NV, Paus R, Fischer A, Lewin GR, Renz H (1999) Abundant production of brain-derived neurotrophic factor by adult visceral epithelia. Am J Pathol 155:1183–1193

[247] Rost B, Hanf G, Ohnemus U, Otto-Knapp R, Groneberg DA, Kunkel G, Noga O (2005) Monocytes of allergics and non-allergics produce, store and release the neurotrophins NGF, BDNF and NT-3. Regul Pept 124:19–25

[248] Noga O, Englmann C, Hanf G, Grutzkau A, Kunkel G (2003) The production, storage and release of the neurotrophins nerve growth factor, brain-derived neurotrophic factor and neurotrophin-3 by human peripheral eosinophils in allergics and non-allergics. Clin Exp Allergy 33:649–654

[249] Rihl M, Kruithof E, Barthel C, De Keyser F, Veys EM, Zeidler H, Yu DT, Kuipers JG, Baeten D (2005) Involvement of neurotrophins and their receptors in spondyloarthritis synovitis: relation to inflammation and response to treatment. Ann Rheum Dis 64:1542–1549

[250] del Porto F, Aloe L, Lagana B, Triaca V, Nofroni I, D'Amelio R (2006) Nerve growth factor and brain-derived neurotrophic factor levels in patients with rheumatoid arthritis treated with TNF-alpha blockers. Ann N Y Acad Sci 1069:438–443

[251] Grimsholm O, Guo Y, Ny T, Forsgren S (2008) Expression patterns of neurotrophins and neurotrophin receptors in articular chondrocytes and inflammatory infiltrates in knee joint arthritis. Cells Tissues Organs 188:299–309

[252] Cai D, Holm JM, Duignan IJ, Zheng J, Xaymardan M, Chin A, Ballard VL, Bella JN, Edelberg JM (2006) BDNF-mediated enhancement of inflammation and injury in the aging heart. Physiol Genomics 24:191–197

[253] Johansson M, Norrgard O, Forsgren S (2007) Study of expression patterns and levels of neurotrophins and neurotrophin receptors in ulcerative colitis. Inflamm Bowel Dis 13:398–409

[254] di Mola FF, Friess H, Zhu ZW, Koliopanos A, Bley T, Di Sebastiano P, Innocenti P, Zimmermann A, Buchler MW (2000) Nerve growth factor and Trk high affinity receptor (TrkA) gene expression in inflammatory bowel disease. Gut 46:670–679

[255] Raap U, Werfel T, Goltz C, Deneka N, Langer K, Bruder M, Kapp A, Schmid-Ott, Wedi B (2006) Circulating levels of brain-derived neurotrophic factor correlate with disease severity in the intrinsic type of atopic dermatitis. Allergy 61:1416–1418

[256] Bajzer M, Seeley RJ (2006) Obesity and gut flora. Nature 444:1009–1010

[257] Backhed F, Ley RE, Sonnenburg JL, Peterson DA, Gordon JI (2005) Host-bacterial mutualism in the human intestine. Science 307:1915–1920

[258] Ley RE, Turnbaugh PJ, Klein S, Gordon JI (2006) Microbial ecology: human gut microbes associated with obesity. Nature 444:1022–1023

[259] Turnbaugh PJ, Ley RE, Mahowald MA, Magrini V, Mardis ER, Gordon JI (2006) An obesity-associated gut microbiome with increased capacity for energy harvest. Nature 444:1027–1031

[260] Ley RE, Backhed F, Turnbaugh P, Lozupone CA, Knight RD, Gordon JL (2005) Obesity alters gut microbial ecology. Proc Natl Acad Sci U S A 102:11070–11075

[261] Backhed F, Ding H, Wang T, Hooper LV, Koh GY, Nagy A, Semenkovich CF, Gordon JI (2004) The gut microbiota as an environmental factor that regulates fat storage. Proc Natl Acad Sci U S A 101:15718–15723

[262] Backhed F, Manchester JK, Semenkovich CF, Gordon JI (2007) Mechanisms underlying the resistance to diet-induced obesity in germ-free mice. Proc Natl Acad Sci U S A 104:979–984

[263] Varel VH, Pond WG, Pekas JC, Yen JT (1982) Influence of high-fibre diet on bacterial populations in gastrointestinal tracts of obese- and lean-genotype pigs. Appl Environ Microbiol 44:107–112

[264] Samuel BS, Shaito A, Motoike T, Rey FE, Backhed F, Manchester JK, Hammer RE, Williams SC, Crowley J, Yanagisawa M, Gordon JI (2008) Effects of the gut microbiota on host adiposity are modulated by the short-chain fatty acid binding G protein-coupled receptor, Gpr41. Proc Natl Acad Sci U S A 105:16767–16772

[265] Le Poul E, Loison C, Struyf S, Springael JY, Lannoy V, Decobecq ME, Brezillon S, Dupriez V, Vassart G, Van Damme J, Parmentier M, Detheux M (2003) Functional characterization of human receptors for short chain fatty acids and their role in polymorphonuclear cell activation. J Biol Chem 278:25481–25489

[266] Cox MA, Jackson J, Stanton M, Rojas-Triana A, Bober L, Laverty M, Yang X, Zhu F, Liu J, Wang S, Monsma F, Vassileva G, Maguire M, Gustafson E, Bayne M, Chou CC, Lundell D, Jenh CH (2009) Short-chain fatty acids act as antiinflammatory mediators by regulating prostaglandin E(2) and cytokines. World J Gastroenterol 15:5549–5557

[267] Cani PD, Bibiloni R, Knauf C, Waget A, Neyrinck AM, Delzenne NM, Burcelin R (2008) Changes in gut microbiota control metabolic endotoxemia-induced inflammation in high-fat diet-induced obesity and diabetes in mice. Diabetes 57:1470–1481

[268] Maslowski KM, Vieira AT, Ng A, Kranich J, Sierro F, Yu D, Schilter HC, Rolph MS, Mackay F, Artis D, Xavier RJ, Teixeira MM, Mackay CR (2009) Regulation of inflammatory responses by gut microbiota and chemoattractant receptor GPR43. Nature 461:1282–1287

[269] Xiong Y, Miyamoto N, Shibata K, Valasek MA, Motoike T, Kedzierski RM, Yanagisawa M (2004) Short-chain fatty acids stimulate leptin production in adipocytes through the G protein-coupled receptor GPR41. Proc Natl Acad Sci U S A 101:1045–1050

[270] Cani PD, Possemiers S, Van de Wiele T, Guiot Y, Everard A, Rottier O, Geurts L, Naslain D, Neyrinck A, Lambert DM, Muccioli GG, Delzenne NM (2009) Changes in gut microbiota control inflammation in obese mice through a mechanism involving GLP-2-driven improvement of gut permeability. Gut 58:1091–1103

[271] Amar J, Burcelin R, Ruiavets JB, Cani PD, Fauvel J, Alessi MC, Chamontin B, Ferrieres J (2008) Energy intake is associated with endotoxemia in apparently healthy men. Am J Clin Nutr 87:1219–1223

[272] Zhang H, DiBaise JK, Zuccolo A, Kudrna D, Braidotti M, Yu Y, Parameswaran P, Crowell MD, Wing R, Rittmann BE, Krajmalnik-Brown R (2009) Human gut microbiota in obesity and after gastric bypass. Proc Natl Acad Sci U S A 106:2365–2370

[273] Whitson BA, D'Cunha J, Hoang CD, Wu B, Ikramuddin S, Buchwald H, Panoskalsis-Mortari A, Kratzke RA, Miller JS, Maddaus MA (2009) Minimally invasive versus open Roux-en-Y gastric bypass: effect on immune effector cells. Surg Obes Relat Dis 5:181–193

[274] Das UN (2008) Is metabolic syndrome X a disorder of the brain? Curr Nutr Food Sci 4:73–108

[275] Middleton FA, Ramos EJB, Xu Y, Diab H, Zhao X, Das UN, Meguid MM (2004) Application of genomic technologies: DNA microarrays and metabolic profiling of obesity in the hypothalamus and in subcutaneous fat. Nutrition 20:14–25

[276] Meguid M, Ramos EJB, Suzuki S, Xu Y, George ZM, Das UN, Hughes K, Quinn R, Chen C, Marx W, Cunningham PRG (2004) A surgical rat model of human Roux-en-Y gastric bypass. J Gastrointest Surg 8:621–630

[277] Bruning JC, Gautam D, Burks DJ, Gillette J, Schubert M, Orban PC, Klein R, Krone W, Muller-Wieland D, Kahn CR (2000) Role of brain insulin receptor in control of body weight and reproduction. Science 289:2122–2125

[278] Wu A, Ying Z, Gomez-Pinilla F (2004) Dietary omega-3 fatty acids normalize BDNF levels, reduce oxidative damage, and counteract learning disability after traumatic brain injury in rats. J Neurotrauma 21:1457–1467

[279] Rao JS, Ertley RN, Lee HJ, DeMar JC Jr, Arnold JT, Rapoport SI, Bazinet RP (2007) n-3 polyunsaturated fatty acid deprivation in rats decreases frontal cortex BDNF via a p38 MAPK-dependent mechanism. Mol Psychiatry 12:36–46

[280] Innis SM, de La Presa Owens S (2001) Dietary fatty acid composition in pregnancy alters neurite membrane fatty acids and dopamine in newborn rat brain. J Nutr 131:118–122

[281] de La Presa Owens S, Innis SM (2000) Diverse, region-specific effects of addition of arachidonic and docosahexaenoic acids to formula with low or adequate linoleic and alpha-linolenic acids on piglet brain monoaminergic neurotransmitters. Pediatr Res 48:125–130

[282] Peters JH, Simasko SM, Ritter RG (2006) Modulation of vagal afferent excitation and reduction of food intake by leptin and cholecystokinin. Physiol Behav 89:477–485

[283] Ueno N, Dube MG, Inui A, Kalra PS, Kalra SP (2004) Leptin modulates orexigenic effects of ghrelin and attenuates adiponectin and insulin levels and selectively the dark-phase feeding as revealed by central leptin gene therapy. Endocrinology 145:4176–4184

[284] Goto M, Arima H, Watanabe M, Hayashi M, Banno R, Sato I, Nagasaki H, Oiso Y (2006) Ghrelin increases neuropeptide Y and agouti-related peptide gene expression in the arcuate nucleus in rat hypothalamic organotypic cultures. Endocrinology 147:5102–5109

[285] Bassil AK, Dass NB, Sanger GJ (2006) The prokinetic-like activity of ghrelin in rat isolated stomach is mediated via cholinergic and tachykininergic motor neurons. Eur J Pharmacol 544:146–152

[286] Borovikova LV, Ivanova S, Zhang M, Yang H, Botchkina GI, Watkins LR, Wang H, Abumrad N, Eaton JW, Tracey KJ (2000) Vagus nerve stimulation attenuates the systemic inflammatory response to endotoxin. Nature 405:458–462

[287] Bernik TR, Friedman SG, Ochani M, DiRaimo R, Ulloa L, Yang H, Sudan S, Czura CJ, Ivanova SM, Tracey KJ (2002) Pharmacological stimulation of the cholinergic antiinflammatory pathway. J Exp Med 195:781–788

[288] Wang H, Yu M, Ochani M, Amella CA, Tanovic M, Susarla S, Li JH, Wang H, Yang H, Ulloa L et al (2003) Nicotinic acetylcholine receptor $\alpha 7$ subunit is an essential regulator of inflammation. Nature 421:384–387

[289] Hersi AI, Kitaichi K, Srivastava LK, Gaudreau P, Quirion R (2000) Dopamine D-5 receptor modulates hippocampal acetylcholine release. Brain Res Mol Brain Res 76:336–340

[290] Das UN (2002) Alcohol consumption and risk of dementia. Lancet 360:490

[291] Minami M, Kimura S, Endo T, Hamaue N, Horafuji M, Togashi H, Matsumoto M, Yaoshioka M, Saito H, Watanabe S, Kobayashi T, Okuyama H (1997) Dietary docosahexaenoic acid increases cerebral acetylcholine levels and improves passive avoidance performance in stroke-prone spontaneously hypertensive rats. Pharmacol Biochem Behav 58:1123–1129

[292] Borkman M, Stolien LH, Pan DA, Jenkins AB, Chisholm DJ, Campbell LV (1993) The relation between insulin sensitivity and the fatty acid composition of skeletal muscle phospholipids. N Engl J Med 328:238–244

[293] Ginsberg BH, Jabour J, Spector AA (1982) Effect of alterations in membrane lipid unsaturation on the properties of the insulin receptor of Ehrlich ascites cells. Biochim Biophys Acta 690:157–164

[294] Somova L, Moodley K, Channa ML, Nadar A (1999) Dose-dependent effect of dietary fish-oil (n-3) polyunsaturated fatty acids on in vivo insulin sensitivity in rat. Methods Find Exp Clin Pharmacol 21:275–278

[295] Huang Y-J, Fang VS, Chou Y-C, Kwok C-F, Ho L-T (1997) Amelioration of insulin resistance and hypertension in a fructose-fed rat model with fish oil supplementation. Metabolism 46:1252–1258

[296] Mori Y, Murakawa Y, Katoh S, Hata S, Yokoyama J, Tajima N, Ikeda Y, Nobukata H, Ishikawa T, Shibutani Y (1997) Influence of highly purified eicosapentaenoic acid ethyl ester on insulin resistance in the Otsuka Long-Evans Tokushima fatty rat, a model of spontaneous non-insulin dependent diabetes mellitus. Metabolism 46:1458–1464

[297] Anthony K, Reed LJ, Dunn JT, Bingham E, Hopkins D, Marsden PK, Amiel SA (2006) Attenuation of insulin-evoked responses in brain networks controlling appetite and reward in insulin resistance. The cerebral basis for impaired control of food intake in metabolic syndrome? Diabetes 55:2986–2992

[298] Flores MBS, Fernandes MFA, Ropello ER, Faria MC, Ueno M, Velloso LA, Saad MJA, Carvalheira JBC (2006) Exercise improves insulin and leptin sensitivity in hypothalamus of Wistar rats. Diabetes 55:2554–2561

[299] Stranahan AM, Lee K, Martin B, Maudsley S, Golden E, Cutler RG, Mattson MP (2009) Voluntary exercise and caloric restriction enhance hippocampal dendritic spine density and BDNF levels in diabetic mice. Hippocampus 19:951–961

[300] Spanswick D, Smith MA, Groppi VE, Logan SD, Ashford MLJ (1997) Leptin inhibits hypothalamic neurons by activation of ATP-sensitive potassium channels. Nature 390:521–525

[301] Harvey J, McKay NG, Walker KS, Van der Kay J, Downes CP, Ashford MLJ (2000) Essential role of phosphoinositide 3-kinase in leptin-induced kATP channel activation in the rat CRI-GI insulinoma cell line. J Biol Chem 275:4660–4669
[302] Spanswick D, Smith MA, Mirshamsi S, Routh VH, Ashford MLJ (2000) Insulin activates ATP-sensitive K+ channels in hypothalamic neuronsof lean, but not obse rats. Nat Neurosci 3:757–762
[303] Loftus TM, Jaworsky DE, Frehywot GL, Townsend CA, Ronnett GV, Lane MD, Kuhajda FP (2000) Reduced food intake and body weight in mice treated with fatty acid synthase inhibitors. Science 288:2379–2381
[304] McGarry GD, Mannaert GP, Foster DW (1977) A possible role for malonyl-CoA in the regulation of hepatic fatty acid oxidation and ketogenesis. J Clin Invest 60:265–270
[305] Ruderman NB, Saha AK, Vavvas D, Witters LA (1999) Malonyl-CoA fuel sensing and insulin resistance. Am J Physiol 276:E1–E18
[306] Ramos EJB, Suzuki S, Meguid MM, Laviano A, Sato T, Chen C, Das UN (2004) Changes in hypothalamic neuropeptide Y and monoaminergic system in tumor-bearing rats: pre- and post-tumor resection and at death. Surgery 136:270–276
[307] Teixeira de Lemos E, Reis F, Baptista S, Pinto R, Sepodes B, Vala H, Rocha-Pereira P, Correia de Silva G, Teixeira N, Silva AS, Carvalho L, Teixeira F, Das UN (2009) Exercise training decreases proinflammatory profile in Zucker diabetic (type 2) fatty rats. Nutrition 25:330–339
[308] Shi X, Stevens GH, Foresman BH, Stern SA, Raven PB (1995) Autonomic nervous system control of the heart: endurance exercise training. Med Sci Sports Exerc 27:1406–1413

Chapter 8
Hypertension

Introduction

Hypertension is common. It is estimated that worldwide, 7.6 million premature deaths (about 13.5 % of the global total) and 92 million deaths and disability-adjusted life years (DALYs) (6.0% of the global total) could be attributed to high blood pressure. About 54% of stroke and 47% of ischemic heart disease worldwide were attributable to high blood pressure. About half this burden was in people with hypertension; the remainder was in those with lesser degrees of high blood pressure. Overall, about 80% of the attributable burden occurred in low-income and middle-income economies, and over half occurred in people aged 45–69 years [1]. Despite the fact that hypertension is common, the exact cause for its occurrence is not clear.

In subjects with secondary hypertension, either atherosclerosis or fibromuscular dysplasia of one or both renal arteries is responsible for renovascular hypertension. The other forms of secondary hypertension such as those due to endocrine causes are rare, but nevertheless should be kept in mind whenever hypertension presents itself in an unusual form. Here, the discussion about the role of various factors in the pathophysiology of hypertension is centered on primary hypertension. But, wherever it is relevant other forms of hypertension is also included.

Hypertension is easily amenable to treatment with the currently available drugs. But, it is anticipated that a better understanding of its cause(s) is expected to lead to newer modes of treatment and development of better drugs that are less expensive, with fewer side effects. The importance of hypertension lays in the fact that it forms one of the risk factors for coronary heart disease (CHD), stroke, atherosclerosis, and peripheral vascular disease (PVD). Recent studies showed that type 2 diabetes mellitus is more common in those with hypertension and vice versa [2, 3], suggesting that there could be some common pathophysiological events in these two diseases. With the increasing incidence of overweight and obesity both in children and adults, it is likely that the incidence of hypertension will also increase. Many pediatricians do not measure blood pressure in their patients, partly with the assumption that hypertension is uncommon in children. But with the increasing incidence of obesity in children, it is probably necessary to measure blood pressure as frequently as

U. N. Das, *Molecular Basis of Health and Disease,*
DOI 10.1007/978-94-007-0495-4_8, © Springer Science+Business Media B.V. 2011

possible, at least, in obese children. The increasing incidence of hypertension in children with obesity suggests that body mass index (BMI) is an important factor that determines the level and degree of blood pressure. But, certainly the development of hypertension in children suggests a rather complex interplay between obesity, uric acid level, dietary sodium intake, inflammation, inheritance and other factors [4]. The detection and monitoring of HTN has significantly improved with the use of ABPM (ambulatory blood pressure monitoring), which allows not only for a more accurate classification and staging of hypertension (HTN), but also for the calculation of more sophisticated parameters such as the AASI (ambulatory arterial stiffness index). Measurement of arterial stiffness enables assessment of arterial dysfunction, which may precede structural vascular changes evaluated by carotid intima media thickness. Sustained HTN eventually leads to end-organ damage HTN eventually leads to end-organ damage [LVH (left ventricular hypertrophy), central nervous system], which in turn increases the risk of cardiovascular morbidity and mortality.

Recent studies revealed that dietary factors, free radicals, nitric oxide (NO), eicosanoids, pro- and anti-inflammatory cytokines, polyunsaturated fatty acids (PUFAs), folic acid, tetrahydrobiopterin (BH_4), and vitamin C play a significant role in the pathobiology of hypertension. These factors/molecules interact with angiotensin converting enzyme (ACE), endothelins, and anti-hypertensive drugs such as calcium antagonists, and β blockers that may have relevance to the prevention and treatment of hypertension [5, 6].

Nutritional Factors in the Pathobiology of HTN

It is generally believed that an excess of salt intake, dietary factors such as cholesterol, psychological factors such as anxiety and stress and some amount of genetic predisposition may contribute to the development of hypertension in an individual.

Several epidemiological and controlled studies have suggested that excessive sodium intake can cause hypertension, possibly by volume expansion and by altering the renin-angiotensin-aldosterone system. In such studies, however, the possible role of other minerals such as calcium, potassium, magnesium, etc., has not received much attention. In addition, the role of other nutrients such as dietary essential fatty acids, vitamins and anti-oxidants and their interaction with free radicals, nitric oxide and angiotensin converting enzyme activity and their relationship to human essential hypertension, was not evaluated well.

Analysis of the first National Health and Nutrition Examination Survey (NHANES) in 1984, NHANES III and NHNES IV revealed that a dietary pattern low in mineral intake, specifically calcium, potassium, and magnesium, was associated with hypertension in American adults [7, 8]. Blood pressure (BP) and nutrient intake data from 10,033 adult participants in NHANES III and 2311 adults in NHANES IV revealed that the association between inadequate mineral consumption and higher BP is valid and has persisted over two decades. It was observed that the BP effect of low mineral intake was most pronounced in those with only systolic hypertension. It was noted that sodium intake was significantly lower in the systolic hypertension group and significantly higher in the diastolic hypertension group compared with the other groups. The nutrient pattern in the combined hypertension group was similar to that of the normotensive group. These findings highlight the possible importance of tailored nutritional recommendations for hypertension based on hypertension category and individual dietary practices and imply that dietary management of hypertension may be more effective if the focus is on the overall nutritional profile rather than single-nutrient intake as currently recommended for most patients.

These results are supported by the fact that an increased dietary potassium intake is protective against the development of hypertensive cardiovascular disease [9–12]. Both calcium and potassium can influence cell membrane stabilization and vascular smooth muscle relaxation (reviewed in [9]). It was also reported that levels of vitamins A and C were low in the hypertensive group compared to normal controls. It is possible that hypertensives have a low intake of these vitamins. This may mean that a subclinical deficiency of these vitamins may have a role in the pathogenesis of hypertension. Hypertensives were found to consume less sodium compared to normotensives [7, 8]. A National Centre for Health Statistics, U.K. study reached a similar conclusion [13]. The relation between sodium consumption and blood pressure is compatible with both nutritional and physiological interactions of nutrients (reviewed in [9]). First, dairy products are a good source of sodium as well as calcium and potassium. Second, the actions of sodium are closely linked to those of potassium and calcium at both the cellular and organ levels. Since sodium intake was significantly lower in the systolic hypertension group and significantly higher in the diastolic hypertension group compared with the other groups is to be taken into account while exploring the role of nutrients in the pathobiology of hypertension.

Interaction(s) Between Minerals, Trace Elements, Vitamins and Essential Fatty Acids

The fact that a deficiency of calcium, potassium, sodium and vitamins A and C may predispose an individual to the development of high blood is interesting. It is possible that when calcium is available in adequate amounts it stabilizes the arterial membranes, blocks its own entry into the cell, and makes the arterial smooth muscles less likely to contract [14]. It is also likely that it is not the calcium alone but calcium in conjunction with other ions such as sodium and potassium that possibly act together to relax the arterial smooth muscles. Thus, a balance between all the ions seems to be more important than calcium or for that matter any one ion in isolation.

Low magnesium levels in serum and other extracellular fluids increase smooth muscle tension and narrow the lumens of arterioles and venules, and thus, can cause vasospasm. Thus, magnesium deficient diets are expected to raise blood pressure in rats and possibly, in humans [9, 15]. Magnesium ions are necessary for the co-operative binding of potassium ions to the cell membrane [9]. Hence, during magnesium deficiency less potassium would be bound leading to increase in extracellular potassium concentration near the cell membrane, which can lead to depolarization and vasoconstriction [9, 16].

Magnesium is essential for the activity of Δ^6 desaturase [9, 17], which converts dietary linoleic acid (LA) to γ-linolenic acid (GLA). GLA can be easily elongated to form dihomo-gamma-linolenic acid the precursor of prostaglandin E_1 (PGE_1), a vasodilator and platelet anti-aggregator [9]. In the Mg^{2+} deficient group the plasma total cholesterol and triglyceride levels were increased while HDL-cholesterol was decreased. In the Mg^{2+}-deficient group the plasma level of lipid peroxides were increased that could be attributed to the increased cytosolic Ca^{2+} in Mg^{2+}-deficiency which can cause: (1) increase of hydro and endoperoxide levels as a consequence of the increase of arachidonic acid release and eicosanoid synthesis in Mg^{2+}-deficiency, and (2) inhibition of the mitochondrial respiratory activity and activation of Ca^{2+}-dependent proteases which may activate the conversion of xanthine dehydrogenase to xanthine oxidase which generates active O_2 species. In the Mg^{2+}-deficient group, the decrease of Δ^6 desaturase activity was attributed to the lower concentration of actual enzyme molecules as a result of the decreased rate of protein synthesis in Mg^{2+}-deficiency. It is also likely that in Mg^{2+}-deficiency states increased catecholamine release may occur that will increase blood pressure [17].

Similarly vitamin C enhances the synthesis of PGE1 from DGLA [9, 18–20]. Vitamin A can block the action of Δ^5 desaturase that is essential for the conversion of DGLA to AA (arachidonic acid). Thus vitamin A can enhance the tissue levels of DGLA which can be utilized to form PGE_1. In a similar fashion, calcium, sodium and potassium also regulate prostaglandin synthesis [9]. It is likely that at optimum physiological concentrations of sodium, potassium and calcium the levels of PGE1 and PGI2 (prostacyclin), which are potent vasodilators and platelet anti-aggregators, remain normal. Thus the, effect of any stimulus given to aggregate platelets and cause vasoconstriction will be abrogated by the formation and liberation of appropriate amounts of GLA, DGLA and AA from the cell membrane lipid pool and by the synthesis of PGE1 and PGI2 [5, 9].

In addition, both salt and calcium seem to modulate the production of endothelial nitric oxide (eNO), a potent vasodilator and platelet anti-aggregator that seem to play a significant role in hypertension. NO production by endothelial cells is essential to prevent the development of hypertension and atherosclerosis [5, 9, 21].

Salt, Calcium, NO, and Hypertension

Inhibition of basal eNO synthesis decreased renal blood flow and sodium excretion [22]. Intrarenal inhibition of eNO synthesis reduced sodium excretion in response to changes in renal arterial pressure without any effect on renal autoregulation, suggesting that eNO has a role in pressure natriuresis. NO released from macula densa affected afferent arteriolar constriction. NO influenced the effects of angiotensin on tubular reabsorption, altered solute transport, and played an active role in the glomerulus. In contrast, in conditions such as glomerulonephritis, enhanced NO generation is from the infiltrating macrophages suggesting a role for iNO (inducible nitric oxide) in proteinuria, mesangial proliferation and other features seen in this condition.

In hypertensive patients, increase in blood pressure by high salt diet (especially in those with salt-sensitive hypertension) correlated with decreased plasma nitrate plus nitrite, and increased asymmetrical dimethylarginine (ADMA) concentrations that were reversed to normalcy following salt restriction [23], suggesting that salt intake modulates eNO synthesis and this could be a mechanism for salt sensitivity in human hypertension via change in ADMA levels.

Dietary calcium reduced blood pressure by enhancing eNO synthesis [24, 25]. Although there is some controversy with regard to the role of NO in salt-induced hypertension [26], it is likely that high salt intake initially stimulates eNO production to maintain blood flow and when the salt intake continues for a prolonged period eNO synthesis falls leading to the development of hypertension. This proposal is supported by the observation that supplementation of L-arginine reduces high salt intake-induced hypertension [27, 28]. Decrease in blood pressure in hypertensives following potassium chloride intake [29] does indicate that potassium enhanced eNO synthesis and release and thus reduced blood pressure. These results imply that at optimum physiologic concentrations of sodium, potassium, calcium, and magnesium the synthesis and release of eNO and other vasodilators such as PGE_1 and PGI_2 remain adequate to maintain normal blood pressure [9, 24].

Asymmetrical Dimethylarginine and Hypertension

One endogenous factor that interferes with eNO synthesis is asymmetrical dimethylarginine (ADMA). Endothelial dysfunction in hypercholesterolemic individuals could be related to plasma concentrations of ADMA [30], which is a strong and independent predictor of overall mortality and cardiovascular outcome in hemodialysis patients [30], and increased plasma concentrations of ADMA have been reported in hypertension and pre-eclampsia [31, 33]. High serum concentrations of ADMA were associated with increased risk of acute coronary events suggesting that endothelial dysfunction as a cause of coronary heart disease (CHD) [34]. This implies that displacing ADMA with excess L-arginine could be useful in hypertension, preeclampsia, and CHD. In offspring of patients with essential hypertension in whom endothelial dysfunction is present could be reverted to normal by intra-brachial L-arginine [35, 36]. These results suggest that impairment in eNO production precedes the onset of hypertension that could be due to increase in the levels of ADMA.

NO, ADMA and Oxidative Stress in Preeclampsia

Hypertension is one of the cardinal features of preeclampsia, a potentially dangerous complication of the second half of pregnancy, labor, or early period after delivery. Preeclampsia is characterized not only by hypertension but these subjects also may have proteinuria, edema of feet and other systemic disturbances. The condition affects about 2.5–3.0% of women. It has the potential to kill either mother or baby or both, even in the developed world (although rarely). Eclampsia is an end stage of the disease characterized by generalized seizures. Preeclampsia cannot be prevented since no clear cut markers to identify potential sufferers have been identified. Risk factors for preeclampsia include: previous history of preeclampsia, primiparity, obesity, family history of preeclampsia, multiple pregnancies, and chronic medical conditions such as long-term hypertension or diabetes. Paradoxically, cigarette smoking reduces the risk. Although, preeclampsia may develop at any time after 20 weeks of gestation, early onset disease is more severe and characterized by a higher rate of small size for gestational age neonates as well as a higher recurrence rate than with later onset disease.

Preeclampsia arises from secondary systemic circulatory disturbances that can be related to maternal endothelial dysfunction. There are two broad categories of preeclampsia: maternal and placental. In placental preeclampsia, the placenta is under hypoxic conditions with oxidative stress, while maternal preeclampsia arises from the interaction between a normal placenta and a maternal constitution that is suffering from, microvascular disease, as is seen with long-term hypertension or diabetes. Mixed presentations, combining maternal and placental contributions, are common. Understanding the pathobiology of preeclampsia may give clues to the pathogenesis of essential hypertension. It can be said that preeclampsia is nature's experiment of reversible hypertension. Since hypertension is completely reversible and disappears after parturition or removal of the placenta and/or the fetus, a better understanding of the pathobiology and/or factors that trigger the onset of preeclampsia may shed light on the pathogenesis of essential hypertension itself. It is possible that same or similar factors that are responsible for preeclampsia may also participate in the pathogenesis of essential hypertension.

Endothelial dysfunction is known to occur in preeclampsia [37] that is associated with decreased NO levels in these patients. Relative messenger RNA expression and protein expression for endothelial nitric oxide synthase were decreased significantly in endothelial cells from preeclampsia compared with cells from normal pregnancies. Horseradish peroxidase leakage (an indication of increased cell permeability in endothelial cells) in preeclamptic endothelial cells was increased > sevenfold compared to control. The inhibition of endothelial nitric oxide synthase with N(G)-Monomethyl-L-arginine, a specific inhibitor of eNO synthase, resulted in an increase in IL-8-induced endothelial cell permeability [38]. These results suggest that increased endothelial permeability is likely to be associated with decreased eNO synthase expression and activity in endothelial cells from preeclampsia.

Pregnancy induced vasodilatation in hypertensive rats was reported to be dependent on endothelial NO than in normotensive Wistar-Kyoto (WKY) rats [39], suggesting that a defect of the endothelial NO generating pathway which promotes vasodilatation in pregnancy may contribute to the predisposition of women with essential hypertension to develop pre-eclampsia [40]. This is supported by the observation that plasma from women with preeclampsia had significantly lower nitrate

/nitrite concentrations and significantly higher lipid peroxide levels than normal pregnant women before the delivery. Lipid peroxide levels were significantly elevated in preeclamptic placenta. After delivery in the preeclamptic group the plasma concentration of nitrate/nitrite was increased and plasma lipid peroxide levels decreased, while these parameters remained unchanged in the normal pregnant women [41]. These results indicate that high levels of lipid peroxides in the circulation, an indication of the increased pro-oxidant state due to enhanced free radical generation, may be the cause of lowered NO synthesis and hypertension observed in preeclamptic women. These results coupled with the observation that in women with preeclampsia 3 weeks of treatment with oral L-arginine (3 gm/day), significantly lowered systolic, diastolic and mean arterial blood pressure as compared with the placebo group (SBP: 134.2 ± 2.9 vs. 143.1 ± 2.8; DBP: 81.6 ± 1.7 vs. 86.5 ± 0.9; MAP: 101.8 ± 1.5 vs. 108.0 ± 1.2 mmHg, $P < 0.01$) lends support to the concept that NO has a significant role in the pathogenesis of preeclampsia. In addition, treatment with exogenous L-arginine significantly elevated 24-h urinary excretion of nitrite/nitrate and mean plasma levels of L-citrulline [42]. It is likely that in women with preeclampsia, prolonged dietary supplementation with L-arginine significantly decreased blood pressure through increased endothelial synthesis and/or bioavailability of NO.

Further studies revealed that plasma nitrite was significantly lower and plasma endothelin levels were significantly higher in pre-eclamptic women than in normotensive pregnant women. Superoxide dismutase activity was decreased (indicating enhanced oxidative stress) and arginase activity was significantly increased in pre-eclamptic patients when compared to normotensive pregnant women. These results suggest that in pre-eclampsia excessive arginase and low superoxide dismutase activity leads to a decrease nitric oxide levels and oxidative stress that may promote microvascular oxidative damage and endothelial dysfunction leading to hypertension [43]. In addition, hydrogen peroxide (H_2O_2), a terminal metabolite of the cellular oxidative stress cascade, was found to be increased in the serum of preeclamptic women at term. H_2O_2 is known to reduce the production of NO by increasing the metabolism of arginases. When the levels of NO and H_2O_2 were simultaneously assessed in the serum of normal and preeclamptic women at 10–15 and 37–40 weeks of pregnancy, and in placentas at delivery, an inverse correlation between increased levels of H_2O_2 and decreased levels of NO early in maternal circulation and at term in placenta was observed. In vitro studies showed that H_2O_2 inhibited NO synthesis of cytotrophoblasts [44]. These results highlight an inverse correlation between H_2O_2 and NO early in maternal circulation and in placenta of women with preeclampsia.

Serum nitrite/nitrate concentration was decreased and creatol (CTL), the oxidized metabolite of creatine, concentration was found to be increased in preeclamptic patients relative to healthy controls during the first trimester of pregnancy, who also exhibited disrupted flow-mediated dilatation (FMD), which was regulated in part by NO. Immunohistochemistry demonstrated strong expression of nitrotyrosine in the vasculature of preeclamptic placentas, while treatment with sera derived from preeclamptic patients increased endothelial expression of inducible NOS (iNOS) mRNA, and this increase was inhibited by angiotensin II (Ang II) receptor type 2 (AT2) blocker. Endothelial NADPH oxidase subunit gp91(phox) expression was increased by treatment with sera from preeclamptic patients and this increase was attenuated by Ang II receptor type 1 (AT1) blocker. These results suggest that NO and ROS play a significant role in the pathogenesis of preeclampsia and that these roles involve Ang II [45].

The studies summarized above clearly establish a role for eNO and free radical and oxidative stress in the pathobiology of pre-eclampsia and possibly, in hypertension, though some studies did not support these observations. For example, concomitant measurement of plasma and urine nitrate and nitrite revealed a reduced nitric oxide in urine but not in the plasma of women with preeclampsia compared with normal pregnant women [46], though these results could be interpreted to mean that the plasma half-life of NO is altered. Another study reported that serum nitrate levels were increased which may reflect either increased production of nitric oxide from an unidentified source or decreased elimination through the kidneys in patients with preeclampsia [47]. These studies imply that the kinetics of NO, free radicals, arginase, ADMA, endothelin, angiotensin II and antioxidants are altered in preeclampsia that may ultimately lead to endothelial dysfunction and decreased vasodilatation and development of hypertension.

VEGF, Endoglin, Placental Growth Factor, TGF-β, Catechol-O-methyltransferase Activity and Preeclampsia

Recent studies suggested that preeclampsia is precipitated by the release of circulating factors from the placenta that induce endothelial dysfunction [48, 49]. Soluble fms-like tyrosine kinase 1 (sFlt1) (also known as soluble vascular endothelial growth factor [VEGF] receptor 1 [sVEGFR1]), a circulating antiangiogenic protein that sequesters the proangiogenic proteins placental growth factor (PlGF) and VEGF, is increased before the onset of clinical disease in the circulation of women with preeclampsia. It was reported that circulating levels of sFlt1 correlate with the severity of preeclampsia and proximity to the onset of hypertension or proteinuria [50–54]. Serum free PlGF and free VEGF levels were found to be decreased before the development of preeclampsia. Overexpression of sFlt1 in pregnant rats resulted in a preeclampsia-like phenotype. Some patients with cancer who were given anti-VEGF therapy have been described to develop hypertension, proteinuria, and the reversible posterior leukoencephalopathy syndrome, which are similar to the clinical picture seen in patients with preeclampsia and eclampsia. These results led to the suggestion that an imbalance in circulating angiogenic factors would lead to vascular endothelial dysfunction and consequently the development of preeclampsia. In a nested case control study, it was reported that circulating soluble endoglin levels increased markedly beginning 2–3 months before the onset of preeclampsia and an increased level of soluble endoglin was usually accompanied by an increased ratio of sFlt1:PlGF. The risk of preeclampsia was greatest among women in the highest quartile of the control distributions for both biomarkers but not for either biomarker alone, suggesting that rising circulating levels of soluble endoglin and ratios of sFlt1:PlGF herald the onset of preeclampsia [55]. It was also reported that soluble endoglin inhibited the formation of capillary tubes in vitro and induced vascular permeability and hypertension in vivo. Its effects in pregnant rats were amplified by coadministration of sFlt1, leading to severe preeclampsia including the HELLP (hemolysis, elevated liver enzymes, low platelets) syndrome and restricted fetal growth. It is important to note that soluble endoglin not only impaired binding of TGF-β1 to its receptors but also interfered with TGF-β1 signaling and eNOS activation in endothelial cells [56]. Thus, ultimately it appears that changes in the concentrations of sFlt1 and soluble endoglin lead to decreased levels of VEGF, PlGF and eNO.

Patients with preeclampsia have higher plasma levels of endothelin-1, a potent vasoconstrictor [57, 58], lipid peroxides and decreased antioxidants suggesting an imbalance between pro and antioxidants [59]. These results coupled with the observation that the levels of sFlt1 and soluble endoglin are enhanced while those of VEGF and PlGF are decreased in patients with preeclampsia suggests that the balance between pro- and anti-angiogenic factors and vasodilator and vasoconstrictor molecules is altered in patients with preeclampsia that ultimately leads to endothelial damage/dysfunction and decreased formation or release of NO, which causes hypertension. This is especially interesting in the light of the observation that eNOS plays a predominant role in VEGF-induced angiogenesis and vascular permeability [60]. Thus, a deficiency of VEGF would ultimately lead to decrease in NO generation.

Several recent studies also showed that patients with preeclampsia possess autoantibodies, termed AT1-AAs that bind and activate the angiotensin II receptor type 1a (AT1 receptor) [61–63]. It was reported that features of preeclampsia, including hypertension, proteinuria, glomerular endotheliosis (a classical renal lesion of preeclampsia), placental abnormalities and small fetus size appeared in pregnant mice after injection with either total IgG or affinity-purified AT1-AAs from women with preeclampsia. These features were prevented by co-injection with losartan, an AT1 receptor antagonist, or by an antibody neutralizing seven–amino-acid epitope peptide [64]. These studies indicate that preeclampsia may be a pregnancy-induced autoimmune disease in which key features of the disease result from autoantibody-induced angiotensin receptor activation, similar to that is seen other autoimmune diseases such as thyrotoxicosis. AT1-AAs can be detected as early in pregnancy as 18 weeks, making them one of the earliest markers to identify women at risk for pre-eclampsia39. Since AT1-AAs can be detected many weeks before the symptoms of preeclampsia, their levels in the plasma could be measured for screening, disease diagnosis and treatment. If it is true that maternal circulating AT1-AAs contribute to preeclampsia, as shown by adoptive transfer animal study results have suggested, the timely identification and removal or inhibition of these autoantibodies from women with preeclampsia may provide considerable therapeutic benefit. If AT1-AAs play a major part in the etiology and pathophysiology of preeclampsia, it may be possible to block autoantibody-mediated AT1 receptor activation and thereby forestall or prevent the onset of the symptoms of preeclampsia. Indeed, it was shown that administration of IgG from individuals with preeclampsia to pregnant mice induced elevations in circulating sFlt1 and soluble endoglin via induction of placental vasculopathy. Thus, these studies suggest that an array of insults have the potential to cause placental damage that is linked to the production of soluble antiangiogenic factors by placenta that, in turn, produce dysfunction of maternal endothelium resulting in preeclampsia.

In another study, it was reported that pregnant mice deficient in catechol-O-methyltransferase (COMT) show a preeclampsia-like phenotype resulting from an absence of 2-methoxyoestradiol (2-ME), a natural metabolite of oestradiol that is elevated during the third trimester of normal human pregnancy. 2-ME ameliorates all pre-eclampsia-like features in the $Comt^{-/-}$ pregnant mice and suppressed placental hypoxia, hypoxia-inducible factor-1α expression and sFLT-1 elevation. In addition, the levels of COMT and 2-ME were found to be significantly lower in women with severe pre-eclampsia [65].

But, it is not yet known whether in patients with essential hypertension the plasma levels of sFlt1 and soluble endoglin, AT1-AAs are enhanced and COMT and 2-ME are abnormal.

Homocysteine and Endothelial Damage

Increased plasma levels of homocysteine are known to be associated with coronary heart disease and atherosclerosis. Homocysteine undergoes auto-oxidation leading to the formation of homocystine, homocysteine-mixed disulfides, and homocysteine thiolactone, during which $O_2^- \cdot$ and hydrogen peroxide (H_2O_2) are generated that could cause endothelial cytotoxicity and dysfunction [66]. These free radicals induce lipid peroxidation that, in turn, oxidizes low-density lipoprotein leading to the formation of oxidized LDL (Ox-LDL) that could initiate and perpetuate atherosclerosis. Homocysteine increases Factor V and Factor XII activity, decreases protein C activation, inhibits thrombomodulin expression, induces tissue factor expression, suppresses heparan sulfate expression, and reduces the binding of tissue-type plasminogen activator to its endothelial cell receptor: annexin II, reducing the production of eNO and PGI_2, events that increase the generation of thrombin and enhances thrombotic tendency and endothelial dysfunction [66–68]. Normal endothelial cells release NO or a related S-nitrosothiol that induces the formation of S-nitroso-homocysteine [69], a potent vasodilator and platelet anti-aggregator. Thus, S-nitroso-homocysteine attenuates sulfhydryl-dependent generation of H_2O_2 and nullifies the prothrombotic actions of homocysteine. However, continued exposure of endothelium to homocysteine reduces the production of eNO and this leads to homocysteine-mediated injury to the endothelium and initiation of atherosclerosis and/or thrombus formation or acceleration of existing atherosclerosis. Hence, methods designed to enhance eNO production could restore the antithrombotic properties of endothelium and prevent atherosclerosis.

Furthermore, homocysteine inhibits glutathione peroxidase (GP) activity *in vitro* and reduces its synthesis in endothelial cells. GP catalyzes the reduction of both H_2O_2 and lipid peroxides to their corresponding alcohols, and thus, prevents inactivation of eNO. Inhibition of GP activity by homocysteine is responsible for its (homocysteine) vascular toxicity [70]. Homocysteine induces smooth muscle cell migration and proliferation [71], upregulated vascular cell adhesion molecule-1 expression, and enhanced monocyte adhesion. Folic acid not only reduced plasma homocysteine levels, but also reduced oxidized LDL-stimulated release of GROα, ENA-78, and interleukin-8 (IL-8), and CC chemokines: monocyte chemoattractant peptide-1 and RANTES in peripheral blood mononuclear cells. Oxidized-LDL-induced release of ENA-78 by peripheral blood mononuclear cells was reduced when cells were incubated with folic acid [72].

Low circulating vitamin B_6 is associated with higher C-reactive protein (CRP), a marker of inflammation, levels independent of plasma homocysteine levels [73, 82]. Vitamin B_6 enhances the production of PGE_1, a potent vasodilator, platelet anti-aggregator, and anti-inflammatory eicosanoid that has anti-inflammatory actions [74, 75]. Thus, vitamin B_6 has anti-inflammatory actions.

Homocysteine enhanced the activity of HMG-CoA reductase in human umbilical vein endothelial cells (HUVECs), and enhanced cellular cholesterol content, whereas simvastatin, an HMG-CoA reductase inhibitor, reduced HUVECs cholesterol content and prevented homocysteine-induced suppression of eNO production in a dose dependent manner [76]. Thus, homocysteine facilitates endothelial dysfunction and atherosclerosis by enhancing cholesterol synthesis. These results indicate that dietary deficiency of folic acid and vitamin B_6 may cause endothelial dysfunction and contribute to the development of hypertension and preeclampsia in a genetically susceptible individual or when other factors that contribute to the development of hypertension are present. It is also evident from the preceding discussion that factors that contribute to the development of oxidative stress have the potential to induce the development of hypertension. In this context, it is interesting to note that several nutritional factors have the ability to modulate oxidative stress and alter endothelial dysfunction.

Nutritional Factors, Oxidant Stress and Endothelial Dysfunction

Folic acid, H_4B (tetrahydrobiopterin), and insulin suppress $O_2^- \cdot$ production and prolong the half-life of NO [9, 24, 66, 74, 75] and thus, preserve endothelium-dependent vasodilation, which also accounts for their anti-inflammatory property [9, 24, 66, 74–77]. Folic acid and H_4B attenuate cholesterol-induced endothelial dysfunction [78] and coronary hyperreactivity to endothelin [79]. Folic acid restores tissue stores of H_4B, whereas H_4B stimulated endothelial cell proliferation. H_4B augments NO generation. Vitamin C stabilizes H_4B and increases its intracellular levels and thus, enhances eNOS activity [80] and vitamin C also enhances NO release by suppressing the formation of total S-nitrosothiols and S-nitrosoalbumin [81]. These results suggest that when L-arginine, folic acid or 5-methyltetrahydrofolic acid, the active form of folic acid, and H_4B when provided together could stimulate eNO synthesis.

Increased Oxidant Stress Occurs in Hypertension

Vascular endothelium produces vasodilators: prostacyclin (PGI_2), PGE_1, NO, and endothelium-derived hyperpolarizing factor (EHF) and other vasoactive factors such as endothelins. Under physiological conditions, a balance is maintained between endothelial vasoconstrictors and vasodilators such that normal blood pressure is maintained. When this balance is altered more in favor of vasoconstrictors, when the concentrations of vasodilators are reduced, or both, hypertension develops. One mechanism by which endothelium-dependent vasodilatation is impaired is due to an increase in the oxidative stress that inactivates NO and PGI_2.

Previously, we showed that in patients with uncontrolled essential hypertension $O_2^- \cdot$, hydrogen peroxide (H_2O_2), and lipid peroxides were produced in significantly large amounts both by unstimulated and stimulated polymorphonuclear leukocytes (PMNs) [82, 83] indicating that there is indeed an increase in oxidative stress in hypertension. The enhanced levels of free radicals and lipid peroxides reverted to normal after the control of hypertension by anti-hypertensive medicines such as calcium antagonists, β blockers and ACE inhibitors [82]. $O_2^- \cdot$ Is known to function as an endothelium-derived vasoconstrictor [84], suggesting that increase in free radical generation seen in untreated hypertensives could be a factor responsible for the heightened peripheral vascular resistance. In addition, decrease in the levels of superoxide dismutase (SOD), catalase, and glutathione peroxidase, and vitamin E was also noted in hypertensives [82] that also reverted to normal after treatment with anti-hypertensive drugs. In this context, it is interesting to note that a fusion protein (HB-SOD) consisting of human Cu–Zn type SOD (superoxide dismutase) and a C-terminal basic peptide with a high affinity for heparan sulfate on endothelial cells not only can localize to the vascular walls but also can effectively prevent the development of hypertension in the spontaneously hypertensive rats (SHR) [85]. These results suggest that the impaired endothelium-dependent vasodilatation seen in human essential hypertension could be due to enhanced generation of free radicals. This is supported by the observation that SOD deficiency is seen in hypertension [82, 86] and that SOD activity decreased with advancing age [86, 87]. Decreased NO bioavailability and increased $O_2^- \cdot$ generation with increasing age could be due to enhanced NAD(P)H oxidase activity that augments $O_2^- \cdot$ generation [87].

In addition, angiotensin II, a potent vasoconstrictor, stimulates free radical generation [82] by up regulating several subunits of membrane bound NADPH oxidases [88, 89]. Furthermore, recent studies reported that reduction of extracellular superoxide dismutase (SOD) in the central nervous system promoted T-cell activation and vascular inflammation, modulated sympathetic outflow and induced hypertension [90]; active oxygen species and thromboxane A_2 reduced angiotensin-II type 2 receptor-induced vasorelaxation in diabetic rats [91]; responsiveness to NO and increased myeloperoxidase-associated tumor necrosis factor-α (TNF-α) plays a role in activation of the PMN NADPH oxidase, thereby contributing to systemic oxidative stress, inflammation, and the development of hypertension [92]; and in healthy middle-aged and older adults, impaired endothelium-dependent dilatation is decreased by higher PMN count mediated by reduced reductions in tetrahydrobiopterin and NO bioavailability [93]. These results emphasize the proposal that free radicals are closely associated with the development of hypertension.

But, this is not without controversy. For example, simvastatin, a hydroxy methyl-glutaryl coenzyme A (HMG-CoA) reductase inhibitor that is used in the management of hyperlipidemias behaves like an anti-oxidant and improves endothelial function [94] and increased SOD and glutathione peroxidase activities. This suggests that SOD by quenching $O_2^- \cdot$ restores endothelial function. Since, simvastatin does not reduce the increased blood pressure this suggests that $O_2^- \cdot$ alone is not responsible for the development of hypertension. Further, anti-oxidants such as vitamin E are not useful to lower elevated blood pressure both in experimental animals and humans.

This is supported by the fact that hypertension that is induced in experimental animals by giving 10% glucose drinking solution showed elevation in the aortic basal $O_2^- \cdot$ production, plasma levels of insulin and glucose, as well as insulin resistance index [95], events that reverted to normal following aspirin administration. An increase in plasma SOD activity was observed in glucose-fed rats but not in aspirin fed rats, suggesting that aspirin prevented the development of hypertension and reduced insulin resistance in glucose-fed rats. Aspirin preferentially blocks the synthesis of thromboxane A_2 (TXA_2) from its precursor arachidonic acid without interfering with the synthesis of PGI_2, a potent vasodilator and platelet anti-aggregator, and enhance the synthesis of anti-inflammatory compounds such as lipoxins, resolvins, protectins maresins and nitrolipids from arachidonic acid (AA), eicosapentaenoic acid (EPA), and docosahexaenoic acid (DHA) [96–98]. It is likely that aspirin increased the production of eNO, PGI_2, lipoxins, resolvins, protectins, maresins and nitrolipids and/or their half-life by decreasing oxidative stress and restoring SOD levels to normal. These studies indicate that the balance between $O_2^- \cdot$ and eNO, PGI_2, lipoxins, resolvins, protectins, maresins and nitrolipids is critical in the prevention of hypertension.

Superoxide Anion Production Is Increased in Hypertension: How and Why?

NAD(P)H oxidase is present in vascular cells and enhanced $O_2^- \cdot$ generation is as a result of its activation in hypertension [99]. NAD(P)H oxidase responds to stimuli such as vasoactive factors, growth factors, and cytokines. Kidneys of adult SHR (spontaneously hypertensive rats) had a significantly greater mRNA for p47phox (a major component of NAD(P)H oxidase). A tenfold up regulation of vascular NAD(P)H oxidase was associated with a threefold increased production of $O_2^- \cdot$ and a concomitant impairment of the eNO signal transduction pathway in hypertension [100]. In experimental animals, chronic inhibition of eNO synthesis increased aortic $O_2^- \cdot$ production, and redox sensitive transcription factors NF-κB and AP-1 [14, 101]. Angiotensin II type 1 receptor antagonists prevented these changes, suggesting that inhibition of eNO synthesis increases vascular oxidative stress and oxidative stress-sensitive signals mediate their actions through angiotensin II type 1 receptors.

Further, angiotensin II stimulates free radical generation by up regulating NAD(P)H oxidase [82, 102, 103] lending support to this view. Thus angiotensin II-induced free radicals play a dominant role in hypertension, endothelial dysfunction, nitrate tolerance, atherosclerosis, and cellular remodeling. This offers an explanation for the beneficial actions of angiotensin receptors antagonists and ACE inhibitors in cardiovascular diseases.

Superoxide Anion and Hypertension

It is logical to suggest that under normal conditions, a balance is maintained between the steady state level of eNO and the $O_2^- \cdot$. NO reacts with $O_2^- \cdot$, producing peroxynitrite. Vascular eNO synthase inhibition increased $O_2^- \cdot$ release, with a corresponding reduction in peroxynitrite formation. Conversely, NO donors and $O_2^- \cdot$ scavengers reduced $O_2^- \cdot$ release, whereas only NO donors enhanced peroxynitrite formation. The changes in the concentrations of NO, peroxynitrite, and $O_2^- \cdot$ were much larger in arteries compared to those seen in veins. A significant correlation was seen between NO bioavailability, peroxynitrite formed and $O_2^- \cdot$ production [104]. Both NO and PGI_2 are inactivated by $O_2^- \cdot$, suggesting that $O_2^- \cdot$ by decreasing the half-life of NO and PGI_2 lowers their circulating concentrations and thus, initiates the development of hypertension.

It is important to know that endothelial cells also produce endothelin, a potent vasoconstrictor and a physiological antagonist of NO. Human endothelial cells and coronary artery smooth muscle cells showed increased endothelin-1 production when exposed to oxidative stress [105], whereas NO reduced the production of endothelin-1 [106], implying that NO and $O_2^- \cdot$ have opposite actions on the production of endothelin-1 by endothelial cells and that the balance struck among these vasoactive molecules is highly relevant in the pathogenesis of hypertension.

NO and Hypertension

NO is synthesized from the semi-essential amino acid, L-arginine. NO is not stored and is formed and released as needed. Three NOS iso-enzymes have been characterized: neuronal or type 1, nNOS; inducible NOS or iNOS; and endothelial NOS or eNOS, each the product of a unique gene, have been identified and well characterized. Type 1 or nNOS is a Ca^{2+}-dependent enzyme found in neuronal tissue and skeletal muscle. Type 2 or iNOS is inducible in a variety of cells and tissues in response to cytokine or endotoxin activation, and this enzyme (iNOS) binds Ca^{2+}/calmodulin so tightly that at normal physiologic levels its activity is functionally Ca^{2+}-independent. Type 3 or eNOS is also Ca^{2+} dependent and is myristoylated and palmitoylated at the N-terminus, modifications, which are needed for localization to the plasmolemmal caveolae of endothelial cells. Though both nNOS and eNOS are constitutive enzymes, all three enzymes can be induced, albeit to different levels and by different stimuli.

Vasodilator response to acetylcholine (which stimulates eNO release from endothelial cells) was significantly reduced in hypertensives [107]. However, the vasodilator response to sodium nitroprusside, a NO donor, was similar in normotensives and hypertensives, suggesting that endothelial dysfunction in essential hypertension are due to selective abnormality of eNO synthesis. An inverse correlation was reported between platelet cytosolic Ca^{2+} and eNO levels indicating a link between hypertension and altered platelet function. This may explain the role of NO in cardiovascular events, suggesting that NO abnormality is not localized to the vascular endothelium but may occur in several other tissues as well.

Oral administration of L-NG-nitro-L-arginine, a NO synthase inhibitor, elevated blood pressure, decreased plasma level and urinary excretion of nitrate ions, and increased peripheral vascular resistance [108]. Infusion of NG-monomethyl1-L-arginine (L-NMMA), an inhibitor of NOS, elevated mean arterial pressure, decreased heart rate and cardiac index, increased total peripheral resistance, urinary sodium and fractional sodium excretions were increased but creatinine clearance remained unchanged [109]. These results suggest that basal generation of eNO regulates peripheral vascular resistance and normal blood pressure.

Cyclosporine Increases Blood Pressure by Augmenting $O_2^- \cdot$ Generation

If it is true that $O_2^- \cdot$ has a role in the pathogenesis of hypertension, it is reasonable to expect that situations that increase blood pressure one should be able to demonstrate enhanced $O_2^- \cdot$ generation. Clinical use of cyclosporine is one such instance. Subcutaneous injections of cyclosporine to experimental animals resulted in impaired vascular response to acetylcholine that was normalized by pre-treatment with SOD [110], supporting the role of $O_2^- \cdot$ generation in the development of hypertension. Cyclosporine increased endothelin-1 and decreased endothelial NOS both in the aorta and the renal cortex lending further support to the involvement of increased $O_2^- \cdot$ generation, decreased eNO synthesis and enhanced endothelin in the pathogenesis of hypertension.

Anti-hypertensive Drugs Enhance eNO Synthesis and Show Antioxidant Property

ACE inhibitors and calcium antagonists not only reduced blood pressure but also increased plasma 6-keto-$PGF_{1\alpha}$, a metabolite of PGI_2, eNO, and normalized cGMP levels [111, 112], indicating that some, if not all, anti-hypertensive drugs act on the eNO-PGI_2-$O_2^- \cdot$ system. Previously, we showed that NO is a potent inhibitor of ACE activity in vitro [113], and both calcium antagonists and β-blockers inhibited free radical generation and formation of lipid peroxide [82]. Thus, anti-hypertensive drugs inhibit $O_2^- \cdot$ generation, increase the half-life of eNO and also enhance eNO production to bring about their antihypertensive action (see Fig. 8.1).

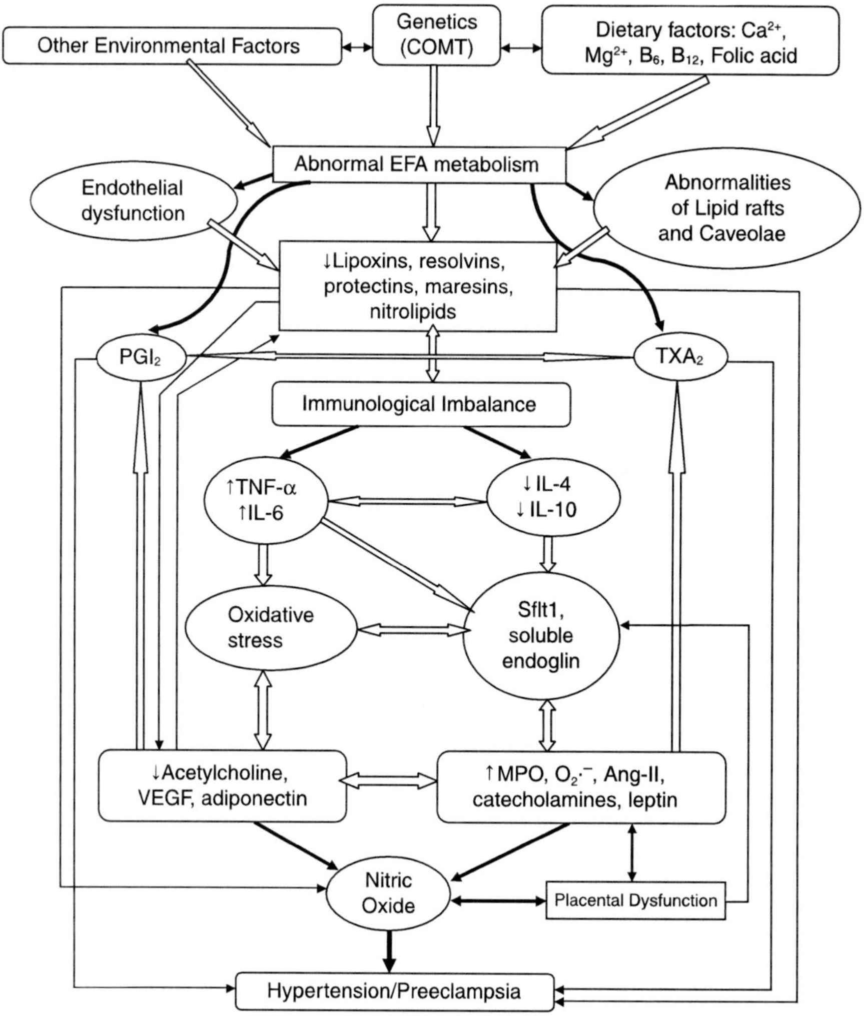

Fig. 8.1 In a genetically predisposed pregnant woman, certain environmental factors such as dietary components may trigger the initiation of preeclampsia One of the genetic factors could be a deficiency of catechol-O-methyltransferase (COMT) activity that results in an absence or deficiency of 2-methoxyoestradiol (2-ME), a natural metabolite of oestradiol that is elevated during the third trimester of normal human pregnancy. 2-ME suppressed placental hypoxia, hypoxia-inducible factor-1α expression and sFLT-1 elevation. In women with severe pre-eclampsia, the levels of COMT and 2-ME were found to be significantly lower. 2-ME enhances the production of PGI_2 (Life Sci 1999; 65: PL167–PL70). 2-ME also has anti-cancer actions (Nature 2000; 407: 390–395). Genetics, dietary and environmental factors may produce immunological imbalance and lead to pro-inflammatory cytokine activation and oxidative stress. Deficiency of dietary factors and genetic factors (such as low activity of Δ^6 and Δ^5 desaturases) leads to decrease in the levels of polyunsaturated fatty acids (such as AA, EPA and DHA) in the cell membrane that causes membrane to be more rigid impairing the production of lipoxins, resolvins, protectins, maresins and activation of the immune system leading to increased production of IL-6, TNF-α that ultimately

Transforming Growth Factor-β (TGF-β) in Hypertension

Circulating levels of TGF-β were significantly higher in hypertensives compared with normotensives [114, 115]. TGF-β1 gene located at chromosome 19q13.1 could be a candidate susceptibility locus for hypertension [116]. Patients with albuminuria in hypertension and diabetic nephropathy showed elevated plasma TGF-β concentration whereas angiotensin receptor blockade and ACE inhibitors decreased its excretion [117–120]. Angiotensin II in association with endothelin and TGF-β activates collagen type I gene resulting in increased formation of extracellular matrix protein in the renal cortex and aorta that leads to renal scarring and end-stage renal disease in patients with hypertension and diabetes mellitus. In contrast, chronic anti-TGF-β antibody significantly reduced blood pressure, proteinuria, and the degree of glomerulosclerosis and renal medullary fibrosis in experimental animals [121]. Increased synthesis of collagen was reported in vitro in the presence of high glucose, and this was reduced by the neutralization of TGF-β indicating that TGF-β enhances collagen synthesis [122]. It is noteworthy that angiotensin II, high salt diet, and cyclosporine stimulated TGF-β expression in the kidney and endothelium [123]. High salt intake and high glucose stimulated eNO synthesis [124, 125] as a compensatory mechanism to abrogate their pro-hypertensive actions. This is so since, NO blocks TGF-β synthesis and thus, suppresses matrix protein synthesis by mesangial cells [126–128] and TGF-β in turn inhibits eNO synthesis [129–131]. Furthermore, TGF-β has angiogenic actions and levels of endoglin, an antagonist of TGF-β, are increased in preeclampsia in which hypertension is known to occur [132]. Unlike in chronic hypertension and long-standing diabetes mellitus in which scarring of kidney is common due to enhanced TGF-β levels, renal sclerosis is uncommon (even if it occurs, it is generally reversible once parturition occurs or abortion is induced) in preeclampsia. This is due to fall in TGF-β levels to normal following parturition. Thus, to start with, elevated TGF-β levels in hypertension and diabetes mellitus is a compensatory phenomenon to enhance angiogenesis and reduce hypertension but, sustained elevations in TGF-β levels ultimately leads to renal failure. Thus, in the

---

causes oxidative stress, increase in sFlt1 and soluble endoglin and decrease in VEGF, acetylcholine, adiponectin levels and increase in MPO activity, free radicals, angiotensin II and catecholamines (increased sympathetic activity). Events that lead to a decrease in the synthesis and release and/or stability and decreased half-life of NO that causes development of hypertension. PUFAs enhance eNO generation, suppress the production of pro-inflammatory cytokines (IL-6 and TNF-α), and form precursors to anti-inflammatory and anti-oxidant lipid molecules lipoxins, resolvins, protectins, and maresins. PUFAs interact with NO to form nitrolipids that possess vasodilator and platelet anti-aggregator actions. PUFAs and their products lipoxins, resolvins, protectins, maresins and nitrolipids suppress MPO and ACE (angiotensin converting enzyme) activity, leukocyte activation and inhibit the production of cytokines IL-6 and TNF-α and enhance the production of IL-4 and IL-10, and thus, are expected to be beneficial in the prevention and management of essential hypertension and preeclampsia. Development of stable synthetic analogues of lipoxins, resolvins, protectins, maresins and nitrolipids that are orally active may prove to be useful in hypertension and preeclampsia short term, increase in TGF-β levels is a protective event but its continued enhanced levels are harmful. These results indicate that a close interaction and feedback regulation exists among TGF-β and NO. Thus, NO seems to play an important role in the pathogenesis of hypertension irrespective of the underlying cause.

Essential Fatty Acids and Blood Pressure

Despite the knowledge that several biologically active molecules (such as superoxide anion, nitric oxide, antioxidants, cytokines, H_2O_2, renin-angiotensin-aldotserone system, TGF-β, nutrients, vitamins, VEGF, endoglin, caveolae, PIGF, autoantibodies against angiotensin receptor-1, polyunsaturated fatty acids, eicosanoids, lipoxins, resolvins, protectins, maresins, nitrolipids) are involved in the regulation of endothelial function, maintenance of vascular tone and blood pressure, and angiogenesis, it is still unclear how and when increase in blood pressure (development of hypertension) occurs and what event(s) and/or molecules initiate the development of hypertension. I present arguments to suggest that the polyunsaturated fatty acid content of the endothelial cells and hypothalamic neurons and their ability to form anti-inflammatory compounds such as lipoxins, resolvins, protectins, maresins and nitrolipids have an important role in the pathobiology of hypertension.

Diets rich in saturated fatty acids elevate blood pressure both in humans and animals and exacerbate spontaneous hypertension [133, 134]. These hypertensive effects have been ascribed to reduced formation of vasodilator prostaglandins such as PGE_1 and PGI_2, and other anti-inflammatory molecules such as lipoxins, resolvins, protectins, maresins and nitrolipids [135]. Dietary LA (linoleic acid; 18:2 n-6) is converted to GLA and DGLA by the action of specific enzymes, which are controlled by genetic, hormonal and nutritional factors (see Chap. 4 for the metabolism of essential fatty acids). Dietary supplementation of DGLA can prevent the increase in the blood pressure induced by feeding saturated fats [136]. This beneficial action of DGLA supplementation is associated with increased synthesis of PGE_1 in these studies [9, 136]. Stress-induced hypertension can also be completely blocked by GLA supplementation [137], suggesting that GLA normalized stress-induced changes in the hypothalamus and in the endocrine organs. This postulated central action for GLA and possibly for its products, which are likely to be eicosanoids, requires that dietary GLA or its products are able to enter the brain bypassing the blood brain barrier and normalize neural and hormonal functions that are associated with stress-induced hypertension.

Since PGs regulate nerve conduction, transmitter release and action and mental function [9, 138–143] it is possible that GLA supplementation to hypertension prone individuals could help to prevent a rise in blood pressure. Previously, McCarron et al. [7, 11, 12] noted that hypertensives consume less LA when compared to the normals. LA can be converted to GLA and DGLA in the body by the action of Δ^6desaturase for the activity of which dietary factors such as magnesium, calcium, potassium, sodium, vitamins A and C are needed as co-factors ([9, 144, 145]; and also see Chap. 4).

Diet rich in fish oil, which contains mainly eicosapentaenoic acid (EPA) and docosahexaenoic acid (DHA), reduce the blood viscosity and lower blood pressure [9, 144–146], by inhibiting the formation of thromboxane A_2 (TXA2), a potent vasoconstrictor and platelet aggregator, and enhancing that of PGI_3, a vasodilator and platelet anti-aggregator [144–147]. Further, EPA blocks the activity of Δ^5desaturase so that the tissue levels of AA will be low and that of DGLA, the precursor of PGE_1, will be high. In addition, EPA enhanced the production of PGI_2 from AA, while AA augmented the synthesis of PGI_3 from EPA, while DGLA increased the metabolism of EPA to form PGI_3 [144, 145, 147–152]. For example, in perfused vascular tissue, DGLA and AA increased the conversion of EPA to PGI_3, a vasodilator and platelet anti-aggregator, whereas orally administered EPA enhanced AA conversion to PGI_2 and inhibited the activity of Δ^6 and Δ^5 desaturases ($\Delta^6 > \Delta^5$), which resulted in enhanced tissue levels of DGLA as a result of decreased conversion of DGLA to AA. The enhanced levels of DGLA could lead to an increase in the formation of PGE_1, a vasodilator and platelet anti-aggregator. This close interaction between DGLA, AA, and EPA implies that optimal levels of DGLA, AA, EPA, and DHA need to be present in the tissues to optimize the formation of various beneficial eicosanoids and lipoxins and resolvins to prevent atherosclerosis (see Fig. 8.2).

Thus, adequate amounts of both n-3 and n-6 fatty acids and other co-factors are necessary for the formation of potent platelet anti-aggregators and vasodilators such as PGE_1, PGI_2 and PGI_3 which can prevent the development of essential hypertension.

Free Radicals, NO, ACE Activity and Essential Hypertension

Previously, we showed that the plasma levels of NO were low in patients with uncontrolled essential hypertension [82] and that they revert to normal after the control of hypertension with various drugs. In the salt sensitive hypertensive rat, the endothelial NO synthase activity is low [153]. Further, ACE inhibitors in addition to their ability to block the synthesis of angiotensin-II, also inhibited the breakdown of bradykinin, a stimulator of the synthesis of NO [154]. Thus, ACE inhibitors elevate the levels of NO and thus, bring about their anti-hypertensive action.

In this study [82], we also noted that the activity of superoxide dismutase (SOD) and the concentration of vitamin E were low and that they revert to normal following the control of hypertension. Both superoxide anion and hydrogen peroxide and the plasma levels of lipid peroxides were higher in patients with uncontrolled essential hypertension which also reverted to normal after the control of hypertension. These results suggest that the changes in free radical generation, NO and anti-oxidants seen in essential hypertension are probably secondary to hypertension. Further, superoxide anion can inactivate PGI_2 and NO and this can lead to an increase in peripheral vascular resistance and hypertension. Also, both calcium antagonists and beta-blockers

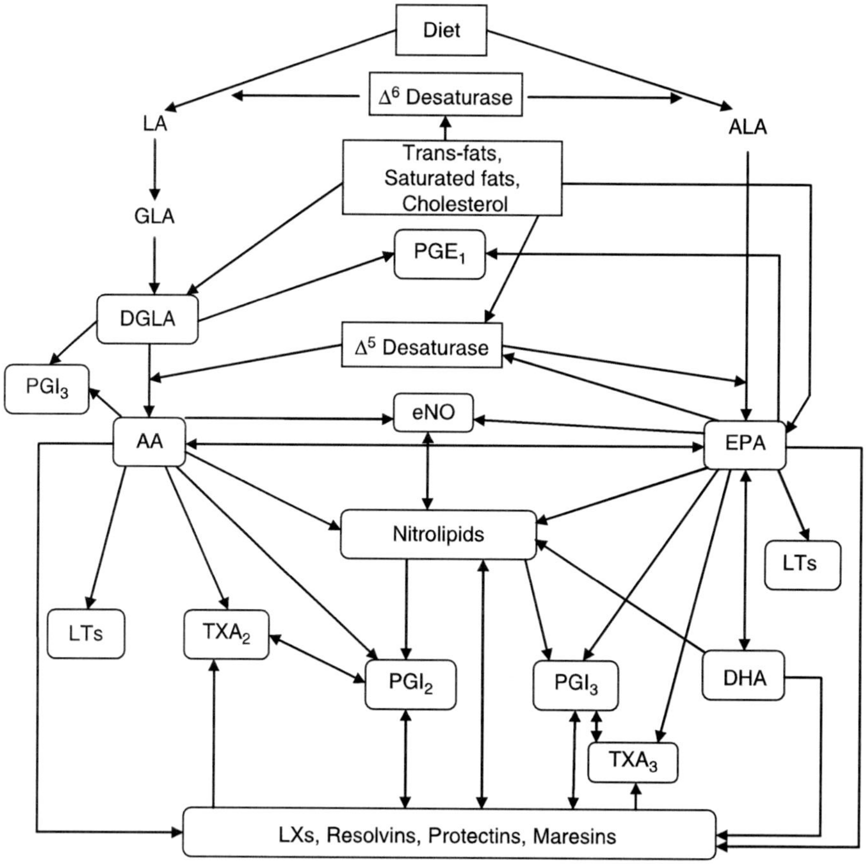

Fig. 8.2 Scheme showing interaction(s) among n-3, n-6 fatty acids and nitrolipids and various eicosanoids derived from PUFAs and lipoxins (LXs), resolvins, protectins and maresins. Trans-fats inhibit the activities of Δ^6 and Δ^5 desaturases and thus, interfere with the formation of AA, EPA, and DHA from their respective precursors LA and ALA, and thus, may reduce the formation and actions of PGE_1, PGI_2, PGI_3, LXs, resolvins, protectins, maresins and nitrolipids and at the same time may augment the formation and/or action of LTs, and TXs. DGLA increases the conversion of EPA to PGI_3, a potent vasodilator and platelet anti-aggregator, while AA augments the conversion of EPA to PGI_3. On the other hand, EPA inhibits the activity of the enzyme Δ^5desaturase that results in an increase in the concentrations of DGLA in the tissues (especially in the endothelial cells), an event that increase the tissue levels of DGLA leading to an increase in the formation of PGE_1, a vasodilator and platelet anti-aggregator. Thus, EPA can indirectly enhance the formation of PGE_1. Statins (HMG-CoA reductase inhibitors) and glitazones (PPARs agonists) may mediate some of their beneficial actions by enhancing the conversion of LA and ALA to DGLA and AA and EPA and DHA and their metabolites such as LXs, resolvins, and protectins and maresins, which are potent anti-inflammatory molecules. Cholesterol and saturated fatty acids also block the activities of both Δ^6 and Δ^5 desaturases and inhibit the conversion of dietary LA and ALA to their respective long-chain metabolites and render cell membrane more rigid. Trans-fats, cholesterol, and saturated fatty acids enhance whereas ω-3 fatty acids decrease the levels of pro-inflammatory cytokines. Thus, trans-fats, cholesterol, and saturated fatty acids have pro-inflammatory actions. AA, EPA and DHA enhance eNO generation while trans-fats, saturated fats and cholesterol inhibit eNO generation.

can inhibit lipid peroxidation *in vitro* to a limited extent [82]. This suggests that the anti-hypertensive drugs currently in use bring about some of their beneficial actions by blocking free radical generation and lipid peroxidation process.

In hypertension, the activity of angiotensin converting enzyme is high compared to the normal controls [113]. Angiotensin-II, a potent vasoconstrictor, can stimulate free radical generation [82]. Prior studies [84] have shown that superoxide anion can cause vasospasm. Hence, the hypertensive action of angiotensin-II can be attributed to its ability to enhance free radical generation.

Further studies showed that the activity of angiotensin converting enzyme (ACE) can be inhibited by NO [113, 155]. This suggests that one possible mechanism by which NO can bring about its anti-hypertensive action could be by its ability to modulate ACE activity in addition to its direct vasodilator action.

Essential Fatty Acids and Hypertension

Epidemiological studies have suggested that people who subsist on vegetarian diet have lower blood pressure than the general population. Vegetarians eat more polyunsaturated fatty acids (PUFAs). Hence, we measured the concentrations of various PUFAs in the plasma phospholipid fraction of patients with hypertension. The levels of linoleic acid (LA), gamma-linolenic acid (GLA), arachidonic acid (AA) and 22:5 n-6 were significantly lower in patients with essential hypertension [83]. The concentrations of dihomoGLA (DGLA) and docosahexaenoic acid (DHA) also tended to be low but not statistically so. Since PUFAs, but not the eicosanoids derived from them, inhibited the activity of ACE and enhanced the synthesis of NO [82, 83, 113], these results suggest that there is a close interaction between ACE activity, PUFAs and NO which may have relevance to the pathobiology of hypertension (Fig. 8.3).

As already discussed above, (a) saturated fatty acids reduce the formation of $PGE_1$and PGI_2, elevate blood pressure, and exacerbate spontaneous hypertension [137], (b) supplementation of LA and DGLA augment the synthesis of PGE_1 and PGI_2 and prevent the increase in blood pressure induced by saturated fats, and (c) fish oil, a rich source of ω-3 fatty acids: eicosapentaenoic acid (EPA, 20:5) and docosahexaenoic acid (DHA, 22:6), reduced blood viscosity and lowered blood pressure [146, 156–162]. In these studies, it was also noted that DHA is more effective than EPA in reducing blood pressure. EPA and DHA inhibit the formation of thromboxane A_2 (TXA_2), a potent vasoconstrictor and platelet aggregator; enhance that of PGI_3,

---

PUFAs interact with NO to form nitrolipids that can release NO and also possess anti-inflammatory and vasodilator actions. Lipoxins, resolvins, protectins and maresins can enhance the synthesis of eNO and PGI_2 and PGI_3. This close interaction between n-3 and n-6 fatty acids, trans-fats, saturated fatty acids, cholesterol and their ability to modify inflammatory markers, production of PGI_2, PGE_1, PGI_3, LXs, resolvins, protectins, maresins, NO, and nitrolipids explains the relationship between various fatty acids, low-grade systemic inflammation, and their role in hypertension (see text for further details)

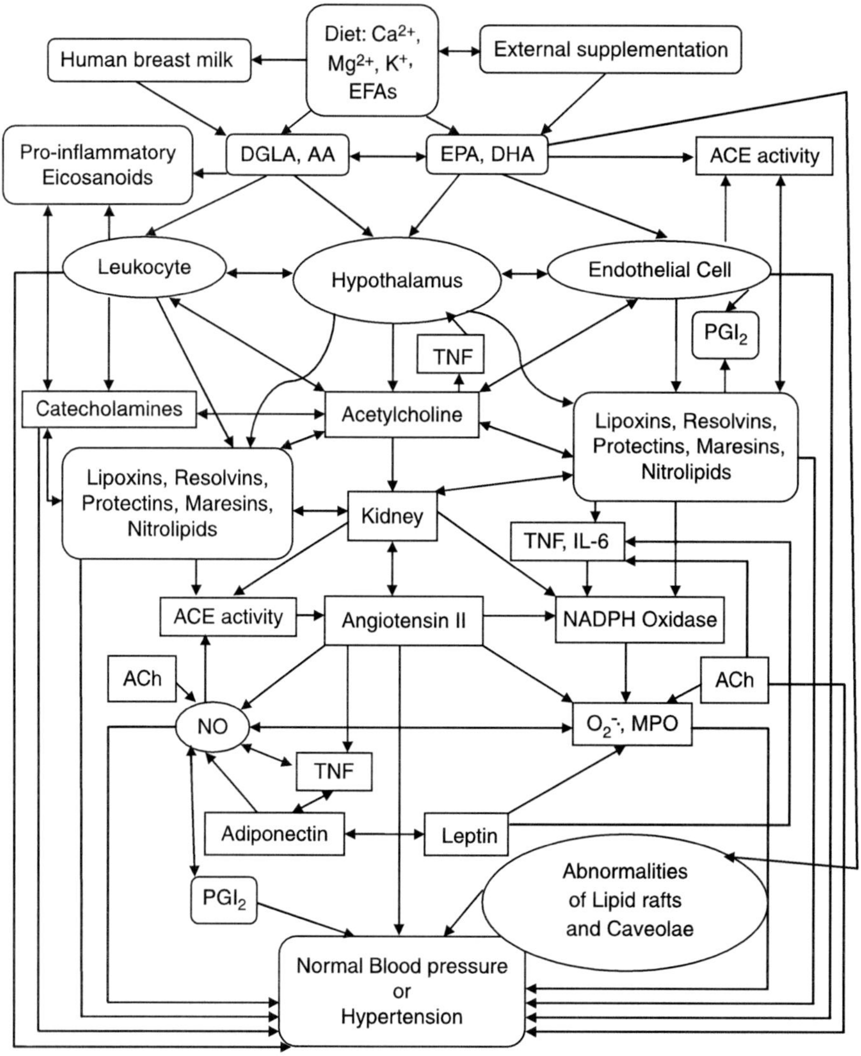

Fig. 8.3 Dietary factors such as EFAs, Mg^{2+}, Ca^{2+}, potassium, sodium, vitamin B_6, and nicotinic acid have a role in the pathobiology of hypertension. Minerals and vitamins may serve as co-factors of the enzymes Δ^6 and Δ^5 desaturases that are essential to metabolize dietary essential fatty acids to their long-chain polyunsaturated fatty acids such as GLA, DGLA, AA, EPA and DHA. AA, EPA and DHA can be metabolized to form anti-inflammatory and anti-oxidant lipid molecules such as lipoxins, resolvins, protectins, maresins and nitrolipids. PUFAs can enhance the production of eNO. A dietary deficiency of EFAs/PUFAs and/or genetic abnormality in the activity of the enzymes Δ^6 and Δ^5 desaturases (the activity of the enzymes may be slow compared to normal) may reduce the formation of AA, EPA and DHA in various tissues especially in the hypothalamus, kidney, endothelial cells (vascular tissue), leukocytes and platelets resulting in reduced formation of lipoxins, resolvins, protectins, maresins, nitrolipids, PGI_2 and PGI_3 that results in an increase in peripheral vascular resistance and enhanced platelet aggregation leading to the initiation and

a vasodilator and platelet anti-aggregator; and lower the tissue levels of AA and enhances those of DGLA, the precursor of PGE_1. Thus, it is expected that provision of adequate amounts of n-3 and n-6 fatty acids in the right proportion may help to prevent the development of hypertension (see Fig. 8.2). It is also important to note that AA, EPA and DHA form precursors to anti-inflammatory compounds such as lipoxins, resolvins, protectins, maresins and nitrolipids. DHA is the direct precursor of protectins and corresponding nitrolipids that have cytoprotective actions and prevent leukocyte activation. It is possible that lipoxins, resolvins, protectins and nitrolipids inhibit leukocyte activation and thus, prevent inappropriate production of superoxide anion, and block the release of pro-inflammatory cytokines such as IL-6, TNF-α, MIF (macrophage migration inhibitory factor) and IL-1 (these cytokines also stimulate the production of free radicals) and thus, block the vasoconstrictor actions of free radicals and reduce blood pressure. Since, DHA is more effective than EPA in reducing blood pressure; it is possible that protectins are the most likely candidates that regulate blood pressure and bring about the anti-hypertensive action of DHA. This implies that development of stable synthetic analogues of protectins may be useful as anti-hypertensive molecules.

In addition, PUFAs, especially DGLA, AA, EPA and DHA, not only form precursors to PGE_1, PGI_2, and PGI_3, but also inhibit ACE activity [113] and augment the synthesis of eNO (reviewed in [144, 145]). L-arginine and eNO, in turn, are known to up regulate the metabolism of PUFAs [163]. This suggests that in the presence of low tissue concentrations of PUFAs the synthesis and release of eNO will be decreased and vice versa. Since endothelial cells are the major source of NO, it is possible that PUFA content of endothelial cells would have a major impact on the synthesis and release of NO. This is supported by the observation that in patients with hypertension the plasma concentrations of LA, AA, and DHA and eNO are low [164]. Normal Asian Indians, who are at high risk of developing insulin resistance and hypertension, have significantly lower concentrations of AA, EPA, and DHA than do normal, healthy Canadians and Americans in their plasma phospholipids

---

progression of atherosclerosis and thrombosis. Decrease formation of PUFAs, lipoxins, resolvins, protectins, maresins and nitrolipids leads to increase in oxidative stress, activation of leukocytes, increase in the formation and release of IL-6 and TNF-α, angiotensin II, enhanced leptin production, events that lead to the development of hypertension. DHA-induced decrease in blood pressure in animal studies could be attributed to increased formation of protectins. DHA can be retroconverted to EPA and both EPA and DHA form precursors to resolvins that have potent anti-inflammatory actions. Lipoxins are anti-inflammatory compounds formed from AA. Lipoxins, resolvins, protectins and maresins inhibit MPO activity and prevent fibrosis. PUFAs and NO inhibit ACE activity and thus, inhibit the production of angiotensin II, a potent vasoconstrictor, pro-hypertensive and pro-oxidant molecule. PUFAs interact with NO to form nitrolipids that donate NO and have vasodilator actions and prevent platelet aggregation. PUFAs, PGI_2, lipoxins, resolvins, protectins and maresins may also possess anti-arrhythmic actions and thus, prevent cardiac arrhythmias (Lipids Health Dis 2008; 7: 37). Altered levels of PUFAs/EFAs can produce changes in the properties and actions of lipid rafts and caveolae of different cells that alters the production of free radicals, NADPH oxidase activity, ACE activity, and production of cytokines, neurotransmitters and other enzymes such as Δ^6 and Δ^5 desaturases, which ultimately influence the development of hypertension

[165]. Further, PUFAs and their metabolites such as some PGs, lipoxins, resolvins and nitrolipids inhibit the synthesis of TNF-α and other pro-inflammatory cytokines [144, 145, 166–168], that have a significant role in insulin resistance and metabolic syndrome [169, 170]. Hence, low plasma concentrations of PUFAs seen in Indian Asians could lead to an increase the production of TNF-α, IL-6 that, in turn, may render them more susceptible to develop insulin resistance, hypertension, and metabolic syndrome. Despite these evidences, it is not clear how, when and why these molecules render an individual to develop hypertension. In this context, it is interesting to note that hypertension could be a low-grade systemic inflammatory condition and seeds of its occurrence in adult life are sown during the perinatal period.

Low-grade Systemic Inflammation Occurs in Hypertension

Elevated plasma IL-6 levels in women with hypertension and insulin resistance in men has been described [171]. A direct correlation between blood pressure and levels of ICAM-1 (intercellular adhesion molecule-1) and IL-6 was noted [172]. A direct relationship between plasma CRP (C-reactive protein) levels and advancing age, BMI (body mass index), systolic blood pressure, HDL, smoking, and hormone replacement therapy was reported in the Women's Health Study [173]. These observations suggest that low-grade systemic inflammation occurs in hypertension. Our earlier observation that in uncontrolled essential hypertension, elevated plasma lipid peroxides and significantly higher levels of leukocyte $O_2^- \cdot$, low eNO, decreased vitamin E and superoxide dismutase (SOD) in RBC membranes occurred lends support to this view [82].

Angiotensin II activates leukocyte NADPH oxidase and enhanced $O_2^- \cdot$ generation [15, 89]. Plasma adiponectin concentrations were enhanced and insulin resistance was decreased after the use of angiotensin converting enzyme (ACE) inhibitors and angiotensin-II receptor blockers [174]. β-blockers and calcium antagonists suppressed $O_2^- \cdot$ generation [82, 113]. This suggests that that β-blockers and calcium antagonists could augment plasma adiponectin levels similar to ACE inhibitors and angiotensin II receptor blockers. It is possible that anti-hypertensive drugs (except β-blockers) reduce peripheral vascular resistance and enhance insulin action by augmenting adiponectin secretion.

Does Adult Hypertension have its Origins in the Perinatal Period?

One of the issues that need to be established is when events that trigger the development of hypertension are initiated? There is reasonable evidence to suggest that adult hypertension has its origins in the perinatal period. For instance, it has been reported that breast milk consumption lowered blood pressure in later life [175]. Previously,

I attributed this beneficial action to the presence of significant amounts of PUFAs in human milk [6, 176–178]. This is supported by the observation that breast-fed infants had a significantly higher percentage of PUFAs in their tissues compared with those of the formula-fed group.

Previously, Weisinger et al. [179] showed that perinatal deficiency of the essential dietary ω-3 fatty acid, α-linolenic acid (ALA), produced a reduction in hypothalamic docosahexaenoic acid (DHA, 22:6 n-3) resulting in hypertension in Sprague-Dawley rats despite restoring the hypothalamic DHA levels to normal in the adult. This suggests that restoring hypothalamic DHA to normal is not sufficient to prevent the development of hypertension. Li et al. [180] reported that animals fed diet rich in ω-6 with very little ALA and then re-fed the control diet rich in ALA for 24 weeks, the DHA levels were still significantly less than the control values in PE (phosphatidylethanolamine), PS (phosphatidylserine) and PI (phosphatidylinositol) fractions, by 9%, 18% and 34%, respectively. These results suggest that n-6/n-3 PUFA imbalance early in life leads to irreversible changes in hypothalamic composition. The increased ALA and reduced DHA proportions in the animals re-fed ALA in later life are consistent with a dysfunction or down-regulation of the conversion of ALA to DHA. Begg et al. [181] report that different sources of ALA (canola or flaxseed oil) are effective in preventing hypertension related to n-3 fatty acid deficiency except that animals which received canola oil had lower body weight, less adiposity, lower plasma leptin levels and consumed less food, whereas animals fed safflower oil + flaxseed oil also had lower but less marked reductions in adiposity and plasma leptin levels compared to those that were given safflower oil only that developed n-fatty acid deficiency. In addition, safflower oil + flaxseed oil fed animals consumed more food and water. These results suggest that body weight, plasma leptin and brain DHA are the main determinants of blood pressure. This study also implies that the interaction between n-3 and n-6 fatty acids influences body weight, plasma leptin and possibly, fatty acid composition and its metabolism in various tissues, especially in tissues that play a role in the pathophysiology of hypertension [175, 176, 182]. But, it is not known whether these DHA-PUFA-deficient animals had any abnormalities in the NO-$O_2^- \cdot$ generation and pro-inflammatory cytokine profile. Based on the current evidence, it is reasonable to predict that there would be a decrease in eNO synthesis/half-life and an increase in $O_2^- \cdot$ generation and IL-6 and TNF-α levels. This implies that availability of adequate amounts of DHA and other PUFAs during the perinatal period prevents the development of hypertension in adulthood. Both EPA and DHA have been reported to inhibit the development of proteinuria and suppressed hypertension in stroke-prone spontaneously hypertensive rats, and prevented the exaggerated growth of vascular smooth muscle cells from these animals through suppression of TGF-β [183, 184]. EPA and DHA products such as lipoxins, resolvins, protectins and maresins are now known to resolve inflammation and act as anti-fibrotic agents, probably by suppressing the action of TGF-β. Hence, the beneficial actions of EPA and DHA both in the prevention of hypertension and fibrosis could be attributed to the increased formation of lipoxins, resolvins, protectins and maresins.

Hence, whenever DHA (and probably other fatty acids such as AA and EPA) levels are low in the brain (especially in the hypothalamus), the production of lipoxins, resolvins, protectins, maresins and nitrolipids will be low resulting in inflammation (as a result of increased production of TNF-α) and induction of hypertension [90, 91]. EPA and possibly, DHA, suppress leptin production [185] that has pro-inflammatory action [186] similar to angiotensin II which may explain the increased plasma leptin levels noted [181] and its role in hypertension since hypertension is a low-grade systemic inflammatory condition [176, 177, 187]. It is noteworthy that DHA is formed from EPA and DHA can be retroconverted to EPA and thus, a dynamic balance might occur between EPA and DHA. AA is also present in the brain but at relatively lower concentrations compared to EPA and DHA. AA forms precursor to pro-inflammatory eicosanoids and anti-inflammatory lipoxins, resolvins and nitrolipids and hence, its role in hypertension needs to be studied.

Based on these results [179–187], it is important to delve more deeply into the role of perinatal deficiency of PUFAs and their role in hypertension. It is likely that EPA/DHA/AA-deficient diets lead to have low levels of AA, EPA and DHA not only in the brain but also in other tissues such as endothelial cells, peripheral leukocytes and kidney that may explain enhanced leukocyte free radical generation in hypertension [188]. High levels of myeloperoxidase generation by activated leukocytes [189] could be secondary to reduced formation of lipoxins, resolvins protectins, maresins and nitrolipids [188]. It is likely (see Figs. 8.1, 8.2 and 8.3) that ALA/EPA/DHA/AA-deficiency leads to the development of hypertension as a result of (a) increased levels of plasma pro-inflammatory cytokines, (b) reduced levels of EPA/DHA, lipoxins, resolvins, protectins, maresins and nitrolipids in various tissues including vascular endothelial cells, hypothalamus, and kidney; (c) high levels of angiotensin II as a result of enhanced activity of ACE in the brain, leukocytes and kidney; (d) augmented production of free radicals due to enhanced NADPH oxidase and release high levels of myeloperoxidase by leucocytes and endothelial cells (e) reduced levels of eNO; (f) decreased plasma levels of adiponectin ((since hypertensives have peripheral insulin resistance and are more prone to develop type 2 diabetes mellitus and metabolic syndrome) [190]); (g) depressed anti-oxidant capacity; (h) enhanced sympathetic tone ((catecholamines have pro-inflammatory actions) [191]) and (i) low acetylcholine levels in the brain and leukocytes ((since acetylcholine is an anti-inflammatory molecule, enhances NO generation and its levels are enhanced by AA/EPA/DHA supplementation) [192, 193]). Furthermore, it is necessary to measure plasma levels of sFlt1 and soluble endoglin and VEGF and correlate with plasma eNO and tissue antioxidants need to be performed. In hypertension, there appears to be an increase in plasma VEGF and sFlt1 levels [194], though this has been disputed in other studies [195], while exercise seems to restore the abnormal levels to normal [196]. Some of these postulations could be performed in humans using peripheral leukocytes and macrophages since they contain the complete intracellular machinery for the generation, release and metabolism of dietary EFAs, lipoxins, resolvins, protectins, maresins, nitrolipids, catecholamines, acetylcholine and serotonin, renin-angiotensin system, and anti-oxidants. Such a study may prove to be interesting. It is recommended that even in preeclampsia similar studies need to be performed, especially with regard to plasma PUFAs and their metabolites. Limited studies evaluated the plasma PUFA levels and the results have been conflicting: some studies showing decrease whole others little or no change [197–199], while increased oxidative stress and imbalance in the PGI_2/TXA_2 levels and enhanced production of pro-inflammatory cytokines have been well documented [200–205]. These results suggest that in preeclampsia there could be defects in the metabolism of PUFAs, imbalance between pro- and anti-inflammatory cytokines and PGI_2/TXA_2, increased oxidative stress, and deficiency of nutritional factors such as Ca^{2+}, vitamin B_6, folic acid, vitamin B_{12}; and possibly, deficiency of lipoxins, resolvins, protectins, maresins and nitrolipids. What is certain is that endothelial dysfunction probably, secondary to alterations in cell membrane fluidity (due to altered PUFA content) and immunological dysfunction leads to the initiation and progression of both hypertension and preeclampsia (see Fig. 8.1).

References

[1] Lawes CMM, Horn SV, Rodgers A; for the International Society of Hypertension (2008) Global burden of blood-pressure-related disease, 2001. Lancet 371:1513–1518

[2] Conen D, Ridker PM, Mora S, Buring JE, Glynn RJ (2007) Blood pressure and risk of developing type 2 diabetes mellitus: the women's health study. Eur Heart J 28:2937–2943

[3] Das UN (2008) Risk of type 2 diabetes mellitus in those with hypertension. Eur Heart J 29:952–953

[4] Feber J, Ahmed M (2010) Hypertension in children: new trends and challenges. Clin Sci (London) 119:151–161

[5] Das UN (2004) Long-chain polyunsaturated fatty acids interact with nitric oxide, superoxide anion, and transforming growth factor-b to prevent human essential hypertension. Eur J Clin Nutr 58:195–203

[6] Das UN (2006) Hypertension as a low-grades systemic inflammatory condition that has its origins in the perinatal period. J Assoc Physicians India 54:133–142

[7] McCarron DA, Morris CD, Henry HJ, Stanton JL (1984) Blood pressure and nutrient intake in the United States. Science 224:1392–1398

[8] Townsend MS, Fulgoni VL 3rd, Stern JS, Adu-Afarwuah S, McCarron DA (2005) Low mineral intake is associated with high systolic blood pressure in the third and fourth national health and nutrition examination surveys: could we all be right? Am J Hypertens 18(2 Pt 1):261–269

[9] Das UN (1985) Minerals, trace elements, and vitamins interact with essential fatty acids and prostaglandins to prevent hypertension, thrombosis, hypercholesterolaemia and atherosclerosis and their attendant complications. IRCS Med Sci 13:684–687

[10] Tannen RL (1983) Effects of potassium on blood pressure control. Ann Intern Med 98:773–780

[11] McCarron DA, Reusser ME (2001) Are low intakes of calcium and potassium important causes of cardiovascular disease? Am J Hypertens 14(6 Pt 2):206S–212S

[12] McCarron DA, Metz JA, Hatton DC (1998) Mineral intake and blood pressure in African Americans. Am J Clin Nutr 68:517–518

[13] Department of Health and Human Services (1983) DHSS Publications. (PH 5), London, pp 83–1676

[14] Kolata G (1984) Does a lack of calcium cause hypertension? Science 225:705–706

[15] Altura BM, Altura BT, Carella A (1983) Magnesium deficiency-induced spasms of umbilical vessels: relation to preeclampsia, hypertension, growth retardation. Science 221:376–378

[16] Siegel G, Walter A, Gustavsson H, Lindman B (1981) Magnesium and membrane function in vascular smooth muscle. Artery 9:232–252
[17] Mahfouz MM, Kummerow FA (1989) Effect of magnesium deficiency on delta 6 desaturase activity and fatty acid composition of rat liver microsomes. Lipids 24:727–732
[18] Horrobin DF, Oka M, Manku MS (1979) The regulation of prostaglandin E1 formation: a candidate for one of the fundamental mechanisms involved in the actions of vitamin C. Med Hypotheses 5:849–858
[19] Srivastava KC (1985) Ascorbic acid enhances the formation of prostaglandin E1 in washed human platelets and prostacyclin in rat aortic rings. Prostaglandins Leukot Med 18:227–233
[20] Manku MS, Oka M, Horrobin DF (1979) Differential regulation of the formation of prostaglandins and related substances from arachidonic acid and from dihomogammalinolenic acid. II. Effects of vitamin C. Prostaglandins Med 3:129–137
[21] Gava AL, Peotta VA, Cabral AM, Vasquez EC, Meyrelles SS (2008) Overexpression of eNOS prevents the development of renovascular hypertension in mice. Can J Physiol Pharmacol 86:458–464
[22] Bachmann S, Mundel P (1994) Nitric oxide in the kidney: synthesis, localization, and function. Am J Kidney Dis 24:112–129
[23] Fujiwara N, Osanai T, Kamada T, Katoh T, Takahashi K, Okumura K (2000) Study on the relationship between plasma nitrite and nitrate level and salt sensitivity in human hypertension: modulation of nitric oxide synthesis by salt intake. Circulation 101:856–861
[24] Das UN (2001) Nutritional factors in the pathobiology of human essential hypertension. Nutrition 17:337–346
[25] Passmore JC, Hatton DC, McCarron DA (1997) Dietary calcium decreases blood pressure without decreasing renal vascular resistance or altering the response to NO blockade. J Lab Clin Med 130:627–634
[26] Schmidt RJ, Beierwaltes WH, Baylis C (2001) Effects of aging and alterations in dietary sodium intake on total nitric oxide production. Am J Kidney Dis 37:900–908
[27] Tan DY, Meng S, Cason GW, Manning RD Jr (2000) Mechanisms of salt-sensitive hypertension: role of inducible nitric oxide synthase. Am J Physiol Regul Comp Physiol 279:R2297–R2303
[28] Bayorh MA, Williams E, Thierry-Palmer M, Sanford G, Emmett N, Harris-Hooker S, Socci RR, Chu TC (1999) Enhanced nitric oxide synthesis reverses salt-induced alterations in blood flow and cGMP levels. Clin Exp Hypertens 21:333–352
[29] Taddei S, Mattei P, Virdis A, Sudano L, Ghiadoni L, Salvetti A (1994) Effect of potassium on vasodilation to acetylcholine in essential hypertension. Hypertension 23:485–490
[30] Boger RH, Bode-Boger SM, Szuba A, Tsao PS, Chan JR, Tangphao O, Blaschke TF, Cooke JP (1998) Asymmetric dimethylarginine: a novel risk factor for endothelial dysfunction. Its role in hypercholesterolemia. Circulation 98:1842–1847
[31] Zoccali C, Bode-Boger SM, Mallamaci F, Benedetto FA, Tripepi G, Malatino LS, Cataliotti A, Bellanuova I, Fermo I, Frolich JC, Boger RH (2001) Plasma concentration of asymmetrical dimethylarginine and mortality in patients with end-stage renal disease: a prospective study. Lancet 358:2113–2117
[32] Vallance P (2001) Importance of asymmetrical dimethylarginine in cardiovascular risk. Lancet 358:2096–2097
[33] Surdacki A, Nowicki M, Sandmann J, Tsikas D, Boeger RH, Bode-Boeger SM, Kruszelnicka-Kwiatkowska O, Kokot F, Dubiel JS, Froelich JC (1999) Reduced urinary excretion of nitric oxide metabolites and increased plasma levels of asymmetric dimethylarginine in men with essential hypertension. J Cardiovasc Pharmacol 33:652–658
[34] Valkonen V-P, Paiva H, Salonen JT, Lakka TA, Lehtimaki T, Laakso J, Laaksonen R (2001) Risk of acute coronary events and serum concentrations of asymmetrical dimethylarginine. Lancet 358:2127–2128
[35] Taddei S, Virdis A, Mattei P, Ghiadoni L, Sudano I, Salvetti A (1996) Defective L-arginine-nitric oxide pathway in offspring of essential hypertensive patients. Circulation 94:1298–1303

[36] McAllister AS, Atkinson AB, Johnston GD, Hadden DR, Bell PM, McCance DR (1999) Basal nitric oxide production is impaired in offspring of patients with essential hypertension. Clin Sci (London) 97:141–147
[37] Var A, Yildirim Y, Onur E, Kuscu NK, Uyanik BS, Goktalay K, Guvenc Y (2003) Endothelial dysfunction in preeclampsia. Increased homocysteine and decreased nitric oxide levels. Gynecol Obstet Invest 56:221–224
[38] Wang Y, Gu Y, Zhang Y, Lewis DF (2004) Evidence of endothelial dysfunction in preeclampsia: decreased endothelial nitric oxide synthase expression is associated with increased cell permeability in endothelial cells from preeclampsia. Am J Obstet Gynecol 190:817–824
[39] Chu ZM, Beilin LJ (1993) Nitric oxide-mediated changes in vascular reactivity in pregnancy in spontaneously hypertensive rats. Br J Pharmacol 110:1184–1188
[40] Seligman SP, Buyon JP, Clancy RM, Young BK, Abramson SB (1994) The role of nitric oxide in the pathogenesis of preeclampsia. Am J Obstet Gynecol 171:944–948
[41] Mutlu-Türkoglu U, Aykaç-Toker G, Ibrahimoglu L, Ademoglu E, Uysal M (1999) Plasma nitric oxide metabolites and lipid peroxide levels in preeclamptic pregnant women before and after delivery. Gynecol Obstet Invest 48:247–250
[42] Rytlewski K, Olszanecki R, Korbut R, Zdebski Z (2005) Effects of prolonged oral supplementation with l-arginine on blood pressure and nitric oxide synthesis in preeclampsia. Eur J Clin Invest 35:32–37
[43] Bernardi F, Constantino L, Machado R, Petronilho F, Dal-Pizzol F (2008) Plasma nitric oxide, endothelin-1, arginase and superoxide dismutase in pre-eclamptic women. J Obstet Gynaecol Res 34:957–963
[44] Aris A, Benali S, Ouellet A, Moutquin JM, Leblanc S (2009) Potential biomarkers of preeclampsia: inverse correlation between hydrogen peroxide and nitric oxide early in maternal circulation and at term in placenta of women with preeclampsia. Placenta 30:342–347
[45] Matsubara K, Matsubara Y, Hyodo S, Katayama T, Ito M (2010) Role of nitric oxide and reactive oxygen species in the pathogenesis of preeclampsia. J Obstet Gynaecol Res 36:239–247
[46] Davidge ST, Stranko CP, Roberts JM (1996) Urine but not plasma nitric oxide metabolites are decreased in women with preeclampsia. Am J Obstet Gynecol 174:1008–1013
[47] Smàrason AK, Allman KG, Young D, Redman CW (1997) Elevated levels of serum nitrate, a stable end product of nitric oxide, in women with pre-eclampsia. Br J Obstet Gynaecol 104:538–543
[48] Roberts JM, Taylor RN, Musci TJ, Rodgers GM, Hubel CA, McLaughlin MK (1989) Preeclampsia: an endothelial cell disorder. Am J Obstet Gynecol 161:1200–1204
[49] Fisher SJ (2004) The placental problem: linking abnormal cytotrophoblast differentiation to the maternal symptoms of preeclampsia. Reprod Biol Endocrinol 2:53
[50] Levine RJ, Maynard Qian C et al (2004) Circulating angiogenic factors and the risk of preeclampsia. N Engl J Med 350:672–683
[51] Chaiworapongsa T, Romero R, Kim YM et al (2005) Plasma soluble vascular endothelial growth factor receptor-1 concentration is elevated prior to the clinical diagnosis of preeclampsia. J Matern Fetal Neonatal Med 17:3–18
[52] Hertig A, Berkane N, Lefevre G et al (2004) Maternal serum sFlt1 concentration is an early and reliable predictive marker of preeclampsia. Clin Chem 50:1702–1703
[53] Maynard SE, Min JY, Merchan J et al (2003) Excess placental soluble fms-like tyrosine kinase 1 (sFlt1) may contribute to endothelial dysfunction, hypertension, and proteinuria in preeclampsia. J Clin Invest 111:649–658
[54] Koga K, Osuga Y, Yoshino O et al (2003) Elevated serum soluble vascular endothelial growth factor receptor 1 (sVEGFR-1) levels in women with preeclampsia. J Clin Endocrinol Metab 88:2348–2351
[55] Levine RJ, Lam C, Qian C, Yu KF, Maynard SE, Sachs BP, Sibai BM, Epstein FH, Romero R, Thadani R, Karumanchi SA, for the CPEP Study Group (2006) Soluble endoglin and other circulating antiangiogenic factors in preeclampsia. N Engl J Med 355:992–1005

[56] Venkatesha S, Toporsian M, Lam C et al (2006) Soluble endoglin contributes to the pathogenesis of preeclampsia. Nat Med 12:642–649
[57] Taylor RN, Varma M, Teng NN, Roberts JM (1990) Women with preeclampsia have higher plasma endothelin levels than women with normal pregnancies. J Clin Endocrinol Metab 71:1675–1677
[58] Nova A, Sibai BM, Barton JR, Mercer BM, Mitchell MD (1991) Maternal plasma level of endothelin is increased in preeclampsia. Am J Obstet Gynecol 165:724–727
[59] Fiore G, Florio P, Micheli L, Nencini C, Rossi M, Cerretani D, Ambrosini G, Giorgi G, Petraglia F (2005) Endothelin-1 triggers placental oxidative stress pathways: putative role in preeclampsia. J Clin Endocrinol Metab 90:4205–4210
[60] Fukumura D, Gohongi T, Kadambi A, Izumi Y, Ang J, Yun C-O, Buerk DG, Huang PL, Jain RK (2001) Predominant role of endothelial nitric oxide synthase in vascular endothelial growth factor-induced angiogenesis and vascular permeability. Proc Natl Acad Sci U S A 98:2604–2609
[61] Luft FC (2000) The pre-eclampsia enigma and the renin-angiotensin system. J Mol Med 78:63–65
[62] Xia Y, Zhou CC, Ramin SM, Kellems RE (2007) Angiotensin receptors, autoimmunity, and pre-eclampsia. J Immunol 179:3391–3395
[63] Xia Y, Ramin SM, Kellems RE (2007) Potential roles of angiotensin receptor-activating autoantibody in the pathophysiology of pre-eclampsia. Hypertension 50:269–275
[64] Zhou CC, Zhang Y, Irani RA, Zhang H, Mi T, Popek EJ, Hicks MJ, Ramin SM, Kellems RE, Xia Y (2008) Angiotensin receptor agonistic autoantibodies induce pre-eclampsia in pregnant mice. Nat Med 14:855–862
[65] Kanasaki K, Palmsten K, Sugimoto H, Ahmad S, Hamano Y, Xie L, Parry S, Augustin HG, Gattone VH, Folkman J, Strauss JF, Kalluri R (2008) Deficiency in catechol-O-methyltransferase and 2-methoxyoestradiol is associated with pre-eclampsia. Nature 453:1117–1122
[66] Loscalzo J (1996) The oxidant stress of hyperhomocyst(e)inemia. J Clin Invest 98:5–7
[67] Heinecke JW, Rosen H, Suzuki LA, Chait A (1987) The role of sulfur-containing amino acids in superoxide production and modification of low-density lipoprotein by arterial smooth muscle cells. J Biol Chem 262:10098–10103
[68] Wall RT, Harlan JM, Harker LA, Striker GE (1986) Homocyst(e)ine-induced endothelial cell injury in vitro: a model for the study of vascular injury. Thromb Res 18:113–121
[69] Stamler JS, Osborne JA, Jaraki O, Rabbani LE, Mullins M, Single D, Loscalzo J (1993) Adverse vascular effects of homocysteine are modulated by endothelium-derived relaxing factor and related oxides of nitrogen. J Clin Invest 91:308–318
[70] Upchurch GR, Welch GN, Freedman JE, Loscalzo J (1995) Homocys(e)ine attenuates endothelial glutathione peroxidase and thereby potentiates peroxide-mediated cell injury. Circulation 92:I–228
[71] Tsai J-C, Perrella MA, Yoshizumi M, Hseih CM, Haber E, Schlegel R, Lee ME (1994) Promotion of vascular smooth muscle cell growth by homocyst(e)ine: a link to atherosclerosis. Proc Natl Acad Sci U S A 91:6369–6373
[72] Holven KB, Aukrust P, Holm T, Ose L, Nenseter MS (2002) Folic acid treatment reduces chemokine release from peripheral blood mononuclear cells in hyperhomocysteinemic subjects. Arterioscler Thromb Vasc Biol 22:699–703
[73] Friso S, Jacques PF, Wilson PWF, Rosenberg IH, Selhub J (2001) Low circulating vitamin B6 is associated with elevation of the inflammation marker C-reactive protein independently of plasma homocysteine levels. Circulation 103:2788–2791
[74] Das UN (1981) Pyridoxine, thrombosis and prostaglandins. Lancet ii:638
[75] Das UN (1982) Hemostatic vitamins and prostaglandins. Med J Aust 2:316
[76] Li H, Lewis A, Brodsky S, Rieger R, Iden C, Goligorsky MS (2002) Homocysteine induces 3-hydroxy-3-methylglutaryl coenzyme A reductase in vascular endothelial cells. Circulation 105:1037–1043

[77] Guidot DM, Hybertson BM, Kitlowski RP, Repine JE (1996) Inhaled nitric oxide prevents IL-1 induced neutrophil accumulation and associated acute edema in isolated rat lungs. Am J Physiol 271:1225–1229

[78] Verhaar MC, Wever RM, Kastelein JJ, van Dam T, Koomans HA, Rabelink TJ (1998) 5-methyltetrahydrofolic acid, the active form of folic acid, restores endothelial function in familial hypercholesterolemia. Circulation 97:237–241

[79] Verma S, Dumont AS, Maitland A (2001) Tetrahydrobiopterin attenuates cholesterol induced coronary hyperreactivity to endothelin. Heart 86:706–708

[80] Huang A, Vita JA, Venema RC, Keaney JF Jr (2000) Ascorbic acid enhances endothelial nitric oxide synthase activity by increasing intracellular tetrahydrobiopterin. J Biol Chem 275:17399–17406

[81] Baker TA, Milstien S, Katusic ZS (2001) Effect of vitamin C on the availability of tetrahydrobiopterin in human endothelial cells. J Cardiovasc Pharmacol 37:333–338

[82] Kumar KV, Das UN (1993) Are free radicals involved in the pathobiology of human essential hypertension? Free Radic Res Commun 19:59–66

[83] Suryaprabha P, Das UN, Koratkar R, Sangeetha P, Ramesh G (1990) Free radical generation, lipid peroxidation and essential fatty acids in uncontrolled hypertension. Prostaglandins Leukot Essent Fatty Acids 41:27–33

[84] Katusic ZS, Vanhoutte PM (1989) Superoxide anion is an endothelium derived contracting factor. Am J Physiol 257:433–437

[85] Nakazono L, Watanabe N, Matsuno K, Sasaki J, Sato T, Inoue M (1991) Does superoxide underlie the pathogenesis of hypertension? Proc Natl Acad Sci U S A 88:10045–10048

[86] Jun T, Ke-Yan F, Catalano M (1996) Increased superoxide anion production in humans: a possible mechanism for the pathogenesis of hypertension. J Hum Hypertens 10:305–309

[87] Hamilton CA, Brosnan MJ, McIntyre M, Graham D, Dominiczak AF (2001) Superoxide excess in hypertension and aging: a common cause of endothelial dysfunction. Hypertension 37 (2 part 2):529–534

[88] Wolf G (2000) Free radical production and angiotensin. Curr Hypertens Rep 2:167–173

[89] Zhang H, Schmeisser A, Garlichs CD, Plotze K, Damme U, Mugge A, Daniel WG (1999) Angiotensin II-induced superoxide anion generation in human vascular endothelial cells: role of membrane-bound NADH-NAD(P)H-oxidases. Cardiovasc Res 44:215–222

[90] Lob HE, Marvar PJ, Guzik TJ, Sharma S, McCann LA, Weyand C, Gordon FJ, Harrison DG (2010) Induction of hypertension and peripheral inflammation by reduction of extracellular superoxide dismutase in the central nervous system. Hypertension 55:277–283

[91] Retailleau K, Belin de Chantemèle EJ, Chanoine S, Guihot AL, Vessières E, Toutain B, Faure S, Bagi Z, Loufrani L, Henrion D (2010) Reactive oxygen species and cyclooxygenase 2-derived thromboxane A2 reduce angiotensin II type 2 receptor vasorelaxation in diabetic rat resistance arteries. Hypertension 55:339–344

[92] Mazor R, Itzhaki O, Sela S, Yagil Y, Cohen-Mazor M, Yagil C, Kristal B (2010) Tumor necrosis factor-alpha: a possible priming agent for the polymorphonuclear leukocyte-reduced nicotinamide-adenine dinucleotide phosphate oxidase in hypertension. Hypertension 55:353–362

[93] Walker AE, Seibert SM, Donato AJ, Pierce GL, Seals DR (2010) Vascular endothelial function is related to white blood cell count and myeloperoxidase among healthy middle-aged and older adults. Hypertension 55:363–369

[94] Carneado J, Alvarez de Sotomayor M, Perez-Guerrero C, Jimmenez L, Herrera MD, Pamies E, Martin-Sanz MD, Stiefel P, Miranda M, Bravo L, Marhuenda E (2002) Simvastatin improves endothelial function in spontaneously hypertensive rats through a superoxide dismutase mediated antioxidant effect. J Hypertens 20:429–437

[95] El Midaoui A, Wu R, De Champlain J (2002) Prevention of hypertension, hyperglycemia and vascular oxidative stress by aspirin treatment in chronically glucose-fed rats. J Hypertens 20:1407–1412

[96] Das UN (2005) Can COX-2 inhibitors-induced increase in cardiovascular disease risk be modified by essential fatty acids? J Assoc Physicians India 53:623–627

[97] Das UN (2006) Clinical laboratory tools to diagnose inflammation. Adv Clin Chem 41:189–229

[98] Serhan CN (2009) Systems approach to inflammation resolution: identification of novel anti-inflammatory and pro-resolving mediators. J Thromb Haemost 7(Suppl 1):44–48

[99] Zalba G, San Jose G, Moreno MU, Fortuno MA, Fortuno A, Beaumont FJ, Diez J (2001) Oxidative stress in arterial hypertension: role of NAD(P)H oxidase. Hypertension 38:1395–1399

[100] Morawietz H, Weber M, Rueckschloss U, Lauer N, Hacker A, Kojda G (2001) Upregulation of vascular NAD(P)H oxidase subunit gp91phox and impairment of the nitric oxide signal transduction pathway in hypertension. Biochem Biophys Res Commun 285:1130–1135

[101] Kitamoto S, Egashira K, Kataoka C, Usui M, Koyanagi M, Takemoto M, Takeshita A (2000) Chronic inhibition of nitric oxide synthesis in rats increases aortic superoxide anion production via the action of angiotensin II. J Hypertens 18:1795–1800

[102] Zhang H, Schmeisser A, Garlichs CD, Plotze K, Damme U, Mugge A, Daniel WG (1999) Angiotensin II-induced superoxide anion generation in human vascular endothelial cells: role of membrane-bound NADH-NAD(P)H-oxidases. Cardiovasc Res 44:215–222

[103] Rey FE, Cifuentes ME, Kiarash A, Quinn MT, Pagano PJ (2001) Novel competitive inhibitor of NAD(P)H oxidase assembly attenuates vascular O(2)(−) and systolic blood pressure in mice. Circ Res 89:408–414

[104] Guzik TJ, West NE, Pillai R, Taggart DP, Channon KM (2002) Nitric oxide modulates superoxide release and peroxynitrite formation in human blood vessels. Hypertension 39:1088–1094

[105] Kaehler J, Sill B, Koester R, Mittmann C, Orzechowski HD, Muenzel T, Meinertz T (2002) Endotelin-1 mRNA and protein in vascular wall cells is increased by reactive oxygen species. Clin Sci (London) 103(Suppl 1):176S–178S

[106] Smith AP, Demoncheaux EA, Higenbottam TW (2002) Nitric oxide gas decreases endothelin-1 mRNA in cultured pulmonary artery endothelial cells. Nitric Oxide 6:153–159

[107] Cardillo C, Kilcoyne CM, Quyyumi AA, Cannon RO 3rd, Panza JA (1998) Selective defect in nitric oxide synthesis may explain the impaired endothelium-dependent vasodilation in patients with essential hypertension. Circulation 97:851–856

[108] Takahashi H, Hara K, Komiyama Y, Masuda M, Murakami T, Nishimura M, Nambu A, Yoshimura M (1995) Mechanism of hypertension induced by chronic inhibition of nitric oxide in rats. Hypertens Res 18:319–324

[109] Haynes WG, Noon JP, Walker BR, Webb DJ (1993) Inhibition of nitric oxide synthesis increases blood pressure in healthy humans. J Hypertens 11:1375–1380

[110] Diederich D, Skopec J, Diederich A, Dai FX (1994) Cyclosporine produces endothelial dysfunction by increased production of superoxide. Hypertension 23(3 Part 2):957–961

[111] Kohno M, Yokokawa K, Minami M, Yasunari K, Maeda K, Kano H, Hanehira T, Yaoshikawa J (1999) Plasma levels of nitric oxide and related vasoactive factors following long-term treatment with angiotensin-converting enzyme inhibitor in patients with essential hypertension. Metabolism 48:1256–1259

[112] Takase H, Sugiyama M, Nakazawa A, Sato K, Ueda R, Dohi Y (2000) Long-term effect of antihypertensive therapy with calcium antagonist or angiotensin converting enzyme inhibitor on serum nitrite/nitrate levels in human essential hypertension. Arzneimittelforschung 50:530–534

[113] Kumar KV, Das UN (1997) Effect of cis-unsaturated fatty acids, prostaglandins, and free radicals on angiotensin converting enzyme activity in vitro. Proc Exp Biol Med 214:374–379

[114] Suthanthiran M, Li B, Song JO, Ding R, Sharma VK, Schwartz JE, August P (2000) Transforming growth factor-beta 1 hyperexpression in African-American hypertensives: a novel mediator of hypertension and/or target organ damage. Proc Natl Acad Sci U S A 97:3479–3484

[115] Derhaschnig U, Shehata M, Herkner H, Bur A, Woisetschlager C, Laggner AN, Hirschl MM (2002) Increased levels of transforming growth factor-beta 1 in essential hypertension. Am J Hypertens 15:207–211

[116] Yamada Y, Fujisawa M, Ando F, Niino N, Tanaka M, Shimokata H (2002) Association of a polymorphism of the transforming growth factor-beta 1 gene with blood pressure in Japanese individuals. J Hum Genet 47:243–248

[117] Houlihan CA, Akdeniz A, Tsalamandris C, Cooper ME, Jerums G, Gilbert RE (2002) Urinary transforming growth factor-beta excretion in patients with hypertension, type 2 diabetes, and elevated albumin excretion rate: effects of angiotensin receptor blockade and sodium restriction. Diabetes Care 25:1072–1077

[118] Russo LM, Osicka TM, Bonnet F, Jerums G, Comper WD (2002) Albuminuria in hypertension is linked to altered lysosomal activity and TGF-beta 1 expression. Hypertension 39:281–286

[119] Wong VY, Laping NJ, Nelson AH, Contino LC, Olson BA, Gygielko E, Campbell WG Jr, Barone F, Brooks DP (2001) Renoprotective effects of carvedilol in hypertensive-stroke prone rats may involve inhibition of TGF beta expression. Br J Pharmacol 134:977–984

[120] Esmatjes E, Flores L, Inigo P, Lario S, Ruilope LM, Campistol JM (2001) Effect of losartan on TGF-beta1 and urinary albumin excretion in patients with type 2 diabetes mellitus and microalbuminuria. Nephrol Dial Transplant 16(Suppl 1):90–93

[121] Dhaly AJ, Hoagland KM, Flasch AK, Jha S, Ledbetter SR, Roman RJ (2002) Antihypertensive effects of chronic anti-TGF-beta antibody therapy in Dahl S rats. Am J Physiol Regul Integr Comp Physiol 283:R757–R767

[122] Riser BL, Cortes P, Yee J, Sharba AK, Asano K, Rodriguez-Barbero A, Narins RG (1998) Mechanical strain- and high-glucose induced alterations in mesangial cell collagen metabolism: role of TGF-beta. J Am Soc Nephrol 9:827–836

[123] Shin GT, Khanna A, Ding R, Sharma VK, Lagman M, Li B, Suthanthiran M (1998) In vivo expression of transforming growth factor-beta1 in humans: stimulation by cyclosporine. Transplantation 65:313–318

[124] Ying WZ, Sanders PW (1999) Dietary salt increases endothelial nitric oxide synthase and TGF-beta in rat aortic endothelium. Am J Physiol 277(4 Part 2):H1293–H1298

[125] Trachtman H, Koss I, Bogart M, Abramowitz J, Futterweit S, Franki N, Singhal PC (1998) High glucose enhances growth factor-stimulated nitric oxide production by cultured rat mesangial cells. Res Commun Mol Pathol Pharmacol 100:213–225

[126] Craven PA, Studer RK, Felder J, Phillips S, DeRubertis FR (1997) Nitric oxide inhibition of transforming growth factor-beta and collagen synthesis in mesangial cells. Diabetes 46:671–681

[127] Studer RK, Georgescu HI, Miller LA, Evans CH (1999) Inhibition of transforming growth factor beta production by nitric oxide-treated chondrocytes: implications for matrix synthesis. Arthritis Rheum 42:248–257

[128] Metukuri MR, Namas R, Gladstone C, Clermont T, Jefferson B, Barclay D, Hermus L, Billiar TR, Zamora R, Vodovotz Y (2009) Activation of latent transforming growth factor-beta1 by nitric oxide in macrophages: role of soluble guanylate cyclase and MAP kinases. Wound Repair Regen 17:578–588

[129] Saura M, Zaragoza C, Herranz B, Griera M, Diez-Marqués L, Rodriguez-Puyol D, Rodriguez-Puyol M (2005) Nitric oxide regulates transforming growth factor-beta signaling in endothelial cells. Circ Res 97:1115–1123

[130] Boutard V, Havouis R, Fouqueray B, Philippe C, Moulinoux JP, Baud L (1995) Transforming growth factor-beta stimulates arginase activity in macrophages. Implications for the regulation of macrophage cytotoxicity. J Immunol 155:2077–2084

[131] Freire-de-Lima CG, Xiao YQ, Gardai SJ, Bratton DL, Schiemann WP, Henson PM (2006) Apoptotic cells, through transforming growth factor-beta, coordinately induce anti-inflammatory and suppress pro-inflammatory eicosanoid and NO synthesis in murine macrophages. J Biol Chem 281:38376–38384

[132] Romero R, Nien JK, Espinoza J, Todem D, Fu W, Chung H, Kusanovic JP, Gotsch F, Erez O, Mazaki-Tovi S, Gomez R, Edwin S, Chaiworapongsa T, Levine RJ, Karumanchi SA (2008) A longitudinal study of angiogenic (placental growth factor) and anti-angiogenic (soluble endoglin and soluble vascular endothelial growth factor receptor-1) factors in normal

pregnancy and patients destined to develop preeclampsia and deliver a small for gestational age neonate. J Matern Fetal Neonatal Med 21:9–23
[133] Tobian L, Ganguli M, Johnson MA, Iwai J (1982) Influence of renal prostaglandins and dietary linoleate on hypertension in Dahl rats. Hypertension 4(Suppl):149–153
[134] Hoffman P, Forster W, Markov ChM (1982) Attenuation of blood pressure increase in spontaneously hypertensive rats by diets enriched with polyunsaturated fatty acids. Prostaglandins Leukot Med 8:151–156
[135] Iacono JM, Judd JT, Marshall MW, Canary JJ, Dougherty RM, Mackin JF et al (1981) The role of dietary essential fatty acids and prostaglandins in reducing blood pressure. Prog Lipid Res 20:349–364
[136] Hassall CH, Kirtland SJ (1984) Dihomo-gamma-linolenic acid reverses hypertension induced in rats by diets rich in saturated fat. Lipids 19:699–703
[137] Mills DE, Ward R (1984) Attenuation of psychosocial stress-induced hypertension by gamma-linolenic acid (GLA) administration in rats. Proc Soc Exp Biol Med 176:32–37
[138] Yasuda H, Maeda K, Sonobe M, Kawabata T, Terada M, Hisanaga T, Taniguchi Y, Kikkawa R, Shigeta Y (1994) Metabolic effect of PGE1 analogue 01206.alpha CD on nerve Na(+)-K(+)-ATPase activity of rats with streptozocin-induced diabetes is mediated via cAMP: possible role of cAMP in diabetic neuropathy. Prostaglandins 47:367–378
[139] Horrobin DF, Durand LG, Manku MS (1977) Prostaglandin E1 modifies nerve conduction and interferes with local anaesthetic action. Prostaglandins 14:103–108
[140] Fu LW, Longhurst JC (2010) Bradykinin and thromboxane A2 reciprocally interact to synergistically stimulate cardiac spinal afferents during myocardial ischemia. Am J Physiol Heart Circ Physiol 298:H235–H244
[141] Wei LL, Shen YD, Zhang YC, Hu XY, Lu PL, Wang L, Chen W (2010) Roles of the prostaglandin E2 receptors EP subtypes in Alzheimer's disease. Neurosci Bull 26:77–84
[142] Yang H, Chen C (2008) Cyclooxygenase-2 in synaptic signaling. Curr Pharm Des 14:1443–1451
[143] Ajmone-Cat MA, Iosif RE, Ekdahl CT, Kokaia Z, Minghetti L, Lindvall O (2006) Prostaglandin E2 and BDNF levels in rat hippocampus are negatively correlated with status epilepticus severity: no impact on survival of seizure-generated neurons. Neurobiol Dis 23:23–35
[144] Das UN (2006) Essential fatty acids – a review. Curr Pharm Biotechnol 7:467–482
[145] Das UN (2006) Essential fatty acids: biochemistry, physiology, and pathology. Biotechnol J 1:420–439
[146] Woodcock BE, Smith E, Lambert WH, Jones WM, Galloway JH, Greaves M et al (1984) Beneficial effect of fish oil on blood viscosity in peripheral vascular disease. Br Med J 288:592–594
[147] Das UN (1987) Biological significance of arachidonic acid. Med Sci Res 24:1485–1490
[148] Bordet JC, Guichardant M, Lagarde M (1986) Arachidonic acid strongly stimulates prostaglandin I3 (PGI3) production from eicosapentaenoic acid in human endothelial cells. Biochem Biophys Res Commun 135:403–410
[149] Juan H, Sametz W (1985) Dihomo-gamma-linolenic acid increases the metabolism of eicosapentaenoic acid in perfused vascular tissue. Prostaglandins Leukot Med 19:79–86
[150] Lagarde M, Véricel E, Croset M, Calzada C, Bordet JC, Guichardant M (1993) Interactions between arachidonic and eicosapentaenoic acids during their dioxygenase-dependent peroxidation. Prostaglandins Leukot Essent Fatty Acids 48:23–25
[151] Bordet JC, Guichardant M, Lagarde M (1988) Hydroperoxides produced by n-6 lipoxygenation of arachidonic and linoleic acids potentiate synthesis of prostacyclin related compounds. Biochim Biophys Acta 958:460–468
[152] Hamazaki T, Hirai A, Terano T, Sajiki J, Kondo S, Fujita T, Tamura Y, Kumagai A (1982) Effects of orally administered ethyl ester of eicosapentaenoic acid (EPA; C20:5, omega-3) on PGI2-like substance production by rat aorta. Prostaglandins 23:557–567
[153] Rees D, Ben-Ishay D, Moncada S (1996) Nitric oxide and the regulation of blood pressure in the hypertension-prone and hypertension-resistant Sabra rat. Hypertension 28:367–371

[154] Cachofeiro V, Sakakibara T, Nasiletti A (1992) Kinins, nitric oxide and the hypertensive effect of captopril and ramipril in hypertension. Hypertension 19:138–145

[155] Persson K, Säfholm AC, Andersson RG, Ahlner J (2005) Glyceryl trinitrate-induced angiotensin-converting enzyme (ACE) inhibition in healthy volunteers is dependent on ACE genotype. Can J Physiol Pharmacol 83:1117–1122

[156] Smith JM, Kopp SJ, Paulson DJ, Daar JT (1993) Effects of dietary fish oil on cardiovascular responsiveness to adrenergic agonists in spontaneously hypertensive rat. Can J Physiol Pharmacol 71:432–438

[157] Mori TA, Bao DQ, Burke V, Puddey IB, Beilin LJ (1999) Docosahexaenoic acid but not eicosapentaenoic acid lowers ambulatory blood pressure and heart rate in humans. Hypertension 34:253–260

[158] Frenoux JM, Prost ED, Belleville JL, Prost JL (2001) A polyunsaturated fatty acid diet lowers blood pressure and improves antioxidant status in spontaneously hypertensive rats. J Nutr 131:39–45

[159] Ait-Yahia D, Madani S, Savelli JL, Prost J, Bouchenak M, Belleville J (2003) Dietary fish protein lowers blood pressure and alters tissue polyunsaturated fatty acid composition in spontaneously hypertensive rats. Nutrition 19:342–346

[160] Paschos GK, Magkos F, Panagiotakos DB, Votteas V, Zampelas A (2007) Dietary supplementation with flaxseed oil lowers blood pressure in dyslipidaemic patients. Eur J Clin Nutr 61:1201–1206

[161] Theobald HE, Goodall AH, Sattar N, Talbot DC, Chowienczyk PJ, Sanders TA (2007) Low-dose docosahexaenoic acid lowers diastolic blood pressure in middle-aged men and women. J Nutr 137:973–978

[162] Rousseau-Ralliard D, Moreau D, Guilland JC, Raederstorff D, Grynberg A (2009) Docosahexaenoic acid, but not eicosapentaenoic acid, lowers ambulatory blood pressure and shortens interval QT in spontaneously hypertensive rats in vivo. Prostaglandins Leukot Essent Fatty Acids 80:269–277

[163] Mohan IK, Das UN (2000) Effect of L-arginine-nitric oxide system on the metabolism of essential fatty acids in chemical-induced diabetes mellitus. Prostaglandins Leukot Essent Fatty Acids 62:35–46

[164] Das UN (1995) Essential fatty acid metabolism in patients with essential hypertension, diabetes mellitus, and coronary heart disease. Prostaglandins Leukot Essent Fatty Acids 52:387–391

[165] Das UN, Kumar KV, Ramesh G (1994) Essential fatty acid metabolism in South Indians. Prostaglandins Leukot Essent Fatty Acids 50:253–256

[166] Kumar GS, Das UN (1994) Effect of prostaglandins and their precursors on the proliferation of human lymphocytes and their secretion of tumor necrosis factor and various interleukins. Prostaglandins Leukot Essent Fatty Acids 50:331–336

[167] Di Battista JA, Martel-Pelletier J, Pelletier J (1999) Suppression of tumor necrosis factor (TNF-alpha) gene expression by prostaglandin E(2). Role of early growth response protein-1 (Egr-1). Osteoarthritis Cartilage 7:395–398

[168] Yoshida T, Cohen S (1982) Biological control of lymphokine function. Fed Proc 41:2480–2483

[169] Kern PA, Saghizadeh M, Ong JM, Bosch RJ, Deem R, Simsolo RB (1995) The expression of tumor necrosis factor in human adipose tissue. Regulation by obesity, weight loss, and relationship to lipoprotein lipase. J Clin Invest 95:2111–2119

[170] Hotamisligil GS, Arner P, Caro JF, Atkinson RL, Spiegelman BM (1995) Increased adipose tissue expression of tumor necrosis factor-alpha in human obesity and insulin resistance. J Clin Invest 95:2409–2415

[171] Das UN (2002) Is metabolic syndrome X an inflammatory condition? Exp Biol Med 227:989–997

[172] Fernandez-Real JM, Vayreda M, Richart C et al (2001) Circulating interleukins 6 levels, blood pressure, and insulin sensitivity in apparently healthy men and women. J Clin Endocrinol Metab 86:1154–1159

[173] Chae CU, Lee RT, Rifai N, Ridker PM (2001) Blood pressure and inflammation in apparently healthy men. Hypertension 38:399–403

[174] Bermudez EA, Rifai N, Buring J, Manson JE, Ridker PM (2002) Interrelationships among circulating interleukins-6, C-reactive protein, and traditional cardiovascular risk factors in women. Arterioscler Thromb Vasc Biol 22:1668–1673

[175] Singhal A, Cole TJ, Lucas A (2001) Early nutrition in preterm infants and later blood pressure: two cohorts after randomized trials. Lancet 357:413–419

[176] Das UN (2010) Metabolic syndrome pathophysiology: the role of essential fatty acids. Wiley-Blackwell, Ames

[177] Das UN (2002) A perinatal strategy for prevebnting adult disease: a role of long-chain polyunsaturated fatty acids. Kluwer, Boston

[178] Das UN (2001) Can perinatal supplementation of long-chain polyunsaturated fatty acids prevent hypertension in adult life? Hypertension 38:e6–e8

[179] Weisinger HS, Armitage JA, Sinclair AJ, Vingrys AJ, Burns PL, Weisinger RS (2001) Perinatal omega-3 fatty acid deficiency affects blood pressure later in life. Nat Med 7:258–259

[180] Li D, Weisinger HS, Weisinger RS, Mathai M, Armitage JA, Vingrys AJ, Sinclair AJ (2006) Omega 6 to omega 3 fatty acid imbalance early in life leads to persistent reductions in DHA levels in glycerophospholipids in rat hypothalamus even after long-term omega 3 fatty acid repletion. Prostaglandins Leukot Essent Fatty Acids 74:391–399

[181] Begg DP, Sinclair AJ, Stah LA, Premaratna SD, Hafandi A, Jois M, Weisinger RS (2010) Hypertension induced by ω-3 polyunsaturated fatty acid deficiency is alleviated by α-linolenic acid regardless of dietary source. Hypertens Res 33:808–813

[182] Das UN (2010) Essential fatty acids and their metabolites and hypertension. Hypertens Res 33:782–785

[183] Mano MT, Bexis S, Abeywardena MY, McMurchie EJ, King RA, Smith RM, Head RJ (1995) Fish oils modulate blood pressure and vascular contractility in the rat and vascular contractility in the primate. Blood Press 4:177–186

[184] Head RJ, Mano MT, Bexis S, Howe PR, Smith RM (1991) Dietary fish oil administration retards the development of hypertension and influences vascular neuroeffector function in the stroke prone spontaneously hypertensive rat (SHRSP). Prostaglandins Leukot Essent Fatty Acids 44:119–122

[185] Priante G, Musacchio E, Valvason C, Baggio B (2009) EPA and DHA suppress AngII- and arachidonic acid induced expression of profibrotic genes in human mesangial cells. J Nephrol 22:137–143

[186] Otero M, Lago R, Lago F, Casanueva FF, Dieguez C, Gómez-Reino JJ, Gualillo O (2005) Leptin, from fat to inflammation: old questions and new insights. FEBS Lett 579:295–301

[187] Das UN (2006) Hypertension as a low-grade systemic inflammatory condition that has its origins in the perinatal period. J Assoc Physicians India 54:133–142

[188] Das UN (2008) Leucocyte activation in coronary heart disease: but how and why? Eur Heart J 29:2317–2318

[189] Walker AE, Seibert SM, Donato AJ, Pierce GL, Seals DR (2010) Vascular endothelial function is related to white blood cell count and myeloperoxidase among healthy middle-aged and older adults. Hypertension 55:363–369

[190] Das UN (2008) Risk of type 2 diabetes mellitus in those with hypertension. Eur Heart J 29:952–953

[191] Flierl MA, Rittirsch D, Nadeau BA, Chen AJ, Sarma JV, Zetoune FS, McGuire SR, List RP, Day DE, Hoesel LM, Gao H, Rooijen NV, Huber-Lang MS, Neubig RR, Ward PA (2007) Phagocyte-derived catecholamines enhance acute inflammatory injury. Nature 449:721–725

[192] Das UN (2008) Beneficial actions of polyunsaturated fatty acids in cardiovascular diseases: but, how and why? Curr Nutr Food Sci 4:2031

[193] Favrelière S, Perault MC, Huguet F, De Javel D, Bertrand N, Piriou A, Durand G (2003) DHA-enriched phospholipid diets modulate age-related alterations in rat hippocampus. Neurobiol Aging 24:233–243

[194] Belgore FM, Blann AD, Li-Saw-Hee FL, Beevers DG, Lip GY (2001) Plasma levels of vascular endothelial growth factor and its soluble receptor (SFlt-1) in essential hypertension. Am J Cardiol 87:805–807

[195] Vyzantiadis T, Karagiannis A, Douma S, Harsoulis P, Vyzantiadis A, Zamboulis C (2006) Vascular endothelial growth factor and nitric oxide serum levels in arterial hypertension. Clin Exp Hypertens 28:603–609

[196] Hansen AH, Nielsen JJ, Saltin B, Hellsten Y (2010) Exercise training normalizes skeletal muscle vascular endothelial growth factor levels in patients with essential hypertension. J Hypertens 28:1176–1185

[197] Wang Y, Walsh SW, Kay HH (2005) Placental tissue levels of nonesterified polyunsaturated fatty acids in normal and preeclamptic pregnancies. Hypertens Pregnancy 24:235–245

[198] Shouk TA, Omar MN, Fayed ST (1999) Essential fatty acids profile and lipid peroxides in severe pre-eclampsia. Ann Clin Biochem 36(Part 1):62–65

[199] Wang YP, Kay HH, Killam AP (1991) Decreased levels of polyunsaturated fatty acids in preeclampsia. Am J Obstet Gynecol 164:812–818

[200] Ouyang YQ, Li SJ, Zhang Q, Cai HB, Chen HP (2009) Interactions between inflammatory and oxidative stress in preeclampsia. Hypertens Pregnancy 28:56–62

[201] Rogers MS, Wang CC, Tam WH, Li CY, Chu KO, Chu CY (2006) Oxidative stress in midpregnancy as a predictor of gestational hypertension and pre-eclampsia. BJOG 113:1053–1059

[202] Chavarría ME, Lara-Gonzàlez L, Gonzàlez-Gleason A, García-Paleta Y, Vital-Reyes VS, Reyes A (2003) Prostacyclin/thromboxane early changes in pregnancies that are complicated by preeclampsia. Am J Obstet Gynecol 188:986–992

[203] Lamarca B (2010) The role of immune activation in contributing to vascular dysfunction and the pathophysiology of hypertension during preeclampsia. Minerva Ginecol 62:105–120

[204] Parrish MR, Murphy SR, Rutland S, Wallace K, Wenzel K, Wallukat G, Keiser S, Ray LF, Dechend R, Martin JN, Granger JP, Lamarca B (2010) The effect of immune factors, tumor necrosis factor-alpha, and agonistic autoantibodies to the angiotensin II type I receptor on soluble fms-like tyrosine-1 and soluble endoglin production in response to hypertension during pregnancy. Am J Hypertens 23:911–916

[205] Guven MA, Coskun A, Ertas IE, Aral M, Zencirci B, Oksuz H (2009) Association of maternal serum CRP, IL-6, TNF-alpha, homocysteine, folic acid and vitamin B12 levels with the severity of preeclampsia and fetal birth weight. Hypertens Pregnancy 28:190–200

Chapter 9
Insulin Resistance, Dyslipidemia, Type 2 Diabetes Mellitus and Metabolic Syndrome

Introduction

Insulin resistance is characterized by diminished ability of tissues to respond to insulin that leads to compensatory hyperinsulinemia. Insulin resistance enhances the risk for type 2 diabetes mellitus, coronary heart disease (CHD), and stroke and is often associated with abdominal obesity, hypertension, hypertriglyceridemia, low high-density lipoprotein (HDL) cholesterol, and hyperglycemia [1]. Although insulin resistance is equated with impaired whole-body insulin-mediated glucose disposal, defective regulation of non-esterified fatty acid and glycerol metabolism occurs even in those in whom glucose tolerance is either normal or only marginally impaired [2]. In view of the fact that insulin resistance is seen in obesity, type 2 diabetes mellitus, CHD, stroke, dyslipidemia and metabolic syndrome, in the present discussion, at times, there could be some repetition of some of the evidences presented in support of certain claims/facts; this is intentional so that a comprehensive picture of the point or evidence under discussion is given at one place instead of asking the reader to go back and forth in search of these evidences or points.

Metabolic syndrome is also known as metabolic syndrome X, syndrome X, insulin resistance syndrome, Reaven's syndrome, and CHAOS. But, one has to make a distinction between insulin resistance *per se* and insulin resistance syndrome. Insulin resistance is simply diminished ability of tissues to respond to insulin and so their requirement of insulin to produce a specific effect is higher compared to normal, while insulin resistance syndrome or metabolic syndrome is characterized by the presence of obesity, insulin resistance, raised blood pressure, atherogenic dyslipidemia, pro-inflammatory state, and prothrombin state and thus, is associated with an increased risk of developing cardiovascular disease and type 2 diabetes mellitus. For ease of understanding, term insulin resistance is used to define a state of diminished ability of tissues to respond to insulin, while metabolic syndrome consists of abdominal obesity, dyslipidemia, raised blood pressure, insulin resistance with or without glucose intolerance, pro-inflammatory state, and prothrombin state. There is considerable evidence to suggest that metabolic syndrome could be an inflammatory condition [3–10].

U. N. Das, *Molecular Basis of Health and Disease*,
DOI 10.1007/978-94-007-0495-4_9, © Springer Science+Business Media B.V. 2011

Metabolic Syndrome

Metabolic syndrome is a risk factor for cardiovascular disease (CVD). The National Cholesterol Education Program's Adult Treatment Panel III report (ATP III) identified six components of the metabolic syndrome that relate to CVD. They are: (a) abdominal obesity, (b) atherogenic dyslipidemia, (c) raised blood pressure, (d) insulin resistance with or without glucose intolerance, (e) pro-inflammatory state, and (f) prothrombin state. Other features of metabolic syndrome include: hyperfibrinogenemia, increased plasminogen activator inhibitor-1 (PAI-1), low tissue plasminogen activator, nephropathy, micro-albuminuria, and hyperuricemia [1].

The incidence of metabolic syndrome is increasing, and the cause(s) for this increasing incidence is not clear. Although genetics could play an important role in the higher prevalence of metabolic syndrome in certain populations, it is not known how genetic factors interact with environmental and dietary factors to increase its incidence. Insulin resistance is present in metabolic syndrome and also in subjects with abdominal obesity, hypertension, type 2 diabetes, hyperlipidemias, CHD, and stroke. Hyperinsulinemia may be a consequence of this. In the early stages of metabolic syndrome, insulin resistance is restricted to muscle tissue whereas adipose tissue is not resistant to insulin [11]. Hence, exercise helps in the prevention and treatment of insulin resistance since; it decreases insulin resistance and enhances glucose utilization in the muscles. In addition, exercise is anti-inflammatory in nature [12, 13]. Exercise not only decreases the levels of pro-inflammatory cytokines CRP, IL-6, and TNF-α but also simultaneously enhances the concentrations of anti-inflammatory cytokines IL-4, IL-10 and TGF-β that also suppress the production of pro-inflammatory cytokines IL-1, IL-2, and TNF-α [13]. Exercise significantly reduced the magnitude of myocardial infarction and this cardioprotective action paralleled the increase in manganese superoxide dismutase (Mn-SOD) activity [14]. Antisense oligo-deoxyribonucleotide administration to Mn-SOD abolished this cardioprotective action implying that ability of exercise to enhance the activity of Mn-SOD is crucial to this protective action. It is likely that an increase in Mn-SOD activity is in response to exercise-induced free radical generation suggesting that free radicals have beneficial actions, especially when they are produced in response to physiological stimulus such as exercise. Administration of antibodies to TNF-α and IL-1 abolished the cardioprotective action of exercise and activation of Mn-SOD, indicating that exercise-induced increase in the levels of IL-6 and TNF-α augment the production of free radicals that, in turn, enhance Mn-SOD activity that elicits the cardioprotective action of exercise. SOD also enhances the half-life of nitric oxide (NO), a potent vasodilator, platelet anti-aggregator, and anti-atherosclerotic molecule. In contrast, anti-oxidant vitamin E counteracted the beneficial effects of exercise, suggesting that endogenous anti-oxidant Mn-SOD is critical to the beneficial actions of exercise and this benefit cannot be imitated by exogenous administration of vitamin E. Thus, regular exercise ensures adequate expression of endogenous anti-oxidants and anti-inflammatory cytokines and thus, brings about its cardioprotective action.

Metabolic Syndrome Is an Inflammatory Condition

Plasma levels of C-reactive protein (CRP), TNF-α, and IL-6, markers of inflammation, are elevated in subjects with obesity, insulin resistance, essential hypertension, type 2 diabetes, and CHD both before and after the onset of these diseases [15–22]. Elevated CRP concentrations were associated with an increased risk of CHD, ischemic stroke, peripheral arterial disease, and ischemic heart disease mortality in healthy men and women. A strong relation between elevated CRP levels and cardiovascular risk factors: fibrinogen, and HDL cholesterol was also reported.

The negative correlation observed between plasma TNF-α and HDL cholesterol, glycosylated hemoglobin, and serum insulin concentrations explain why CHD is more frequent in obese compared to healthy or lean subjects [15]. Subjects with elevated CRP levels were two times more likely to develop diabetes at 3–4 years of follow-up period [23]. Dietary glycemic load is significantly and positively associated with plasma CRP in healthy middle-aged women [24] suggesting that hyperglycemia induces inflammation.

TNF-α has a role in insulin resistance and type 2 diabetes mellitus. Acute raise in plasma glucose levels in normal and impaired glucose tolerance (IGT) subjects increased plasma IL-6, TNF-α, and IL-18 levels and these increases were much larger and lasted longer in IGT subjects compared to control [25]. TNF-α secretion was suppressed in younger subjects in response to glucose challenge, but not in the older subjects [26]. Furthermore, hyperglycemia induced the production of acute phase reactants from the adipose tissue [27]. Hence, the increased incidence of type 2 diabetes and metabolic syndrome in the elderly could attributed to alterations in the homeostatic mechanisms that control TNF-α, IL-6, and CRP levels. This evidence suggests that low-grade systemic inflammation has a role in the development of type 2 diabetes.

Elevated plasma IL-6 levels in women with hypertension and insulin resistance in men was noted [28]. A graded positive relationship between blood pressure and levels of ICAM-1 (intercellular adhesion molecule-1) and IL-6 was noted in healthy men [29] indicating that plasma CRP and IL-6 are elevated in insulin resistance and hypertension. These evidences suggest that various components of the metabolic syndrome are associated with an increase in the concentrations of markers of inflammation and hence, metabolic syndrome can be considered as an inflammatory condition.

Why Abdominal Obesity Occurs?

Abdominal obesity is the most common and dominant component of the metabolic syndrome and it is likely that visceral adipose tissue accumulation is one the main culprits in the development of metabolic syndrome and insulin resistance [30]. There is reasonable evidence to believe that enhanced expression of 11β-hydroxysteroid dehydrogenase type 1 (11β-HSD-1) enzyme selectively in adipose tissue causes abdominal obesity and induces insulin-resistant diabetes, hyperlipidemia, hyperphagia and hyperleptinemia [31, 32] that implies that abdominal obesity is like localized Cushing's syndrome. Adipocyte 11β-HSD-1 mRNA concentrations are associated with adiposity, and genetic variations in the 11β-HSD-1 gene are associated with Type 2 diabetes mellitus, plasma insulin concentrations and insulin action, independent of obesity indicating that 11β-HSD-1 gene is under tissue-specific regulation, and has tissue-specific consequences [32]. Obese men had no difference in their whole-body rate of regenerating cortisol but showed a more rapid conversion of ^{3}H cortisone to ^{3}H cortisol in abdominal subcutaneous adipose tissue. Insulin infusion produced a marked decrease in adipose 11β-HSD-1 activity in lean but not in obese men, suggesting that cortisol generation is increased selectively within adipose tissue in obesity, and this increase in 11β-HSD-1 activity is resistant to insulin-mediated down regulation [33]. Thus, inhibitors of 11β-HSD-1 enzyme in adipose tissue could enhance insulin sensitivity. The observation that 11β-HSD-1 deficiency protects against the development of diet-induced abdominal obesity and remain insulin sensitive is in support of this idea. 11β-HSD-1$^{(-/-)}$ mice expressed lower resistin and TNF-α, but higher PPAR-γ, adiponectin, and uncoupling protein-2 (UCP-2) mRNA levels in adipose tissue, and 11β-HSD-1$^{(-/-)}$ adipocytes showed higher basal and insulin-stimulated glucose uptake, and β-HSD-1$^{(-/-)}$ mice did show reduced visceral fat accumulation upon high-fat feeding [34]. Thus, manipulation of 11β-HSD-1 prevents the development of features of metabolic syndrome and an increase in 11β-HSD-1 activity suppresses adiponectin, PPAR-γ, and UCP-2 activities.

Furthermore, TNF-α and IL-1β produced a dose-dependent increase in 11β-HSD-1 activity only in the subcutaneous and omental adipose cells, but had no effect on 11β-HSD-1 activity in hepatocytes. Insulin-like growth factor I (IGF-I), similar to insulin, caused a dose-dependent inhibition of 11β-HSD-1 activity in subcutaneous and omental stromal cells, but not in human hepatocytes. PPAR-γ ligands significantly increased 11β-HSD-1 activity in omental and subcutaneous adipose cells [35]. These results suggest that tissue-specific regulation of 11β-HSD-1 occurs and suggests that the response of omental adipose cells differs from that seen in subcutaneous adipocytes. Glucocorticoids, which induce abdominal obesity, insulin resistance and show anti-inflammatory actions, inhibit TNF-α synthesis [36], whereas in subcutaneous adipocytes from lean subjects, TNF-α inhibited adiponectin, an endogenous insulin sensitizer and anti-inflammatory molecule; release but had no effect on adiponectin release from subcutaneous or omental adipocytes from obese subjects. In contrast, dexamethasone inhibited adiponectin release [37]. Thus, the interaction among glucocorticoids, TNF-α, 11β-HSD-1 activity, adiponectin, insulin, and PPARs is complex.

Glucose Is Pro-inflammatory in Nature

Glucose is not only the principle source of energy for cells but it also has pro-inflammatory actions. Glucose challenge stimulates generation of reactive oxygen species by leukocytes and decreases vitamin E levels. High calorie diet rich in carbohydrates, fats (especially saturated and trans-fats) or protein stimulates the production of reactive oxygen species [38–41]. Increase in plasma pro-inflammatory cytokines IL-6, TNF-α, and IL-18 triggered by acutely raised plasma glucose levels both in normal and IGT subjects can be prevented by simultaneous administration of glutathione [27, 42], suggesting that an oxidative mechanism is responsible for the increase in circulating cytokines. IL-6 and TNF-α activate NADPH oxidase and thus enhance the generation of reactive oxygen species [43], suggesting that saturated and trans-fats or protein-rich diet stimulated oxidative stress is due to an increase in IL-6, TNF-α, IL-18, and CRP. NF-κB is stimulated by CRP, IL-6, TNF-α. NF-κB, in turn, stimulates superoxide anion generation, induces monocyte chemoattractant peptide (MCP-1) and iNOS gene expression and activates vascular smooth muscle cells [44], and increases plasminogen activator inhibitor-1 expression [45]. Thus, CRP, IL-6, and TNF-α show pro-inflammatory and pro-atherogenic actions by enhancing oxidative stress by activating NF-κB and/or NADPH oxidase. Since, plasma levels of lipid peroxides, a marker of increase in free radical generation, are elevated in patients with type 2 diabetes mellitus, hypertension, and CHD (which are all components of the metabolic syndrome) it is reasonable to assume that an augmented expression of NADPH oxidase and decreased expression of antioxidative enzymes occurs in these conditions that, in turn, could lead to a decrease in the production of adiponectin and an increase in plasminogen activator inhibitor-1, IL-6, and monocyte chemotactic protein-1 [46–52]. The oxidative stress induced by high calorie diet (or even normal diet) and subsequent increase in plasma glucose is completely prevented in the presence of adequate tissue concentrations of anti-oxidants vitamin E and glutathione. In contrast, dietary restriction, exercise, and weight loss suppress oxidative stress by inhibiting the generation and release of pro-inflammatory cytokines [53–55].

Insulin Is Anti-inflammatory in Nature

Insulin suppresses the production of TNF-α, IL-6, IL-1, IL-2, and macrophage migration inhibitory factor (MIF), and enhances the production of IL-4 and IL-10 [56–61]. Thus, insulin has anti-inflammatory actions. This suggests that one of the functions of hyperinsulinemia could be to prevent or arrest low-grade inflammation that is seen in type 2 diabetes mellitus, hyperglycemia and metabolic syndrome. On the other hand, leptin has pro-inflammatory actions [53, 60–63]. Since hyperinsulinemia and hyperleptinemia are present in obese children [64, 65], it is likely that low-grade systemic inflammation seen is initiated early in life.

Endothelial Nitric Oxide in Metabolic Syndrome

Of all the free radicals that are responsible for oxidative stress, superoxide anion has a dominant role in metabolic syndrome. Superoxide anion interacts with nitric oxide (NO) and inactivates it producing peroxynitrite radical. Enhanced superoxide anion production is responsible for the NO deficit seen in diabetes and its associated vascular dysfunction [66, 67]. Reduced eNO (endothelial nitric oxide) and increased superoxide anion generation is seen in insulin resistance, obesity, hypertension, and CHD [48–55]. Reduced eNO generation could occur as a result of enzymatic uncoupling of L-arginine oxidation, deficiency of L-arginine, increased plasma concentrations of asymmetrical dimethylarginine (ADMA), decreased concentrations of co-factors of NO synthesis such as tetrahydrobiopterin, folic acid, and vitamin C [66, 68]. Elevated plasma CRP, IL-6, and TNF-α levels have been associated not only with obesity and insulin resistance but also with hypertriglyceridemia, glucose intolerance, and hyperleptinemia, and negatively correlated with HDL cholesterol [69–71]. HDL stimulates endothelial nitric oxide (eNO) synthesis [72] and NO, in turn, inhibits LDL oxidation [73, 74]. Increase in the levels of oxidized LDL and superoxide anion, and reduced levels of eNO are in support of the idea that low grade inflammation is present and also suggests the reasons for the enhanced risk of atheroslcerosis and thrombosis, and CHD in subjects with metabolic syndrome. We observed increased plasma lipid peroxides and decreased NO concentrations in type 2 diabetes, hypertension, and CHD [48–50, 75–77] lending support to this view.

As already discussed in the previous chapter on hypertension (see Chap. 8), patients with uncontrolled essential hypertension have higher plasma lipid peroxides, low NO, and their leukocytes generated significantly higher levels of superoxide anion, and RBC membranes contained low vitamin E and superoxide dismutase [51], which reverted to normal following control of blood pressure with various anti-hypertensive drugs. Even physiological concentrations of angiotensin II activate NADPH oxidase [78] and enhance generation of free radicals. Both angiotensin converting enzyme (ACE) inhibitors and angiotensin-II receptor blockers increased adiponectin and eNO concentrations in patients with hypertension [79, 80] suggesting that insulin resistance in hypertension is due to low adiponectin and eNO levels. Since a positive correlation exists between plasma adiponectin levels and insulin sensitivity, and blockade of the rennin-angiotensin system increases adiponectin and eNO concentrations, this may explain the beneficial actions of ACE inhibitors and angiotensin-II receptor blockers in diabetes, hypertension and CHD. This also suggests that anti-hypertensive drugs increase insulin sensitivity and enhance insulin action; suppress free radical generation, augment eNO generation, and stimulate adiponectin synthesis and thus, show limited anti-inflammatory actions.

Insulin resistance is accompanied by increase in peripheral vascular resistance due to decreased eNO generation that is also responsible for endothelial dysfunction. High-fructose-fed rats showed decrease in metabolic clearance rate of glucose compared to control that was reverted to normal by sodium nitroprusside infusion, a donor of NO [81], suggesting that NO improves insulin resistance. Sustained hyperinsulinemia causes impairment of NO production that contributes to insulin resistance and hypertension [82]. Insulin resistant experimental animals have depleted tetrahydrobiopterin (H_4B) and elevated 7, 8-dihydrobiopterin (7, 8 H_2B) (activating and inactivating cofactors of nitric oxide synthase respectively) that leads to reduced eNOS activity and increased superoxide anion generation that leads to impaired NO-dependent vasodilation. A stepwise decrease in the maximal acetylcholine-induced vasodilation (that is due to eNO) and plasma $H_4B/7,8\text{-}H_2B$ ratio, and increase in coronary lipid peroxide production as insulin sensitivity decreased was reported. The acetylcholine-induced vasodilation was positively correlated with insulin sensitivity, whereas $H_4B/7$, $8\text{-}H_2B$ ratio was inversely correlated with insulin sensitivity, indicating that both abnormal pteridine metabolism and vascular oxidative stress are linked to coronary endothelial dysfunction and reduced NO generation in insulin-resistant subjects [83]. In addition, subjects with insulin resistance showed elevated plasma concentrations of ADMA, an endogenous inhibitor of NO [84]. These results emphasize that insulin resistance is accompanied by decrease in eNO production that could be due to increase in ADMA levels and oxidant stress.

Mice with targeted disruption of eNOS are hypertensive and insulin resistant and also had a 1.5 to 2-fold elevation of the cholesterol, triglyceride, and free fatty acid plasma concentration, and elevated plasma leptin, uric acid and fibrinogen levels and glucose intolerance on a high fat diet but were not obese [85]. These evidences indicate eNOS deficiency could trigger many of the abnormalities of metabolic syndrome. These eNOS knockout mice are similar to the neuron-specific disruption of the insulin receptor gene (NIRKO) mice that also develop obesity, hyperglycemia, hypertriglyceridemia, insulin resistance, hyperleptinemia and hyperphagia [86]. Since insulin stimulated the production of eNO [87, 88] and inhibited TNF-α production [89–91], and disruption of insulin receptor in the brain produced features of metabolic syndrome, it reasonable to propose that a decrease in the number of insulin receptors, defect in the function of insulin receptors, insulin lack or resistance in the neuronal cells leads to the development of metabolic syndrome even when pancreatic β cells are normal.

The well-known increase in plasma leptin levels seen in obese subjects has been attributed to the increased body fat mass and a relationship between fasting concentrations of leptin and insulin has also been described. In moderately overweight men with type 2 diabetes with a mean body mass index (BMI) of 26.8 kg/m^2, fasting leptin level was significantly and positively correlated with BMI and with fasting insulin while it negatively correlated with the glucose disposal rate, while leptin was inversely correlated with HDL-cholesterol [92]. In the study subjects, the highest fasting leptin levels were observed in those patients with the most expressed insulin resistance independent of body composition. Leptin is known to augment eNO generation [93–96] suggesting that hyperleptinemia may be a defensive mechanism to reduce insulin resistance since NO is known to enhance tissue sensitivity to insulin [81, 82] as described above. But, continued exposure to leptin decreased eNO synthesis, increased $O_2^- \cdot$ and $ONOO^-$ levels and depleted intracellular L-arginine of endothelial cells. The increased eNOS expression and a reduced L-arginine content led to eNOS uncoupling, a reduction in bioavailable NO, and an elevated concentration of $O_2^- \cdot$ and $ONOO^-$ that was partially reversed by L-arginine supplementation [97, 98]. These results suggest that long-standing hyperleptinemia induces an endothelial NO/$ONOO^-$ imbalance that leads to endothelial dysfunction in obesity and diabetes mellitus.

Despite the fact that obesity, insulin resistance, dyslipidemia, diabetes mellitus and metabolic syndrome are associated with low-grade systemic inflammation and cause endothelial dysfunction, it is still not clear as to how and when they are initiated. There is now reasonable evidence to suggest that they may have their origins in the perinatal period and hypothalamus and neurotransmitters and hypothalamic peptides play a significant role in their pathogenesis. If this proposal is true, it implies that preventive and therapeutic measures need to be instituted during the perinatal period to stem the current epidemic of insulin resistance, obesity, type 2 diabetes mellitus and metabolic syndrome.

Perinatal Origins of Metabolic Syndrome

Epidemiological, experimental and clinical studies revealed that the fetal nutritional environment can pattern adult obesity. A higher prevalence of obesity was reported in those who were of either low or high birth weight. Heavier mothers had heavier babies and these babies went on to have a high BMI in adult life [99]. The offspring of women who had type 2 diabetes mellitus, gestational diabetes or impaired glucose tolerance is at high risk of developing obesity and type 2 diabetes mellitus and other features of metabolic syndrome [100, 101]. Subjects who were small babies tend to have a more abdominal distribution of adipose tissue, a significantly reduced muscle mass, and a high overall body fat content in adult life [102–104]. Results from the Avon Longitudinal Study of Pregnancy and Childhood (ALSPAC) showed [105] that fetal growth is influenced by both fetal genes and maternal-uterine-placental factors. Important maternal-placental factors included parity, smoking and weight gain, but also maternal genetic factors in the mother or fetal placenta, including the mitochondrial DNA 16189 variant and H19. These maternal genetic factors influenced smaller, growth-restrained infants. Fetal genes included the insulin gene (INS) VNTR (variable number of tandem repeat), which were found to be associated with birth size and cord blood IGF-II levels. During postnatal life, the INS VNTR III/III genotype remained associated with body size, including body mass index and waist circumference, and also lowered insulin sensitivity among girls. However, Rapid "catch-up" early postnatal weight gain strongly predicted later childhood obesity and insulin resistance; among these children, those with INS VNTR class I alleles were more obese. These results suggest that human metabolism/genes are more adopted to famine conditions and hence, with the availability of abundant and calorie-rich diet these genetic factors and their interactions with maternal and childhood environmental factors are contributing to both childhood and adult obesity and its consequent metabolic abnormalities such as type 2 diabetes mellitus.

This is supported by a study performed in young indigenous women of childbearing age in rural communities in north Queensland [106], which revealed that 41% of Aboriginal and 69% of Torres Strait Islander (TSI) women had central obesity, 62% were smokers, 71% drank alcohol regularly and of those, 60% did so at harmful levels. One third of Aboriginal and 16% of TSI women had very low red cell folate (RCF) levels. In the group followed up, there was a mean annual waist gain of 1.6 cm in Aboriginal women and 1.2 cm in TSI, 0.5 kg/m^2 in BMI and 1.5 kg in weight. Incidence of new type 2 diabetes mellitus was 29.1 per 1,000 person-years in Aboriginal women and 13.9 per 1,000 person-years among TSI. High prevalence and incidence of central obesity and diabetes, poor nutrition, high rates of alcohol use and tobacco smoking together with young maternal age, could provide a poor intra-uterine environment for many indigenous Australian babies, and contribute to high perinatal morbidity including childhood obesity. These results indicate that community level interventions are needed to improve pre-pregnancy and perinatal nutrition.

In a review of the incidence of obesity, type 2 diabetes mellitus and metabolic syndrome in urban South Asian/Asian Indian adults and children but also in economically disadvantaged people residing in urban slums and rural areas revealed similar pattern [107]. It was opined that rapid nutrition, lifestyle, and socioeconomic transitions, consequent to increasing affluence, urbanization, mechanization, and rural-to-urban migration, psychological stress in urban setting, genetic predisposition, adverse perinatal environment, and childhood "catch up" obesity could be contributing to the high prevalence and incidence of obesity, type 2 diabetes mellitus and metabolic syndrome in this population. Atherogenic dyslipidemia, glucose intolerance, thrombotic tendency, subclinical inflammation, and endothelial dysfunction were higher in South Asians than Caucasians and these manifestations were more severe and were seen since childhood in South Asians than Caucasians. Metabolic syndrome and cardiovascular risk in South Asians was also heightened by their higher body fat, truncal subcutaneous fat, intra-abdominal fat and ectopic fat deposition (such as liver fat). This study reemphasized the high degree of cardiometabolic risk in South Asians, starting at an early age [108].

The impact of perinatal nutrition on the development of obesity, type 2 diabetes and metabolic syndrome in adult life is supported by the report that when the normal litter size of Wistar rats (n = 10; controls) was reduced from day 3 to day 21 of life to only 3 pups per mother (small litters, SL; overnutrition), throughout life, SL rats displayed significant hyperphagia, overweight, hyperinsulinemia, impaired glucose tolerance, elevated triglycerides, and an increased systolic blood pressure. In adulthood, an increase of galanin-neurons in the arcuate hypothalamic nucleus (ARC) that positively correlated to body weight; and hyperinsulinemia and increased hypothalamic insulin levels in SL rats during early postnatal life was observed. By the end of the critical hypothalamic differentiation period (which is day 21 of life in mice), SL rats had increased number of GAL-neurons in the ARC, showing a positive correlation to body weight and insulin. Thus, these results indicated that neonatally acquired persisting malformation of hypothalamic galaninergic neurons, induced by early overfeeding and hyperinsulinism, promoted the development of overweight and metabolic syndrome-like alterations during life [109].

These results suggest that nutrient supply early in pregnancy and perinatal and childhood influences the development of obesity, type 2 diabetes mellitus and metabolic syndrome in adult life probably, by inducing changes in the expression, localization, and action of specific neuropeptides in the appetite regulatory network within the brain.

Hypothalamic Neuropeptides and Food Intake

Appetite is controlled by several neuropeptides: appetite stimulating neuropeptide Y (NPY) and agouti-related peptide (AgRP), and the appetite inhibitory molecules pro-opiomelanocortin (POMC), the precursor for α-melanocyte stimulating hormone (α-MSH), and cocaine and amphetamine-regulated transcript (CART), which are expressed within the hypothalamus and act together to regulate energy balance. These neuropeptides interact and show positive and negative feedback control among themselves. Thus, an increase in appetite stimulating peptides during fasting will suppress the levels of satiety molecules and vice versa. This feedback regulation may be lost or altered in subjects with obesity, metabolic syndrome and cancer. NPY neurons respond to alterations in the concentrations of plasma glucose, insulin, and leptin. Increased food intake results in increases in circulating concentrations of leptin that are sensed by the leptin receptors expressed on arcuate (ARC) and DMN (dorsomedial nucleus) neurons leading to a fall in hypothalamic NPY mRNA that results in decreased food intake. AgRP that is coexpressed with NPY in the ARC is an endogenous antagonist of anorexigenic melanocortin receptors MC3-R and MC4-R in the PVN (paraventricular nucleus) and other hypothalamic regions. α-MSH is an endogenous anorexigenic peptide that acts on the melanocortin receptors to suppress food intake. CART, localized within the POMC (proopiomelanocortin) neurons in the hypothalamus, also suppresses food intake [110, 111].

Appetite Regulatory Centers Are in Place During Perinatal Period and Fine-tuned/Programmed by Maternal and Perinatal Factors

NPY is present within the fetal ARC from as early as 14.5 days gestation; NPY/AgRP projections between the ARC and DMN develop around 10–11 days after birth whereas NPY containing projections to the PVN develop around 15–16 days [111, 112]. In the rodent, arcuate nucleus of the hypothalamus (ARH)-derived neuropeptide Y (NPY) and proopiomelanocortin (POMC) neurons have efferent projections throughout the hypothalamus that do not fully mature until the second and third postnatal weeks. In the fetal Japanese macaque, NPY mRNA was expressed in the ARH, paraventricular nucleus (PVH), and dorsomedial nucleus of the hypothalamus (DMH) as early as G100. ARH-derived NPY projections to the PVH were initiated at G100 but were limited and variable; however, there was a modest increase in density and number by G130. ARH-NPY/agouti-related peptide (AgRP) fiber projections to efferent target sites were completely developed by G170, but the density continued to increase in the postnatal period. In contrast to NPY/AgRP projections, α-MSH fibers were minimal at G100 and G130 but were moderate at G170. Several significant species differences between rodent and the nonhuman primate (NHP) were also reported. There were few NPY/catecholamine projections to the PVH and ARH prior to birth, while projections were increased in the adult. A substantial proportion of the catecholamine fibers did not coexpress NPY. In addition, cocaine and amphetamine-related transcript (CART) and α-melanocyte stimulating hormone (α-MSH) were not colocalized in fibers or cell bodies. As a consequence of the prenatal development of these neuropeptide systems in the non-human primate, the maternal environment may critically influence these circuits. Additionally, because differences exist in the neuroanatomy of NPY and melanocortin circuitry the regulation of these systems may be different in primates than in rodents [113].

Lactation induced an increase in NPY activity in two neuronal populations in the hypothalamus: the arcuate nucleus (ARH) and the dorsomedial nucleus (DMH) area. Injection of the retrograde tracer, fluorogold (FG), into the PVH of lactating females revealed that NPY cells (identified by in situ hybridization for NPY mRNA) were observed throughout the ARH; however, a greater number of double-labeled cells were found in the caudal portion than the rostral portion of the ARH. Thus, NPY neurons in the caudal portion of the ARH provide the major ARH NPY input into the PVH area. These results coupled with the observation that activation of NPY neurons in the ARH during lactation is confined to the caudal portion suggest that the lactation-activated NPY neurons project to the PVH. FG-labeled NPY cells were also identified in the DMH area, providing the evidence that the NPY neurons in the DMH area activated during lactation also project to the PVH, indicating that the increase in NPY activity is important in modulating some of the physiological alterations occurring during lactation, such as the increase in food intake, in part through modulating PVH neuronal activity [114]. Thus, during lactation, the levels of NPY, which plays an important role in mediating food intake, are significantly elevated in a number of hypothalamic areas, including the arcuate nucleus (ARH). Additional studies revealed that the expression of agouti-related protein (AGRP), a homologue of the skin agouti protein, mRNA signal was found mostly in the ventromedial portion of the ARH, which contained a high density of NPY neurons. A significant increase in AGRP mRNA content was observed in the mid- to caudal portion of the ARH of lactating animals compared with diestrous females. No difference was found in the rostral portion of the ARH. Double-label in situ hybridization for AGRP and NPY showed that almost all of the NPY-positive neurons throughout the ARH also expressed AGRP mRNA signal. Furthermore, AGRP expression was confined almost exclusively to NPY-positive neurons. Thus, it is clear that during lactation, AGRP gene expression was significantly elevated in a subset of the AGRP neurons in the ARH. The high degree of colocalization of AGRP and NPY, coupled with observation that increased NPY expression occurred in the ARH in response to suckling, suggests that AGRP and NPY are coordinately regulated and are involved in the increase in food intake during lactation [115].

These results imply that factors that influence brain growth and development during which the expression of various neurotransmitters in the hypothalamic nuclei are being coordinately developed will have substantial impact on the development of appetite regulatory centers that, in turn, determine subsequent food intake in later life. For instance, the amount and type of food consumed during suckling in the rat plays an important role in determining subsequent food intake and preferences in later life. Thus, postnatal over nutrition in rats led to an increased early weight gain and fat deposition, hyperphagia, obesity, hyperleptinemia, hyperglycemia, hyperinsulinemia and insulin resistance, indices of metabolic syndrome that is accompanied by alterations in the concentrations and actions of the appetite regulatory neuropeptides in the form of decreased mean areas of neuronal nuclei and cytoplasm within the PVN, VMN, and ARC and a significant increase in the number of NPY containing neurons within the ARC and decreased immunostaining for both POMC and α-MSH [111–117]. In contrast, when rats are undernourished during the perinatal period the offspring develop significant hyperphagia and obesity when maintained on a high fat diet and showed an increase in the relative mass of retroperitoneal fat. For example, pregnant Wistar rats fed an 8% protein diet during pregnancy and lactation (low-protein group; LP) while control mothers (CO) received a 17% protein isocaloric standard diet, LP offspring displayed underweight at birth and during suckling, while leptin levels were not altered. At weaning, under basal conditions cholecystokinin-8S (CCK-8S) was decreased in LP offspring in the paraventricular hypothalamic nucleus and arcuate hypothalamic nucleus, as well as in the dorsomedial hypothalamic nucleus, lateral hypothalamic area and ventromedial hypothalamic nucleus. These data indicate that (a) an inhibition of the satiety peptide CCK-8S in main regulators of body weight and food intake occurred in low-protein malnourished newborn rats; (b) no direct relationship existed between hypothalamic CCK-8S and circulating leptin at this age; and (c) no neurochemical signs of hypothalamic CCKergic dysregulation could be seen at the age of weaning [118] but may manifest only in the adult life.

These results suggest that perinatal malnutrition and growth retardation at birth that are important risk factors for the development of overweight and metabolic syndrome in later life could alter the hypothalamic neuropeptidergic systems which are seem to be highly vulnerable to persisting malorganization due to perinatal malnutrition. Thus, the neuropeptides that regulate appetite centers and their responses to stimuli such as glucose, insulin, leptin, and other environmental stimuli such as bisphenol are "programmed" in the fetal and perinatal stages of development. Hence, factors that govern the growth and development of brain and biochemical stimuli such as glucose, insulin and fatty acids (both saturated and unsaturated fatty acids that may include both short chain and long chain fats) that influence the development of various hypothalamic neurons may have long-lasting impression or programming affects on the appetite regulating centers. This ultimately could influence the dietary preferences and the development of obesity and metabolic syndrome in later life [111, 119–123].

Ventromedial Hypothalamus may have a Role in the Development of Type 2 Diabetes Mellitus

There is reasonable evidence to suggest that insulin resistance, obesity, type 2 diabetes mellitus, hypertension, and consequently metabolic syndrome are disorders of the brain, especially due to hypothalamic dysfunction. Ventromedial hypothalamic

(VMH) lesion in rats induces hyperphagia and excessive weight gain, fasting hyperglycemia, hyperinsulinemia, hypertriglyceridemia and impaired glucose tolerance [124, 125]. Intraventricular administration of antibodies to neuropeptide Y (NPY) abolished the hyperphagia and *ob* mRNA (leptin mRNA) in these animals, suggesting that increased release of NPY plays a role in hyperphagia and obesity and *ob* gene is up regulated even in non-genetically obese animals [126, 127]. Increased NPY concentrations were noted in the paraventricular, ventromedial (VMH), and lateral hypothalamic areas of streptozotocin-induced diabetic rats [128]. Streptozotocin-induced diabetes produced significant decreases in extracellular concentrations of norepinephrine (NA/NE), serotonin (5-HT), and their metabolites, a pronounced increase in extracellular GABA, in the VMH [129]. Long-term infusion of norepinephrine plus serotonin into the VMH impairs pancreatic islet function in as much as VMH norepinephrine and serotonin levels are elevated in hyperinsulinemic and insulin-resistant animals [130]. Streptozotocin-induced diabetes caused an increase in NE/NA concentrations in the PVN with a concurrent increase in serum corticosterone, increased the concentrations of NE/NA, dopamine and serotonin in the ARC and NE/NA concentrations in the lateral hypothalamus, VMH and suprachiasmatic nucleus [131]. Treatment with insulin completely reversed these effects, while leptin treatment was ineffective. The restoration of serotonergic activity and other hypothalamic peptide abnormalities to normal by insulin therapy suggests that there is a cross-talk between insulin and hypothalamic peptides and monoaminergic neurotransmitters. These results imply that dysfunction of VMH impairs pancreatic β cell function and induces metabolic abnormalities seen in metabolic syndrome.

Long-term effect of VMH lesions (16 weeks after making VMH lesion) on glucose metabolism, pancreatic insulin content, abdominal fat distribution and vascular complications in male Goto-Kakizaki (GK) rats revealed that food intake increased, plasma glucose, insulin and triglyceride levels were markedly elevated in VMH-lesioned GK (GK-VMH) rats compared with that in sham-operated GK (GK-sham) rats. The insulin content of pancreas at 16 weeks after operation was markedly decreased in GK-VMH rats, a significant 1.2-fold increase in mesenteric fat weight and a 1.3-fold higher ratio of mesenteric fat weight to subcutaneous fat weight in GK rats compared with sham-operated rats was noticed. The urinary excretions of protein and albumin in GK-VMH rats were greater than those in GK-sham rats. Histological examinations of the kidneys in GK-VMH rats revealed that the glomerular basement membranes were thicker than those of GK-sham rats, the descending aorta in GK-VMH rats showed morphologic changes in the intima characteristic of an early stage of atherosclerosis. These results suggest that VMH lesioned rats show visceral fat accumulation, develop typical diabetic complications, including both microangiopathy and macroangiopathy [132]. Thus, hypothalamic neurons and neurotransmitters seem to play a crucial role in the regulation of insulin secretion and metabolic syndrome and hence, it (metabolic syndrome) could very well be a disorder of the brain [133].

Furthermore, the Goto-Kakizaki (GK) rat, a nonobese strain in which a spontaneous type of non-insulin-dependent diabetes mellitus develops without apparent macroangiopathy, when induced with ventromedial hypothalamic (VMH) lesion showed hyperphagia and a further deterioration in glucose metabolism. Male GK rats with VMH lesions showed marked increase in plasma glucose levels 16 weeks after the operation: VMH lesion GK rats, 19.3 ± 2.0 mmol/l, vs sham-operated GK rats, 10.1 ± 1.3 mmol/l; $p < 0.0001$. In addition, these VMH lesion GK rats 16 weeks after the surgery revealed intimal thickening and significantly increased infiltrating cells into the intima as compared with sham-operated GK rats. Electron microscopic examination showed an increased number of microvilli and lysosomes in endothelial cells, infiltration of macrophages and lymphocytes into the intima, and migration of medial smooth muscle cells into the intima that are early events in atherosclerosis. These results indicate that VMH lesion produced significant hyperglycemia that led to the development of metabolic syndrome and atherosclerosis [134].

Insulin Receptors in the Brain and the Metabolic Syndrome

Brain is rich in insulin receptors especially in the olfactory bulb, the hypothalamus, and the pituitary. Insulin signaling seems to have a significant role in the regulation of food intake, neuronal growth and differentiation, and in modulating neurotransmitter release and synaptic plasticity in the CNS [135–137]. Insulin administration into the VMN and PVN reduced food intake [138, 139]. Infusion of insulin specific antibodies or anti-sense oligonucleotides directed against insulin receptor in to the third ventricle reduced hepatic sensitivity to circulating insulin, increased hepatic glucose production, suggesting that the action of insulin in the brain regulates liver glucose metabolism [140]. ICV insulin infusion blocked the effects of both fasting and streptozotocin-induced diabetes to increase expression of NPY mRNA in the arcuate nucleus [141]. Conversely, insulin increased hypothalamic POMC mRNA content [142]. Both insulin and leptin suppressed NPY/AgRP neurons in the arcuate nucleus, while activating POMC/CART neurons. This suggests that a cross-talk exists between insulin and leptin apart from their ability to share their common ability to suppress anabolic, while activating catabolic, regulatory neurocircuitry [143].

The possibility that resistance to the action of insulin in the brain contributing to the development of obesity, type 2 diabetes mellitus and metabolic syndrome received further support from the observation that brain glucose metabolism, using [^{18}F]fluorodeoxyglucose positron emission tomography, in seven insulin-sensitive [(HOMA-IR) = 1.3] and seven insulin-resistant [(HOMA-IR) = 6.3] men showed that during suppression of endogenous insulin by somatostatin, with and without an insulin infusion that elevated insulin to 24.6 ± 5.2 and 23.2 ± 5.8 mU/l, concentrations similar to fasting levels of the resistant subjects and approximately threefold above those of the insulin-sensitive subjects, insulin-evoked change in global cerebral metabolic rate for glucose was reduced in insulin resistance (+7 vs. +17.4%, $P = 0.033$). Insulin was associated with increased metabolism in ventral striatum and prefrontal cortex and with decreased metabolism in right amygdala/hippocampus and cerebellar vermis ($P < 0.001$), relative to global brain. Insulin's effect was less in ventral striatum and prefrontal cortex in the insulin-resistant subjects (mean ± SD for right ventral striatum 3.2 ± 3.9 vs. 7.7 ± 1.7, $P = 0.017$). These results indicated that brain insulin resistance exists in peripheral insulin resistance, especially in regions subserving appetite and reward, implying that diminished link between control of food intake and energy balance may contribute to development of obesity in insulin resistance [144].

Mechanism of Action of Insulin Receptors in the Brain and Elsewhere

If it is true that insulin acts on its receptors located at specific areas, especially hypothalamus, of the brain, how does it bring about its action?

NIRKO Mice

Insulin acts on ATP-sensitive K^+ channels (K_{ATP} channels) of hypothalamic neurons and acts downstream of NPY and POMC neurons and integrates the signals of peripheral and central energy homeostasis. Leptin, like insulin, also activates K_{ATP} channels in glucose-responsive hypothalamic neurons [145, 146]. Glucose-responsive neurons from Zucker fatty (*fa/fa*) rats that develop obesity, which have a leptin receptor mutation, are insensitive to both insulin and leptin, explaining as to why ICV insulin inhibits neither food intake nor NPY gene expression in these *fa/fa* rats [147, 148]. Thus, insulin seems to interact with neuropeptides and regulate food intake. For example, neuron-specific disruption of the insulin receptor gene (NIRKO) in mice does not interfere with brain development and neuronal survival but these mice showed increased food intake, both male and female mice developed diet-sensitive obesity with increases in body fat and plasma leptin levels, insulin resistance, hyperinsulinemia and hypertriglyceridemia, features that are seen in metabolic syndrome [86, 149]. This study indicates that a decrease in the number of insulin receptors, defect in the function of insulin receptors, insulin lack or insulin resistance in the brain could lead to the development of metabolic syndrome even when pancreatic β cells are normal [149].

FIRKO Mice

In contrast, mice with fat-specific disruption of the insulin receptor gene (FIRKO mice) had low fat mass, showed loss of the normal relationship between plasma leptin and body weight, were protected against age-related and hypothalamic lesion-induced obesity, obesity-related glucose intolerance, had an extended life span and were protected from age-related obesity [150, 151]. These FIRKO mice exhibited polarization of adipocytes into populations of large and small cells, which differ in expression of fatty acid synthase, C/EBP alpha, and SREBP-1. White adipose tissue of FIRKO mice is also characterized by a polarization into two major populations of adipocytes, one small (<50 μm) and one large (>100 μm), which differ with regard to basal triglyceride synthesis and lipolysis, as well as in the expression of fatty acid synthase, sterol regulatory element-binding protein 1c, CCAAT/enhancer-binding protein alpha (C/EBP-alpha) and GLUT-1. In adipocytes from FIRKO mice, basal glucose uptake was unchanged compared with the controls, but insulin-stimulated glucose uptake was reduced by more than 90% in both large and small cells. On the other hand, basal incorporation of glucose into triglycerides and basal rates of lipolysis in large FIRKO adipocytes were increased significantly compared with the control, indicating increased lipid turnover in large adipocytes of FIRKO mice. Insulin inhibited lipolysis in both large and small cells of control mice, but not in the FIRKO mice.

Gene expression analysis using RNA isolated from large and small adipocytes of FIRKO and control (IR lox/lox) mice showed that of the 12,488 genes analyzed, 111 genes were expressed differentially in the four populations of adipocytes studied. These alterations exhibited ten defined patterns and occurred in response to two distinct regulatory effects. 63 genes were identified as changed in expression depending primarily upon adipocyte size, including C/EBP-alpha, C/EBP-delta, superoxide dismutase 3, and the platelet-derived growth factor receptor. 48 genes were regulated primarily by impairment of insulin signaling, including TGF-β, IFN-γ, IGF-I receptor, activating transcription factor 3, aldehyde dehydrogenase 2, and protein kinase C delta. These data suggest an intrinsic heterogeneity of adipocytes with differences in gene expression related to adipocyte size and insulin signaling, indicating that insulin signaling in adipocytes is critical for development of obesity and associated metabolic abnormalities, and abrogation of insulin signaling in fat prevents the development of obesity and metabolic syndrome [150–153].

MIRKO Mice

Skeletal muscle insulin resistance is among the earliest detectable defects in humans with type 2 diabetes. Disruption of the insulin receptor gene in mouse skeletal muscle (MIRKO mice) that exhibited a muscle-specific >95% reduction in receptor content and early signaling events, showed elevated fat mass, serum triglycerides, and free fatty acids, but blood glucose, serum insulin, and glucose tolerance were normal [154]. Thus, insulin resistance in muscle contributes to the altered fat metabolism associated with type 2 diabetes, but tissues other than muscle appear to be more involved in insulin-regulated glucose disposal. On the other hand, Moller et al. [155] showed that transgenic mice that overexpress dominant-negative insulin receptors specifically in striated muscle have a severe defect in muscle insulin receptor-mediated signaling and modest hyperinsulinemia. Hind limb perfusion studies in such transgenic mice revealed that maximal rates of insulin-stimulated muscle 3-O-methylglucose transport were reduced by 32–40% with proportional defects in[^{14}C]glucose uptake, lactate production, and muscle glycogen synthesis. It was also noted that though body weights were normal, transgenic mice had a 22–38% increase in body fat; with a reciprocal decrease (10–15%) in body protein and an increase in mean gonadal fat pad weight. Oral glucose tolerance test in 6-month-old transgenic mice revealed an increase in fasting glucose levels by 25%, and peak values were 22–40% higher. Transgenic mice also had a 37% decrease in plasma lactate levels and modest increases in levels of plasma triglycerides and FFA (29% and 15%, respectively). These results led to the conclusion that (a) severe defects in muscle insulin receptor function resulted in impaired insulin-stimulated glucose uptake and metabolism in the muscle; (b) muscle-specific insulin resistance contributed to the development of obesity; and (c) a defect in insulin-mediated muscle glucose disposal is sufficient to result in impaired glucose tolerance and other features of the insulin resistance syndrome, including hyperinsulinemia and dyslipidemia. Similar results were obtained when MIRKO mice were studied under hyperinsulinemic-euglycemic conditions [156]. It was found that insulin-stimulated muscle glucose transport and glycogen synthesis were decreased by about 80%, whereas insulin-stimulated fat glucose transport was increased threefold in MIRKO mice, demonstrating that selective insulin resistance in muscle promotes redistribution of substrates to adipose tissue thereby contributing to increased adiposity and development of the prediabetic syndrome.

LIRKO Mice

Liver plays a central role in the control of glucose homeostasis and is subject to complex regulation by substrates, insulin, and other hormones. Liver-specific insulin receptor knockout (LIRKO) mice exhibited dramatic insulin resistance, severe glucose intolerance, and a failure of insulin to suppress hepatic glucose production and to regulate hepatic gene expression and were paralleled by marked hyperinsulinemia due to a combination of increased insulin secretion and decreased insulin clearance. With aging, the LIRKO mice exhibited surprisingly, progressive decline in fasting blood glucose levels such that by 6 months of age, LIRKO mice showed fasting hypoglycemia rather than hyperglycemia. Furthermore, the severe impairment in glucose tolerance that was observed at 2 months of age was no longer apparent at 6 months. This change in phenotype was not associated with reexpression of insulin receptor in the aged LIRKO liver, and the IGF-1 receptor was not detectable in LIRKO liver. The expression of both glycolytic and gluconeogenic enzymes in the liver did show some normalization as the LIRKO mice aged. Thus, the appearance of hypoglycemia in fasted LIRKO mice suggested the development of an acquired liver failure that affected glucose production rather than morphological and functional changes, and the metabolic phenotype became less severe. Thus, insulin signaling in liver is critical in regulating glucose homeostasis and maintaining normal hepatic function. At 6–12 months of age, the livers of LIRKO mice were not only smaller than those of age- and gender-matched controls, multiple pale nodules throughout the liver. At 12-months of age, the liver now showed dysplastic nodules that disrupted the lobular architecture and there was increased lipid accumulation. The most characteristic ultrastructural feature of the LIRKO hepatocytes was the presence of enlarged mitochondria similar to those observed when there is increased oxidative stress such as in alcoholic liver disease [157]. These morphological abnormalities are reminiscent of non-alcoholic fatty liver disease (NAFLD) that is common in subjects with metabolic syndrome. These data provide evidence for a direct role of insulin signaling in the liver in the regulation of postprandial glucose homeostasis and whole-body insulin sensitivity and for the important role of the insulin receptor in liver for insulin clearance in vivo. Insulin signaling in liver is also essential for the maintenance of normal hepatocyte morphology and function. Furthermore, these findings suggest that liver insulin resistance may promote β cell hyperplasia and hyperinsulinemia observed in type 2 diabetes mellitus, but that the progression to full-blown diabetes with fasting hyperglycemia requires defects in tissues other than liver.

βIRKO Mice

Though the exact relationship between pancreatic β cell defect in the secretion of insulin and type 2 diabetes mellitus is not clear, it is likely that a dysfunction of the β cell function could be present in this disease. Mice in which the insulin receptor gene in the β cells is specifically inactivated (βIRKO mice) exhibited a selective loss of insulin secretion in response to glucose and a progressive impairment of glucose tolerance [158]. βIRKO mice showed a loss of first-phase insulin secretion in response to glucose, but not to arginine, similar to that observed in humans with type 2 diabetes mellitus. These mice also showed a progressively impaired glucose tolerance over 6 months. The alterations in the islet function of βIRKO mice was not associated with any significant changes in morphology, changes in the size of the pancreatic β cells, in the ratio of β to non-β cells but, at the age of 4 months βIRKO mice showed 35% lower insulin content compared with controls. Electron microscopic study of β cells revealed well-preserved cells with no apparent differences in the cell membrane, endoplasmic reticulum, Golgi apparatus or electron-dense insulin-containing granules within the β cells in the βIRKO versus the controls. Although, the islets were somewhat smaller in the 4-month old βIRKO mice, the distribution of GLUT-2 appeared to be comparable with that in the controls, while the *ob*/*ob* mice showed barely detectable GLUT-2 in the β cells.

The βIRKO mice provided the first in vivo model demonstrating the consequences of a lack of functional insulin receptors in the islet cell. These mice showed a selective loss of glucose-stimulated early insulin release and age-dependent inability to handle glucose challenge, defects that are very similar to those seen in patients with type 2 diabetes mellitus [158]. The observations made in the βIRKO mice provide direct evidence of a functional role for the insulin receptor in the islet β cell function in the maintenance of glucose homeostasis and suggest that insulin resistance at the β cell may be a significant factor in the development of a loss of glucose-stimulated insulin secretion by the β cells. Based on these observations, it can be proposed that in type 2 diabetes mellitus in which insulin receptor-specific resistance at the β cell level coupled with insulin resistance at the periphery could result in the classical pathophysiological findings seen in this disease.

Brown Adipose-specific Insulin Receptor Knockout (BATIRKO) Mice

Brown adipose tissue plays a significant role in determining peripheral insulin sensitivity [159], as well as thermal adaptation. When brown adipose tissue-specific insulin receptor knockout mice were developed, the brown adipose tissue developed normally in these animals, but surprisingly brown adipose tissue undergoes atrophy as the mice age. Paradoxically, the age-dependent loss of brown adipose tissue was found to be associated with deterioration of β-cell function, decrease in β-cell mass that ultimately produced hyperglycemia [160]. These results suggest that the maintenance of adequate β-cell mass and function for brown adipose tissue seems to be necessary. It is likely that brown adipose tissue produces some soluble factors {? adiponectin or similar factor(s)} that have a broader metabolic action. This cross-talk between pancreatic β-cells and the brown adipose tissue is somewhat similar to the connection between obesity and accelerated cancer progression. It was reported that stromal cells from white adipose tissue (WAT) cooperate with the endothelium to promote blood vessel formation through the secretion of soluble trophic factors. It was observed that tumors recruit WAT-derived cells. Adipose stromal and endothelial cells that enter into systemic circulation home to and engraft into tumor stroma and vasculature, respectively. It was also reported that recruitment of adipose stromal cells by tumors is sufficient to promote tumor growth. Furthermore, migration of stromal and vascular progenitor cells from WAT grafts to tumors accelerated cancer progression, providing evidence for a direct relationship between obesity and cancer [161].

Thus, the development of study of the BATIRKO mice revealed two important aspects of insulin resistance. The first one is the fact that insulin plays an important role in development or maintenance of brown adipose tissue since, mice lacking insulin receptor exhibited brown fat atrophy in an age-dependent manner. Though the exact signaling pathway by which insulin receptor signaling plays a role in brown fat adipogenesis is not clear, the possibility that the expression of C/EBP-α { (CCAAT/enhancer-binding protein-α is a protein that in humans is encoded by the *CEBPA* gene [162, 163]. The protein encoded by this intronless gene is a bZIP transcription factor which can bind as a homodimer to certain promoters and enhancers. It forms heterodimers with the related proteins CEBP-β and CEBP-γ. The encoded protein binds to the promoter and modulates the expression of the gene encoding leptin. Also, the encoded protein can interact with CDK2 and CDK4, thereby inhibiting these kinases and causing growth arrest in cultured cells. CEBP-α interacts with Cyclin-dependent kinase 2 and Cyclin-dependent kinase 4 [164]} in brown adipose tissue is strongly dependent on the insulin receptor throughout development [161]. The second important aspect observed is that the BATIRKO mice have an altered regulation of glucose homeostasis since, the lack of insulin receptor in the brown adipose tissue led to reduced β cell mass, a significant decrease in basal insulin, and a marked insulin-secretion defect in response to glucose in vivo and in isolated islets, events that led to a diabetic phenotype with fasting hyperglycemia and impaired glucose tolerance [161]. This phenotype became apparent in an age-dependent manner, suggesting that the diabetic phenotype is related to brown fat atrophy, though the exact mechanism that accounts for the moderate decrease in β cell mass and insulin levels is unclear.

VENIRKO Mice

Insulin receptors on vascular endothelial cells have been suggested to participate in insulin-regulated glucose homeostasis by facilitating transcytosis of insulin from the intravascular to extravascular space, by promoting vasodilation and enhancing blood flow, and by generation of signaling mediators [165, 166]. Mice with a vascular endothelial cell insulin receptor knockout (VENIRKO) generated using the Cre-*loxP* system showed blood glucose and insulin concentrations, glucose and insulin tolerance tests, the time of course of insulin action on glucose disposal during a euglycemic-hyperinsulinemic clamp; were found to be similar to those seen in the control. But these VENIRKO mice showed almost 30–60% reduction in endothelial nitric oxide synthase (eNOS) and endothelin-1 mRNA levels in endothelial cells, aorta, and heart, while expression of VEGF was normal. Surprisingly, VENIRKO mice showed lower systolic, diastolic and mean blood pressures than controls, but responded normally to high and low salt diets. These VENIRKO mice showed insulin resistance when put on low-salt diet [167, 168]. These results suggest that inactivation of the insulin receptor on endothelial cells has no major consequences on vascular development or glucose homeostasis under basal conditions, but had altered expression of VEGF that may play alower calcium intake role in maintaining vascular tone and regulation of insulin sensitivity to dietary salt intake. In this context, the work of McCarron et al. [169] is particularly relevant who showed that lower calcium intake was the most consistent factor in hypertensive individuals and across the population studied (based on HANES I survey) higher intakes of calcium, potassium, and sodium were associated with lower mean systolic blood pressure and lower absolute risk of hypertension. These results led to the suggestion that nutritional deficiencies and not excesses are what distinguish overweight or hypertensive individuals from normal subjects in the USA. In fact, it was suggested that caloric restriction increases the risk of further reducing an individual's exposure to nutrients that may be essential for maintaining normal mean arterial pressures [169]. The results seen with VENIRKO mice that showed insulin resistance when put on low-salt diet [167] is reminiscent of the suggestions made by McCarron et al. [169]. Since, hypertension and type 2 diabetes mellitus can occur together [170], insulin resistance is common in these two conditions, and as VENIRKO mice showed insulin resistance when put on low-salt diet, it is reasonable to suggest that salt intake could alter endothelial function and play a significant role in insulin resistance. Furthermore, both salt and calcium seem to modulate the production of endothelial nitric oxide (eNO), a potent vasodilator and platelet anti-aggregator, which seem to play a significant role in hypertension (see Chap. 8 for further discussion on this aspect). Based on the results of the studies done in VENIRKO mice [167, 168] and the observation that these mice developed insulin resistance when put on low-salt diet suggests that minerals (such as salt, Ca^{2+}, Mg^{2+}) have a regulatory role in the maintenance of vascular tone and in the pathogenesis of hypertension by modulating eNO generation and possibly, minerals posses ability to regulate insulin resistance either through eNO generation or other pathways. It is likely that dietary sodium and Ca^{2+} may regulate voltage-gated sodium channels, sodium-calcium exchanger (NCX1) and Na^{+}-K^{+}-ATPase activity and thus, modulate insulin secretion from β cells [171–173] and eNO generation.

Insulin, GLUT-4 and Glucose Transport

Insulin-stimulated glucose uptake by adipose tissue and muscle is facilitated by Glucose transporter type 4 (GLUT-4), a protein that is encoded in humans by the *GLUT-4* gene, and hence, is also referred to as insulin-sensitive glucose transporter. GLUT-4 is found in adipose tissues and striated muscle (skeletal and cardiac) and is responsible for insulin-regulated glucose translocation into the cell. GLUT-4 is expressed only in muscle and fat cells, the major tissues in the body that respond to insulin [174–177]. In the absence of insulin, GLUT-4 is sequestered in the interior of muscle and fat cells within lipid bilayers of vesicles. Insulin induces the translocation of GLUT-4 from intracellular storage sites to the plasma membrane. Insulin binds to the insulin receptor in its dimeric form. The receptor phosphorylates and subsequently activates IRS-1, which in turn binds the enzyme PI-3 kinase which converts the membrane lipid PIP2 to PIP3. PIP3 generates a binding site for PKB (protein kinase B), and also for PDK1 which, being localized together with PKB, can phosphorylate and activate PKB. Once phosphorylated, PKB becomes active and phosphorylates other targets that stimulate GLUT-4 to be expressed on the plasma membrane. At the cell surface, GLUT-4 permits the facilitated diffusion of circulating glucose down its concentration gradient into muscle and fat cells. Glucose is rapidly phosphorylated by glucokinase in the liver and hexokinase in other tissues to form glucose-6-phosphate, which then enters glycolysis or is polymerized into glycogen. Glucose-6-phosphate cannot diffuse back out of cells, which also serves to maintain the concentration gradient for glucose to passively enter cells [178]. Contraction also stimulates the cell to translocate GLUT-4 receptors to the surface. This is especially true in cardiac muscle, where continuous contraction can be relied upon; but is observed to a lesser extent in skeletal muscle [179]. GLUT-4 interacts with Death-associated protein 6 [180].

Muscle-specific GLUT-4 Knockout Mice (MG4KO)

In view of the importance of GLUT-4 in glucose transport, it is important to determine if different types of insulin resistance in a single tissue might produce different alterations in glucose homeostasis. To answer this question, a muscle-specific GLUT-4 knockout mouse was developed and studied [181, 182]. These animals showed a profound reduction in basal glucose transport, had severe insulin resistance and glucose intolerance from an early age, suggesting that GLUT-4-mediated glucose transport in muscle is essential to the maintenance of normal glucose homeostasis. Muscle-specific GLUT-4 knockout mice had a 55% decrease in insulin-stimulated whole body glucose uptake, an effect that could be attributed to a 92% decrease in insulin-stimulated muscle glucose uptake. In addition, surprisingly, these animals also showed that insulin's ability to stimulate adipose tissue glucose uptake and suppress hepatic glucose production was significantly impaired that seems to be secondary to a primary defect in muscle glucose transport indicating that secondary defects in insulin action in adipose tissue and liver developed due to glucose toxicity. These secondary defects probably contribute to insulin resistance and to the development of diabetes. Despite the fact that mice with muscle-specific knockout of the Glut-4 glucose transporter were insulin resistant and mildly diabetic, muscle glycogen content in the fasted state was increased. These animals showed increased basal glycogen synthase activity (by 34%) and had decreased glycogen phosphorylase activity (by 17%) in the muscle, while contraction-induced glycogen breakdown was found to be normal. The increased glycogen synthase activity occurred in spite of decreased signaling through the insulin receptor substrate 1 (IRS-1)-phosphoinositide (PI) 3-kinase-Akt pathway and increased glycogen synthase kinase 3β (GSK3β) activity in the basal state. Hexokinase II activity was increased, leading to an approximately twofold increase in glucose-6-phosphate levels. In addition, the levels of two scaffolding proteins that are glycogen-targeting subunits of protein phosphatase 1 (PP1), the muscle-specific regulatory subunit (RGL) and the protein targeting to glycogen (PTG), were increased by 3.2- to 4.2-fold in muscle compared to wild-type mice. The catalytic activity of PP1, which dephosphorylates and activates glycogen synthase, was also increased that dominated over the GSK3 effects, since glycogen synthase phosphorylation on the GSK3-regulated site was decreased. Thus, the markedly reduced glucose transport in muscle resulted in increased glycogen synthase activity due to increased hexokinase II, glucose-6-phosphate, and RGL and PTG levels and enhanced PP1 activity. This, combined with decreased glycogen phosphorylase activity, resulted in increased glycogen content in muscle in the fasted state when glucose transport was reduced explaining why muscle glycogen content in the fasted state was increased [183].

In contrast to these results, mouse model of type 2 diabetes generated by genetic disruption of one allele of GLUT-4 (GLUT-4$^{+/-}$) exhibited decreased GLUT-4 expression and glucose uptake in muscle that accompanied impaired whole-body glucose utilization, hyperinsulinemia, hyperglycemia, hypertension and heart histopathology but did not have obesity [184]. In order to prevent the onset of impaired muscle GLUT-4 expression and glucose utilization, a fast-twitch muscle-specific GLUT-4 transgene – the myosin light chain (MLC) promoter – driven transgene MLC-GLUT4 – into GLUT-4$^{+/-}$ mice (MLC-GLUT-4$^{+/-}$ mice) was introduced that resulted in normalization of GLUT-4 expression and 2-deoxyglucose uptake levels in fast-twitch muscles of MLC-GLUT 4$^{+/-}$ mice. In contrast to GLUT-4$^{+/-}$ mice, MLC-GLUT 4$^{+/-}$ mice exhibited normal whole-body glucose utilization, and were not hyperinsulinemic and hyperglycemic. Even the occurrence of diabetic heart histopathology in MLC-GLUT-4$^{+/-}$ mice was reduced to control levels. These results, suggest that the onset of a diabetic phenotype in GLUT-4$^{+/-}$ mice can be avoided by preventing decreases in muscle GLUT-4 expression and glucose uptake [185].

In another study, it was reported that disruption of the GLUT-4 gene in the mice (GLUT-4$^{-/-}$) were growth-retarded and exhibited decreased longevity associated with cardiac hypertrophy and severely reduced adipose tissue deposits. Blood glucose levels in female *GLUT-4*-null mice were not significantly elevated in either the fasting or fed state; while the male *GLUT-4*-null mice had moderately reduced glycemia in the fasted state and increased glycemia in the fed state. But, both female and male *GLUT-4*-null mice exhibited postprandial hyperinsulinemia, indicating the presence of insulin resistance. These animals showed increased expression of GLUT-2 in the liver and GLUT-1 in the heart but not skeletal muscle. Oral glucose tolerance tests show that both female and male *GLUT-4*-null mice have the ability to clear glucose as efficiently as controls, but insulin tolerance tests indicated that these mice were less sensitive to insulin action. These results suggest that functional GLUT-4 protein is not needed for maintaining nearly normal glycemia but that GLUT-4 is essential for sustained growth, normal cellular glucose and fat metabolism, and expected longevity [186].

Based on the preceding discussion, it is evident that in contrast to the MIRKO mice [154–156], MG4KO mice [181–186] have severe whole body insulin resistance, fasting hyperglycemia and glucose intolerance, but show no increase in body fat content or associated hyperlipidemia. These results indicate that, perhaps, GLUT-4 receptor is more important to maintain glucose homeostasis than insulin receptor.

Fat-specific GLUT-4 Knockout Mice

Similar to the MG4KO mice, fat-specific GLUT-4 knockout mice were created to study the importance of GLUT-4 in adipose tissue [187]. These (G4A$^{-/-}$) mice showed normal growth and adipose mass despite markedly impaired insulin-stimulated glucose uptake in adipocytes. Even though GLUT-4 expression was well preserved in muscle, these (G4A$^{-/-}$) mice developed insulin resistance in muscle and liver and showed decreased biological responses and impaired activation of phosphoinositide-3-OH kinase. G4A$^{-/-}$ mice developed glucose intolerance and hyperinsulinemia. The insulin resistance seen in this animal model could not be accounted for by changes in circulating free fatty acids, triglycerides or leptin, or changes in TNF-α expression in adipose tissue. It is interesting to note that the degree of glucose intolerance and insulin resistance in G4A$^{-/-}$ mice was found to be similar to that in mice with muscle-selective ablation of GLUT-4 (MG4KO) [181], thereby suggesting distinct and complementary roles for adipose tissue and skeletal muscle GLUT-4 in mediating glucose disposal in vivo. These evidences reaffirm the importance of glucose transport in adipose tissue and muscle tissue and their critical role in glucose homeostasis. Thus, the adipose and muscle-selective down regulation of GLUT-4 seen in human obesity and type 2 diabetes mellitus may contribute to the development of insulin resistance seen in obesity, type 2 diabetes, polycystic ovary syndrome (PCOS) and hypertension [188–193].

PUFAs, Expression of Insulin Receptors and GLUTs and Diabetes Mellitus

In this context, it is important to note that the expression and function of both the insulin receptor and GLUTs depend on the cell membrane architecture. Previously, I proposed that alterations in the cell membrane fluidity is a critical factor that determines the expression of number of insulin receptors and possibly, GLUTs and their affinity and/or response to insulin [59, 60, 111, 194–199]. This proposal is supported by the observation that subjects with obesity, insulin resistance, type diabetes mellitus and hypertension, conditions that are closely associated with insulin resistance and in which GLUT-4 and insulin receptor expression are decreased in several tissues especially in the adipose tissue and muscle [154–156, 181–186, 188–193], have low plasma and tissue concentrations of various polyunsaturated fatty acids (PUFAs) such as γ-linolenic acid (GLA), dihomo-GLA (DGLA), arachidonic acid (AA), eicosapentaenoic acid (EPA), and docosahexaenoic acid (DHA) [200, 201]. Furthermore, in animal studies, we showed that oral feeding of oils rich in PUFAs (such as GLA, AA, EPA and DHA) and pure fatty acids prevented the development of chemical-induced diabetes mellitus [202–206] suggesting that presence of adequate amounts of PUFAs protect pancreatic β cells from toxic actions of various chemicals. It was also observed that both cyclooxygenase and lipoxygenase inhibitors did not block the protective action of PUFAs against alloxan-induced diabetes mellitus, though some individual eicosanoids were found to be effective in preventing alloxan-induced β-cell cytotoxicity in vitro and diabetes mellitus in vivo [207, 208]. Thus, it is possible that the cytoprotective and anti-diabetic action of PUFAs observed in our studies could be due to the formation of anti-inflammatory products such as lipoxins, resolvins, protectins, maresins and nitrolipids [60, 194].

Recently, Gonzalez-Periz et al. [209] showed that increased intake of n-3-PUFAs alleviates obesity-induced insulin resistance and advanced hepatic steatosis in obese mice. These beneficial effects were associated with up-regulation of PPAR-γ, GLUT-2 and GLUT-4, and insulin receptor signaling (i.e., IRS-1 and IRS-2) in both adipose tissue and liver. In addition, n-3-PUFAs induced the expression and the production of the potent anti-inflammatory, antisteatotic, and insulin-sensitizing adipokine, adiponectin, and induced the phosphorylation of AMPK. These beneficial actions were found to be associated with a decrease in the formation of n-6-PUFA-derived eicosanoids such as PGE_2 and 5-HETE and a concomitant increase in the generation of beneficial molecules protectins and resolvins that are derived principally from n-3 PUFAs. But, it should be noted here that AA can give rise to both pro- and anti-inflammatory molecules such as PGEs, TXs (thromboxanes) and Leukotrienes and lipoxins and resolvins. In patients with obesity, type 2 diabetes mellitus and hypertension, plasma levels of AA have been found to be low [60, 194] and at the same time AA can prevent chemical-induced cytotoxicity to β cells in vitro and diabetes mellitus in vivo [203]. This protective action could be attributed to the formation of beneficial lipoxins and resolvins. But, it is not yet known how the formation of PGs, TXs, LTs and lipoxins and resolvins is modulated under physiological and pathological conditions. It is possible that when excess of PGs, LTs, TXs are formed from AA cytotoxicity occurs whereas when lipoxins and resolvins are formed in adequate amounts cytoprotection is seen.

One factor that regulates the expression of insulin receptors and GLUTs on the cell membrane could be the quality and quantity of PUFAs present in the cell membrane phospholipids. In the presence of adequate amounts of PUFAs (especially AA, EPA and DHA), the number of insulin receptors and GLUTS will be normal or adequate and the affinity of these receptors to insulin will be optimal, whereas when the phospholipid content of AA, EPA and DHA is low and/or the cholesterol and other saturated fatty acids are present in excess amounts in the cell membrane it leads to increase in the rigidity of the membrane and decreased expression insulin receptors and GLUTs and decreased affinity of insulin to these receptors as has been shown with oleic acid treatment [210]. This would ultimately lead to insulin resistance. Hence, it is likely that when adequate amounts of PUFAs are provided in the diet or formed from their precursors and incorporated in the cell membrane lipids of muscle, liver and adipose and pancreatic β tissues/cells, the insulin sensitivity will be normal.

Yet another mechanism by which PUFAs are able to regulate insulin secretion, insulin receptor and GLUTs expression is by increasing the concentration of caveolin-1 and caveolin-3 in caveolae and by their effects on the Ras/Raf-1/extracellular signal regulated kinase (ERK)/mitogen-activated protein kinase pathway [211].

It is known that insulin stimulation of GLUT-4 translocation requires at least two distinct insulin receptor-mediated signals: one leading to the activation of phosphatidylinositol 3 (PI-3) kinase and the other to the activation of the small GTP binding protein TC10. TC10 is processed through the secretory membrane trafficking system and localizes to caveolin-enriched lipid raft microdomains. Disruption of lipid raft microdomains inhibited the insulin stimulation of GLUT-4 translocation and TC10 lipid raft localization and activation without affecting PI-3 kinase signaling. These data suggest that insulin stimulation of GLUT-4 translocation in adipocytes requires the spatial separation and distinct compartmentalization of the PI-3 kinase and TC10 signaling pathways [212]. Since PUFAs have the ability to change the composition of caveolae, it is expected that when adequate amounts of PUFAs are present in the caveolae the translocation of GLUT-4 in response to insulin stimulation will be optimum and thus, PUFAs are able to decrease insulin resistance. This proposal is supported by the work of Gonzalez-Periz et al. [209] who showed that increased intake of n-3-PUFAs alleviates obesity-induced insulin resistance in obese mice by the up-regulation of GLUT-2 and GLUT-4 in both adipose tissue and liver.

In contrast to these results, studies showed that when male weanling Wistar/NIN rats were given trans-fatty acids (TFAs) and saturated fatty acids (SFAs) in the diet upregulated the mRNA levels of resistin, and downregulated adiponectin and GLUT-4 suggesting that insulin resistance in TFA- and SFAs-fed rats is due to decrease in the expression of GLUT-4 in adipose tissue [213]. These beneficial actions of

PUFAs (especially n-3 PUFAs) on GLUT-4 are in addition to their favorable action on PPARs and ability to suppress the production of IL-6, TNF-α and resistin and preventing infiltration of adipose tissue by macrophages that generally induce insulin resistance and by enhancing the formation of anti-inflammatory lipid molecules such as lipoxins, resolvins, protectins and maresins [214–218].

Polygenic Knockout Models

Type 2 diabetes mellitus is known to be polygenic in origin and hence, it is likely that knocking out one gene may not give clues to all the abnormalities seen in this condition. This led to the study of a heterozygous double-knockout mouse model of insulin receptor and IRS-1. It was observed that insulin receptor and heterozygote IRS-1 knockout mice exhibited only mild sub-clinical insulin resistance with 1.5 to 2-fold elevation in insulin levels and mild pancreatic β-cell hyperplasia [167]. On the other hand, the IR.IRS-1 double-heterozygous knockout mice showed marked insulin resistance with 10-fold increase in pancreatic β-cell mass. These mice developed diabetes by 4–6 months of age that too only $\sim$50% of them. These compound heterozygote animals developed diabetes with delayed onset, showed a marked synergism between insulin receptor defect and the IRS-1 defect and only $\sim$50% of the mice developed diabetes. These results suggest that an additional gene(s) contribute to or protect these animals from the development of diabetes.

Phenotype of male mice that are double heterozygous for the insulin receptor and IRS-1 showed marked variation depending on the genetic background. This difference appears to be due to differences in insulin resistance but not as a result of pancreatic β-cell failure [167]. It was also reported that the susceptibility or resistance of different types of knockout mice varied depending on the F2 intercross between mice based on their genetic background (either they are B6 or 129Sv). The incidence of diabetes in double heterozygous male intercross mice was found to be 60% at 6 months of age and their responses to glucose showed a wide variation and a bimodal distribution as has been reported in humans with type 2 diabetes mellitus [219]. These results suggest that the insulin receptor/IRS-1 knockout mouse is somewhat similar to human type 2 diabetes mellitus that has a polygenic etiology, probably genetically programmed insulin resistance, a delayed age of onset and a biphasic relationship between insulin and glucose levels.

Triple Heterozygous Knockouts (IR/IRS-1/p85)

In an extension of the double knockout studies, Kido et al. produced even more complex compound heterozygous animals such as mice with three partial defects in insulin signaling (IR/IRS-1/IRS-2 or IR/IRS-1/p85) [220]. The IR/IRS-1/IRS-2 triple heterozygous mouse showed severely impaired glucose tolerance and a doubling of the incidence of diabetes compared to the double heterozygotes. On the other hand, the IR/IRS-1/p85 knockout was found to be less severely affected than the IR/IRS-$1^{+/-}$ mouse. It was also noted that heterozygosity for the p85 allele protected mice form becoming diabetic [221]. These results indicate that p85 may represent a novel therapeutic target for enhancing insulin signaling [222].

It is evident from the preceding discussion how genetic predisposition plays itself out in the oligogenic and heterogeneous pathogenesis of type 2 diabetes mellitus. It is also clear that balance and interaction(s) among various proteins can affect the efficiency of signaling both positively and negatively. The IRS knockout and tissue specific knockouts have clearly demonstrated the contribution of different tissues to the pathogenesis of type 2 diabetes mellitus, though at times their contribution and roles are so unpredictable in the whole scheme of things. It is also evident that insulin has important effects in tissues such as brain and pancreatic β-cells that may have relevance to their role in the pathogenesis of type 2 diabetes mellitus. But, unfortunately in all these studies the changes in the levels of various cytokines, nitric oxide, free radicals, antioxidants, and PUFAs and their metabolites have not been studied. Nevertheless, these studies emphasize the important functional role for the insulin receptor in glucose sensing by the liver, adipose tissue, brain, muscle and pancreatic beta cell and suggest that defects in insulin signaling at the level of the beta cell and other cells may contribute to the observed alterations in insulin secretion in type 2 diabetes.

It would have been interesting had some studies pertaining to the changes in the levels of various neurotransmitters, gut peptides, and hypothalamic peptides and neurotransmitters was studied in these knockout models. It is not clear whether in these knockout animal models, the changes in plasma glucose, insulin and glucose homeostasis observed are solely due to the knockout of the specific gene or such knockout of gene(s) also has unexpected consequences elsewhere such as changes in the hypothalamic neurotransmitters, peptides, gut peptides and hormones, etc. It is not unreasonable to expect such changes in the whole organism since, in general, homeostatic mechanisms are expected to produce changes in other tissues and organs despite the induced genetic manipulation is supposed to have specific action(s).

For instance, food deprivation induced increase in NPY levels in the paraventricular nucleus (PVN) returned to the control range following insulin injections, which did not alter blood glucose levels. This change *in vivo* NPY release in the PVN of food-deprived rats also decreased in response to peripheral insulin injections. Both insulin and insulin-like growth factor-II (IGF-II) decreased the release of NPY in a dose dependent fashion from the PVN *in vitro*, suggesting that the site of insulin action on the hypothalamic NPY network is at the level of NPY nerve terminals and that both insulin and IGF-II decrease NPY release from the PVN [223]. Since NPY is a potent orexigenic signal and as insulin and IGF-II decrease hypothalamic NPY, it is suggested that presence of adequate amounts of insulin, insulin receptors and IGF-II in the brain can reduce appetite, and thus, control obesity and hyperglycemia.

It is evident from the preceding discussion that insulin when used at a dose that does not produce any change in the plasma glucose has the ability to alter hypothalamic NPY levels, an unexpected observation. In a similar fashion, it is not unlikely that specific gene knockout manipulations could produce some far reaching changes in other tissues and organs. We are simply missing such changes since they were never looked for. The best example of such far reaching changes occurring in a distant organ such as brain when the manipulation is performed in the gut is detailed below. In general and in the early days of gastric bypass surgery, it was thought that weight loss is due to physical limitation on the size of the stomach and reduced length of the small intestine. But, now there is reasonable evidence to suggest that the weight loss noted in these patients who underwent gastric bypass surgery is, at least, in part due to changes in the levels of neurotransmitters in the hypothalamic nuclei.

Weight Loss After Gastric Bypass and Changes in Hypothalamic Neuropeptides and Monoamines

Roux-en-gastric bypass (RYGB) and other bariatric operations are employed as a therapeutic procedure to induce weight loss in subjects with extreme obesity. RYGB produces on an average 49–65% weight loss within 2–5 years [224, 225]. Besides weight loss, RYGB ameliorates diabetes, hyperlipidemia, and other obesity-related metabolic abnormalities [224–226]. We developed a surgical rat model of human RYGB to study the molecular mechanisms involved in weight loss and amelioration of metabolic abnormalities in diet-induced obese animals [227]. These studies revealed that gastric bypass surgery produced significant weight loss due to reduced caloric intake with a reduction in meal size and meal number, accompanied by a decrease in serum glucose, insulin, leptin, triglyceride concentrations, and subcutaneous abdominal fat [228]. In addition, RYGB-induced weight loss was associated with a decrease in NPY in ARC, pPVN (parvocellular part of paraventricular nucleus of hypothalamus), and mPVN (magnocellular part of PVN) and an increase in α-MSH in ARC, pPVN, and mPVN compared with controls. 5HT-$_{1B}$-receptor in pPVN and mPVN increased in RYGB compared to control [229]. These results suggest that weight loss seen after RYGB and diet control could be due to specific changes in hypothalamic peptides. Serotonin innervations are widely distributed in the hypothalamus and it innervates NPY neurons both in the ARC and PVN. Serotonin suppresses food intake. Thus, weight loss seen in RYGB and PF groups could be related to alterations in the concentrations of specific hypothalamic signaling peptides that regulate appetite, food intake and satiety. In addition, weight loss induced after RYGB procedure decreased plasma lipids, insulin resistance, leptin, sTNFR1 (soluble tumor necrosis factor receptor 1), and IL-6 and adiponectin and ghrelin increased significantly and simultaneously insulin resistance improved after weight loss and correlated with high adiponectin levels [230]. These results suggest that the proinflammatory molecules that are enhanced in obesity are suppressed after RYGB procedure. Though reduction in weight and amelioration of pro-inflammatory environment in those who underwent RYGB procedure is accompanied by changes in the concentrations of hypothalamic peptides and monoamines, it is not clear whether the latter (hypothalamic peptides and monoamines) modulate inflammation. If it is true that metabolic syndrome is a low-grade systemic inflammatory condition in which hypothalamic peptides and monoamines play a significant role, it is anticipated that there could exist a relationship between hypothalamus and inflammation.

Monoaminergic Amines and Hypothalamic and Gut Peptides and Inflammation

Dopamine

Dopamine is a neurotransmitter and also has cardiovascular properties. It is used in patients with systemic inflammatory response syndrome (SIRS) to maintain hemodynamic stability. Polymorphonuclear leukocytes (PMNLs) isolated from healthy volunteers and patients with SIRS and treated with varying doses of dopamine and a dopamine D-1 receptor agonist showed a significant delay in PMNL apoptosis in patients with SIRS compared with controls. Treatment of isolated PMNLs from both healthy controls and patients with SIRS with dopamine induced apoptosis. PMNL ingestive and cytocidal capacity were both decreased in patients with SIRS compared with controls. Treatment with dopamine significantly increased phagocytic function [231]. These data indicate that dopamine induces PMNL apoptosis and modulates its function both in healthy controls and in patients with SIRS.

PMN obtained from healthy subjects stimulated with lipopolysaccharide (LPS) and TNF-α showed a significant increase in transendothelial migration and upregulation of CD11b/CD18. Similarly, HUVEC (human umbilical vein endothelial cells) stimulated with LPS and TNF-α showed upregulation of E-selectin/ ICAM-1 expression compared with normal EC (endothelial cells). Dopamine significantly decreased PMN transmigration, attenuated PMN CD11b/CD18 and the endothelial molecules E-selectin and ICAM-1 expression and in addition, decreased chemoattractant effect of IL-8 [232]. Thus, dopamine seems to have anti-inflammatory actions by attenuating the initial interaction between PMN and the endothelium, and modulating PMN exudation.

Infusion of dopamine in septic mice increased splenocyte apoptosis and decreased splenocyte proliferation and IL-2 release of septic mice without any effect on sepsis-induced changes in leukocyte distribution. Furthermore, dopamine inhibited splenocyte proliferation and the release of the TH1-cytokines IL-2 and IFN-γ that paralleled a decrease in the survival of dopamine-treated septic animals [233], suggesting that dopamine modulates cellular immune functions in a murine model of sepsis.

It is interesting to note that obese subjects have decreased dopamine receptors and decreased dopamine levels in the brain [234] and hence, are believed to have "reward deficiency syndrome". Since dopamine has anti-inflammatory actions, and so the decrease in the dopamine receptor number or content in the brain of subjects with obesity could facilitate low-grade inflammation that may affect hypothalamus and eventually lead to hypothalamic dysfunction and the development of metabolic syndrome.

Serotonin (5-hydroxytryptamine)

Administration of 5-hydroxytryptamine (serotonin) or its precursor, 5-hydroxy-L-tryptophan (5-HTPH), produced marked depression of T cell dependent, humoral, hemolytic, primary immune response in mice. Serotonin caused a marked reduction of the thymus weight [235]. It is noteworthy that serotonin content decreased in ventral part of the anterior hypothalamus within 20 min after immunization of rats with sheep red blood cells [236] suggesting that hypothalamic serotonin content could influence immune response. It was reported that elevation of active serotonin level resulted in the inhibition of immune response and the nuclei raphe serotoninergic system provided an inhibitory mechanism of the immune response modulation, which is realized via the hypothalamus-hypophysis-adrenals axis. This inhibitory action of serotonin on immune response is brought about by its ability to attenuate suppressor cell function [237]. In addition, serotonin inhibited oxidative burst of human phagocytes and exerted a dose dependent inhibition of the myeloperoxidase activity, suggesting that serotonin modulates the oxidative burst of phagocytes and decreases the generation of reactive oxygen species [238]. Serotonin significantly inhibited the production of TNF and IL-12, whereas IL-10, NO and PGE_2 production were increased. These immunomodulatory actions of serotonin were mimicked by 5-HT(2) receptor agonist but were not abrogated by 5-HT(2) receptor antagonist, suggesting the possible involvement of other 5-HT receptors. Inhibitors of cyclooxygenase and antibody to PGE_2 abrogated the inhibitory and stimulatory effect of serotonin on TNF and IL-10 production, respectively, whereas NO synthase inhibitor eliminated serotonin-stimulated IL-10 increase. Furthermore, PGE_2 significantly increased alveolar macrophage IL-10 and NO production. These results suggest that serotonin alters the cytokine network through the production of PGE_2 and NO [239]. In addition, serotoninergic receptors (5-HTR) are expressed by a broad range of inflammatory cell types, including dendritic cells (DCs). 5-HT induced oriented migration in immature but not in LPS-matured DCs via activation of 5-HTR_1 and 5-HTR_2 receptor subtypes. 5-HT increased migration of pulmonary DCs to draining lymph nodes in vivo. By binding to 5-HTR_3, 5-HTR_4 and 5-HTR_7 receptors, 5-HT up-regulated production of the pro-inflammatory cytokine IL-6. 5-HT influenced chemokine release by human monocyte-derived DCs and 5-HT induced maturation of DCs and enabled them to secrete high amounts of IL-10 from low IL-12p70 secreting phenotype. Furthermore, 5-HT favored the outcome of a Th2 immune response both in vitro and in vivo [240]. These and other results suggest that 5-HT is a potent regulator of human dendritic cell function and immune response and has pro-inflammatory actions. The ability of serotonin to enhance inflammatory reactions in the skin, lung and gastrointestinal tract seems to be, in part, mediated by its action on mast cells. For instance, mouse bone marrow-derived mast cells (mBMMC) and human CD34(+)-derived MC (huMC) expressed mRNA for multiple 5-HT receptors. Though serotonin did not induce degranulation of mBMMC and huMC, it did induce mBMMC and huMC adherence to fibronectin; immature and mature mBMMC and huMC migration and their chemotaxis. 5-HT induced accumulation of MC in the dermis of 5-HT(1A)R(+/+) mice, but not in 5-HT(1A) receptor knockout mouse[5-HT(1A)R(-/-)]. Thus, it is clear that both mouse and human MC respond to 5-HT through the 5-HT(1A) receptor and 5-HT promotes inflammation by increasing MC at the site of tissue injury [241].

Neuropeptide Y

Neuropeptide Y (NPY) is a sympathetic comediator that can regulate immunological functions including T cell activation and migration of blood leukocytes. Leukocytes expressed high amounts of NPY mRNA and peptide, similar to expression levels in sympathetic ganglia. During acute allograft rejection, leukocyte NPY expression drastically dropped to 1% of control levels suggesting that it modulates immune response and inflammation [242]. NPY and NPY-related receptor specific peptides reduced granulocyte accumulation into carrageenan-induced air pouch (an experimental model of inflammation), attenuated phagocytosis attained via Y1 receptor, decreased peroxide production mediated via Y2 and Y5 receptors activation and increased nitric oxide production via Y1 receptor [243]. These results emphasize the fact that NPY has anti-inflammatory actions. It is noteworthy that NPY-induced modulation of the immune and inflammatory responses is regulated by tissue-specific expression of different receptor subtypes (Y1–Y6) and the activity of the enzyme dipeptidyl peptidase 4 (DP4, CD26) that terminates the action of NPY on Y1 receptor subtype. It was noted that NPY suppressed paw edema in adult and aged, but not in young rats. Furthermore, plasma DP4 activity decreased, while macrophage DP4 activity, as well as macrophage CD26 expression increased with aging. Further studies showed that anti-inflammatory effect of NPY is mediated via Y1 and Y5 receptors. In contrast to the in vivo situation, NPY stimulated macrophage nitric oxide production in vitro only in young rats, and this effect was mediated via Y1 and Y2 receptors. Thus, age-dependant modulation of inflammatory reactions by NPY is determined by plasma, but not macrophage DP4 activity at different ages [244]. It is known that with age the production of TNF-α and IL-6 increase, appetite decreases and the tendency to develop metabolic syndrome is increased. Thus, the decrease in DP4 activity with age could be compensatory phenomena to counteract age-associated pro-inflammatory process. But, NPY is an orexigenic peptide and thus, may overcompensate age-associated decline in appetite and paradoxically promote the development of metabolic syndrome.

In an animal model of colitis, an increase in enteric neuronal NPY and nNOS expression in WT (wild-type) mice was noted. WT mice showed more inflammation compared to $NPY^{(-/-)}$ as indicated by higher clinical and histological scores, and myeloperoxidase (MPO) activity ($p < 0.01$). WT mice had increased nitrite, decreased glutathione (GSH) levels and increased catalase activity indicating enhanced oxidative stress. The lower histological scores, MPO and chemokine KC in S.T.-treated nNOS$^{(-/-)}$ and NPY$^{(-/-)}$/nNOS$^{(-/-)}$ mice support the contention that loss of NPY-induced nNOS attenuated inflammation. NPY-treated rat enteric neurons in vitro exhibited increased nitrite and TNF-α production [245]. These results indicate that NPY mediated increase in nNOS is a determinant of oxidative stress and subsequent inflammation. These results emphasize the close interaction among NPY, NOS and pro-inflammatory cytokine TNF-α and their modulatory influence on inflammation and metabolic syndrome.

Gastrin-releasing peptide (GRP, 10^{-10} M), NPY (10^{-10} M), somatostatin (10^{-10} M) and vasoactive intestinal peptide (VIP, 10^{-9} M) modulated the production of IL-1β, IL-6 and TNF-α by peripheral whole blood cells from healthy young and old people. GRP, NPY, somatostatin and VIP stimulated the production of IL-1β in old subjects, and NPY, somatostatin and VIP in young ones. The production of IL-6 was enhanced by GRP, NPY and VIP in young and old people. The TNF-α production was stimulated by NPY and somatostatin in young subjects, and by NPY, somatostatin and VIP in old ones, whereas GRP produced a decrease of TNF-α in young persons. GRP in old subjects and VIP in young and old subjects stimulated LPS-induced IL-6 production by whole blood cells. On the contrary, GRP and VIP inhibited LPS-induced TNF-α production in young controls [246]. Thus, neuropeptides have the ability to modulate the production of pro-inflammatory cytokines by peripheral blood cells at physiological concentrations indicating the close relationship among appetite and food intake regulating neuropeptides and inflammation. Paradoxically, cytokines IL-1β, IL-6, and TNF-α did not alter either basal or stimulated NPY release from the hypothalamic slices [247] suggesting that, at least, in some instances of anorexia such as cancer cachexia wherein the concentrations of these cytokines are increased, anorexia is not due to their effect on NPY levels. Since NPY is present in human adipose tissue, insulin increases NPY secretion, and adipocyte treatment with rh-NPY downregulated leptin secretion but had no effect on adiponectin and TNF-α secretion [248], it can be suggested that anti-lipolytic action of NPY promotes an increase in adipocyte size in hyperinsulinemic conditions and adipocyte-derived NPY mediates reduction of leptin secretion that may have implications for central feedback of adiposity signals.

Ghrelin

Ghrelin, an orexigenic peptide produced by the gut, produces specific inhibition of fatty acid biosynthesis induced by AMP-activated protein kinase (AMPK) resulting in decreased hypothalamic levels of malonyl-CoA and increased carnitine palmitoyltransferase 1 (CPT1) activity. In addition, fasting downregulated fatty acid synthase (FAS) in a region-specific manner, an effect that is mediated by an AMPK and ghrelin-dependent mechanisms. Thus, decreasing AMPK activity in the ventromedial nucleus of the hypothalamus (VMH) is sufficient to inhibit ghrelin's effects on FAS expression and feeding. Modulation of hypothalamic fatty acid metabolism specifically in the VMH in response to ghrelin is a physiological mechanism that controls feeding [249]. In addition, ghrelin that acts as an endogenous ligand for the growth hormone secretagogue receptor (GHS-R), are expressed in human T lymphocytes and monocytes, where ghrelin acts via GHS-R to specifically inhibit the expression of proinflammatory anorectic cytokines such as IL-1β, IL-6, and TNF-α. Ghrelin inhibited while leptin upregulated GHS-R expression on human T lymphocytes suggesting a reciprocal regulatory network by which ghrelin and leptin control immune cell activation and inflammation. Ghrelin exerts potent anti-inflammatory effects and attenuated endotoxin-induced anorexia in a murine endotoxemia model [250]. Thus, ghrelin functions as a key signal, coupling the metabolic axis to the immune system.

It was reported that pretreatment of phagocytic leukocytes with a GHS-R antagonist,[D-Lys3]-GHRP-6, abolished the stimulatory effects of trout ghrelin and des-VRQ-trout ghrelin on superoxide production. Ghrelin increased mRNA levels of superoxide dismutase and GH expressed in trout phagocytic leukocytes. Immunoneutralization of GH by addition of anti-salmon GH serum to the medium blocked the stimulatory effect of ghrelin on superoxide production [251], suggesting that ghrelin stimulates phagocytosis in fish leukocytes through a GHS-R-dependent pathway, and also that the effect of ghrelin is mediated, at least in part, by GH secreted by leukocytes. Furthermore, when the serum levels of ghrelin and its relationship with disease activity and nutritional status were evaluated in patients with inflammatory bowel disease (IBD), it was noted that serum ghrelin levels were significantly higher in patients with active ulcerative colitis and Crohn's disease than in those in remission (108 ± 11 pg/ml vs. 71 ± 13 pg/ml for ulcerative colitis patients, $P < 0.001$; 110 ± 10 pg/ml vs. 75 ± 15 pg/ml for Crohn's disease patients, $P < 0.001$). Circulating ghrelin levels in patients with these two diseases were positively correlated with sedimentation, fibrinogen and CRP and were negatively correlated with IGF-1, BMI, fat mass (%), and fat free mass (%). These results indicate that ghrelin levels may be important in determination of the activity in IBD patients and evaluation of nutritional status [252].

Ghrelin and GH secretagogue receptor 1b were expressed in PBMCs (peripheral blood mononuclear cells) of subjects with metabolic syndrome. Ghrelin gene expression correlated positively with the expressions of TNF-α ($P < 0.001$), IL-1β ($P < 0.001$) and IL-6 ($P = 0.026$), but was not associated with the plasma ghrelin concentration. At baseline, the plasma ghrelin levels were associated with fasting serum insulin concentrations, insulin sensitivity index and high-density lipoprotein cholesterol. Weight, BMI or waist circumference and acute insulin response in intravenous glucose tolerance test were found to best strong predictors of the ghrelin concentration. These results indicate an autocrine role for ghrelin within an immune microenvironment in view of its expression in PBMCs [253]. Thus, ghrelin expression in PBMCs could be used as a marker of low-grade systemic inflammation seen in metabolic syndrome along with plasma TNF-α and IL-6.

TNF-α is a glycoprotein hormone with important functions in inflammation and apoptosis, serves as a pro-inflammatory cytokine in the defense against viral, bacterial and parasitic infections and autoimmune disorders, and influences energy homeostasis and has an anorexigenic effect on the hypothalamus. TNF-α is also involved in the pathogenesis of psychiatric disorders such as depression or narcolepsy. On the other hand, ghrelin is a peptide hormone that primarily regulates eating behavior through modulation of expression of orexigenic peptides in the hypothalamus. Ghrelin administration increases food intake and body weight, while weight loss in turn increases ghrelin levels. Ghrelin possesses anti-inflammatory properties and has antidepressant and anxiolytic properties. Therefore, it is suggested that TNF-α and ghrelin seem to have opposite effects regarding the hypothalamic regulation of eating behavior, modulation of the immune response and the state of mental health. In a similar fashion, hypothalamic monoamines serotonin, dopamine, and acetylcholine and peptides such as NPY, BDNF, and melanocortins not only modulate eating behavior but also participate in the regulation of immune response and inflammation.

Melanocortin

The proopiomelanocortin (POMC) gene is transcribed in several tissues including the corticotroph of the anterior pituitary, neurons of the arcuate nucleus of the hypothalamus, and cells in the dermis and the lymphoid system. In these cells, POMC propeptide is processed posttranslationally to result in a series of small peptides. Thus, pituitary corticotrophs express prohormone convertase 1 (PC1) but not PC2, resulting in the production of N-terminal peptide, joining peptide, ACTH and lipotropin. Expression of PC2 in the hypothalamus leads to the production of α-, β-, and γ-MSH (melanocyte stimulating hormone) but not ACTH. The action of these melanocortin peptides is mediated by five G protein-coupled seven transmembrane domain receptors[melanocortin receptor type 1 (MC3R to MC5R)]. Both MC3R and MC4R are highly expressed in the central nervous system and play an important role in the control of food intake and energy balance. In particular, there are two distinct subsets of neurons in the arcuate nucleus of the hypothalamus that express MC3R and MC4R that together with their downstream target sites make up the central melanocortin system. POMC neurons produce the anorectic peptide α-MSH together with cocaine- and amphetamine-related transcript (CART), whereas a separate group expresses the orexigenic neuropeptide Y (NPY) and agouti-related protein (AgRP). AgRP is a potent MC3R and MC4R antagonist. Activation of the NPY/AgRP neurons increases food intake and decreases energy expenditure, whereas activation of POMC neurons decreases food intake and increases energy expenditure. The long isoform of the leptin receptor is highly expressed on the arcuate neurons, and leptin regulates these two neuronal populations in a reciprocal manner: suppressed levels of leptin after a fast decrease POMC mRNA and increase AgRP mRNA in the hypothalamus. From the arcuate nucleus, POMC and AgRP have extensive projections to several hypothalamic regions, including the lateral hypothalamus and the paraventricular nucleus. Cell bodies within the lateral hypothalamus contain the orexigenic peptide melanin concentrating hormone, and neurons of the paraventricular nucleus express TRH (thyrotropin releasing hormone). Thus, via this second order signaling, the melanocortin peptides exert their effects. In addition, melanocortins have potent anti-inflammatory effects that are mediated by direct effects on cells of the immune system as well as indirectly by affecting the function of resident non-immune cells and suppress NF-κB activation, expression of adhesion molecules and chemokine receptors, production of pro-inflammatory cytokines and other mediators. Thus α-MSH modulates inflammatory cell proliferation, activity and migration [254, 255].

Acetylcholine

Acetylcholine (Ach), the principal vagal neurotransmitter, suppresses inflammation and is termed as the "cholinergic anti-inflammatory pathway". These neural signals transmitted via the vagus nerve inhibit cytokine release through a mechanism that requires the alpha7 subunit-containing nicotinic acetylcholine receptor (alpha7nAChR). Vagus nerve regulation of peripheral functions is controlled by brain nuclei and neural networks. Studies showed that brain acetylcholinesterase activity controls systemic and organ specific TNF production during endotoxemia. Peripheral administration of the acetylcholinesterase inhibitor galantamine reduced serum TNF levels through vagus nerve signaling, and protected against lethality during murine endotoxemia. Administration of a centrally-acting muscarinic receptor antagonist abolished the suppression of TNF by galantamine, indicating that suppressing acetylcholinesterase activity, coupled with central muscarinic receptors, controls peripheral cytokine responses. Administration of galantamine to alpha7nAChR knockout mice failed to suppress TNF levels, indicating that the alpha7nAChR-mediated cholinergic anti-inflammatory pathway is required for the anti-inflammatory effect of galantamine. Thus, inhibition of brain acetylcholinesterase suppresses systemic inflammation through a central muscarinic receptor-mediated and vagal- and alpha7nAChR-dependent mechanism [256–258]. Ach modulates the production and actions of other hypothalamic monoamines serotonin, dopamine, and acetylcholine and peptides: NPY, BDNF, and melanocortins and thus, participates in the regulation of energy homeostasis.

Adrenaline and Noradrenaline

Patients with stress hyperglycemia and type 2 diabetes mellitus have increase in noradrenaline and adrenaline and decrease in serotonin and its metabolites [259–266] in the brain and increased production and release of catecholamines from the phagocytes in the peripheral circulation. This assumes importance in the light of the observation that sympathetic activation is associated with metabolic syndrome and increased risk of cardiovascular disease. In a study of 104 type 2 diabetic patients it was observed that blood concentrations of hs-CRP, IL-6 and plasminogen activator inhibitor-1 were higher in diabetic patients with than in those without metabolic syndrome. Both the 24-h mean LF (low frequency, both sympathetic and parasympathetic activities) and the low frequency to high frequency (LF-to-HF ratio) were also significantly higher in diabetic patients with than in those without metabolic syndrome. The LF-to-HF ratio was significantly higher in diabetic patients with a high CRP concentration (>3.0 mg/l) than in those with a low (<1.0 or = or) or moderate CRP (≤3.0 mg/l) concentration ($P < 0.001$ and $P < 0.01$, respectively). These results suggest that type 2 diabetic patients with metabolic syndrome have elevated markers of inflammation and evidence of cardiac sympathetic predominance [267]. Since adrenaline and noradrenaline have pro-inflammatory actions it is reasonable to suggest that the existence of low-grade systemic inflammation in metabolic syndrome could be due to sympathetic over activity. Since normally a balance is maintained between sympathetic and parasympathetic nervous systems, it implies that in metabolic syndrome plasma or tissue and leukocyte Ach levels will be low that has anti-inflammatory action whereas the production and release of catecholamines and consequent sympathetic over activity will be present.

Gut Peptides

Incretins are gastrointestinal hormones that enhance insulin release from pancreatic β cells after eating before blood glucose levels become elevated. Incretins also slow the rate of absorption of nutrients into the blood stream by reducing gastric emptying and may directly reduce food intake. Incretins inhibit glucagon release from α cells of the Islets of Langerhans. The two main incretins are glucagon-like peptide-1 (GLP-1) and gastric inhibitory peptide (GIP). Both GLP-1 and GIP are rapidly inactivated by the enzyme dipeptidyl peptidase-4 (DPP-4).

GLP-1 that has a half-life of less than 2 min is derived from the transcription product of the proglucagon gene and is produced by the intestinal L cell. The low half-life of GLP-1 is due to its rapid degradation by the enzyme dipeptidyl pepidase-4. Some of the known physiological functions of GLP-1 are it:

- increases insulin secretion from the pancreas in a glucose-dependent manner
- decreases glucagon secretion from the pancreas
- increases β cell mass and insulin gene expression
- inhibits acid secretion and gastric emptying in the stomach
- decreases food intake by increasing satiety
- promotes insulin sensitivity

In addition, GLP-1 also has immunomodulatory and anti-inflammatory actions.

GLP-1 binding and GLP-1 receptor mRNA expression is detected in both astrocytes and microglia. GLP-1 treatment induced morphological changes in microglia from the ramified type to the amoeboid type, suggesting an increase in the production and release of endogenous GLP-1. GLP-1 prevented the LPS-induced IL-1β mRNA expression, increased cAMP concentration and cAMP response element-binding protein phosphorylation in astrocytes indicating that it is a modulator of inflammation in the central nervous system [268].

Pro-inflammatory cytokines IL-1β, IFN-γ, and TNF-α inhibited the proliferation of pancreatic β cells in vitro through the extracellular signal-regulated kinase 1/2 (ERK1/2) activation, the signaling pathway involved in β cell replication. GLP-1 completely reversed the cytokine-induced inhibition of ERK phosphorylation and increased β cell proliferation threefold in cytokine-treated cultures. While pro-inflammatory cytokines reduced islet cell ERK1/2 activation and β cell proliferation in pancreatic islet culture, GLP-1 was capable of reversing this effect [269], suggesting that GLP-1 not only has anti-inflammatory actions but is also capable of preventing the loss of pancreatic β cells and may, in fact, enhance their proliferation and thus, preserve insulin secreting ability of β cells.

In addition, inhibition of DPP-4 that increases the circulating levels of incretins GLP-1 and GIP preserved islet mass in rodent models of type 1 diabetes. DPP-4 inhibitor, sitagliptin, treatment of NOD mice before and after islet transplantation resulted in prolongation of islet graft survival by decreasing insulitis and reducing migration of isolated splenic CD4+ T-cells, possibly, by the activation of protein kinase A and Rac1. These results indicate that both GLP-1 and GIP enhance graft survival through a pathway involving cAMP/PKA/Rac1 activation [270] and thus show immunosuppressive and anti-inflammatory properties. Furthermore, studies performed in DP4 deficient rats revealed a phenotype involving reduced diet-induced body weight gain and improved glucose tolerance associated with increased levels of GLP-1 and bound leptin as well as decreased aminotransferases and triglycerides. These experimental animals also showed anxiolytic-like and reduced stress-like responses, and several immune alterations, such as differential leukocyte subset composition at baseline, blunted natural killer cell and T-cell functions, and altered cytokine levels, indicating that incretins might modulate central nervous system and immune functions in vivo [271].

Leptin

Leptin is not only involved in the pathobiology of obesity and metabolic syndrome but also has pro-inflammatory actions. In inflammatory condition such as ankylosing spondylitis, leptin, IL-6 and TNF-α mRNA expressions of PMBCs were significantly higher than controls. Stimulation of PBMCs by exogenous leptin significantly increased the production of IL-6 and TNF-α in patients with ankylosing spondylitis in a dose-dependent fashion compared to controls [272] implying its pro-inflammatory effect in pathogenesis of ankylosing spondylitis.

These results are interesting in the light of the fact that consumption of dietary fats is an important factor contributing to obesity. Both in rodents and humans, the consumption of fat-rich diets blunts leptin and insulin anorexigenic signaling in the hypothalamus by a mechanism dependent on the in situ activation of inflammation. Consumption of dietary fats induced apoptosis of neurons and a reduction of synaptic inputs in the arcuate nucleus and lateral hypothalamus. This effect is dependent upon diet composition, and not on caloric intake. The presence of an intact TLR4 receptor protects cells from further apoptotic signals. In diet-induced inflammation of the hypothalamus, activation of pro-inflammatory pathways occurs that play a central role in the development of resistance to leptin and insulin [273]. The increase in the concentrations of leptin in response to high fat diet [274] may aggravate inflammation that, in turn, induces apoptosis of hypothalamic nuclei leading to the initiation and progression of the metabolic syndrome. Insulin resistance and hyperinsulinemia seen in obesity and other related conditions may, in fact, be a protective event since insulin has anti-inflammatory actions. Thus, hyperinsulinemia is beneficial though its presence implies the beginning of the metabolic syndrome. Both hyperleptinemia and hyperinsulinemia lead to reduced sympathetic activity [274] that also contributes to the pathophysiology of obesity and development of metabolic syndrome.

Cholecystokinin

The autonomic nervous system plays an important role in sensing luminal contents in the gut by way of hard-wired connections and chemical messengers, such as cholecystokinin (CCK). Ingestion of dietary fat stimulates CCK receptors, and leads to attenuation of the inflammatory response by way of the efferent vagus nerve and nicotinic receptors. Vagotomy and administration of antagonists for CCK and nicotinic receptors blunted the inhibitory effect of high-fat enteral nutrition on hemorrhagic shock-induced TNF-α and IL-6 release. Furthermore, the protective effect of high-fat enteral nutrition on inflammation-induced intestinal permeability was abrogated by vagotomy and administration of antagonists for CCK and nicotinic receptors, suggesting that there exists a neuroimmunologic pathway, controlled by nutrition [275]. This anti-inflammatory action of CCK could be a protective pathway developed to prevent inflammation that occurs due to the consumption of high fat diet.

Thus there is both pro- and anti-inflammatory actions exhibited by various hypothalamic monoaminergic and peptide molecules and those produced by the gut that not only regulate appetite, satiety and food intake but also modulate immune response (see Fig. 9.1). Based on these findings, it is no surprise that obesity, insulin resistance, hypertension, dyslipidemia and metabolic syndrome are low-grade systemic inflammatory conditions [226].

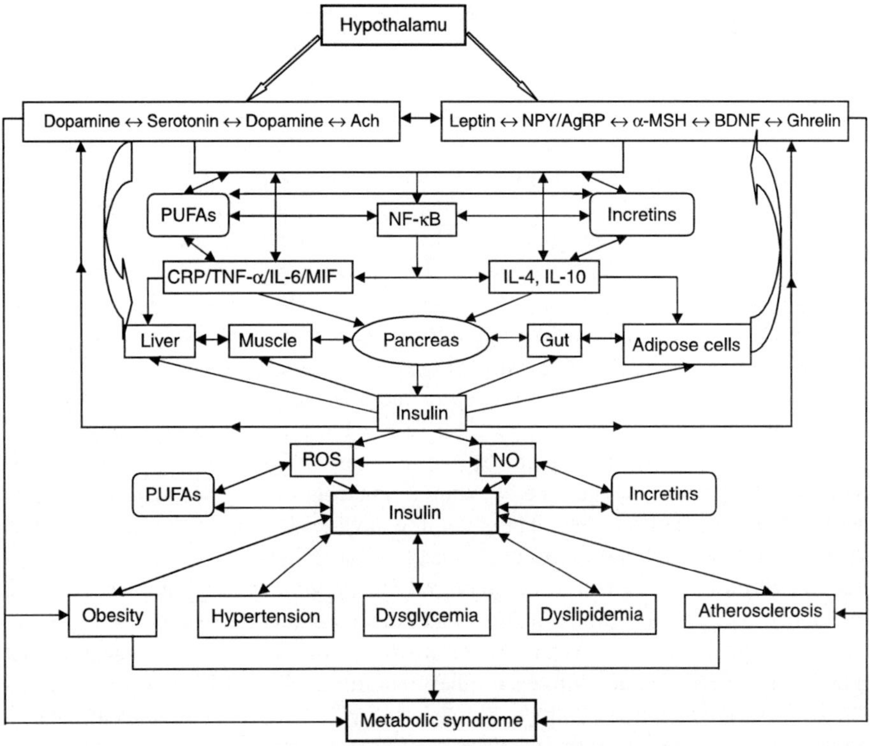

Fig. 9.1 Scheme showing interaction(s) among hypothalamus, liver, gut, adipose tissue, muscle and cytokines, gut and hypothalamic peptides and neurotransmitters, PUFAs and various components of metabolic syndrome. Following normal food intake an increase in the production of TNF-α and IL-6 and consequently enhanced plasma CRP levels and a decrease in anti-inflammatory cytokines IL-4 and IL-10 occur. TNF-α and IL-6 cause oxidative stress, decrease eNO and PGI_2 (prostacyclin) and adiponectin levels. This causes insulin resistance. As a result of this increase in insulin secretion occurs. Insulin not only normalizes plasma glucose, lipid and amino acid concentrations but also functions as an anti-inflammatory molecule by suppressing TNF-α and IL-6 and enhancing IL-4 and IL-10 synthesis and secretion. Following this, adiponectin levels raise and insulin sensitivity and the balance between pro- and anti-inflammatory cytokines is restored to normal

Neurotransmitters and Gut Peptides as Modulators of Inflammation and Immune Response

It is likely that peripheral tissues (such as muscle, adipose cells, etc.), pancreatic β cells and hypothalamic neurons communicate with each other to maintain energy homeostasis. For example, food intake prompts the release of gut peptides such as ghrelin, cholecystokinin (CCK), GLP-1 and GIP that could interact with hypothalamic neurons and signal hunger and satiety sensations. CCK reduces food intake by acting at CCK-1 receptors on vagal afferent neurons. Leptin mRNA has been reported in vagal afferent neurons, some of which also express CCK-1 receptor, suggesting that leptin, alone or in cooperation with CCK, might activate vagal afferent neurons, and influence food intake via a vagal route. A much higher prevalence of CCK and leptin sensitivity amongst cultured vagal afferent neurons that innervate stomach or duodenum than there was in the overall vagal afferent population was reported. Almost all leptin-responsive gastric and duodenal vagal afferents also were sensitive to CCK. Leptin, infused into the upper GI tract arterial supply, reduced meal size, and enhanced satiation evoked by CCK, indicating that vagal afferent neurons are activated by leptin and that this activation is likely to participate in meal termination by enhancing vagal sensitivity to CCK [276]. Injection of leptin increased hypothalamic leptin expression in *ob*/*ob* mice; suppressed body weight and adiposity; voluntarily decreased dark-phase food intake; suppressed plasma levels of adiponectin, TNF-α, free fatty acids and insulin, concomitant with normoglycemia; and elevated ghrelin levels for extended period. Leptin administration rapidly decreased plasma gastric ghrelin and adipocyte adiponectin but not TNF-α level, thereby demonstrating a peripheral restraining action of leptin on the secretion of hormones of varied origins. Ghrelin administration stimulated feeding in controls but was found to be ineffective in leptin-treated wt mice. Thus, leptin expressed locally in the hypothalamus counteracted the central orexigenic effects of peripheral ghrelin, suggesting that leptin and ghrelin interact with each other and thus, regulate energy homeostasis and metabolism [277]. In addition, incubation of the hypothalamic explants with ghrelin significantly increased NPY and AGRP mRNA expression [278], suggesting that ghrelin and NPY interact with each other. Ghrelin facilitated both cholinergic and tachykininergic excitatory pathways, consistent with activity within the enteric nervous system and possibly the vagus nerve [279], indicating that sympathetic and parasympathetic (especially vagus nerve) nerves carry messages from the peripheral tissues and β cells to the hypothalamus and *vice versa*. Thus, ultimately all the messages that reach hypothalamus are integrated, codified and relayed to the target tissues to maintain overall energy balance (see Fig. 9.2).

This is supported by the recent report that adenovirus-mediated expression of PPAR-γ2 in the liver induces acute hepatic steatosis while markedly reducing peripheral adiposity, changes that were accompanied by increased energy expenditure and improved systemic insulin sensitivity. It was noted that hepatic vagotomy and selective afferent blockage of the hepatic vagus reversed, whereas thiazolidinedione, a PPAR-γ agonist, enhanced these changes [280]. These results emphasize that there is a neuronal pathway consisting of the afferent vagus from the liver and efferent sympathetic nerves to adipose tissues that is involved in the regulation of energy expenditure, systemic insulin sensitivity, glucose metabolism, and fat distribution between the liver and the periphery. In this context, it is important to note that pro-inflammatory cytokine production is regulated by the efferent vagus nerve. This "cholinergic anti-inflammatory pathway" mediated by acetylcholine (ACh), when stimulated, inhibited the production of TNF, IL-1, MIF, and HMGB1 and activation of NF-κB expression [281–283]. Thus, the effects of PPAR-γ agonist and vagus nerve stimulation are similar in that both improved systemic insulin sensitivity, reduced TNF-α production and showed anti-inflammatory actions [280, 282]. Since, ACh is a neurotransmitter and regulates the secretion and actions of serotonin,

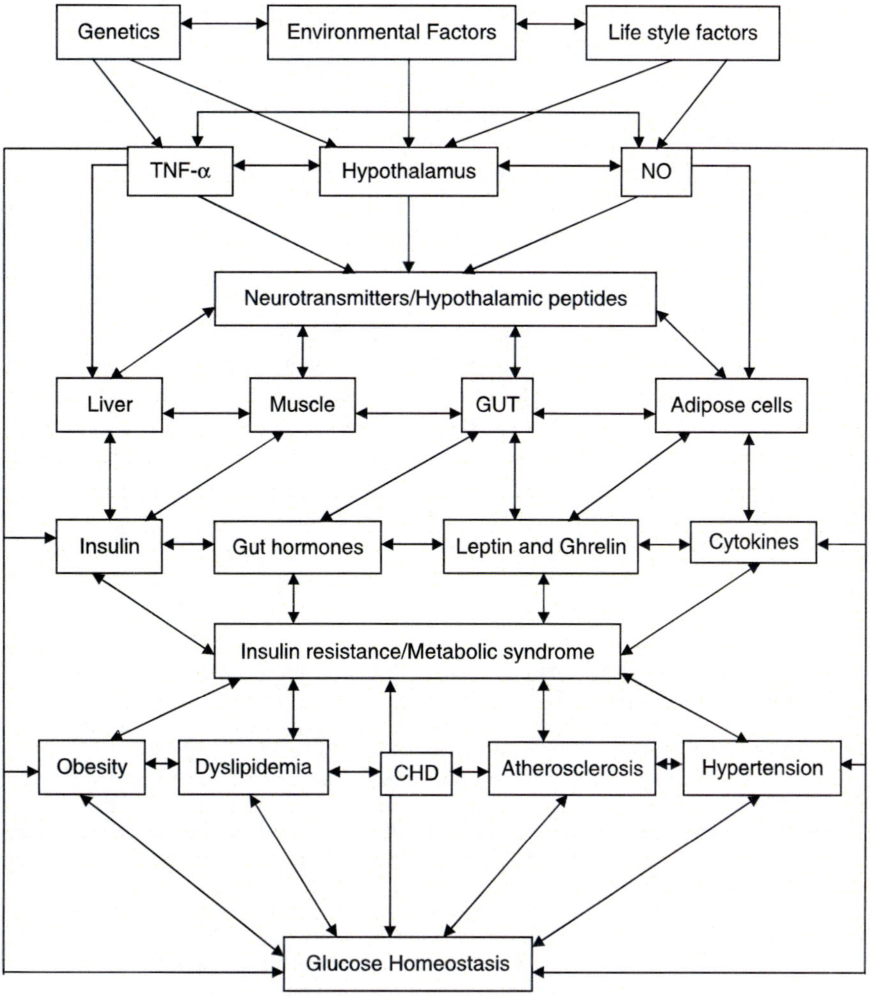

Fig. 9.2 Scheme showing relationship among genetic and environmental factors and target organs involved in the development of insulin resistance/metabolic syndrome

dopamine and other neuropeptides [284], it is evident that a complex network of interaction(s) exists between these molecules in the regulation of energy homeostasis. In this context, it is pertinent to note that brain insulin resistance exists in peripheral insulin resistance, especially in regions subserving appetite and reward [285]; and exercise enhanced the sensitivity of hypothalamus to the actions of leptin and insulin and the appetite-suppressive actions of exercise are mediated by the hypothalamus [286]. These evidences emphasize the significant role of hypothalamus and inflammation in the maintenance of energy balance and pathobiology of metabolic syndrome. Furthermore, recent findings that alterations in the composition of adipose tissue T cells occur early in obesity and shape the relationship between immunity and metabolism [287–290] lends support to the original proposals that obesity and metabolic syndrome are indeed inflammatory conditions [3, 5, 9, 10, 59–61].

Based on these evidences, the sequence of events that could lead to the initiation and perpetuation of the metabolic syndrome could be as follows: food intake increases the production of TNF-α and IL-6 and decreases those of anti-inflammatory cytokines IL-4 and IL-10, and adiponectin. TNF-α and IL-6 induce oxidative stress and activate NF-κB leading to insulin resistance and consequent hyperinsulinemia. Insulin secreted in response to food intake not only normalizes plasma glucose, lipid and amino acid concentrations but also suppresses TNF-α and IL-6 and enhance IL-4 and IL-10 synthesis resulting in the restoration of balance between pro- and anti-inflammatory cytokines and suppression of oxidative stress. On the other hand, continued consumption of energy rich diet leads to a state of low-grade systemic inflammation and chronic oxidative stress. Dietary restriction, exercise, and weight loss suppress free radical generation and oxidative stress [55], decrease the production of TNF-α and IL-6 and enhance IL-4 and IL-10, and adiponectin synthesis.

References

[1] Reaven GM (1993) Role of insulin resistance in human disease (syndrome X): an expanded definition. Annu Rev Med 44:121–131

[2] Krentz AJ, Nattrass M (1996) Insulin resistance: a multifaceted metabolic syndrome. Insights gained using a low-dose insulin infusion technique. Diabet Med 13:30–39

[3] Das UN (2002) Is metabolic syndrome X an inflammatory condition? Exp Biol Med 227:989–997

[4] Ford ES (2003) The metabolic syndrome and C-reactive protein, fibrinogen, and leukocyte count: findings from the third national health and nutrition examination survey. Atherosclerosis 168:351–358

[5] Das UN (2002) Obesity, metabolic syndrome X, and inflammation. Nutrition 18:430–432

[6] Albert MA, Glynn RJ, Ridker PM (2003) Plasma concentration of C-reactive protein and the calculated Framingham coronary heart disease risk score. Circulation 108:161–165

[7] Luc G, Bard J-M, Juhan-Vague I et al (2003) C-reactive protein, interleukins-6, and fibrinogen as predictors of coronary heart disease. The PRIME study. Arterioscler Thromb Vasc Biol 23:1255–1261

[8] Ridker PM, Cushman M, Stampfer MJ, Tracy RP, Hennekens CH (1997) Inflammation, aspirin, and the risk of cardiovascular disease in apparently healthy men. N Engl J Med 336:973–979

[9] Ramos EJB, Ramos EJB, Xu Y, Romanova I, Chen C, Quinn R, Inui A, Das UN, Meguid MM (2003) Is obesity an inflammatory disease? Surgery 134:329–335

[10] Das UN (2004) Metabolic syndrome X: an inflammatory condition? Curr Hypertens Rep 6:66–73

[11] Grundy SM, Brewer B Jr, Cleema JI, Smith SC, Lenfant C for the conference participants (2004) Definition of metabolic syndrome. NHLBI/AHA conference proceedings. Circulation 109:433–438

[12] Das UN (2006) Exercise and inflammation. Eur Heart J 27:1385–1386

[13] Das UN (2004) Anti-inflammatory nature of exercise. Nutrition 20:323–326

[14] Yamashita N, Hoshida S, Otsu K et al (1999) Exercise provides direct biphasic Cardioprotection via manganese superoxide dismutase activation. J Exp Med 189:1699–1706

[15] Das UN (2002) A perinatal strategy to prevent coronary heart disease. Nutrition 19:1022–1027

[16] Albert MA, Glynn RJ, Ridker PM (2003) Plasma concentration of C-reactive protein and the calculated Framingham coronary heart disease risk score. Circulation 108:161–165

[17] van der Meer IM, de Maat MPM, Hak AE et al (2002) C-reactive protein predicts progression of atherosclerosis measured as various sites in the arterial tree. The Rotterdam study. Stroke 33:2750–2755

[18] Luc G, Bard J-M, Juhan-Vague I et al (2003) C-reactive protein, interleukins-6, and fibrinogen as predictors of coronary heart disease. The PRIME study. Arterioscler Thromb Vasc Biol 23:1255–1261

[19] Engstrom G, Hedblad B, Stavenow L, Lind P, Janzon L, Lindgarde F (2003) Inflammation-sensitive plasma proteins are associated with future weight gain. Diabetes 52:2097–2101

[20] Mosca L (2002) C-reactive protein-to screen or not to screen. N Engl J Med 347:1615–1617

[21] Castell JV, Gomez-Lechon MJ, David M, Horano T, Kishimoto T, Heinrich PC (1988) Recombinant human interleukins-6 (IL-6/BSF-2/HSF) regulates the synthesis of acute phase proteins in human hepatocytes. FEBS Lett 232:347–350

[22] Barzilay JI, Abraham L, Heckbert SR, Cushman M, Kuller LH, Resnick HE, Tracy RP (2001) The relation of markers of inflammation to the development of glucose disorders in the elderly. Diabetes 50:2384–2389

[23] Kim MJ, Yoo KH, Park HS, Chung SM, Jin CJ, Lee Y, Shin YG, Chung CH (2005) Plasma adiponectin and insulin resistance in Korean type 2 diabetes mellitus. Yonsei Med J 46:42–50

[24] Liu S, Manson JE, Buring JE, Stampfer MJ, Willett WC, Ridker PM (2002) Relation between a diet with a high glycemic load and plasma concentrations of high-sensitivity C-reactive protein in middle-aged women. Am J Clin Nutr 75:492–498

[25] Esposito K, Nappo F, Marfella R et al (2002) Inflammatory cytokine concentrations are acutely increased by hyperglycemia in humans. Role of oxidative stress. Circulation 106:2067–2072

[26] Kirwan JP, Krishnan RK, Weaver JA, Del Aguila LF, Evans WJ (2001) Human aging is associated with altered TNF-α production during hyperglycemia and hyperinsulinemia. Am J Physiol 281:E1137–E1143

[27] Lin Y, Rajala MW, Berger JP, Moller DE, Barzilai N, Scherer PE (2001) Hyperglycemia-induced production of acute phase reactants in adipose tissue. J Biol Chem 276:42077–42083

[28] Fernandez-Real JM, Vayreda M, Richart C et al (2001) Circulating interleukins 6 levels, blood pressure, and insulin sensitivity in apparently healthy men and women. J Clin Endocrinol Metab 86:1154–1159

[29] Chae CU, Lee RT, Rifai N, Ridker PM (2001) Blood pressure and inflammation in apparently healthy men. Hypertension 38:399–403

[30] Brochu M, Tchernof A, Dionne IJ, Sites CK, Eltabbakh GH, Sims EA, Poehlman ET (2001) What are the physical characteristics associated with a normal metabolic profile despite a high level of obesity in postmenopausal women? J Clin Endocrinol Metab 86:1020–1025

[31] Masuzaki H, Paterson J, Shinyama H et al (2001) A transgenic model of visceral obesity and the metabolic syndrome. Science 294:2166–2170

[32] Nair S, Lee YH, Lindsay RS, Walker BR, Tataranni PA, Bogardus C, Baier LJ, Permana PA (2004) 11beta-Hydroxysteroid dehydrogenase type 1: genetic polymorphisms are associated with type 2 diabetes in Pima Indians independently of obesity and expression in adipocyte and muscle. Diabetologia 47:1088–1095

[33] Sandeep TC, Andrew R, Homer N, Andrews RC, Smith K, Walker BR (2005) Increased in vivo regeneration of cortisol in adipose tissue in human obesity and effects of the 11beta-hydroxysteroid dehydrogenase type 1 inhibitor carbenoxolone. Diabetes 54:872–879

[34] Morton NM, Paterson JM, Masuzaki H, Holmes MC, Staels B, Fievet C, Walker BR, Flier JS, Mullins JJ, Seckl JR (2004) Novel adipose tissue-mediated resistance to diet-induced visceral obesity in 11 beta-hydroxysteroid dehydrogenase type 1-deficient mice. Diabetes 53:931–938

[35] Tomlinson JW, Moore J, Cooper MS, Bujalska I, Shahmanesh M, Burt C, Strain A, Hewison M, Stewart PM (2001) Regulation of expression of 11beta-hydroxysteroid dehydrogenase type 1 in adipose tissue: tissue-specific induction by cytokines. Endocrinology 142:1982–1989
[36] Han J, Thompson P, Beutler B (1990) Dexamethasone and pentoxifylline inhibit endotoxin-induced cachectin/tumor necrosis factor synthesis at separate points in the signaling pathway. J Exp Med 172:391–394
[37] Degawa-Yamauchi M, Moss KA, Bovenkerk JE, Shankar SS, Morrison CL, Lelliott CJ, Vidal-Puig A, Jones R, Considine RV (2005) Regulation of adiponectin expression in human adipocytes: effects of adiposity, glucocorticoids, and tumor necrosis factor alpha. Obes Res 13:662–669
[38] Mohanty P, Hamouda W, Garg R, Aljada A, Ghanim H, Dandona P (2000) Glucose challenge stimulates reactive oxygen species (ROS) generation by leucocytes. J Clin Endocrinol Metab 85:2970–2973
[39] Mohanty P, Ghanim H, Hamouda W, Aljada A, Garg R, Dandona P (2002) Both lipid and protein intakes stimulate increased generation of reactive oxygen species by polymorphonuclear leukocytes and mononuclear cells. Am J Clin Nutr 75:767–772
[40] Schulze PC, Yoshioka J, Takahashi T, He Z, King GL, Lee RT (2004) Hyperglycemia promotes oxidative stress through inhibition of thioredoxin function by thioredoxin-interacting protein. J Biol Chem 279:30369–30374
[41] Lin Y, Berg AH, Iyengar P, Lam TK, Giacca A, Combs TP, Rajala MW, Du X, Rollman B, Li W, Hawkins M, Barzilai N, Rhodes CJ, Fantus IG, Brownlee M, Scherer PE (2005) The hyperglycemia-induced inflammatory response in adipocytes: the role of reactive oxygen species. J Biol Chem 280:4617–4626
[42] Das UN (2002) Is insulin an endogenous cardioprotector? Crit Care 6:389–393
[43] Fan J, Frey RS, Rahman A, Malik AB (2002) Role of neutrophil NADPH oxidase in the mechanism of tumor necrosis factor-α-induced NF-kB activation and intercellular adhesion molecule-1 expression in endothelial cells. J Biol Chem 277:3404–3411
[44] Hattori Y, Matsumura M, Kasai K (2003) Vascular smooth muscle cell activation by C-reactive protein. Cardiovasc Res 58:186–195
[45] Devaraj S, Xu DY, Jialal I (2003) C-reactive protein increases plasminogen activator inhibitor-1 expression and activity in human aortic endothelial cells. Circulation 107:398–404
[46] Furukawa S, Fujita T, Shimabukuro M, Iwaki M, Yamada Y, Nakajima Y, Nakayama O, Makishima M, Matsuda M, Shimomura I (2004) Increased oxidative stress in obesity and its impact on metabolic syndrome. J Clin Invest 114:1752–1761
[47] Fortuno A, San Jose G, Moreno MU, Beloqui O, Diez J, Zalba G (2006) Phagocytic NADPH oxidase overactivity underlies oxidative stress in metabolic syndrome. Diabetes 55:209–215
[48] Mohan IK, Das UN (1997) Oxidant stress, anti-oxidants and nitric oxide in non-insulin dependent diabetes mellitus. Med Sci Res 25:55–57
[49] Kumar KV, Das UN (1994) Lipid peroxides and essential fatty acids in patients with coronary heart disease. J Nutr Med 4:33–37
[50] Das UN, Kumar KV, Mohan IK (1994) Lipid peroxides and essential fatty acids in patients with diabetes mellitus and diabetic nephropathy. J Nutr Med 4:149–155
[51] Kumar KV, Das UN (1993) Are free radicals involved in the pathobiology of human essential hypertension? Free Radic Res Commun 19:59–66
[52] Das UN (2003) Pathobiology of metabolic syndrome X in obese and non-obese south Asian Indians: further discussion and some suggestions. Nutrition 19:560–562
[53] Das UN (2001) Is obesity an inflammatory condition? Nutrition 17:953–966
[54] Pitsavos C, Panagiotakos DB, Chrysohoou C, Tzima N, Das UN, Stefanadis C (2007) Diet, exercise and C-reactive protein levels in people with abdominal obesity: the ATTICA study. Angiology 58:225–233
[55] Dandona P, Mohanty P, Ghanim H et al (2001) The suppressive effect of dietary restriction and weight loss in the obese on the generation of reactive oxygen species by leukocytes, lipid peroxidation, and protein carbonylation. J Clin Endocrinol Metab 86:355–362

[56] Das UN (2001) Is insulin an anti-inflammatory molecule? Nutrition 17:409–413
[57] Jeschke MG, Einspanier R, Klein D et al (2002) Insulin attenuates the systemic inflammatory response. Mol Med 8:443–450
[58] Dandona P, Aljada A, Mohanty P (2002) The anti-inflammatory and potential anti-atherogenic effect of insulin: a new paradigm. Diabetologia 45:924–930
[59] Das UN (2002) A perinatal strategy for preventing adult diseases: the role of long-chain polyunsaturated fatty acids. Kluwer Academic, Boston, MA
[60] Das UN (2010) Metabolic syndrome pathophysiology: the role of essential fatty acids fatty acids and their metabolites. Wiley-Blackwell, Ames, IA
[61] Das UN (2010) Metabolic syndrome as a low-grade systemic inflammatory condition. Expert Rev Endocrinol 4:577–592
[62] Otero M, Lago R, Lago F, Casanueva FF, Dieguez C, Gómez-Reino JJ, Gualillo O (2005) Leptin, from fat to inflammation: old questions and new insights. FEBS Lett 579:295–301
[63] Otero M, Lago R, Gomez R, Lago F, Gomez-Reino JJ, Gualillo O (2006) Leptin: a metabolic hormone that functions like a proinflammatory adipokine. Drug News Perspect 19:21–26
[64] Whincup PH, Gilg JA, Papacosta O et al (2002) Early evidence of ethnic differences in cardiovascular risk: cross sectional comparison of British South Asian and white children. BMJ 324:1–6
[65] Kalhan R, Puthawala K, Agarwal S, Amini SB, Kalhan SC (2001) Altered lipid profile, leptin, insulin, and anthropometry in offspring of South Asian immigrants in the United States. Metabolism 50:1197–1202
[66] Das UN (2003) Folic acid says NO to vascular diseases. Nutrition 19:686–692
[67] Barbato JE, Zuckerbraun BS, Overhaus M, Raman KG, Tzeng E (2005) Nitric oxide modulates vascular inflammation and intimal hyperplasia in insulin resistance and the metabolic syndrome. Am J Physiol Heart Circ Physiol 289:H228–H236
[68] Leif A, Mather K (2004) Insulin resistance, metabolic syndrome and vascular diseases: update on mechanistic linkages. Can J Cardiol 20(Suppl B):66B–76B
[69] Jovinge S, Hamsten A, Tomvall P et al (1998) Evidence for a role of tumor necrosis factor alpha in disturbances of triglycerides and glucose metabolism predisposing to coronary heart disease. Metabolism 47:113–118
[70] Corica F, Allegra A, Corsonello A et al (1999) Relationship between plasma leptin levels and the tumor necrosis factor alpha system in obese subjects. Int J Obes Metab Disord 23:355–360
[71] Das UN (2005) Long-chain polyunsaturated fatty acids, endothelial lipase and atherosclerosis. Prostaglandins Leukot Essent Fatty Acids 72:173–179
[72] Denicola A, Batthyany C, Lissi E et al (2002) Diffusion of nitric oxide into low density lipoprotein. J Biol Chem 277:932–936
[73] Ji Y, Diao J, Han Y, Huang Y, Bai H, Chen Q, Fan L, Ferro A (2006) Pyridoxine prevents dysfunction of endothelial cell nitric oxide production in response to low-density lipoprotein. Atherosclerosis 188:84–94
[74] Das UN (2001) Nutritional factors in the pathobiology of human essential hypertension. Nutrition 17:337–346
[75] Vanizor B, Orem A, Karahan SC, Kiran E, Erem C, Aliyazicioğlu R, Uydu HA (2001) Decreased nitric oxide end-products and its relationship with high density lipoprotein and oxidative stress in people with type 2 diabetes without complications. Diabetes Res Clin Pract 54:33–39
[76] Barbato JE, Zuckerbraun BS, Overhaus M, Raman KG, Tzeng E (2005) Nitric oxide modulates vascular inflammation and intimal hyperplasia in insulin resistance and the metabolic syndrome. Am J Physiol Heart Circ Physiol 289:H228–H236
[77] Konukoğlu D, Serin O, Turhan MS (2005) Plasma total homocysteine concentrations in obese and non-obese female patients with type 2 diabetes mellitus; its relations with plasma oxidative stress and nitric oxide levels. Clin Hemorheol Microcirc 33:41–46
[78] Seshiah PN, Weber DS, Rocic P, Valppu L, Taniyama Y, Griendling KK (2002) Angiotensin II stimulation of NAD(P)H oxidase activity. Circ Res 91:406–413

[79] Furuhashi M, Ura N, Higashiura K et al (2003) Blockade of the rennin-angiotensin system increases adiponectin concentrations in patients with essential hypertension. Hypertension 42:76–81
[80] Zhang H, Zhang B, Tang BY, Chen YY, Zhu L, Shen YL (2005) Angiotensin-converting enzyme inhibitors potentiate subthreshold preconditioning through NO and mitoK(ATP) channel. Sheng Li Xue Bao 57:453–460
[81] Oshida Y, Tachi Y, Morishita Y, Kitakoshi K, Fuku N, Han YQ, Ohsawa I, Sato Y (2000) Nitric oxide decreases insulin resistance induced by high-fructose feeding. Horm Metab Res 32:339–342
[82] Fang TC, Wu CC, Huang WC (2001) Inhibition of nitric oxide synthesis accentuates blood pressure elevation in hyperinsulinemic rats. J Hypertens 19:1255–1262
[83] Shinozaki K, Hirayama A, Nishio Y et al (2001) Coronary endothelial dysfunction in the insulin-resistant state is linked to abnormal pteridine metabolism and vascular oxidative stress. J Am Coll Cardiol 38:1821–1828
[84] Stuhlinger MC, Abbasi F, Chu JW et al (2002) Relationship between insulin resistance and an endogenous nitric oxide synthase inhibitor. JAMA 287:1420–1426
[85] Cook S, Hugli O, Egli M, Vollenweder P, Burcelin R, Nicod P, Thorens B, Scherrer U (2003) Clustering of cardiovascular risk factors mimicking the human metabolic syndrome X in eNOS null mice. Swiss Med Wkly 133:360–363
[86] Bruning JC, Gautam D, Burks DJ, Gillette J, Schubert M, Orban PC, Klein R, Krone W, Muller-Wieland D, Kahn CR (2000) Role of brain insulin receptor in control of body weight and reproduction. Science 289:2122–2125
[87] Cardillo C, Nambi SS, Kilcoyne CM, Choucair WK, Katz A, Quon MJ, Panza JA (1999) Insulin stimulates both endothelin and nitric oxide activity in the human forearm. Circulation 100:820–825
[88] Das UN (2004) Long-chain polyunsaturated fatty acids interact with nitric oxide, superoxide anion, and transforming growth factor-β to prevent human essential hypertension. Eur J Clin Nutr 58:195–203
[89] Li J, Zhang H, Wu F, Nan Y, Ma H, Guo W, Wang H, Ren J, Das UN, Gao F (2008) Insulin inhibits tumor necrosis factor-alpha induction in myocardial ischemia/reperfusion: role of Akt and endothelial nitric oxide synthase phosphorylation. Crit Care Med 36:1551–1558
[90] Das UN (2000) Possible beneficial action(s) of glucose-insulin-potassium regimen in acute myocardial infarction and inflammatory conditions: a hypothesis. Diabetologia 43:1081–1082
[91] Das UN (2002) Insulin and the critically ill. Crit Care 6:262–263
[92] Fischer S, Hanefeld M, Haffner SM, Fusch C, Schwanebeck U, Köhler C, Fücker K, Julius U (2002) Insulin-resistant patients with type 2 diabetes mellitus have higher serum leptin levels independently of body fat mass. Acta Diabetol 39:105–110
[93] White V, Gonzàlez E, Capobianco E, Pustovrh C, Soez C, Romanini MC, Jawerbaum A (2004) Modulatory effect of leptin on nitric oxide production and lipid metabolism in term placental tissues from control and streptozotocin-induced diabetic rats. Reprod Fertil Dev 16:363–372
[94] White V, Gonzàlez E, Capobianco E, Pustovrh C, Martínez N, Higa R, Baier M, Jawerbaum A (2006) Leptin modulates nitric oxide production and lipid metabolism in human placenta. Reprod Fertil Dev 18:425–432
[95] Procopio C, Andreozzi F, Laratta E, Cassese A, Beguinot F, Arturi F, Hribal ML, Perticone F, Sesti G (2009) Leptin-stimulated endothelial nitric-oxide synthase via an adenosine 5′-monophosphate-activated protein kinase/Akt signaling pathway is attenuated by interaction with C-reactive protein. Endocrinology 150:3584–3593
[96] Lampiao F, du Plessis SS (2008) Insulin and leptin enhance human sperm motility, acrosome reaction and nitric oxide production. Asian J Androl 10:799–807
[97] Blanquicett C, Graves A, Kleinhenz DJ, Hart CM (2007) Attenuation of signaling and nitric oxide production following prolonged leptin exposure in human aortic endothelial cells. J Investig Med 55:368–377

[98] Korda M, Kubant R, Patton S, Malinski T (2008) Leptin-induced endothelial dysfunction in obesity. Am J Physiol Heart Circ Physiol 295:H1514–H1521

[99] Parsons TJ, Power C, Manor O (2001) Fetal and early life growth and body mass index from birth to early adulthood in 1958 birth cohort: longitudinal study. BMJ 323:1331–1335

[100] Dorner G, Plagemann A (1994) Perinatal hyperinsulinism as possible predisposing factor for diabetes mellitus, obesity and enhanced cardiovascular risk in later life. Horm Metab Res 26:213–221

[101] Buchanan TA, Kjos SL (1999) Gestational diabetes: risk or myth? J Clin Endocrinol Metab 84:1854–1857

[102] Singhal A, Wells J, Cole TJ, Fewtrell M, Lucas A (2003) Programming lean body mass: a link between birth weight, obesity, and cardiovascular disease? Am J Clin Nutr 77:726–730

[103] Loos RJ, Beunen G, Fagard R, Derom C, Vlietinck R (2001) Birth weight and body composition in young adult men-a prospective twin study. Int J Obes Relat Metab Disord 25:1537–1545

[104] Loos RJ, Beunen G, Fagard R, Derom C, Vlietinck R (2002) Birth weight and body composition in young women: a prospective twin study. Am J Clin Nutr 75:676–682

[105] Ong KK, Dunger DB (2004) Birth weight, infant growth and insulin resistance. Eur J Endocrinol 151(Suppl 3):U131–U139

[106] McDermott R, Campbell S, Li M, McCulloch B (2009) The health and nutrition of young indigenous women in north Queensland -intergenerational implications of poor food quality, obesity, diabetes, tobacco smoking and alcohol use. Public Health Nutr 12:2143–2149

[107] Misra A, Khurana L (2009) The metabolic syndrome in South Asians: epidemiology, determinants, and prevention. Metab Syndr Relat Disord 7:497–514

[108] Bhardwaj S, Misra A, Khurana L, Gulati S, Shah P, Vikram NK (2008) Childhood obesity in Asian Indians: a burgeoning cause of insulin resistance, diabetes and sub-clinical inflammation. Asia Pac J Clin Nutr 17(Suppl 1):172–175

[109] Plagemann A, Harder T, Rake A, Voits M, Fink H, Rohde W, Dörner G (1999) Perinatal elevation of hypothalamic insulin, acquired malformation of hypothalamic galaninergic neurons, and syndrome x-like alterations in adulthood of neonatally overfed rats. Brain Res 836:146–155

[110] McMillen IC, Adam CL, Muhlhausler BS (2005) Early origins of obesity: programming the appetite regulatory system. J Physiol 565:9–17

[111] Das UN (2010) Obesity: genes, brain, gut and environment. Nutrition 26:459–473

[112] Grove KL, Smith MS (2003) Ontogeny of the hypothalamic neuropeptide Y system. Physiol Behav 79:47–63

[113] Grayson BE, Allen SE, Billes SK, Williams SM, Smith MS, Grove KL (2006) Prenatal development of hypothalamic neuropeptide systems in the nonhuman primate. Neuroscience 143:975–986

[114] Li C, Chen P, Smith MS (1998) Neuropeptide Y (NPY) neurons in the arcuate nucleus (ARH) and dorsomedial nucleus (DMH), areas activated during lactation, project to the paraventricular nucleus of the hypothalamus (PVH). Regul Pept 75–76:93–100

[115] Chen P, Li C, Haskell-Luevano C, Cone RD, Smith MS (1999) Altered expression of agouti-related protein and its colocalization with neuropeptide Y in the arcuate nucleus of the hypothalamus during lactation. Endocrinology 140:2645–2650

[116] Davidowa H, Li Y, Plagemann A (2003) Altered responses to orexigenic (AGRP, MCH) and anorexigenic (alpha-MSH, CART) neuropeptides of paraventricular hypothalamic neurons in early postnatally overfed rats. Eur J Neurosci 18:613–621

[117] Fahrenkrog S, Harder T, Stolaczyk E, Melchior K, Franke K, Dudenhausen JW, Plagemann A (2004) Cross-fostering to diabetic rat dams affects early development of mediobasal hypothalamic nuclei regulating food intake, body weight, and metabolism. J Nutr 134:648–654

[118] Plagemann A, Rittel F, Waas T, Harder T, Rohde W (1999) Cholecystokinin-8S levels in discrete hypothalamic nuclei of weanling rats exposed to maternal protein malnutrition. Regul Pept 85(2–3):109–113

[119] Plagemann A (2004) 'Fetal programming' and 'functional teratogenesis': on epigenetic mechanisms and prevention of perinatally acquired lasting health risks. J Perinat Med 32:297–305

[120] López M, Seoane LM, Tovar S, García MC, Nogueiras R, Diéguez C, Searís RM (2005) A possible role of neuropeptide Y, agouti-related protein and leptin receptor isoforms in hypothalamic programming by perinatal feeding in the rat. Diabetologia 48:140–148

[121] Picó C, Oliver P, Sànchez J, Miralles O, Caimari A, Priego T, Palou A (2007) The intake of physiological doses of leptin during lactation in rats prevents obesity in later life. Int J Obes (Lond) 31:1199–1209

[122] Parente LB, Aguila MB, Mandarim-de-Lacerda CA (2008) Deleterious effects of high-fat diet on perinatal and postweaning periods in adult rat offspring. Clin Nutr 27:623–634

[123] Ryan KK, Haller AM, Sorrell JE, Woods SC, Jandacek RJ, Seeley RJ (2010) Perinatal exposure to bisphenol-a and the development of metabolic syndrome in CD-1 mice. Endocrinology 151:2603–2612

[124] Axen KV, Li X, Fung K, Sclafani A (1994) The VMH-dietary obese rat: a new model of non-insulin-dependent diabetes mellitus. Am J Physiol 266(3 Pt 2):R921–R928

[125] Keno Y, Tokunaga K, Fujioka S, Kobatake T, Kotani K, Yoshida S, Nishida M, Shimomura I, Matsuo T, Odaka H et al (1994) Marked reduction of pancreatic insulin content in male ventromedial hypothalamic-lesioned spontaneously non-insulin-dependent diabetic (Goto-Kakizaki) rats. Metabolism 43:32–37

[126] Dube MG, Kalra PS, Crowley WR, Kalra SP (1995) Evidence of a physiological role for neuropeptide Y in ventromedial hypothalamic lesion-induced hyperphagia. Brain Res 690:275–278

[127] Funahashi T, Shimomura I, Hiraoka H, Arai T, Takahashi M, Nakamura T, Nozaki S, Yamashita S, Takemura K, Tokunaga K et al (1995) Enhanced expression of rat obese (ob) gene in adipose tissues of ventromedial hypothalamus (VMH)-lesioned rats. Biochem Biophys Res Commun 211:469–475

[128] Williams G, Gill JS, Lee YC, Cordoso HM, Okpere BE, Bloom SR (1989) Increased neuropeptide Y concentrations in specific hypothalamic regions of streptozotocin-induced diabetic rats. Diabetes 38:321–327

[129] Ohtani N, Ohta M, Sugano T (1997) Microdialysis study of modification of hypothalamic neurotransmitters in streptozotocin-diabetic rats. J Neurochem 69:1622–1628

[130] Takahashi A, Ishimaru H, Ikarashi Y, Maruyama Y (1994) Aspects of hypothalamic neuronal systems in VMH lesion-induced obese rats. J Auton Nerv Syst 48:213–219

[131] Barber M, Kasturi BS, Austin ME, Patel KP, Mohan Kumar SM, Mohan Kumar PS (2003) Diabetes-induced neuroendocrine changes in rats: role of brain monoamines, insulin and leptin. Brain Res 964:128–135

[132] Yoshida S, Yamashita S, Tokunaga K, Yamane M, Shinohara E, Keno Y et al (1996) Visceral fat accumulation and vascular complications associated with VMH lesioning of spontaneously non-insulin-dependent diabetic GK rat. Int J Obes Relat Metab Disord 20:909–916

[133] Das UN (2008) Is metabolic syndrome X a disorder of the brain? Curr Nutr Food Sci 4:73–108

[134] Nishida M, Miyagawa JI, Tokunaga K, Yamamoto K, Keno Y, Kobatake T, Yoshida S, Nakamura T, Odaka H, Ikeda H, Hanafusa T, Yamashita S, Kameda-Takemura K, Matsuzawa Y (1997) Early morphologic changes of atherosclerosis induced by ventromedial hypothalamic lesion in the spontaneously diabetic Goto-Kakizaki rat. J Lab Clin Med 129:200–207

[135] Wan Q, Xiong ZG, Man HY, Ackerley CA, Braunton J, Lu WY, Becker LE, MacDonald JF, Wang YT (1997) Recruitment of functional GABAA receptors to postsynaptic domains by insulin. Nature 388:686–690

[136] Bruning JC, Gautam D, Burks DJ, Gillette J, Schubert M, Orban PC, Klein R, Krone W, Muller-Wieland D, Kahn CR (2000) Role of brain insulin receptor in control of body weight and reproduction. Science 289:2122–2125

[137] Hill JM, Lesniak MA, Pert CB, Roth J (1986) Autoradiographic localization of insulin receptors in rat brain: prominence in olfactory and limbic areas. Neuroscience 17:1127–1138

[138] Menendez JA, Atrens DM (1991) Insulin and the paraventricular hypothalamus: modulation of energy balance. Brain Res 555:193–201
[139] McGowan MK, Andrews KM, Grossman SP (1992) Chronic intrahypothalamic infusions of insulin or insulin antibodies alter body weight and food intake in the rat. Physiol Behav 51:753–766
[140] Obici S, Feng Z, Karkanias G, Baskin DG, Rossetti L (2002) Decreasing hypothalamic insulin receptors causes hyperphagia and insulin resistance in rats. Nat Neurosci 5:566–572
[141] Sipols AJ, Baskin DG, Schwartz MW (1995) Effect of intracerebroventricular insulin infusion on diabetic hyperphagia and hypothalamic neuropeptide gene expression. Diabetes 44:147–151
[142] Benoit SC, Air EL, Coolen LM, Strauss R, Jackman A, Clegg DJ, Seeley RJ, Woods SC (2002) The catabolic action of insulin in the brain is mediated by melanocortins. J Neurosci 22:9048–9052
[143] Schwartz MW (2000) Central nervous system control of food intake. Nature 404:661–671
[144] Anthony K, Reed LJ, Dunn JT, Bingham E, Hopkins D, Marsden PK, Amiel SA (2006) Attenuation of insulin-evoked responses in brain networks controlling appetite and reward in insulin resistance: the cerebral basis for impaired control of food intake in metabolic syndrome? Diabetes 55:2986–2992
[145] Harvey J, Ashford ML (2003) Leptin in the CNS: much more than a satiety signal. Neuropharmacology 44:845–854
[146] Mirshamsi S, Laidlaw HA, Ning K, Anderson E, Burgess LA, Gray A, Sutherland C, Ashford ML (2004) Leptin and insulin stimulation of signalling pathways in arcuate nucleus neurones: PI3K dependent actin reorganization and KATP channel activation. BMC Neurosci 5:54
[147] Ikeda H, West DB, Pustek JJ, Figlewicz DP, Greenwood MRC, Porte D Jr, Woods SC (1986) Intraventricular insulin reduces food intake and body weight of lean but not obese Zucker rats. Appetite 7:381–386
[148] Schwartz MW, Marks J, Sipols AJ, Baskin DG, Woods SC, Kahn SE, Porte D Jr (1991) Central insulin administration reduces neuropeptide Y mRNA expression in the arcuate nucleus of food-deprived lean (Fa/Fa) but not obese (fa/fa) Zucker rats. Endocrinology 128:2645–2647
[149] Das UN (2007) Metabolic syndrome X is a low-grade systemic inflammatory condition with its origins in the perinatal period. Curr Nutr Food Sci 3:277–295
[150] Bluher M, Patti M-E, Gesta S, Kahn BB, Kahn CR (2004) Intrinsic heterogeneity in adipose tissue of fat-specific insulin receptor knock-out mice is associated with differences in patterns of gene expression. J Biol Chem 279:31891–31901
[151] Blüher M, Michael MD, Peroni OD, Ueki K, Carter N, Kahn BB, Kahn CR (2002) Adipose tissue selective insulin receptor knockout protects against obesity and obesity-related glucose intolerance. Dev Cell 3:25–38
[152] Blüher M, Kahn BB, Kahn CR (2003) Extended longevity in mice lacking the insulin receptor in adipose tissue. Science 299:572–574
[153] Bluher M (2008) Fat tissue and long life. Obes Facts 1:176–182
[154] Brüning JC, Michael MD, Winnay JN, Hayashi T, Hörsch D, Accili D, Goodyear LJ, Kahn CR (1998) A muscle-specific insulin receptor knockout exhibits features of the metabolic syndrome of NIDDM without altering glucose tolerance. Mol Cell 2:559–569
[155] Moller DE, Chang PY, Yaspelkis BB 3rd, Flier JS, Wallberg-Henriksson H, Ivy JL (1996) Transgenic mice with muscle-specific insulin resistance develop increased adiposity, impaired glucose tolerance, and dyslipidemia. Endocrinology 137:2397–2405
[156] Kim JK, Michael MD, Previs SF, Peroni OD, Mauvais-Jarvis F, Neschen S, Kahn BB, Kahn CR, Shulman GI (2000) Redistribution of substrates to adipose tissue promotes obesity in mice with selective insulin resistance in muscle. J Clin Invest 105:1791–1797
[157] Michael MD, Kulkarni RN, Postic C, Previs SF, Shulman GI, Magnuson MA, Kahn CR (2000) Loss of insulin signaling in hepatocytes leads to severe insulin resistance and progressive hepatic dysfunction. Mol Cell 6:87–97
[158] Kulkarni RN, Brüning JC, Winnay JN, Postic C, Magnuson MA, Kahn CR (1999) Tissue-specific knockout of the insulin receptor in pancreatic beta cells creates an insulin secretory defect similar to that in type 2 diabetes. Cell 96:329–339

[159] Lowell BB, Susulic V, Hamann A, Lawitts JA, Himms-Hagen J, Boyer BB, Kozak LP, Flier JS (1993) Development of transgenic mice after genetic ablation of brown adipose tissue. Nature 366:740–742
[160] Guerra C, Navarro P, Valverde AM, Arribas M, Bruning J, Kozak LP, Kahn CR, Benito M (2001) Brown adipose tissue-specific insulin receptor knockout shows diabetic phenotype without insulin resistance. J Clin Invest 108:1205–1213
[161] Zhang Y, Daquinag A, Traktuev DO, Amaya-Manzanares F, Simmons PJ, March KL, Pasqualini R, Arap W, Kolonin MG (2009) White adipose tissue cells are recruited by experimental tumors and promote cancer progression in mouse models. Cancer Res 69:5259–5266
[162] Szpirer C, Riviere M, Cortese R, Nakamura T, Islam MQ, Levan G, Szpirer J (1992) Chromosomal localization in man and rat of the genes encoding the liver-enriched transcription factors C/EBP, DBP, and HNF1/LFB-1 (CEBP, DBP, and transcription factor 1, TCF1, respectively) and of the hepatocyte growth factor/scatter factor gene (HGF). Genomics 13:293–300
[163] Cao Z, Umek RM, McKnight SL (1991) Regulated expression of three C/EBP isoforms during adipose conversion of 3T3-L1 cells. Genes Dev 5:1538–1552
[164] Wang H, Iakova P, Wilde M, Welm A, Goode T, Roesler WJ, Timchenko NA (2001) C/EBPalpha arrests cell proliferation through direct inhibition of Cdk2 and Cdk4. Mol Cell 8:817–828
[165] Kuboki K, Jiang ZY, Takahara N, Ha SW, Igarashi M, Yamauchi T, Feener EP, Herbert TP, Rhodes CJ, King GI (2000) Regulation of endothelial constitutive nitric oxide synthase gene expression in endothelial cells and in vivo: a specific vascular action on insulin. Circulation 101:676–681
[166] Zeng K, Nystrom FH, Ravichandran LV, Cong LN, Kirby M, Mostowski H, Quon MJ (2000) Roles for insulin receptor, PI3-kinase, and Akt in insulin-signaling pathways related to production of nitric oxide in human vascular endothelial cells. Circulation 101:1539–1545
[167] Kahn CR (2003) Lessons about the control of glucose homeostasis and the pathogenesis of diabetes from knockout mice. Trans Am Clin Climatol Assoc 114:125–148
[168] Kondo T, Vicent D, Suzuma K, Yanagisawa M, King GL, Holzenberger M, Kahn RC (2003) Knockout of insulin and IGF-1 receptors on vascular endothelial cells protects against retinal neovascularization. J Clin Invest 111:1835–1842
[169] McCarron DA, Morris CD, Henry HJ, Stanton JL (1984) Blood pressure and nutrient intake in the United States. Science 224:1392–1398
[170] Das UN (2008) Risk of type 2 diabetes mellitus in those with hypertension. Eur Heart J 29:952–953
[171] Ernst SJ, Aguilar-Bryan L, Noebels JL (2009) Sodium channel beta1 regulatory subunit deficiency reduces pancreatic islet glucose-stimulated insulin and glucagon secretion. Endocrinology 150:1132–1139
[172] Hamming KS, Soliman D, Webster NJ, Searle GJ, Matemisz LC, Liknes DA, Dai XQ, Pulinilkunnil T, Riedel MJ, Dyck JR, Macdonald PE, Light PE (2010) Inhibition of {beta}-cell sodium-calcium exchange enhances glucose-dependent elevations in cytoplasmic calcium and insulin secretion. Diabetes 59:1686–1693
[173] Owada S, Larsson O, Arkhammar P, Katz AI, Chibalin AV, Berggren PO, Bertorello AM (1999) Glucose decreases Na^+, K^+ -ATPase activity in pancreatic beta-cells. An effect mediated via Ca^{2+} -independent phospholipase A2 and protein kinase C-dependent phosphorylation of the alpha-subunit. J Biol Chem 274:2000–2008
[174] James DE, Brown R, Navarro J, Pilch PF (1988) Insulin-regulatable tissues express a unique insulin-sensitive glucose transport protein. Nature 333:183–185
[175] James DE, Strube M, Mueckler M (1989) Molecular cloning and characterization of an insulin-regulatable glucose transporter. Nature 338:83–87
[176] Birnbaum MJ (1989) Identification of a novel gene encoding an insulin-responsive glucose transporter protein. Cell 57:305–315
[177] Bell GI, Murray JC, Nakamura Y, Kayano T, Eddy RL, Fan YS, Byers MG, Shows TB (1989) Polymorphic human insulin-responsive glucose-transporter gene on chromosome 17p13. Diabetes 38:1072–1075

[178] Watson RT, Kanzaki M, Pessin JE (2004) Regulated membrane trafficking of the insulin-responsive glucose transporter 4 in adipocytes. Endocr Rev 25:177–204
[179] Lund S, Holman GD, Schmitz O, Pedersen O (1995) Contraction stimulates translocation of glucose transporter GLUT4 in skeletal muscle through a mechanism distinct from that of insulin. Proc Natl Acad Sci U S A 92:5817–5821
[180] Lalioti VS, Vergarajauregui S, Pulido D, Sandoval IV (2002) The insulin-sensitive glucose transporter, GLUT4, interacts physically with Daxx. Two proteins with capacity to bind Ubc9 and conjugated to SUMO1. J Biol Chem 277:19783–19791
[181] Zisman A, Peroni OD, Abel D, Michael MD, Mauvais-Jarvis F, Lowell BB, Wojtaszewski JFP, Hirshman MF, Virkamaki A, Goodyear LJ, Kahn CR, Kahn BB (2000) Targeted disruption of the glucose transported 4 selectively in muscle causes insulin resistance and glucose intolerance. Nat Med 6:924–928
[182] Kim JK, Zisman A, Fillmore JJ, Peroni OD, Kotani K, Perret P, Zong H, Dong J, Kahn CR, Kahn BB, Shulman GI (2001) Glucose toxicity and the development of diabetes in mice with muscle-specific inactivation of GLUT4. J Clin Invest 108:153–160
[183] Kim YB, Peroni OD, Aschenbach WG, Minokoshi Y, Kotani K, Zisman A, Kahn CR, Goodyear LJ, Kahn BB (2005) Muscle-specific deletion of the Glut4 glucose transporter alters multiple regulatory steps in glycogen metabolism. Mol Cell Biol 25:9713–9723
[184] Stenbit AE, Tsao TS, Li J, Burcelin R, Geenen DL, Factor SM, Houseknecht K, Katz EB, Charron MJ (1997) GLUT4 heterozygous knockout mice develop muscle insulin resistance and diabetes. Nat Med 3:1096–1101
[185] Tsao TS, Stenbit AE, Factor SM, Chen W, Rossetti L, Charron MJ (1999) Prevention of insulin resistance and diabetes in mice heterozygous for GLUT4 ablation by transgenic complementation of GLUT4 in skeletal muscle. Diabetes 48:775–782
[186] Katz EB, Stenbit AE, Hatton K, DePinhot R, Charron MJ (1995) Cardiac and adipose tissue abnormalities but not diabetes in mice deficient in GLUT4. Nature 377:151–155
[187] Abel ED, Peroni O, Kim JK, Kim YB, Boss O, Hadro E, Minnemann T, Shulman GI, Kahn BB (2001) Adipose-selective targeting of the GLUT4 gene impairs insulin action in muscle and liver. Nature 409:729–733
[188] Garvey WT, Maianu L, Hueckstead TP, Birnbaum MJ, Molina JM, Ciaraldi TP (1991) Pretranslational suppression of a glucose transporter protein causes insulin resistance in adipocytes from patients with non-insulin-dependent diabetes mellitus and obesity. J Clin Invest 87:1072–1081
[189] Dohm GL, Elton CW, Friedman JE, Pilch PF, Pories WJ, Atkinson SM Jr, Caro JF (1991) Decreased expression of glucose transporter in muscle from insulin-resistant patients. Am J Physiol 260(3 Pt 1):E459–E463
[190] Rosenbaum D, Haber RS, Dunaif A (1993) Insulin resistance in polycystic ovary syndrome: decreased expression of GLUT-4 glucose transporters in adipocytes. Am J Physiol 264(2 Pt 1):E197–E202
[191] Giacchetti G, Faloia E, Taccaliti A, Morosini PP, Arnaldi G, Soletti F, Mantero F, Accili D, De Pirro R (1994) Decreased expression of insulin-sensitive glucose transporter mRNA (GLUT-4) in adipose tissue of non-insulin-dependent diabetic and obese patients: evaluation by a simplified quantitative PCR assay. J Endocrinol Invest 17:709–715
[192] Pedersen O, Kahn CR, Flier JS, Kahn BB (1991) High fat feeding causes insulin resistance and a marked decrease in the expression of glucose transporters (Glut 4) in fat cells of rats. Endocrinology 129:771–777
[193] Friedman JE, Ishizuka T, Liu S, Farrell CJ, Bedol D, Koletsky RJ, Kaung HL, Ernsberger P (1997) Reduced insulin receptor signaling in the obese spontaneously hypertensive Koletsky rat. Am J Physiol 273(5 Pt 1):E1014–E1023
[194] Das UN (2010) Metabolic syndrome pathophysiology: the role of essential fatty acids. Wiley-Blackwell, Ames, IA
[195] Das UN (1994) Insulin resistance and hyperinsulinemia: are they secondary to an alteration in the metabolism of essential fatty acids? Med Sci Res 22:243–245

[196] Das UN (1999) GLUT-4, tumor necrosis factor, essential fatty acids and daf-genes and their role in insulin resistance and non-insulin dependent diabetes mellitus. Prostaglandins Leukot Essent Fatty Acids 60:13–20
[197] Das UN (1999) GLUT-4, tumor necrosis factor, essential fatty acids and daf-genes and their role in glucose homeostasis, insulin resistance, non-insulin dependent diabetes mellitus and longevity. J Assoc Physicians India 47:431–435
[198] Das UN (2005) A defect in the activity of Δ^6 and Δ^5 desaturases may be a factor predisposing to the development of insulin resistance syndrome. Prostaglandins Leukot Essent Fatty Acids 72:343–350
[199] Das UN (2002) The lipids that matter from infant nutrition to insulin resistance. Prostaglandins Leukot Essent Fatty Acids 67:1–12
[200] Das UN, Vijay Kumar K, Krishna Mohan I (1994) Lipid peroxides and essential fatty acids in patients with diabetes mellitus and diabetic nephropathy. J Nutr Med 4:149–155
[201] Das UN (1995) Essential fatty acid metabolism in patients with essential hypertension, diabetes mellitus and coronary heart disease. Prostaglandins Leukot Essent Fatty Acids 52:387–391
[202] Krishna Mohan I, Das UN (2001) Prevention of chemically-induced diabetes mellitus in experimental animals by polyunsaturated fatty acids. Nutrition 17:126–151
[203] Suresh Y, Das UN (2001) Protective action of arachidonic acid against alloxan-induced cytotoxicity and diabetes mellitus. Prostaglandins Leukot Essent Fatty Acids 64:37–52
[204] Suresh Y, Das UN (2003) Long-chain polyunsaturated fatty acids and chemically-induced diabetes mellitus: effect of ω-6 fatty acids. Nutrition 19:93–114
[205] Suresh Y, Das UN (2003) Long-chain polyunsaturated fatty acids and chemically-induced diabetes mellitus: effect of ω-3 fatty acids. Nutrition 19:213–228
[206] Suresh Y, Das UN (2006) Differential effect of saturated, monounsaturated, and polyunsaturated fatty acids on alloxan-induced diabetes mellitus. Prostaglandins Leukot Essent Fatty Acids 74:199–213
[207] Sailaja Devi MM, Das UN (2004) Effect of prostaglandins against alloxan-induced cytotoxicity to insulin secreting insulinoma RIN cells in vitro. Prostaglandins Leukot Essent Fatty Acids 71:309–318
[208] Sailaja MMS, Das UN (2006) Effect of prostaglandins against alloxan-induced diabetes mellitus. Prostaglandins Leukot Essent Fatty Acids 74:39–60
[209] Gonzàlez-Périz A, Horrillo R, Ferré N, Gronert K, Dong B, Moràn-Salvador E, Titos E, Martínez-Clemente M, López-Parra M, Arroyo V, Joan Clària J (2009) Obesity-induced insulin resistance and hepatic steatosis are alleviated by omega-3 fatty acids: a role for resolvins and protectins. FASEB J 23:1946–1957
[210] Xie W, Hamilton JA, Kirkland JL, Corkey BE, Guo W (2006) Oleate-induced formation of fat cells with impaired insulin sensitivity. Lipids 41:267–271
[211] Bousserouel S, Raymondjean M, Brouillet A, Bèrèziat G, Andrèani M (2004) Modulation of cyclin D1 and early growth response factor-1 gene expression in interleukin-1b-treated rat smooth muscle cells by n-6 and n-3 polyunsaturated fatty acids. Eur J Biochem 271:4462–4473
[212] Watson RT, Shigematsu S, Chiang SH, Mora S, Kanzaki M, Macara IG, Saltiel AR, Pessin JE (2001) Lipid raft microdomain compartmentalization of TC10 is required for insulin signaling and GLUT4 translocation. J Cell Biol 154:829–840
[213] Saravanan N, Haseeb A, Ehtesham NZ (2005) Ghafoorunissa. Differential effects of dietary saturated and trans-fatty acids on expression of genes associated with insulin sensitivity in rat adipose tissue. Eur J Endocrinol 153:159–165
[214] Pérez-Matute P, Pérez-Echarri N, Martínez JA, Marti A, Moreno-Aliaga MJ (2007) Eicosapentaenoic acid actions on adiposity and insulin resistance in control and high-fat-fed rats: role of apoptosis, adiponectin and tumour necrosis factor-alpha. Br J Nutr 97:389–398
[215] Todoric J, Löffler M, Huber J, Bilban M, Reimers M, Kadl A, Zeyda M, Waldhäusl W, Stulnig TM (2006) Adipose tissue inflammation induced by high-fat diet in obese diabetic mice is prevented by n-3 polyunsaturated fatty acids. Diabetologia 49:2109–2119

[216] Duque-Guimarães DE, de Castro J, Martinez-Botas J, Sardinha FL, Ramos MP, Herrera E, do Carmo MG (2009) Early and prolonged intake of partially hydrogenated fat alters the expression of genes in rat adipose tissue. Nutrition 25:782–789
[217] Taouis M, Dagou C, Ster C, Durand G, Pinault M, Delarue J (2002) N-3 polyunsaturated fatty acids prevent the defect of insulin receptor signaling in muscle. Am J Physiol Endocrinol Metab 282:E664–E671
[218] Poulsen RC, Gotlinger KH, Serhan CN, Kruger MC (2008) Identification of inflammatory and proresolving lipid mediators in bone marrow and their lipidomic profiles with ovariectomy and omega-3 intake. Am J Hematol 83:437–445
[219] DeFronzo RA (1992) Pathogenesis of type 2 (non-insulin-dependent) diabetes mellitus: a balanced overview. Diabetologia 35:389–397
[220] Kido Y, Burks DJ, Withers D, Bruning JC, Kahn CR, White MF, Accili D (2000) Tissue-specific insulin resistance in mice with mutations in the insulin receptor, IRS-1, and IRS-2. J Clin Invest 105:199–205
[221] Mauvais-Jarvis F, Ueki K, Fruman DA, Hirshman MF, Sakamoto K, Goodyear LJ, Iannacone M, Accili D, Cantley LC, Kahn CR (2002) Reduced expression of the murine p85alpha subunit of phosphoinositide 3-kinase improves insulin signaling and ameliorates diabetes. J Clin Invest 109:141–149
[222] Ueki K, Fruman DA, Brachmann SM, Tseng YH, Cantley LC, Kahn CR (2002) Molecular balance between the regulatory and catalytic subunits of phosphoinositide 3-kinase regulates cell signaling and survival. Mol Cell Biol 22:965–977
[223] Sahu A, Dube MG, Phelps CP, Sninsky CA, Kalra PS, Kalra SP (1995) Insulin and insulin-like growth factor II suppress neuropeptide Y release from the nerve terminals in the paraventricular nucleus: a putative hypothalamic site for energy homeostasis. Endocrinology 136:5718–5724
[224] Fisher BL, Schauer P (2002) Medical and surgical options in the treatment of severe obesity. Am J Surg 184:9S–16S
[225] Pories WJ, Swanson MS, MacDonald KG, Long SB, Morris PG, Brown BM, Barakat HA, de Ramon RA, Israel G, Dolezal JM (1995) Who would have thought it? An operation proves to be the most effective therapy for adult-onset diabetes mellitus. Ann Surg 222:339–350
[226] Panagiotakos DB, Pitsavos C, Das UN, Skoumas Y, Stefanadis C (2007) The implications of anthropometric, inflammatory and glycaemic control indices in the epidemiology of the metabolic syndrome given by different definitions: a classification analysis. Diabetes Obes Metab 9:660–668
[227] Meguid MM, Ramos EJB, Suzuki S, Xu Y, George ZM, Das UN, Hughes K, Quinn R, Chen C, Marx W, Cunningham PRG (2004) A surgical model of human Roux-en-Y gastric bypass. J Gastrointest Surg 8:621–630
[228] Xu Y, Ramos EJB, Middleton F, Romanova I, Quinn R, Chen C, Das UN, Inui A, Meguid MM (2004) Gene expression profiles post Roux-en-Y gastric bypass. Surgery 136:246–252
[229] Romanova I, Ramos EJB, Xu Y, Quinn R, Chan C, George ZM, Inui A, Das UN, Meguid MM (2004) Neurobiologic changes in the hypothalamus associated with weight loss after gastric bypass. J Am Coll Surg 199:887–895
[230] Vendrell J, Broch M, Vilarrasa N, Molina A, Gomez JM, Gutierrez C, Simon I, Soler J, Richart C (2004) Resistin, adiponectin, ghrelin, leptin, and proinflammatory cytokines: relationships in obesity. Obes Res 12:962–971
[231] Sookhai S, Wang JH, McCourt M, O'Connell D, Redmond HP (1999) Dopamine induces neutrophil apoptosis through a dopamine D- 1 receptor-independent mechanism. Surgery 126:314–322
[232] Sookhai S, Wang JH, Winter D, Power C, Kirwan W, Redmond HP (2000) Dopamine attenuates the chemoattractant effect of interleukin-8: a novel role in the systemic inflammatory response syndrome. Shock 14:295–299
[233] Oberbeck R, Schmitz D, Wilsenack K, Schuler M, Husain B, Schedlowski M, Exton MS (2006) Dopamine affects cellular immune functions during polymicrobial sepsis. Intensive Care Med 32:731–739

[234] Wang G-J, Volkow ND, Logan J, Pappas NR, Wong CT, Zhu W, Netusil N, Fowler JS (2001) Brain dopamine and obesity. Lancet 357:354–357

[235] Bliznakov EG (1980) Serotonin and its precursors as modulators of the immunological responsiveness in mice. J Med 11:81–105

[236] Mashek K, Devoino LV, Kadletsova O, Idova GV, Morozova NB (1985) Changes in the level of serotonin in the brain and immunocompetent organs during the formation of the immune response. Fiziol Zh SSSR Im I M Sechenova 71:992–997

[237] Devoino L, Morozova N, Cheido M (1988) Participation of serotoninergic system in neuroimmunomodulation: intraimmune mechanisms and the pathways providing an inhibitory effect. Int J Neurosci 40:111–128

[238] Ciz M, Komrskova D, Pracharova L, Okenkova K, Cizova H, Moravcova A, Jancinova V, Petrikova M, Lojek A, Nosal R (2007) Serotonin modulates the oxidative burst of human phagocytes via various mechanisms. Platelets 18:583–590

[239] Menard G, Turmel V, Bissonnette EY (2007) Serotonin modulates the cytokine network in the lung: involvement of prostaglandin E2. Clin Exp Immunol 150:340–348

[240] Muller T, Durk T, Blumental B, Grimm M, Cicko S, Panther E, Sorichter S, Herouy Y, Di Virgilio F, Ferrari D, Norgauer J, Idzko M (2009) 5-hydroxytryptamine modulates migration, cytokine and chemokine release and T-cell priming capacity of dendritic cells in vitro and in vivo. PLoS One 4:e6453

[241] Kushnir-Sukhov NM, Gilfillan AM, Coleman JW, Brown JM, Bruening S, Toth M, Metcalfe DD (2006) 5-hydroxytryptamine induces mast cell adhesion and migration. J Immunol 177:6422–6432

[242] Holler J, Zakrzewicz A, Kaufmann A, Wilhelm J, Fuchs-Moll G, Dietrich H, Padberg W, Kuncova J, Kummer W, Grau V (2008) Neuropeptide Y is expressed by rat mononuclear blood leukocytes and strongly down-regulated during inflammation. J Immunol 181:6906–6912

[243] Dimitrijevic M, Stanojevic S, Mitic S, Vujic V, Kovacevic- Jovanovic V, Mitic K, von Horsten S, Kosec D (2006) Neuropeptide Y (NPY) modulates oxidative burst and nitric oxide production in carrageenan-elicited granulocytes from rat air pouch. Peptides 27:3208–3215

[244] Dimitrijevic M, Stanojevic S, Mitic S, Kustriovic N, Vujic V, Miletic T, Kovacevic-Jovanovic V (2008) The anti-inflammatory effect of neuropeptide Y (NPY) in rats is dependent on dipeptidyl peptidase 4 (DP4) activity and age. Peptides 29:2179–2187

[245] Chandrasekharan B, Bala V, Kolachala VL, Vijaykumar M, Jones D, Gewirtz AT, Sitaraman SV, Srinivasan S (2008) Targeted deletion of neuropeptide Y (NPY) modulates experimental colitis. PLoS One 3:e3304

[246] Hernanz A, Tato E, De la Fuente M, de Miguel E, Arnalich F (1996) Differential effects of gastrin-releasing peptide, neuropeptide Y, somatostatin and vasoactive intestinal peptide on interleukin-1 beta, interleukin-6 and tumor necrosis factor-alpha production by whole blood cells from healthy young and old subjects. J Neuroimmunol 71:25–30

[247] King PJ, Widdowson PS, Doods H, Williams G (2000) Effect of cytokines on hypothalamic neuropeptide Y release in vitro. Peptides 21:143–146

[248] Kos K, Harte AL, James S, Snead DR, O'Hare JP, McTernan PG, Kumar S (2007) Secretion of neuropeptide Y in human adipose tissue and its role in maintenance of adipose tissue mass. Am J Physiol Endocrinol Metab 293:E1335–E1340

[249] Lopez M, Lage R, Saha AK et al (2008) Hypothalamic fatty acid metabolism mediates the orexigenic action of ghrelin. Cell Metab 7:389–399

[250] Dixit VD, Schaffer EM, Pyle RS, Collins GD, Sakthivel SK, Palaniappan R, Lillard JW Jr, Taub DD (2004) Ghrelin inhibits leptin- and activation-induced proinflammatory cytokine expression by human monocytes and T cells. J Clin Invest 114:57–66

[251] Yada, Mutoh K, Azuma T, Hyodo S, Kangawa K (2006) Ghrelin stimulates phagocytosis and superoxide production in fish leukocytes. J Endocrinol 189:57–65

[252] Ates Y, Degertekin B, Erdil A, Yaman H, Dagalp K (2008) Serum ghrelin levels in inflammatory bowel disease with relation to disease activity and nutritional status. Dig Dis Sci 53:2215–2221

[253] Mager U, Kolehmainen M, d Mello VD, Schwab U, Laaksonen DE, Rauramaa R, Gylling H, Atalay M, Pulkkinen L, Uusitupa M (2008) Expression of ghrelin gene in peripheral blood mononuclear cells and plasma ghrelin concentrations in patients with metabolic syndrome. Eur J Endocrinol 158:499–510

[254] Luger TA, Brzoska T (2007) alpha-MSH related peptides: a new class of anti-inflammatory and immunomodulating drugs. Ann Rheum Dis 66(Suppl 3):52–55

[255] Taylor AW, Kitaichi N (2008) The diminishment of experimental autoimmune encephalomyelitis (EAE) by neuropeptide alpha- melanocyte stimulating hormone (alpha-MSH) therapy. Brain Behav Immun 22:639–646

[256] Pavlov VA, Parrish WR, Rosas-Ballina M, Ochani M, Puerta M, Ochani K, Chavan S, Al-Abed Y, Tracey KJ (2009) Brain acetylcholinesterase activity controls systemic cytokine levels through the cholinergic anti-inflammatory pathway. Brain Behav Immun 23:41–45

[257] Das UN (2007) Acetylcholinesterase and butyrylcholinesterase as possible markers of low-grade systemic inflammation. Med Sci Monit 13:RA214–RA221

[258] Van Maanen MA, Vervoordeldonk MJ, Tak PP (2009) The cholinergic anti-inflammatory pathway: towards innovative treatment of rheumatoid arthritis. Nat Rev Rheumatol 5:229–232

[259] Fuenmayor LD (1979) The effect of fasting on the metabolism of 5-hydroxytryptamine and dopamine in the brain of the mouse. J Neurochem 33:481–485

[260] Agardh CD, Carlsson A, Linqvist M, Siesjo BK (1979) The effect of pronounced hypoglycemia on monoamine metabolism in rat brain. Diabetes 28:804–809

[261] MacKenzie RG, Trulson ME (1978) Effects of insulin and streptozotocin-induced diabetes on brain tryptophan and serotonin metabolism in rats. J Neurochem 30:205–211

[262] Shimizu H, Bray GA (1990) Effects of insulin on hypothalamic monoamine metabolism. Brain Res 510:251–258

[263] Baranov VG, Propp MV, Sokoloverova IM, Savchenko ON, Onegova RF (1980) Dopamine, norepinephrine and serotonin content in various parts of the hypothalamus in alloxan diabetes. Probl Endokrinol (Mosk) 26:43–48

[264] Lackovic Z, Salkovic M, Relia M (1990) Effect of long-lasting diabetes mellitus on rat and human brain monoamines. J Neurochem 54:143–147

[265] Bhattacharya SK, Saraswati M (1991) Effect of intracerebroventricularly administered insulin on brain monoamines and acetylcholine in euglycaemic and alloxan-induced hyperglycaemic rats. Indian J Exp Biol 29:1095–1100

[266] Di Giulio AM, Tenconi B, La Croix R, Mantegazza P, Abbracchio MP, Cattabeni F, Gorio A (1989) Denervation and hyperinnervation in the nervous system of diabetic animals and diabetic encephalopathy. J Neurosci Res 24:362–368

[267] Aso Y, Wakabayashi S, Nakano T, Yamamoto R, Takebayashi K, Inukai T (2006) High serum high-sensitivity C-reactive protein concentrations are associated with relative cardiac sympathetic overactivity during the early morning period in type 2 diabetic patients with metabolic syndrome. Metabolism 55:1014–1021

[268] Iwai T, Ito S, Tanimitsu K, Udagawa S, Oka J (2006) Glucagon-like peptide-1 inhibits LPS-induced IL-1beta production in cultured rat astrocytes. Neurosci Res 55:352–360

[269] Blandino-osano M, Perez-Arana G, Mellado-Gil JM, Segundo C, Aguilar-Diosdado M (2008) Anti-proliferative effect of pro-inflammatory cytokines in cultured beta cells is associated with extracellular signal-regulated kinase 1/2 pathway inhibition: protective role of glucagon-like peptide-1. J Mol Endocrinol 41:35–44

[270] Kim SJ, Nian C, Doudet DJ, McIntosh CH (2009) Dipeptidyl peptidase IV inhibition with MK0431 improves islet graft survival in diabetic NOD mice partially via T-cell modulation. Diabetes 58:641–651

[271] Frerker N, Raber K, Bode F, Skripuletz T, Nave H, Klemann C, Pabst R, Stephan M, Schade J, Brabant G, Wedekind D, Jacobs R, Jörns A, Forssmann U, Straub RH, Johannes S, Hoffmann T, Wagner L, Demuth HU, von Hörsten S (2009) Phenotyping of congenic dipeptidyl peptidase 4 (DP4) deficient Dark Agouti (DA) rats suggests involvement of DP4 in neuro-, endocrine, and immune functions. Clin Chem Lab Med 47:275–287

[272] Park MC, Chung SJ, Park YB, Lee SK (2009) Pro-inflammatory effect of leptin on peripheral blood mononuclear cells of patients with ankylosing spondylitis. Joint Bone Spine 76:170–175

[273] Moraes JC, Coope A, Morari J, Cintra DE, Roman EA, Pauli JR, Romanatto T, Carvalheir JB, Oliveira AL, Saad MJ, Velloso LA (2009) High-fat diet induces apoptosis of hypothalamic neurons. PLoS One 4:e5045

[274] Matsumoto T, Miyatsuji A, Miyawaki T, Yanagimoto Y, Moritani T (2003) Potential association between endogenous leptin and sympatho-vagal activities in young obese Japanese women. Am J Hum Biol 15:8–15

[275] Luyer MD, Greve JW, Hadfoune M, Jacobs JA, Dejong CH, Buurman WA (2005) Nutritional stimulation of cholecystokinin receptors inhibits inflammation via the vagus nerve. J Exp Med 202:1023–1029

[276] Peters JH, Simasko SM, Ritter RG (2006) Modulation of vagal afferent excitation and reduction of food intake by leptin and cholecystokinin. Physiol Behav 89:477–485

[277] Ueno N, Dube MG, Inui A, Kalra PS, Kalra SP (2004) Leptin modulates orexigenic effects of ghrelin and attenuates adiponectin and insulin levels and selectively the dark-phase feeding as revealed by central leptin gene therapy. Endocrinology 145:4176–4184

[278] Goto M, Arima H, Watanabe M, Hayashi M, Banno R, Sato I, Nagasaki H, Oiso Y (2006) Ghrelin increases neuropeptide Y and agouti-related peptide gene expression in the arcuate nucleus in rat hypothalamic organotypic cultures. Endocrinolog 147:5102–5109

[279] Bassil AK, Dass NB, Sanger GJ (2006) The prokinetic-like activity of ghrelin in rat isolated stomach is mediated via cholinergic and tachykininergic motor neurones. Eur J Pharmacol 544:146–152

[280] Uno K, Katagiri H, Yamada T, Ishigaki Y, Ogihara T, Imai J, Hasegawa Y, Gao J, Kaneko K, Iwasaki H, Ishihara H, Sasano H, Inukai K, Mizuguchi H, Asano T, Shiota M, Nakazato M, Oka Y (2006) Neuronal pathway from the liver modulates energy expenditure and systemic insulin sensitivity. Science 312:1656–1659

[281] Borovikova LV, Ivanova S, Zhang M, Yang H, Botchkina GI, Watkins LR, Wang H, Abumrad N, Eaton JW, Tracey KJ (2000) Vagus nerve stimulation attenuates the systemic inflammatory response to endotoxin. Nature 405:458–462

[282] Bernik TR, Friedman SG, Ochani M, DiRaimo R, Ulloa L, Yang H, Sudan S, Czura CJ, Ivanova SM, Tracey KJ (2002) Pharmacological stimulation of the cholinergic antiinflammatory pathway. J Exp Med 195:781–788

[283] Wang H, Yu M, Ochani M, Amella CA, Tanovic M, Susarla S, Li JH, Wang H, Yang H, Ulloa L et al (2003) Nicotinic acetylcholine receptor $\alpha 7$ subunit is an essential regulator of inflammation. Nature 421:384–387

[284] Hersi AI, Kitaichi K, Srivastava LK, Gaudreau P, Quirion R (2000) Dopamine D-5 receptor modulates hippocampal acetylcholine release. Brain Res Mol Brain Res 76:336–340

[285] Anthony K, Reed LJ, Dunn JT, Bingham E, Hopkins D, Marsden PK, Amiel SA (2006) Attenuation of insulin-evoked responses in brain networks controlling appetite and reward in insulin resistance. The cerebral basis for impaired control of food intake in metabolic syndrome? Diabetes 55:2986–2992

[286] Flores MBS, Fernandes MFA, Ropello ER, Faria MC, Ueno M, Velloso LA, Saad MJA, Carvalheira JBC (2006) Exercise improves insulin and leptin sensitivity in hypothalamus of Wistar rats. Diabetes 55:2554–2561

[287] Nishimura S, Manabe I, Nagasaki M et al (2009) CD8+ effector T cells contribute to macrophage recruitment and adipose tissue inflammation in obesity. Nat Med 15:914–921

[288] Winer S, Chan Y, Paltser G et al (2009) Normalization of obesity-associated insulin resistance through immunotherapy. Nat Med 15:921–930

[289] Feurer M, Herrero L, Cipolletta D et al (2009) Lean, but not obese, fat is enriched for a unique population of regulatory T cells that affect metabolic parameters. Nat Med 15:930–940

[290] Lumeng CN, Maillard I, Saltiel AR (2009) T-ing up inflammation in fat. Nat Med 15:846–847

Chapter 10
Atherosclerosis

Introduction

Atherosclerosis, the major underlying cause for coronary heart disease (CHD), is a dynamic process. In majority of the instances, hyperlipidemia, diabetes mellitus, hypertension, obesity, hyperhomocysteinemia and smoking are the main risk factors for the development of atherosclerosis and CHD. Several studies revealed that in CHD, hypertension, diabetes mellitus, hyperlipidemias, and obesity, EFA (essential fatty acids) metabolism is abnormal such that plasma and tissue concentrations of γ-linolenic acid (GLA), dihomo-GLA (DGLA), arachidonic acid (AA), eicosapentaenoic acid (EPA), and docosahexaenoic acid (DHA) in the phospholipid fraction are low [1–8]. Increased intake of polyunsaturated fatty acids (PUFAs especially in the form of GLA, DGLA, EPA and DHA) protects against the development of these diseases both in experimental animals [9–12] and humans [13], though the exact mechanism of this protective action is unclear. GLA, DGLA, AA, EPA, and DHA form precursors to prostaglandin E_1 (PGE_1), prostacyclin (PGI_2), PGI_3, lipoxins (LXs), resolvins, neuroprotectin D1 (NPD1), enhance NO generation, and interact with NO to form nitrolipids that have anti-inflammatory actions, prevent platelet aggregation, inhibit leukocyte activation and augment wound healing and resolve inflammation that may account for their beneficial actions. This implies that an altered EFA metabolism in the form of a block in the activity of Δ^6 and Δ^5 desaturases, which are essential for the formation of long-chain metabolites from dietary linoleic acid (LA, 18:2 ω-6) and α-linolenic acid (ALA, 18:3 ω-3), and inadequate formation of anti-inflammatory lipoxins, resolvins, protectins, maresins and nitrolipids from their precursor PUFAs could lead to the initiation, progression and aggravation of atherosclerosis.

Lipoxins and their aspirin-triggered carbon-15 epimers are key mediators of endogenous anti-inflammation and resolution. Aspirin-triggered lipoxin A_4 analog (ATL-1) have been shown to modulate reactive oxygen species (ROS) generation in endothelial cells. Pre-treatment of endothelial cells with ATL-1 completely blocked ROS production triggered by different agents, inhibited the phosphorylation and translocation of the cytosplamic NAD(P)H oxidase subunit p47 (phox) to the cell

U. N. Das, *Molecular Basis of Health and Disease*,
DOI 10.1007/978-94-007-0495-4_10, © Springer Science+Business Media B.V. 2011

membrane as well as NAD(P)H oxidase activity and impaired the redox-sensitive activation of the transcriptional factor NF-κB, suggesting that lipoxins play a protective role against the development and progression of atherosclerosis and various cardiovascular diseases in which endothelial dysfunction is known to exist [14]. These results are supported in experiments performed with apolipoprotein E-deficient mice with (a) global leukocyte 12/15-lipoxygenase deficiency, (b) normal enzyme expression, or (c) macrophage-specific 12/15-lipoxygenase overexpression in which it was noted that 12/15-lipoxygenase expression protected mice against atherosclerosis via its role in the biosynthesis of lipoxin A_4, resolvin D1, and protectin D1. These lipid mediators showed potent agonist actions on macrophages and vascular endothelial cells that reduced the magnitude of the local inflammatory response suggesting that a failure of local resolution mechanisms may underlie the unremitting inflammation that fuels atherosclerosis [15]. The evidence that lipoxins, resolvins and protectins are anti-inflammatory compounds and pro-inflammatory cytokines are elevated in atherosclerosis lends support to the belief that atherosclerosis is an inflammatory condition.

Atherosclerosis Is a Low-grade Systemic Inflammatory Condition

Atherosclerosis is common both in the developed and developing countries. An increase in the plasma concentrations of C-reactive protein (CRP), tumor necrosis factor-α (TNF-α), interleukin-6 (IL-6), myeloperoxidase (MPO), lipoprotein associated phospholipase A_2 (Lp-PLA$_2$), and lipid peroxides occurs in atherosclerosis suggesting that it is a low-grade systemic inflammatory condition [16–19]. Myeloperoxidase (MPO), a leukocyte enzyme; CRP, produced by endothelial cells and liver; lipoprotein-associated phospholipase A_2 (Lp-PLA$_2$), produced by macrophages, are known to be expressed in greater concentrations in atherosclerotic lesions [20]. Elevated Lp-PLA$_2$ is significantly and independently associated with a twofold higher risk for CHD events including myocardial infarction, neovascularization, and death from cardiac disease [21–24].

It is known that patients with atherosclerosis have low circulating endothelial nitric oxide (eNO) levels due to its decreased production by endothelial cells [25–27], increased generation of reactive oxygen species (ROS) by infiltrating leukocytes and macrophages, and decreased anti-oxidant content of the endothelial cells at atheroslcerosis-prone areas of the blood vessels due to their exposure to increased ROS leads to an imbalance between the pro- and anti-oxidant status that is tilted more in favor of pro-oxidants leads to endothelial damage [28–32]. The decrease in the production of eNO by endothelial cells may, in part, be due to enhanced levels of asymmetrical dimethylarginine (ADMA) that inhibits eNO generation that may lead to increased mortality due to cardiovascular diseases [33–35]. This increase in ADMA, an endogenous inhibitor of eNOS that is considered to be a risk factor for atherosclerosis was found to be associated with decreased vasorelaxation of saphenous veins to acetylcholine and bradykinin, chemicals mediators that are known to enhance NO generation. In addition, high serum ADMA was associated with higher total $O_2^{-\cdot}$ production in both saphenous veins and internal mammary arteries. These results indicate an association between ADMA, eNO and $O_2^{-\cdot}$ generation and vascular function and implies that increase in serum ADMA levels could produce eNOS uncoupling in the human vascular endothelium of patients with coronary artery disease [36]. What is more significant is the observation that plasma ADMA concentrations were found to be positively related to internal carotid/bulb intimal-media thickness, suggesting that ADMA promotes subclinical atherosclerosis in a site-specific manner, with a greater proatherogenic influence at known vulnerable sites in the arterial tree [37]. In addition to ADMA, homocysteine also augments the formation of superoxide anion and reduces the synthesis and release of eNO [38, 39]. Homocysteine markedly reduced the increase in haem oxygenase (HO) activity and HO-1 protein expression induced by sodium nitroprusside. High levels of homocysteine also abolished hypoxia-mediated HO-1 expression [40]. Thus, hyperhomocysteinemia accelerates the atherosclerotic process by reducing the formation of NO and carbon monoxide that are vasodilators and inhibitors of atherosclerosis and enhancing the production of harmful reactive oxygen species.

Mediators of Inflammation in Atherosclerosis

Some of the important mediators of inflammation include: histamine, serotonin, lysosomal enzymes, prostaglandins (PGs), leukotrienes (LTs), thromboxanes (TXs), platelet activating factors (PAFs), ROS, NO, HOCL, various cytokines, kinin system, coagulation/fibrinolysis system, and complement system. NO has both pro- and anti-inflammatory actions depending on the source and the local concentration. Histamine, serotonin, bradykinin, complement system and coagulation cascade are well known for their involvement in infections, inflammatory process and sepsis and septic shock.

The major cellular sources of these mediators are platelets, neutrophils, monocytes/macrophages, mast cells, and mesenchymal cells such as endothelium, smooth muscle cells, fibroblasts, and most epithelia. It is likely that one mediator triggers the release of another mediator that acts on the target tissue. These secondary mediators either potentiate the action of the initial mediator or paradoxically abrogate its action. Thus, the ultimate degree of inflammation depends on the balance between such pro- and anti-inflammatory mediators. In some instances, anti-inflammatory chemicals or signals initiated may not only act on the target tissue but also on other tissues to suppress inflammation. Once released or activated, most of these mediators are inactivated or decay quickly. For instance, AA and its metabolites have a short half-life, whereas specific or non-specific enzymes inactivate kinins. On the other hand, ROS and NO are scavenged by specific or non-specific antioxidants [16].

Under normal physiological conditions, a balance is maintained between pro- and anti-inflammatory molecules. This delicate balance is tilted more towards the pro-inflammatory molecules in atherosclerosis leading to its initiation and progression. If this altered balance between pro- and anti-inflammatory molecules is restored to normal, then it is likely that atheroslcerosis process could be prevented. Of the several molecules that influence atherosclerotic process, the most important factors are the following: CRP, fibrinogen, serum amyloid A, TNF-α, monocyte chemoattractant protein-1 (MMP-1), IL-6, IL-8, IL-1, IL-4, IL-10, IL-12, IL-15, IL-18, IL-33, ICAM-1 (intercellular adhesion molecule-1), VCAM-1 (vascular cell adhesion molecule-1), MMPs (metalloproteinases), IFN-γ (interferon-γ), reactive oxygen species and nitric oxide [41]. The various lipids such as HDL, LDL, VLDL, triglycerides and cholesterol seem to have the ability to modulate the synthesis, release and/or action of the various factors that have been outlined above and thus, participate in the initiation and progression and prevent or arrest the process of atherosclerosis. For example, cholesterol can function as a pro-inflammatory molecule by interfering with the metabolism of essential fatty acids (EFAs) [1]; enhancing the production of IL-6 and TNF-α [42, 43] and superoxide anion [44, 45]. In addition, it is possible that by interfering with the metabolism of essential fatty acids and thus, decreasing the formation of their long-chain metabolites arachidonic acid, eicosapentaenoic acid and docosahexaenoic acid, cholesterol may suppress the formation of anti-inflammatory lipid molecules such as lipoxins, resolvins, protectins and maresins that have anti-atherosclerotic properties [15]. IL-12 is an anti-inflammatory cytokine that can arrest or prevent atherosclerosis, while all other factors enumerated above: CRP, TNF, IL-6, IL-1, IL-4, fibrinogen, MCP-1, serum amyloid A, IL-15, IL-8, IL-18, ICAM-1, VCAM-1 and MMPs seem to enhance the progression of atherosclerosis [41]. Interestingly, knockout of interferon-γ (IFN-γ) [46] and interleukin-18 (IL-18) [47, 48] seem to retard atherosclerosis.

It is noteworthy that NO reacts with PUFAs to yield their respective nitroalkene derivatives that can be detected in plasma. These nitroalkene derivatives, termed as nitrolipids, produce vascular relaxation, inhibit neutrophil degranulation and superoxide formation, and inhibit platelet activation [49–51]. Nitrolipids have endogenous PPAR-γ ligand activity and release NO [51]. These actions of nitrolipids prevent platelet aggregation, thrombus formation and atherosclerosis, and prevent inflammation. This implies that nitrolipids have a significant role in low-grade systemic inflammatory conditions such as atherosclerosis. Thus, PUFAs not only form precursors to various eicosanoids, resolvins, lipoxins, resolvins, protectins and maresins but also react with NO to form nitrolipids that have platelet anti-aggregatory action, prevent thrombus formation and thus, arrest atherosclerosis.

Thus, PUFAs and their metabolites such as eicosanoids, lipoxins, resolvins, protectins, maresins and nitrolipids; various pro- and anti-inflammatory cytokines, free radicals, nitric oxide (including ADMA) and various antioxidants seem to play critical role in the pathobiology of atherosclerosis. As most, if not all, of these molecules are involved in the inflammatory process at one stage or the other, it is quite but natural that atherosclerosis is an inflammatory process. Thus, whenever pro-inflammatory molecules are produced in excess compared to the anti-inflammatory molecules, atherosclerosis is initiated. Since endothelial cells are at the centre of atherosclerosis, it is important that endothelial cells remain healthy and are able to produce adequate amounts of anti-inflammatory molecules so that atherosclerosis is kept under check. In this context, the interaction among endothelial cells, platelets and leukocytes assumes great significance.

Cross Talk Among Platelets, Leukocytes and Endothelial Cells

It is reasonable to believe that there is a close interaction among platelets, leukocytes and endothelial cells, cells that are critical in the pathobiology of atherosclerosis. The cross-talk among these three types of cells could ultimately determine the initiation and progression of atherosclerosis and thrombosis. For instance, under normal conditions, endothelial cells produce adequate amounts of PGE_1 from DGLA; PGI_2 from AA; LXs, resolvins, protectins and maresins from AA, EPA and DHA; formation of adequate amounts of nitrolipids, which have anti-inflammatory and platelet anti-aggregatory actions, due to an interaction between PUFAs and NO and NO from L-arginine such that the pro-inflammatory and pro-atherosclerotic events are successfully abrogated. Some of these pro-inflammatory and pro-atherogenic stimuli include: hemodynamic forces, hyperlipidemia, hypertension, hyperglycemia, smoking, etc. These factors induce the expression of pro-inflammatory genes that initiate and accelerate atherosclerosis at the points of shear stress, enhance infiltration of intima by leukocytes and macrophages, cause low-level activation of NF-κB and elevated expression of VCAM-1 and ICAM-1, IL-1, IL-6, MCP-1, as well as antioxidant genes glutathione peroxidase and glutathione-S-transferase 2, and pro-inflammatory eicosanoids such as TXA_2, PGE_2, $PGF_{2\alpha}$, LTs, and other PGs, TXs, and LTs, and increased production and release of free radicals and UCP (uncoupling proteins) expression occurs in endothelial cells, platelets, and leukocytes in atherosclerosis-susceptible regions, and endothelial cells themselves may show changes in cell shape and proliferation. These events can be prevented and atherosclerosis process is arrested by the production of adequate amounts of PGE_1, PGI_2, PGI_3, LXs, resolvins, protectins, maresins, nitrolipids, NO, and anti-inflammatory cytokines such as IL-4, IL-10, TGF-β by endothelial cells, provided there are adequate stores of respective precursors of various PUFAs and L-arginine and their respective enzymes. This suggests that under physiological conditions a delicate balance is maintained between pro- and anti-inflammatory and pro and anti-atherosclerotic factors. When this delicate balance is tilted more towards pro-atherosclerotic and pro-inflammatory factors, atherosclerosis occurs [52]. Since, hyperlipidemia, hypertension, hyperglycemia and smoking have pro-atherogenic properties and atherosclerosis is a low-grade systemic inflammatory process and is associated with decreased formation anti-atherosclerotic molecules such as lipoxins, resolvins, protectins, maresins, nitrolipids, PGE_1, PGI_2, PGI_3, and anti-inflammatory cytokines, it is implied that even in these diseases there could occur a deficiency of these beneficial molecules. In the light of this, it will be interesting to study the role of lipoxins, resolvins, protectins, maresins, nitrolipids, PGE_1, PGI_2, PGI_3, and anti-inflammatory cytokines in the pathobiology of hypertension, hyperlipidemias and diabetes mellitus (especially type 2).

Lipoxins in Rheumatoid Arthritis

Recent studies revealed that premature atherosclerosis could occur in patients with rheumatoid arthritis, lupus, scleroderma and other collagen vascular diseases. It is possible that in these inflammatory conditions, the formation of anti-inflammatory molecules such as lipoxins, resolvins, protectins, maresins and nitrolipids and anti-inflammatory cytokines are inadequate which could explain chronic unresolved inflammation and persistence of the disease process.

This proposal is supported by the observation that the conversion of endogenous arachidonic acid (AA) by polymorphonuclear cells (PMN) from patients with rheumatoid arthritis (RA) before (D0) and 1 day (D1) after antiinflammatory drug therapy revealed that large amounts of 5,15-diHETE and significant levels of lipoxins (from 2 to 20 ng/10^7 PMN) were produced with individual differences between donors. The production of lipoxins after treatment may be related to long-term clinical improvement of some patients [53]. In human fibroblast like synoviocytes TGF-β acted synergistically with IL-1β to stimulate IL-6 protein levels, whereas LXA_4 (lipoxin A_4) inhibited IL-6 expression in dose- and time-dependent manner. LXA_4 at nanomolar concentrations altered the MMP-1 and MMP-3 expression levels of IL-1β and TGF-β2- stimulated fibroblast like synoviocytes, while both IL-1β and TGF-β2 up-regulated LXA_4R (lipoxin A_4 receptor) mRNA. These results demonstrated that LXA_4Rs mediate the effects of LXA_4 on inflammatory responses upon stimulation of human fibroblast like synoviocytes with IL-1β and TGF-β2, suggesting that production of LXA_4 might constitute an important mechanism by which human synovial fibroblast activation is regulated [54]. It was reported that serum amyloid A (SAA), an acute phase reactant with cytokine-like properties, binds to the same seven transmembrane G protein-coupled receptor ligated by LXA_4. LXA_4 induced stimulation of tissue inhibitors of metalloproteinase-2, whereas SAA induced IL-8 and matrix metalloproteinase-3 production. SAA up-regulated NF-κB and AP-1 DNA binding activity, while LXA_4 markedly inhibited these responses after IL-1β stimulation and the NF-κB pathway proved to be the preeminent for the biological responses elicited by both ligands. These findings suggest that two endogenous molecules, SAA and LXA_4 targeting a common receptor, could participate in the pathogenesis of inflammatory arthritis by differentially regulating inflammatory responses in tissues expressing the lipoxin A receptor [55].

It was also reported that both LXA_4 and 15-epi-LXA_4 were present at significantly higher levels in rheumatoid arthritis synovial fluid (10.34 ± 14.12 ng/ml for LXA_4) compared to osteoarthritis synovial fluid (0.66 ± 0.77 ng/ml for LXA_4). Interestingly, logarithmic concentration of LXA_4 was significantly correlated with that of leukotriene B_4 (LTB_4) and prostaglandin E_2 in rheumatoid and osteoarthritis synovial fluids. Similarly, expressions of LX receptor and 15-LOX (lipoxygenase) mRNA were stronger in rheumatoid arthritis synovium than osteoarthritis synovium, while the expression of mRNA for IL-13, which induces 15-LOX, was significantly stronger in rheumatoid synovium than osteoarthritis synovium. These results suggest that 15-LOX induced by IL-13 might regulate the production of LXA_4 [56] to have an antiinflammatory effect against proinflammatory lipid mediators in inflamed joints and implies that production of adequate amounts of LXA_4 might suppress rheumatoid arthritis. In this context, it is interesting to note that in murine collagen-induced arthritis model, COX-2 and its metabolite, proinflammatory PGE_2, are present in the joints during the resolution phase of arthritis, and blocking COX-2 activity and PGE_2 production within this period perpetuated, instead of attenuated, inflammation. On the other hand, repletion with PGE_2 analogs restored homeostasis and resolved inflammation by the formation of LXA_4. Thus, it appears that there is a close link between the cyclooxygenase-lipoxygenase pathways and PGE_2 serves as a feedback inhibitor essential for limiting chronic inflammation in autoimmune arthritis such as rheumatoid arthritis. These results may also explain as to why COX-2 inhibitors are palliative rather than curative in humans, because blocking PGE_2 may inhibit the production of PGE_2-stimulated production of LXA_4 that seems to be essential for resolution of arthritis [57]. It is paradoxical that production of PGE_2, a pro-inflammatory eicosanoid, is essential for the induction of anti-inflammatory LXA_4. These results also indicate that inflammation and its resolution are two sides of the same coin. Thus, understanding the interaction between pro- and anti-inflammatory eicosanoids is essential to develop meaningful therapeutic strategies for the management of rheumatoid arthritis. This may hold good for other collagen vascular diseases such as lupus, systemic sclerosis and Wegener's granulomatosus. It is possible that resolvins, protectins, maresins and nitrolipids, which are also anti-inflammatory compounds, are present in the synovial fluid and tissues and are needed for resolution of the inflammatory process in various diseases. Based on these studies, it is reasonable to propose that plasma, tissue or synovial levels of LXA_4, resolvins, protectins, maresins and nitrolipids could be used as prognostic markers of progression and resolution of various inflammatory conditions.

The possible role of anti-inflammatory lipoxins, resolvins, protectins, maresins and nitrolipids is further strengthened by the observation that P-selectin knockout mice had higher neutrophil infiltration and more severe glomerulonephritis than the wild-type mice raising the possibility that P-selectin has anti-inflammatory function and these mice also had lower renal lipoxin A_4 [58, 59]. These results indicate that P-selectin may serve as an anti-inflammatory molecule by enhancing lipoxin A_4 generation. In a similar fashion, it is possible that even in collagen vascular diseases (such as rheumatoid arthritis and lupus) steroids and other drugs used bring about their anti-inflammatory actions at least, in part, by augmenting the generation of lipoxin A_4. This argument is supported by the observation that inhibition of leukocyte infiltration by aspirin and dexamethasone is due to their action at the lipoxin A_4 receptor [60, 61]. These results also indicate that leukocyte function is regulated by lipoxins. Thus, leukocyte activation and, possibly, those of eosinophils is suppressed by lipoxins and a deficiency of lipoxins may lead to excess activation of leukocytes due to the absence of negative feed-back control exerted by lipoxins and possibly, other anti-inflammatory lipid mediators such as resolvins, protectins, maresins and nitrolipids [62, 63].

Leukocytes and Atherosclerosis

Leukocytosis is a marker of inflammation. High leukocyte count is associated with a greater cardiovascular risk and adverse outcome [64–68]. Infiltration of intima by leukocytes and macrophages is one of the earliest events to occur in atherosclerosis. Elevated LDL, hypertension, hyperglycemia, shear stress and other systemic factors initiate and accelerate atherosclerosis. Despite the fact that the entire vascular endothelium is exposed to these systemic factors, atherosclerotic lesions occur in a patchy manner and develop preferentially at bifurcations, branch points, and inner curvatures of arteries, suggesting that local factors play a critical role in the development of atherosclerosis. Hemodynamic forces to these regions may induce the expression of pro-inflammatory genes [69–71] that initiate and accelerate atherosclerosis at these points of shear stress. Experiments performed in normocholesterolemic C57BL/6 mice and rabbits showed that low-level activation of NF-κB and elevated expression of VCAM-1 and ICAM-1 occurs in endothelial cells in atherosclerosis-susceptible regions of the ascending aorta [72–74]. Gene expression profiling studies revealed that at the sites of atheroslcerosis-prone regions endothelial cells showed upregulation of pro-inflammatory genes IL-1, IL-6, MCP-1, as well as antioxidant genes glutathione peroxidase and glutathione-S-transferase 2, and endothelial cells themselves demonstrated changes in cell shape and proliferation [72, 75]. Endothelial cells in these atherosclerosis-prone regions of aorta showed increases in LDL and cholesterol transport and retention [76–78]. Furthermore, intimal accumulation of LDL and its oxidation products preceded monocyte recruitment into early atherosclerotic lesions, suggesting that lipid accumulation triggers inflammatory response characterized by upregulation of the expression of chemokines and adhesion molecules in the lesion-prone areas in the intima that contributes to leukocyte accumulation and atherosclerotic lesion formation [79–82]. These evidences imply that at atheroslcerosis-prone regions of the normal intima inflammatory response is triggered on introduction of atherosclerotic risk factors (such as hyperlipidemia, hypertension, shear stress, etc.) that leads to upregulation of several proinflammatory genes including various adhesion molecules and chemokines, which mediate accumulation of leukocytes and initiation and perpetuation of atherosclerosis. Jongstra-Bilen et al. [82] showed that significantly lower numbers of intimal CD68$^+$ leukocytes were found in inbred atheroslcerosis-resistant mice compared to wild type, and the predominant mechanism for the accumulation of these leukocytes was due to continued recruitment of bone marrow-derived blood monocytes, suggestive of low-grade inflammation. In contrast, intimal CD68$^+$ leukocytes were reduced in VCAM-1-deficient mice suggesting that in the intima of atherosclerosis-predisposed regions increased expression of proinflammatory genes occurs. Based on these results, it can be said that healthy endothelial cells despite exposure to hemodynamic factors are able to resist induction of excess expression of adhesion molecules, resist increases in LDL and cholesterol transport and retention, abrogate activation of NF-κB and the induction of expression of pro-inflammatory genes at bifurcations, branch points, and inner curvatures of arteries, regions that are prone for atherosclerosis due to enhanced infiltration by monocytes, $CD68^+$ leukocytes, and macrophages. This implies that under normal physiological conditions healthy endothelial cells produce factors/molecules that successfully counter pro-atherosclerotic events. Essential fatty acids and their beneficial metabolites such as lipoxins, resolvins, protectins, maresins and nitrolipids could be such factors that appear to be critical for maintaining endothelial cell structural integrity and function.

Uncoupling Protein-1, Essential Fatty Acids, and Atherosclerosis

The patchy manner in which atherosclerosis occurs suggests that arterial walls undergo regional disturbances of metabolism that include the uncoupling of respiration and oxidative phosphorylation, which may be characteristic of blood vessels being predisposed to the development of atheroslcerosis [83]. Oxidative stress is implicated in atherosclerosis. Mitochondrial electron transport accounts for most reactive oxygen species (ROS) production [84]. ROS production occurs during mitochondrial respiration that also produces energy in the form of ATP, resulting from ADP phosphorylation, as electrons at complex I and III react with molecular oxygen to form superoxide [85]. Uncoupling proteins (inner mitochondrial membrane anion transporters) allow protons to leak back into the mitochondrial matrix, thereby decreasing the potential energy available for ADP phosphorylation and ROS generation. Superoxide anion activates uncoupling proteins [86, 87] that, in turn, limit further superoxide generation by dissipating proton motive force and thus, decreases oxidative stress. This is supported by the observation that uncoupling decreases glucose-induced ROS formation and abrogates pathways associated with vascular damage in endothelial cells in vitro [88]. In contrast, UCP-2 in macrophages decreases ROS and atheroslcerosis [89]. Although, these results appear to be in conflict with the proposal that inefficient vascular metabolism is detrimental, it is known that uncoupling agents produce smooth muscle contraction and cause hypertension [90], and it was reported that respiratory uncoupling is increased in the aortae of experimental animals that are susceptible to atherosclerosis [83]. These results imply that the efficiency of vascular wall energy metabolism could be a determinant of atherosclerotic lesion development. It was found that [91] UCP-1 expression in aortic smooth muscle cells causes hypertension and increases atheroslcerosis without affecting cholesterol levels [84]. This increase in UCP-1 expression also enhanced superoxide anion production and decreased the availability of NO, suggesting that oxidative stress has been elevated. These results led to the proposal that inefficient metabolism in blood vessels causes atherosclerosis.

As discussed above, it is evident that atherosclerosis is a low-grade systemic inflammatory condition in which infiltrating leukocytes and macrophages play a critical role. One of the earliest signs of atherosclerosis is the development of abnormal mitochondria in smooth muscle cells [92], suggesting that mitochondrial dysfunction triggers the disease. The results of Bernal-Mizrachi et al. [91] discussed above lend support to this view. Arteries have marginal oxygenation [93] and hypoxia reduces the respiratory control ratio [94]. Uncoupled respiration precedes atherosclerosis at lesion-prone sites but not at the sites that are resistant to atherosclerosis [83]. Disease-free aortae have abundant concentrations of the essential fatty acid-linoleate, whereas fatty streaks (an early stage of atherosclerosis) are deficient in EFAs [91, 95, 96]. EFA deficiency promotes respiratory uncoupling [97, 98] and atherosclerosis [1, 99, 100]. Bernal-Mizrachi et al. [91] showed that oxidative stress increases ROS generation and decreases NO formation and/or availability to be associated with smooth muscle expression of UCP-1. These results [91] and other studies [69–78] emphasize the importance of local disturbances of metabolism in the arterial wall are responsible for atherosclerosis and vascular disease, suggesting that strategies designed to revert to normal EFA, ROS, leukocyte and endothelial cell function and their mitochondrial function and restoring the balance between pro- and anti-inflammatory cytokines could prevent or postpone vascular diseases including CHD. In this context, the many beneficial actions of EFAs and their products such as lipoxins, resolvins, protectins, maresins and nitrolipids especially, with regard to their ability to enhance NO generation, regulate UCP expression, suppress the production of pro-inflammatory cytokines and superoxide anion and maintain the integrity of endothelial cells is particularly interesting.

PUFAs of ω-3 and ω-6 Series, Trans-fats, Saturated Fats, Cholesterol and Their Role in Atherosclerosis

Atherosclerotic plaque rupture is known to be responsible for sudden coronary events. Felton et al. [101] reported that the concentrations of all fatty acids were increased at the edge of disrupted plaques compared with the center, but as a proportion of total fatty acids, ω-6 were lower. These results suggest that ω-6 fatty acids have a significant role in atherosclerosis and it is likely that some of the inconsistent results obtained in some studies with EPA and DHA could be attributed to inadequate provision or utilization of ω-6 fatty acids, especially DGLA and AA. It is possible that there is a close interaction between ω-3 and ω-6 fatty acids, which could influence one's susceptibility or resistance to atherosclerosis. In this context, it is interesting to note that EPA/DHA readily get incorporated into the atheromatous plaque, and patients treated with fish oil had more thick fibrous caps and no signs of inflammation compared with plaques in patients in the control and sunflower oil groups. Furthermore, the number of macrophages in plaques from patients receiving fish oil was lower than in the other two groups, suggesting that atherosclerotic plaques readily incorporate ω-3 PUFAs from fish-oil supplementation, inducing changes that can enhance stability of atherosclerotic plaques [102]. In contrast, trans-fatty acids may render atheromatous plaques unstable, partly by displacing ω-3 fatty acids, interfering with

ω-3 fatty acid metabolism and activating inflammatory responses and endothelial dysfunction [103, 104]. These results suggest that trans-fats not only enhance the risk of CAD [105, 106] but also induce plaque instability. In addition, trans-fats interfere with the activity of Δ^6 and Δ^5 desaturases [1, 100, 107] that are essential for the conversion of dietary LA and ALA to their respective long-chain metabolites. Thus, there is a close interaction between ω-3, ω-6 fatty acids and trans-fats.

In this context, the interaction between ω-3 and ω-6 fatty acids is particularly significant as already discussed in Chap. 4 which showed that DGLA increases the conversion of EPA to PGI_3, AA augmented the conversion of EPA to PGI_3, EPA enhances the tissue levels of DGLA leading to increase in the formation of PGE_1, events that prevent atherosclerosis. In contrast, trans-fats interfere with the formation of DGLA, AA, EPA, and DHA and thus, prevent the formation of anti-atherosclerotic molecules: PGE_1, PGI_2, PGI_3, lipoxins, resolvins, protectins, maresins and nitrolipids and at the same time may augment the formation and/or action of LTs, and TXs that promote atherosclerosis. Furthermore, even the beneficial action of statins (HMG-CoA reductase inhibitors) and glitazones (PPARs agonists) seem to be mediated by EFAs and their metabolites such as LXs, resolvins, and protectins [108–114], which are potent anti-inflammatory molecules [1, 115–117]. On the other hand, cholesterol and saturated fatty acids similar to trans-fats block the activities of both Δ^6 and Δ^5 desaturases and inhibit the conversion of dietary LA and ALA to their respective long-chain metabolites and render cell membrane more rigid [1]. Studies did suggest that increase in the consumption of trans-fats, cholesterol, and saturated fatty acids and increase in their plasma concentrations enhanced [118–120], whereas consumption of ω-3 fatty acids decreased the levels of inflammatory markers especially pro-inflammatory cytokines [121]. These results imply that trans-fats, cholesterol, and saturated fatty acids have pro-inflammatory actions, while ω-3 fatty acids possess anti-inflammatory actions. The ability of trans-fats, saturated fatty acids and cholesterol to enhance plasma levels of pro-inflammatory cytokines may, in part, be due to their ability to suppress the production of ω-3 EPA and DHA since both EPA and DHA have been shown to inhibit T cell proliferation and their ability to produce IL-6 and TNF-α, which are pro-inflammatory cytokines [122, 123]. These evidences suggest that ω-3 and ω-6 fatty acids, trans-fats, saturated fatty acids and cholesterol modulate inflammation and thus, influence the pathobiology of atherosclerosis, CAD and stroke.

Atheroprotective Actions of ω-3 and ω-6 Fatty Acids

It is evident from the preceding discussion that both ω-3 and ω-6 PUFAs interact with each other to prevent atherosclerosis, CAD, CVD, and stroke, though ω-3 EPA and DHA seem to be having a more dominant role compared to ω-6 in this beneficial action. PUFAs display a multitude of actions (such as ability to lower plasma triglycerides, cholesterol and apolipoprotein B and alter hemostatic system; see

Table 10.1 Summary of effects of PUFAs on nuclear receptors involved in the regulation of lipogenesis

Nuclear receptor	Effects on gene regulation	Expected changes		
		TG	HDL	LDL
PPAR-α	↑	↓↓	↑	↓
LXR	↓	↓↓	↓	↓
FXR	↑	↓↓	↑	↑
HNF-4α	↓	↓↓	↓	↔

FXR Farnesol X receptor, *HDL* High-density lipoprotein, *HNF-4α* Hepatocyte nuclear factor-4α, *LDL* Low-density lipoprotein, *LXR* Liver X receptor, *PPAR-α* Peroxisome proliferator-activated receptor, ↑ Increase, ↓ Decrease, ↔ Neutral effect

Table 10.1 also for the actions of PUFAs on lipid metabolism) to prevent atherosclerosis which have been outlined in Chap. 4 and elsewhere [1, 107]. Here the actions of PUFAs only on endothelial cells and inflammation are highlighted that are relevant to atherosclerosis.

Effects on Endothelial Function

PUFAs, especially GLA, DGLA, AA, EPA, and DHA are necessary for endothelial health and normal function. Endothelial cells need to produce adequate amounts of NO, PGI_2, PGE_1, LXs, resolvins, protectins, maresins, and nitrolipids to prevent adhesion of platelets, leukocytes and macrophages to their surface that are known to produce ROS and pro-inflammatory cytokines and induce endothelial dysfunction. For endothelial cells to prevent platelet, leukocyte and macrophage adhesion and infiltration, they not only should be capable of producing adequate amounts of NO, PGI_2, PGI_3, LXs, resolvins, protectins, maresins and nitrolipids but also suppress the expression of adhesion molecules on their surface and prevent the synthesis and release of IL-6, TNF-α, and MIF (macrophage migration inhibitory factor). EPA and DHA reduce adhesion and migration of monocytes and inhibit leukocyte-endothelial cell interactions that involve increased endothelial expression of leukocyte adhesion molecules or endothelial activation [124–128]. Consumption of DHA/EPA was found to reduce endothelial expression of vascular cell adhesion molecile-1 (VCAM-1), E-selectin, intercellular adhesion molecule-1 (ICAM-1), IL-6 and IL-8 in response to IL-1, IL-4, TNF-α, and bacterial endotoxin [124–129]. Johansen et al. [130] reported that ω-3 fatty acids decreased both tissue plasminogen activator antigen and soluble thrombomodulin, whereas in the placebo group soluble E-selectin and soluble VCAM-1 increased. These studies [124–130] suggest that ω-3 fatty acids decrease hemostatic markers of inflammation, show anti-inflammatory properties and inhibit endothelial activation.

Smooth muscle cell proliferation plays a significant role in the pathogenesis of atheroslcerosis and restenosis. Cornwell et al. [131, 132] showed that both ω-3 and ω-6 fatty acids (especially AA, EPA, and DHA) inhibited smooth muscle cell proliferation and that this is related to the amount of lipid peroxides formed in the cells. Several other investigators have confirmed these findings [133, 134]. These studies imply that intracellular deficiency of PUFAs could lead to the initiation and progression of atherosclerosis. Pakala et al. [135] showed that smooth muscle cell proliferation induced by serotonin at the sites of vascular injury can be blocked by EPA and DHA, whereas Nakayama et al. [136] demonstrated that EPA inhibited TGF-β1 mRNA and cdk2 activity in vascular smooth muscle cells from spontaneously hypertensive rats. Others have confirmed these results [137, 138] suggesting that EPA and DHA, and possibly other PUFAs prevent endothelial activation, smooth muscle cell proliferation, and thus, prevent atheroslcerosis [139, 140].

PUFAs Inhibit Angiotensin-converting Enzyme (ACE) Activity and Augment Endothelial Nitric Oxide Generation

PUFAs inhibited leukocyte ACE activity [141, 142] suggesting that they could function as endogenous regulators of ACE activity, and thus regulate the formation of (angiotensin-II) Ang-II. PUFAs enhance endothelial NO generation [107, 143]. Hence, when tissue/cell concentrations of PUFAs are low the formation of Ang-II will be high whereas that of eNO will be low. Plasma concentrations of PUFA and eNO are low in hypertension, diabetes mellitus, lupus, atherosclerosis, insulin resistance, and obesity [1, 144, 145]. Furthermore, a 25-nucelotide ACE deletion polymorphism increases ACE activity and such individuals showed a higher risk of developing stroke, obesity, emphysema, bipolar affective disorders, and cancers [146, 147]. This suggests that an altered ACE activity and EFA metabolism play a role in many diseases. Furthermore, angiotensin II has pro-inflammatory actions [141] and thus, PUFAs by suppressing the formation of angiotensin II could function as anti-inflammatory molecules.

In addition, transgenic rats overexpressing both human renin and angiotensinogen genes (dTGR) develop hypertension, inflammation, and renal failure and showed decreased formation epoxy-eicosatrienoic acids (5,6-, 8,9-, 11,12- and 14,15-EETs) and hydroxyeicosa-tetraenoic acids (19- and 20-HETEs) from AA. These EETs and HETEs inhibited IL-6 and TNF-α-induced activation of NF-κB and prevented vascular inflammation [148] suggesting that AA and other PUFAs not only regulate ACE activity and Ang-II levels but also possess anti-inflammatory properties.

EPA and AA stimulate eNO synthesis [1, 143]. NO has potent anti-atherosclerotic and anti-inflammatory actions. Aspirin enhances the formation of eNO through the generation of epi-lipoxins that may explain its anti-inflammatory action [149]. Epi-lipoxins that have potent anti-inflammatory actions enhance the generation of NO that, in turn, prevents the interaction between leukocytes and the vascular endothelium. NO stimulates the formation of PGI_2 from AA [149] and lipoxins are derived from AA, EPA, and DHA. Aspirin inhibits TXA_2 formation, a potent platelet aggregator and vasoconstrictor, and enhances PGI_2 formation, a platelet anti-aggregator and vasodilator, and thus brings about its anti-atherosclerotic actions.

PUFAs Suppress the Production of Pro-inflammatory Cytokines

ALA, DGLA, EPA, and DHA, LXs (lipoxins), resolvins and possibly, protectins, maresins and nitrolipids suppress pro-inflammatory IL-1, IL-2, IL-6, macrophage migration inhibitory factor (MIF), HMGB1 (high mobility group box 1) and TNF-α production by T cells and other cells [1, 122, 123, 150, 151], and thus could function as endogenous anti-inflammatory molecules. PGE_2, $PGF_{2\alpha}$, TXA_2 and LTs derived from AA also modulate IL-6 and TNF-α production. These results imply levels of IL-6 and TNF-α at the sites of inflammation and injury may depend on the local levels of various PUFAs and eicosanoids formed from them. In particular, the suppressive action of DHA on IL-1β and TNF-α production by stimulated human retinal vascular endothelial cells [152] is interesting since this suggests that it (DHA) and possibly other PUFAs may be important to prevent atherosclerosis, macular degeneration, and diabetic retinopathy. EPA and DHA suppress the production of pro-inflammatory cytokines and bring about their anti-inflammatory actions by increasing PPAR-γ mRNA and protein activity [153].

Effects on HMG-CoA Reductase Enzyme

The two sterol regulatory element-binding proteins (SREBPs): SREBP-1 and SREBP-2, each ~1,150 amino acids in length, control the transcription of the genes for the low-density lipoprotein (LDL) receptor and 3-hdroxy-3-methylglutaryl coenzyme A (HMG-CoA) synthase. The proteolytic processing of both SREBPs is blocked by sterol overloading and enhanced when sterols are depleted by statins, the HMG-CoA synthesis inhibitors [154]. Cholesterol depletion that occurs due to the use of statins leads to proteolytic activation of transcription factors of the SREBPs and also induces PPAR-γ expression [155], suggesting that PPAR-γ expression is controlled by SREBPs. Similar to statins, AA, EPA, and DHA are useful in the treatment of hyperlipidemias, have anti-proliferative action on tumor cells both *in vitro* and *in vivo*, bind to DNA and regulate the expression of genes and oncogenes. More importantly, PUFAs are also potent inhibitors of the HMG-CoA reductase enzyme [108, 156–158] and interact with SREBPs. Statins enhance plasma AA concentrations and decrease the ratio of EPA to AA significantly. The beneficial actions of PUFAs in atherosclerosis can be attributed to the formation of anti-inflammatory compounds such as lipoxins and resolvins.

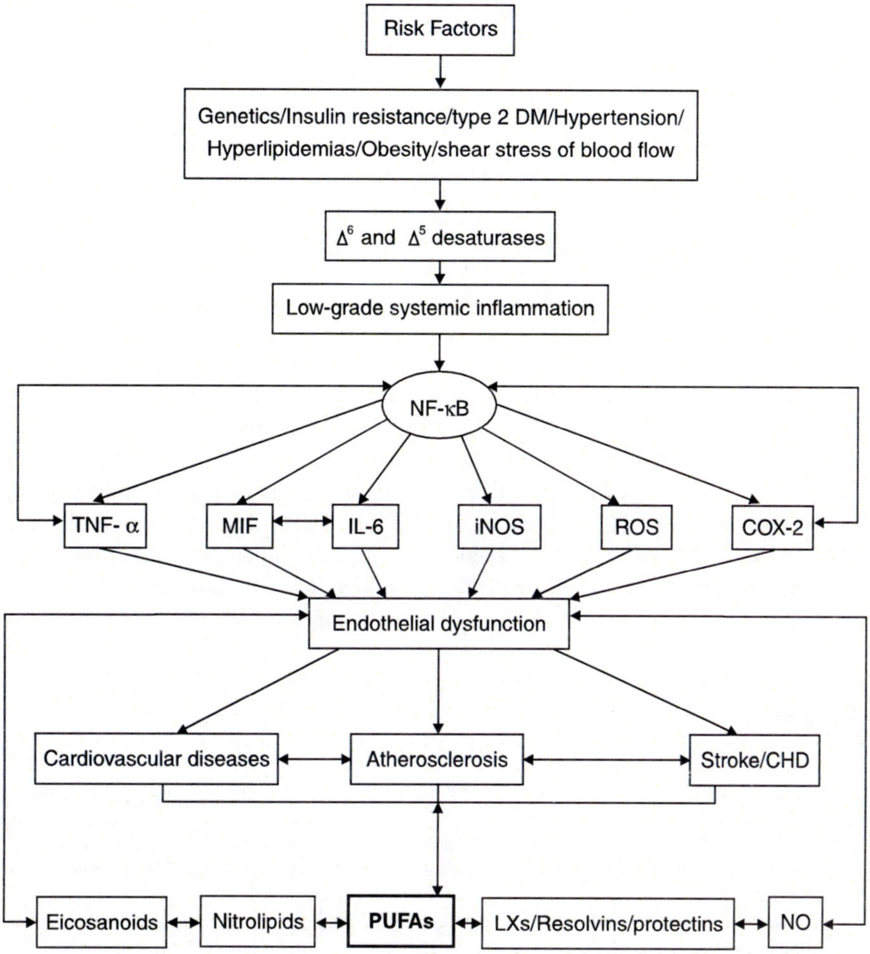

Fig. 10.1 Scheme showing the relationship among various mediators of endothelial dysfunction and CHD/stroke and the role of PUFAs and their metabolites in these processes

PUFAs have inhibitory effects on SREBP-1a and SREBP-1c [159]. In CaCo-2 cells, PUFAs decreased gene and protein expression of SREBP-1 and FAS mRNA by interfering with LXR activity [160], and in rats PUFAs enhanced cholesterol losses via bile acid synthesis [161]. In the intestine, dietary PUFAs suppress SREBP-1c mRNA without altering expression of its target genes, fatty acid synthase, acetyl-CoA carboxylase, or ATP citrate lyase and decreased intestinal fatty acid synthesis by a posttranscriptional mechanism independent of the SREBP pathway [162]. Feeding mice on fish oil diet for 2 weeks decreased serum cholesterol and triacylglycerol levels; by 50% and 60% respectively, hepatic FPP (farnesyl diphosphate synthase, a SREBP target enzyme that is subject to negative-feedback regulation by sterols in co-ordination with HMG-CoA reductase) synthase and HMG-CoA reductase mRNAs were decreased by 70% and 40% respectively. PUFAs down regulate hepatic cholesterol synthesis by impairing the SREBP pathway [163]. PUFAs reduce SREBP-mediated gene transcription by increasing intracellular cholesterol content through the hydrolysis of cellular sphingomyelin, and the lipid second messenger ceramide, a product of sphingomyelin hydrolysis, decreased SRE-mediated gene transcription of SREBP-1 and SREBP-2 [164].

Based on the preceding discussion, it is clear that atherosclerosis is a low-grade inflammatory condition and PUFAs (especially ω-3 EPA and DHA) are useful in its prevention and management. PUFAs also inhibit ACE and HMG-CoA reductase activities and behave as endogenous ACE inhibitors. Statins similar to PUFAs and their products such as lipoxins, resolvins, protectins, maresins and nitrolipids suppress the production of pro-inflammatory cytokines, modulate SREBP pathway and thus, inhibit atheroslcerosis both by lowering plasma triglycerides and cholesterol levels (see Table 10.1), and modulating inflammatory events.

These evidences suggest that atherosclerosis can be prevented/arrested if endothelial cells are able to produce adequate amounts of various PUFAs such that they in turn lead to the formation of beneficial PGE_1, PGI_2, PGI_3, LXs, resolvins, protectins, maresins and nitrolipids that are capable of suppressing inflammation, expression of various adhesion molecules on the surface of endothelial cells, and prevent leukocyte, monocyte and macrophage infiltration of endothelial cells (see Fig. 10.1).

References

[1] Das UN (2006) Essential fatty acids: biochemistry, physiology and pathology. Biotechnol J 1:420–439
[2] Das UN (1995) Essential fatty acid metabolism in patients with essential hypertension, diabetes mellitus and coronary heart disease. Prostaglandins Leukot Essent Fatty Acids 52:387–391
[3] Kumar KV, Das UN (1994) Lipid peroxides and essential fatty acids in patients with coronary heart disease. J Nutr Med 4:33–37
[4] Das UN (2001) Nutritional factors in the pathobiology of human essential hypertension. Nutrition 17:337–346
[5] Das UN (2001) Can perinatal supplementation of long chain polyunsaturated fatty acids prevent hypertension in adult life? Hypertension 38:e6–e8
[6] Das UN (2003) Can perinatal supplementation of long-chain polyunsaturated fatty acids prevent diabetes mellitus? Eur J Clin Nutr 57:218–226
[7] Das UN (2005) A defect in the activity of Δ^6 and Δ^5 desaturases may be a factor predisposing to the development of insulin resistance syndrome. Prostaglandins Leukot Essent Fatty Acids 72:343–350
[8] Wang L, Folsom AR, Eckfeldt JH (2003) Plasma fatty acid composition and incidence of coronary heart disease in middle aged adults: the Atherosclerosis Risk in Communities (ARIC) study. Nutr Metab Cardiovasc Dis 13:256–266
[9] Zheng ZJ, Folsom AR, Ma J, Arnett DK, McGovern PG, Eckfeldt JH (1999) Plasma fatty acid composition and 6-year incidence of hypertension in middle-aged adults: the Atherosclerosis Risk in Communities (ARIC) study. Am J Epidemiol 150:492–500
[10] Suresh Y, Das UN (2006) Differential effect of saturated, monounsaturated, and polyunsaturated fatty acids on alloxan-induced diabetes mellitus. Prostaglandins Leukot Essent Fatty Acids 74:199–213

[11] Suresh Y, Das UN (2003) Long-chain polyunsaturated fatty acids and chemically-induced diabetes mellitus: effect of ω-6 fatty acids. Nutrition 19:93–114
[12] Suresh Y, Das UN (2003) Long-chain polyunsaturated fatty acids and chemically-induced diabetes mellitus: effect of ω-3 fatty acids. Nutrition 19:213–228
[13] Mozaffarian D, Ascherio A, Hu FB, Stampfer MJ, Willett WC, Siscovick DS, Rimm EB (2005) Interplay between different polyunsaturated fatty acids and risk of coronary heart disease in men. Circulation 111:157–164
[14] Nascimento-Silva V, Arruda MA, Barja-Fidalgo C, Fierro IM (2007) Aspirin-triggered lipoxin A4 blocks reactive oxygen species generation in endothelial cells: a novel antioxidative mechanism. Thromb Haemost 97:88–98
[15] Merched AJ, Ko K, Gotlinger KH, Serhan CN, Chan L (2008) Atherosclerosis: evidence for impairment of resolution of vascular inflammation governed by specific lipid mediators. FASEB J 22:3595–3606
[16] Das UN (2006) Clinical laboratory tools to diagnose inflammation. Adv Clin Chem 41:189–229
[17] Brennan ML, Penn MS, Lente FV, Nambi V, Shishehbor MH, Aviles RJ, Goormastic M, Pepoy ML, McErlean ES, Topol EJ, Nissen SE, Hazen SL (2003) Prognostic value of myeloperoxidase in patients with chest pain. N Engl J Med 349:1595–1604
[18] Ballantyne CM, Hoogeveen RC, Bang H, Coresh J, Folsom AR, Heiss G, Sharrett AR (2004) Lipoprotein-associated phospholipase A_2, high-sensitivity C-reactive protein, and risk for incident coronary heart disease in middle-aged men and women in the Atherosclerosis Risk in Communities (ARIC) study. Circulation 109:837–842
[19] Oei HHS, van der Meer IM, Hofman A, Koudstaal PJ, Stijnen T, Breteler MMB, Witteman JCM (2005) Lipoprotein-associated phospholipase A_2 activity is associated with risk of coronary heart disease and ischemic stroke. The Rotterdam study. Circulation 111:570–575
[20] Häkkinen T, Luoma JS, Hiltunen MO, Macphee CH, Milliner KJ, Patel L, Rice SQ, Tew DG, Karkola K, Ylä-Herttuala S (1999) Lipoprotein-associated phospholipase A(2), platelet-activating factor acetylhydrolase, is expressed by macrophages in human and rabbit atherosclerotic lesions. Arterioscler Thromb Vasc Biol 19:2909–2917
[21] Lp-PLA(2) Studies Collaboration, Thompson A, Gao P, Orfei L, Watson S, Di Angelantonio E, Kaptoge S, Ballantyne C, Cannon CP, Criqui M, Cushman M, Hofman A, Packard C, Thompson SG, Collins R, Danesh J (2010) Lipoprotein-associated phospholipase A(2) and risk of coronary disease, stroke, and mortality: collaborative analysis of 32 prospective studies. Lancet 375:1536–1544
[22] Anuurad E, Ozturk Z, Enkhmaa B, Pearson TA, Berglund L (2010) Association of lipoprotein-associated phospholipase A2 with coronary artery disease in African-Americans and Caucasians. J Clin Endocrinol Metab 95:2376–2383
[23] Hatoum IJ, Hu FB, Nelson JJ, Rimm EB (2010) Lipoprotein-associated phospholipase A2 activity and incident coronary heart disease among men and women with type 2 diabetes. Diabetes 59:1239–1243
[24] Brilakis ES, Khera A, Saeed B, Banerjee S, McGuire DK, Murphy SA, de Lemos JA (2008) Association of lipoprotein-associated phospholipase A2 mass and activity with coronary and aortic atherosclerosis: findings from the Dallas heart study. Clin Chem 54:1975–1981
[25] Wilcox JN, Subramanian RR, Sundell CL, Tracey WR, Pollock JS, Harrison DG, Marsden PA (1997) Expression of multiple isoforms of nitric oxide synthase in normal and atherosclerotic vessels. Arterioscler Thromb Vasc Biol 17:2479–2488
[26] Garlichs CD, Beyer J, Zhang H, Schmeisser A, Plötze K, Mügge A, Schellong S, Daniel WG (2000) Decreased plasma concentrations of L-hydroxy-arginine as a marker of reduced NO formation in patients with combined cardiovascular risk factors. J Lab Clin Med 135:419–425
[27] Napoli C, Ignarro LJ (2001) Nitric oxide and atherosclerosis. Nitric Oxide 5:88–97
[28] Loffredo L, Pignatelli P, Cangemi R, Andreozzi P, Panico MA, Meloni V, Violi F (2006) Imbalance between nitric oxide generation and oxidative stress in patients with peripheral arterial disease: effect of an antioxidant treatment. J Vasc Surg 44:525–530

[29] Aygul R, Kotan D, Demirbas F, Ulvi H, Deniz O (2006) Plasma oxidants and antioxidants in acute ischaemic stroke. J Int Med Res 34:413–418
[30] Collino M, Aragno M, Mastrocola R, Gallicchio M, Rosa AC, Dianzani C, Danni O, Thiemermann C, Fantozzi R (2006) Modulation of the oxidative stress and inflammatory response by PPAR-gamma agonists in the hippocampus of rats exposed to cerebral ischemia/reperfusion. Eur J Pharmacol 530:70–80
[31] Kumar KV, Das UN (1994) Lipid peroxides and essential fatty acids in patients with coronary heart disease. J Nutr Med 4:33–37
[32] Das UN (2000) Free radicals, cytokines and nitric oxide in cardiac failure and myocardial infarction. Mol Cell Biochem 215:145–152
[33] Zoccali C, Bode-Böger S, Mallamaci F, Benedetto F, Tripepi G, Malatino L, Cataliotti A, Bellanuova I, Fermo I, Frölich J, Böger R (2001) Plasma concentration of asymmetrical dimethylarginine and mortality in patients with end-stage renal disease: a prospective study. Lancet 358:2113–2117
[34] Maas R, Quitzau K, Schwedhelm E, Spieker L, Rafflenbeul W, Steenpass A, Lüscher TF, Böger RH (2007) Asymmetrical dimethylarginine (ADMA) and coronary endothelial function in patients with coronary artery disease and mild hypercholesterolemia. Atherosclerosis 191:211–219
[35] Juonala M, Viikari JS, Alfthan G, Marniemi J, Kähönen M, Taittonen L, Laitinen T, Raitakari OT (2007) Brachial artery flow-mediated dilation and asymmetrical dimethylarginine in the cardiovascular risk in young Finns study. Circulation 116:1367–1373
[36] Antoniades C, Shirodaria C, Leeson P, Antonopoulos A, Warrick N, Van-Assche T, Cunnington C, Tousoulis D, Pillai R, Ratnatunga C, Stefanadis C, Channon KM (2009) Association of plasma asymmetrical dimethylarginine (ADMA) with elevated vascular superoxide production and endothelial nitric oxide synthase uncoupling: implications for endothelial function in human atherosclerosis. Eur Heart J 30:1142–1150
[37] Maas R, Xanthakis V, Polak JF, Schwedhelm E, Sullivan LM, Benndorf R, Schulze F, Vasan RS, Wolf PA, Böger RH, Seshadri S (2009) Association of the endogenous nitric oxide synthase inhibitor ADMA with carotid artery intimal media thickness in the Framingham Heart Study offspring cohort. Stroke 40:2715–2719
[38] Ungvari Z, Pacher P, Rischàk K, Szollàr L, Koller A (1999) Dysfunction of nitric oxide mediation in isolated rat arterioles with methionine diet-induced hyperhomocysteinemia. Arterioscler Thromb Vasc Biol 19:1899–1904
[39] Das UN (2003) Folic acid says NO to vascular diseases. Nutrition 19:686–692
[40] Sawle P, Foresti R, Green CJ, Motterlini R (2001) Homocysteine attenuates endothelial haem oxygenase-1 induction by nitric oxide (NO) and hypoxia. FEBS Lett 508:403–406
[41] Abbasi SH, Boroumand MA (2010) Expanded network of inflammatory markers of atherogenesis: where are we now? Open Cardiovasc Med J 4:38–44
[42] Zhu X, Lee JY, Timmins JM, Brown JM, Boudyguina E, Mulya A, Gebre AK, Willingham MC, Hiltbold EM, Mishra N, Maeda N, Parks JS (2008) Increased cellular free cholesterol in macrophage-specific Abca1 knock-out mice enhances pro-inflammatory response of macrophages. J Biol Chem 283:22930–22941
[43] Li Y, Schwabe RF, DeVries-Seimon T, Yao PM, Gerbod-Giannone MC, Tall AR, Davis RJ, Flavell R, Brenner DA, Tabas I (2005) Free cholesterol-loaded macrophages are an abundant source of tumor necrosis factor-alpha and interleukin-6: model of NF-kappaB- and map kinase-dependent inflammation in advanced atherosclerosis. J Biol Chem 280:21763–21772
[44] Sanguigni V, Pignatelli P, Caccese D, Pulcinelli FM, Lenti L, Magnaterra R, Martini F, Lauro R, Violi F (2002) Increased superoxide anion production by platelets in hypercholesterolemic patients. Thromb Haemost 87:796–801
[45] Pignatelli P, Sanguigni V, Lenti L, Loffredo L, Carnevale R, Sorge R, Violi F (2007) Oxidative stress-mediated platelet CD40 ligand upregulation in patients with hypercholesterolemia: effect of atorvastatin. J Thromb Haemost 5:1170–1178
[46] Gupta S, Pablo AM, Jiang X, Wang N, Tall AR, Schindler C (1997) IFN-γ potentiates atherosclerosis in ApoE knock-out mice. J Clin Invest 99:2752–2761

[47] Elhage R, Jawien J, Rudling M et al (2003) Reduced atherosclerosis in interleukin-18 deficient apolipoprotein E-knockout mice. Cardiovasc Res 59:234–240
[48] Tenger C, Sundborger A, Jawien J, Zhou X (2005) IL-18 accelerates atherosclerosis accompanied by elevation of IFN-γ and CXCL16 expression independently of T cells. Arterioscler Thromb Vasc Biol 25:791–796
[49] Baker PRS, Lin Y, Schopfer FJ, Woodcock SR, Groeger AL, Batthyany C, Swooney S, Long MH, Iles KE, Baker LMS, Branchaud BP, Chen Y, Freeman BA (2005) Fatty acid transduction of nitric oxide signaling: multiple nitrated unsaturated fatty acid derivatives exist in human blood and urine and serve as endogenous peroxisome proliferator-activated receptor ligands. J Biol Chem 280:42464–42475
[50] Coles B, Bloodsworth A, Clark SR, Lewis MJ, Cross AR, Freeman BA, O'Donnell VB (2002) Nitrolinoleate inhibits superoxide generation, degranulation, and integrin expression by human neutrophils. Circ Res 91:375–381
[51] Lima ES, Bonim MG, Augusto O, Barbeiro HV, Souza HP, Abdalla DSP (2005) Nitrated lipids decompose to nitric oxide and lipid radicals and cause vasorelaxation. Free Radic Biol Med 39:532–539
[52] Das UN (2009) Cross talk among leukocytes, platelets, and endothelial cells and its relevance to atherosclerosis and coronary heart disease. Curr Nutr Food Sci 5:75–93
[53] Thomas E, Leroux JL, Blotman F, Chavis C (1995) Conversion of endogenous arachidonic acid to 5,15-diHETE and lipoxins by polymorphonuclear cells from patients with rheumatoid arthritis. Inflamm Res 44:121–124
[54] Sodin-Semrl S, Spagnolo A, Barbaro B, Varga J, Fiore S (2004) Lipoxin A4 counteracts synergistic activation of human fibroblast-like synoviocytes. Int J Immunopathol Pharmacol 17:15–25
[55] Sodin-Semrl S, Spagnolo A, Mikus R, Barbaro B, Varga J, Fiore S (2004) Opposing regulation of interleukin-8 and NF-kappaB responses by lipoxin A4 and serum amyloid A via the common lipoxin A receptor. Int J Immunopathol Pharmacol 17:145–156
[56] Hashimoto A, Hayashi I, Murakami Y, Sato Y, Kitasato H, Matsushita R, Iizuka N, Urabe K, Itoman M, Hirohata S, Endo H (2007) Antiinflammatory mediator lipoxin A4 and its receptor in synovitis of patients with rheumatoid arthritis. J Rheumatol 34:2144–2153
[57] Chan MM, Moore AR (2010) Resolution of inflammation in murine autoimmune arthritis is disrupted by cyclooxygenase-2 inhibition and restored by Prostaglandin E(2)-mediated lipoxin A(4) production. J Immunol 184:6418–6426
[58] Mayadas TN, Mendrick DL, Brady HR et al (1996) Acute passive antiglomerular basement membrane nephritis in P-selectin- deficient mice. Kidney Int 49:1342–1349
[59] Tam FWK (2002) Role of selectins in glomerulonephritis. Clin Exp Immunol 129:1–3
[60] Perretti M, Chiang N, La M, Fierro IM, Marullo S, Getting SJ, Solito E, Serhan CN (2002) Endogenous lipid- and peptide-derived anti-inflammatory pathways generated with glucocorticoid and aspirin treatment activate the lipoxin A4 receptor. Nat Med 8:1296–1302
[61] Hashimoto A, Murakami Y, Kitasato H, Hayashi I, Endo H (2007) Glucocorticoids co-interact with lipoxin A4 via lipoxin A4 receptor (ALX) up-regulation. Biomed Pharmacother 61:81–85
[62] Planaguma A, Levy BD (2008) Uncontrolled airway inflammation in lung disease represents a defect in counter-regulatory signaling. Future Lipidol 3:697–704
[63] Starosta V, Pazdrak K, Boldogh I, Svider T, Kurosky A (2008) Lipoxin A4 counterregulates GM-CSF signaling in eosinophilic granulocytes. J Immunol 181:8688–8699
[64] Danesh J, Collins R, Appleby P, Peto R (1998) Association of fibrinogen, C-reactive protein, albumin, or leukocyte count with coronary heart disease: meta-analyses of prospective studies. JAMA 279:1477–1482
[65] Modan B, Schor S, Shani M (1975) Acute myocardial infarction. Prognostic value of white blood cell count and blood glucose level. JAMA 233:266–267
[66] Bovill EG, Bild DE, Heiss G, Kuller LH, Lee MH, Rock R, Wahl PW (1996) White blood cell counts in persons aged 65 years or more from the cardiovascular health study. Correlations with baseline clinical and demographic characteristics. Am J Epidemiol 143:1107–1115

[67] Brown DW, Giles WH, Croft JB (2001) White blood cell count: an independent predictor of coronary heart disease mortality among a national cohort. J Clin Epidemiol 54:316–322
[68] Kruk M, Przyłuski J, Kalińczuk L, Pregowski J, Kadziela J, Kaczmarska E, Petryka J, Kepka C, Klopotowski M, Chmielak Z, Ciszewski A, Demkow M, Karcz M, Witkowski A, Ruzyłło W (2009) Hemoglobin, leukocytosis and clinical outcomes of ST-elevation myocardial infarction treated with primary angioplasty: ANIN myocardial infarction registry. Circ J 73:323–329
[69] Brooks AR, Lelkes PI, Rubanyi GM (2002) Gene expression profiling of human aortic endothelial cells exposed to disturbed flow and steady laminar flow. Physiol Genomics 9:27–41
[70] Garcia-Cardena G, Comander J, Anderson KR, Blackman BR, Gimbrone MA Jr (2001) Biomechanical activation of vascular endothelium as a determinant of its functional phenotype. Proc Natl Acad Sci U S A 98:4478–4485
[71] Gimbrone MA Jr, Nagel T, Topper JN (1997) Biomechanical activation: an emerging paradigm in endothelial adhesion biology. J Clin Invest 100:S61–S65
[72] Hajra L, Evans AI, Chen M, Hyduk SJ, Collins T, Cybulsky MI (2000) The NF-kappaB signal transduction pathway in aortic endothelial cells is primed for activation in regions predisposed to atherosclerotic lesion formation. Proc Natl Acad Sci U S A 97:9052–9057
[73] Nakashima Y, Raines EW, Plump AS, Breslow JL, Ross R (1998) Upregulation of CAM-1 and ICAM-1 at atheroslcerosis- prone sites on the endothelium in the ApoE-deficient mouse. Arterioscler Thromb Vasc Biol 18:842–851
[74] Iiyama K, Hajra L, Iiyama M, Li H, DiChiara M, Medoff BD, Cybulsky MI (1999) Patterns of vascular cell adhesion molecule-1 and intercellular adhesion molecule-1 expression in rabbit and mouse atherosclerotic lesions and at sites predisposed to lesion formation. Circ Res 85:199–207
[75] Passerini AG, Polacek DC, Shi C, Francesco NM, Manduchi E, Grant GR, Pritchard WF, Powell S, Chang GY, Stoeckert CJ Jr, Davies PF (2004) Coexisting proinflammatory and antioxidative endothelial transcription profiles in a disturbed flow region of the adult porcine aorta. Proc Natl Acad Sci U S A 101:2482–2487
[76] Schwenke DC, Carew TE (1989) Initiation of atherosclerotic lesions in cholesterol-fed rabbits. I. Focal increases in arterial LDL concentration precede development of fatty streak lesions. Arteriosclerosis 9:895–907
[77] Schwenke DC, Carew TE (1989) Initiation of atherosclerotic lesions in cholesterol-fed rabbits. II. Selective retention of LDL vs. selective increases in LDL permeability in susceptible sites of arteries. Arteriosclerosis 9:908–918
[78] Schwenke DC (1995) Selective increase in cholesterol at atherosclerosis-susceptible aortic sites after short- term cholesterol feeding. Arterioscler Thromb Vasc Biol 15:1928–1937
[79] Napoli C, D'Armiento FP, Mancini FP, Postiglione A, Witzturn JL, Palumbo G, Palinski W (1997) Fatty streak formation occurs in human fetal aortas and is greatly enhanced by maternal hypercholesterolemia. Intimal accumulation of low density lipoprotein and its oxidation precede monocytic recruitment into early atherosclerotic lesions. J Clin Invest 100:2680–2690
[80] Shi W, Haberland ME, Jien ML, Shih DM, Lusis AJ (2000) Endothelial responses to oxidized lipoproteins determine genetic susceptibility to atherosclerosis in mice. Circulation 102:75–81
[81] Cybulsky MI, Won D, Haidari M (2004) Leukocyte recruitment to Atherosclerotic lesion. Can J Cardiol 20(Suppl B):24B–28B
[82] Jongstra-Bilen J, Haidari M, Zhu SN, Chen M, Guha D, Cybulsky MI (2006) Low-grade chronic inflammation in regions of the normal mouse arterial intima predisposed to atherosclerosis. J Exp Med 203:2073–2083
[83] Santerre RF, Nicolosi RJ, Smith SC (1974) Respiratory control in preatherosclerotic susceptible and resistant pigeon aortas. Exp Mol Pathol 20:397–406
[84] Droge W (2002) Free radicals in the physiological control of cell function. Physiol Rev 82:47–95

[85] Nohl H (1994) Generation of superoxide radicals as byproducts of cellular respiration. Ann Biol Clin (Paris) 52:199–204
[86] Echtay KS, Roussel D, St-Pierre J, Jekabsons MB, Cadenas S, Stuart JA, Harper JA, Roebuck SJ, Morrison A, Pickering S, Clapham JC, Brand MD (2002) Superoxide activates mitochondrial uncoupling proteins. Nature 415:96–99
[87] Murphy MP, Echtay KS, Blaikie FH, Asin-Cayuela J, Cocheme HM, Green K, Buckingham JA, Taylor ER, Hurrell F, Hughes G, Miwa S, Cooper CE, Svistunenko DA, Smith RA, Brand MD (2003) Superoxide activates uncoupling proteins by generating carbon-centered radicals and initiating lipid peroxidation: studies using a mitochondria- targeted spin trap derived from alpha-phenyl-N-tert- butylnitrone. J Biol Chem 278:48534–48545
[88] Nishikawa T, Edelstein D, Du XL, Yamagishi S, Matsumura T, Kaneda Y, Yorek MA, Beebe D, Oates PJ, Hammes HP, Giardino I, Brownlee M (2000) Normalizing mitochondrial superoxide production blocks three pathways of hyperglycaemic damage. Nature 404:787–790
[89] Blanc J, Alves-Guerra MC, Esposito B, Rousset S, Gourdy P, Rixquier D, Tedgui A, Miroux B, Mallat Z (2003) Protective role of uncoupling protein 2 in atherosclerosis. Circulation 107:388–390
[90] Petterson G (1985) Effect of dinitrophenol and anoxia on isometric tension in rabbit colon smooth muscle. Acta Pharmacol Toxicol (Copenh) 57:184–189
[91] Bernal-Mizrachi C, Gates AC, Weng S, Imamura T, Knutsen RH, DeSantis P, Coleman T, Townsend RR, Muglia LJ, Semenkovich CF (2006) Vascular respiratory uncoupling increases blood pressure and atherosclerosis. Nature 435:502–506
[92] Watts H (1963) In: Jones RJ (ed) Evolution of the atherosclerotic plaque. University of Chicago, Chicago, IL, p 117
[93] Levin M, Leppanen O, Evaldsson M, Wiklund O, Bondjers G, Bjornheden T (2003) Mapping of ATP, glucose, glycogen, and lactate concentrations within the arterial wall. Arterioscler Thromb Vasc Biol 25:1801–1807
[94] Jennings RB, Kaltenbach JP, Sommens HM (1967) Mitochondrial metabolism in ischemic injury. Arch Pathol 84:15–19
[95] Smith EB (1962) The effects of age and of early atheromata on the intimal lipids in men. Biochem J 84:49p
[96] Smith EB (1962) Lipids carried by S1 0-12 lipoprotein in normal and hypercholesterolaemic serum. Lancet 2:530–534
[97] Klein PD, Johnson RM (1954) Phosphorous metabolism in unsaturated fatty acid-deficient rats. J Biol Chem 211:103–110
[98] Hayashida T, Portman OW (1960) Swelling of liver mitochondria from rats fed diets deficient in essential fatty acids. Proc Soc Exp Biol Med 103:656–659
[99] Cornwell DG, Panganamala RV (1981) Atherosclerosis an intracellular deficiency in essential fatty acids. Prog Lipid Res 20:365–376
[100] Das UN (2006) Essential fatty acids-a review. Curr Pharm Biotechnol 7:467–482
[101] Felton CV, Crook D, Davies MJ, Oliver MF (1997) Relation of plaque lipid composition and morphology to the stability of human aortic plaques. Arterioscler Thromb Vasc Biol 17:1337–1345
[102] Thies F, Garry JM, Yaqoob P, Rerkasem K, Williams J, Shearman CP, Gallagher PJ, Calder PC, Grimble RF (2003) Association of n-3 polyunsaturated fatty acids with stability of atherosclerotic plaques: a randomised controlled trial. Lancet 361:477–485
[103] Mozaffarian D (2006) Trans fatty acids – effects on systemic inflammation and endothelial function. Atheroscler Suppl 7:29–32
[104] Mozaffarian D, Rimm EB, King IB, Lawler RL, McDonald GB, Levy WC (2004) Trans fatty acids and systemic inflammation in heart failure. Am J Clin Nutr 80:1521–1525
[105] Willett WC, Stampfer MJ, Manson JE, Colditz GA, Speizer FE, Rosner BA, Sampson LA, Hennekens CH (1993) Intake of trans fatty acids and risk of coronary heart disease among women. Lancet 341:581–585

[106] Ascherio A, Hennekens CH, Buring JE, Master C, Stampfer MJ, Willett WC (1994) Trans-fatty acids intake and risk of myocardial infarction. Circulation 89:94–101
[107] Das UN (2000) Beneficial effect(s) of n-3 fatty acids in cardiovascular diseases: but, why and how? Prostaglandins Leukot Essent Fatty Acids 63:351–362
[108] Das UN (2001) Essential fatty acids as possible mediators of the actions of statins. Prostaglandins Leukot Essent Fatty Acids 65:37–40
[109] Levine L (2003) Statins stimulate arachidonic acid release and prostaglandin I2 production in rat liver cells. Lipids Health Dis 2:1
[110] Jula A, Marniemi J, Ronnemaa T, Virtanen A, Huupponen R (2005) Effects of diet and simvastatin on fatty acid composition in hypercholesterolemic men: a randomized controlled trial. Arterioscler Thromb Vasc Biol 25:1952–1959
[111] Harris JI, Hibbeln JR, Mackey RH, Muldoon MF (2004) Statin treatment alters serum n-3 and n-6 fatty acids in hypercholesterolemic patients. Prostaglandins Leukot Essent Fatty Acids 71:263–269
[112] Bellini MJ, Polo MP, de Alaniz MJ, de Bravo MG (2003) Effect of simvastatin on the uptake and metabolic conversion of palmitic, dihomo-gamma-linoleic and alpha-linolenic acids in A549 cells. Prostaglandins Leukot Essent Fatty Acids 69:351–367
[113] Rise P, Pazzucconi F, Sirtori CR, Galli C (2001) Statins enhance arachidonic acid synthesis in hypercholesterolemic patients. Nutr Metab Cardiovasc Dis 11:88–94
[114] Birnbaum Y, Ye Y, Lin Y, Freeberg SY, Nishi SP, Martinez JD, Huang MH, Uretsky BF, Perez-Polo JR (2006) Augmentation of myocardial production of 15-epi- lipoxin-A_4 by pioglitazone and atorvastatin in the rat. Circulation 114:929–935
[115] Morris T, Stables M, Gilroy DW (2006) New perspectives on aspirin and the endogenous control of acute inflammatory resolution. Scientific WorldJournal 6:1048–1065
[116] Schwab JM, Serhan CN (2006) Lipoxins and new lipid mediators in the resolution of inflammation. Curr Opin Pharmacol 6:414–420
[117] Serhan CN (2005) Novel omega – 3-derived local mediators in anti-inflammation and resolution. Pharmacol Ther 105:7–21
[118] Mozaffarian D, Pischon T, Hankinson SE, Rifai N, Joshipura K, Willett WC, Rimm EB (2004) Dietary intake of trans fatty acids and systemic inflammation in women. Am J Clin Nutr 79:606–612
[119] Baer DJ, Judd JT, Clevidence BA, Tracy RP (2004) Dietary fatty acids affect plasma markers of inflammation in healthy men fed controlled diets: a randomized crossover study. Am J Clin Nutr 79:969–973
[120] Lopez-Garcia E, Schulze MB, Meigs JB, Manson JE, Rifai N, Stampfer MJ, Willett WC, Hu FB (2005) Consumption of trans fatty acids is related to plasma biomarkers of inflammation and endothelial dysfunction. J Nutr 135:562–566
[121] Lopez-Garcia E, Schultze MB, Manson JE, Meigs JB, Albert CM, Rifai N, Willett WC, Hu FB (2004) Consumption of (n-3) fatty acids is related to plasma biomarkers of inflammation and endothelial activation in women. J Nutr 134:1806–1811
[122] Kumar GS, Das UN, Kumar KV, Madhavi, Das NP, Tan BKH (1992) Effect of n-6 and n-3 fatty acids on the proliferation and secretion of TNF and IL-2 by human lymphocytes in vitro. Nutr Res 12:815–823
[123] Kumar GS, Das UN (1994) Effect of prostaglandins and their precursors on the proliferation of human lymphocytes and their secretion of tumor necrosis factor and various interleukins. Prostaglandins Leukot Essent Fatty Acids 50:331–334
[124] Prichard BNC, Smith CCT, Ling KLE, Betteridge DJ (1995) Fish oils and cardiovascular disease. BMJ 310:819–820
[125] Ambring A, Johansson M, Axelsen M, Gan L, Strandvik B, Friberg P (2006) Mediterranean-inspired diet lowers the ratio of serum phospholipid n-6 to n-3 fatty acids, the number of leukocytes and platelets, and vascular endothelial growth factor in healthy subjects. Am J Clin Nutr 83:575–581
[126] De Caterina R, Massaro M (2005) Omega-3 fatty acids and the regulation of expression of endothelial pro-atherogenic and pro-inflammatory genes. J Membr Biol 206:103–116

[127] Chen W, Esselman WJ, Jump DB, Busik JV (2005) Anti-inflammatory effect of docosahexaenoic acid on cytokine-induced adhesion molecule expression in human retinal vascular endothelial cells. Invest Ophthalmol Vis Sci 46:4342–4347

[128] Weber C, Erl W, Pietsch A, Danesch U, Weber PC (1995) Docosahexaenoic acid selectively attenuates induction of vascular cell adhesion molecule-1 and subsequent monocytic cell adhesion to human endothelial cells stimulated by tumor necrosis factor-alpha. Arterioscler Thromb Vasc Biol 15:622–628

[129] Eschen O, Christensen JH, De Caterina R, Schmidt EB (2004) Soluble adhesion molecules in healthy subjects: a dose- response study using n-3 fatty acids. Nutr Metab Cardiovasc Dis 14:180–185

[130] Johansen O, Seljeflot I, Hostmark AT, Arnesen H (1999) The effect of supplementation with omega-3 fatty acids on soluble markers of endothelial function in patients with coronary heart disease. Arterioscler Thromb Vasc Biol 19:1681–1686

[131] Morisaki N, Sprecher H, Milo GE, Cornwell DG (1982) Fatty acid specificity in the inhibition of cell proliferation and its relationship to lipid peroxidation and prostaglandin biosynthesis. Lipids 17:893–899

[132] Cornwell DG, Panganamala RV (1981) Atherosclerosis: an intracellular deficiency in essential fatty acids. Prog Lipid Res 20:365–376

[133] Hirafuji M, Machida T, Hamaue N, Minami M (2003) Cardiovascular protective effects of n-3 polyunsaturated fatty acids with special emphasis on docosahexaenoic acid. J Pharmacol Sci 92:308–316

[134] Abeyawardena MY, Head RJ (2001) Long chain n-3 polyunsaturated fatty acids and blood vessel function. Cardiovasc Res 52:361–371

[135] Pakala R, Pakala R, Sheng WL, Benedict CR (1999) Eicosapentaenoic acid and docosahexaenoic acid block serotonin-induced smooth muscle cell proliferation. Arterioscler Thromb Vasc Biol 19:2316–2322

[136] Nakayama M, Fukuda N, Watanabe Y, Soma M, Hu WY, Kishioka H, Satoh C, Kubo A, Kamatsuse K (1999) Low dose of eicosapentaenoic acid inhibits exaggerated growth of vascular smooth muscle cells from spontaneously hypertensive rats through suppression of transforming growth factor-beta. J Hypertens 17:1421–1430

[137] Nobukata H, Ishikawa T, Obata M, Shibutani Y (2000) Long-term administration of highly purified eicosapentaenoic acid ethyl ester improves the dysfunction of vascular endothelial and smooth muscle cells in male WBN/Kob rats. Metabolism 49:1588–1591

[138] Mori Y, Nobukata H, Harada T, Kasahara T, Tajima N (2003) Long- term administration of highly purified eicosapentaenoic acid ethyl ester improves blood coagulation abnormalities and dysfunction of vascular endothelial cells in Otsuka Long-Evans Tokushima fatty rats. Endocr J 50:603–611

[139] Huttner JJ, Gwebu ET, Panganamala RV, Milo GE, Cornwell DC, Sharma HM, Geer JC (1977) Fatty acids and their prostaglandin derivatives: inhibitors of proliferation in aortic smooth muscle cells. Science 197:289–291

[140] Fan YY, Ramos KS, Chapkin RS (2001) Dietary gamma-linolenic acid suppresses aortic smooth muscle cell proliferation and modifies atherosclerotic lesions in apolipoprotein E knockout mice. J Nutr 131:1675–1681

[141] Das UN (2005) Is angiotensin II an endogenous pro-inflammatory molecule? Med Sci Monit 11:RA155–RA162

[142] Kumar KV, Das UN (1997) Effect of cis-unsaturated fatty acids, prostaglandins, and free radicals on angiotensin-converting enzyme activity in vitro. Proc Soc Exp Biol Med 214:374–379

[143] Okuda Y, Kawashima K, Sawada T, Tsurumaru K, Asano M, Suzuki S, Soma M, Nakajima T, Yamashita K (1997) Eicosapentaenoic acid enhances nitric oxide production by cultured human endothelial cells. Biochem Biophys Res Commun 232:487–491

[144] Das UN (2002) A perinatal strategy for preventing adult diseases: the role of long-chain polyunsaturated fatty acids. Kluwer Academic, Boston, MA

[145] Das UN, Mohan IK, Raju TR (2001) Effect of corticosteroids and eicosapentaenoic acid/docosahexaenoic acid on pro-oxidant and anti-oxidant status and metabolism of essential fatty acids in patients with glomerular disorders. Prostaglandins Leukot Essent Fatty Acids 65:197–203
[146] Bunk S (2002) ACEs wild. Scientist 16:22–24
[147] Moskowitz DW (2002) Is angiotensin I-converting enzyme a 'master' disease gene? Diabetes Technol Ther 4:683–711
[148] Kaergel E, Muller DN, Honeck H, Theuer J, Shagdarsuren E, Mullally A, Luft FC, Schunck W-H (2002) P450-dependent arachidonic acid metabolism and angiotensin-II-induced renal damage. Hypertension 40:273–279
[149] Gilroy DW (2005) New insights into the anti-inflammatory actions of aspirin- induction of nitric oxide through the generation of epi-lipoxins. Mem Inst Oswaldo Cruz 100(Suppl 1):49–54
[150] Wang W, Diamond SL (1997) Does elevated nitric oxide production enhance the release of prostacyclin from shear stressed aortic endothelial cells? Biochem Biophys Res Commun 233:748–751
[151] Arita M, Bianchini F, Aliberti J, Sher A, Chiang N, Hong S, Yang R, Petasis NA, Serhan CN (2005) Stereochemical assignment, antiinflammatory properties, and receptor for the omega-3 lipid mediator resolvin E1. J Exp Med 201:713–722
[152] Dooper MM, van Riel B, Graus YM, M'Rabet L (2003) Dihomo- gamma-linolenic acid inhibits tumour necrosis factor-alpha production by human leucocytes independently of cyclooxygenase activity. Immunology 110:348–357
[153] Chen W, Esselman WJ, Jump DB, Busik JV (2005) Anti-inflammatory effect of docosahexaenoic acid on cytokine-induced adhesion molecule expression in human retinal vascular endothelial cells. Invest Ophthalmol Vis Sci 46:4342–4347
[154] Li H, Ruan XZ, Powis SH, Fernando R, Mon WY, Wheeler DC, Moohead JF, Varghese Z (2005) EPA and DHA reduce LPS-induced inflammation responses in HK-2 cells: evidence for a PPAR- gamma-dependent mechanism. Kidney Int 67:867–874
[155] Sheng Z, Otani H, Brown MS, Goldstein JL (1995) Independent regulation of sterol regulatory element-binding proteins 1 and 2 in hamster liver. Proc Natl Acad Sci U S A 92:935–938
[156] Fajas L, Schoonjans K, Gelman L, Kim JB, Najib J, Martin G, Fruchart JC, Briggs M, Spiegelman BM, Auwerx J (1999) Regulation of peroxisome proliferator-activated receptor gamma expression by adipocyte differentiation and determination factor 1/sterol regulatory element binding protein 1: implications for adipocyte differentiation and metabolism. Mol Cell Biol 19:5495–5503
[157] El-Sohemy A, Archer MC (1999) Regulation of mevalonate synthesis in low density lipoprotein receptor knockout mice fed n-3 or n-6 polyunsaturated fatty acids. Lipids 34:1037–1043
[158] Nakamura N, Hamazaki T, Jokaji H, Minami S, Kobayashi M (1998) Effect of HMG-CoA reductase inhibitors on plasma polyunsaturated fatty acid concentration in patients with hyperlipidemia. Int J Clin Lab Res 28:192–195
[159] Das UN (2000) Essential fatty acids and osteoporosis. Nutrition 16:286–290
[160] Hannah VC, Ou J, Luong A, Goldstein JL, Brown MS (2001) Unsaturated fatty acids down-regulate srebp isoforms 1a and 1c by two mechanisms in HEK-293 cells. J Biol Chem 276:4365–4372
[161] Field FJ, Born E, Murthy S, Mathur SN (2002) Polyunsaturated fatty acids decrease the expression of sterol regulatory element-binding protein-1 in CaCo-2 cells: effect on fatty acid synthesis and triacylglycerol transport. Biochem J 368(Pt 3):855–864
[162] Xu J, Cho H, O'Malley S, Park JH, Clarke SD (2002) Dietary polyunsaturated fats regulate rat liver sterol regulatory element binding proteins-1 and -2 in three distinct stages and by different mechanisms. J Nutr 132:3333–3339
[163] Field FJ, Born E, Mathur SN (2003) Fatty acid flux suppresses fatty acid synthesis in hamster intestine independently of SREBP-1 expression. J Lipid Res 44:1199–1208

[164] Le Jossic-Corcos C, Gonthier C, Zaghini I, Logette E, Shechter I, Bournot P (2005) Hepatic farnesyl diphosphate synthase expression is suppressed by polyunsaturated fatty acids. Biochem J 385(Pt 3):787–794

[165] Worgall TS, Johnson RA, Seo T, Gierens H, Deckelbaum RJ (2002) Unsaturated fatty acid-mediated decreases in sterol regulatory element-mediated gene transcription are linked to cellular sphingolipid metabolism. J Biol Chem 277:3878–3885

Chapter 11
Osteoporosis

Introduction

Osteoporosis in which bone mineral density (BMD) is reduced leads to an increased risk of fracture. In osteoporosis not only the BMD is reduced but even the bone microarchitecture is disrupted, and the amount and variety of proteins in bone is altered. WHO (world health organization) defined osteoporosis in women as a bone mineral density 2.5 standard deviations below peak bone mass (20-year-old healthy female average) as measured by DXA; the term "established osteoporosis" includes the presence of a fragility fracture [1].

Osteoporosis is most common in women after menopause, called postmenopausal osteoporosis, but may also develop in men, and may occur in anyone in the presence of particular hormonal disorders and other chronic diseases or as a result of medications, specifically glucocorticoids, when the disease is called steroid- or glucocorticoid-induced osteoporosis. Given its influence in the risk of fragility fracture, osteoporosis may significantly affect life expectancy and quality of life.

Disease of the parathyroid glands (hyperparathyroidism) is also a major cause of osteoporosis. Hyperparathyroidism should be considered in any patient with severe osteoporosis, osteoporosis occurring at a young age, or osteoporosis in a male.

Dietary Protein and Osteoporosis

Dietary protein increases urinary calcium losses and has been associated with higher rates of hip fracture in cross-cultural studies. In a prospective study, a cohort of 85,900 women, aged 35–59 years, who were participants in the Nurses' Health Study an increased risk of forearm fracture was noted for women who consumed more than 95 g/day compared with those who consumed less than 68 g/day. A similar increase in risk was observed for animal protein, but no association was found for consumption of vegetable protein [2]. In contrast to these results, it was reported that the risk of hip fracture was negatively associated with total protein intake. Animal rather than vegetable sources of protein appeared to account for this association. Thus, intake of dietary protein, especially from animal sources, may be associated

U. N. Das, *Molecular Basis of Health and Disease*,
DOI 10.1007/978-94-007-0495-4_11,

with a reduced incidence of hip fractures in postmenopausal women [3]. But, it is important to note that protein deficiency contributes to the occurrence of osteoporotic fractures not only by decreasing bone mass but also by altering muscle function. Furthermore, malnutrition is associated with increased morbidity in patients with osteoporotic fractures. A low IGF-1 (insulin-like growth factor-1) level is a risk factor for hip fracture. In subjects with appropriate intakes of vitamin D and calcium, giving protein supplements to correct an inadequate spontaneous protein intake increases circulating IGF-1 levels, improves clinical outcomes after hip fracture, and prevents bone mineral density loss at the proximal femur. Supplemental protein also significantly reduces the length of inpatient rehabilitation care. These data emphasize the importance of adequate nutrient intake in the prevention and treatment of osteoporotic fractures [4]. This is supported by the observation that elderly persons who have osteoporotic hip fracture are often undernourished, particularly with respect to protein. Protein malnutrition may contribute to the occurrence and outcome of hip fracture. In a 6-month, randomized, double-blind, placebo-controlled trial with a 6-month post-treatment follow-up performed in 82 patients (mean age, 80.7 ± 7.4 years) with recent osteoporotic hip fracture, when were given calcium supplementation, 550 mg/day, and one dose of vitamin D, 200,000 IU (at baseline) and protein supplementation, 20 g/day, or isocaloric placebo (among controls), patients who received protein supplements had significantly greater increases in serum levels of IGF-1 ($85.6\% \pm 14.8\%$ and $34.1\% \pm 7.2\%$ at 6 months); and an attenuation of the decrease in proximal femur bone mineral density at 12 months. Median stay in rehabilitation wards was shorter for patients who received protein supplements than for controls. These results suggest that protein repletion after hip fracture was associated with increased serum levels of IGF-1, attenuation of proximal femur bone loss, and shorter stay in rehabilitation hospitals [5]. It was noted that for every 15-g/day increase in animal protein intake, BMD increased by 0.016 g/cm^2 at the hip, 0.012 g/cm^2 at the femoral neck, 0.015 g/cm^2 at the spine, and 0.010 g/cm^2 for the total body. Conversely, a negative association between vegetable protein and BMD was observed in both males and females. These results are in support of the argument that dietary animal protein has a protective role in the skeletal health of elderly women [6]. These studies [4–6] are in support of the hypothesis that high calcium intake combined with adequate protein intake based on a high ratio of vegetable to animal protein may be protective against osteoporosis [7].

It is interesting to note that elderly women with a high dietary ratio of animal to vegetable protein intake have more rapid femoral neck bone loss and a greater risk of hip fracture than do those with a low ratio, suggesting that an increase in vegetable protein intake and a decrease in animal protein intake may decrease bone loss and the risk of hip fracture [8]. Animal studies clearly showed that both energy and protein deficiencies may contribute to age-related bone loss, highlighting the importance of sustaining adequate energy and protein provision to preserve skeletal integrity in the elderly [9].

These and other studies indicate that osteoporosis can be prevented with lifestyle changes and sometimes medication; in people with osteoporosis, treatment may involve both. Lifestyle change includes exercise and preventing falls as well as adequate protein intake (neither excess nor low). Medication includes calcium, vitamin D, and bisphosphonates.

Magnesium and Osteoporosis

Magnesium is one of the essential nutrients for optimal bone mineralization. Magnesium-deprived rats had diminished bone volume and abnormal histological features consistent with disorganized and chaotic bone remodeling; suggesting that low-magnesium intake during growth can alter the quality and quantity of bone. Thus, magnesium deprivation may contribute to the development of osteoporosis [10].

In this context, the anticonvulsive and antihypertensive values of magnesium in eclampsia, and its antiarrhythmic applications in a variety of cardiac diseases is noteworthy. It is well recognized that Mg^{2+} is an essential nutrient and its deficiency elicits neuromuscular manifestations. Mg^{2+} is useful in the management of pre-eclampsia and eclampsia, delays preterm birth, influences premenstrual syndrome, and ameliorates migraine headaches, disorders that almost exclusively or largely afflict women. Estrogen enhances Mg^{2+} utilization and uptake by soft tissues and bone that has been reported to be the reason for the resistance of young women to heart disease and osteoporosis, as well as increased prevalence of these diseases in postmenopausal women [11].

Low dietary Mg^{2+} may be a risk factor for osteoporosis, partly, by lowering serum parathyroid hormone concentrations even though serum calcium remained essentially normal. Mg^{2+} depletion also resulted in a significantly lower serum $1,25(OH)_2$-vitamin D levels. Histomorphometry and micro-computerized tomography demonstrated decreased bone volume and trabecular thickness in Mg^{2+} deficient animals with no difference in osteoclast or osteoblast number. It is noteworthy that Mg^{2+} resulted in almost 138–150% increase in TNF-α level in osteoclasts with a simultaneous increase in substance P by 179–432% [12, 13]. In addition, it was reported that low dietary Mg^{2+} induced low bone mineral density and osteoporosis resulted in a decrease in Osteoprotegerin (OPG) and an increase in receptor activator of nuclear factor kB ligand (RANKL) that could be responsible for increased osteoclastogenesis [14]. The relationship between Mg^{2+} deficiency and the resultant increase in TNF-α in osteoclasts is supported by the observation that a low magnesium diet resulted in a greater increase in osteoclast number and decrease in osteoblast number in the wild-type mice with a trend toward greater eroded bone perimeter, as compared to TNF-r-KO (TNF-α receptor knockout) mice confirming that TNF-α plays a significant role in Mg^{2+} deficiency-induced bone loss [15]. Thus, Mg^{2+} deficiency results in an increase in inflammatory cytokines which indicates that osteoporosis could be an inflammatory condition.

Osteoporosis Is a Low-grade Systemic Inflammatory Condition

It is evident from the preceding discussion [10–15] that osteoporosis seen in Mg^{2+} deficiency state is associated with enhanced TNF-α in osteoclasts suggesting that proinflammatory state and proinflammatory cytokines have a role in osteoporosis. This concept is supported by the observation that the proinflammatory cytokines IL-1β and IL-6 play a central role in the acceleration of postmenopausal bone loss. Estrogen has anti-inflammatory actions and suppresses the production of pro-inflammatory cytokines: IL-1, IL-6, TNF-α and MIF [16–19], though one in vitro study reported that estrogen did not affect normal macrophage migration and also failed to suppress production of MIF by guinea pig lymphocytes stimulated with antigen [20]. In a 5-year longitudinal study performed in 165 perimenopausal women, who were randomized to receive hormone-replacement therapy (HRT) or no treatment, it was noted that serum IL-6 increased with age, but cytokines did not correlate with baseline BMD. HRT led to small increases in IL-1ra (IL-1 receptor antagonist, $p < 0.001$) and IL-6 ($p < 0.05$), with a decrease in sIL-6R (soluble IL-6 receptor, $p < 0.01$) and no change in IL-1β while no changes were observed in the control group. IL-1ra was inversely correlated with bone loss. In addition, a weak positive correlation between sIL-6R and bone loss was noted. High IL-6 levels were associated with slower bone loss. In summary these studies showed that serum IL-1ra and sIL-6R are influenced by HRT and are associated with the rate of bone loss in perimenopausal women [21]. These results are in support of the concept that the decline in ovarian function with menopause is associated with spontaneous increases in proinflammatory cytokines IL-1, IL-6, and TNF-α. The exact mechanisms by which estrogen interferes with cytokine activity are still incompletely known but may potentially include interactions of the ER (estrogen receptor) with other transcription factors, modulation of nitric oxide activity, antioxidative effects, plasma membrane actions, and changes in immune cell function. Experimental and clinical studies strongly support a link between the increased state of proinflammatory cytokine activity and postmenopausal bone loss [22–24] that may also be relevant to vascular homeostasis and the development of atherosclerosis. This is especially so as atherosclerosis is also a low-grade systemic inflammatory condition.

Furthermore, raloxifene hydrochloride, a selective oestrogen receptor modulator that increases bone mineral density and decrease biochemical markers of bone turnover in postmenopausal women, without stimulatory effects on breast or uterus was found to decrease serum osteocalcin and parathyroid hormone, and urine deoxypyridinoline levels to normal levels with treatment. Serum 25-OH vitamin D levels after treatment in the patient group were higher than those in the control group. Serum IL-6, TNF-alpha and TGF-β1 levels did not change significantly with treatment. However, serum levels of IL-6 and TGF-β1 in the patient group after treatment, decreased to levels lower than those found in the control group. Thus, raloxifene treatment reduced bone turnover biochemical markers, parathyroid hormone and induces 25-OH vitamin D in postmenopausal women and decrease serum proinflammatory cytokine levels in the postmenopausal period [25].

In a clinical study of patients with and without osteoporosis in which the concentrations of the cytokines such as adiponectin, leptin, Osteoprotegerin (OPG), soluble receptor activator of nuclear factor kappaB ligand (s-RANKL), TNF-α, and IL-6 in the extracellular fluid from the human bone marrow when evaluated it was observed that osteoporotic women had higher content of proinflammatory and adipogenic cytokines, and decreased leptin bioavailability [26]. The later observation of decreased bioavailability of leptin in the bone marrow supernatant fluid is interesting since, it is known that leptin inhibits bone formation by the osteoblasts, while leptin deficiency results in a high bone mass phenotype [27] despite hypogonadism in experimental animals. This indicates that leptin deficiency seen in the bone marrow supernatant fluid in osteoporosis could a compensatory mechanism in response to osteoporosis. Leptin-deficient and leptin receptor-deficient mice that are obese and hypogonadic showed increased bone formation leading to high bone mass despite hypogonadism and hypercortisolism. This phenotype was found to be dominant, independent of the presence of fat, and specific for the absence of leptin signaling. Osteoblasts were found to be devoid of leptin signaling but intracerebroventricular infusion of leptin caused bone loss in leptin-deficient and wild-type mice, suggesting that leptin is a potent inhibitor of bone formation acting through the central nervous system [28]. It was reported that leptin deficiency results in low sympathetic tone and genetic or pharmacological ablation of adrenergic signaling led to a leptin-resistant high bone mass. In addition, β-adrenergic receptors on osteoblasts regulated their proliferation, and a β-adrenergic agonist decreased bone mass in leptin-deficient and wild-type mice while a β-adrenergic antagonist increased bone mass in wild-type and ovariectomized mice. None of these manipulations affected body weight. These results led to the suggestion that leptin-dependent neuronal regulation of bone formation is mediated through the sympathetic nervous system [29]. It is interesting in this context to note that both leptin and catecholamines have proinflammatory actions [30–37]. Thus, leptin-induced bone loss and sympathetic nervous system involvement in decreasing bone mass can be linked to their proinflammatory actions since, IL-6, TNF-α that are proinflammatory cytokines produce osteoporosis. These results imply that activation of parasympathetic nervous system may protect against osteoporosis. It is important to note that acetylcholine the principal neurotransmitter of the parasympathetic nervous system is a potent anti-inflammatory molecule [38–40]. Thus, there seems to be a close interaction(s) among leptin, osteoblasts and osteoclasts, sympathetic nervous system and pro- and anti-inflammatory molecules. Furthermore, exercise that is beneficial in the prevention and treatment of osteoporosis is known to have anti-inflammatory actions, enhance parasympathetic tone and suppress the production of IL-6 and TNF-α [41–47].

In a population-based sample of 188 home-dwelling, middle-aged and older adults (104 women, mean age 59 years) in whom high-frequency heart rate variability (HF) and pre-ejection period (PEP) served as markers of cardiac parasympathetic and sympathetic tone, respectively, it was noted that an inverse relationship exists between HF and CRP [48]. This suggests that the higher the parasympathetic tone the lower the inflammatory marker hs-CRP. This is in line with the evidence that higher parasympathetic tone exerts anti-inflammatory action since acetylcholine; the principal parasympathetic neurotransmitter has anti-inflammatory actions. Thus, leptin, catecholamines and sympathetic nervous system and β-adrenergic agonists decrease bone mass at least, in part, by activating the inflammatory cascade while exercise is beneficial in the management of osteoporosis by enhancing parasympathetic tone that, in turn, suppresses production of inflammatory cytokines such as IL-6, TNF-α and MIF. In addition, ageing and menopause are associated with an increase in the production of pro-inflammatory cytokines IL-6 and TNF-α that may explain as to why osteoporosis is common in these instances. Based on these evidences, it is clear that osteoporosis could be considered as a low-grade systemic inflammatory condition. Since, low-grade systemic inflammation is also a feature of endothelial dysfunction, insulin resistance, type 2 diabetes mellitus, hypertension and metabolic syndrome, it is reasonable to propose that osteoporosis is likely to be present in all these conditions. Furthermore, plasma and tissue levels of pro-inflammatory cytokines IL-6, TNF-α, MIF and HMGB1 (high mobility group box 1) are known to be elevated in rheumatoid arthritis, lupus and systemic sclerosis that may also explain the frequent presence of osteoporosis in these diseases. What is more important is the presence of endothelial dysfunction and insulin resistance in rheumatoid arthritis, lupus and systemic sclerosis that may also explain the development of premature atherosclerosis and increased frequency of cardiovascular diseases in these rheumatological conditions [49–52]. Since endothelial cells produce nitric oxide (NO) that is important to prevent atherosclerosis and cardiovascular diseases, and endothelial dysfunction could be associated with osteoporosis, it is likely that NO may have a modulatory function on osteoblasts and osteoclasts.

Nitric Oxide in Osteoporosis

Excessive osteoclastic activity has been implicated in osteoporosis, Paget disease of bone, rheumatoid arthritis, and the growth of metastases in bone. Nitric oxide (NO) appears to be an important modulator of osteoporosis. NO produced by the vascular endothelium and nervous system is involved in both neurotransmission and the regulation of blood pressure. NO seems to be a potent inhibitor of osteoclast function. NO when used at 30 μM produced a decrease in the osteoclast spread area that was also associated with a reduction of bone resorption. These actions of NO are produced, probably, by stimulating guanylate cyclase, with a consequent increase in cyclic GMP. But a different mode of action is likely in the osteoclast since dibutyryl or 8-bromo cyclic GMP have no effect. The abundance of NO-producing endothelial cells in bone marrow and their proximity to osteoclasts suggests that marrow endothelial cells play a physiological role in the regulation of osteoclastic activity and in the prevention of osteoporosis [53]. This is supported by the observation that (a) osteoclasts exhibit substantial NOS (nitric oxide synthase) activity that may account for basal NO production; (b) nitroprusside, a NO donor, markedly decreased the number of bone pits and the average pit area in comparison with control in vitro; (c) while NOS inhibition by N-nitro-L-arginine methyl ester or aminoguanidine dramatically increased the number of bone pits and the average resorption area per pit; and (d) inhibition of NOS activity in vitro and in vivo resulted in potentiation of osteoclast activity. These findings suggest that endogenous NO production in osteoclast cultures may regulate resorption activity and modulation of NOS and NO levels by cells within the bone microenvironment may be a sensitive mechanism for local control of osteoclast bone resorption.

Both constitutive and inducible forms of NO synthase are expressed by bone-derived cells, while interleukin-1 (IL-1), tumor necrosis factor (TNF), and interferon-γ (IFN-γ) are potent stimulators of NO production. In combination with other cytokines, IFN-γ markedly induces NO production, which leads to the suppression of osteoclast formation and activity of mature osteoclasts that could be responsible for the selective inhibitory effect of IFN-γ on cytokine-induced bone resorption. Constitutive production of NO (that is produced by endothelial cells) promotes the proliferation of osteoblast-like cells and modulates osteoblast function. On the other hand, high concentrations of NO that are produced by the inducible NOS in response to pro-inflammatory cytokines such as IL-1, TNF-α and IL-6 have inhibitory action on osteoblast lineage, and possibly, enhance osteoclast proliferation and lead to the onset of osteoporosis. Thus, NO production appears to have varied actions on resorption: at lower concentrations, NO prevents bone resorption while moderate to high induction of NO potentiates bone resorption. Thus, NO is an important regulatory molecule in bone with effects on cells of the osteoblast and osteoclast lineage and represents one of the molecules produced by osteoblasts which directly regulate osteoclastic activity. Stimulation of NO production in bone by proinflammatory cytokines may be involved in the onset and progression of osteoporosis with cytokine activation seen in rheumatoid arthritis, tumor associated osteolysis, and postmenopausal osteoporosis [54].

These results are supported by the observation that osteoclasts exhibit substantial NOS activity that may account for basal NO production. Chicken osteoclasts when exposed to nitroprusside that donates NO markedly decreased the number of bone pits and the average pit area in comparison with control cultures. On the other hand, NOS inhibition by N-nitro-L-arginine methyl ester or aminoguanidine dramatically increased the number of bone pits and the average resorption area per pit. In ovariectomized rats, a model of osteoporosis, aminoguanidine potentiated the loss of bone mineral density and aminoguanidine also caused a loss of bone mineral density in the sham-operated rats suggesting that inhibition of NOS activity potentiates osteoclast activity [55]. Hence, it can be said that modulation of NOS and NO levels by cells within the bone microenvironment controls osteoclast and osteoblast activity and thus, NO plays a significant role in the pathogenesis of osteoporosis. These results led to the suggestion that NO or NO donors could be of benefit in the prevention and treatment of osteoporosis that is supported by the report that administration of nitroglycerine (NO donor) alone prevented ovariectomy-induced bone loss, while the combination of 17-beta-estradiol + nitroglycerine did not further enhance the bone mass or femur weight, and the OVX-induced bone loss was not further aggravated by L-NAME, a potent inhibitor of NO synthesis. But, surprisingly it was noted that L-NAME completely inhibited the ability of 17-beta-estradiol to prevent or reverse ovariectomy induced bone loss, suggesting that the protective effect of estrogens against bone loss is mediated through NO [56]. In a similar fashion, even corticosteroid-induced osteoporosis was also prevented by NO donors and NO precursor- L-arginine [57]. These results are supported by the report that data from the Study of Osteoporotic Fractures showed that women using nitrates intermittently had substantially greater hip and heel BMD (bone mineral density) than nonusers. Surprisingly, it was noted that women taking daily nitrates had slightly greater hip BMD but the same heel BMD as nonusers, suggesting that intermittent administration of nitrates may enhance BMD [58]. These results indicate that intermittent nitrate is of greater value in the prevention of osteoporosis than daily nitrate use since daily use of nitrates may result in tachyphylaxis.

Post-menopausal Osteoporosis, Cytokines and NO

In this context, it may be noted that decline in ovarian function with menopause is associated with spontaneous increases in proinflammatory cytokines IL-1, IL-6, and TNF-α. The exact mechanisms by which estrogen interferes with cytokine activity are still incompletely known but may potentially include interactions of the ER with other transcription factors, modulation of nitric oxide activity, antioxidative effects, plasma membrane actions, and changes in immune cell function. Both experimental and clinical studies support a link between the increased state of proinflammatory cytokine activity and postmenopausal bone loss indicating that excess production of cytokines could initiate and perpetuate osteoporosis not only in post-menopausal women [59] but also in various inflammatory conditions. These results imply that estrogen has anti-inflammatory actions, enhances NO generation and thus, hormone replacement therapy is able to prevent osteoporosis. It is likely that changes in the NO activity and cytokine profile seen in post-menopausal women may also be relevant to vascular homeostasis and the development of atherosclerosis seen in them. Thus, increased incidence of atherosclerosis, cardiovascular diseases, Alzheimer's disease, insulin resistance and metabolic syndrome (including hypertension) seen with ageing and post-menopausal women could be attributed to changes in the interactions of the ER with other transcription factors, modulation of nitric oxide activity, antioxidative effects, and changes in immune cell. In other words, this implies that methods designed to enhance eNO generation and suppress the enhanced pro-inflammatory cytokine concentrations could be of benefit in all these conditions.

Dose Dependent Action of NO on Bone

But, it is important to note that the affect of NO in osteoporosis and other diseases depends on the local concentrations of NO. Hao et al. [60] reported that when ovariectomized rats were treated with different doses of nitroglycerine (a donor of NO) treatment with low-dose nitroglycerin, middle-dose nitroglycerin, and 17-beta-estradiol maintained bone mineral density and reversed the effects of ovariectomy on dry weight of the bones, ash weight and calcium content when compared with those in the control group. Paradoxically, there were no differences in the bone mineral density, dry weight, ash weight, or calcium concentration between the ovariectomized-only rats and the rats treated with high-dose nitroglycerin. These results suggest that NO treatment can counteract bone loss in ovariectomized rats, while excess NO is not of benefit in the treatment of osteoporosis. Thus, there appears to be a dose response relationship between NO and osteoporosis-lower and moderate doses prevent osteoporosis whereas excess NO could be harmful. This has important therapeutic implications. For instance, eNOS may be beneficial in preventing osteoporosis whereas NO generated due to the activation of iNOS (that leads to excess production of NO) may actually cause osteoporosis. Activation of iNOS occurs during infections, rheumatoid arthritis, osteomyelitis and other inflammatory conditions that could be responsible for osteoporosis seen in these diseases. On the other hand, osteoporosis seen in postmenopausal women and ageing is related to reduced eNOS activity. Since eNOS reflects endothelial integrity and function, this may also explain the coexistence of atherosclerosis, cardiovascular diseases and osteoporosis in the elderly. Thus, it is important to enhance the activity of eNOS in postmenopausal women and the elderly, whereas iNOS needs to be tamed in inflammatory conditions to suppress the occurrence of osteoporosis. In this context, it is interesting to note that subjects who used 5 or 20 mg of isosorbide mononitrate, a donor of NO, for 12 weeks showed a decrease in urine N-telopeptide, a marker of bone resorption, and an increase in serum bone-specific alkaline phosphatase, a marker of bone formation, suggesting that commonly used isosorbide mononitrate may be useful in the prevention of post-menopausal osteoporosis [61]. These results are supported by a nationwide population-based pharmaco-epidemiological case-control study that included 124,655 subjects who had sustained a fracture during the year 2000 (cases) and 373,962 age- and sex-matched controls that showed an approximately 15% reduced risk of fractures in users of organic nitrates [62].

Since under normal conditions a balance is maintained between reactive oxygen species and eNO, it is expected that in postmenopausal women and the elderly there could occur an increase in free radical generation as a result of reduced eNOS activity. Such an assumption is supported by the work of Ozgocmen et al. [63] who showed that in postmenopausal women aged 40–65 had significantly lower erythrocyte catalase enzyme activity and higher erythrocyte malondialdehyde (MDA) and erythrocyte nitric oxide levels in comparison to controls whereas erythrocyte superoxide dismutase and glutathione peroxidase enzyme activity were similar. In plasma, postmenopausal osteoporotic women had significantly higher SOD (superoxide dismutase) enzyme activity and higher MDA levels whereas similar glutathione peroxidase enzyme activity and NO levels compared to non-porotic controls. Significant correlation was found between erythrocyte SOD, CAT enzyme activities and erythrocyte NO levels with proximal femur BMD. Thus, these results suggest that oxidative stress markers may be an important indicator for bone loss in postmenopausal women.

The involvement of NO in osteoporosis is further strengthened by the report that in women aged 65 years and older the heterozygous Glu/Asp genotype had a borderline statistically significantly lower rate of hip fracture than either the Glu/Glu genotype (HR = 0.87, 95% CI: 0.74, 1.01) or the Asp/Asp genotype (HR = 0.78, 95% CI: 0.62, 0.98) suggesting that allelic variation at the NOS3 locus maybe associated with hip fracture risk [64].

Anti-diabetic Drug Metformin, NO and Osteoporosis

Metformin, a widely used drug in the management of type 2 diabetes mellitus, lowers blood glucose levels by decreasing hepatic glucose production and increasing glucose utilization. Metformin has beneficial actions on circulating lipids that has been linked to its ability to reduce fatty liver. Metformin activates AMPK in hepatocytes; as a result, acetyl-CoA carboxylase (ACC) activity is reduced, fatty acid oxidation is induced, and expression of lipogenic enzymes is suppressed. Activation of AMPK by metformin suppresses expression of SREBP-1, a key lipogenic transcription factor. Hepatic expression of SREBP-1 (and other lipogenic) mRNAs and protein is reduced in metformin-treated animals; activity of the AMPK target, ACC, is also reduced. AMPK inhibition inhibited metformin's inhibitory effect on glucose production by hepatocytes. In isolated rat skeletal muscles, metformin stimulated glucose uptake in parallel with AMPK activation. Thus, AMPK activation seems to be at the core of the beneficial effects of metformin [65, 66].

When the effects of metformin on the differentiation and mineralization of osteoblastic MC3T3-E1 cells as well as the intracellular signal transduction mechanism were studied, it was observed that metformin (50 μM) significantly increased collagen-I and osteocalcin mRNA expression, stimulated alkaline phosphatase activity, and enhanced cell mineralization. Moreover, metformin significantly activated AMPK in dose- and time-dependent manners, and induced endothelial nitric oxide synthase (eNOS) and bone morphogenetic protein-2 (BMP-2) expressions, actions that were blocked in the presence of AMPK inhibitor. These results emphasize the fact that metformin-induced eNOS and BMP-2 expressions and its ability to induce differentiation and mineralization of osteoblasts via activation of the AMPK signaling pathway might explain its beneficial effects not only in diabetes but also osteoporosis by promoting bone formation [67].

The adenosine monophosphate-activated protein kinase (AMPK) that is a conserved regulator of the cellular response to low energy is activated when intracellular adenosine triphosphate (ATP) concentrations decrease and AMP concentrations increase in response to nutrient deprivation and pathological stresses. Thus, AMPK is activated in response to glucose limitation. Furthermore, AMPK has a critical role in many metabolic processes, including glucose uptake and fatty acid oxidation in muscle, fatty acid synthesis and gluconeogenesis in the liver, and the regulation of food intake in the hypothalamus [68]. In the liver, AMPK is regulated in response to adipokines adiponectin and resistin, which serve to stimulate and inhibit AMPK activation, respectively [69, 70]. Exercise activate AMPK in muscle and in liver and thus, is believed to bring about some of its actions on blood glucose regulation and thought to therapeutically act in part through stimulation of this pathway in those tissues [71, 72]. Recent studies reported that the kinase LKB1, a protein-threonine kinase, is sufficient to activate AMPK since LKB1 phosphorylates and activates AMPK. This is supported by the observation that deletion of LKB1 in the liver of adult mice resulted in a nearly complete loss of AMPK activity. This loss of LKB1 function led to the development of hyperglycemia with increased gluconeogenesis and lipogenesis.

In LKB1-deficient livers, TORC2, a transcriptional coactivator of CREB (cAMP response element–binding protein), was dephosphorylated and entered the nucleus, driving the expression of peroxisome proliferator-activated receptor-γ coactivator 1α (PGC-1α), which in turn induced gluconeogenesis. Inhibition of TORC2 led to reduced PGC-1α expression and normalized blood glucose levels in mice with deleted liver LKB1, indicating that TORC2 is a critical target of LKB1/AMPK signals in the regulation of gluconeogenesis [73]. No reduction in blood glucose was noted in the experimental animals in which LKB1 had been deleted in the liver providing direct evidence that the LKB1 tumor suppressor is the major upstream activating kinase for AMPK in liver. Thus, the activation of LKB1 appears to regulate downstream kinases such as AMPK that phosphorylate a transcriptional coactivator TORC2, resulting in its inactivation through sequestration in the cytoplasm. This pathway under normal conditions integrates cellular (AMPK) and hormonal (SIK) inputs to negatively regulate transcriptional events that promote synthesis of gluconeogenic enzymes. In the absence of LKB1, no kinase is active to phosphorylate TORC2, and hence, gluconeogenesis occurs. These data also provided proof that AMPK activation is absolutely required for the glucose-lowering action of metformin in intact animals. Since, the deletion of LKB1 in the liver did not impair AMPK activation in muscle, yet it eliminated the effect of metformin on serum glucose levels suggesting that metformin primarily decreases blood glucose concentrations by decreasing hepatic gluconeogenesis. LKB1 also acts as a tumor suppressor. Thus, increased CREB-dependent or SREBP-1-dependent transcription could have a role in LKB1-dependent tumorigenesis, implying that metformin may have a role in the inhibition of tumor growth [74–76].

In addition, metformin increases ONOO(−) (reactive nitrogen species) generation and exposure of eNOS-deficient hepatocytes to metformin did not suppress gluconeogenesis, activate AMPK or increase ONOO(−) generation. Furthermore, metformin lowered fasting blood glucose levels in wild-type diabetic mice, but not in eNOS-deficient diabetic mice, suggesting that activation of AMPK by metformin is dependent on ONOO(−) and the action of metformin in liver is dependent on intra-hepatocellular eNOS generation [77].

Polyunsaturated Fatty Acids and Osteoporosis

Based on the preceding discussion it is likely that strategies designed to enhance eNO synthesis could prevent osteoporosis. Similar to estrogen, statins and polyunsaturated fatty acids (PUFAs) also enhance eNO synthesis and hence, are likely to be useful in osteoporosis [78–81]. Essential fatty acids and their metabolites can prevent osteoporosis [80–83]. Thus, these three structurally different agents: estrogen, statins and PUFAs have the same beneficial action namely prevention of osteoporosis by their ability to augment constitutional (or endothelial) nitric oxide generation, which is known to be beneficial in osteoporosis. Estrogen, statins and PUFAs suppress the production of pro-inflammatory cytokines that could also be responsible

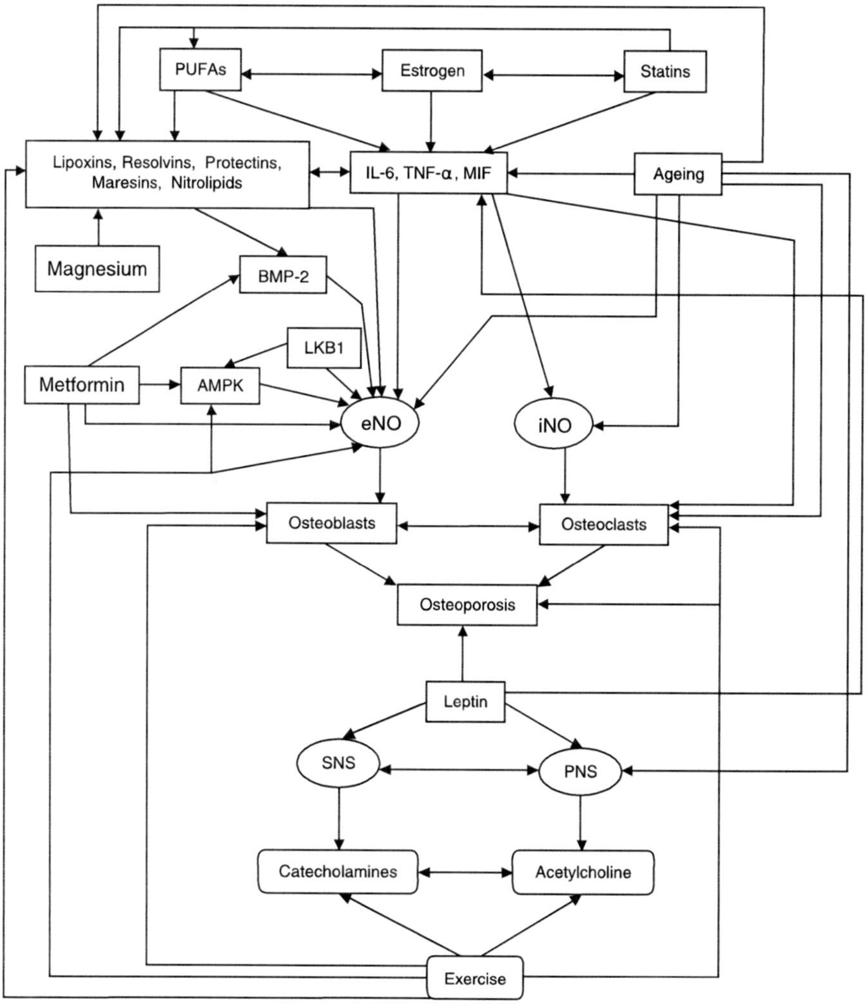

Fig. 11.1 Scheme showing the interaction(s) among various factors involved in the pathobiology of osteoporosis. Osteoporosis is common in post-menopausal state, ageing and inflammatory conditions. Estrogen inhibits the production of pro-inflammatory cytokines such as IL-6 and TNF-α and ageing may also be associated with increase in the production of these cytokines. Both post-menopausal state and ageing cause decrease in eNO generation while eNO inhibits osteoporosis by enhancing osteoblast activity and decreasing osteoclast differentiation and proliferation. Polyunsaturated fatty acids and their products such as lipoxins, resolvins, protectins and nitrolipids posses anti-inflammatory actions and thus, they may prevent osteoporosis. PUFAs form an important constituent of cell membranes and regulate cell membrane fluidity and the expression of various receptors. PUFAs may regulate the expression of estrogen receptors on the cell membrane and thus, an interaction could exist between estrogen and PUFAs. Ageing may decrease the activity of Δ^6 and Δ^5 desaturases, enzymes that are essential for the conversion of dietary essential fatty acids to their long-chain metabolites such as AA, EPA and DHA that form precursors to anti-inflammatory compounds lipoxins, resolvins and protectins. Hence, ageing could lead to decreased formation of AA,

for their beneficial action in osteoporosis. One possibility is that PUFAs especially arachidonic acid, eicosapentaenoic acid and docosahexaenoic acid may enhance the formation of anti-inflammatory compounds such as lipoxins, resolvins and protectins that may inhibit osteoporosis [84, 85].

Thus, understanding the molecular mechanisms of glucose regulation and the actions of metformin suggest that this anti-diabetic drug enhances eNO generation, activates AMPK and LKB1 that form the basis of its beneficial actions in the treatment of hyperglycemia, osteoporosis and cancer. Similar to metformin, even PUFAs and their products such as lipoxins, resolvins and protectins also enhance eNO synthesis and posses anti-inflammatory actions and thus, could be of significant benefit in osteoporosis, insulin resistance and cancer. Exercise also has actions similar to metformin. Exercise enhances eNO generation, activates AMPK and LKB1 and hence, the beneficial actions of exercise in the prevention of osteoporosis, ability to lower blood glucose levels and prevention of cancer can also be ascribed to the same molecular actions as those seen with metformin. Estrogen also stimulates eNO generation [78, 79] and this could be one mechanism by which it is beneficial in osteoporosis. Raloxifene, a selective estrogen receptor modulator that is effective for the prevention of post-menopausal osteoporosis, also stimulates nitric oxide (NO) synthesis [86]. Thus, eNO plays a significant role in the prevention of osteoporosis (see Fig. 11.1).

It is evident from the preceding discussion that post-menopausal osteoporosis is an inflammatory condition and that eNO, AMPK and LKB1 participate not only in glucose homeostasis but also regulate inflammation either directly or indirectly and control tumor cell growth. Thus, under certain conditions metformin may show

---

EPA and DHA and as a result their products: lipoxins, resolvins and protectins. Deficiency of AA, EPA, DHA and lipoxins, resolvins and protectins could lead to low-grade inflammation due to the absence of negative control exerted by PUFAs and their anti-inflammatory products. Magnesium enhances the activities of desaturases and thus, could potentially enhance the formation of PUFAs and their anti-inflammatory products lipoxins, resolvins and protectins. Endothelial NO enhances the activity and differentiation and proliferation of osteoblasts while inducible NO enhances the activity and proliferation of osteoclasts. In inflammatory conditions such as rheumatoid arthritis and lupus, osteoporosis could be as a result of enhanced iNOS. Leptin enhances osteoporosis by augmenting sympathetic activity, while β-adrenergic antagonists increase bone mass. Both leptin and catecholamines have pro-inflammatory actions, while acetylcholine has anti-inflammatory actions. It is likely that catecholamines and leptin could inhibit the formation of anti-inflammatory compounds lipoxins, resolvins and protectins from PUFAs whereas acetylcholine could enhance their formation. But this needs to be established. Anti-diabetic drug metformin enhances eNO generation, activates AMPK, and augmented the production of BMP-2 (bone morphogenetic protein-2) and induced the proliferation of osteoblasts and thus, it may prevent osteoporosis. Metformin may also have anti-inflammatory actions but it is not known whether it can also augment the formation of lipoxins, resolvins and protectins. Exercise augments the formation of eNO, enhances the generation of lipoxins (and possibly that of resolvins and protectins), increases the activity of AMPK and LKB1, enhances parasympathetic tone and brain acetylcholine levels and shows anti-inflammatory actions by decreasing the formation of IL-6 and TNF-α. Thus, exercise can prevent osteoporosis. Exercise may also enhance the activity of osteoblasts and their proliferation while decreasing the activity and proliferation of osteoclasts. For further details see text anti-inflammatory actions [87–90]. These findings reinforce the emerging intimate relationship that exists between physiological control of metabolism (especially of glucose and fat) and cancer. The mTOR {there is a cross-talk between mTOR, mammalian target of rapamycin, and AMPK, [91]}, insulin, and LKB1 pathways represent a fundamental eukaryotic network governing cell growth in response to environmental nutrients. Dysregulation of each contributes to diabetes, cancer and other diseases.

References

[1] WHO (1994) Assessment of fracture risk and its application to screening for postmenopausal osteoporosis. Report of a WHO Study Group. World Health Organ Tech Rep Ser 843:1–129
[2] Feskanich D, Willett WC, Stampfer MJ, Colditz GA (1996) Protein consumption and bone fractures in women. Am J Epidemiol 143:472–479
[3] Munger RG, Cerhan JR, Chiu BC (1999) Prospective study of dietary protein intake and risk of hip fracture in postmenopausal women. Am J Clin Nutr 69:147–152
[4] Rizzoli R, Ammann P, Chevalley T, Bonjour JP (2001) Protein intake and bone disorders in the elderly. Joint Bone Spine 68:383–392
[5] Schürch MA, Rizzoli R, Slosman D, Vadas L, Vergnaud P, Bonjour JP (1998) Protein supplements increase serum insulin-like growth factor-I levels and attenuate proximal femur bone loss in patients with recent hip fracture. A randomized, double-blind, placebo-controlled trial. Ann Intern Med 128:801–809
[6] Promislow JH, Goodman-Gruen D, Slymen DJ, Barrett-Connor E (2002) Protein consumption and bone mineral density in the elderly: the Rancho Bernardo Study. Am J Epidemiol 155:636–644
[7] Weikert C, Walter D, Hoffmann K, Kroke A, Bergmann MM, Boeing H (2005) The relation between dietary protein, calcium and bone health in women: results from the EPIC-Potsdam cohort. Ann Nutr Metab 49:312–318
[8] Sellmeyer DE, Stone KL, Sebastian A, Cummings SR (2001) A high ratio of dietary animal to vegetable protein increases the rate of bone loss and the risk of fracture in postmenopausal women. Study of Osteoporotic Fractures Research Group. Am J Clin Nutr 73:118–122
[9] Mardon J, Habauzit V, Trzeciakiewicz A, Davicco MJ, Lebecque P, Mercier S, Tressol JC, Horcajada MN, Demigné C, Coxam V (2008) Influence of high and low protein intakes on age-related bone loss in rats submitted to adequate or restricted energy conditions. Calcif Tissue Int 82:373–382
[10] Carpenter TO, Mackowiak SJ, Troiano N, Gundberg CM (1992) Osteocalcin and its message: relationship to bone histology in magnesium-deprived rats. Am J Physiol 263(1 Pt 1):E107–E114
[11] Seelig MS (1993) Interrelationship of magnesium and estrogen in cardiovascular and bone disorders, eclampsia, migraine and premenstrual syndrome. J Am Coll Nutr 12:442–458
[12] Rude RK, Gruber HE, Norton HJ, Wei LY, Frausto A, Kilburn J (2005) Dietary magnesium reduction to 25% of nutrient requirement disrupts bone and mineral metabolism in the rat. Bone 37:211–219
[13] Rude RK, Gruber HE, Norton HJ, Wei LY, Frausto A, Mills BG (2004) Bone loss induced by dietary magnesium reduction to 10% of the nutrient requirement in rats is associated with increased release of substance P and tumor necrosis factor-alpha. J Nutr 134:79–85
[14] Rude RK, Gruber HE, Wei LY, Frausto A (2005) Immunolocalization of RANKL is increased and OPG decreased during dietary magnesium deficiency in the rat. Nutr Metab (Lond) 2:24
[15] Rude RK, Wei L, Norton HJ, Lu SS, Dempster DW, Gruber HE (2009) TNFalpha receptor knockout in mice reduces adverse effects of magnesium deficiency on bone. Growth Factors 27:370–376

[16] Ito A, Bebo BF Jr, Matejuk A, Zamora A, Silverman M, Fyfe-Johnson A, Offner H (2001) Estrogen treatment down-regulates TNF-alpha production and reduces the severity of experimental autoimmune encephalomyelitis in cytokine knockout mice. J Immunol 167:542–552

[17] Salem ML (2004) Estrogen, a double-edged sword: modulation of TH1- and TH2-mediated inflammations by differential regulation of TH1/TH2 cytokine production. Curr Drug Targets Inflamm Allergy 3:97–104

[18] Giraud SN, Caron CM, Pham-Dinh D, Kitabgi P, Nicot AB (2010) Estradiol inhibits ongoing autoimmune neuroinflammation and NFkappaB-dependent CCL2 expression in reactive astrocytes. Proc Natl Acad Sci U S A 107:8416–8421

[19] Houdeau E, Moriez R, Leveque M, Salvador-Cartier C, Waget A, Leng L, Bueno L, Bucala R, Fioramonti J (2007) Sex steroid regulation of macrophage migration inhibitory factor in normal and inflamed colon in the female rat. Gastroenterology 132:982–993

[20] Kitzmiller JL, Rocklin RE (1980) Lack of suppression of lymphocyte MIF production by estradiol, progesterone and human chorionic gonadotropin. J Reprod Immunol 1:297–306

[21] Abrahamsen B, Bonnevie-Nielsen V, Ebbesen EN, Gram J, Beck-Nielsen H (2000) Cytokines and bone loss in a 5-year longitudinal study – hormone replacement therapy suppresses serum soluble interleukin-6 receptor and increases interleukin-1-receptor antagonist: the Danish osteoporosis prevention study. J Bone Miner Res 15:1545–1554

[22] Pfeilschifter J, Köditz R, Pfohl M, Schatz H (2002) Changes in proinflammatory cytokine activity after menopause. Endocr Rev 23:90–119

[23] Clowes JA, Riggs BL, Khosla S (2005) The role of the immune system in the pathophysiology of osteoporosis. Immunol Rev 208:207–227

[24] Ginaldi L, Di Benedetto MC, De Martinis M (2005) Osteoporosis, inflammation and ageing. Immun Ageing 2:14

[25] Ozmen B, Kirmaz C, Aydin K, Kafesciler SO, Guclu F, Hekimsoy Z (2007) Influence of the selective oestrogen receptor modulator (raloxifene hydrochloride) on IL-6, TNF-alpha, TGF-beta1 and bone turnover markers in the treatment of postmenopausal osteoporosis. Eur Cytokine Netw 18:148–153

[26] Pino AM, Ríos S, Astudillo P, Fernàndez M, Figueroa P, Seitz G, Rodríguez JP (2010) Concentration of adipogenic and proinflammatory cytokines in the bone marrow supernatant fluid of osteoporotic women. J Bone Miner Res 25:492–498

[27] Takeda S, Karsenty G (2001) Central control of bone formation. J Bone Miner Metab 19:195–198

[28] Ducy P, Amling M, Takeda S, Priemel M, Schilling AF, Beil FT, Shen J, Vinson C, Rueger JM, Karsenty G (2000) Leptin inhibits bone formation through a hypothalamic relay: a central control of bone mass. Cell 100:197–207

[29] Takeda S, Elefteriou F, Levasseur R, Liu X, Zhao L, Parker KL, Armstrong D, Ducy P, Karsenty G (2002) Leptin regulates bone formation via the sympathetic nervous system. Cell 111:305–317

[30] Das UN (2001) Is obesity an inflammatory condition? Nutrition 17:953–966

[31] Lord GM, Matarese G, Howard JK, Bloom SR, Lechler RI (2002) Leptin inhibits the anti-CD3-driven proliferation of peripheral blood T cells but enhances the production of proinflammatory cytokines. J Leukoc Biol 72:330–338

[32] Otero M, Lago R, Gómez R, Lago F, Gómez-Reino JJ, Gualillo O (2006) Leptin: a metabolic hormone that functions like a proinflammatory adipokine. Drug News Perspect 19:21–26

[33] Lappas M, Permezel M, Rice GE (2005) Leptin and adiponectin stimulate the release of proinflammatory cytokines and prostaglandins from human placenta and maternal adipose tissue via nuclear factor-kappaB, peroxisomal proliferator-activated receptor-gamma and extracellularly regulated kinase 1/2. Endocrinology 146:3334–3342

[34] Aleffi S, Petrai I, Bertolani C, Parola M, Colombatto S, Novo E, Vizzutti F, Anania FA, Milani S, Rombouts K, Laffi G, Pinzani M, Marra F (2005) Upregulation of proinflammatory and proangiogenic cytokines by leptin in human hepatic stellate cells. Hepatology 42:1339–1348

[35] Flierl MA, Rittirsch D, Huber-Lang M, Sarma JV, Ward PA (2008) Catecholamines – crafty weapons in the inflammatory arsenal of immune/inflammatory cells or opening pandora's box? Mol Med 14:195–204
[36] Flierl MA, Rittirsch D, Nadeau BA, Sarma JV, Day DE, Lentschz AB, Huber-Lang MS, Ward PA (2009) Upregulation of phagocyte-derived catecholamines augments the acute inflammatory response. PLoS One 4:e4414
[37] Flierl MA, Rittirsch D, Nadeau BA, Chen AJ, Sarma JV, Zetoune FS, McGuire SR, List RP, Day DE, Hoesel LM, Gao H, Rooijen NV, Huber-Lang MS, Neubig RR, Ward PA (2007) Phagocyte-derived catecholamines enhance acute inflammatory injury. Nature 449:721–726
[38] Pavlov VA, Wang H, Czura CJ, Friedman SG, Tracey KJ (2003) The cholinergic anti-inflammatory pathway: a missing link in neuroimmunomodulation. Mol Med 9:125–134
[39] Matthay MA, Ware LB (2004) Can nicotine treat sepsis? Nat Med 10:1161–1162
[40] Wang H, Yu M, Ochani M, Amella CA, Tanovic M, Susarla S, Li HJ, Wang H, Yang H, Ulloa L, Al-Abed Y, Czura CJ, Tracey KJ (2003) Nicotinic acetylcholine receptor a7 subunit is an essential regulator of inflammation. Nature 421:384–388
[41] Das UN (2004) Anti-inflammatory nature of exercise. Nutrition 20:323–326
[42] Das UN (2006) Exercise and inflammation. Eur Heart J 27:1385–1386
[43] Lemos TL, Reis F, Baptista S, Pinto R, Sepodes B, Vala H, Rocha-Pereira P, Correia de Silva G, Teixeira N, Silva SA, Carvalho L, Teixeira F, Das UN (2009) Exercise training decreases proinflammatory profile in Zucker diabetic (type 2) fatty rats. Nutrition 25:330–339
[44] Seals DR, Chase PB (1989) Influence of physical training on heart rate variability and baroreflex circulatory control. J Appl Physiol 66:1886–1895
[45] O'Leary DS, Seamans DP (1993) Effect of exercise on autonomic mechanisms of baroreflex control of heart rate. J Appl Physiol 75:2251–2257
[46] Charlton GA, Crawford MH (1997) Physiologic consequences of training. Cardiol Clin 15:345–354
[47] Levy WC, Cerqueira MD, Harp GD, Johannessen KA, Abrass IB, Schwartz RS, Stratton JR (1998) Effect of endurance exercise training on heart rate variability at rest in healthy young and older men. Am J Cardiol 82:1236–1241
[48] Singh P, Hawkley LC, McDade TW, Cacioppo JT, Masi CM (2009) Autonomic tone and C-reactive protein: a prospective population-based study. Clin Auton Res 19:367–374
[49] Meyer O (2001) Atherosclerosis and connective tissue diseases. Joint Bone Spine 68:564–575
[50] Bijl M (2003) Endothelial activation, endothelial dysfunction and premature atherosclerosis in systemic autoimmune diseases. Neth J Med 61:273–277
[51] Tyrrell PN, Beyene J, Feldman BM, McCrindle BW, Silverman ED, Bradley TJ (2010) Rheumatic disease and carotid intima-media thickness: a systematic review and meta-analysis. Arterioscler Thromb Vasc Biol 30:1014–1026
[52] MacIntyre I, Zaidi M, Alam AS, Datta HK, Moonga BS, Lidbury PS, Hecker M, Vane JR (1991) Osteoclastic inhibition: an action of nitric oxide not mediated by cyclic GMP. Proc Natl Acad Sci U S A 88:2936–2940
[53] Kasten TP, Collin-Osdoby P, Patel N, Osdoby P, Krukowski M, Misko TP, Settle SL, Currie MG, Nickols GA (1994) Potentiation of osteoclast bone-resorption activity by inhibition of nitric oxide synthase. Proc Natl Acad Sci U S A 91:3569–3573
[54] Evans DM, Ralston SH (1996) Nitric oxide and bone. J Bone Miner Res 11:300–305
[55] Kasten TP, Collin-Osdoby P, Patel N, Osdoby P, Krukowski M, Misko TP, Settle SL, Currie MG, Nickols GA (1994) Potentiation of osteoclast bone-resorption activity by inhibition of nitric oxide synthase. Proc Natl Acad Sci U S A 91:3569–3573
[56] Wimalawansa SJ, De Marco G, Gangula P, Yallampalli C (1996) Nitric oxide donor alleviates ovariectomy-induced bone loss. Bone 18:301–304
[57] Wimalawansa SJ, Chapa MT, Yallampalli C, Zhang R, Simmons DJ (1997) Prevention of corticosteroid-induced bone loss with nitric oxide donor nitroglycerin in male rats. Bone 21:275–280
[58] Jamal SA, Browner WS, Bauer DC, Cummings SR (1998) Intermittent use of nitrates increases bone mineral density: the study of osteoporotic fractures. J Bone Miner Res 13:1755–1759

[59] Pfeilschifter J, Köditz R, Pfohl M, Schatz H (2002) Changes in proinflammatory cytokine activity after menopause. Endocr Rev 23:90–119
[60] Hao YJ, Tang Y, Chen FB, Pei FX (2005) Different doses of nitric oxide donor prevent osteoporosis in ovariectomized rats. Clin Orthop Relat Res 435:226–231
[61] Jamal SA, Cummings SR, Hawker GA (2004) Isosorbide mononitrate increases bone formation and decreases bone resorption in postmenopausal women: a randomized trial. J Bone Miner Res 19:1512–1517
[62] Rejnmark L, Vestergaard P, Mosekilde L (2006) Decreased fracture risk in users of organic nitrates: a nationwide case-control study. J Bone Miner Res 21:1811–1817
[63] Ozgocmen S, Kaya H, Fadillioglu E, Aydogan R, Yilmaz Z (2007) Role of antioxidant systems, lipid peroxidation, and nitric oxide in postmenopausal osteoporosis. Mol Cell Biochem 295:45–52
[64] Taylor BC, Schreiner PJ, Zmuda JM, Li J, Moffett SP, Beck TJ, Cummings SR, Lee JM, Walker K, Ensrud KE; for the SOF Research Group (2006) Association of endothelial nitric oxide synthase genotypes with bone mineral density, bone loss, hip structure, and risk of fracture in older women: the SOF study. Bone 39:174–180
[65] Zhou G, Myers R, Li Y, Chen Y, Shen X, Fenyk-Melody J, Wu M, Ventre J, Doebber T, Fujii N, Musi N, Hirshman MF, Goodyear LJ, Moller DE (2001) Role of AMP-activated protein kinase in mechanism of metformin action. J Clin Invest 108:1167–1174
[66] Hawley SA, Gadalla AE, Olsen GS, Hardie DG (2002) The antidiabetic drug metformin activates the AMP-activated protein kinase cascade via an adenine nucleotide-independent mechanism. Diabetes 51:2420–2425
[67] Kanazawa I, Yamaguchi T, Yano S, Yamauchi M, Sugimoto T (2008) Metformin enhances the differentiation and mineralization of osteoblastic MC3T3-E1 cells via AMP kinase activation as well as eNOS and BMP-2 expression. Biochem Biophys Res Commun 375:414–419
[68] Xue B, Kahn BB (2006) AMPK integrates nutrient and hormonal signals to regulate food intake and energy balance through effects in the hypothalamus and peripheral tissues. J Physiol 574:73–83
[69] Banerjee RR, Rangwala SM, Shapiro JS, Rich AS, Rhoades B, Qi Y, Wang J, Rajala MW, Pocai A, Scherer PE, Steppan CM, Ahima SA, Obici S, Rossetti L, Lazar MA (2004) Regulation of fasted blood glucose by resistin. Science 303:1195–1198
[70] Yamauchi T, Kamoni J, Minokoshi Y, Ito Y, Waki H, Uchida S, Yamashita S, Noda M, Kita S, Ueki K, Eto K, Akanuma Y, Froguel P, Foufelle F, Ferre P, Carling D, Kimurai S, Nagai R, Kahn BB, Kadowaki T (2002) Adiponectin stimulates glucose utilization and fatty-acid oxidation by activating AMP-activated protein kinase. Nat Med 8:1288–1295
[71] Chen ZP, Stephens TJ, Murthy S, Canny BJ, Hargreaves M, Witters LA, Kemp BE, McConell GK (2003) Effect of exercise intensity on skeletal muscle AMPK signaling in humans. Diabetes 52:2205–2212
[72] Ho-Jin K, Josef B, Goodyear LJ (2008) LKB1 and AMPK and the regulation of skeletal muscle metabolism. Curr Opin Clin Nutr Metab Care 11:227–232
[73] Shaw RJ, Lamia KA, Vasquez D, Koo S-H, Bardeesy N, DePinho RA, Montminy M, Cantley LC (2005) The Kinase LKB1 mediates glucose homeostasis in liver and therapeutic effects of metformin. Science 310:1642–1646
[74] Huang X, Wullschleger S, Shpiro N, McGuire VA, Sakamoto K, Woods YL, McBurnie W, Fleming S, Alessi DR (2008) Important role of the LKB1-AMPK pathway in suppressing tumorigenesis in PTEN-deficient mice. Biochem J 412:211–221
[75] Zakikhani M, Dowling RJ, Sonenberg N, Pollak MN (2008) The effects of adiponectin and metformin on prostate and colon neoplasia involve activation of AMP-activated protein kinase. Cancer Prev Res (Phila Pa) 1:369–375
[76] Kisfalvi K, Eibl G, Sinnett-Smith J, Rozengurt E (2009) Metformin disrupts crosstalk between G protein-coupled receptor and insulin receptor signaling systems and inhibits pancreatic cancer growth. Cancer Res 69:6539–6545

[77] Fujita Y, Hosokawa M, Fujimoto S, Mukai E, Abudukadier A, Obara A, Ogura M, Nakamura Y, Toyoda K, Nagashima K, Seino Y, Inagaki N (2010) Metformin suppresses hepatic gluconeogenesis and lowers fasting blood glucose levels through reactive nitrogen species in mice. Diabetologia 53:1472–1481

[78] Sheng JZ, Arshad F, Braun JE, Braun AP (2008) Estrogen and the Ca2+-mobilizing agonist ATP evoke acute NO synthesis via distinct pathways in an individual human vascular endothelium-derived cell. Am J Physiol Cell Physiol 294:C1531–C1541

[79] Hamilton CA, Groves S, Carswell HV, Brosnan MJ, Graham D, Dominiczak AF (2006) Estrogen treatment enhances nitric oxide bioavailability in normotensive but not hypertensive rats. Am J Hypertens 19:859–866

[80] Das UN (2002) Nitric oxide as the mediator of the antiosteoporotic actions of estrogen, statins, and essential fatty acids. Exp Biol Med (Maywood) 227:88–93

[81] Das UN (2002) Estrogen, statins, and polyunsaturated fatty acids: similarities in their actions and benefits-is there a common link? Nutrition 18:178–188

[82] Bhattacharya A, Rahman M, Banu J, Lawrence RA, McGuff HS, Garrett IR, Fischbach M, Fernandes G (2005) Inhibition of osteoporosis in autoimmune disease prone MRL/Mpj-Fas(lpr) mice by N-3 fatty acids. J Am Coll Nutr 24:200–209

[83] Högström M, Nordström P, Nordström A (2007) n-3 Fatty acids are positively associated with peak bone mineral density and bone accrual in healthy men: the NO_2 study. Am J Clin Nutr 85:803–807

[84] Serhan CN (2004) Clues for new therapeutics in osteoporosis and periodontal disease: new roles for lipoxygenases? Expert Opin Ther Targets 8:643–652

[85] Kühn H, O'Donnell VB (2006) Inflammation and immune regulation by 12/15-lipoxygenases. Prog Lipid Res 45:334–356

[86] Leitzbach D, Weckler N, Madajka M, Malinski T, Wiemer G, Linz W (2005) Restoration of endothelial function via enhanced nitric oxide synthesis after long-term treatment of raloxifene in adult hypertensive rats. Arzneimittelforschung 55:86–92

[87] Xavier DO, Amaral LS, Gomes MA, Rocha MA, Campos PR, Cota BD, Tafuri LS, Paiva AM, Silva JH, Andrade SP, Belo AV (2010) Metformin inhibits inflammatory angiogenesis in a murine sponge model. Biomed Pharmacother 64:220–225

[88] Ersoy C, Kiyici S, Budak F, Oral B, Guclu M, Duran C, Selimoglu H, Erturk E, Tuncel E, Imamoglu S (2008) The effect of metformin treatment on VEGF and PAI-1 levels in obese type 2 diabetic patients. Diabetes Res Clin Pract 81:56–60

[89] Takemura Y, Osuga Y, Yoshino O, Hasegawa A, Hirata T, Hirota Y, Nose E, Morimoto C, Harada M, Koga K, Tajima T, Yano T, Taketani Y (2007) Metformin suppresses interleukin (IL)-1beta-induced IL-8 production, aromatase activation, and proliferation of endometriotic stromal cells. J Clin Endocrinol Metab 92:3213–3218

[90] Hattori Y, Suzuki K, Hattori S, Kasai K (2006) Metformin inhibits cytokine-induced nuclear factor kappaB activation via AMP-activated protein kinase activation in vascular endothelial cells. Hypertension 47:1183–1188

[91] Vila-Bedmar R, Lorenzo M, Fernàndez-Veledo S (2010) Adenosine 5′-monophosphate-activated protein kinase-mammalian target of rapamycin cross talk regulates brown adipocyte differentiation. Endocrinology 151:980–992

Chapter 12
Alzheimer's Disease, Schizophrenia and Depression

Introduction

Alzheimer's disease (AD), the most common form of dementia, is named after German physician Alois Alzheimer, who first described it in 1906. AD is a progressive neurodegenerative disorder characterized by amyloid plaques composed of aggregated amyloid beta plaques, neurofibrillary tangles (NFT) that are composed of hyperphosphorylated tau and synaptic defects resulting in neuritic dystrophy and neuronal death [1]. It is now believed that AD is the most common form of dementia in the ageing population especially in the USA. AD produces loss of memory and problems with thinking and behavior severe enough to affect work, lifelong hobbies or social life. Alzheimer's gets worse over time, and it is fatal. Today it is the seventh-leading cause of death in the United States. The severity of AD may be significant enough to eventually interfere with daily life and thus, these patients may need constant family support to survive. It is estimated that about 5.3 million Americans now have Alzheimer's disease. It is important to note, however, that AD is not a normal part of aging. The duration of Alzheimer's disease can vary from 3 to 20 years, but many die an average of 4–6 years after diagnosis. The disease initially presents with mild cognitive impairment such as memory lapses, especially in forgetting familiar words or names or the location of keys, eyeglasses or other everyday objects. As the disease progresses, they are unable to recognize and remember spouse and children, do not respond to the environment, and ultimately lose ability to speak and control movement.

Pathobiology of Alzheimer's Disease

Plaques and tangles are believed to play a significant role in the pathogenesis of Alzheimer's disease. Plaques that build up between nerve cells contain deposits of the protein fragment *β-amyloid*. Tangles are twisted fibers of the protein *tau* and form inside dying cells. The plaques and tangles when form in areas that are important in learning and memory could lead to memory loss. Plaques and tangles found in

U. N. Das, *Molecular Basis of Health and Disease*,
DOI 10.1007/978-94-007-0495-4_12, © Springer Science+Business Media B.V. 2011

Alzheimer's disease block communication among nerve cells and interfere with their function and ultimately induce their apoptosis that lead to loss of memory and other mental abilities. In AD, the levels of the neurotransmitter acetylcholine (Ach) are lower that may explain some if not all features of Alzheimer's disease. Hence, efforts are being made to develop drugs that either enhance its levels or interfere with its catabolism so that the levels of Ach are maintained near normal levels with the hope that this would prevent or stabilize the disease process.

It has been suggested that missense mutations in amyloid precursor protein (APP), presenillin-1 (PS-1), presenillin-2 (PS-2) genes alter the proteolysis of APP and increase the generation of Aβ42 (amyloid β 42). The accumulation of Aβ42 could trigger the inflammatory responses by causing microglial activation that leads to the release of pro-inflammatory cytokines such as interleukin-6 (IL-6), IL-1, tumor necrosis factor-α (TNF-α) and macrophage migration inhibitory factor (MIF) [2–4]. These evidences add to the increasing amounts of evidence that suggests that inflammatory processes are involved in the neurotoxicity of AD. The pro-inflammatory cytokines released by the activated microglia may lead to neuronal death and dysfunction by (a) enhancing glutamate-induced excitotoxicity [5]; (b) inhibition of long-term potentiation, which limits functional plasticity after neuronal injury [6, 7]; and (c) inhibition of hippocampal neurogenesis [8]. Of all the pro-inflammatory cytokines that are believed to have a role in AD, TNF-α appears to be particularly important as a potential intermediary in AD. Several recent studies reported elevated TNF-α level in the cerebrospinal fluid (CSF) and serum of AD patients [9–11], and it was reported that a single nucleotide polymorphism in the TNF-α gene is associated with earlier onset of AD [12]. These observations suggested that efforts made to suppress neuroinflammation could be a rational approach to halt the AD disease process. With this belief, therapeutic attempts to suppress or prevent microglial activation or proinflammatory cytokine release in addition to anti-amyloid therapies are being attempted.

Several genes that are likely to increase susceptibility for AD include: apolipoprotein E (ApoE 4) variant [13], 2-macroglobulin [14], the K-variant of butyrylcholinesterase [15], and several mitochondrial genes [16]. Some of the other factors that also play a role in the aetiopathogenesis of AD include: brain metabolic abnormalities, environmental factors, and age related decrease in neuronal membrane fluidity that by increasing the formation of amyloid beta plaques and hyperphosphorylation of tau protein may cause death of the neurons [17].

Mutations in presenilins lead to dominant inheritance of Familial Alzheimer's disease (FAD). These mutations alter the cleavage of γ-secretase of the amyloid precursor protein, resulting in the increased ratio of Aβ42/Aβ40 and accelerated amyloid plaque pathology in transgenic mouse models [18]. Proteolytic processing of APP by β-secretase, γ-secretase, and caspases could lead to the generation of A-beta peptide and carboxyl-terminal fragments (CTF) of APP, which may also participate in the pathogenesis of Alzheimer's disease [19]. Missense mutations in the gene encoding APP, as well as those in the genes encoding PS-1 and PS-2, share the common feature of altering the γ-secretase cleavage of APP to increase the production of the amyloidogenic Aβ42, a primary component of amyloid plaques in both familial and sporadic AD.

Our bioinformatics study revealed that presenilin-1 (PS-1), presenilin-2 (PS-2), and amyloid precursor protein (APP) out of 73 proteins that were studied appeared to be the key pathological proteins in the evolution of Alzheimer's disease (unpublished data).

Thus, factors that influence the initiation and progression and play a role in the pathogenesis of AD include: (i) Aβ42/ Aβ40 ratio and oligomers of these peptides; (ii) oxidative stress; (iii) proinflammatory cytokines produced by activated glial cells, (iv) alterations in cholesterol homeostasis, and (v) alterations in cholinergic nervous system [20].

Amyloid β in AD

Amyloid beta (Aβ) is a major component of amyloid plaques characterizing AD and various factors that could contribute to its accumulation include increased rates of its production and/or impaired clearance. Insulin degrading enzyme is responsible for the degradation and clearance of amyloid beta in the brain thus, may also have a role in the pathogenesis of AD [21].

Several studies showed that Aβ is toxic to cultured neuronal cells and induces tau phosphorylation [17, 22]. Tau being a microtubule-associated protein that is essential for the stabilization of neuronal microtubules under normal physiological conditions, when it undergoes modifications, mainly through phosphorylation, it can result in the generation of aberrant aggregates that are toxic to neurons [23]. Based on this evidence, amyloid vaccine (both passive and active immunization against amyloid) has been attempted to arrest and even reverse both plaque pathology and behavioral phenotypes in the transgenic animals [24]. But, human clinical trials gave negative results.

Aβ42 fibrils significantly accelerate neurofibrillary tangles formation in mice providing support to the role of amyloid beta in AD. Mutations in tau give rise to neurofibrillary tangles but not plaques. In contrast, mutations in APP or in the probable APP proteases give rise to both plaques and tangles indicating that amyloid pathology occurs upstream of tau pathology. Although the exact mechanism(s) is not clear, it appears that amyloid β enhance free radical generation and induce inflammation that results in loss in the cholinergic system including decrease in choline acetyltransferase level, choline uptake, and decrease in acetylcholine (ACh) level which may cause cognitive deficits seen in AD.

Oxidative Stress Causes Neuronal Death

Aβ causes hydrogen peroxide (H_2O_2) accumulation in cultured hippocampal neurons [25] that results in oxidative damage to cellular phospholipid membranes indicating that lipid peroxidation participates in the pathogenesis of AD [26, 27]. The loss of membrane integrity produced by Aβ-induced free-radical damage leads to cellular dysfunction, including inhibition of ion-motive ATPase, loss of calcium homeostasis, inhibition of glial cell Na^+-dependent glutamate uptake system that results in NMDA receptors mediated delayed neurodegeneration, loss of protein transporter function, disruption of signaling pathways, and activation of nuclear transcription factors and apoptotic pathways.

In a recent study, it was observed that manipulation to genetically boost the ability to quench free radicals of specific mitochondrial origin in mice resulted in protection against vascular as well as neuronal deficits that are typically seen in AD [27]. It was also noted that the vascular deficits were improved via antioxidant modulation of the endothelial nitric oxide synthase, while neuronal deficits were improved via modulation of the phosphorylation status of the protein tau, a neuronal cytoskeletal stabilizer. These findings directly link free radicals of specific mitochondrial origin to AD-associated vascular and neuronal pathology.

Alzheimer's is an Inflammatory Condition

There is mounting evidence to suggest that inflammation plays a significant role in the pathobiology of Alzheimer's disease [28]. In general, it is thought that a chronic inflammatory response mediated by amyloid-beta is a major factor in the pathology of AD. In vivo administration of Aβ produced an inflammatory response and vascular disruption as seen in the brains of AD patients. Increased levels of IL-1β have been detected in brain tissue, cerebrospinal fluid, and blood/serum from AD patients. Aβ stimulated the production of TNF-α in brain astrocytes and murine monocytes. When adult male rats were perfused via an intra-aortic cannula with either Aβ alone, interleukin-1 receptor antagonist (IL-1 ra) plus Aβ, tumor necrosis factor-binding protein (TNF-bp) plus Aβ or sterile saline, serum TNF-α, IL-1β, Aβ and NO showed a significant increase in TNF-α and Aβ but not in IL-1β or NO after the injection of Aβ. In rats given Aβ alone there was extensive vascular disruption, including endothelial and smooth muscle damage with leukocyte adhesion and migration, while animals receiving either IL-1ra or TNF-bp before Aβ showed no in vivo leukocyte extravasation or vascular damage. These results suggested that pro-inflammatory cytokines TNF-α and IL-1β seem to mediate the vascular disruption and inflammatory response initiated by Aβ, implying that inflammation plays a role in the pathogenesis of AD [29].

Lombardi et al. [30] observed a significant increase in the blood mononuclear cell CD4, CD25 and CD28 antigen expression in the AD group (55.3%, 24.8% and 65.1%) compared to healthy subjects (44.5%, 10.3% and 54.3%). In this study, lipopolysaccharide-stimulated in vitro production of IL-1β, TNF-α, IL-6 and IL-10 measured by a whole blood culture system was found to be significantly higher in AD patients compared with control. Furthermore, the observed differences were more evident at late stages of the disease, suggesting that immune activation is present in AD. This enhanced immune activation and/or inflammatory activity seen in AD is supported by the observation that such activation is also present in the brains of AD patients compared with age-matched control patients [31]. Continuous neuroinflammatory processes including glial activation is seen in AD [32]. Microglia and astrocytes would be activated, perceiving Aβ oligomers and fibrils as foreign material, since Aβ assemblies are apparently never observed during the development of brain and in the immature nervous system [1].

β-Amyloid fibrils have been shown to activate parallel mitogen-activated protein kinase pathways in microglia and THP1 monocytes [33]. It was reported that microglia from human AD brain exposed to Aβ produced and secreted a wide range of inflammatory mediators, including cytokines, chemokines, growth factors, complements, and reactive oxygen intermediates [34]. Significant dose-dependent increase in the production of prointerleukin-1, IL-6, TNF-α, monocyte chemoattractant protein-1, macrophage inflammatory peptide-1, IL-8, and macrophage colony-stimulating factor were observed after exposure to preaggregated Amyloid β-42. These and several other evidences emphasize the role of inflammation in the pathogenesis of AD [35–43].

Based on these studies, it is clear that inflammation plays a significant role in the pathobiology of Alzheimer's disease. Hence, methods designed to suppress inflammation and more importantly, the levels of TNF-α could form a new approach in its management. Such efforts could involve inhibition of TNF-α synthesis or neutralizing its actions by using specific monoclonal antibodies [44]. In this context, it is interesting to note that cholinergic upregulation leads to suppression of inflammation. It was shown that the use of acetylcholinesterase (AChE)inhibitors that leads to an increase in acetylcholine levels suppressed lymphocyte proliferation and pro-inflammatory cytokine production, as well as extracellular esterase activity. Anti-inflammatory activity was mediated by the alpha7 nicotinic acetylcholine receptor (neuronal); while the muscarinic receptor had the opposite effect. Treatment of experimental autoimmune encephalomyelitis (EAE), with an anti- sense oligodeoxynucleotide, targeted to AChE mRNA, reduced the clinical severity of the disease and inflammation intensity.

These results suggest that AChEI (acetylcholinesterase inhibition) increases the concentration of extracellular acetylcholine (ACh), rendering it available for interaction with a nicotinic receptor expressed on lymphocytes that leads to suppression of inflammation [45] and emphasize the importance of cholinergic balance in neurological disorders, such as Alzheimer's disease. These results are particularly interesting since, AD has been linked to a deficiency in the brain neurotransmitter acetylcholine [46].

Cholinergic System in Alzheimer's Disease

A primary clinical symptom of Alzheimer's dementia is the progressive deterioration in learning and memory ability that could be attributed to profound loss in the cholinergic system of brain, including dramatic loss of choline acetyltransferase level, choline uptake, and acetylcholine (ACh) level in the neocortex and hippocampus and reduced number of the cholinergic neurons in basal forebrain and nucleus basalis of Meynert, areas that are closely associated with cognitive deficits in AD [47]. Hence, interventions that enhance acetylcholine levels or block further fall in ACh levels may improve cholinergic neurotransmission that, in turn, lead to improvement in learning and memory in AD [48]. Since acetylcholine has anti-inflammatory actions, it is reasonable to predict that a decrease in the levels of ACh may aggravate the inflammatory process and progression of AD. This "cholinergic anti-inflammatory pathway" mediated by ACh acts by inhibiting the production of TNF, IL-1, MIF, and HMGB1 and suppresses the activation of NF-κB expression [49].

Both plasma and cerebrospinal fluid levels of pro-inflammatory cytokines: IL-1 and TNF-α are increased in patients with AD [50, 51]. Systemic injection of IL-1 decreased extracellular acetylcholine in the hippocampus suggesting that increased concentrations of IL-1 in patients with AD could be responsible for lowered cerebral acetylcholine levels seen. In addition, IL-1 stimulates the beta-amyloid precursor protein promoter that is found in the form of amyloid plaques in the brains of AD diseased patients. Furthermore, receptors of IL-1 are on APP (amyloid precursor protein) mRNA positive cells and its ability to promote APP gene expression suggests that IL-1 plays an important role in AD [52]. The involvement of inflammatory process in the pathogenesis of AD is further supported by the observation that inhibition or neutralizing the actions of TNF-α could be of benefit to these patients [44, 53].

In addition, pro-inflammatory cytokines seem to have the ability to interfere with the synthesis and actions of neurotrophic factors that are essential for the survival of neurons.

Neurotrophic Factors and AD

Neurotrophic factors such as nerve growth factor (NGF) and brain-derived neurotrophic factor (BDNF) are known to exert neurotrophic actions on the cholinergic neurons of the basal forebrain nuclei. These neurotrophic factors are synthesized by hippocampal and cortical neurons that are located in the projection field of the basal forebrain cholinergic neurons. Maintenance of the normal levels of NGF- and BDNF-mRNAs is mediated predominantly by NMDA receptors. Synthesis of BDNF and NGF in neurons of the hippocampus is regulated by neuronal activity, while the glutamate system is responsible for their upregulation and the GABAergic system for down regulation. During early postnatal development, the activity dependent regulation of NGF and BDNF is mediated by NMDA receptors that are also influenced by the cholinergic system [54]. In the postmortem samples of hippocampus from AD and control donors, it was noticed that BDNF was decreased in AD, suggesting the possibility that deceased expression of BDNF may contribute to the progression of cell death in AD [55–57]. These results coupled with the observation that

IL-1β decreases BDNF messenger RNA expression in the hippocampus indicates that inflammatory cytokines initiate and enhance the progression of AD by decreasing BDNF production. Hence, it is likely that enhanced levels of IL-1 in seen in AD may cause alterations in the hippocampal neuron synaptic plasticity by stimulating β-amyloid production and by decreasing BDNF production [58]. In addition, IL-1β could render neurons vulnerable to degeneration by interfering with BDNF-induced neuroprotection by compromising the PI3-K/Akt pathway-mediated protection by BDNF and suppressing Akt and MAPK/ERK activation. Activation of CREB, a target of these signaling pathways, was severely depressed by IL-1 β. It is presumed that IL-1 β interferes with BDNF signaling at the docking step that conveys activation to the PI3-K/Akt and Ras/MAPK pathways since it was noted that IL-1β suppressed the activation of the respective scaffolding proteins IRS-1 and Shc, an effect that involves ceramide generation. IL-1-induced interference with BDNF neuroprotection and signal transduction was corrected, in part, by ceramide production inhibitors and mimicked by the cell-permeable C2-ceramide, suggesting that IL-1 β interferes with BDNF signaling involving a ceramide-associated mechanism [59].

Exercise enhances the levels of BDNF in the brain providing support to the concept that physical exercise is beneficial in AD [60–62]. The importance of BDNF in AD is further supported by the observation that in amyloid-transgenic mice, BDNF gene delivery, when administered after disease onset, reversed synapse loss, partially normalized aberrant gene expression, improved cell signaling and restored learning and memory. These outcomes occurred independently of effects on amyloid plaque load. In aged rats, BDNF infusion reversed cognitive decline, improved age-related perturbations in gene expression and restored cell signaling. In adult rats and primates, BDNF prevented lesion-induced death of cortical neurons. In aged primates, BDNF reversed neuronal atrophy and ameliorated age-related cognitive impairment. These and other observations indicate that BDNF exerts substantial protective effects on crucial neuronal circuitry involved in AD [63, 64] and suggests that BDNF has therapeutic potential in the management of AD [65, 66].

Recent studies suggested that lithium decreases Amyloid β peptide production and inhibits the activity of glycogen synthase kinase-3 which induces aggregation of tau protein into tangles, and tau hyperphosphorylation. This is supported by the observation that the incidence of AD is lower in patients on chronic lithium treatment, which also increases BDNF activity [67]. In a randomized, single-blinded, placebo-controlled, parallel-group multicenter 10-week study, lithium treatment in AD patients produced a significant increase of BDNF serum levels, and additionally a significant diminution of cognitive impairment was found. This suggests that upregulation of BDNF might be part of a neuroprotective effect of lithium in AD patients [68]. These results once again support the concept that BDNF has a significant role in AD and methods designed to enhance brain BDNF levels are of significant benefit in this condition.

Yet another method of enhancing brain BDNF levels is the use of polyunsaturated fatty acids (PUFAs).

PUFAs in Alzheimer's Disease

There is reasonable evidence to suggest that PUFAs, especially ω-3 fatty acids, could be useful in the prevention and treatment of AD. PUFAs, eicosapentaenoic acid (EPA) and docosahexaenoic acid (DHA) of ω-3 series and arachidonic acid of ω-6 series are essential for neurocognitive development and normal brain functioning [69], and DHA improved memory performance in aged mice [70–72]. A reduction in dietary DHA in an Alzheimer's mouse model showed loss of postsynaptic proteins associated with increased oxidation, increased caspase-cleaved actin, which was localized in dendrites. In contrast, when DHA-restricted mice were given DHA, the fatty acid protected them against dendritic pathology and behavioral deficits and increased anti-apoptotic BAD phosphorylation. These results suggest that DHA is useful in preventing Alzheimer's disease in which synaptic loss is critical [73].

DHA attenuated amyloid-β secretion accompanied by the formation of neuroprotectin D1 (NPD1), a DHA-derived 10,17S-docosatriene [74, 75]. In addition, DHA inhibited IL-6 and TNF-α production that are neurotoxic and increased the synthesis of endothelial nitric oxide (eNO), a neurotransmitter. In Alzheimer's disease, hippocampal DHA and NPD1 were reduced including the expression of enzymes involved in NPD1 synthesis, cytosolic phospholipase A_2 and 15-ipoxygenase [75, 76]. NPD1 repressed amyloid β-induced activation of pro-inflammatory genes and upregulated the antiapoptotic genes encoding Bcl-2, Bcl-xl and Bfl-1 (A1) indicating its (NPD1) anti-inflammatory nature. Soluble amyloid precursor protein-α stimulates NPD1 synthesis from DHA [75] that, in turn, could prevent neuronal death.

Presenilin, a major component of γ-secretase, generates amyloid-β. Overexpression of phospholipase D1 decreases the catalytic activity of γ-secretase [77], and releases PUFAs as evidenced by increased formation of prostaglandin E_2 [78], suggesting that PUFAs may regulate the activity of γ-secretase. AA and DHA enhance acetylcholine levels in the brain [79, 80] that may account for their beneficial effects in AD. Furthermore, EPA and DHA enhance eNO generation [71], suppress production of pro-inflammatory cytokines [71] that may also be responsible for the beneficial effects of PUFAs in AD. In addition, DHA inhibited Aβ(1-42)-fibril formation with a concomitant reduction in the levels of soluble Aβ(1-42) oligomers. The polymerization (into fibrils) of preformed oligomers treated with DHA was inhibited, indicating that DHA not only obstructs their formation but also inhibits their transformation into fibrils. DHA inhibited Aβ(1-42)-induced toxicity in SH-S5Y5 cells. These evidences suggest that by restraining Aβ(1-42) toxic tri/tetrameric oligomers, DHA may limit amyloidogenic neurodegenerative disease AD [81, 82]. But not all studies are in support of such beneficial action of DHA. For instance, Johansson et al. [83] reported that DHA and AA at micellar concentrations stabilized soluble Aβ42 wild-type protofibrils, thereby hindering their conversion to insoluble fibrils. As a consequence, DHA sustained amyloid-β-induced toxicity in PC12 cells over time, whereas Aβ without DHA stabilization resulted in reduced toxicity, as Aβ formed fibrils. These results are in contradiction to both epidemiologic and animal studies wherein DHA was found to be beneficial for cognition and reducing the risk for AD. DHA and AA exert a number of biological effects on cells, including activation of transcriptional factors and signal transduction systems implicated in AD. DHA-enriched diet affects APP processing and trafficking, lowering $A\beta$ levels in AD mice. These properties of DHA and AA would be beneficial, and can outweigh the presumed detrimental effect of PUFAs on protofibrils stabilization.

PUFAs and Neurogenesis and Neurite Outgrowth

It is known that raise in intracellular cyclic AMP (cAMP) levels promote differentiation of neuroblastoma cells. Activation of adenylate cyclase that increased intracellular cAMP levels were accompanied by a decrease in cell proliferation and an increase in neurite outgrowth, an effect that were exaggerated when combined with phosphodiesterase enzyme inhibitors. Increasing cAMP levels not only resulted in decreased proliferation but also increased morphological differentiation and enhanced their acetylcholinesterase activity. On the other hand, prostaglandin E_1 (PGE_1) promoted differentiation of neuroblastoma cells with little effect on cAMP levels, suggesting that elevation of cAMP is sufficient for inhibiting proliferation and promoting neurite outgrowth of neuroblastoma cells, but is not a necessary condition for inducing differentiation [84]. This may have relevance to the differentiation of neuronal stem cells to adult neuronal cells. It is interesting to note that neurite outgrowth response can be directly induced by arachidonic acid (10 μM), a response that is inhibited by N- and L-type calcium channel antagonists. PGE_1 is derived from its precursor dihomo-γ-linolenic acid (DGLA) that, in turn, can be elongated to form arachidonic acid (AA, 20:4 ω-6). In cells, AA can be generated by phospholipase A_2 (PLA_2) or by the sequential activities of a phospholipase C (to generate diacylglycerol) and diacylglycerol lipase. It was reported that the neurite outgrowth stimulated by cell adhesion molecules (CAMs)-NCAM, N-cadherin, and L1 is abolished by an inhibitor of diacylglycerol lipase acting at a site upstream from calcium channel activation [85]. These results suggest that AA and/or one of its metabolites is the second messenger that activates calcium channels in the CAM signalling pathway leading to axonal growth, an evidence that is supported by the observation that AA can increase voltage-dependent calcium currents in cardiac myocytes. In fact, it has been shown that LTB_4 and LXA_4 that are derived from AA have a regulatory role in the proliferation and differentiation of murine neural stem cells (NSCs) that were isolated from embryo brains [86]. Proliferation of NSCs was stimulated by LTB_4 (3-100 nM) that was blocked by its receptor antagonist, while LXA_4, and its aspirin-triggered-15-epi-LXA_4 stable analog attenuated growth of NSCs. Both lipoxygenase inhibitors and LTB_4 receptor antagonists caused apoptosis and cell death. Further studies showed that growth-related gene expressions such as epidermal growth factor (EGF) receptor, cyclin E, p27, and caspase 8 were tightly regulated by LTB_4(4) and

LXA_4. LTB_4 not only stimulated the proliferation of NSCs but also induced their differentiation as monitored by neurite outgrowth and microtubule-associated protein 2 (MAP2) expressions. These results suggest that LTB_4 and LXA_4 that are derived from the same precursor AA regulate proliferation and differentiation of NSCs [86]. Since human brain is rich in AA this implies that one of the functions of AA in the brain is to regulate the differentiation and proliferation of NSCs and thus, control the size of the brain and function(s) of the neurons.

Immature cerebellar granule neurons express large amounts of 5-lipoxygenase whose inhibition effectively reduced cell proliferation suggesting that neuronal expression of 5-lipoxygenase is crucial for neurogenesis in vitro, and possibly also in vivo [87]. It is important to note here that 5-lipoxygenase is necessary for the conversion of AA to LTB_4 and LXA_4.

The involvement of AA and its products in neuronal stem cell differentiation, neurogenesis and neurite outgrowth may have implications for mental disorders including schizophrenia. It was noted that Fabp7, a fatty acid binding protein 7, is one of the genes controlling prepulse inhibition (PPI) is a compelling endophenotype (biological markers) for mental disorders including schizophrenia [88]. PPI gene was found to be associated with schizophrenia. Disruption of FABP7 dampened hippocampal neurogenesis. PUFAs are ligands for FABP members and are abundantly expressed in neural stem/progenitor cells in the hippocampus. Administration of AA for 4 weeks after birth promoted neurogenesis in wild type rats; and raising Pax6 (+/-) pups on an AA-containing diet enhanced neurogenesis and partially improved PPI in adult animals, suggesting the potential benefit of AA in ameliorating PPI deficits relevant to psychiatric disorders and implied that the effect may be correlated with augmented postnatal neurogenesis [89]. These results coupled with the observation that positive modulators of PLA_2 (especially of $cPLA_2$ and $iPLA_2$) or supplementation with AA in combination with cognitive training could be a valuable therapeutic strategy for cognitive enhancement in early-stage AD [90] reaffirms the role of PUFAs in brain development and growth and some neurological conditions.

Similar to AA, DHA also stimulated neurite outgrowth [91] by activating syntaxin 3 that is specifically involved in fast calcium-triggered exocytosis of neurotransmitters. SNAP25 (synaptosomal-associated protein of 25 kDa), a syntaxin partner implicated in neurite outgrowth, interacted with syntaxin 3 only in the presence of AA that allowed the formation of the binary syntaxin 3-SNAP 25 complex. AA stimulated syntaxin 3 to form the ternary SNARE complex (soluble N-ethylmaleimide-sensitive factor attachment protein receptor), which is needed for the fusion of plasmalemmal precursor vesicles into the cell surface membrane that leads to membrane fusion that facilitates neurite outgrowth. Dietary omega-3 linolenic acid (ALA) and DHA were found to be capable of efficiently substitute for arachidonic acid in activating syntaxin 3. It was reported that DHA deficiency states could lead to altered SNARE complex binding or disassembly [92]. On the other hand, supplementation of DHA to experimental animals was found to be trigger a neuronal program that enhanced synaptogenesis [93]. These results imply that EPA, DHA, and AA when given in optimal amounts could be of benefit in the prevention and treatment of Alzheimer's disease and other neurological conditions such as depression and anxiety.

In addition, DHA has been shown to promote neuronal survival by facilitating membrane translocation/activation of Akt, whereas DHA deficiency increased hippocampal neuronal susceptibility to apoptosis [94]. Retinal pigment epithelium and synaptic membranes contain the highest content of DHA of all cell membranes, and DHA is required for retinal pigment epithelium functional integrity. NPD_1 (neuroprotectin D_1) that is formed from DHA counteracted H_2O_2/TNF-α-induced apoptosis by upregulating anti-apoptotic proteins Bcl-2 and Bcl-$_{XL}$ and decreasing pro-apoptotic Bax and Bad proteins. In addition, NPD_1 inhibited oxidative stress-induced caspase-3 activation, IL-1β-stimulated expression of cyclo-oxygenase-2 and thus, protected against oxidative stress-induced apoptosis and inflammation [95–97].

Furthermore, PUFAs can modulate the expression, properties, and action of dopamine, serotonin, and acetylcholine [98–100], especially during the perinatal period during which the growth and development of brain is maximum. Thus, ω-3 PUFAs and ω-6 AA are essential for neural function, including neurotransmission, membrane fluidity, ion channel and enzyme regulation and gene expression, prevent inflammation, and thus, are of benefit in Alzheimer's disease. These results also imply that preparation of stable analogues of lipoxins, resolvins and protectins could be of benefit in the management Alzheimer's disease.

In addition, EFA deficiency can promote respiratory uncoupling that increases oxidative stress and decreases nitric oxide availability or biological action [101–104]. PUFAs modulate the activities of uncoupling proteins, and under certain specific conditions may function as anti-oxidants and prevent free radical-induced damage to cells. Thus, ω-3 PUFAs modulate neural function, neurotransmission, membrane fluidity, ion channel and enzyme regulation and gene expression, prevent inflammation, and are essential for brain growth and development that may explain their role and beneficial actions in Alzheimer's disease.

Since both BDNF and PUFAs have a significant role in the pathobiology of AD, it is imperative that they interact with each other.

Interaction(s) Between PUFAs and BDNF

In animal experiments, inhibition of COX enzymes blocked increases in BDNF and PGE2 following spatial learning, inhibited induction of long-term potentiation (LTP; the major contemporary model of synaptic plasticity), and induced substantial and sustained deficits in spatial learning. Surprisingly, prior exercise increased endogenous BDNF levels sufficiently to reverse the effects of COX inhibition, suggesting that COX enzymes play a permissive role in synaptic plasticity and spatial learning via a BDNF-associated mechanism [105]. Contrary to these results, it was reported that PGs cause memory deficits in contextual fear conditioning that occur following IL-1β injection [106], whereas COX inhibition blocked these disruptions. These results imply that physiological levels of PGs could enhance memory while high or low concentrations may cause memory deficits. PGs injected directly into the dorsal hippocampus impaired context memory and significantly reduced post-conditioning levels of BDNF within the hippocampus, suggesting that PGs interact with BDNF to produce their memory-impairing effects.

It was reported that DHA-enriched diet significantly increased spatial learning ability, and these effects were enhanced by exercise. The DHA-enriched diet increased levels of pro-brain-derived neurotrophic factor (P-BDNF), mature BDNF, activated forms of CREB and synapsin I whereas the additional application of exercise boosted their levels further. DHA supplementation reduced hippocampal oxidized protein level, increased the levels of activated forms of hippocampal Akt and CaMKII whereas a combination of a DHA diet and exercise resulted in a much greater reduction and increase respectively [107]. Thus, both exercise and BDNF complement each other's actions.

In addition, omega-3 enriched dietary supplements can provide protection against reduced plasticity and impaired learning ability after traumatic brain injury by normalizing BDNF levels and reducing oxidative damage indicating that these fatty acids have a direct stimulatory action on the production of BDNF [108]. These and other studies clearly demonstrate that EPA/DHA/AA have a significant role in neuronal growth, synaptic plasticity, memory improvement and reduction in oxidative stress in the brain; enhance BDNF levels in the brain and thus, bring about their beneficial actions. It is likely that BDNF may need the cooperation of PUFAs to bring about their actions and for its own stabilization in the brain. The beneficial actions of exercise in memory formation and improvement in synaptic plasticity may also attributed to increased formation and utilization of PUFAs and enhanced levels of BDNF, especially since it is known that exercise augments the formation of lipoxins that are anti-inflammatory compounds and have neuroprotective actions. PUFAs and BDNF seem to act together to prevent Alzheimer's disease.

In conclusion, EPA, DHA and AA, are not only essential for the growth and development of brain but also have an important role in improving memory and protection against dendritic pathology and behavioral deficits and increased anti-apoptotic BAD phosphorylation and thus, prevent Alzheimer's disease. PUFAs attenuate amyloid-β secretion, decrease the catalytic activity of γ-secretase, inhibit IL-6 and TNF-α production, increase the synthesis of eNO, stimulate neurite outgrowth by activating syntaxin 3 and enhance the production of acetylcholine in the brain actions that explain their ability both in the prevention and treatment of Alzheimer's disease (see Fig. 12.1). It is likely that some of the beneficial actions of PUFAs could be due to increased formation of lipoxins, resolvins, protectins, maresins and nitrolipids that are anti-inflammatory and neuroprotective molecules. PUFAs augment the production of BDNF that is essential for growth, differentiation and survival of neurons in the brain. These results suggest that a combination of PUFAs and BDNF could be of significant benefit in the prevention and treatment of Alzheimer's disease. Since AA/EPA/DHA form precursors to potent anti-inflammatory and neuroprotective molecules such as lipoxins, resolvins, protectins and maresins, it is likely that the plasma, cerebrospinal fluid and tissue levels of various PUFAs and their metabolites could serve as markers both to predict and prognosticate the development and progression of Alzheimer's and other neurodegenerative diseases including other types of dementia.

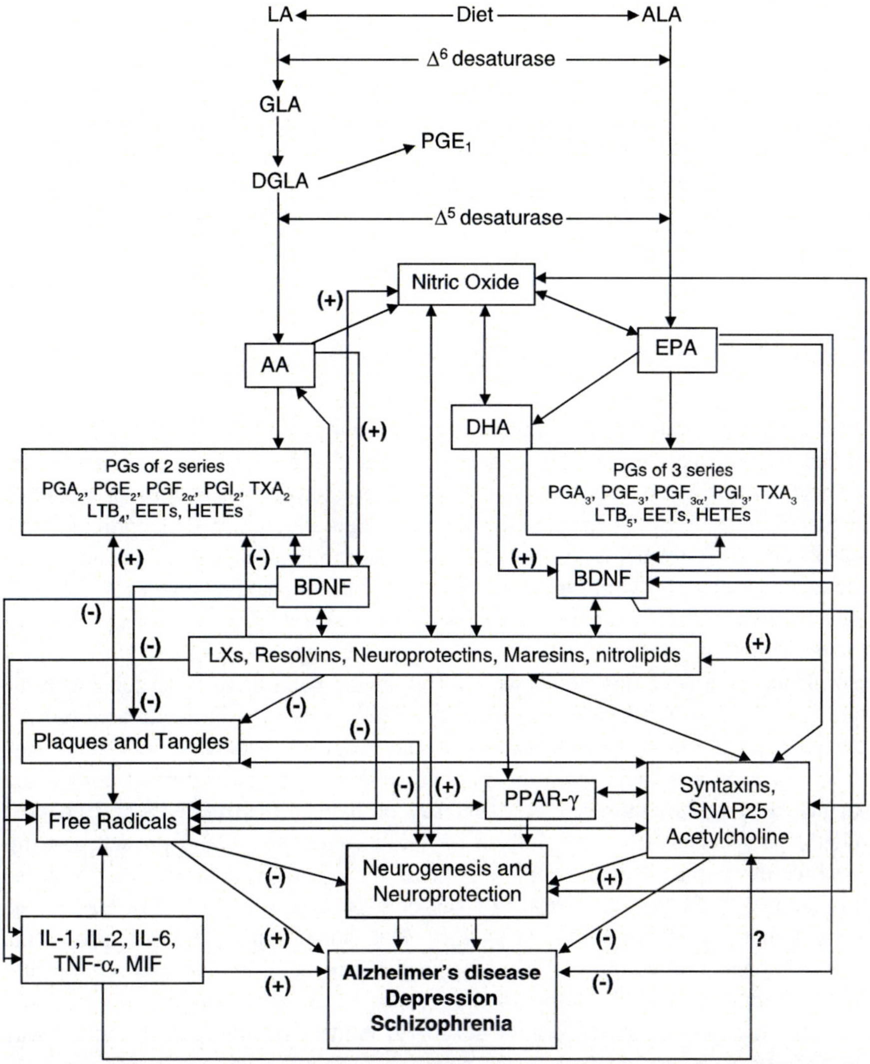

Fig. 12.1 Scheme showing the relationship among PUFAs and their products, BDNF, neurotransmitters such as acetylcholine and cytokines and their role in Alzheimer's disease. (+) Indicates initiation and/or progression of disease or increase in the synthesis or action; (−) Indicates protection from disease and better prognosis or decrease in the synthesis or action; **?** Indicates that possibly, cytokines inhibit syntaxin, SNAP25 and acetylcholine formation or interfere with their action. Though the role of PPAR-γ in Alzheimer's disease/depression/schizophrenia is not discussed in detail, it is believed that PPAR-γ or its agonists prevent AD/depression/schizophrenia and show neuroprotective and enhance neurogenesis properties (see references [109–112]). PUFAs and their products such as lipoxins, resolvins, protectins, maresins and nitrolipids enhance PPAR-γ activity and have PPAR-γ agonistic activity

Schizophrenia

Schizophrenia is characterized by delusions, hallucinations, disorganized speech (e.g., frequent derailment or incoherence), grossly disorganized or catatonic behavior, negative symptoms, i.e., affective flattening, alogia, or avolition, and social/occupational dysfunction. Schizophrenia may result from altered neuronal membrane structure and metabolism, and/or dysregulation of the inflammatory response system.

Prenatal and Perinatal Factors on Psychopathology

Since brain growth and development occurs predominantly from 2nd trimester of pregnancy to 5 years of age, it has been proposed that prenatal and perinatal factors may play a role in the pathobiology of various neuropsychiatric conditions including schizophrenia. It has been thought that major depression and schizophrenia were associated with not being breast-fed, maternal emotional problems, cannabis use, trauma and maternal viral infections [113–116]. Breast-feeding enhances cognitive development [117], though some studies did not support this conclusion [118, 119]. Schizophrenia is preceded by childhood cognitive impairments that led to the proposal that breast-feeding could be a factor in the pathobiology of schizophrenia [120–122]. Human breast milk is rich in PUFAs such as γ-linolenic acid (GLA), DGLA, AA, EPA, and DHA that led to the proposal that these fatty acids may have a role in schizophrenia especially since they form an important component of neuronal cell membranes. A deficiency of PUFAs (especially AA, EPA, and DHA) may have an adverse impact on brain development and growth and impact the development of schizophrenia [123–128]. Thus, it is likely that decreased availability of PUFAs as a result of sub-optimal breast-feeding may lead to the development of schizophrenia [127–130]. But, some studies did not support the possible beneficial action of breast-feeding in protecting against the risk of later schizophrenia [131–133]. Some of the reasons for these conflicting reports could be: the duration of breast-feeding, the PUFA content of breast milk and the way these fatty acids are handled by the newborn. It is possible that longer the duration of breast-feeding the greater will be its impact on the newborn. Thus, subjects who received breast-feeding for more than 6–12 months are likely to have obtained significantly larger amounts of PUFAs compared those who were breast-fed for much shorter duration. It is known that the fatty acid composition of the breast milk depends on the diet and large variations in the PUFA content of the breast milk have been reported in women of different regions and countries. Hence, it is not surprising that there were both positive and negative reports with regard to the effect of breast-feeding in protection against the development of schizophrenia in later life.

Early Fetal Environment and Development and Schizophrenia

It is believed that inadequate fetal nutrition can alter the body's structure, physiology, and metabolism (fetal programming) that predispose them to develop chronic illnesses in adulthood including schizophrenia [134]. Responses to the environmental influences, especially when they occur during pregnancy, may be expressed by the offspring during their adult life, and may not be seen/expressed by the mother. One such environmental factor that can have a life-long impact on the offspring could be the nutrition during pregnancy. Changes triggered by environmental factors may have long term consequences in the offspring, especially if these events are induced during sensitive, often brief, periods of development. It is noteworthy that maternal infections are one such environmental factor that may be expressed in the fetus rather than in the mother. For instance, maternal exposure to glucocorticoids in pregnancy induces hypertension, insulin resistance, obesity and altered muscle mass as well as alterations in the hypothalamic-pituitary-adrenal axis in the adult progeny [135]. It is likely that maternal infection induces excess production of pro-inflammatory cytokines by the infiltrating macrophages, T cells and the neurons themselves may induce adverse effects on the developing fetal neurons (brain) resulting in the development of schizophrenia in adult life [136]. Exposure of fetal neurons to such noxious stimulus may render them more susceptible to further damage even by suboptimal doses of pro-inflammatory cytokines later that may predispose the fetus to develop schizophrenia in adult life.

Higher frequencies of obstetric, neonatal, and maternal complications have been recorded in adolescents who had psychiatric disorders suggesting that complications during pregnancy and at birth may render the newborn more vulnerable to environmental events precipitating psychiatric conditions [137–142]. These findings suggest the importance of maternal and perinatal factors in the development of central nervous system and psychiatric conditions such as schizophrenia in adulthood.

Disruption of the serotonin transporter (5-HTT) early in brain development affects the development of brain circuits that deal with stress response and a polymorphism that reduced their 5-HTT activities were more likely than others to become depressed in response to stressful experiences [143, 144]. Serotonin (5-HT) is a trophic factor that modulates developmental processes such as neuronal division, differentiation, migration, and synaptogenesis. Inhibition of 5-HTT functions can alter brain development in subtle ways. For example, genetic inactivation of 5-HTT affects barrel formation in the somatosenory cortex and alters segregation of retinal axons; mice lacking the 5-HTT gene showed reduced dorsal raphe' firing rates and fewer serotonergic neurons [145], suggesting that alterations in the structure and function of serotonergic nuclei can contribute to the altered behavioral responses noted. Developmental and/or genetic factors affecting anxiety- or depression-related behaviors alter hippocampal structure, amygdala function and receptor expression in the prefrontal cortex, structures that are known to receive significant serotonergic innervation [146–149]. 5-HTT function modulates the development of brain systems involved in emotional and stress related responses and low expressing 5-HTT variants may act during fetal brain development to modify brain circuits or gene expression that may predispose the carriers of these alleles to schizophrenia. It is likely that low-expressing 5-HTT variants act during development of brain, especially during perinatal period, and modify brain circuits or gene expression that predisposes carriers of these alleles to emotional disorders such as schizophrenia. Thus, nutritional factors acting during fetal, perinatal, and infancy periods may influence not only the growth and development of brain but also ability of the neurons to synthesize, secrete, and express receptors for various neurotransmitters in such a way that these early life events would ultimately have an impact in the development of schizophrenia in adult life. This is supported by the observation that early fetal environment and parental factors have a major impact on suicidal behavior in adolescents and young adults [138–143, 150].

Maternal Infections and Schizophrenia

Maternal viral infection could increase the risk for schizophrenia in the offspring. Mice born to mothers who had respiratory tract infection at mid-gestation showed features of schizophrenia [151]. Prenatal immune challenge disrupted sensorimotor gating in adult rats [152] and these animals showed increased serum levels of IL-2 and IL-6, suggesting that prenatal immune events have a role in the pathogenesis of schizophrenia [153], implying that schizophrenia could be a low-grade systemic inflammatory condition.

There is evidence that specific immunological abnormalities do occur in schizophrenia. One of the causes for restricted fetal growth could be maternal infection. The increased plasma concentrations of ILs observed in the maternal plasma secondary to infections do cause anorexia that may lead to decreased intake of balanced food by the mother that ultimately leads to fetal growth restriction. In addition, the pro-inflammatory cytokines may have deleterious actions on the growth and development of fetal brain. It is likely that maternal malnutrition itself may render them more susceptible to infections that, in turn, interferes with normal feeding behavior and thus, exacerbates malnutrition.

Large population studies showed that the incidence of schizophrenia is higher among people who were born in urban settings that are associated with a higher risk of maternal influenza infection during pregnancy. Mice born to mothers who were exposed to respiratory infection at mid gestation showed abnormal behavior as adults which resemble those seen in schizophrenics. Since no virus could be detected in the affected offspring, it suggests that maternal immune system itself could lead to the brain changes in their offspring. Similar changes in the offspring were observed following stimulation of maternal immune system using a synthetic double-stranded RNA that evoked an anti-viral immune response lends support to this view [153]. Mice born to infected mothers showed changes in cortex and hippocampus, loss of Purkinje cells in the cerebellum; changes that have also been described in schizophrenic patients.

Is Schizophrenia an Inflammatory Condition?

Serum and cerebrospinal fluid (CSF) IL-2, IL-6, IL-8 and tumor necrosis factor-α (TNF-α) levels were reported to be elevated in patients with schizophrenia [154–158]. Relapse-prone patients had significantly higher levels of CSF IL-2 than patients who did not relapse and the use of risperidone decreased interferon-γ (IFN-γ) production and enhanced IL-10 (a suppressor of Th_1 response) [159]. Haloperidol and perazine decreased the release of IL-1β and TNF-α from monocytes of schizophrenic patients [157]. IL-2 treatment-induced behavioral changes that could be blocked by a selective dopamine D1 receptor antagonist or by a relatively high dose of a D2 antagonist [160] indicating that IL-2 induces and/or increases psychiatric abnormalities by causing aberrations in central dopaminergic transmission. Rat cortical cultures exposed to IL-1β, IL-6 and TNF-α showed decreased neuronal survival [161]. Pro-inflammatory cytokines interfere with the actions of various neurotransmitters and induce features of schizophrenia [162]. These data suggest that schizophrenia could be an inflammatory condition as a result of increased pro-inflammatory cytokines during gestational period which cause injury to fetal neurons that in turn increases the risk of schizophrenia in adult life. If this is true, suppression of production of pro-inflammatory cytokines could form a new approach in the prevention and management of schizophrenia. In this context, it is important to note that PUFAs not only inhibit the production of pro-inflammatory cytokines but also possess neuroprotective properties indicating that they may function as endogenous anti-inflammatory, neuroprotective and anti-schizophrenic molecules.

PUFAs and Their Metabolites and Schizophrenia

As human brain is rich in PUFAs and forms an important component of the neuronal cell membranes and is essential for fetal growth and development, it is reasonable to expect that they would have a significant role in schizophrenia [123]. Newborn and pre-term infants have limited capacity to synthesize EPA, DHA and AA from their precursor essential fatty acids: ALA and LA. Dietary AA improved first year growth of pre-term infants [163], whereas EPA and DHA increase birth weight by prolonging gestation and by increasing the fetal growth rate [164, 165]. These evidences indicate that low-birth weight infants are likely to have decreased tissue and plasma concentrations of various PUFAs. EPA, DHA, and AA, inhibit TNF-α and IL-6 production; enhance eNO generation, inhibit 3-hydroxy-3-methylglutaryl coenzyme A (HMG-CoA) reductase and angiotensin converting enzyme activities, function as endogenous ligands for PPARs, and suppress leptin gene expression [166–170]. Thus PUFAs suppress inflammation, regulate cholesterol metabolism, and control appetite and food intake. PUFAs augment brain acetylcholine levels that, in turn regulate the synthesis, release and actions of other neurotransmitters such as serotonin, dopamine, catecholamines and other hypothalamic peptides such as neuropeptide Y, melanocortins, etc. and regulate autonomic nervous system [169,

170]. In view of these various actions, it is no surprise that PUFAs also have an important role in schizophrenia.

Reduced levels of membrane DHA, EPA, and AA, and increased levels of peroxidation products have been reported in schizophrenics [171–173]. However, the reductions in levels of both AA and DHA were much smaller in medicated versus never-medicated patients and especially those of AA and DHA levels were much higher in chronic medicated patients than drug-naïve first-episode patients [174]. It is likely that many, if not all, the antipsychotics modulate metabolism of PUFAs and thus, brings about their beneficial actions in schizophrenia. This argument is supported by the observation that chronic valproate and lithium treatment decrease AA turnover in brain phospholipids [175–177]. EPA and DHA inhibit protein kinase C, inositol monophosphatase, and inositolpolyphosphatase that leads to a decrease in inositol-1,4,5-triphosphate ($InsP_3$) response activity [178–181] that could be responsible for their beneficial actions in schizophrenia.

PUFAs suppress the production of pro-inflammatory cytokines IL-2, IL-6 and TNF-α both in vitro and in vivo [182–185]. Oral supplementation of EPA is useful in schizophrenia [186–188]. Thus, the PUFAs are beneficial in schizophrenia at least, in part, due to their ability to suppress the production of pro-inflammatory cytokines, which are elevated in this condition. EPA, DHA and AA have neuroprotective and cytoprotective actions and prevent apoptosis of neurons [189–194]. Some of the beneficial actions of various PUFAs in schizophrenia could be attributed to the formation of anti-inflammatory and neuroprotective compounds lipoxins, resolvins, protectins, maresins and nitrolipids. It is likely that schizophrenics who do not respond to PUFAs and anti-schizophrenic drugs are unable to form adequate amounts of lipoxins, resolvins, protectins and nitrolipids. If this is true, it will be interesting to measure plasma and CSF levels of various PUFAs, lipoxins, resolvins, protectins and nitrolipids and correlate their concentrations to the response to treatment. It is possible that plasma levels of PUFAs, lipoxins, resolvins, protectins and nitrolipids could be used as markers of response to treatment and as predictors of relapse and for predicting the prognosis.

It is likely that injury, infection and inflammation increase the production of pro-inflammatory cytokines during pregnancy both in the mother and the fetus that, in turn, interfere with the growth and development of the growing fetal brain by inducing apoptosis of developing neurons, altering the balance between dopaminergic and serotoninergic neurons and predispose them to develop schizophrenia in adult life. DHA, EPA and AA and their products lipoxins, resolvins, protectins and nitrolipids could prevent these events by virtue of their neuroprotective action and inhibitory action on the production of pro-inflammatory cytokines. It is proposed that sub-clinical deficiency of PUFAs and reduced formation of lipoxins, resolvins, protectins and nitrolipids may lead to enhanced production of pro-inflammatory cytokines due to the absence of the negative feed-back control exerted by these lipid molecules. In view of this, it will be interesting to study whether supplementation of PUFAs to high-risk pregnant and lactating mothers prevents/postpones the development of schizophrenia in their progeny. The possible beneficial effect of breast-feeding in the prevention of schizophrenia could be due to the presence of significant amounts of PUFAs in the human breast milk. Decreased formation of lipoxins, resolvins, protectins and nitrolipids from various PUFAs may render neuronal tissue susceptible to the cytotoxic actions of pro-inflammatory cytokines. Increased consumption of PUFAs may enhance the formation of lipoxins, resolvins, protectins and nitrolipids that may protect the neuronal tissue from environmental insults.

Depression

Similar to AD, even depression appears to be an inflammatory condition in which PUFAs play a significant role.

The exact cause(s) of depression is not clear but, several theories have been proposed some of which include: psychological, psychosocial, hereditary, evolutionary and biological factors. Most biological theories focus on the monoamines serotonin, norepinephrine and dopamine.

Depressed persons may have cognitive symptoms of recent onset, such as forgetfulness. Depression often coexists with physical disorders such as stroke, other cardiovascular diseases, Parkinson's disease, and chronic obstructive pulmonary disease.

Patients with depression may have smaller hippocampal volumes [195–197]. A comprehensive meta-analysis revealed that patients with depression showed large volume reductions in frontal regions, especially in the anterior cingulate and orbitofrontal cortex with smaller reductions in the prefrontal cortex. The hippocampus, the putamen and caudate nucleus showed moderate volume reductions. Thus, major depressive disorder is characterized by structural brain abnormalities, particularly in those brain areas that are involved in emotion processing and stress-regulation [198]. It has also been reported that smaller volume of the caudate nucleus may be related to the pathophysiology of major depressive disorder and may account for abnormalities of the cortico-striatal-pallido-thalamic loop in this disease [199]. In fact, it was observed that compared with individuals at low familial risk of the development of depression, high-risk individuals have reduced hippocampal volume, indicating that neuroanatomic anomalies associated with depression may precede the onset of a depressive episode and influence the development and course of depression [200].

Several studies showed higher incidence of white matter hyperintense lesions identified by brain MRI in patients with geriatric depression than in healthy elderly subjects, which correlated with cognitive impairment and clinician-rated depressive symptoms than those with none and/or mild lesions. These results suggest that subcortical/frontal type cognitive impairment and the persistence of depressive symptoms in geriatric depression is related to moderate deep white matter lesions more often complicated in the late-onset group [201]. Among the unipolar patients, length of illness and presence of mood disorder in a first-degree relative were found to be related to deep and periventricular white matter lesions, respectively, suggesting that these brain lesions are more directly related to late-life and more severe cases of these illnesses [202]. Based on a meta-analysis study, it was concluded that hyperintensities are not specific to bipolar disorder, but appear at similar rates in unipolar depression and schizophrenia. Thus, the role of hyperintensities in the pathogenesis, pathophysiology, and treatment of bipolar disorder remained unclear [203] but it led to the theory of vascular depression.

Depression May Be Associated with Low BDNF Levels

There may be a link between depression and neurogenesis of the hippocampus, a center for both mood and memory. Loss of hippocampal neurons is found in some depressed individuals and correlates with impaired memory and dysthymic mood. Drugs may increase serotonin levels in the brain, stimulating neurogenesis and thus increasing the total mass of the hippocampus. This increase may help to restore mood and memory. Similar relationships have been observed between depression and an area of the anterior cingulate cortex implicated in the modulation of emotional behavior. One of the neurotrophins responsible for neurogenesis is BDNF. The level of BDNF in the blood plasma of depressed subjects is reduced compared to the normal [204, 205]. It is interesting to note that low BDNF levels were found even in healthy humans who showed depressive personality traits. These results provide further support for the hypothesis that BDNF may be central to the development of depressive mood states [206]. It is imperative to note that serum and plasma BDNF levels are low in depressed patients compared with control subjects, while in whole blood, BDNF levels remain unaltered in the depressed subjects compared with control subjects. The serum/blood BDNF ratio was lower in patients with major depression. These results suggest that an alteration of serum or plasma BDNF is not due to the change in blood BDNF but could be related to mechanisms of BDNF release [207].

Antidepressant treatment increases the blood level of BDNF [208–210]. Although decreased plasma BDNF levels have been found in many other disorders, there is strong evidence that BDNF is involved in the cause of depression and the mechanism of action of antidepressants.

There is also evidence that major depression may be caused in part by an overactive hypothalamic-pituitary-adrenal axis (HPA axis) that results in an effect similar to the neuro-endocrine response to stress. Investigations revealed increased levels of the hormone cortisol and enlarged pituitary and adrenal glands, suggesting disturbances of the endocrine system may play a role in some psychiatric disorders, including major depression. Oversecretion of corticotropin-releasing hormone from the hypothalamus is thought to drive this, and is implicated in the cognitive and arousal symptoms. But, some studies found no significant differences in terms of HPA-axis function, but lowered serum BDNF levels in burnout group as compared to healthy controls was reported. Logistic regression analysis revealed that emotional exhaustion, depersonalization and depression were significantly associated with burnout, whereas serum BDNF levels correlated negatively with emotional exhaustion, depersonalization and correlated positively with competence. However, no significant relationships between cortisol levels and serum BDNF levels, depression, anxiety, psychosomatic complaints and burnout inventory was noted. These results suggest that low BDNF might contribute to the neurobiology of burnout syndrome including altered mood and cognitive functions [211]. On the other hand, increased plasma glucocorticoid levels are known to render the hippocampus, a structure important for learning and memory, susceptible to neuronal damage suggesting that HPA axis dysregulation and cognitive deficits seen in depression could be related. It was reported that high stress reactive mice exhibited hippocampus-dependent memory deficits along with decreased hippocampal, but not plasma BDNF levels providing evidence that HPA axis interacts with BDNF secretion in major depression [212].

In fact, it was noted that interferon-α (IFN-α) therapy, used for hepatitis C infection, induced depressive symptoms in these patients was found to be associated with decreased serum levels of BDNF. Furthermore, pro-inflammatory cytokine levels predicted lower BDNF levels, whereas low BDNF levels, as well as increased cytokine levels, were independently associated with the development of depressive symptoms during IFN-α treatment. These findings suggest that IFN-α-induced immune activation on depression may be related to decrease in serum BDNF levels [213], supporting the concept that BDNF plays a major role in depression.

BDNF and Serotonin Interact with Each Other

Studies done in BDNF-heterozygous (+/−) mice, which exhibit abnormal behaviors such as obesity, anxiety and aggression, revealed that dietary restriction significantly ameliorated obesity, anxiety and aggression in these mice that also showed that the concentrations of 5-HT and 5-HIAA in the frontal cortex, and 5-HT in the hippocampus were significantly lower than those of wild-type mice. Dietary restriction significantly increased the levels of 5-HT and 5-HIAA in the frontal cortex of BDNF (+/−) mice. These observations suggest that the benefits seen with dietary restriction are associated with changes in the serotonergic system that is possibly influenced by BDNF [214]. It was reported also that chronic corticosterone decreased BDNF protein (−16.6%) in whole hippocampus and BDNF mRNA (−19%) in the hypothalamic CA3 area. In both the frontal cortex and hippocampus, tissue levels of 5-HT and 5-HIAA were increased and decreased, respectively, suggesting that chronic corticosterone impairs hippocampal BDNF function (hippocampal atrophy is seen in depression) and alters the brain tissue levels of 5-HT and 5-HIAA [215]. These results indicate an intricate relationship among corticosterone (and thus, HPA axis), BDNF and serotonin system with particular reference to hippocampus.

Several studies showed that chronic but not acute treatment with antidepressant drugs targeting the central 5-HT system, enhances the expression BDNF. In a study designed to evaluate the relationship between 5-HT(6)-receptor activation on hippocampal and cortical levels of mRNA expression of BDNF and Arc in the rat, it was noted that the administration of selective 5-HT(6)-receptor agonist caused a bell-shaped dose response on hippocampal BDNF mRNA expression, having no effect at 0.1 mg/kg, a significant up-regulation at 1 mg/kg and no effect at 10 mg/kg. The up-regulation in BDNF expression observed at 1 mg/kg was completely blocked by pre-treatment with the selective 5-HT(6)-receptor antagonist. These results provide evidence for the involvement of the 5-HT(6) receptor in regulating BDNF expression, suggesting that as opposed to more general 5-HT receptor activation by antidepressants, direct 5-HT(6)-receptor activation results in a more rapid rise in BDNF [216].

These evidences are in line with the monoamine hypothesis that suggests that depression is due to the underactivity in the brain of monoamines, such as dopamine, serotonin, and norepinephrine. The fact that the monoamine oxidase inhibitors (MAOIs) and tricyclic antidepressants are effective in the treatment of depression lends support to this concept. This hypothesis also led to the development of SSRIs (selective serotonin reuptake inhibitors).

Interestingly, Both BDNF and serotonin have a modulatory influence on inflammation.

BDNF and Inflammation

Peripheral tissue inflammation produced by an intraplantar injection of Freund's adjuvant into the rat paws produced a significant increase in the percentage of BDNF-immunoreactive (IR) neuron profiles in the L5 dorsal root ganglion and marked elevation in the expression of BDNF-IR terminals in the spinal dorsal horn. These results suggest that peripheral tissue inflammation induced an increase in BDNF synthesis in the dorsa root ganglion and elevated anterograde transport of BDNF to the spinal dorsal horn [217]. Two hours of induction of bladder inflammation, a significant increase in the levels of NGF, BDNF and neurotrophin-3 mRNAs was noted [218] suggesting that neurotrophic factors have a role in inflammation. Patients with asthma showed enhanced levels of BDNF in their bronchoalveolar fluid [219]. Activated human T cells, B cells, monocytes, and, T helper: T_{H1} and T_{H2}-type $CD4^+$ T cell lines secrete bioactive BDNF upon antigen stimulation. BDNF is present in inflammatory infiltrates in patients with acute disseminated encephalitis and multiple sclerosis [220] where BDNF is believed to have neuroprotective properties. Thus, BDNF and other neurotrophins have two functions: to protect the brain neurons from inflammatory events [221, 222] whereas in the respiratory tract, peripheral nerves and urinary bladder may function as pro-inflammatory molecules [223–225]. BDNF is present in several tissues such as brain, salivary glands, stomach, duodenum, ileum, colon, lung, heart, liver, pancreas, kidney, oviduct, uterus, bladder, and monocytes and eosinophils [226–228] and may have a role in inflammatory conditions such as collagen vascular diseases [229–231], myocardial injury [232], inflammatory bowel disease [233, 234], and atopic dermatitis [235]. Since BDNF plays a significant role in depression and it modulates inflammation and inflammatory events [236], it raises the interesting possibility that depression could be an inflammatory condition.

Serotonin and Catecholamines Modulate Inflammation

Similar to BDNF, serotonin and catecholamines that have a role in the pathogenesis of depression also participate in the inflammatory process. Serotoninergic receptors (5-HTR) are expressed by a broad range of inflammatory cell types, including dendritic cells (DCs). Serotonin inhibited oxidative burst of human phagocytes and exerted a dose dependent inhibition of the myeloperoxidase activity [237] and has a significant influence on the production of TNF, IL-12, IL-10, NO and PGE_2 [238]. These and other results suggest that 5-HT is a potent regulator of human dendritic cell function and immune response and has pro-inflammatory actions [239, 240]. The ability of serotonin to enhance inflammatory reactions in the skin, lung and gastrointestinal tract seems to be, in part, mediated by its action on mast cells.

Similarly, even catecholamines have been shown to have potent pro-inflammatory actions [241–244], lending further support to the concept that depression could be an inflammatory condition.

Depression is an Inflammatory Condition

There is evidence to suggest that pro-inflammatory cytokines may act as neuromodulators and modulate the behavioral, neuroendocrine and neurochemical features of depressive disorders. Diseases in which inflammation is the key feature such as rheumatoid arthritis and lupus are often accompanied by depression. Administration of pro-inflammatory cytokines such as IFN-γ induces depressive symptomatology. In animal studies, it was reported that administration of pro-inflammatory cytokines induces "sickness behaviour" that has a pattern of behavioral alterations that is very similar to the behavioral symptoms of depression in humans. Cytokines may also act on the central nervous system and cause the hypothalamic-pituitary-adrenal (HPA) axis hyperactivity that is seen in depressive disorders. Pro-inflammatory cytokines may cause HPA axis hyperactivity by disturbing the negative feedback inhibition of circulating corticosteroids (CSs) on the HPA axis. Deficiency in serotonergic (5-HT) neurotransmission that is seen in major depression could be due to the cytokine(s)-induced reduction in 5-HT levels by lowering the availability of its precursor tryptophan (TRP) through activation of the TRP-metabolising enzyme indoleamine-2,3-dioxygenase (IDO) [245].

Pro-inflammatory cytokines might cause depressive illness [245] based on the observations that: (a) activation of the immune system, and administration of endotoxin (LPS) or interleukin-1 (IL-1) to experimental animals induces sickness behavior, which resembles depression [246]; (b) activation of the immune system is observed in many depressed patients [247]; (c) depression is frequent in those with immune dysfunction [248]; (d) chronic treatment with antidepressants inhibits sickness behavior induced by LPS [249]; (f) pro-inflammatory cytokines activate the hypothalamo-pituitary-adrenocortical axis (HPAA), which is activated in depressed patients [245]; (g) cytokines activate cerebral noradrenergic systems that is known to occur in depressed patients [248]; and (h) several pro-inflammatory cytokines activate brain serotonergic systems, which have been implicated in major depressive illness and its treatment [250, 251]. These results suggest that depression could be a low-grade systemic inflammatory condition.

Central nervous system regulates the production of pro-inflammatory cytokines: TNF, IL-1, HMGB1, IL-6, and MIF through the efferent vagus nerve [252, 253]. Acetylcholine, the principal vagus neurotransmitter, inhibits the production of pro-inflammatory cytokines through a mechanism dependent on the $\alpha 7$ nicotinic acetylcholine receptor subunit. Since vagal nerve stimulation (VNS) is of benefit in depression, I previously proposed that the beneficial effect of VNS in depression is due to its (VNS) inhibitory action on the production of pro-inflammatory cytokines [254].

But, it remains to be seen as to why the inflammatory process is activated in depression. It is possible that the activation of the immune system and inflammation in depression is secondary to the failure of the negative feed-back exerted by PUFAs that are present in large amounts in the brain. Thus, it is likely that one of the primary functions of PUFAs in the brain is to protect it from the cytotoxic actions of an activated immune system and the resultant inflammatory process.

Depression and PUFAs

A significant decrease of ω-3 fatty acids in plasma and/or in the membranes of red blood cells in subjects with depression has been reported [255–257]. PUFAs, especially ω-3 fatty acids suppress the production of IL-1β, IL-2, IL-6, TNF-α and MIF (macrophage migration inhibitory factor). If this is true, it implies that ω-3 fatty acids play a significant role in major depression, through their roles in maintaining membrane fluidity that influences neurotransmission and by modulating the production of pro-inflammatory cytokines [167]. In addition, antidepressants exhibit an immunoregulating effect by reducing the release of pro-inflammatory cytokines, by increasing the release of endogenous antagonists of pro-inflammatory cytokines like IL-10 and, finally, by acting like inhibitors of cyclo-oxygenase [258]. Double blind placebo-controlled and other studies [259–261] showed that administration of ω-3 fatty acids EPA and DHA was associated with a longer period of remission among depressed patients. Thus, epidemiological, experimental and clinical data favor the idea that PUFAs could play a role in the pathogenesis and/or the treatment of depression.

PUFAs by themselves or by giving rise to their anti-inflammatory products such as lipoxins, resolvins, protectins, maresins and nitrolipids are able to suppress unwarranted immune activation and inflammation. For instance, induction of LXA_4 accompanied the in vivo suppression of IL-12 responsiveness of murine splenic dendritic cells (DCs) after microbial stimulation. This paralysis of DC function was found to be dependent on the presence of lipoxygenase involved in LXA_4 biosynthesis. DCs pre-treated with LXA_4 were found to be refractory to microbial stimulation for IL-12 production in vitro and mice injected with a stable LXA_4 analog showed reduced splenic DC mobilization and IL-12 responses in vivo [262]. These findings indicate that the induction of lipoxins provide a potent mechanism for regulating DC function during the innate response to pathogens and other stimuli. In addition, it is likely that PUFAs and several of their products such as prostaglandins, leukotrienes, thromboxanes, lipoxins, resolvins, protectins and nitrolipids are able to regulate the proliferation and differentiation of embryonic stem cells in addition to their capacity to suppress inflammation and thus, play a major role in coronary heart disease, stroke, diabetes mellitus, hypertension, atherosclerosis, cancer, depression, schizophrenia, Alzheimer's disease, and collagen vascular diseases [263]. Hence, it will be interesting to study not only the plasma levels of various PUFAs but also of their products, especially lipoxins, resolvins, protectins and nitrolipids in the plasma, CSF and other body fluids, their metabolites in the urine and various tissues (such as biopsy specimens) before and after the administration of PUFAs and correlate their levels with the progression and response of the diseases to various therapies employed and in the natural history of the diseases. It is possible that those subjects who are able to generate adequate amounts of the anti-inflammatory compounds lipoxins, resolvins, protectins, maresins and nitrolipids (in addition to the formation of PGI_2, PGI_3, PGE_1) show a better response to the therapies employed or show faster recovery and have less aggressive disease(s) and demonstrate good prognosis. Thus, development of more stable analogues of lipoxins, resolvins, protectins, maresins and nitrolipids that possess the same anti-inflammatory activities as that of the natural compounds may find a niche place in the management of several inflammatory conditions including depression (See Fig. 12.1 that gives a general scheme as to the role of PUFAs and their products in AD, schizophrenia and depression).

References

[1] Selkoe DJ (2001) Alzheimer's disease: genes, proteins, and therapy. Physiol Rev 81:741–766

[2] Rosenberg PB (2005) Clinical aspects of inflammation in Alzheimer's disease. Int Rev Psychiatry 17:503–514

[3] McGeer PL, McGeer EG (2002) Local neuroinflammation and the progression of Alzheimer's disease. J Neurovirol 8:529–538

[4] Mrak RE, Griffin WST (2005) Glia and their cytokines in progression of neurodegeneration. Neurobiol Aging 26:349–354

[5] Chao CC, Hu S, Ehrlich L, Peterson PK (1995) Interleukin-1 and tumor necrosis factor-alpha synergistically mediate neurotoxicity: involvement of nitric oxide and of N-methyl-D-aspartate receptors. Brain Behav Immun 9:355–365

[6] Bellinger FP, Madamba SG, Campbell IL, Siggins GR (1995) Reduced long-term potentiation in the dentate gyrus of transgenic mice with cerebral overexpression of interleukin-6. Neurosci Lett 198:95–98
[7] Tancredi V, D'Arcangelo G, Grassi F et al (1992) Tumor necrosis factor alters synaptic transmission in rat hippocampal slices. Neurosci Lett 146:176–178
[8] Vallieres L, Campbell IL, Gage FH et al (2002) Reduced hippocampal neurogenesis in adult transgenic mice with chronic astrocytic production of interleukin-6. J Neurosci 22:486–492
[9] Tarkowski E, Andreasen N, Tarkowski A et al (2003) Intrathecal inflammation precedes development of Alzheimer's disease. J Neurol Neurosurg Psychiatry 74:1200–1205
[10] Alvarez A, Cacabelos R, Sanpedro C, Garcia-Fantini M, Aleixandre M (2007) Serum TNF-alpha levels are increased and correlate negatively with free IGF-I in Alzheimer disease. Neurobiol Aging 28:533–536
[11] Zuliani G, Ranzini M, Guerra G et al (2007) Plasma cytokines profile in older subjects with late onset Alzheimer's disease or vascular dementia. J Psychiatr Res 22:305–311
[12] Lio D, Annoni G, Licastro F, Crivello A, Forte GI, Scola L, Colonna-Romano G, Candore G, Arosio B, Galimberti L, Vergani C, Caruso C (2006) Tumor necrosis factor-alpha-308A/G polymorphism is associated with age at onset of Alzheimer's disease. Mech Ageing Dev 127:567–571
[13] Poierier J, Minnich A, Davignon J (1995) Apolipoprotein E, synaptic plasticity and Alzheimer's disease. Ann Med 27:663–670
[14] Blacker D, Wilcox MA, Laird NM, Rodes L, Horvath SM, Go RC, Perry R, Watson B, Bassett SS, McInnis MG, Albert MS, Hyman BT, Tanzi RE (1998) Alpha-2 macroglobulin is genetically associated with Alzheimer disease. Nat Genet 19:357–360
[15] Sridhar GR, Thota H, Allam AA, Babu CS, Prasad AS, Divakar Ch (2006) Alzheimer's disease and type 2 diabetes mellitus: the cholinesterase connection? Lipids Health Dis 5:28
[16] Law A, Gauthier S, Quirion R (2001) Say NO to Alzheimer's disease: the putative links between nitric oxide and dementia of the Alzheimer's type. Brain Res Brain Res Rev 35:73–96
[17] Iqbal K, Grundke-Iqbal I (2005) Metabolic/signal transduction hypothesis of Alzheimer's disease and other tauopathies. Acta Neuropathol (Berl) 109:25–31
[18] Wang R, Wang B, He W, Zheng H (2006) Wild-type presenilin 1 protects against Alzheimer disease mutation-induced amyloid pathology. J Biol Chem 281:15330–15336
[19] Selkoe DJ (1999) Translating cell biology into therapeutic advances in Alzheimer's disease. Nature 399(6738 Suppl):A23–A31
[20] Rojo L, Sjoberg MK, Hernandez P, Zambrano C, Maccioni RB (2006) Roles of cholesterol and lipids in the etiopathogenesis of Alzheimer's disease. J Biomed Biotechnol 2006:73976
[21] Edland SD (2004) Insulin-degrading enzyme, apolipoprotein E, and Alzheimer's disease. J Mol Neurosci 23:213–217
[22] Avila J, Lucas JJ, Perez M, Hernandez F (2004) Role of tau protein in both physiological and pathological conditions. Physiol Rev 84:361–384
[23] Morgan D, Diamond DM, Gottschall PE, Ugen KE, Dickey C, Hardy J, Duff K, Jantzen P, DiCarlo G, Wilcock D, Connor K, Hatcher J, Hope C, Gordon M, Arendash GW (2000) A beta peptide vaccination prevents memory loss in an animal model of Alzheimer's disease. Nature 408:982–985
[24] Mattson MP, Lovell MA, Furukawa K, Markesbery WR (1995) Neurotrophic factors attenuate glutamate-induced accumulation of peroxides, elevation of intracellular calcium concentration and neurotoxicity and increase antioxidant enzyme activities in hippocampal neurons. J Neurochem 65:1740–1751
[25] Koppaka V, Axelsen PH (2000) Accelerated accumulation of amyloid beta proteins on oxidatively damaged lipid membranes. Biochemistry 39:10011–10016
[26] Toda T, Nakamura M, Morisawa H, Hirota M, Nishigaki R, Yoshimi Y (2010) Proteomic approaches to oxidative protein modifications implicated in the mechanism of aging. Geriatr Gerontol Int 10(Suppl 1):S25–S31

[27] Massaad CA, Amin SK, Hu L, Mei Y, Klann E, Pautler RG (2010) Mitochondrial superoxide contributes to blood flow and axonal transport deficits in the Tg2576 mouse model of Alzheimer's disease. PLoS One 5:e10561

[28] Gupta A, Pansari K (2003) Inflammation and Alzheimer's disease. Int J Clin Pract 57:36–39

[29] Sutton ET, Thomas T, Bryant MW, Landon CS, Newton CA, Rhodin JA (1999) Amyloid-beta peptide induced inflammatory reaction is mediated by the cytokines tumor necrosis factor and interleukin-1. J Submicrosc Cytol Pathol 31:313–323

[30] Lombardi VR, García M, Rey L, Cacabelos R (1999) Characterization of cytokine production, screening of lymphocyte subset patterns and in vitro apoptosis in healthy and Alzheimer's Disease (AD) individuals. J Neuroimmunol 97:163–171

[31] Dumery L, Bourdel F, Soussan Y, Fialkowsky A, Viale S, Nicolas P, Reboud-Ravaux M (2001) Beta-Amyloid protein aggregation: its implication in the physiopathology of Alzheimer's disease. Pathol Biol (Paris) 49:72–85

[32] Calingasan NY, Erdely HA, Altar AC (2002) Identification of CD40 ligand in Alzheimer's disease and in animal models of Alzheimer's disease and brain injury. Neurobiol Aging 23:31–39

[33] McDonald DR, Bamberger ME, Combs CK, Landreth GE (1998) Beta-Amyloid fibrils activate parallel mitogen-activated protein kinase pathways in microglia and THP1 monocytes. J Neurosci 18:4451–4460

[34] Lue LF, Rydel R, Brigham EF, Yang LB, Hampel H, Murphy GM, Brachova L, Yan SD, Walker DG, Shen Y, Rogers J (2001) Inflammatory repertoire of Alzheimer's disease and nondemented elderly microglia in vitro. Glia 35:72–79

[35] Lorton D, Kocsis JM, King L, Madden K, Brunden KR (1996) beta-Amyloid induces increased release of interleukin-1 beta from lipopolysaccharide-activated human monocytes. J Neuroimmunol 67:21–29

[36] Szczepanik AM, Funes S, Petko W, Ringheim GE (2001) IL-4, IL-10 and IL-13 modulate A beta(1-42)-induced cytokine and chemokine production in primary murine microglia and a human monocyte cell line. J Neuroimmunol 113:49–62

[37] Patel NS, Paris D, Mathura V, Quadros AN, Crawford FC, Mullan MJ (2005) Inflammatory cytokine levels correlate with amyloid load in transgenic mouse models of Alzheimer's disease. J Neuroinflammation 2:9

[38] Rainero I, Bo M, Ferrero M, Valfre W, Vaula G, Pinessi L (2004) Association between the interleukin-1alpha gene and Alzheimer's disease: a meta-analysis. Neurobiol Aging 25:1293–1298

[39] Grammas P, Ovase R (2001) Inflammatory factors are elevated in brain microvessels in Alzheimer's disease. Neurobiol Aging 22:837–842

[40] Remarque EJ, Weverling-Rijnsburger AW, Laterveer JC, Blauw GJ, Westendorp RG (2001) Patients with Alzheimer's disease display a pro-inflammatory phenotype. Exp Gerontol 36:171–176

[41] Luterman JD, Haroutunian V, Yemul S, Ho L, Purohit D, Aisen PS, Mohs R, Pasinetti GM (2000) Cytokine gene expression as a function of the clinical progression of Alzheimer disease dementia. Arch Neurol 57:1153–1160

[42] Akama KT, Van Eldik LJ (2000) Beta-amyloid stimulation of inducible nitric-oxide synthase in astrocytes is interleukin-1beta- and tumor necrosis factor-alpha (TNFalpha)-dependent, and involves a TNFalpha receptor-associated factor- and NFkappaB-inducing kinase-dependent signaling mechanism. J Biol Chem 275:7918–7924

[43] Licastro F, Pedrini S, Caputo L, Annoni G, Davis LJ, Ferri C, Casadei V, Gimaldi LM (2000) Increased plasma levels of interleukin-1, interleukin-6 and alpha-1-antichymotrypsin in patients with Alzheimer's disease: peripheral inflammation or signals from the brain? J Neuroimmunol 103:97–102

[44] Rosenberg PB (2006) Cytokine inhibition for treatment of Alzheimer's disease. MedGenMed 8:24

[45] Nizri E, Hamra-Amitay Y, Sicsic C, Lavon I, Brenner T (2006) Anti-inflammatory properties of cholinergic up-regulation: a new role for acetylcholinesterase inhibitors. Neuropharmacology 50:540–547

[46] Tabet N (2006) Acetylcholinesterase inhibitors for Alzheimer's disease: anti-inflammatories in acetylcholine clothing! Age Ageing 35:336–338
[47] Cummings JL, Kaufer D (1996) Neuropsychiatric aspects of Alzheimer's disease: the cholinergic hypothesis revisited. Neurology 47:876–883
[48] Giacobini E (2004) Cholinesterase inhibitors: new roles and therapeutic alternatives. Pharmacol Res 50:433–440
[49] Borovikova LV, Ivanova S, Zhang M, Yang H, Botchkina GI, Watkins LR, Wang H, Abumrad N, Eaton JW, Tracey KJ (2000) Vagus nerve stimulation attenuates the systemic inflammatory response to endotoxin. Nature 405:458–462
[50] Cacabelos R, Barquero M, Garcia P, Alvaez XA, Varela de Seijas E (1991) Cerebrospinal fluid interleukin-1 beta (IL-1 beta) in Alzheimer's disease and neurological disorders. Methods Find Exp Clin Pharmacol 13:455–458
[51] Fillit H, Ding WH, Buee L, Kalman J, Altstiel L, Lawlor B, Wolf-Klein G (1991) Elevated circulating tumor necrosis factor levels in Alzheimer's disease. Neurosci Lett 129:318–320
[52] Donnelly RJ, Friedhoff AJ, Beer B, Blume AJ, Vitek MP (1990) Interleukin-1 stimulates the beta-amyloid precursor protein promoter. Cell Mol Neurobiol 10:485–495
[53] Tobinick E, Gross H, Weinberger A, Cohen H (2006) TNF-alpha modulation for treatment of Alzheimer's disease: a 6-month pilot study. MedGenMed 8:25
[54] Berzaghi MP, Cooper J, Castren E, Zafra F, Sofroniew M, Thoenen H, Lindholm D (1993) Cholinergic regulation of brain-derived neurotrophic factor (BDNF) and nerve growth factor (NGF) but not neurotrophin-3 (NT-3) mRNA levels in the developing rat hippocampus. J Neurosci 13:3818–3826
[55] Phillips HS, Hains JM, Armanini M, Laramee GR, Johnson SA, Winslow JW (1991) BDNF mRNA is decreased in the hippocampus of individuals with Alzheimer's disease. Neuron 7:695–702
[56] Connor B, Young D, Yan Q, Faull RL, Synek B, Dragunow M (1997) Brain-derived neurotrophic factor is reduced in Alzheimer's disease. Brain Res Mol Brain Res 49:71–81
[57] Soontornniyomkii V, Wang G, Pittman CA, Hamilton RL, Wiley CA, Achim CL (1999) Absence of brain-derived neurotrophic factor and trkB receptor immunoreactivity in glia of Alzheimer's disease. Acta Neuropathol 98:345–348
[58] Lapchak PA, Araujo DM, Hefti F (1993) Systemic interleukin-1 beta decreases brain-derived neurotrophic factor messenger RNA expression in the rat hippocampal formation. Neuroscience 53:297–301
[59] Tong L, Balazs R, Soiampornkul R, Thangnipon W, Cotman CW (2008) Interleukin-1 beta impairs brain derived neurotrophic factor-induced signal transduction. Neurobiol Aging 29:1380–1393
[60] Russo-Neustadt A, Beard RC, Cotman CW (1999) Exercise, antidepressant medications, and enhanced brain derived neurotrophic factor expression. Neuropsychopharmacology 21:679–682
[61] Seifert T, Brassard P, Wissenberg M, Rasmussen P, Nordby P, Stallknecht B, Adser H, Jakobsen AH, Pilegaard H, Nielsen HB, Secher NH (2010) Endurance training enhances BDNF release from the human brain. Am J Physiol Regul Integr Comp Physiol 298:R372–R377
[62] Griffin EW, Bechara RG, Birch AM, Kelly AM (2009) Exercise enhances hippocampal-dependent learning in the rat: evidence for a BDNF-related mechanism. Hippocampus 19:973–980
[63] Gustad J, Benitez A, Smith J, Glickman E, Spitznagel MB, Alexander T, Juvancic-Heltzel J, Murray L (2008) Serum brain-derived neurotrophic factor is associated with cognitive function in healthy older adults. J Geriatr Psychiatry Neurol 21:166–170
[64] Nagahara AH, Merrill DA, Coppola G, Tsukada S, Schroeder BE, Shaked GM, Wang L, Blesch A, Kim A, Conner JM, Rockenstein E, Chao MV, Koo EH, Geschwind D, Masliah E, Chiba AA, Tuszynski MH (2009) Neuroprotective effects of brain-derived neurotrophic factor in rodent and primate models of Alzheimer's disease. Nat Med 15:331–337
[65] Lindsay RM (1994) Neurotrophic growth factors and neurodegenerative diseases: therapeutic potential of the neurotrophins and ciliary neurotrophic factor. Neurobiol Aging 15:249–251

[66] Knusel B, Gao H (1996) Neurotrophins and Alzheimer's disease: beyond the cholinergic neurons. Life Sci 58:2019–2027
[67] Yeh HL, Tsai SJ (2008) Lithium may be useful in the prevention of Alzheimer's disease in individuals at risk of presenile familial Alzheimer's disease. Med Hypotheses 71:948–951
[68] Leyhe T, Eschweiler GW, Stransky E, Gasser T, Annas P, Basun H, Laske C (2009) Increase of BDNF serum concentration in lithium treated patients with early Alzheimer's disease. J Alzheimers Dis 16:649–656
[69] Gamoh S, Hashimoto M, Hossain S, Masumura S (2001) Chronic administration of docosahexaenoic acid improves the performance of radial arm maze task in aged rats. Clin Exp Pharmacol Physiol 28:266–270
[70] Lim GP, Calon F, Morihara T, Yang F, Teter B, Ubeda O, Salem N Jr, Frautschy SA, Cole GM (2005) A diet enriched with the omega-3 fatty acid docosahexaenoic acid reduces amyloid burden in an aged Alzheimer mouse model. J Neurosci 25:3032–3040
[71] Das UN (2008) Folic acid and polyunsaturated fatty acids improve cognitive function and prevent depression, dementia, and Alzheimer's disease – but how and why? Prostaglandins Leukot Essent Fatty Acids 78:11–19
[72] Cole GM, Frautschy SA (2010) DHA may prevent age-related dementia. J Nutr 140:869–874
[73] Hashimoto M, Hossain S, Shimada T, Shido O (2006) Docosahexaenoic acid-induced protective effect against impaired learning in amyloid beta-infused rats is associated with increased synaptosomal membrane fluidity. Clin Exp Pharmacol Physiol 33:934–939
[74] Lukiw WJ, Cui JG, Marcheselli VL, Bodker M, Botkjaer A, Gotlinger K, Serhan CN, Bazan NG (2005) A role for docosahexaenoic acid-derived neuroprotectin D1 in neural cell survival and Alzheimer disease. J Clin Invest 115:2774–2783
[75] Marcheselli VL, Hong S, Lukiw WJ, Tian XH, Gronet K, Musto A, Hardy M, Gimenez JM, Chiang N, Serhan CN, Bazan NG (2003) Novel docosanoids inhibit brain ischemia-reperfusion-mediated leukocyte infiltration and pro-inflammatory gene expression. J Biol Chem 278:43807–43817
[76] Mukherjee PK, Marcheselli VL, Serhan CN, Bazan NG (2004) Neuroprotectin D1: a docosahexaenoic acid-derived docosatriene protects human retinal pigment epithelial cells from oxidative stress. Proc Natl Acad Sci U S A 101:8491–8496
[77] Cai D, Netzer WJ, Zhong M, Lin Y, Du G, Frohman M, Foster DA, Sisodia SS, Xu H, Gorelick FS, Greengard P (2006) Presenilin-1 uses phospholipase D1 as a negative regulator of beta-amyloid formation. Proc Natl Acad Sci U S A 103:1941–1946
[78] Kim SY, Ahn BH, Min KJ, Lee YH, Joe EH, Min DS (2004) Phospholipase D isozymes mediate epigallocatechin gallate-induced cyclooxygenase-2 expression in astrocyte cells. J Biol Chem 279:38125–38133
[79] Almeida T, Cunha RA, Ribeiro JA (1999) Facilitation by arachidonic acid of acetylcholine release from the rat hippocampus. Brain Res 826:104–111
[80] Aid S, Vancassel S, Linard A, Lavialle M, Guesnet P (2005) Dietary docosahexaenoic acid[22: 6(n-3)] as a phospholipid or a triglyceride enhances the potassium chloride-evoked release of acetylcholine in rat hippocampus. J Nutr 135:1008–1013
[81] Hossain S, Hashimoto M, Katakura M, Miwa K, Shimada T, Shido O (2009) Mechanism of docosahexaenoic acid-induced inhibition of in vitro Abeta1-42 fibrillation and Abeta1-42-induced toxicity in SH-S5Y5 cells. J Neurochem 111:568–579
[82] Hashimoto M, Shahdat HM, Katakura M, Tanabe Y, Gamoh S, Miwa K, Shimada T, Shido O (2009) Effects of docosahexaenoic acid on in vitro amyloid beta peptide 25–35 fibrillation. Biochim Biophys Acta 1791:289–296
[83] Johansson AS, Garlind A, Berglind-Dehlin F, Karlsson G, Edwards K, Gellerfors P, Ekholm-Pettersson F, Palmblad J, Lannfelt L (2007) Docosahexaenoic acid stabilizes soluble amyloid-beta protofibrils and sustains amyloid-beta-induced neurotoxicity in vitro. FEBS J 274:990–1000
[84] Lando M, Abemayor E, Verity MA, Sidell N (1990) Modulation of intracellular cyclic adenosine monophosphate levels and the differentiation response of human neuroblastoma cells. Cancer Res 50:722–727

[85] Williams EJ, Walsh FS, Doherty P (1994) The production of arachidonic acid can account for calcium channel activation in the second messenger pathway underlying neurite outgrowth stimulated by NCAM, N-cadherin, and L1. J Neurochem 62:1231–1234
[86] Wada K, Arita M, Nakajima A, Katayama K, Kudo C, Kamisaki Y, Serhan CN (2006) Leukotriene B4 and lipoxin A4 are regulatory signals for neural stem cell proliferation and differentiation. FASEB J 20:1785–1792
[87] Uz T, Manev R, Manev H (2001) 5-Lipoxygenase is required for proliferation of immature cerebellar granule neurons in vitro. Eur J Pharmacol 418:15–22
[88] Watanabe A, Toyota T, Owada Y, Hayashi T, Iwayama Y, Matsumata M, Ishitsuka Y, Nakaya A, Maekawa M, Ohnishi T, Arai R, Sakurai K, Yamada K, Kondo H, Hashimoto K, Osumi N, Yoshikawa T (2007) Fabp7 maps to a quantitative trait locus for a schizophrenia endophenotype. PLoS Biol 5:e297
[89] Maekawa M, Takashima N, Matsumata M, Ikegami S, Kontani M, Hara Y, Kawashima H, Owada Y, Kiso Y, Yoshikawa T, Inokuchi K, Osumi N (2009) Arachidonic acid drives postnatal neurogenesis and elicits a beneficial effect on prepulse inhibition, a biological trait of psychiatric illnesses. PLoS One 4:e5085
[90] Schaeffer EL, Forlenza OV, Gattaz WF (2009) Phospholipase A2 activation as a therapeutic approach for cognitive enhancement in early-stage Alzheimer disease. Psychopharmacology (Berl) 202:37–51
[91] Darios F, Davletov B (2006) Omega-3 and omega-6 fatty acids stimulate cell membrane expansion by acting on syntaxin 3. Nature 440:813–817
[92] Pongrac JL, Slack PJ, Innis SM (2007) Dietary polyunsaturated fat that is low in (n-3) and high in (n-6) fatty acids alters the SNARE protein complex and nitrosylation in rat hippocampus. J Nutr 137:1852–1856
[93] Wurtman RJ, Cansev M, Ulus IH (2009) Synapse formation is enhanced by oral administration of uridine and DHA, the circulating precursors of brain phosphatides. J Nutr Health Aging 13:189–197
[94] Akbar M, Calderon F, Wen Z, Kim HY (2005) Docosahexaenoic acid: a positive modulator of Akt signaling in neuronal survival. Proc Natl Acad Sci U S A 102:10858–10863
[95] Lukiw WJ, Cui JG, Marcheselli VL, Bodker M, Botkjaer A, Gotlinger K, Serhan CN, Bazan NG (2005) A role for docosahexaenoic acid-derived neuroprotectin D1 in neural cell survival and Alzheimer disease. J Clin Invest 115:2774–2783
[96] Marcheselli VL, Hong S, Lukiw WJ, Tian XH, Gronet K, Musto A, Hardy M, Gimenez JM, Chiang N, Serhan CN, Bazan NG (2003) Novel docosanoids inhibit brain ischemia-reperfusion-mediated leukocyte infiltration and pro-inflammatory gene expression. J Biol Chem 278:43807–43817
[97] Mukherjee PK, Marcheselli VL, Serhan CN, Bazan NG (2004) Neuroprotectin D1: a docosahexaenoic acid-derived docosatriene protects human retinal pigment epithelial cells from oxidative stress. Proc Natl Acad Sci U S A 101:8491–8496
[98] Almeida T, Cunha RA, Ribeiro JA (1999) Facilitation by arachidonic acid of acetylcholine release from the rat hippocampus. Brain Res 826:104–111
[99] Aid S, Vancassel S, Linard A, Lavialle M, Guesnet P (2005) Dietary docosahexaenoic acid[22: 6(n-3)] as a phospholipid or a triglyceride enhances the potassium chloride-evoked release of acetylcholine in rat hippocampus. J Nutr 135:1008–1013
[100] de La Presa Owens S, Innis SM (1999) Docosahexaenoic and arachidonic acid prevent a decrease in dopaminergic and serotoninergic neurotransmitters in frontal cortex caused by a linoleic and alpha-linolenic acid deficient diet in formula-fed piglets. J Nutr 129:2088–2093
[101] Bernal-Mizrachi C, Gates AC, Weng S et al (2005) Vascular respiratory uncoupling increases blood pressure and atherosclerosis. Nature 435:502–506
[102] Cha SH, Fukushima A, Sakuma K, Kagawa Y (2001) Chronic docosahexaenoic acid intake enhances expression of the gene for uncoupling protein 3 and affects pleiotropic mRNA levels in skeletal muscle aged C57BL/6NJcl mice. J Nutr 131:2636–2642
[103] Klein PD, Johnson RM (1954) Phosphorous metabolism in unsaturated fatty acid-deficient rats. J Biol Chem 211:103–110

[104] Hayashida T, Portman OW (1960) Swelling of liver mitochondria from rats fed diets deficient in essential fatty acids. Proc Soc Exp Biol Med 103:656–659
[105] Shaw KN, Commins S, O'Mara SM (2003) Deficits in spatial learning and synaptic plasticity induced by the rapid and competitive broad-spectrum cyclooxygenase inhibitor ibuprofen are reversed by increasing endogenous brain-derived neurotrophic factor. Eur J Neurosci 17:2438–2446
[106] Hein AM, Stutzman DL, Bland ST, Barrientos RM, Watkins LR, Rudy JW, Maier SF (2007) Prostaglandins are necessary and sufficient to induce contextual fear learning impairments after interleukin-1 beta injections into the dorsal hippocampus. Neuroscience 150:754–763
[107] Wu A, Ying Z, Gomez-Pinilla F (2008) Docosahexaenoic acid dietary supplementation enhances the effects of exercise on synaptic plasticity and cognition. Neuroscience 155:751–759
[108] Wu A, Ying Z, Gomez-Pinilla F (2004) Dietary omega-3 fatty acids normalize BDNF levels, reduce oxidative damage, and counteract learning disability after traumatic brain injury in rats. J Neurotrauma 21:1457–1467
[109] Inestrosa NC, Godoy JA, Quintanilla RA, Koenig CS, Bronfman M (2005) Peroxisome proliferator-activated receptor gamma is expressed in hippocampal neurons and its activation prevents beta-amyloid neurodegeneration: role of Wnt signaling. Exp Cell Res 304:91–104
[110] Rosa AO, Kaster MP, Binfaré RW, Morales S, Martín-Aparicio E, Navarro-Rico ML, Martinez A, Medina M, García AG, López MG, Rodrigues AL (2008) Antidepressant-like effect of the novel thiadiazolidinone NP031115 in mice. Prog Neuropsychopharmacol Biol Psychiatry 32:1549–1556
[111] Yaffe K, Kanaya AM, Lindquist K, Hsueh WC, Cummings SR, Beamer B, Newman A, Rosano C, Li R, Harris T, Health ABC Study (2008) PPAR-gamma Pro12Ala genotype and risk of cognitive decline in elders. Neurobiol Aging 29:78–83
[112] d'Abramo C, Ricciarelli R, Pronzato MA, Davies P (2006) Troglitazone, a peroxisome proliferator-activated receptor-gamma agonist, decreases tau phosphorylation in CHOtau4R cells. J Neurochem 98:1068–1077
[113] Allen NB, Lewinsphn PM, Seeley JR (1998) Prenatal and perinatal influences on risk for psychopathology in childhood and adolescence. Dev Psychopathol 10:513–529
[114] McCreadie RG (1997) The Nothsdale Schizophrenia Surveys. 16. Breast-feeding and schizophrenia: preliminary results and hypotheses. Br J Psychiatry 170:334–337
[115] Shi L, Fatemi SH, Sidwell RW, Patterson PH (2003) Maternal influenza infection causes marked behavioral and pharmacological changes in the offspring. J Neurosci 23:297–302
[116] Harley M, Kelleher I, Clarke M, Lynch F, Arseneault L, Connor D, Fitzpatrick C, Cannon M (2010) Cannabis use and childhood trauma interact additively to increase the risk of psychotic symptoms in adolescence. Psychol Med 40(10):1627–1634
[117] Lucas A, Morley R, Cole TJ et al (1992) Breast milk and subsequent intelligence quotient in children born pre-term. Lancet 239:261–264
[118] Gale CR, Martyn CN (1996) Breast feeding, dummy use, and adult intelligence. Lancet 347:1057
[119] Malloy MH, Berendes H (1998) Does breast feeding influence intelligence quotients at 9 and 10 years of age. Early Hum Dev 50:209–217
[120] Jones P, Rodgers B, Murray R et al (1994) Child developmental risk factors for schizophrenia in the 1946 birth cohort. Lancet 344:1398–1402
[121] Crow TJ, Done DJ, Sacker A (1995) Childhood precursors of psychosis as clues to its evolutionary origins. Eur Arch Psychiatry Clin Neurosci 245:61–69
[122] Peet M, Poole J, Laugharne J (1997) Infant feeding and the development of schizophrenia. Schizophr Res 24:255–256
[123] Das UN (2002) The lipids that matter from infant nutrition to insulin resistance. Prostaglandins Leukot Essent Fatty Acids 67:1–12
[124] Das UN (2003) Long-chain polyunsaturated fatty acids in the growth and development of the brain and memory. Nutrition 19:62–65

[125] Das UN (2003) Can memory be improved? A discussion on the role of ras, GABA, acetylcholine, NO, insulin, TNF-α, and long-chain polyunsaturated fatty acids in memory formation and consolidation. Brain Dev 25:251–261
[126] Makrides M, Neumann M, Simmer K, Pater J, Gibson R (1995) Are long-chain polyunsaturated fatty acids essential nutrients in infancy? Lancet 345:1463–1468
[127] Hedelin M, Löf M, Olsson M, Lewander T, Nilsson B, Hultman CM, Weiderpass E (2010) Dietary intake of fish, omega-3, omega-6 polyunsaturated fatty acids and vitamin D and the prevalence of psychotic-like symptoms in a cohort of 33,000 women from the general population. BMC Psychiatry 10:38
[128] Watari M, Hamazaki K, Hirata T, Hamazaki T, Okubo Y (2010) Hostility of drug-free patients with schizophrenia and n-3 polyunsaturated fatty acid levels in red blood cells. Psychiatry Res 177:22–26
[129] Amminger GP, Schäfer MR, Papageorgiou K, Klier CM, Cotton SM, Harrigan SM, Mackinnon A, McGorry PD, Berger GE (2010) Long-chain omega-3 fatty acids for indicated prevention of psychotic disorders: a randomized, placebo-controlled trial. Arch Gen Psychiatry 67:146–154
[130] Ohara K (2007) The n-3 polyunsaturated fatty acid/dopamine hypothesis of schizophrenia. Prog Neuropsychopharmacol Biol Psychiatry 31:469–474
[131] Leask SJ, Done DJ, Crow TJ, Richards M, Jones PB (2000) No association between breastfeeding and adult psychosis in two national birth cohorts. Br J Psychiatry 177:218–221
[132] Sasaki T, Okazaki Y, Akaho R et al (2000) Type of feeding during infancy and later development of schizophrenia. Schizophr Res 42:79–82
[133] Amore M, Balista C, McCreadie RG, Cimmino C, Pisani F, Bevilacqua G, Ferrari G (2003) Can breast-feeding protect against schizophrenia? Case-control study. Biol Neonate 83:97–101
[134] Bateson P, Barker D, Clutton-Brock T, Deb D, D'Udine B, Foley RA, Cluckman P, Godfrey K, Kirkwood T, Lahr MM, McNamara J, Metcalfe NB, Monaghan P, Spencer HG, Sultan SE (2004) Developmental plasticity and human health. Nature 430:419–421
[135] Waterland RA, Garza C (1999) Potential mechanisms of metabolic imprinting that lead to chronic disease. Am J Clin Nutr 69:179–197
[136] Fricchione GL, Bilfinger TV, Stefano GB (1996) The macrophage and neuropsychiatric disorders. Neurobiology 9:16–29
[137] Salk L, Lipsitt LP, Sturner WQ, Reilly BM, Levat RH (1985) Relationship of maternal and perinatal conditions to eventual adolescent suicide. Lancet 1:624–627
[138] Neugebauer R, Reuss ML (1998) Association of maternal, antenatal and perinatal complications with suicide in adolescence and young adulthood. Acta Psychiatr Scand 97:412–418
[139] Barker DJP, Osmond C, Rodin J, Fall CHD, Winter PD (1995) Low weight gain in infancy and suicide in adult life. BMJ 311:1203
[140] Mittendorfer-Rutz E, Rasmussen F, Wasserman D (2004) Restricted fetal growth and adverse maternal psychosocial and socioeconomic conditions as risk factors for suicidal behaviour of offspring: a cohort study. Lancet 364:1135–1140
[141] Allebeck P, Allgulander C, Henninggsohn I, Jakobsson SW (1991) Causes of death in a cohort of 50465 young men-validity of recorded suicide as underlying cause of death. Scand J Soc Med 19:242–247
[142] Nilsson PM, Nyberg P, Ostergren PO (2001) Increased susceptibility to stress at a psychological assessment of stress tolerance is associated with impaired fetal growth. Int J Epidemiol 30:75–80
[143] Ansorge MS, Zhou M, Lira A, Hen R, Gingrich JA (2004) Early-life blockade of the 5-HT transporter alters emotional behavior in adult mice. Science 306:879–881
[144] Caspi A, Sugden K, Moffitt TE, Taylor A, Craig IW, Harrington H, McClay J, Mill J, Martin J, Braithwaite A, Poulton R (2003) Influence of life stress on depression: moderation by a polymorphism in the 5-HT gene. Science 301:386–390
[145] Rumajogee P, Verge D, Hanoun N, Brisorgueil MJ, Hen R, Lesch KP, Hamon M, Miquel MC (2004) Phenotype in the absence of 5-HT autoreceptors or the 5-HT transporter: involvement of BDNF and camp. Eur J Neurosci 19:937–944

[146] Liu D, Diorio J, Tannenbaum B, Caldji C, Francis D, Freedman A, Sharma A, Pearson D, Plotsky PM, Meaney MJ (1997) Maternal care, hippocampal glucocorticoid receptors, and hypothalamic-pituitary-adrenal responses to stress. Science 277:1659–1662

[147] Gross C, Zhuang X, Stark K, Ramboz S, Oosting R, Kirby L, Santarelli L, Beck S, Hen R (2002) Serotonin A1 receptor acts during development to establish normal anxiety-like behavior in the adult. Nature 416:396–400

[148] Hariri AR, Mattay VS, Tessitore A, Kolachana B, Fera F, Goldman D, Egan MF, Weinberger DR (2002) Serotonin transporter genetic variation and the response of the human amygdala. Science 297:400–403

[149] Smythe JW, Rowe WB, Meaney MJ (1994) Neonatal handling alters serotonin (5-HT) turnover and 5-HT2 receptor binding in selected brain regions: relationship to the handling effect on glucocorticoid receptor expression. Brain Res Dev Brain Res 80:183–189

[150] Oquendo MA, Baca-Garcia E (2004) Nurture versus nature: evidence of intrauterine effects on suicidal behaviour. Lancet 364:1102–1104

[151] Shi L, Fatemi SH, Sidwell RW, Patterson PH (2003) Maternal influenza infection causes marked behavioral and pharmacological changes in the offspring. J Neurosci 23:297–302

[152] Borrell J, Vela JM, Arevalo-Martin A, Molina-Holgado E, Guaza C (2002) Prenatal immune challenge disrupts sensorimotor gating in adult rats. Implications for the etiopathogenesis of schizophrenia. Neuropsychopharmacology 26:204–211

[153] Editorial (2003) Of mice and mental illness. Nat Neurosci 6:323

[154] Zhang XY, Zhou DF, Zhang PY, Wu GY, Cao LY, Shen YC (2002) Elevated interleukins-2, interleukins-6 and interleukins-8 serum levels in neuroleptic-free schizophrenia: association with psychopathology. Schizophr Res 57:247–258

[155] Ebrinc S, Top C, Oncul O, Basoglu C, Cavuslu S, Cetin M (2002) Serum interlukin 1 alpha and interleukins 2 levels in patients with schizophrenia. J Int Med Res 30:314–317

[156] McAllister CG, van Kammen DP, Rehn TJ, Miller AL, Gurklis J, Kelley ME, Yao J, Peters JL (1995) Increases in CSF levels of interleukins-2 in schizophrenia: effects of recurrence of psychosis and medication status. Am J Psychiatry 152:1291–1297

[157] Kowalski J, Blada P, Kucia K, Madej A, Herman ZS (2001) Neuroleptics normalize increased release of interleukins-1 beta and tumor necrosis factor-alpha from monocytes in schizophrenia. Schizophr Res 50:169–175

[158] Maes M, Bocchio Chiavetto L, Bignotti S, Battisa Tura GJ, Pioli R, Boin F, Kenis G, Bosmans E, de Jongh R, Altamura CA (2002) Increased serum interleukins-8 and interleukins-10 in schizophrenic patients resistant to treatment with neuroleptics and the stimulatory effects of clozapine on serum leukemia inhibitory factor receptor. Schizophr Res 54:281–291

[159] Cazzullo CL, Sacchetti E, Galluzzo A, Adorni A, Pegoraro M, Bosis S, Colombo F, Trabattoni D, Zagliani A, Clerici M (2002) Cytokine profiles in schizophrenic patients treated with risperidone: a 3-month follow-up study. Prog Neuropsychopharmacol Biol Psychiatry 26:33–39

[160] Zalcman SS (2002) Interleukin-2-induced increases in climbing behavior: inhibition by dopamine D-1 and D-2 receptor antagonists. Brain Res 944:157–164

[161] Marx CE, Jarskog LF, Lauder JM, Lieberman JA, Gilmore JH (2001) Cytokine effects on cortical neuron MAP-2 immunoreactivity: implications for schizophrenia. Biol Psychiatry 50:743–749

[162] Kim YK, Kim L, Lee MS (2000) Relationships between interleukins, neurotransmitters and psychopathology in drug-free male schizophrenics. Schizophr Res 44:165–175

[163] Carlson SE, Werkman SH, Peeples JM, Cooke RJ, Tolley EA (1993) Arachidonic acid status correlates with first year growth in preterm infants. Proc Natl Acad Sci U S A 90:1073–1077

[164] Olsen SF, Olsen J, Frische G (1990) Does fish consumption during pregnancy increase fetal growth? A study of the size of the newborn, placental weight and gestational age in relation to fish consumption during pregnancy. Int J Epidemiol 19:971–977

[165] Baguma-Nibasheka M, Brenna JT, Nathaniesz PW (1999) Delay of preterm delivery in sheep by omega-3 long-chain polyunsaturates. Biol Reprod 60:698–701

[166] Das UN (2001) Essential fatty acids as possible mediators of the actions of statins. Prostaglandins Leukot Essent Fatty Acids 65:37–40
[167] Kumar KV, Das UN (1997) Effect of cis-unsaturated fatty acids, prostaglandins, and free radicals on angiotensin-converting enzyme activity in vitro. Proc Soc Exp Biol Med 214:374–379
[168] Reseland JE, Haugen F, Hollung K, Solvoll K, Halvorsen B, Brude I R, Nenseter MS, Christiansen EN, Drevon CA (2001) Reduction of leptin gene expression by dietary polyunsaturated fatty acids. J Lipid Res 42:743–750
[169] Das UN (2002) A perinatal strategy for preventing adult disease. Kluwer Academic, Norwell, MA
[170] Das UN (2010) Metabolic syndrome pathophysiology: the role of essential fatty acids. Wiley-Blackwell, Ames, IA
[171] Hibbeln JR, Makino KK, Martin CE, Dickerson F, Boronow J, Fenton WS (2003) Smoking, gender, and dietary influences on erythrocyte essential fatty acid composition among patients with schizophrenia or schizoaffective disorder. Biol Psychiatry 53:431–441
[172] Aravindakshan M, Sitasawad S, Debsikdar V, Ghate M, Evans D, Horrobin DF, Bennett C, Ranjekar PK, Mahadik SP (2003) Essential polyunsaturated fatty acid and lipid peroxide levels in never-medicated and medicated schizophrenia patients. Biol Psychiatry 53:56–64
[173] Assies J, Lieverse R, Vreken P, Wanders RJ, Dinhemans PM, Linszen DH (2001) Significantly reduced docosahexaenoic and docosapentaenoic acid concentrations in erythrocyte membranes from schizophrenic patients compared with a carefully matched control group. Biol Psychiatry 49:510–522
[174] Khan MM, Evans DR, Gunna V, Scheffer RE, Parikh VV, Mahadik SP (2002) Reduced erythrocyte membrane essential fatty acids and increased lipid peroxides in schizophrenia at the never-medicated first-episode of psychosis and after years of treatment with antipsychotics. Schizophr Res 58:1–10
[175] Chang MC, Contreras MA, Rosenberger TA, Rintala JJ, Bell JM, Rapoport SI (2001) Chronic valproate treatment decreases the in vivo turnover of arachidonic acid in brain phospholipids: a possible common effect of mood stabilizers. J Neurochem 77:796–803
[176] Chang MC, Bell JM, Purdon AD, Chikhale EG, Grange E (1999) Dynamics of docosahexaenoic acid metabolism in the central nervous system: lack of effect of chronic lithium treatment. Neurochem Res 24:399–406
[177] Chang MC, Grange E, Rabin O, Bell JM, Allen DD, Rapoport SI (1996) Lithium decreases turnover of arachidonate in several brain phospholipids. Neurosci Lett 220:171–174
[178] Kim SHF, Weeber EJ, Sweatt JD, Stoll AL, Marangell LB (2001) Inhibitory effects of omega-3 fatty acids on protein kinase C activity in vitro. Mol Psychiatry 6:246–248
[179] Chaudhry A, Laychock SG, Rubin RP (1987) The effects of fatty acids on phosphoinositide synthesis and myo-inositol accumulation in exocrine pancreas. J Biol Chem 262:17426–17431
[180] Sperling RI, Benincaso AI, Knoell CT, Larkin JK, Austen KF, Robinson DR (1993) Dietary omega-3 polyunsaturated fatty acids inhibit phosphoinositide formation and chemotaxis in neutrophils. J Clin Invest 91:651–660
[181] Williams RSB, Cheng L, Mudge AW, Harwood AJ (2002) A common mechanism of action for three mood-stabilizing drugs. Nature 417:292–295
[182] Kumar SG, Das UN, Kumar KV, Madhavi N, Das NP, Tan BKH (1992) Effect of n-6 and n-3 fatty acids on the proliferation and secretion of TNF and IL-2 by human lymphocytes in vitro. Nutr Res 12:815–823
[183] Kumar SG, Das UN (1994) Effect of prostaglandins and their precursors on the proliferation of human lymphocytes and their secretion of tumor necrosis factor and various interleukins. Prostaglandins Leukot Essent Fatty Acids 50:331–334
[184] Das UN (1994) Beneficial effect of eicosapentaenoic acid and docosahexaenoic acid in the management of systemic lupus erythematosus and its relationship to the cytokine network. Prostaglandins Leukot Essent Fatty Acids 51:207–213

[185] Endres S, Ghorbani R, Kelley VE et al (1989) The effect of dietary supplementation with n-3 polyunsaturated fatty acids on the synthesis of interleukins-1 and tumor necrosis factor by mononuclear cells. N Engl J Med 320:265–271

[186] Peet M, Horrobin DF, E-E Multicentre Study Group (2002) A dose-ranging exploratory study of the effects of ethyl-eicosapentaenoate in patients with persistent schizophrenic symptoms. J Psychiatr Res 36:7–18

[187] Peet M, Brind J, Ramchand CN, Shah S, Vankar GK (2001) Two double-blind placebo-controlled pilot studies of eicosapentaenoic acid in the treatment of schizophrenia. Schizophr Res 49:243–251

[188] Puri BK, Richardson AJ, Horrobin DF, Easton T, Saeed N, Oatridge A, Hajnal JV, Bydder GM (2000) Eicosapentaenoic acid treatment in schizophrenia associated with symptom remission, normalization of blood fatty acids, reduced neuronal membrane phospholipid turnover and structural brain changes. Int J Clin Pract 54:57–63

[189] Lonergan PE, Martin DSD, Horrobin DF, Lynch MA (2002) Neuroprotective effect of eicosapentaenoic acid in hippocampus of rats exposed to γ-radiation. J Biol Chem 277:20804–20811

[190] Martin DSD, Lonergan PE, Boland B, Fogarty MP, Brady M, Horrobin DF, Campbell VA, Lynch MA (2002) Apoptotic changes in the aged brain are triggered by interleukin-1β-induced activation of p38 and reversed by treatment with eicosapentaenoic acid. J Biol Chem 277:34239–34246

[191] Kim H-Y, Akbar M, Lau A, Edsall L (2000) Inhibition of neuronal apoptosis by docosahexaenoic acid (22:6n-3). Role of phosphatidylserine in antiapoptotic effect. J Biol Chem 275:35215–35223

[192] Kim H-Y, Akbar M, Kim K-Y (2001) Inhibition of neuronal apoptosis by polyunsaturated fatty acids. J Mol Neurosci 16:223–227

[193] Das UN, Devi GR, Rao KP, Rao MS (1985) Prostaglandins and their precursors can modify genetic damage induced by benzo (a) pyrene and gamma-radiation. Prostaglandins 29:911–920

[194] Das UN, Devi GR, Rao KP, Rao MS (1989) Prostaglandins can modify gamma-radiation and chemical-induced cytotoxicity and genetic damage both in vitro and in vivo. Prostaglandins 38:689–699

[195] Sheline YI, Wang PW, Gado MH, Csernansky JG, Vannier MW (1996) Hippocampal atrophy in recurrent major depression. Proc Natl Acad Sci U S A 93:3908–3913

[196] Sheline YI, Sanghavi M, Mintun MA, Gado MH (1999) Depression duration but not age predicts hippocampal volume loss in medically healthy women with recurrent major depression. J Neurosci 19:5034–5043

[197] Lange C, Irle E (2004) Enlarged amygdala volume and reduced hippocampal volume in young women with major depression. Psychol Med 34:1059–1064

[198] Koolschijn PC, van Haren NE, Lensvelt-Mulders GJ, Hulshoff Pol HE, Kahn RS (2009) Brain volume abnormalities in major depressive disorder: a meta-analysis of magnetic resonance imaging studies. Hum Brain Mapp 30:3719–3735

[199] Kim MJ, Hamilton JP, Gotlib IH (2008) Reduced caudate gray matter volume in women with major depressive disorder. Psychiatry Res 164:114–122

[200] Chen MC, Hamilton JP, Gotlib IH (2010) Decreased hippocampal volume in healthy girls at risk of depression. Arch Gen Psychiatry 67:270–276

[201] Murata T, Kimura H, Omori M, Kado H, Kosaka H, Iidaka T, Itoh H, Wada Y (2001) MRI white matter hyperintensities, (1)H-MR spectroscopy and cognitive function in geriatric depression: a comparison of early- and late-onset cases. Int J Geriatr Psychiatry 16:1129–1135

[202] Sassi RB, Brambilla P, Nicoletti M, Mallinger AG, Frank E, Kupfer DJ, Keshavan MS, Soares JC (2003) White matter hyperintensities in bipolar and unipolar patients with relatively mild-to-moderate illness severity. J Affect Disord 77:237–245

[203] Taylor WD, MacFall JR, Payne ME, McQuoid DR, Steffens DC, Provenzale JM, Krishnan RR (2005) Greater MRI lesion volumes in elderly depressed subjects than in control subjects. Psychiatry Res 139:1–7

[204] Karege F, Perret G, Bondolfi G, Schwald M, Bertschy G, Aubry JM (2002) Decreased serum brain-derived neurotrophic factor levels in major depressed patients. Psychiatry Res 109:143–148

[205] Shimizu E, Hashimoto K, Okamura N, Koike K, Komatsu N, Kumakiri C, Nakazato M, Watanabe H, Shinoda N, Okada S, Iyo M (2003) Alterations of serum levels of brain-derived neurotrophic factor (BDNF) in depressed patients with or without antidepressants. Biol Psychiatry 54:70–75

[206] Lang UE, Hellweg R, Gallinat J (2004) BDNF serum concentrations in healthy volunteers are associated with depression-related personality traits. Neuropsychopharmacology 29:795–798

[207] Karege F, Bondolfi G, Gervasoni N, Schwald M, Aubry JM, Bertschy G (2005) Low brain-derived neurotrophic factor (BDNF) levels in serum of depressed patients probably results from lowered platelet BDNF release unrelated to platelet reactivity. Biol Psychiatry 57:1068–1072

[208] Sen S, Duman R, Sanacora G (2008) Serum brain-derived neurotrophic factor, depression, and antidepressant medications: meta-analyses and implications. Biol Psychiatry 64:527–532

[209] Piccinni A, Marazziti D, Catena M, Domenici L, Del Debbio A, Bianchi C, Mannari C, Martini C, Da Pozzo E, Schiavi E, Mariotti A, Roncaglia I, Palla A, Consoli G, Giovannini L, Massimetti G, Dell'Osso L (2008) Plasma and serum brain-derived neurotrophic factor (BDNF) in depressed patients during 1 year of antidepressant treatments. J Affect Disord 105:279–283

[210] Huang TL, Lee CT, Liu YL (2008) Serum brain-derived neurotrophic factor levels in patients with major depression: effects of antidepressants. J Psychiatr Res 42:521–525

[211] Onen Sertoz O, Tolga Binbay I, Koylu E, Noyan A, Yildirim E, Elbi Mete H (2008) The role of BDNF and HPA axis in the neurobiology of burnout syndrome. Prog Neuropsychopharmacol Biol Psychiatry 32:1459–1465

[212] Knapman A, Heinzmann JM, Hellweg R, Holsboer F, Landgraf R, Touma C (2010) Increased stress reactivity is associated with cognitive deficits and decreased hippocampal brain-derived neurotrophic factor in a mouse model of affective disorders. J Psychiatr Res 44:566–575

[213] Kenis G, Prickaerts J, van Os J, Koek GH, Robaeys G, Steinbusch HW, Wichers M (2010) Depressive symptoms following interferon-alpha therapy: mediated by immune-induced reductions in brain-derived neurotrophic factor? Int J Neuropsychopharmacol 29:1–7

[214] Koizumi H, Hashimoto K, Iyo M (2006) Dietary restriction changes behaviours in brain-derived neurotrophic factor heterozygous mice: role of serotonergic system. Eur J Neurosci 24:2335–2344

[215] Jacobsen JP, Mørk A (2006) Chronic corticosterone decreases brain-derived neurotrophic factor (BDNF) mRNA and protein in the hippocampus, but not in the frontal cortex, of the rat. Brain Res 1110:221–225

[216] de Foubert G, O'Neill MJ, Zetterström TS (2007) Acute onset by 5-HT(6)-receptor activation on rat brain brain-derived neurotrophic factor and activity-regulated cytoskeletal-associated protein mRNA expression. Neuroscience 147:778–785

[217] Cho HJ, Kim JK, Zhou XF, Rush RA (1997) Increased brain-derived neurotrophic factor immunoreactivity in rat dorsal root ganglia and spinal cord following peripheral inflammation. Brain Res 764:269–272

[218] Oddiah D, Anand P, McMahon SB, Rattray M (1998) Rapid increase of NGF, BDNF and NT-3 mRNAs in inflamed bladder. Neuroreport 9:1455–1458

[219] Virchow JC, Julius P, Lommatzsch M, Luttmann W, Renz H, Braun A (1998) Neurotrophins are increased in bronchoalveolar lavage fluid after segmental allergen provocation. Am J Respir Crit Care Med 158:2002–2005

[220] Kerschensteiner M, Gallmeier E, Behrens L, Leal VV, Misgeld T, Klinkert WE, Kolbeck R, Hoppe E, Oropeza-Wekerle RL, Bartke I, Stadelmann C, Lassmann H, Wekerle H, Hohlfeld R (1999) Activated human T cells, B cells, and monocytes produce brain-derived neurotrophic

factor in vitro and in inflammatory brain lesions: a neuroprotective role of inflammation? J Exp Med 189:865–870
[221] Tabakman R, Lecht S, Sephanova S, Arien-Zakay H, Lazarovici P (2004) Interactions between the cells of the immune and nervous system: neurotrophins as neuroprotection mediators in CNS injury. Prog Brain Res 146:387–401
[222] Makar TK, Trisler D, Sura KT, Sultana S, Patel N, Bever CT (2008) Brain derived neurotrophic factor treatment reduces inflammation and apoptosis in experimental allergic encephalomyelitis. J Neurol Sci 270:70–76
[223] Ricci A, Mariotta S, Saltini C, Falasca C, Giovagnoli MR, Mannino F, Graziano P, Sciacchitano S, Amenta F (2005) Neurotrophin system activation in bronchoalveolar lavage fluid immune cells in pulmonary sarcoidosis. Sarcoidosis Vasc Diffuse Lung Dis 22:186–194
[224] Hahn C, Islamian AP, Renz H, Nockher WA (2006) Airway epithelial cells produce neurotrophins and promote the survival of eosinophils during allergic airway inflammation. J Allergy Clin Immunol 117:787–794
[225] Bennedich Kahn L, Gustafsson LE, Olgart Hoglund C (2008) Brain-derived neurotrophic factor enhances histamine-induced airway responses and changes levels of exhaled nitric oxide in guinea pigs in vivo. Eur J Pharmacol 595:78–83
[226] Lommatzsch M, Braun A, Mannsfeldt A, Botchkarev VA, Botchkarev NV, Paus R, Fischer A, Lewin GR, Renz H (1999) Abundant production of brain-derived neurotrophic factor by adult visceral epithelia. Am J Pathol 155:1183–1193
[227] Rost B, Hanf G, Ohnemus U, Otto-Knapp R, Groneberg DA, Kunkel G, Noga O (2005) Monocytes of allergics and non-allergics produce, store and release the neurotrophins NGF, BDNF and NT-3. Regul Pept 124:19–25
[228] Noga O, Englmann C, Hanf G, Grutzkau A, Kunkel G (2003) The production, storage and release of the neurotrophins nerve growth factor, brain-derived neurotrophic factor and neurotrophin-3 by human peripheral eosinophils in allergics and non-allergics. Clin Exp Allergy 33:649–654
[229] Rihl M, Kruithof E, Barthel C, De Keyser F, Veys EM, Zeidler H, Yu DT, Kuipers JG, Baeten D (2005) Involvement of neurotrophins and their receptors in spondyloarthritis synovitis: relation to inflammation and response to treatment. Ann Rheum Dis 64:1542–1549
[230] del Porto F, Aloe L, Lagana B, Triaca V, Nofroni I, D'Amelio R (2006) Nerve growth factor and brain-derived neurotrophic factor levels in patients with rheumatoid arthritis treated with TNF-alpha blockers. Ann N Y Acad Sci 1069:438–443
[231] Grimsholm O, Guo Y, Ny T, Forsgren S (2008) Expression patterns of neurotrophins and neurotrophin receptors in articular chondrocytes and inflammatory infiltrates in knee joint arthritis. Cells Tissues Organs 188:299–309
[232] Cai D, Holm JM, Duignan IJ, Zheng J, Xaymardan M, Chin A, Ballard VL, Bella JN, Edelberg JM (2006) BDNF-mediated enhancement of inflammation and injury in the aging heart. Physiol Genomics 24:191–197
[233] Johansson M, Norrgard O, Forsgren S (2007) Study of expression patterns and levels of neurotrophins and neurotrophin receptors in ulcerative colitis. Inflamm Bowel Dis 13:398–409
[234] di Mola FF, Friess H, Zhu ZW, Koliopanos A, Bley T, Di Sebastiano P, Innocenti P, Zimmermann A, Buchler MW (2000) Nerve growth factor and Trk high affinity receptor (TrkA) gene expression in inflammatory bowel disease. Gut 46:670–679
[235] Raap U, Werfel T, Goltz C, Deneka N, Langer K, Bruder M, Kapp A, Schmid-Ott, Wedi B (2006) Circulating levels of brain-derived neurotrophic factor correlate with disease severity in the intrinsic type of atopic dermatitis. Allergy 61:1416–1418
[236] Das UN (2010) Obesity: genes, brain, gut, and environment. Nutrition 26:459–473
[237] Ciz M, Komrskova D, Pracharova L, Okenkova K, Cizova H, Moravcova A, Jancinova V, Petrikova M, Lojek A, Nosal R (2007) Serotonin modulates the oxidative burst of human phagocytes via various mechanisms. Platelets 18:583–590
[238] Menard G, Turmel V, Bissonnette EY (2007) Serotonin modulates the cytokine network in the lung: involvement of prostaglandin E2. Clin Exp Immunol 150:340–348

[239] Muller T, Durk T, Blumental B, Grimm M, Cicko S, Panther E, Sorichter S, Herouy Y, Di Virgilio F, Ferrari D, Norgauer J, Idzko M (2009) 5-hydroxytryptamine modulates migration, cytokine and chemokine release and T-cell priming capacity of dendritic cells in vitro and in vivo. PLoS One 4:e6453
[240] Das UN (2010) Metabolic syndrome is a low-grade systemic inflammatory condition. Expert Rev Endocrinol Metab 5:577–592
[241] Aso Y, Wakabayashi S, Nakano T, Yamamoto R, Takebayashi K, Inukai T (2006) High serum high-sensitivity C-reactive protein concentrations are associated with relative cardiac sympathetic overactivity during the early morning period in type 2 diabetic patients with metabolic syndrome. Metabolism 55:1014–1021
[242] Maestroni GJ (2000) Dendritic cell migration controlled by alpha 1b-adrenergic receptors. J Immunol 165:6743–6747
[243] Rassler B (2007) The role of catecholamines in formation and resolution of pulmonary oedema. Cardiovasc Hematol Disord Drug Targets 7:27–35
[244] Flierl MA, Rittirsch D, Nadeau BA, Chen AJ, Sarma JV, Zetoune FS, McGuire SR, List RP, Day DE, Hoesel LM, Gao H, Van Rooijen N, Huber-Lang MS, Neubig RR, Ward PA (2007) Phagocyte-derived catecholamines enhance acute inflammatory injury. Nature 449:721–725
[245] Schiepers OJ, Wichers MC, Maes M (2005) Cytokines and major depression. Prog Neuropsychopharmacol Biol Psychiatry 29:201–217
[246] Das UN (2007) Is depression a low-grade systemic inflammatory condition? Am J Clin Nutr 85:1665–1666
[247] O'Brien SM, Scott LV, Dinan TG (2004) Cytokines: abnormalities in major depression and implications for pharmacological treatment. Hum Psychopharmacol 19:397–403
[248] Dunn AJ, Swiergiel AH, de Beaurepaire R (2005) Cytokines as mediators of depression: what can we learn from animal studies? Neurosci Biobehav Rev 29:891–909
[249] Castanon N, Leonard BE, Neveu PJ, Yirmiya R (2002) Effects of antidepressants on cytokine production and actions. Brain Behav Immun 16:569–574
[250] Sluzewska A, Rybakowski J, Bosmans E, Sobieska M, Berghmans R, Maes M, Wiktorowicz K (1996) Indicators of immune activation in major depression. Psychiatry Res 64:161–167
[251] Basterzi AD, Aydemir C, Kisa C, Aksaray S, Tuzer V, Yazici K, Goka E (2005) IL-6 levels decrease with SSRI treatment in patients with major depression. Hum Psychopharmacol 20:473–476
[252] Borovikova LV, Ivanova S, Zhang M, Yang H, Botchkina GI, Watkins LR, Wang H, Abumrad N, Eaton JW, Tracey KJ (2000) Vagus nerve stimulation attenuates the systemic inflammatory response to endotoxin. Nature 405:458–462
[253] Ulloa L (2005) The vagus nerve and the nicotinic anti-inflammatory pathway. Nat Rev Drug Discov 4:673–683
[254] Das UN (2007) Vagus nerve stimulation, depression and inflammation. Pscychoneuropharmacology 32:2053–2054
[255] Colin A, Reggers J, Castronovo V, Anssean M (2003) Lipids, depression, and suicide. Encephale 29:49–58
[256] Su KP, Huang SY, Chiu CC, Shen WW (2003) Omega-3 fatty acids in major depressive disorder: a preliminary double-blind, placebo-controlled trial. Eur Neuropscychopharmacol 13:267–271
[257] Ranjekar PK, Hinge A, Hegde MV et al (2003) Decreased antioxidant enzymes and membrane essential polyunsaturated fatty acids in schizophrenia and bipolar mood disorder patients. Psychiatry Res 121:109–122
[258] Denkins Y, Kempf D, Ferniz M, Nileshwar S, Marchetti D (2005) Role of omega-3 polyunsaturated fatty acids on cyclooxygenase-2 metabolism in brain-metastatic melanoma. J Lipid Res 46:1278–1284
[259] Parker G, Gibson NA, Brotchie H, Heruc G, Rees AM, Hadzi-Pavlovic D (2006) Omega-3 fatty acids and mood disorders. Am J Psychiatry 163:969–978
[260] De Vriese SR, Christophe AB, Maes M (2004) In humans, the seasonal variation in polyunsaturated fatty acids is related to the seasonal variation in violent suicide and serotonergic markers of violent suicide. Prostaglandins Leukot Essent Fatty Acids 71:13–18

[261] Arvindakshan M, Ghate M, Ranjekar PK, Evans DR, Mahadik SP (2003) Supplementation with a combination of omega-3 fatty acids and antioxidants (vitamins E and C) improves the outcome of schizophrenia. Schizophr Res 62:195–204

[262] Aliberti J, Hieny S, Reis e Sousa C, Serhan CN, Sher A (2002) Lipoxin-mediated inhibition of IL-12 production by DCs: a mechanism for regulation of microbial immunity. Nat Immunol 3:76–82

[263] Das UN (2011) Influence of polyunsaturated fatty acids and their metabolites on stem cell biology. Nutrition 27:21–25

Chapter 13
Rheumatological Conditions

Introduction

Rheumatological conditions (also termed as collagen vascular diseases) are a group of disorders that affect mainly the joints (small or large or both types). Though the terms rheumatological conditions and collagen vascular diseases are used interchangeably, it may be mentioned here that it is better to use the term rheumatological conditions or systemic autoimmune diseases for conditions such as rheumatoid arthritis (RA), systemic lupus erythematosus (SLE or simply called as lupus), systemic sclerosis, and dermatomyositis The terms "collagen vascular disease" and "collagen-vascular disease," has been in use since 1962 (and possibly earlier), are synonyms for systemic autoimmune disease. The term "collagen vascular disease" is a misnomer: these diseases affect many structures in addition to vascular structures, and they affect many molecules in addition to the collagen molecule. They are also referred to as connective tissue diseases. However, although the systemic autoimmune diseases affect connective tissue, they also affect many other tissue types, including muscle tissue and neural tissue. In addition, many connective tissue diseases (such as scurvy and Marfan's syndrome) are not autoimmune in nature. Systemic lupus erythematosus and rheumatoid arthritis can cause vasculitis. However, these diseases affect many structures other than blood vessels.

Rheumatoid arthritis (RA) and lupus are the most important of all systemic autoimmune diseases both in terms of their frequency of occurrence in the population and the devastating course they can take in some, if not all, patients. Though clinically RA and lupus have substantial differences, the underlying pathophysiology appears to be similar, if not identical. For the sake of brevity and easy understanding a major portion of the discussion will be centered on RA and lupus and where it is necessary specific mention or discussion about each of them will be made.

U. N. Das, *Molecular Basis of Health and Disease*,
DOI 10.1007/978-94-007-0495-4_13,

Autoimmunity

It is believed that autoimmunity is the failure of an organism to recognize its own constituent parts as *self*, which allows an immune response against its own cells and tissues and thus to disease(s). Almost any tissue/organ/system of the body could be the target of such an immunological attack though certain tissues/organs are more commonly involved. The most common autoimmune diseases are: celiac disease, type 1 diabetes mellitus, rheumatoid arthritis (RA), systemic lupus erythematosus (SLE), Sjogren's, Hashimoto's thyroiditis, Graves' disease, idiopathic thrombocytopenic purpura (ITP), etc. Paul Ehrlich, at the beginning of the twentieth century, was the first to propose the concept of **horror autotoxicus**, wherein a "normal" body does not mount an immune response against its own tissues. Thus, any autoimmune response was perceived to be abnormal and postulated to be connected with human disease. At present, autoimmune responses are considered an integral part of vertebrate immune systems, normally prevented from causing disease by the phenomenon of immunological tolerance to self-antigens.

In general, it is considered that certain amount of low level of autoimmunity does exist in the healthy body and is, in fact, beneficial. Such low-level autoimmunity might aid in the recognition of neoplastic cells by CD8+ T cells, and thus reduce the incidence of cancer. Low-level of autoimmunity may allow a rapid immune response in the early stages of an infection when the availability of foreign antigens limits the response (i.e., when there are few pathogens present).

Self and Non-self and Immunological Tolerance

It has been proposed that loss of immunological tolerance to "self antigens" could trigger the development of auto-antibodies to body's own tissues that may ultimately lead to initiation and elaboration of specific immune response against self determinants. The exact genesis of immunological tolerance is still elusive, but several theories have been proposed. Hypotheses that have gained widespread attention are

1. **Clonal Deletion theory:** Originally proposed by Burnet, which states that self-reactive lymphoid cells are destroyed during the development of the immune system in an individual [1]; failure to do so would trigger the elaboration of self-reacting antibodies that could specifically destroy the target tissues/organs culminating in a specific disease.
2. **Clonal Anergy theory:** This theory proposed by Nossal states that self-reactive T- or B-cells become inactivated in the normal individual and cannot amplify the immune response [2] and when such activation for unexplained reasons fails it could lead to the production of autoantibodies.
3. **Idiotype Network theory:** Jerne proposed that a network of antibodies capable of neutralizing self-reactive antibodies exists naturally within the body [3].

4. **Clonal Ignorance theory,** according to which host immune responses are directed to ignore self-antigens [4], while
5. **The "Suppressor population" or "Regulatory T cell"** theories suggest that regulatory T-lymphocytes (commonly $CD4^{+}FoxP3^{+}$ cells, among others) function to prevent, downregulate, or limit autoaggressive immune responses in the immune system.

In essence, these theories propose that in one way or the other, self is recognized and non-self is attacked. But, when this delicate balance is upset leading to the recognition of the self as non-self it would lead to the production of antibodies against the self and destruction of the specific tissues/organs/system leading to the onset and progression of the specific disease.

In addition, self-tolerance is needed so that self is not recognized as non-self even by error. Tolerance can be differentiated into "Central" and "Peripheral" tolerance. It is not clear whether or not the above-stated checking mechanisms operate in the central lymphoid organs (Thymus and Bone Marrow) or the peripheral lymphoid organs (lymph node, spleen, etc., where self-reactive B-cells may be destroyed). It is likely that these theories are not mutually exclusive, and evidence has been mounting suggesting that all of these mechanisms may actively contribute to vertebrate immunological tolerance.

A puzzling feature of the documented loss of tolerance seen in spontaneous human autoimmunity is that it is almost entirely restricted to the autoantibody responses produced by B lymphocytes. Loss of tolerance by T cells has been extremely hard to demonstrate, and where there is evidence for an abnormal T cell response it is usually not to the antigen recognized by autoantibodies. Thus, in rheumatoid arthritis there are autoantibodies to IgG Fc but apparently no corresponding T cell response. In lupus there are autoantibodies to DNA, which cannot evoke a T cell response, and limited evidence for T cell responses implicates nucleoprotein antigens. In celiac disease there are autoantibodies to tissue transglutaminase but the T cell response is to the foreign protein gliadin. This disparity has led to the idea that human autoimmune disease is in most cases (with probable exceptions including type I diabetes) based on a loss of B cell tolerance which makes use of normal T cell responses to foreign antigens in a variety of aberrant ways [4].

In addition, there appears to be many other factors that render an individual to develop autoimmune disorders/diseases some of which include: genetic factors, gender, and environmental factors.

Genetic Factors

Certain individuals are genetically susceptible to developing autoimmune diseases. This susceptibility is associated with multiple genes plus other risk factors. Genetically-predisposed individuals do not always develop autoimmune diseases.

Three main sets of genes are suspected in many autoimmune diseases. These genes are related to: immunoglobulins, T-cell receptors and the major histocompatibility complexes (MHC).

The first two that are involved in the recognition of antigens, are inherently variable and susceptible to recombination. These variations enable the immune system to respond to a very wide variety of invaders, but may also give rise to lymphocytes capable of self-reactivity.

Strong evidence to suggest that certain MHC class II allotypes are strongly correlated with autoimmune diseases is evident from the observations that (a) HLA DR2 is strongly positively correlated with lupus and multiple sclerosis, and negatively correlated with type1 diabetes mellitus. A very strong association between lupus and HLA-DR3 and anti-La seemed to account for any associations with TNF-α alleles on an extended DR3 haplotype. In a study performed to know the MHC class III TNF-lymphotoxin (TNF-LT) region (6p21.3) as a possible susceptibility locus for RA, it was noted that TNF-LT region appears to influence susceptibility to RA, distinct from HLA-DR [5–7]. No significant difference in the frequency and carriage rate of IL-1 α polymorphisms between RA patients and the controls was noted. The β 2/2 genotype of IL-1 β was more common in female RA patients compared with controls. A lower carriage rate of IL-1β 2 occurred in male RA patients. A higher carriage rate of IL-1 α 2 was associated with a higher ESR, HAQ score, and vit-D3, but conversely a lower SJC, a lower RF and a lower BMD at the lumbar spine. A higher frequency of IL-1 α 1 was found to be associated with a lower CRP value, while an increased IL-1 β 2 carriage with active rheumatoid disease as indicated by a higher CRP, ESR and pain score and a higher BMD at the lumbar spine and lower vit-D3. Thus, polymorphisms of the IL-β gene affects RA occurrence and carriage of IL-1β 2 polymorphisms with more active disease in RA, whereas the presence of both the IL-1 α 2 and the IL-1 β 1 allele influences bone resorption [8]. An association between RA and a polymorphic IL-4 gene sequence located in 5q31–33, and a prognostic value of a polymorphism in IL-1β exon 5, which allowed prediction of erosive disease with a specificity of 91.8% in 42.1% of patients was reported [9]. Such association between various cytokine polymorphism and RA and lupus have been described but their clinical significance remains to be determined [10–15]; (b) HLA DR3 is correlated strongly with Sjögren's syndrome, myasthenia gravis, lupus and type 1 diabetes mellitus; (c) HLA DR4 is correlated with the genesis of rheumatoid arthritis, Type 1 diabetes mellitus, and pemphigus vulgaris. Fewer correlations exist with MHC class I molecules. The most notable and consistent is the association between HLA B27 and ankylosing spondylitis. The contributions of genes outside the MHC complex with autoimmune diseases remain to be established.

Gender

Nearly 75% of the more than 23.5 million Americans who suffer from autoimmune disease are women, although it is less-frequently acknowledged that millions of men also suffer from these diseases. In general, autoimmune diseases that develop in men tend to be more severe. A few autoimmune diseases that men are just as or more likely to develop as women, include: ankylosing spondylitis, Wegener's granulomatosus, type 1 diabetes mellitus, Crohn's disease and psoriasis.

The reasons for the sex role in autoimmunity are unclear. Women appear to generally mount larger inflammatory responses than men when their immune systems are triggered, increasing the risk of autoimmunity. Involvement of sex steroids is indicated by that many autoimmune diseases tend to fluctuate during pregnancy, in the menstrual cycle or when using oral contraception. It has been suggested that the slight exchange of cells between mothers and their children during pregnancy may induce autoimmunity. This would tip the gender balance in the direction of the female.

A preponderance of 16-α-hydroxylated estrogens, as observed in rheumatoid arthritis synovial fluids, is an unfavorable sign in synovial inflammation. 17β-estradiol administered during hormone replacement therapy will rapidly increase estrone sulfate after conversion in adipose tissue by aromatases, hormone replacement therapy can have proinflammatory effects by providing estrone sulfate to the inflamed synovial tissue. In addition, the use of combined oral contraceptives is associated with an increased risk of lupus. Estrogens are generally considered as enhancers of cell proliferation and humoral immune response [16].

Another theory suggests the female high tendency to get autoimmunity is due to an imbalanced X chromosome inactivation. The X-inactivation skew theory has recently been confirmed experimentally in scleroderma and autoimmune thyroiditis [17]. This theory suggests that autoreactive T cells may fail to be tolerized by self antigens encoded by one of the two X chromosomes. In the periphery, these autoreactive T cells may stimulate B cells expressing the target X-encoded antigen. Alternatively, the X-encoded genes cause autoimmunity by affecting B or T cells directly. An attractive feature of X-inactivation hypotheses is that the discordance rate between monozygotic twins may readily be explained, because otherwise-identical twins may have different X-inactivation patterns [18–20].

Recent evidence indicates that lupus could be an epigenetic disease characterized by impaired T cell DNA methylation. Women have two X chromosomes; one is inactivated by mechanisms including DNA methylation. It was suggested that demethylation of sequences on the inactive X may cause gene overexpression uniquely in women, predisposing them to lupus. This suggestion has been verified by observing the expression and methylation of CD40LG, a B cell costimulatory molecule encoded on the X chromosome, in experimentally demethylated T cells from men and women and in men and women with lupus. Bisulfite sequencing revealed that CD40LG is unmethylated in men, while women have one methylated and one unmethylated gene. 5-Azacytidine, a DNA methyltransferase inhibitor, demethylated CD40LG and doubled its expression on CD4(+) T cells from women but not men, while increasing TNFSF7 expression equally between sexes. Similar studies demonstrated that CD40LG demethylates in CD4(+) T cells from women with lupus, and that women but not men with lupus overexpress CD40LG on CD4(+) T cells, while both overexpress TNFSF7. These studies demonstrated that regulatory sequences on the inactive X chromosome demethylate in T cells from women with lupus, contributing to CD40LG overexpression uniquely in women, suggesting that demethylation of CD40LG and perhaps other genes on the inactive X may contribute to the striking female predilection of this disease [20]. Similar Skewed X chromosome inactivation may be a risk factor for the occurrence of other autoimmune disorders, including juvenile idiopathic arthritis [21].

Environmental Factors

An interesting inverse relationship exists between infectious diseases and autoimmune diseases. In areas where multiple infectious diseases are endemic, autoimmune diseases are quite rarely seen. The hygiene hypothesis attributes these correlations to the immune manipulating strategies of pathogens. For example, parasitic infections are associated with reduced activity of autoimmune disease [22–24]. The putative mechanism is that the parasite attenuates the host immune response in order to protect itself. This may provide a serendipitous benefit to a host that also suffers from autoimmune disease. It is likely that parasites exert bystander immunosuppression of pathogenic T cells that mediate autoimmune diseases. It was observed that infection of mice with *Fasciola hepatica* resulted in recruitment of dendritic cells, macrophages, eosinophils, neutrophils, and CD4(+) T cells into the peritoneal cavity. The dendritic cells and macrophages in infected mice expressed IL-10 and they had low surface expression of costimulatory molecules and/or MHC class II. Furthermore, most CD4(+) T cells in the peritoneal cavity of infected mice secreted IL-10, but not IFN-gamma or IL-4. A less significant expansion of CD4(+)Foxp3(+) T cells was noted. *Fasciola hepatica*-specific Tr1-type clones generated from infected mice suppressed proliferation and IFN-γ production by Th1 cells and suppression of parasite-specific Th1 and Th2 responses was reversed in IL-10-defective mice. Infection with *Fasciola hepatica* suppressed immune responses to autoantigens and attenuated the clinical signs of experimental autoimmune encephalomyelitis that was found to be associated with suppression of autoantigen-specific IFN-γ and IL-17 production. The suppression of Th1 and Th17 responses and attenuation of experimental autoimmune encephalomyelitis by *Fasciola hepatica* was maintained in IL-10(-/-) mice but was reversed by neutralization of TGF-β in vivo, suggesting that Fasciola*hepatica*-induced IL-10 subverts parasite-specific Th1 and Th2 responses, and that *Fasciola hepatica*-induced TGF-β plays a critical role in bystander suppression of autoantigen-specific Th1 and Th17 responses that mediate autoimmune diseases [24]. These results imply that adequate production of TGF-β is necessary to prevent or suppress autoimmune diseases or alternatively a defect in the production of TGF-β could be considered as a factor that predisposes to the development and progression of autoimmune diseases.

Despite these interesting results, it is important to note that an equally strong association of certain microbial organisms with autoimmune diseases has been documented. For example, *Klebsiella pneumoniae* and coxsackievirus B have been strongly correlated with ankylosing spondylitis and type 1 diabetes mellitus, respectively. It has been postulated that the infecting organism produces super-antigens that are capable of polyclonal activation of B-lymphocytes, and production of large amounts of antibodies of varying specificities, some of which may be self-reactive.

Certain chemical agents and drugs can also be associated with the genesis of autoimmune conditions, or conditions that simulate autoimmune diseases. The most striking of these is the drug-induced lupus erythematosus. Usually, withdrawal of the offending drug cures the symptoms in a patient.

Cigarette smoking is an established risk factor for both incidence and severity of rheumatoid arthritis. This may relate to abnormal citrullination of proteins, since the effects of smoking correlate with the presence of antibodies to citrullinated peptides [25–27].

Pathogenesis of Autoimmunity

Despite many years of research, the exact pathogenesis of autoimmune diseases is not clear. Several mechanisms are thought to be operative in the pathogenesis of autoimmune diseases, against a backdrop of genetic predisposition and environmental modulation. A summary of some of the important mechanisms are detailed below:

1. **T-Cell Bypass:** A normal immune system requires the activation of B-cells by T-cells before the former can produce antibodies in large quantities. This requirement of a T-cell can be bypassed in rare instances, such as infection by organisms producing super-antigens, which are capable of initiating polyclonal activation of B-cells, or even of T-cells, by directly binding to the β-subunit of T-cell receptors in a non-specific fashion. Thus, three models that have been suggested for the early events leading to autoimmunity are: (a) polyclonal activation of competent B cells, (b) direct activation of competent T cells, and (c) bypass of specifically tolerant T cells and activation of competent B cells [28–32]. For example, neonatal islet-specific expression of TNF-α in nonobese diabetic mouse promotes diabetes by provoking islet-infiltrating antigen-presenting cells to present islet peptides to autoreactive T cells. It was observed that TNF-α promotes autoaggression of both effector CD4(+) and CD8(+) T cells. Whereas CD8(+) T cells are critical for diabetes progression, CD4(+) T cells play a lesser role. TNF-α-mediated diabetes development was not dependent on CD154-CD40 signals or activated CD4(+) T cells. Instead, TNF-α promoted cross-presentation of islet antigen to CD8(+) T cells using a unique CD40-CD154-independent pathway. These data indicates that inflammatory stimuli can bypass CD154-CD40 immune regulatory signals and cause activation of autoreactive T cells [33].
2. **T-Cell-B-Cell discordance:** A normal immune response is assumed to involve B and T cell responses to the same antigen, even though B cells and T cells recognize very different things: conformations on the surface of a molecule for B cells and pre-processed peptide fragments of proteins for T cells. However, the exact mechanism of this recognition and how B and T cells recognize different

antigens very specifically and distinctly is not clear. But, what is known is that a B cell recognizing a specific antigen endocytoses the same and processes it and presents it in a presentable from to a T cell. B cells recognizing IgG Fc could get help from any T cell responding to an antigen co-endocytosed with IgG by the B cell as part of an immune complex. For example, in coeliac disease it appears that B cells recognizing tissue transglutaminase are helped by T cells recognizing gliadin.

3. **Aberrant B cell receptor-mediated feedback:** A feature of human autoimmune disease is that it is largely restricted to a small group of antigens, several of which have known signaling roles in the immune response (DNA, C1q, IgG Fc, Ro, Con. A receptor, Peanut agglutinin receptor (PNAR)). This implies that spontaneous autoimmunity may result when the binding of antibody to certain antigens leads to aberrant signals being fed back to parent B cells through membrane bound ligands. These ligands include B cell receptor (for antigen), IgG Fc receptors, CD21, which binds complement C3d, Toll-like receptors 9 and 7 (which can bind DNA and nucleoproteins) and PNAR. More indirect aberrant activation of B cells can also be envisaged with autoantibodies to acetyl choline receptor and hormone and hormone binding proteins. Thus, the concept of T-cell-B-cell discordance envisages that such discordance leads to the development of self-perpetuating autoreactive B cells [34–36]. Autoreactive B cells in spontaneous autoimmunity are seen as surviving because of subversion both of the T cell help pathway and of the feedback signal through B cell receptor, thereby overcoming the negative signals responsible for B cell self-tolerance without necessarily requiring loss of T cell self-tolerance.
4. **Molecular mimicry:** An exogenous antigen may happen to share structural similarities with certain host antigens; thus, any antibody produced against this antigen (which mimics the self-antigens) can bind to the host antigens, and amplify the immune response. The idea of molecular mimicry arose in the context of rheumatic fever that follows infection with Group A beta-haemolytic streptococci. Several autoantigens that have been identified include cardiac myosin epitopes, vimentin, and other intracellular proteins. In the heart tissue, antigen-driven oligoclonal T cell expansions were probably the effectors of the rheumatic heart lesions. These cells are CD4(+) and produced inflammatory cytokines (TNF-α and IFN-γ) [37–39]. It is also likely that the disease is due to e.g., an unusual interaction between immune complexes, complement components and endothelium.
5. **Idiotype cross reaction in autoimmunity:** Idiotypes are antigenic epitopes found in the antigen-binding portion (Fab) of the immunoglobulin molecule. Autoimmunity can arise as a result of a cross-reaction between the idiotype on an antiviral/anti-bacterial antibody and a host cell receptor for the organism in question. In this case, the host-cell receptor is envisioned as an internal image of the virus/bacteria, and the anti-idiotype antibodies can react with the host cells [40–42].
6. **Cytokine dysregulation:** Cytokines are divided into two groups according to the population of cells whose functions they promote: helper T-cells type 1 or type

2 (T_H1 and T_H2 respectively). Typically, T_H1 cytokines are: IL-2, TNF-α and IFN-γ, while T_H2 cytokines are IL-4, IL-5, IL-6, IL-10, and IL-13 [43].

The interactions between cytokines from the T_H1/T_H2 model are not a simple straightforward interaction and are more complicated than is originally thought. For example, the T_H2 cytokine IL-10 inhibits cytokine production of both T_H subsets in humans. Human IL-10 (h-IL-10) suppresses the proliferation and cytokine production of all T cells and the activity of macrophages, but continues to stimulate plasma cells, ensuring that antibody production still occurs. As such, h-IL-10 is not truly a promoter of the T_H2 response in humans, but acts to prevent over-stimulation of helper T cells while still maximizing the production of antibodies.

There are also other types of T cells that can influence the expression and activation of helper T cells, such as natural regulatory T cells [44], along with less common cytokine profiles such as the T_H3 subset of helper T cells. Terms such as "regulatory" and "suppression" have become ambiguous after the discovery that helper $CD4^+$ T cells that are capable of regulating (and suppressing) their own responses outside of dedicated regulatory T cells have been described [45–48].

One major difference between regulatory T cells and effector T cells is that regulatory T cells typically serve to modulate and deactivate the immune response, while effector T cell groups usually begin with immune-promoting cytokines and then switch to inhibitory cytokines later in their life cycle. The latter is a feature of T_H3 cells, which transform into a regulatory subset after its initial activation and cytokine production.

Both regulatory T cells and T_H3 cells produce the cytokine TGF-β and IL-10. Both cytokines are inhibitory to helper T cells; TGF-β suppresses the activity of most of the immune system. TGF-β may not suppress activated T_H2 cells as effectively as it might suppress naive cells, but it is not typically considered a T_H2 cytokine [49–51].

Recently, the characterization of another novel T helper subtype, T helper 17 cells (T_H17) has cast further doubt on the basic T_H1/T_H2 model. These IL-17 producing cells are now thought to have their own distinct effector and regulatory functions [52–54]. These T_H17 cells are a novel CD4(+) subset that preferentially produces IL-17, IL-17F, and IL-22 cytokines. T_H17 cells appear to play a critical role in sustaining the inflammatory response and their presence is closely associated with autoimmune disease and may play an essential role in protection against certain extracellular pathogens. However, T_H17 cells with specificity for self-antigens are highly pathogenic and lead to the development of inflammation and severe autoimmunity. A combination of TGF-β plus IL-6 and the transcription factors STAT3 and RORgammat were recently described to be essential for initial differentiation of T_H17 cells and IL-23 for the later stabilization of the T_H17 cell subset. IL-21 produced by T_H17 themselves plays an important role in the amplification of Th17 cells. Thus, Th17 cells may undergo three distinct steps of development: differentiation, amplification and stabilization in which distinct cytokines play a role [55]. The specific effector functions of T_H17 cells expand beyond previously described effects of T_H1 and T_H2 immunity, with specific roles in host defense against certain pathogens and in organ-specific autoimmunity. The potential dynamics of T_H17 cell populations and their interplay with other inflammatory cells in the induction of tissue inflammation in host defense and organ-specific autoimmunity is especially interesting with particular reference to autoimmune diseases [56–64].

For example, the frequencies of IL-17-positive and IL-22-positive CD4+ T cells were increased in peripheral blood mononuclear cells (PBMCs) from patients with ankylosing spondylitis and patients with RA, resulting in secretion of higher quantities of IL-17 by PBMCs following stimulation, supporting the possibility that Th17 cells, particularly when present in excess of IL-10-producing cells, are involved in the pathogenesis of inflammatory arthritis [56].

Many of the cytokines are also expressed by other immune cells, and it is becoming clear that while the original T_H1/T_H2 model is interesting and gives a broad view of the role of various cytokines in inflammatory and autoimmune diseases, it is far too simple to define its entire role or actions. In some *in vivo* studies, individual helper T cells usually do not match the specific cytokine profiles of the T_H model, and many cells express cytokines from both profiles. But, yet the T_H model has played an important role in developing our understanding of the roles and behaviour of helper T cells and the cytokines they produce during an immune response. Recent studies showed that another T helper subset may exist called as T_H9 cells. These T_H9 cells are believed to elaborate IL-9 that defends against helminth infections [65–67]. Though, the secretion of IL-9, initially recognized as a T_H2 cytokine, has been attributed to a novel CD4 T cell subset T_H9 in the murine system, it can also be secreted by mouse T_H17 cells and may mediate aspects of the proinflammatory activities of T_H17 cells. IL-9 is secreted by human naive CD4 T cells in response to differentiation by T_H9 (TGF-β and IL-4) or T_H17 polarizing conditions. Surprisingly, these differentiated naive cells did not coexpress IL-17 and IL-9, unless they were repeatedly stimulated under T_H17 differentiation-inducing conditions. In contrast to the naive cells, memory CD4 T cells could be induced to secrete IL-9 by simply providing TGF-β during stimulation without the requirement of neither IL-4 nor proinflammatory cytokines. Addition of TGF-β to the T_H17-inducing cytokines (IL-1β, IL-6, IL-21, IL-23) that induce memory cells to secrete IL-17, resulted in the marked coexpression of IL-9 in IL-17 producing memory cells. The proinflammatory cytokine mediating TGF-β-dependent coexpression of IL-9 and IL-17 was identified to be IL-1β. Circulating monocytes are potent costimulators of IL-9 production by T_H17 cells via their capacity to secrete IL-1β. Patients with autoimmune diabetes exhibited a higher frequency of memory CD4 cells with the capacity to transition into IL-9(+)IL-17(+) cells, suggesting that the presence of IL-17(+)IL-9(+) CD4 cells induced by IL-1β may play a role in human autoimmune disease [67]. These results indicate the plasticity of T cells in their ability to secrete different types of cytokines. Thus, it is likely that depending on the local conditions and the presence or absence of various cytokines the same type of T cells could be stimulated to produce different types of cytokines.

7. **Dendritic cell apoptosis:** Dendritic cells present antigens to lymphocytes. Hence, it is proposed that dendritic cells that are defective in apoptosis can lead to inappropriate systemic lymphocyte activation and consequent decline in self-tolerance [68–71].
8. **Epitope spreading or epitope drift:** when the immune reaction changes from targeting the primary epitope to also targeting other epitopes, it could lead to autoimmune diseases [72, 73]. In contrast to molecular mimicry theory, the other epitopes need not be structurally similar to the primary one.

It is evident from the preceding discussion that a delicate balance is maintained between pro- and anti-inflammatory cytokines and when this balance is upset more in favor of pro-inflammatory cytokines it could lead to autoimmune diseases such as RA and lupus. Such an imbalance in the immune system could be triggered by various environmental factors including bacteria, viruses, hormones and other factors such as vitamin D in an individual who has the required genetic background that renders them susceptible to develop the autoimmune diseases. In addition to cytokines, other molecules that are secreted by macrophages, dendritic cells, T and B cells, endothelial cells, mast cells and other cells involved in the pathobiology of inflammation and immune response such as free radicals, NO, eicosanoids, and the anti-oxidant defenses such as catalase, superoxide dismutase, glutathione etc., all seem to play a significant role in the pathobiology of various autoimmune diseases. A brief description of the pathobiology of lupus is described below with special emphasis on the role of essential fatty acids and their metabolites in autoimmune diseases.

Systemic Lupus Erythematosus

Systemic lupus erythematosus (SLE, also called as lupus), a disease of unknown aetiology that is more common in women than in men, is characterized by non-destructive arthritis/arthralgias, a cutaneous rash, vasculitis, involvement of the central nervous system (CNS) and renal and cardiopulmonary manifestations. Although genetic, environmental and sex hormonal factors have been implicated in the pathogenesis of lupus, it is known that several cytokines, nitric oxide (NO), free radicals, a deranged immune system, a deficient anti-oxidant defenses, and Toll-like receptors have a significant role both in the initiation and perpetuation of the inflammatory process observed. The fundamental process in lupus appears to be rendering DNA and RNA antigenic that leads to the production of anti-DNA and ant-RNA antibodies and the formation of immune complexes. These antibodies and immune complexes, in turn, trigger both a local and a systemic inflammatory response that ultimately leads to target organ/tissue damage seen. The susceptibility to develop lupus in a given individual seems to have, at least, partly a genetic basis though this is still not very clear. Once the inflammatory process is triggered, this leads to the production of a variety of pro-inflammatory cytokines such as interleukin-1 (IL-1), IL-6, tumor necrosis factor-α (TNF-α), interferons (IFNs), macrophage migration inhibitory factor (MIF),

HMGB1 (high mobility group B1) and possibly, a reduction in the elaboration of anti-inflammatory cytokines such as IL-10, IL-4, and TGF-β. This imbalance between the pro- and anti-inflammatory cytokines coupled with increased secretion of free radicals such as superoxide anion ($O_2^{-\cdot}$), hydrogen peroxide (H_2O_2), singlet oxygen, inducible nitric oxide (iNO), and other reactive oxygen species (ROS) by activated monocytes, macrophages, polymorphonuclear leukocytes (PMNL), T cells, Kupffer cells, glial cells in the brain, and other organ specific reticuloendothelial cells would ultimately cause target tissue/organ damage seen in lupus [74–83]. Not all patients of lupus have the same manifestations and the clinical presentation of the same patient at different time periods is different. Thus, a patient my initially present with cutaneous manifestations and over a period of time the involvement of joints, kidneys, and other organs becomes apparent. In yet another, the initial presentation may be proteinuria and after a while other features of the disease become evident. This type of varied presentation(s) is at times baffling and suggests that the involvement of various organs and tissues due to the underlying inflammatory process is varied, the degree of involvement may differ both in time and extent and more importantly is unpredictable. To understand and devise new methods of treatment that are appropriate for a given lupus patient, it calls for a thorough understanding of the inflammatory process itself.

Pathobiology of Inflammation with Emphasis on Chronic Inflammation

Inflammation, which may be local or systemic and/or chronic or acute, is a reaction to injurious agents, either external or internal, that consists of both vascular and cellular responses. During inflammation, the reaction of blood vessels is unique that leads to the accumulation of fluid and leukocytes in extravascular tissues. This reaction can be in the form of vasodilatation that is seen as hyperemia at the site(s) of injury, which serves the essential function of increasing the blood supply to the injured tissue/organ so that adequate elimination of the inflammation-inducing agent is achieved and/or repair process can occur after the inflammation subsides. Thus, both injury and repair are two faces of the inflammatory process and it is difficult to separate these two processes. In fact, in majority of the instances, both inflammation to injury and repair occur almost simultaneously. It is possible that in majority of the instances especially, in rheumatological conditions, the injury-inducing agent such as immune complexes may have regressed to some extent but the tissue repair process or resolution of the inflammatory process may not set in that leads to tissue or organ dysfunction. Thus, failure of the resolution of the injury adequately by itself could perpetuate the disease process. Hence, understanding the relationship between injury and resolution of inflammation and the molecules that take part in these events is essential not only to understand the disease process but also to develop newer therapeutic strategies. In this context, a brief discussion of the various components of the inflammation and molecules involved the same is mandatory.

Components and Mediators of the Inflammatory Response

The inflammatory response mainly consists of two components: a vascular response and a cellular response. The vascular and cellular reactions of both acute and chronic inflammation are mediated by chemical factors secreted by various cells that take part in the inflammatory process and/or responding to the inflammatory stimulus. Once the inflammatory process is initiated, tissues/organs try to elaborate certain anti-inflammatory chemicals and signals that try to minimize tissue damage and eliminate the harmful effects of inflammation and induce resolution of the inflammatory process and restore the structure and function of the damage tissues/organs to normal. Thus, ultimately recovery of a tissue/organ from the inflammatory process and regaining of its function depends on the balance between pro- and anti-inflammatory chemicals and events that occur as a result of these mutually antagonistic processes. Once inflammation is terminated either by endogenous mediators/repair processes and/or by modern medical techniques (that may include antibiotics, anti-inflammatory drugs, chemical and surgical measures) and the offending agent is successfully removed, all the secreted mediators and the cellular responses are either broken down or dissipated and the tissues/organs in question revert to their natural physiological state, to the extent possible, depending on the degree of damage and repair that has occurred. Both vascular and cellular components of inflammation are regulated by a variety of mediators such as cytokines, free radicals, nitric oxide (NO), carbon monoxide (CO), eicosanoids, lipoxins, resolvins, protectins, nitrolipids, and their ability to interact with various chemokines, toll-like receptors (TLRs), anti-oxidants, adhesion molecules, etc. [84].

Histamine, thrombin, and platelet activating factor (PAF) stimulate the redistribution of P-selectin from its intracellular stores to the cell surface; whereas macrophages, mast cells, and endothelial cells secrete pro-inflammatory cytokines IL-1, TNF-α, and chemokines that act on endothelial cells and induce the expression of several adhesion molecules. This results in the expression of E-selectin on the surface of endothelial cells. Simultaneously, leukocytes express carbohydrate ligands for the selectins that allow them to bind to the endothelial selectins. This binding of leukocytes to endothelium is a low-affinity interaction that results in rolling of leukocytes on the surface of endothelium [84].

On the other hand, IL-1 and TNF-α and other pro-inflammatory cytokines induce the expression of ligands for integrins such as VCAM-1 and ICAM-1. Chemokines produced at the sites of inflammation or injury act on endothelial cells such that proteoglycans (such as heparan sulfate glycosaminoglycans) are expressed at high concentrations on their surface, whereas they activate leukocytes to convert low-affinity integrins such as VLA4 and LFA-1 to high-affinity state. These events ultimately lead to firm binding of activated leukocytes to activated endothelial cells such that cytoskeleton of leukocytes is reorganized, and they spread out on the endothelial surface. Binding of activated leukocytes to endothelial surface induces endothelial dysfunction and damage due to ROS, iNO produced by leukocytes. These adherent leukocytes migrate through inter-endothelial spaces towards the site of injury by binding to PECAM-1 (platelet endothelial cell adhesion molecule) or CD31. Leukocytes pierce the basement membrane by secreting collagenases that digest collagen.

Leukocytes emigrate towards the sites of inflammation by chemotaxis. Some of the endogenous chemoattractants include (but not limited to): components of the complement system such as C5a, lipoxygenase pathway products such as leukotriene B_4 (LTB_4), and some cytokines such as IL-8. Chemoattractants bind to specific seven transmembrane G-protein-coupled receptors (GPCRs) on the surface of leukocytes that, in turn, activate phospholipase C (PLC), phosphoinositol-3-kinase (PI3K) and protein kinases. PLC and PI3K act on cell membrane phospholipids to generate lipid second messengers such as inositol triphosphate (IP3) that increase cytosolic calcium (Ca^{2+}) and activate small GTPases of the Rac/Rho/cdc2 family as well as numerous kinases. GTPases induce polymerization of actin that helps in the motility of the leukocytes. In this context, it is interesting to note that eNO synthase activation is critical for vascular leakage during acute inflammation [84].

In order to produce inflammation, leukocytes generate ROS on activation. Products of necrotic cells, antigen-antibody complexes, cytokines, and chemokines also induce leukocyte activation. Different classes of leukocyte cell surface receptors recognize different stimuli. For instance, chemokines, lipid mediators, and N-formyl-methionyl peptides increase integrin avidity, and produce cytoskeletal changes that aids leukocyte chemotaxis; microbial lipopolysaccharide (LPS) binds to toll-like receptors (TLRs) on leukocyte membrane leading to their activation and production of cytokines and ROS that are essential for the killing of microbes; and binding of microbial products to mannose receptor augments leukocyte phagocytic process that aids in the elimination of the invading organisms. Activation of leukocytes by various stimuli triggers several signaling pathways that result in increases in cytosolic Ca^{2+} and activation of protein kinase C (PKC) and phospholipase A_2 (PLA_2). PLA_2 activation leads to the release of lipids such as arachidonic acid (AA, 20:4 ω-6), eicosapentaenoic acid (EPA, 20:5 ω3), and docosahexaenoic acid (DHA, 22:6 ω-3) from the cell membrane lipid pools. AA, and possibly EPA and DHA themselves could increase cytosolic Ca^{2+} and PKC concentrations in various cells [85–87]. Furthermore, AA by itself has the ability to activate leukocytes [87]. These results suggest that dietary lipids have the ability to modulate leukocyte responses and the inflammatory process. Products of AA, EPA, and DHA such as prostaglandins (PGs), leukotrienes (LTs), lipoxins (LXs), and resolvins have both positive and negative influences on leukocyte activation, chemotaxis, inflammation and its resolution [88–90]. Some of the products that are released by activated leukocytes include: AA and its metabolites, lysosomal enzymes, ROS, NO, various cytokines, various leukocyte adhesion molecules and other surface receptors such as TLRs, GPCRs, receptors for opsonins, etc.

Leukocytes use mannose receptors and scavenger receptors to bind and ingest bacteria, though they can engulf other particles without attachment to specific receptors. Opsonins enhance the efficiency of phagocytosis. Killing and degradation of the ingested bacteria or particles both by leukocytes and macrophages is accomplished by ROS, NO, myeloperoxidase (MPO), ozone and proteases [84].

Some of the important mediators of inflammation include: histamine, serotonin, lysosomal enzymes, PGs, LTs, PAFs, ROS, NO, HOCL, various cytokines, kinin system, coagulation and fibrinolysis system, and complement system.

The role of arachidonic acid and other polyunsaturated fatty acids and their products in inflammation calls for a special mention in view of their diverse role in inflammation and in particular in lupus. The metabolism of essential fatty acids and various products formed from them and their actions has been discussed in details in Chap. 4. It is suffice to mention here that AA, EPA and DHA also give rise to anti-inflammatory molecules such as lipoxins, resolvins, protectins, maresins and nitrolipids that suppress the production of pro-inflammatory cytokines such as IL-6, TNF-α, MIF, and possibly, HMGB-1, and inhibit the activation of MPO (myeloperoxidase), generation of superoxide anion and other free radicals, enhance the release of eNO, events that lead to suppression of inflammation, resolution of the inflammatory events, wound healing and thus, restoration of function of the target tissues and organs. It is conceivable that in conditions such as rheumatoid arthritis and lupus and other rheumatological conditions there could be a deficiency of these potent anti-inflammatory molecules that may be responsible for the continued inflammatory process and hence, persistence of the disease.

Cytokines in Inflammation

Cytokines regulate cellular immune responses and participate in both acute and chronic inflammation. TNF-α, IL-1, IL-6, MIF, IL-17, IL-23 and HMGB-1 (high mobility group B-1) have pro-inflammatory actions, whereas IL-4 and IL-10 have anti-inflammatory actions, and antagonize the actions of IL-1, IL-6 and TNF-α [84, 91]. Recent studies showed that endothelial cells, adipose tissue, Kupffer cells, and glial cells are capable of producing both pro- and anti-inflammatory cytokines. Endotoxin and other microbial products, immune complexes, physical injury, and other inflammatory stimuli activate endothelial cells, leukocytes, and fibroblasts, and induce systemic acute-phase reactions. TNF-α, IL-6, and IL-1 activate endothelial cells and induce the synthesis of endothelial adhesion molecules, other cytokines, chemokines, growth factors, eicosanoids, and nitric oxide (NO), events that increase the thrombotic tendency on the surface of the endothelium [84, 92]. TNF primes neutrophils, leading to augmented responses of these cells to other mediators, and stimulates neutrophils to produce ROS. IL-1, IL-6, and TNF-α induce the systemic acute-phase responses such as fever, loss of appetite, slow-wave sleep, features that may be seen in patients with lupus and other rheumatological conditions; the release of neutrophils into the circulation, release of corticotropin and corticosteroids. Sustained and increased production of TNF-α that occurs during chronic intracellular infections such as tuberculosis and neoplastic diseases causes cachexia. Increased production of IL-1, IL-6, and TNF-α is seen in rheumatoid arthritis and lupus, and other collagen vascular diseases. This discovery led to the development anti-TNF-α

antibodies and TNF-α receptor blockers that found their use in the treatment of these conditions though are not always very effective.

Inflammatory and Anti-inflammatory Molecules and Antioxidants in Lupus

It is evident from the preceding discussion that many molecules are involved in the pathobiology of inflammation. Lupus, which is an autoimmune collagen vascular disease, is characterized by increased production of IL-1, IL-6, TNF-α, IFN-γ, MIF, HMGB1, iNO, ROS, various chemokines, MPO, GM-CSF, G-CSF, endothelin, and hs-CRP [74, 93–98]. In contrast, the concentrations of PGI_2, PGE_1, eNO, and anti-oxidants such as superoxide dismutase (SOD) and glutathione peroxidase are decreased whereas those of lipid peroxides are increased [99–101]. Pro-inflammatory cytokines such as IL-1, IL-6, TNF-α, IFN-γ, HMGB1, and MIF that are released in large amounts (possibly, due to the loss of feed-back control exerted by T_H2 cells, and/or deficiency of their cytokines as depicted in Fig. 13.1) in lupus by activated neutrophils, macrophages, T cells, synovial cells, fibroblasts, and endothelial cells not only initiate the inflammatory process but also perpetuate the inflammation since these cytokines, in turn, stimulate neutrophils, macrophages, T cells, synovial cells, fibroblasts, and endothelial cells to produce free radicals, various eicosanoids, and cytokines in an autocrine fashion [74, 84]. In addition, IL-1, and possibly other pro-inflammatory cytokines, increases the production of endothelin-1 by endothelial cells that is a potent vasoconstrictor. Increased basal and stimulated endothelin-1 concentrations are associated with the enhanced and prolonged vasospasm (Raynaud's phenomena) seen in lupus and other collagen vascular conditions such as scleroderma [102]. Furthermore, IL-2 stimulates the production of autoantibodies and worsens immune-mediated disease [103]. For instance, treatment with human recombinant IL-2 (rhIL-2) augmented the severity of T-cell mediated experimental allergic encephalomyelitis in Lewis rats [104] and accelerated the appearance of autoimmune insulin-dependent diabetes mellitus in the BB rat model [105]. Administration of IL-2 in the setting of lupus has the potential to expand B cell populations and increase production of pathogenic autoantibodies. Activated T cells and macrophages induce the formation of new blood vessels (angiogenesis), and this induction is mediated by cytokines: IL-1, IFN-γ, and TNF-α [106]. These cytokines including IL-6, CSF-1 (colony stimulating factor-1) initiate immune response, induces cell proliferation, augment matrix-degrading protease activity, and cause resorption of bone (osteoporosis). Proteases released by activated neutrophils are responsible for bony erosions seen that are more common in rheumatoid arthritis than in lupus.

On the other side of the spectrum, certain anti-inflammatory molecules are produced that try to contain the inflammatory process and induce resolution of the disease process. For instance, transforming growth factor-β (TGF-β) down regulates inflammation. Various cells including monocytes, fibroblasts, platelets and synovial tissue produce it. TGF-β stimulates collagen transcription and inhibits collagenase

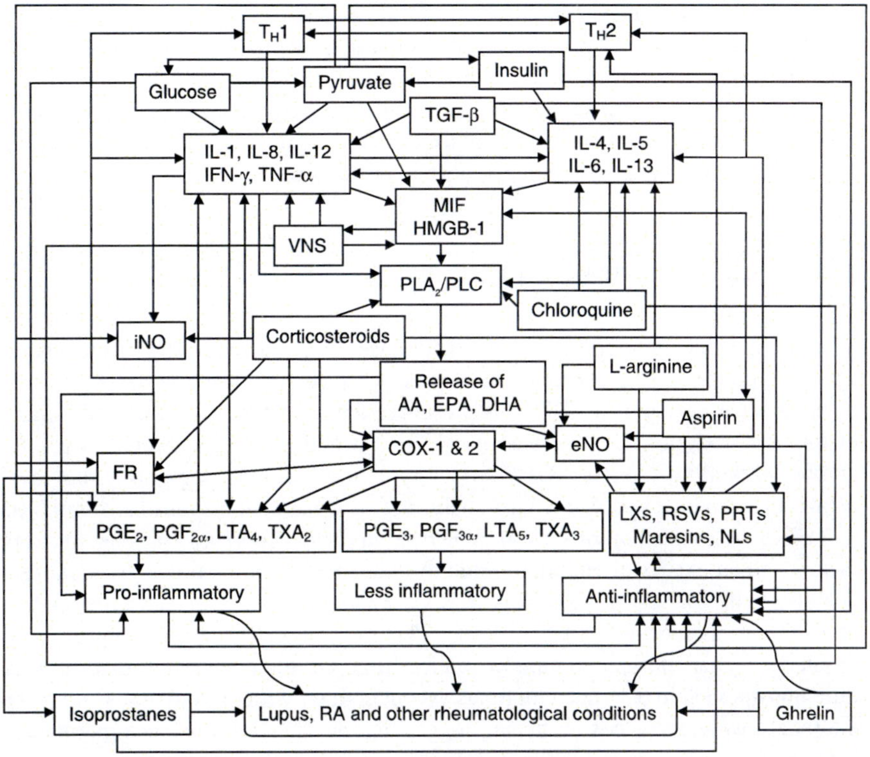

Fig. 13.1 Scheme showing possible interaction between PUFAs (AA, EPA, DHA), their products such as PGs, LTs, TXs, LXs, resolvins, protectins and maresins and T_H1 and T_H2 and their respective cytokines. PUFAs have direct actions on T_H1 and T_H2 responses and cytokines by themselves without being converted to their products. Ghrelin, isoprostanes (formed due to the action of free radicals on PUFAs), insulin and pyruvate also have anti-inflammatory actions. For further details see text

transcription in fibroblasts and suppresses IL-1-stimulated collagenase transcription (reviewed in [74]). TGF-β regulates immune response both in vitro and in vivo, and regulates its own production through an autocrine amplification pathway, especially in synoviocytes. TGF-β antagonizes serum or PDGF (platelet derived growth factor)-stimulated synoviocyte growth, induces a more differential phenotype in immature synovial fibroblasts, induces collagen and fibronectin formation, and glycosaminoglycans by articular chondrocytes and thus, counters the degradation of cartilage induced by IL-1 and other cytokines [107]. In addition, TGF-β inhibits the growth of capillary endothelial cells, suppresses IL-1 and IL-2-dependent T cell proliferation, inhibits free radical generation by human monocytes, and participates in wound healing and fracture repair [102]. This suggest that TGF-β negatively regulates all the destructive and pro-inflammatory actions of IL-1, IL-2, TNF-α, HMGB1, and MIF that are important to initiate the repair process and restore normalcy in lupus and other collagen vascular conditions. In contrast, excess production of TGF-β

provokes and perpetuates fibroblast proliferation leading to abnormal sclerosis in the skin and internal organs in scleroderma and kidney in lupus that leads to the late stage complications in these conditions. Thus, TGF-β is a double-edged sword: when present in sub-normal amounts inflammation may go unchecked and higher amounts may provoke abnormal sclerosis. Hence, it is important to maintain normal amounts of TGF-β at a given site for normal physiology.

TGF-β in Scleroderma and Lupus

Scleroderma, one of the collagen vascular diseases, is characterized by vascular damage and fibrosis of skin and internal organs. No systemic elevation of TGF-β was noted in scleroderma. But, TGF-β is present in the subcutaneous tissue and inflammatory infiltrate of early scleroderma [108]. Normal fibroblasts express a greater number of binding sites for TGF-β under hypoxic conditions, suggesting that local factors that produce hypoxic environment may modulate the response of fibroblasts to TGF-β. A deficiency of NO or EDRF (endothelium derived vascular relaxing factor) and an excess of endothelin due to injury to or dysfunction of vascular endothelial cells not only produce Raynaud's phenomena but also enhance the binding of TGF-β to its receptors on fibroblasts. This may augment the proliferation of fibroblasts, a characteristic feature of scleroderma. On the other hand, there is very little fibrosis seen in patients with lupus except in those with discoid lupus and when the renal involvement is significant, suggesting that there is a balance maintained between inflammation and fibrosis that may depend on the efficiency of activating latent TGF-β in lupus.

TGF-β is the most potent naturally occurring immunosuppressant [109], produced by all cells of the immune system. TGF-β controls the proliferation and the fate of cells through apoptosis. In TGF-β knockout mice, lack of TGF-β_1 initiates indiscriminate loss of self-tolerant T cells leading to dysregulation of B cell activity that causes the production of autoantibodies and development of a lupus-like illness [110, 111]. In patients with lupus, lymphocyte TGF-β_1 production is decreased, whereas spontaneous polyclonal IgG and autoantibody production can be abrogated by treatment with IL-2 and TGF-β_1 [112, 113]. This is supported by the observation that in patients with lupus, low normal TGF-β_1 activation was associated with increased lymphocyte apoptosis, irreversible organ damage, and increased atherosclerosis [109–114]. There is evidence to suggest that an imbalance between T_H1 and T_H2 cells plays a significant role in the pathobiology of lupus [109–118].

Immune Dysfunction in Lupus

One method of determining the balance between T_H1 and T_H2 cells is by measuring their respective cytokines. Studies focused on cytokines and T-helper cells in the peripheral blood in patients with lupus. However, such studies have revealed inconsistent results that led to confusion as to the exact role of T_H1 and T_H2 cells in lupus [116–119]: some studies suggested that a predominance of T_H1 cytokines occurs in lupus, whereas others failed to support this observation. On the other hand, measurement of peripheral cytokine profile led to the suggestion that lupus could be a disease mediated by T_H2 dominance [117]. Some studies did suggest that T_H1 cells may replace T_H2 pathway and may aid progression of lupus to active nephritis [120]. In this context, it should be noted that activation of T cells occurs at the site of disease involvement and so peripheral plasma measurement of cytokines may not reflect the actual type of T cells that are actively participating in the disease, though this is useful at times. One of the major complications of lupus is the involvement of kidney resulting in renal failure that may prove fatal in many patients. In lupus, kidney biopsy is ideal to study intrarenal lymphocyte activation. Although, renal biopsy is performed often in patients with lupus whenever renal involvement is suspected, it is not without complications. Recently, measurement of messenger RNA (mRNA) expression in urinary sediment has been described [121]. In a study wherein the mRNA expression in the urinary sediment of lupus patients was performed and compared with their urinary and intra-renal protein expression, it was found that urinary mRNA and protein expressions of T-bet were significantly higher in lupus with active nephritis compared to those with inactive disease. In contrast, the urinary and protein expressions of GATA-3 were significantly lower in lupus patients with active nephritis. Furthermore, in those in whom kidney biopsy was done, tubular expressions of T-bet and GATA-3 significantly correlated with the histological activity index [122]. These results suggest that active lupus nephritis is associated with increased T-bet and decreased GATA-3 expression in the urinary sediment and kidney tissue indicating a predominant T_H1 type of T-lymphocyte activation. In this context, it is relevant to note that T-bet promotes T_H1 lineage commitment and forms an autoregulatory positive-feedback loop with IFN-γ to maintain a T_H1-mediated immune response [123], whereas GATA-3 promotes T_H2 differentiation and induces T_H2 cytokine production [124]. Thus, the relative expression of T-bet and GATA-3, resulting in a swing in the T_H1 and T_H2 expressions, would ultimately determine the type of T helper cell expression. Based on the results of this study [81], it is evident that measurement of T-helper cell transcription factor gene expression is feasible and probably aids in the assessment and risk stratification of lupus patients.

Loss of Self-tolerance in Lupus

Diverse autoantibodies directed against a variety of intra- and extracellular components exist in lupus for years before the development of the disease [125], suggesting that normal physiologic mechanisms that maintain tolerance to self-antigens have been breached. A subpopulation of T cells known as Tregs establishes and preserves self-tolerance [126]. It is opined that abnormal T cell clones exist that mediate defective helper and suppressor functions, which result in autoantibody generation by forbidden B cell clones. Defective signaling cascades give rise to a primary T cell disorder that result in impaired effector functions in lupus [127]. These effector dysfunctions are a result of skewed expression of various effector molecules including CD40 ligand (e.g., CD154) and various cytokines and may reflect an imbalance of gene expression. Impaired effector T cell function as a result of skewed cytokine production creates a microenvironment that facilitates a strong T_H2 response relative to T_H1 and Treg activity, which leads to overproduction of IL-4, IL-6, and IL-10 by T_H2 and underproduction of IL-2, IL-12, TGF-β, and IFN-γ by T_H1 and Tregs that results in imbalanced autocrine and paracrine effects on T and B cells in the microenvironment. This imbalance in the cytokine production and reduced numbers of $CD4^+CD25^+$ Tregs results in insufficient suppressor activity in lupus that results in dysregulated immune response driving both physiologic and forbidden B cell clones to overproduce antibodies and autoantibodies, which results in hypergammaglobulinemia. These events occur despite the existence of other counter regulatory mechanisms, including expression of the cell surface molecule cytotoxic T lymphocyte antigen 4 (CTLA4) [128]. IL-2 is mainly produced by activated $CD4^+$ and $CD8^+$ T cells and binds to high-affinity cell surface IL-2 receptors (IL-2Rs) expressed by T cells, B cells, NK cells and APCs (antigen presenting cells). Originally, it was believed that IL-2 is a growth factor. Studies with IL-$2^{-/-}$ and IL-$2R^{-/-}$ knockout mice revealed that IL-2 is not a growth factor *in vivo* but serves as a third signal that stimulates clonal expansion of effector cells to promote tolerogenic responses and to regulate development and function of $CD4^+CD25^+$ Tregs and $CD8^+$ Tregs to maintain tolerance [129, 130].

These evidences are supported by the recent observation that the frequency of $CD4^+$ $CD25^+$ Tregs were significantly decreased in patients with active pediatric lupus patients compared with patients with inactive lupus and controls and was inversely correlated with disease activity and serum anti-double-stranded DNA levels [131]. Furthermore, an elevated surface expression of GITR in $CD4^+$ $CD25^+$ T cells, elevated mRNA expression of CTLA-4 in $CD4^+$ T cells and higher amounts of mRNA expression for FOXP3 in $CD4^+$ cells in patients with active lupus disease compared with patients with inactive disease and control was noted. These results indicate a defective Treg population in paediatric lupus implying a role for FOXP3, CTLA-4 and GITR and $CD4^+$ Tregs in the pathogenesis of lupus. These results are in agreement of those reported by Valencia et al. [132] who reported a significant decrease in the suppressive function of $CD4^+CD25^+$ Tregs from peripheral blood of patients with active lupus as compared with normal donors and patients with inactive lupus. More importantly, $CD4^+CD25^+$ Tregs isolated from patients with active lupus expressed reduced levels of FoxP3 mRNA and protein and poorly suppressed the proliferation and cytokine secretion of $CD4^+$ effector T cells in vitro. In contrast, the expression of FoxP3 mRNA and protein and in vitro suppression of the proliferation of $CD4^+$ effector T cells by Tregs isolated from inactive lupus patients was comparable to that of normal individuals. It is interesting to note that in vitro activation of $CD4^+CD25^+$ Tregs from patients with active lupus increased FoxP3 mRNA and protein expression and restored their suppressive function demonstrating that the defect in $CD4^+CD25^+$ Treg function in patients with active lupus is reversible. Similar results were not only reported by Lyssuk et al. [133] but they also showed that both in newly admitted patients with the first manifestations of the disease, and those treated with cytostatics and steroids the coexpression of FoxP3 on $CD4^+CD25$ T cells was significantly reduced in both groups regardless of the therapy. The ability of Tregs to suppress proliferation of autologous $CD8^+$ and $CD4^+$ T cells was significantly reduced in both groups of patients compared to healthy donors though impaired production of Tregs in lupus patients could be partly restored by conventional treatments. These results imply that measurement of FoxP3 on $CD4^+CD25$ T cells and Tregs in lupus could form a marker of response to therapy and prognostic indicator and can be a new therapeutic strategy in lupus. It may be noted here that these results are not without controversy. For instance, Lin et al. [134] reported that lupus patients had a higher $FOXP3^+$ T-cell frequency and absolute $CD4^+CD25\text{-}FOXP3^+$ cell count than normal individuals, and the frequencies of $CD4^+CD25^+FOXP3^+$ and $CD4^+FOXP3^+$ cells were positively correlated with the disease activity. On the other hand, the differences in frequencies and absolute counts of $FOXP3^+$ T cells between normal controls and rheumatoid arthritis (RA) patients were found to be insignificant. Moreover, lupus and RA patients appear to express two FOXP3 transcript variants in peripheral blood mononuclear cells at the levels similar to normal individuals. Despite these controversial results, it is opined that analysis on peripheral blood $FOXP3^+$ T cells may be useful for the evaluation of lupus disease activity. It is possible that both a decrease and an increase in $FOXP3^+$ T cells could be hallmark of active lupus [131–134].

It is interesting to note that patients with B-non-Hodgkin's lymphoma (B-NHL), who usually have a poor immune response, had higher percentages of $CD4^+CD25^+$-Tregs in their peripheral blood with or without chemotherapy compared to the healthy controls [135], which may be one of the important reasons of immunosuppression in B-NHL. Since $CD4^+CD25^+$ regulatory T cells (Tregs) mediate immune suppression through cell-cell contact with surface molecules, particularly cytotoxic T lymphocyte-associated antigen 4 (CTLA-4), glucocorticoid-induced tumor necrosis factor receptor family-related protein (GITR), and transforming growth factor β (TGF-β), Zhang et al. [136] characterized the expression of surface markers on peripheral blood mononuclear cells-derived Tregs in patients with atopic asthma and healthy subjects, and the effect of inhaled corticosteroid on them. Their study revealed that equivalent numbers of peripheral Tregs were found in patients with atopic asthma (stable and acute) and healthy subjects. Tregs in the study subjects preferentially expressed CTLA-4, GITR, toll-like receptor 4 (TLR4), latency-associated peptide (LAP/TGF-beta1), and forkhead box P3 (FOXP3). Patients with acute asthma showed decreased numbers of $CD4^+CD25^+LAP^+$ T cells compared to healthy subjects and stable asthmatics. It is noteworthy that inhaled corticosteroid enhanced the percentage of Tregs expressing LAP in vivo and in vitro dose-dependently. Furthermore, the percentages of Tregs expressing LAP were negatively correlated with total serum IgE levels and severity of asthma. These results suggest that corticosteroids have the ability to enhance the percentage of Tregs that could explain their immunosuppressive properties.

It is noteworthy that adjunction of high-dose cyclosporine (50 mg/kg cyclosporine) to pre-transplant donor-specific blood transfusion abrogated Tregs generation, whereas a lower dose (10 mg/kg) of cyclosporine promoted Tregs development either in synergy with peri-operative donor-specific blood transfusion or by its own effect [137]. These data suggests that at times lower dose of cyclosporine is to be preferred to induce immunosuppression especially in patients with lupus and rheumatoid arthritis aimed at inducing Tregs.

UV Radiation, Immune Response, Mast Cells and its Role in Lupus

The ability of UV radiation to suppress immune system is particularly interesting since understanding the molecular mechanisms of its action could pave way to develop newer therapeutic strategies for immunological disorders. The beauty of immunosuppression induced by UV radiation lies in the fact that in contrast to conventional immunosuppression by immunosuppressive drugs, UV radiation does not compromise the immune system in a general but rather in an antigen-specific fashion via induction of immunotolerance. This effect is mostly mediated via regulatory T cells (Treg) induced by UV radiation. Though it is possible that several subtypes of UV-induced Treg may exist, the best characterized are those which inhibit contact hypersensitivity. Induction of these Tregs by UV radiation is an active process which requires antigen presentation by UV-damaged but still alive Langerhans cells (LC) in the lymph nodes. UV-induced Treg have been characterized as expressing $CD4^+$ and $CD25^+$ and as releasing upon activation the immunosuppressive cytokine interleukin-10 (IL-10). Once activated in an antigen-specific manner, they suppress immune responses in a general fashion via the release of IL-10, a phenomenon called bystander suppression [138]. These evidences suggest that it is possible to use UV radiation to induce immunosuppression.

On the other hand, patients with lupus are sensitive to UV radiation and sun light (including artificial light) that is known to exacerbate their skin lesions. Hence developing a method of giving UV radiation that induces immunosuppression without the exacerbation of skin lesions in patients with lupus will be a real challenge. The absence of immunosuppression to UV radiation in patients with lupus is rather puzzling given its immunosuppressive nature. The ability of UV radiation to worsen skin lesions in patients with lupus suggests that the responses of keratinocytes to UV radiation in normal vs lupus are entirely different: in the normal UV radiation induces the generation of $CD4^+CD25^+$-Tregs that, in turn, release immunosuppressive cytokine interleukin-10 (IL-10) to produce immunosuppression; while in the subjects with lupus this phenomena is not only absent but, in fact, may induce inflammation. The capacity of UV radiation to reproduce skin lesions in patients with lupus has been attributed to UV-induced keratinocytes apoptosis, aberrant expression of inducible nitric oxide synthase, and an abnormality in the role of regulatory T cells (Tregs) and chemokines for lymphocyte recruitment [139, 140]. For instance, the sensitivity of Treg to CD95L-mediated apoptosis seems to be responsible for the increased loss and consequent decrease in the number of these cells in patients with active lupus [141, 142].

The paradoxical role of UV radiation in inducing immunosuppression in the normal but worsening the skin lesion in the lupus seems to be related to the interaction between UV radiation and mast cells. A direct correlation between dermal mast cell prevalence in dorsal skin of different mouse strains and susceptibility to UVB-induced systemic immuno-suppression has been observed. For instance, highly UV-susceptible C57BL/6 mice have high dermal mast cell prevalence while BALB/c mice, which require considerable UV radiation for 50% immunosuppression, have low mast cell prevalence. There is also a functional link between the prevalence of dermal mast cells and susceptibility to UVB- and cis-urocanic acid (UCA)-induced systemic immunosuppression. Mast cell-depleted mice are unresponsive to UVB or cis-UCA-induced systemic immunosuppression unless they are previously reconstituted at the irradiated or cis-UCA-administered site with bone marrow-derived mast cell precursors. Cis-UCA do not stimulate mast cell degranulation directly but it stimulated neuropeptide release from sensory c-fibers which, in turn, efficiently degranulate mast cells. It was noted that histamine, and not TNF-α was the product from mast cells that stimulated downstream immunosuppression since, histamine receptor antagonists reduced UVB and cis-UCA-induced systemic immunosuppression. Histamine stimulated keratinocyte prostanoid production and indomethacin, a potent prostaglandin synthesis inhibitor, inhibited UVB induced immunosuppression an effect that was not cumulative with the histamine receptor antagonists [143]. Thus, both histamine and prostaglandin E_2 are important mediators in downstream immunosuppression brought about by UV radiation and both (histamine and PGE_2) regulate the development of T_H2 cells and reduced expression of T_H1 immune responses such as a contact hypersensitivity reaction.

Mast Cells in Rheumatological Conditions

In contrast to this, there is evidence to suggest that mast cells play an important role in the pathogenesis of lupus, rheumatoid arthritis and other collagen vascular diseases. In a study [144] that set out to evaluate the role of mast cells in immuno complex-mediated injury to skin, as immune complex-induced injury is an important pathogenic factor in antibody-mediated nephritis, lupus, RA, and other similar diseases, it was noted that in mast cell-deficient WBB6F1-W/Wv mice induction of reverse Arthus reaction the neutrophil influx was only 40% and edema 60% of that in their congenic controls (WBB6F1$^{(-)+/+}$). Hemorrhage and mast cell release were also significantly reduced in the mast cell-deficient mice. On the other hand, mast cell reconstitution restored the magnitude of the reaction in WBB6F1-W/Wv equivalent to that in WBB6F1$^{(-)+/+}$ mice. The 5-lipoxygenase inhibitor A-63162 significantly decreased neutrophil accumulation, edema, and hemorrhage in WBB6F1$^{(-)+/+}$, but not in mast cell-deficient mice, whereas mast cell reconstitution of WBB6F1-W/Wv mice restored the effect of A-63162. These results clearly showed that mast cells and their mediators, including leukotrienes, contribute to reverse Arthus reaction and thus play a significant role in collagen vascular diseases. Furthermore, morphometric studies in renal biopsy specimens revealed the presence of large number of interstitial mast cells in lupus nephropathy with a significant positive correlation between interstitial mast cell count and relative interstitial volume in the development of interstitial fibrosis [145] further attesting to the role of mast cells in lupus. These results are in agreement with the belief that mast cells may contribute to the pathogenesis of connective tissue diseases [146, 147], scleroderma, vasculitic syndromes, and lupus. Inhibition of the growth factor receptor of human mast cells, c-Kit, by the selective tyrosine kinase inhibitor imatinib mesylate, induces apoptosis of synovial tissue mast cells and preliminary findings suggest that inhibition of c-Kit could have anti-rheumatic activity in the treatment of patients with RA and spondyloarthropathies. These results emphasize the fact that agents that are immunosuppressive in normal subjects (such as UV radiation) may, in fact, serve as pro-inflammatory stimuli in lupus.

Based on the preceding discussion, it is evident that efforts made to enhance NO generation, enhance the number of Treg cells, block pro-inflammatory eicosanoid synthesis, and stabilize mast cells could be of benefit in lupus and other collagen vascular diseases. In this context, the role of polyunsaturated fatty acids (PUFAs) and their pro- and anti-inflammatory metabolites and their role in inflammation are interesting.

PLA_2, TNF-α, MIF and Pro- and Anti-inflammatory Lipids

Inflammation need to be considered not as a single event but as a process with multiple checkpoints where each phase of cellular influx, persistence, and resolution of inflammation is controlled by several endogenous stop and go signals. It is evident from the discussion in Chap. 4 (the section on metabolism of essential fatty acids) that products from AA, EPA, and DHA not only form precursors to various pro-inflammatory molecules: PGE_2, $PGF_{2\alpha}$, TXs, and LTs but also give rise to lipoxins (LXs) and aspirin-triggered 15-epimer LXs (ATLs) that are anti-inflammatory in nature. For these pro- and anti-inflammatory compounds to form, initial activation of PLA_2 is essential so that the necessary precursors could be released from the cell membrane lipid pool. Thus, there are two phases of release of PUFAs (especially that of AA): one at onset of the generation of pro-inflammatory PGs, TXs, and LTs and one at the time of resolution for the synthesis of anti-inflammatory LXs and aspirin-triggered 15 epimer LXs (ATLs).

Three classes of phospholipases (PLAs), based on cellular disposition and calcium dependence, control the release of AA and other PUFAs from membrane phospholipids. These are: calcium-independent PLA_2 ($iPLA_2$), secretory PLA_2 ($sPLA_2$), and cytosolic PLA_2 ($cPLA_2$) [148]. Each class of PLA_2 is further divided into isoenzymes for which there are ten for mammalian $sPLA_2$, at least three for $cPLA_2$, and two for $iPLA_2$. During the early phase of inflammation, COX-derived PGs and lipoxygenase-derived LTs initiate exudate formation and inflammatory cell influx [149]. In the murine air pouch model, injection of TNF-α caused an immediate influx of neutrophils concomitant with PGE_2 and LTB_4 production, and during the phase of resolution of inflammation an increase in LXA_4 (lipoxin A_4) production occurred that stopped neutrophil influx and enhanced phagocytosis of debris [150]. Similarly, in a carrageenin-induced pleurisy model, the onset of inflammation is characterized by rapid PMN influx mediated by PGE_2 synthesis, and as inflammation starts subsiding COX-2-derived anti-inflammatory PGs: PGD_2 and its product 15deoxy$\Delta^{12-14}PGJ_2$ formation occurs that induced resolution of inflammation with a simultaneous decrease in PGE_2 synthesis [151]. These results suggest that there are two waves of release of AA and other PUFAs: one at the onset of inflammation that causes the synthesis and release of PGE_2 and a second at resolution for the synthesis of anti-inflammatory PGD_2, 15deoxy$\Delta^{12-14}PGJ_2$, and lipoxins that are essential for the suppression of inflammation. Thus, COX-2 enzyme has both harmful and useful actions by virtue of its ability to give rise to pro-inflammatory and anti-inflammatory PGs and LXs. But, what is not clear is the mechanism by which these specific pro- and anti-inflammatory lipids are directed to form from the same precursor namely, AA, EPA and/or DHA at the precise time. In this context, PLA_2 isoforms responsible for the phenomenon of lipid class in acute resolving inflammation needs close attention.

Induction of acute pleurisy in rats by injecting carrageenin, increased type VI $iPLA_2$ protein expression was the principal isoforms expressed from the onset of inflammation up to 24 h. In contrast, type IIa and V $sPLA_2$ was expressed from the beginning of 48 h till 72 h, whereas type IV $cPLA_2$ was not detectable during the early phase of acute inflammation but increased progressively during resolution peaking at 72 h. This increase in type IV $cPLA_2$ was mirrored by a parallel increase in COX-2 expression [152]. In fact, it was observed that the increase in $cPLA_2$ and COX-2 occurred in parallel, suggesting a close functional interaction or enzymatic coupling between these two. This is supported by the observation that during the resolution phase of inflammation, PGD_2, 15deoxy$\Delta^{12-14}PGJ_2$, LXs, and resolvins are formed [149–153] with a simultaneous increase in the expression of COX-2 during the late stages of inflammation, especially when it is resolving. In summary, type VI $iPLA_2$ was highly expressed at the onset of inflammation, whereas types IIa and V sPLA2 and type IV cPLA2 were the predominant isoforms of PLA_2 expressed during the resolution phases of acute inflammation. Thus, there is a clear-cut role for different types of PLA_2 in distinct and different phases of inflammation. Glucocorticoids, which are potent anti-inflammatory compounds, suppress $cPLA_2$ and $sPLA_2$ expression [154, 155]. In addition, dexamethasone inhibited IL-1β-induced increase in $cPLA_2$ expression *in vitro* with both dexamethasone and IL-1β having little or no effect on $iPLA_2$ expression [152]. This suggests that type VI $iPLA_2$ expression is refractory to the suppressive effects of glucocorticoids suggesting that it is critical to acute inflammation. Selective inhibition of $cPLA_2$ resulted in the reduction of pro-inflammatory molecules PGE_2, LTB_4, IL-1β, and platelet-activating factor (PAF). It is important to note that COX-2 expression increased in tandem with type IV $cPLA_2$. Since increase in the expression of $cPLA_2$ is associated with resolution of inflammation, it implies that this calcium dependent isoforms of PLA_2 is responsible for the release of AA, EPA, and DHA that, in turn, is utilized by COX-2 enzyme to generate anti-inflammatory compounds such as PGD_2, 15deoxy$\Delta^{12-14}PGJ_2$, LXs, and resolvins. This was confirmed by the observation that when specific $cPLA_2$ and $sPLA_2$ inhibitors were used, the inflammatory process continued with little or no resolution. Furthermore, inhibition of types IIa and V $sPLA_2$ not only decreased PAF and LXA_4 (lipoxin A_4) but also resulted in a reduction in $cPLA_2$ and COX-2 activities. These results suggest that $sPLA_2$-derived PAF and LXA_4 induce COX-2 and type IV $cPLA_2$. Even IL-1β was found to induce $cPLA_2$ expression. This indicates that one of the functions of IL-1 is not only to induce inflammation but also to induce $cPLA_2$ expression to initiate resolution of inflammation [156, 157]. Both PAF and IL-1β increased type IV $cPLA_2$ in A549 cells but was without effect on COX-2. On the other hand, LXA_4 increased COX-2 expression in macrophages and fibroblasts. Taking these results together, it suggests that types IIa and V sPLA2 are necessary for the induction of type IV $cPLA_2$ and COX-2 enzymes to produce resolution of inflammation [152]. It is noteworthy that synthetic glucocorticoid dexamethasone inhibits both $cPLA_2$ and $sPLA_2$ expression, whereas type IV $iPLA_2$ expression is refractory to its suppressive actions [152–154]. Studies revealed that IL-1β more than doubled type IV $cPLA_2$ expression whereas dexamethasone reversed this effect by ~40%. On the other hand, both IL-1β and dexamethasone had little or no effect on type VI $iPLA_2$ expression [152]. It is known that long-term use of glucocorticoids is harmful to the patient in that it may suppress momentarily but once the drug is stopped it leads to intense flare up of inflammation. This can be attributed to the fact that inhibition of $cPLA_2$ and $sPLA_2$ (and so decrease in the release of $cPLA_2$ and $sPLA_2$-induced AA, EPA, and DHA release) leads to a reduction in the formation of PAF and LXA_4 that are needed to enhance macrophage phagocytosis of apoptotic cells and removal of debris to resolve inflammation [149, 158]. Activated $iPLA_2$ contributes to the conversion of inactive proIL-1β to active IL-1β, which in turn induces $cPLA_2$ expression that is necessary for resolution of inflammation. Both TNF-α and MIF are pro-inflammatory molecules and are known to activate $cPLA_2$ and induce COX-2. But, paradoxically, both TNF-α and MIF perpetuate inflammation, despite activating $cPLA_2$ and COX-2. This suggests that both TNF-α and MIF suppress the synthesis of LXs, PGD_2, 15deoxy$\Delta^{12-14}PGJ_2$ from $cPLA_2$-induced release of AA/EPA/DHA. On the other hand, LXs, especially LXA_4 inhibit TNF-α-induced production of ILs; promote TNF-α mRNA decay, TNF-α secretion, and leukocyte trafficking and thus attenuated inflammation. This close interaction between various PLA_2s, COX-2, PGD_2, LXA_4, and PAF in the initiation, maintenance, and resolution of inflammation suggests that any imbalance in this complex interplay between various PLA_2s, IL-1, TNF-α, MIF, COX-2, PGD_2, LXA_4, and local and systemic production of glucocorticoids during the various phases of inflammation could lead to either less optimal inflammation or persistence of inflammation. In addition, the association between $cPLA_2$ and COX enzymes seems to be cell/tissue specific. For instance, in COS-1 cells, COX-1 and COX-2 couple specifically with $cPLA_2$ for PGE_2 production, but not with $sPLA_2$. In contrast, in mast cells $sPLA_2$ is coupled closely with COX-1 for the immediate production of PGD_2 whereas to facilitate delayed synthesis and release of PGD_2, $sPLA_2$ closely interacts with COX-1 [159, 160]. On the other hand, as discussed above, in rat carrageenin pleurisy model, $cPLA_2$ couples preferentially with COX-2 to synthesize and release anti-inflammatory PGs. These results attest to the fact that there is a differential association between various PLA_2s and COX enzymes and this may depend on the cells/tissues that are under examination and the stimuli that is involved.

It is evident from the preceding discussion that type VI $iPLA_2$ expression dominates the initial phase of inflammation that leads to the production of PGE_2, LTB_4, PAF, and IL-1β with concomitant lower levels of expression of type IIa and V $sPLA_2$ and type IV $cPLA_2$. Once the acute phase of inflammation subsides, the resolution phase is characterized by the sequential expression of $sPLA_2$ (types IIa and V) that leads to the synthesis of PAF and LXA_4 that, in turn, induces the expression of type IV $cPLA_2$ which in association with COX-2 synthesizes PGD_2 and paves the way for the resolution of inflammation. It is evident that the local levels of endogenous glucocorticoids play a major role in the resolution of the inflammatory process. In experimental animals, corticosterone is released very early in the course of inflammation by stimulating the hypothalamic-pituitary-adrenal axis by pro-inflammatory mediators such as TNF-α, IL-1β, and IL-6, an event that is critical to the resolution of inflammation [161]. Studies showed that $iPLA_2$ is resistant to the inhibitory actions of dexamethasone whereas both $cPLA_2$ and $sPLA_2$ are inhibited. During the normal course of an inflammatory process, the local concentrations of endogenous corticosterone are high, whereas at the time of resolution they are low so that both $cPLA_2$ and $sPLA_2$ are expressed to enhance the production of LXs, PGD_2, and 15deoxy$\Delta^{12-14}PGJ_2$ to help in the resolution of inflammation. Continued use of corticosteroids for an extended period of time, suppresses $sPLA_2$ and $cPLA_2$ expression that are essential for the production of LXs, PGD_2 and 15deoxy$\Delta^{12-14}PGJ_2$ to resolve inflammation. This may explain why long-term use of steroids leads to non-healing of inflammatory lesions and a flare up of the inflammatory process as soon as steroids are stopped.

It is important to note that $iPLA_2$ has other important actions in inflammation that include: (a) enhancing the conversion of pro-IL-1β to IL-1β by IL-converting enzyme [162]; (b) eicosanoid biosynthesis; and (c) clearance of debris. In contrast, high concentrations of $cPLA_2$ suppress the conversion of pro-IL-1β to IL-1β. During the resolution of inflammation, LXA_4 that is formed as a result of increased expression of $sPLA_2$ is not only essential to switch off inflammatory cell infiltration but also to enhance macrophage phagocytosis to remove the debris. Similarly, PAF also augments macrophage phagocytosis. The formation of both LXA_4 and PAF is maximal at the initiation of resolution of inflammation. Furthermore, both LXA_4 and PAF have the ability to upregulate COX-2 and $cPLA_2$ expression, and COX-2 brings about the synthesis of PGD_2 and 15deoxy$\Delta^{12-14}PGJ_2$ that have anti-inflammatory actions. These results suggest that there is a very close and inter-connected loop of events that, under normal physiological conditions, act in a concerted manner to resolve inflammation.

These findings have important implications both to acute and chronic inflammation. In chronic inflammatory conditions such as RA and lupus the flares and remissions are somewhat similar to onset and resolution respectively of acute inflammatory process described above in as much as the cell profile and mediators that initiate the response are similar. In chronic inflammatory conditions such as RA and lupus, NSAIDs (non-steroidal anti-inflammatory drugs) are prescribed from months to years despite which the disease progression or destruction of target tissues (in RA joints and bones) continues. It is evident from the preceding discussion that COX-2 has an important role in resolving inflammation [163] and hence, the failure of NSAIDs to halt the progression of disease(s) could be due to the inhibition of COX-2 (see Fig. 13.2).

Glucocorticoids, COX Enzymes, LTs, Cytokines, NO, LXs, and Inflammation

Both oral and parenteral corticosteroids are widely used in the treatment of various inflammatory conditions. Although corticosteroids are very effective anti-inflammatory compounds in view of their pleiotropic actions, they also have significant side effects such as growth retardation in children, hypertension, immunosuppression, peripheral insulin resistance, delayed or impaired wound healing, osteoporosis, and other metabolic disturbances. Glucocorticoids bring about their anti-inflammatory actions by (i) the induction and activation of annexin 1 (also called as lipocortin-1) [164], (ii) the induction of mitogen-activated protein kinase (MAPK) phosphatase 1 [165], and (iii) the inhibition of COX-2 [166]. Annexin 1 or Lipocortin-1 physically interacts with and inhibits $cPLA_{2\alpha}$ so that AA is not released in adequate amounts to form precursor to various pro-inflammatory eicosanoids. As already discussed above, there is reasonable evidence to suggest that increased expression of $cPLA_2$ is necessary to give rise to anti-inflammatory molecules such as PGD_2 and $15deoxy\Delta^{12-14}PGJ_2$, and LXs. Thus, the timing and quality and quantity of expression (perhaps a pulsatile expression) of $cPLA_2$ and the local concentrations of glucocorticoids could be one important factor that determines the progression and/or resolution of inflammation. The selective inhibition of COX-2 and iNOS expression by glucocorticoids could explain their potent anti-inflammatory actions [166, 167]. Glucocorticoids also inhibit the production of pro-inflammatory cytokines such as IL-1, IL-6, TNF-α, and MIF [168–170]. Glucocorticoids mediate their inhibitory action on iNOS and COX enzymes through lipocortin-1 (annexin1) [164]. On the other hand, eNO activates constitutive (COX-1) resulting in optimal release of PGE_2, whereas iNO activates COX-2 resulting in markedly increased release of PGE_2 that results in inflammation [171]. This implies that constitutive production of NO and PGE_2 are anti-inflammatory in nature whereas inducible production of NO and PGE_2 are pro-inflammatory, simply because the quantities of NO and PGE_2 are extremely high in the later instance. It may be noted here that low concentrations of glucocorticoids enhance MIF synthesis that, in turn, overrides glucocorticoid-mediated inhibition of secretion of other pro-inflammatory cytokines. MIF induces

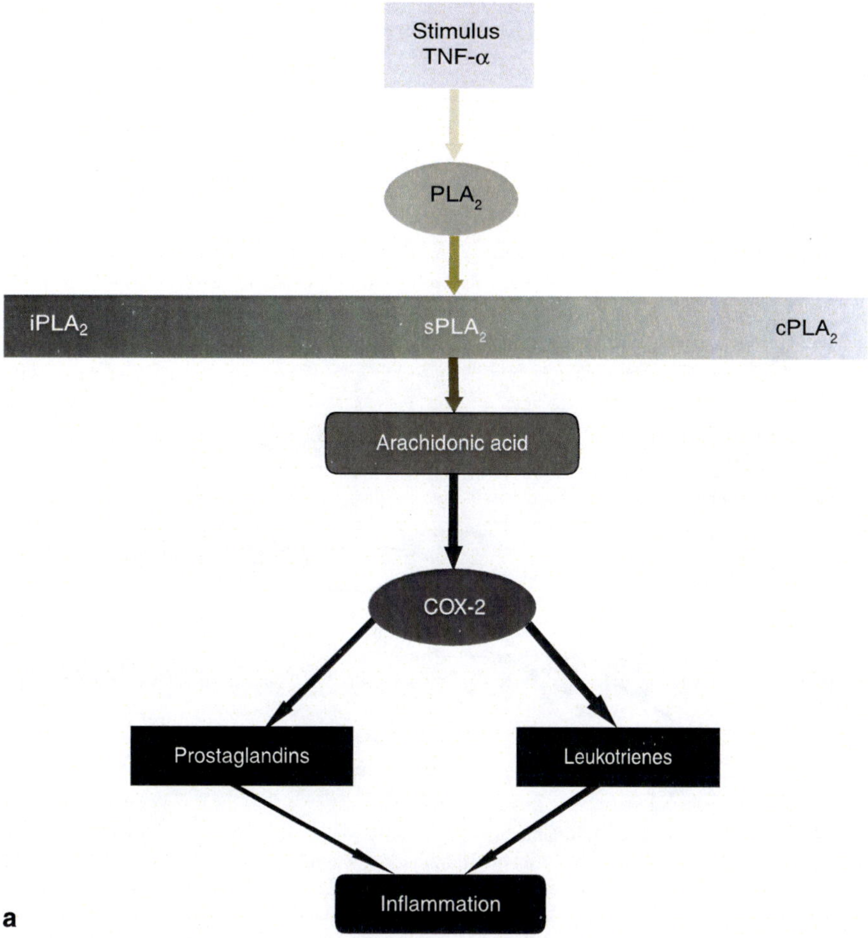

Fig. 13.2 **a** Scheme showing the role of eicosanoids in inflammation. Three classes of phospholipases control the release of AA and other PUFAs from membrane phospholipids. These are: calcium-independent PLA_2 ($iPLA_2$), secretory PLA_2 ($sPLA_2$), and cytosolic PLA_2 ($cPLA_2$). Each class of PLA_2 is further divided into isoenzymes for which there are ten for mammalian $sPLA_2$, at least three for $cPLA_2$, and two for $iPLA_2$. During the early phase of inflammation, $iPLA_2$ is activated for the release of AA and subsequently COX-2-derived PGs and lipoxygenase-derived LTs initiate exudate formation and inflammatory cell influx leading to the onset of inflammation. **b** Scheme showing the role of various prostaglandins and lipoxins in inflammation and its resolution. There are two waves of release of AA and other PUFAs: one at the onset of inflammation that causes the synthesis and release of PGE_2 and LTB_4 and a second at resolution for the synthesis of anti-inflammatory PGD_2, 15deoxy$\Delta^{12-14}PGJ_2$, and lipoxins that are essential for the suppression of inflammation. Increased type VI $iPLA_2$ protein expression was the principal isoforms expressed from the onset of inflammation up to 24 h. In contrast, type IIa and V $sPLA_2$ was expressed from the beginning of 48 h till 72 h whereas type IV $cPLA_2$ was not detectable during the early phase of acute inflammation but increased progressively during resolution peaking at 72 h. This increase in type IV $cPLA_2$ was mirrored by a parallel increase in COX-2 expression. The increase in $cPLA_2$ and COX-2 occurred in parallel, suggesting a close functional interaction or enzymatic coupling between these two. During the resolution phase of inflammation, PGD_2, 15deoxy$\Delta^{12-14}PGJ_2$, LXs, and resolvins are formed with a simultaneous increase in the expression of COX-2 during the late

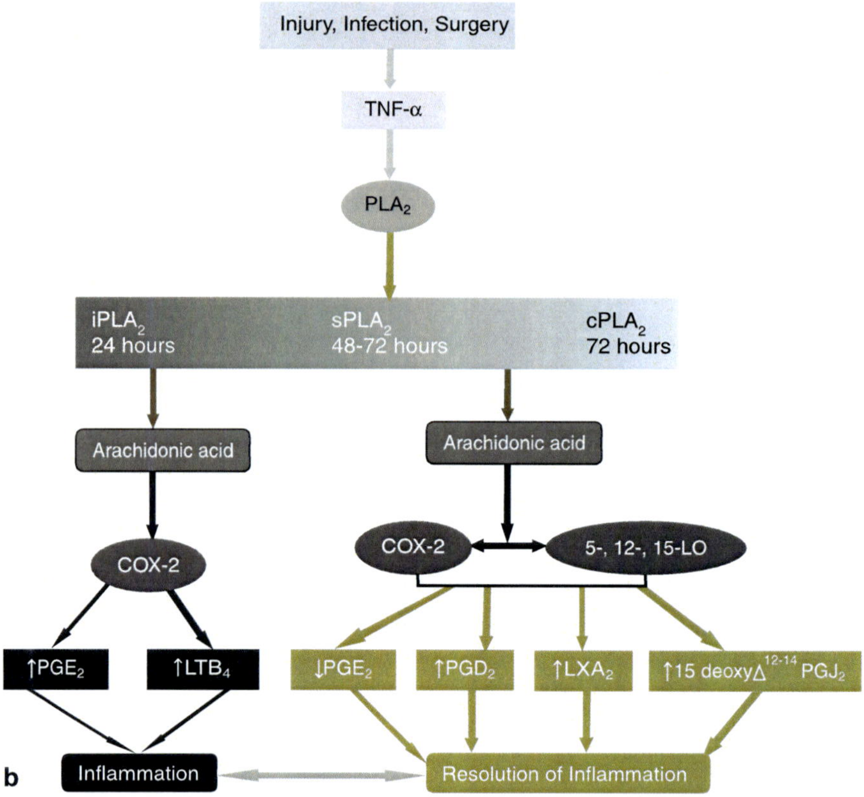

Fig. 13.2 (Continued) stages of inflammation, especially when it is resolving. In summary, type VI $iPLA_2$ was highly expressed at the onset of inflammation, whereas types IIa and V sPLA2 and type IV cPLA2 were the predominant isoforms of PLA_2 expressed during the resolution phases of acute inflammation. **c** Scheme showing the role of various prostaglandins and lipoxins in inflammation and its resolution. (−) Indicates inhibition or suppression of action; (+) Indicates activation or enhancement of action. Type VI $iPLA_2$ is expressed at the onset of inflammation, whereas types IIa and V sPLA2 and type IV cPLA2 are predominantly expressed during the resolution phases of acute inflammation. Glucocorticoids, the potent anti-inflammatory compounds, suppress $cPLA_2$ and $sPLA_2$ expression. In addition, dexamethasone inhibits IL-1β-induced increase in $cPLA_2$ expression, while both dexamethasone and IL-1β have little or no effect on $iPLA_2$ expression, suggesting that type VI $iPLA_2$ expression is refractory to the suppressive effects of glucocorticoids. Selective inhibition of $cPLA_2$ results in the reduction of pro-inflammatory PGE_2, LTB_4, IL-1β, and platelet-activating factor (PAF). COX-2 expression is increased in tandem with type IV $cPLA_2$, suggesting that this enzyme is responsible for the release of AA, EPA, and DHA that, in turn, is utilized by COX-2 enzyme to generate anti-inflammatory compounds PGD_2, 15deoxy$\Delta^{12-14}PGJ_2$, LXs, and resolvins. Specific inhibition of $cPLA_2$ and $sPLA_2$ leads to continuation of the inflammatory process with no resolution. Inhibition of types IIa and V $sPLA_2$ not only decrease PAF and LXA_4 but also results in a reduction in $cPLA_2$ and COX-2 activities, indicating that $sPLA_2$-derived PAF and LXA_4 induce COX-2 and type IV $cPLA_2$. Even IL-1β induces $cPLA_2$ expression. Thus, one of the functions of IL-1 is not only to induce inflammation but also to induce $cPLA_2$ expression to initiate resolution of inflammation. Both PAF and IL-1β increased type IV $cPLA_2$ in A549 cells but was without effect on COX-2. On the other hand, LXA_4 increased COX-2 expression in macrophages and fibroblasts. These results suggest that types IIa and V sPLA2 are necessary for the induction of type IV $cPLA_2$ and COX-2 enzymes to produce resolution of inflammation. Dexamethasone inhibits both $cPLA_2$

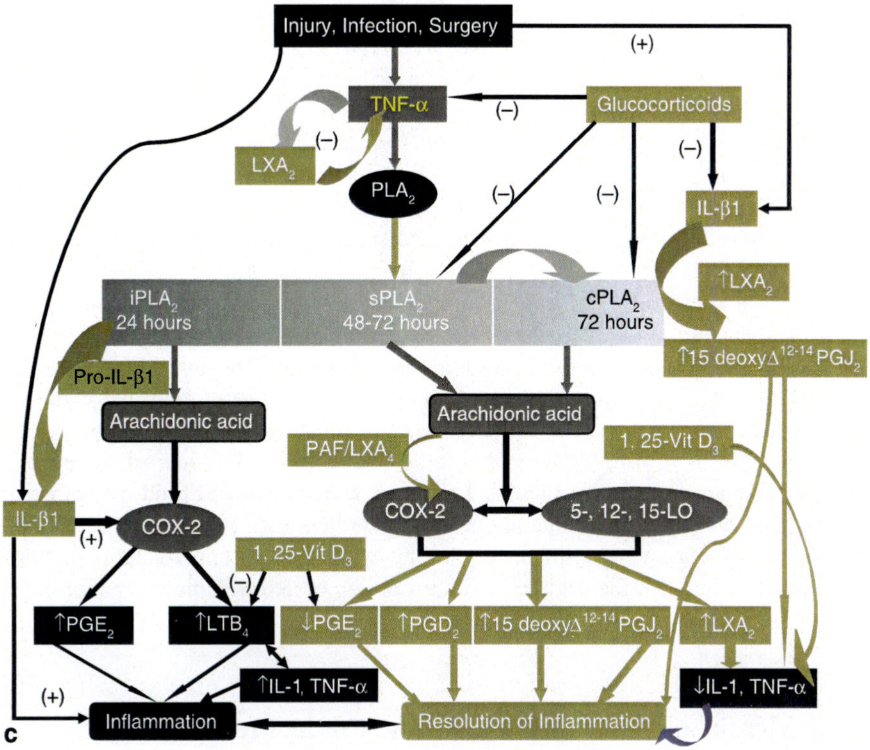

Fig. 13.2 (Continued) and $sPLA_2$ expression, whereas type IV $iPLA_2$ expression is refractory to its suppressive actions. IL-1β enhances type IV $cPLA_2$ expression whereas dexamethasone reverses this effect. Both IL-1β and dexamethasone have no effect on type VI $iPLA_2$ expression. Long-term use of glucocorticoids is harmful to the patient since its ability to inhibit $cPLA_2$ and $sPLA_2$ (and so decrease in the release of $cPLA_2$ and $sPLA_2$-induced AA, EPA, and DHA release) leads to a reduction in the formation of PAF and LXA_4 that are needed to enhance macrophage phagocytosis of apoptotic cells and removal of debris to resolve inflammation. Activated $iPLA_2$ contributes to the conversion of inactive proIL-1β to active IL-1β, which in turn induces $cPLA_2$ expression that is necessary for resolution of inflammation. Both TNF-α and MIF suppress the synthesis of LXs, PGD_2, 15deoxy$\Delta^{12-14}PGJ_2$ from $cPLA_2$-induced release of AA/EPA/DHA. On the other hand, LXs, especially LXA_4 inhibit TNF-α-induced production of ILs; promote TNF-α mRNA decay, TNF-α secretion, and leukocyte trafficking and thus attenuated inflammation

the production of TNF-α and vice versa. Thus, there is a close interaction between glucocorticoids, MIF, TNF-α, NO, and eicosanoids.

Glucocorticoids have been shown to accelerate the catabolism of LTC_4 (leukotriene C_4), a pro-inflammatory molecule, by enhancing the activity of γ-glutamyl transpeptidase [172]. Oral prednisone reduced PGD_2, 15-HETE, and enhanced those of 5-HETE and LTE_4 (a less pro-inflammatory metabolite of LTC_4). LTB_4, a potent pro-inflammatory molecule, and TXB_2 (a metabolite of pro-inflammatory molecule TXA_2) levels fell significantly in macrophages following treatment with prednisone [173], suggesting that glucocorticoids alter and macrophage eicosanoid synthesis such that the inflammatory process is dampened. In this context, it is interesting to note that 15-HPETE, an anti-inflammatory eicosanoid formed via lipoxygenase pathway, causes a significant increase in the rate of TNF degradation [174], an action that may also be seen with LXs. On the other hand, LXA_4 inhibited not only the secretion of TNF-α [175], but also prevented TNF-α-induced production of IL-1β, IL-6, cyclin E expression, and NF-κB activation [176]. Thus, glucocorticoids and lipoxins have similar actions on inflammation, both are anti-inflammatory, but their mechanisms of action seem to be different. In this context, it is important to note that both TNF-α and glucocorticoids have opposite actions on PLA_2: the former stimulates [177] while the later inhibits [167]. In fact, there is reasonable evidence to suggest that activation of $cPLA_2$ is crucial to the actions of TNF-α. This indicates that $cPLA_2$ and other PLA_2s play a central role in the pathobiology of inflammation and its resolution that could be attributed to the fact that LCPUFAs released by PLA_2 form precursors to several pro- and anti-inflammatory compounds as discussed above. The interaction between pro-inflammatory cytokines such as TNF-α, ILs, and MIF and lipid mediators of inflammation and anti-inflammatory process and repair indicates that there is a close feedback regulation among them and that the cytokines that initiate and perpetuate inflammation are also the triggers of the formation of anti-inflammatory molecules and repair process (see Figs. 13.1 and 13.2).

Cell Membrane Fatty Acid Content Could Modulate Inflammation and Repair

The amount and type of PUFA(s) released in response to inflammatory stimuli depends on the cell membrane phospholipid fatty acid content that, in turn, determines the quality and quantity of pro- and anti-inflammatory lipids formed. Since EFAs: LA and ALA that are desaturated and elongated to form LCPUFAs, are obtained direct from diet, this suggests that dietary content EFAs could be one factor that determines the degree of inflammation. As dietary intake of PUFAs can modify cell membrane fatty acid composition, this has the potential to modulate cell/tissue response to infection, injury and inflammatory events as a result of their conversion to various pro- and anti-inflammatory lipid molecules. Increased dietary intake of GLA, DGLA, and EPA/DHA substantially decreases inflammatory response [178–183] as a result of decreased formation of pro-inflammatory eicosanoids and cytokines, and an increase in the production of beneficial eicosanoids: LXs, resolvins, protectins, PGE_1, PGI_2, PGI_3, HPETEs, and eNO that resolve inflammation [163, 184–189]. A cell membrane that is rich in GLA/DGLA/EPA/DHA and contains appropriate amounts of AA, there could occur specific activation of $sPLA_2$ and $cPLA_2$ in response to an inflammatory stimulus leading to the formation of increased amounts of LXs, PGD_2 and 15deoxy$\Delta^{12-14}PGJ_2$, eNO, GSNO, PGE_1, PGI_2, PGI_3, and HPETEs that dampen inflammatory process. This is supported by the observation that human embryonic kidney cells, in the presence of exogenous PUFAs (fatty acids that were used in this study were AA, LA, and oleic acid), on exposure to IL-1, preferentially released AA due to the activation of $sPLA_2$-IIA, type IV $cPLA_2$, and type VI $iPLA_2$. The degree of activation of these PLA_2 was as follows: $sPLA_2$-IIA > type IV $cPLA_2$ > type VI $iPLA_2$, indicating that exogenous PUFAs preferentially activate type IIA $sPLA_2$-mediated AA release from IL-1 stimulated cells and the order of release was AA > LA > oleic acid [190, 191]. This is interesting since, it is evident from the preceding discussion that activation of $cPLA_2$ and $sPLA_2$ would lead to the formation of anti-inflammatory LXs, PGD_2 and 15deoxy$\Delta^{12-14}PGJ_2$ lending support to the hypothesis that lipid composition of the cell membrane can potentially modulate response to inflammation.

Since PUFAs form precursors to both pro- and anti-inflammatory compounds, the severity and persistence of inflammation depends on the balance between pro- and anti-inflammatory molecules formed from AA, EPA, and DHA. This implies that failure to generate adequate amounts of LXs and resolvins could lead to chronic inflammatory conditions such as RA, lupus, glomerulonephritis, and other conditions. The persistence of inflammation in these conditions could be due to continued synthesis and secretion of inflammatory cytokines such as IL-1, IL-2, IL-6, IL-8, TNF-α, and MIF. On the other hand, IFN-γ and IL-13 could trigger production of LXs and resolvins such that resolution of inflammation is initiated. When this delicate balance between pro- and anti-inflammatory cytokines and PGs and LXs and resolvins is disregulated, it will lead to persistence of inflammation (see Figs. 13.1 and 13.2). It is possible that the balance (or ratio) between the concentrations of AA and EPA in the cell membrane could be one factor that determines the amount of LXs, resolvins, PGD_2 and 15deoxy$\Delta^{12-14}PGJ_2$ formed. For instance, the presence of large amounts of AA in the membrane phospholipids could preferentially lead to the formation of pro-inflammatory lipids. On the other hand, when adequate amounts of EPA/DHA are present it could lead to the formation of PGD_2 and 15deoxy$\Delta^{12-14}PGJ_2$, LXs, and resolvins so that resolution of inflammation occurs early.

Nitric Oxide, Lipid Peroxides, and Antioxidant Status in Lupus

Both atherosclerosis and lupus are inflammatory disorders and so atherosclerosis may have an accelerated progression in lupus. Nitric oxide (NO) is an important mediator of inflammation including the inflammation associated with atherosclerosis and lupus. Endothelial nitric oxide synthase (eNOS)-mediated constitutive expression of NO promotes endothelial integrity and normal vascular function, whereas inducible nitric oxide synthase (iNOS) mediated expression of NO promotes endothelial dysfunction and atherogenesis. Hence, the balance between normal vascular function and atherogenesis may be mediated by differences in the quantity, location, and timing of NO production within vessel walls. In this context, it is noteworthy that statins have anti-inflammatory properties and reverse many of the deleterious effects associated with NO metabolism in atherosclerosis. They do so by augmenting eNOS and inhibiting iNOS expression. It is possible that even in lupus it is important to augment eNOS and inhibit iNOS to prevent vascular dysfunction.

Serum nitrite levels were reported to be significantly elevated in patients with lupus (mean $\pm$ SEM = 37 $\pm$ 6 μM/l) compared with controls (15 $\pm$ μM/l; $P < 0.01$), and were elevated in patients with active lupus compared with those with inactive disease (46 $\pm$ 7 μM/l versus 30 $\pm$ 7 μM/l; $P < 0.01$) and serum nitrite levels correlated with disease activity and with levels of antibodies to double-stranded DNA. Endothelial cell expression of iNOS in lupus patients was significantly greater compared with controls, and higher in patients with active disease compared with those with inactive lupus. Even keratinocyte expression of iNOS was also significantly elevated in lupus compared with controls, whereas eNOS activity was similar in patients with active lupus and inactive lupus patients and normal controls in either the vascular endothelium or the keratinocytes [192]. These and other results suggest that iNO production is enhanced in active lupus including neuropsychiatric lupus [193–195] that could be responsible for vascular and cutaneous inflammation seen in these patients.

Both glomerular endothelial cells and endothelium of cortical vessels contain eNOS. On the other hand, iNOS was localized in mesangial cells, glomerular epithelial cells and infiltrating cells in the diseased glomeruli, whereas iNOS was hardly detected in control kidneys. In addition, the expression pattern of eNOS in each glomerulus was the reverse of that of iNOS. In IgAN (IgA nephropathy) and lupus nephropathy, eNOS correlated negatively with the degree of glomerular injury, while the expression of iNOS correlated positively with the degree of glomerular injury in the same tissues [196], suggesting that eNOS and iNOS may have diametrically opposite actions on glomerular function and nephropathy: eNOS being beneficial whereas iNOS induces inflammation and glomerular injury. Furthermore, it was noted that immunosuppressant mycophenolate mofetil and PGJ_2 have been shown to suppress iNOS that may explain their beneficial actions in lupus [197, 198]. The practical implication of enhanced iNOS may lie in the fact native human DNA can be modified by peroxynitrite ($ONOO^-$) as a result of increased production of NO and this modified DNA was found to be a better antigen due to the formation of neo-epitopes that could incite the generation of anti-DNA antibodies that may induce autoimmune response in lupus [199]. Since eNOS is a constitutional enzyme whereas iNOS is an inducible enzyme, the amount of NO released due to the activation of eNOS and iNOS are different. Small amounts of NO are generated during the activation of eNOS whereas the amount of NO released is large as a result of iNOS activation. This suggests that physiological amounts of NO are anti-inflammatory whereas excess NO has pro-inflammatory actions.

Oxidant Stress, Anti-oxidants, NO and PUFAs in Lupus

The increase in pro-inflammatory cytokines that occurs in patients with active lupus and RA are known to be capable of attracting and stimulating neutrophils, macrophages and T cells that, in turn, produce free radicals, eicosanoids and cytokines in an autocrine fashion [74, 200]. The activated neutrophils release IL-1 which activates macrophages, T cells and synovial cells to produce prostaglandins, IL-2, TNF and free radicals. In addition, IL-1 enhances the production of endothelin-1 in cultured endothelial cells [201] and this could contribute to vasospasm [102] seen in lupus. Endothelial cells also produce PGI_2 and NO that are potent vasodilators and platelet anti-aggregators and are natural antagonists of endothelin-1. Hence, enhancing the production of NO and/or decreasing endotehlin-1 could be beneficial in Raynaud's phenomenon. Increased basal and stimulated serum endothelin concentrations has been described in patients with Raynaud's disease [102] that may explain, at least in part, the association of Raynaud's phenomenon in lupus.

Previously, I reported that drug-resistant Raynaud's phenomenon in lupus responds to oral L-arginine therapy [202] that was attributed to an increase in the generation of NO, since L-arginine is its precursor. Subsequently, I described the beneficial action of L-arginine and EPA/DHA in lupus and their ability to increase plasma NO levels [100, 101]. In these studies, plasma endothelin-1 concentrations were not measured to know whether there was a simultaneous decrease in its levels that also could have contributed to the response seen. But, it is interesting to note that all these patients are doing well for the past 23 years.

It is likely that under normal conditions, a balance is maintained between eNO and endothelin-1; and when this balance is altered it leads to Raynaud's phenomena. Hence, restoring endothelial integrity in lupus is important. My previous studies [203, 204] studies and those of Fries et al. [203] suggest that this could be achieved by providing L-arginine, ω-3 fatty acids and adequate immunosuppressive drugs and donors of NO [204]. Based on these data, it is suggested that serial measurements of plasma concentrations of various cytokines, pro- and anti-oxidants, PGI_2, NO, asymmetrical dimethyl arginine, endothelin-1, and indices of endothelial integrity and function in lupus could help in predicting the development and response of Raynaud's phenomenon in lupus.

In our studies [100, 101, 202], it was observed that patients with lupus have decreased plasma levels of EPA, DHA, NO and anti-oxidants glutathione peroxidase and enhanced levels of lipid peroxides that could be restored to normal by providing oral EPA/DHA and L-arginine.

Since enhanced levels of IL-1 seems to be the reason for the enhanced production of free radicals, lipid peroxides that, in turn, decrease the half-life of NO (since superoxide anion can inactivate NO), it is also worthwhile to devise methods to suppress IL-1 production. IL-1 production can be inhibited by 1,25-dihydroxyvitamin D_3, which is known to have immunoregulatory actions [205].

1,25-dihydroxyvitamin D_3 Suppresses Autoimmunity

The active form of vitamin D3, 1-α,25-dihydroxyvitamin D_3 (1,25-vitamin D_3), suppresses in vitro immunoglobulin production by activated peripheral blood mononuclear cells (PBM) from normal human subjects by inhibiting T helper/inducer T_H cell activity. It was reported that 1,25-vitamin D_3 abrogated the inducing effect of T_H cells on immunoglobulin synthesis by B cells. In addition, 1,25-(OH)2-D_3 produced a dramatic inhibition of IL-2 production by activated PBM, but did not inhibit IL-2 receptor generation by these cells, suggesting that the T_H lymphocyte is the specific cellular target for the immunoinhibitory effects of 1,25-vitamin D_3 [205]. It was also reported that 1,25-vitamin D_3 blocks IL-2 production and interferes with IL-2-thymocyte interaction, which results in inhibition of thymocyte proliferation [206]. 1,25-vitamin D_3 specifically inhibited, in a time- and dose-dependent fashion, the generation of cytotoxic activity from cultured CD16+ peripheral blood NK cells and IL-2 production by PHA-activated peripheral blood lymphocytes but did not interfere with the cytotoxic function of NK cells. Interestingly, exogenous IL-2 reversed this suppressive effect [208]. In addition, it was reported that 1,25-vitamin D_3 inhibited not only IL-12-generated IFN-γ production, but also suppressed IL-4 and IL-13 expression induced by IL-4 [209]. Studies done with experimental autoimmune uveitis (EAU) as a model for human cell-mediated autoimmunity, it was reported that oral 1,25-dihydroxyvitamin D_3 prevented as well as partly reversed disease and suppressed immunological responses by directly suppressing IL-17 induction in purified naive CD4(+) T cells without inhibiting Th-17 lineage commitment, as reflected by unaltered RORgammat, STAT3, and FoxP3 expression. Innate immune response parameters in draining lymph nodes of treated mice were suppressed, as was production of IL-1, IL-6, TNF-α, and IL-12/IL-23p40, but not IL-10, by explanted splenic dendritic cells (DC). Supernatants of calcitriol-conditioned bone marrow-derived dendritic cells had reduced ability to support Th-17 polarization of naive CD4(+) T cells in vitro and in vivo. Thus, 1,25-dihydroxyvitamin D_3 appears to suppress autoimmunity by inhibiting the Th-17 response [210].

In view of these evidences, it is reasonable to propose that 1,25-dihydroxyvitamin D_3 could be given to patients with lupus, RA and other rheumatological conditions to suppress the activity of the disease [74].

ADMA is Useful in Lupus and Other Rheumatological Conditions

It is known that plasma eNO levels are low in patients with lupus and RA [100, 101]. These low levels of NO could trigger vasospasm and cause Raynaud's phenomenon seen in lupus and other rheumatological conditions. The low levels of NO could be due to several reasons: (a) substrate deficiency; (b) low activity of the eNOS enzyme; and (c) rapid inactivation of eNO.

Arginase and nitric oxide synthase (NOS) compete for the same substrate, L-arginine. The reciprocal regulation of arginase and NOS in L-arginine-metabolizing pathways is known to occur. It was observed that both serum arginase activity and protein levels were significantly higher in patients with RA than in patients with lupus or osteoarthritis (OA) or in healthy controls. A significant correlation between the serum concentrations of arginase protein and rheumatoid factor was noted in RA patients. These data indicate that increased arginase production occurs in RA and this may have an important role in its pathogenesis [211].

Asymmetric dimethylarginine (ADMA) is an endogenous nitric oxide synthesis inhibitor and independent risk factor for endothelial dysfunction and cardiovascular disease. Mean plasma ADMA levels were significantly higher in patients with lupus with a history of cardiovascular episodes than in patients without such a history. In multiple regression analysis a high SLEDAI (SLE disease activity index) score, high titre of anti-dsDNA antibodies, and low serum HDL were significantly associated with high plasma ADMA levels. These results suggest that in patients with lupus, plasma ADMA levels are significantly associated with cardiovascular episodes, measures of disease activity, and organ damage, independently of an unfavourable lipid profile [212]. These results have been supported by several other studies [213, 214]. In addition, it was also noted that high TNF-α levels and other factors such as high levels of sVCAM-1 that reflect endothelial damage are indicators of high risk of cardiovascular disease in lupus and other rheumatological conditions [215, 216]. These results indicate that pro-inflammatory molecules are able to reduce the production of eNO and induce damage to the endothelial cells and thus, enhance the risk of cardiovascular disease in lupus and RA. One method of reducing the plasma ADMA levels is by giving oral L-arginine that displaces ADMA and enhances plasma eNO levels.

In summary, it is clear that under physiological conditions a delicate balance is maintained between pro- and anti-inflammatory molecules and this balance is tilted more in favor of pro-inflammatory molecules in lupus and RA. Efforts made to restore this balance to normal by suppressing the pro-inflammatory events could be of significant therapeutic value in rheumatological conditions. One such approach could be in using naturally occurring anti-inflammatory lipids such as lipoxins, resolvins and protectins or their stable synthetic analogues; insulin that has anti-inflammatory actions; ethylpyruvate, lipid (especially DHA/lipoxin)-enriched albumin infusions and ghrelin that has anti-inflammatory actions [217–225] (see Figs. 13.1 and 13.2).

References

[1] Burnet FM (1957) A modification of Jerne's theory of antibody production using the concept of Clonal selection. Aust J Sci 20:67–69

[2] Pike B, Boyd A, Nossal G (1982) Clonal anergy: the universally anergic B lymphocyte. Proc Natl Acad Sci U S A 79:2013–2017

[3] Jerne N (1974) Towards a network theory of the immune system. Ann Immunol (Paris) 125C:373–389

[4] Edwards JC, Cambridge G, Abrahams VM (1999) Do self perpetuating B lymphocytes drive human autoimmune disease? Immunology 97:1868–1876

[5] Graham RR, Ortmann W, Rodine P, Espe K, Langefeld C, Lange E, Williams A, Beck S, Kyogoku C, Moser K, Gaffney P, Gregersen PK, Criswell LA, Harley JB, Behrens TW (2007) Specific combinations of HLA-DR2 and DR3 class II haplotypes contribute graded risk for disease susceptibility and autoantibodies in human SLE. Eur J Hum Genet 15:823–830

[6] Mulcahy B, Waldron-Lynch F, McDermott MF, Adams C, Amos CI, Zhu DK, Ward RH, Clegg DO, Shanahan F, Molloy MG, O'Gara F (1996) Genetic variability in the tumor necrosis factor-lymphotoxin region influences susceptibility to rheumatoid arthritis. Am J Hum Genet 59:676–683

[7] Newton J, Brown MA, Milicic A, Ackerman H, Darke C, Wilson JN, Wordsworth BP, Kwiatkowski D (2003) The effect of HLA-DR on susceptibility to rheumatoid arthritis is influenced by the associated lymphotoxin alpha-tumor necrosis factor haplotype. Arthritis Rheum 48:90–96
[8] Zhang X, Llamado L, Pillay I, Price P, Will R (2002) Interleukin-1 gene polymorphism disease activity and bone mineral metabolism in rheumatoid arthritis. Chin Med J (Engl) 115:46–49
[9] Cantagrel A, Navaux F, Loubet-Lescoulié P, Nourhashemi F, Enault G, Abbal M, Constantin A, Laroche M, Mazières B (1999) Interleukin-1beta, interleukin-1 receptor antagonist, interleukin-4, and interleukin-10 gene polymorphisms: relationship to occurrence and severity of rheumatoid arthritis. Arthritis Rheum 42:1093–1100
[10] Paradowska-Gorycka A, Wojtecka-Lukasik E, Trefler J, Wojciechowska B, Lacki JK, Maslinski S (2010) Association between IL-17F gene polymorphisms and susceptibility to and severity of rheumatoid arthritis (RA). Scand J Immunol 72:134–141
[11] Paradowska-Gorycka A, Trefler J, Maciejewska-Stelmach J, Lacki JK (2010) Interleukin-10 gene promoter polymorphism in Polish rheumatoid arthritis patients. Int J Immunogenet 37:225–231
[12] Ying B, Shi Y, Pan X, Song X, Huang Z, Niu Q, Cai B, Wang L (2011) Association of polymorphisms in the human IL-10 and IL-18 genes with rheumatoid arthritis. Mol Biol Rep 38:379–385
[13] Santos LL, Morand EF (2009) Macrophage migration inhibitory factor: a key cytokine in RA, SLE and atherosclerosis. Clin Chim Acta 399:1–7
[14] Zou YQ, Lu LJ, Li SJ, Zeng T, Wang XD, Bao CD, Chen SL, Yang CD (2008) The levels of macrophage migration inhibitory factor as an indicator of disease activity and severity in adult-onset Still's disease. Clin Biochem 41:519–524
[15] Ayoub S, Hickey MJ, Morand EF (2008) Mechanisms of disease: macrophage migration inhibitory factor in SLE, RA and atherosclerosis. Nat Clin Pract Rheumatol 4:98–105
[16] Cutolo M, Brizzolara R, Atzeni F, Capellino S, Straub RH, Puttini PC (2010) The immunomodulatory effects of estrogens: clinical relevance in immune-mediated rheumatic diseases. Ann N Y Acad Sci 1193:36–42
[17] Stewart JJ (1999) Theory and treatment of the X-inactivation chimera in female-prevalent autoimmune disease. Arch Immunol Ther Exp (Warsz) 47:355–359
[18] Brix TH, Knudsen GP, Kristiansen M, Kyvik KO, Orstavik KH, Hegedüs L (2005) High frequency of skewed X-chromosome inactivation in females with autoimmune thyroid disease: a possible explanation for the female predisposition to thyroid autoimmunity. J Clin Endocrinol Metab 90:5949–5953
[19] Selmi C, Invernizzi P, Gershwin ME (2006) The X chromosome and systemic sclerosis. Curr Opin Rheumatol 18:601–605
[20] Lu Q, Wu A, Tesmer L, Ray D, Yousif N, Richardson B (2007) Demethylation of CD40LG on the inactive X in T cells from women with lupus. J Immunol 179:6352–6358
[21] Uz E, Mustafa C, Topaloglu R, Bilginer Y, Dursun A, Kasapcopur O, Ozen S, Bakkaloglu A, Ozcelik T (2009) Increased frequency of extremely skewed X chromosome inactivation in juvenile idiopathic arthritis. Arthritis Rheum 60:3410–3412
[22] Saunders K, Raine T, Cooke A, Lawrence C (2007) Inhibition of autoimmune type 1 diabetes by gastrointestinal helminth infection. Infect Immun 75:397–407
[23] Wallberg M, Harris R (2005) Co-infection with Trypanosoma brucei brucei prevents experimental autoimmune encephalomyelitis in DBA/1 mice through induction of suppressor APCs. Int Immunol 17:721–728
[24] Walsh KP, Brady MT, Finlay CM, Boon L, Mills KH (2009) Infection with a helminth parasite attenuates autoimmunity through TGF-beta-mediated suppression of Th17 and Th1 responses. J Immunol 183:1577–1586
[25] Bang SY, Lee KH, Cho SK, Lee HS, Lee KW, Bae SC (2010) Smoking increases rheumatoid arthritis susceptibility in individuals carrying the HLA-DRB1 shared epitope, regardless

of rheumatoid factor or anti-cyclic citrullinated peptide antibody status. Arthritis Rheum 62:369–377

[26] Linn-Rasker SP, van der Helm-van Mil AH, van Gaalen FA, Kloppenburg M, de Vries RR, le Cessie S, Breedveld FC, Toes RE, Huizinga TW (2006) Smoking is a risk factor for anti-CCP antibodies only in rheumatoid arthritis patients who carry HLA-DRB1 shared epitope alleles. Ann Rheum Dis 65:366–371

[27] Criswell LA, Merlino LA, Cerhan JR, Mikuls TR, Mudano AS, Burma M, Folsom AR, Saag KG (2002) Cigarette smoking and the risk of rheumatoid arthritis among postmenopausal women: results from the Iowa Women's Health Study. Am J Med 112:465–471

[28] Weigle WO (1981) Self-nonself recognition by T and B lymphocytes and their roles in autoimmune phenomena. Arthritis Rheum 24:1044–1053

[29] Blossom S, Gilbert KM (1999) Antibody production in autoimmune BXSB mice. I. CD40L-expressing B cells need fewer signals for polyclonal antibody synthesis. Clin Exp Immunol 118:147–153

[30] Bouzahzah F, Jung S, Craft J (2003) CD4+ T cells from lupus-prone mice avoid antigen-specific tolerance induction in vivo. J Immunol 170:741–748

[31] Choe JY, Crain B, Wu SR, Corr M (2003) Interleukin 1 receptor dependence of serum transferred arthritis can be circumvented by toll-like receptor 4 signaling. J Exp Med 197:537–542

[32] Le Saout C, Mennechet S, Taylor N, Hernandez J (2008) Memory-like CD8+ and CD4+ T cells cooperate to break peripheral tolerance under lymphopenic conditions. Proc Natl Acad Sci U S A 105:19414–19419

[33] Green EA, Wong FS, Eshima K, Mora C, Flavell RA (2000) Neonatal tumor necrosis factor alpha promotes diabetes in nonobese diabetic mice by CD154-independent antigen presentation to CD8(+) T cells. J Exp Med 191:225–238

[34] Hillion S, Garaud S, Devauchelle V, Bordron A, Berthou C, Youinou P, Jamin C (2007) Interleukin-6 is responsible for aberrant B-cell receptor-mediated regulation of RAG expression in systemic lupus erythematosus. Immunology 122:371–380

[35] Daridon C, Devauchelle V, Hutin P, Le Berre R, Martins-Carvalho C, Bendaoud B, Dueymes M, Saraux A, Youinou P, Pers JO (2007) Aberrant expression of BAFF by B lymphocytes infiltrating the salivary glands of patients with primary Sjögren's syndrome. Arthritis Rheum 56:1134–1144

[36] Liossis SN, Kovacs B, Dennis G, Kammer GM, Tsokos GC (1996) B cells from patients with systemic lupus erythematosus display abnormal antigen receptor-mediated early signal transduction events. J Clin Invest 98:2549–2557

[37] Guilherme L, Kalil J (2010) Rheumatic fever and rheumatic heart disease: cellular mechanisms leading autoimmune reactivity and disease. J Clin Immunol 30:17–23

[38] Root-Bernstein R, Vonck J, Podufaly A (2009) Antigenic complementarity between coxsackie virus and streptococcus in the induction of rheumatic heart disease and autoimmune myocarditis. Autoimmunity 42:1–16

[39] Faé KC, Diefenbach da Silva D, Bilate AM, Tanaka AC, Pomerantzeff PM, Kiss MH, Silva CA, Cunha-Neto E, Kalil J, Guilherme L (2008) PDIA3, HSPA5 and vimentin, proteins identified by 2-DE in the valvular tissue, are the target antigens of peripheral and heart infiltrating T cells from chronic rheumatic heart disease patients. J Autoimmun 31:136–141

[40] Chen QY, Rowley MJ, Mackay IR (1999) Anti-idiotypic antibodies to anti-PDC-E2 in primary biliary cirrhosis and normal subjects. Hepatology 29:624–631

[41] Stefanescu M, Onu A, Matache C, Ramos-Morales F, Fischer S, Szegli G (1995) Naturally occurring anti-idiotypic antibodies to anti-phosphotyrosine in systemic lupus erythematosus interact with SRC-homology 2 domains. Autoimmunity 22:81–86

[42] Nordling C, Kleinau S, Klareskog L (1992) Down-regulation of a spontaneous arthritis in male DBA/1 mice after administration of monoclonal anti-idiotypic antibodies to a cross-reactive idiotope on anti-collagen antibodies. Immunology 77:144–146

[43] Romagnani S (1991) Type 1 T helper and type 2 T helper cells: functions, regulation and role in protection and disease. Int J Clin Lab Res 21:152–158

[44] Dorshkind K, Klimpel GR, Rosse C (1980) Natural regulatory cells in murine bone marrow: inhibition of in vitro proliferative and cytotoxic responses to alloantigens. J Immunol 124:2584–2588
[45] Poojary KV, Kong YC, Farrar MA (2010) Control of th2-mediated inflammation by regulatory T cells. Am J Pathol 177:525–531
[46] Murai M, Krause P, Cheroutre H, Kronenberg M (2010) Regulatory T-cell stability and plasticity in mucosal and systemic immune systems. Mucosal Immunol 3:443–449
[47] Shalaby KH, Martin JG (2010) Overview of asthma; the place of the T cell. Curr Opin Pharmacol 10:218–225
[48] Gandhi R, Farez MF, Wang Y, Kozoriz D, Quintana FJ, Weiner HL (2010) Cutting edge: human latency-associated peptide+ T cells: a novel regulatory T cell subset. J Immunol 184:4620–4624
[49] Carrier Y, Yuan J, Kuchroo VK, Weiner HL (2007) Th3 cells in peripheral tolerance. I. Induction of Foxp3-positive regulatory T cells by Th3 cells derived from TGF-beta T cell-transgenic mice. J Immunol 178:179–185
[50] Carrier Y, Yuan J, Kuchroo VK, Weiner HL (2007) Th3 cells in peripheral tolerance. II. TGF-beta-transgenic Th3 cells rescue IL-2-deficient mice from autoimmunity. J Immunol 178:172–178
[51] Weiner HL (2001) Oral tolerance: immune mechanisms and the generation of Th3-type TGF-beta-secreting regulatory cells. Microbes Infect 3:947–954
[52] Korn T, Bettelli E, Oukka M, Kuchroo VK (2009) IL-17 and Th17 cells. Annu Rev Immunol 27:485–517
[53] Ghilardi N, Ouyang W (2007) Targeting the development and effector functions of TH17 cells. Semin Immunol 19:383–393
[54] Bettelli E, Korn T, Kuchroo VK (2007) Th17: the third member of the effector T cell trilogy. Curr Opin Immunol 19:652–657
[55] Korn T, Oukka M, Kuchroo V, Bettelli E (2007) Th17 cells: effector T cells with inflammatory properties. Semin Immunol 19:362–371
[56] Toh ML, Kawashima M, Hot A, Miossec P, Miossec P (2010) Role of IL-17 in the Th1 systemic defects in rheumatoid arthritis through selective IL-12Rbeta2 inhibition. Ann Rheum Dis 69:1562–1567
[57] Shen H, Goodall JC, Hill Gaston JS (2009) Frequency and phenotype of peripheral blood Th17 cells in ankylosing spondylitis and rheumatoid arthritis. Arthritis Rheum 60:1647–1656
[58] Li X, Yuan FL, Lu WG, Zhao YQ, Li CW, Li JP, Xu RS (2010) The role of interleukin-17 in mediating joint destruction in rheumatoid arthritis. Biochem Biophys Res Commun 397:131–135
[59] Paradowska-Gorycka A, Grzybowska-Kowalczyk A, Wojtecka-Lukasik E, Maslinski S (2010) IL-23 in the pathogenesis of rheumatoid arthritis. Scand J Immunol 71:134–145
[60] Li Q, Cong B, Shan B, Zhang J, Chen H, Wang T, Ma C, Qin J, Wen D, Yu F (in press) Cholecystokinin octapeptide exerts its therapeutic effects on collagen-induced arthritis by suppressing both inflammatory and Th17 responses. Rheumatol Int
[61] Wei B, Pei G (2010) microRNAs: critical regulators in Th17 cells and players in diseases. Cell Mol Immunol 7:175–181
[62] Crispín JC, Tsokos GC (2010) Interleukin-17-producing T cells in lupus. Curr Opin Rheumatol 22:499–503
[63] Ma J, Yu J, Tao X, Cai L, Wang J, Zheng SG (2010) The imbalance between regulatory and IL-17-secreting CD4(+) T cells in iupus patients. Clin Rheumatol 29:1251–1258
[64] Henriques A, Ines L, Couto M, Pedreiro S, Santos C, Magalhães M, Santos P, Velada I, Almeida A, Carvalheiro T, Laranjeira P, Morgado JM, Pais ML, da Silva JA, Paiva A (2010) Frequency and functional activity of Th17, Tc17 and other T-cell subsets in systemic lupus erythematosus. Cell Immunol 264:97–103
[65] Staudt V, Bothur E, Klein M, Lingnau K, Reuter S, Grebe N, Gerlitzki B, Hoffmann M, Ulges A, Taube C, Dehzad N, Becker M, Stassen M, Steinborn A, Lohoff M, Schild H, Schmitt E,

Bopp T (2010) Interferon-regulatory factor 4 is essential for the developmental program of T helper 9 cells. Immunity 33:192–202

[66] Nowak EC, Noelle RJ (2010) Interleukin-9 as a T helper type 17 cytokine. Immunology 131:169–173

[67] Beriou G, Bradshaw EM, Lozano E, Costantino CM, Hastings WD, Orban T, Elyaman W, Khoury SJ, Kuchroo VK, Baecher-Allan C, Hafler DA (2010) TGF-beta induces IL-9 production from human Th17 cells. J Immunol 185:46–54

[68] Ozaki Y, Ito T, Son Y, Amuro H, Shimamoto K, Sugimoto H, Katashiba Y, Ogata M, Miyamoto R, Murakami N, Amakawa R, Fukuhara S (2010) Decrease of blood dendritic cells and increase of tissue-infiltrating dendritic cells are involved in the induction of Sjögren's syndrome but not in the maintenance. Clin Exp Immunol 159:315–326

[69] Khan S, Greenberg JD, Bhardwaj N (2009) Dendritic cells as targets for therapy in rheumatoid arthritis. Nat Rev Rheumatol 5:566–571

[70] Baldwin HM, Ito-Ihara T, Isaacs JD, Hilkens CM (2010) Tumour necrosis factor alpha blockade impairs dendritic cell survival and function in rheumatoid arthritis. Ann Rheum Dis 69:1200–1207

[71] Marti L, Golmia R, Golmia AP, Paes AT, Guilhen DD, Moreira-Filho CA, Scheinberg M (2009) Alterations in cytokine profile and dendritic cells subsets in peripheral blood of rheumatoid arthritis patients before and after biologic therapy. Ann N Y Acad Sci 1173:334–342

[72] Voynova E, Tchorbanov A, Prechl J, Nikolova M, Baleva M, Erdei A, Vassilev T (2008) An antibody-based construct carrying DNA-mimotope and targeting CR1(CD35) selectively suppresses human autoreactive B-lymphocytes. Immunol Lett 116:168–173

[73] Singh RR, Hahn BH (1998) Reciprocal T-B determinant spreading develops spontaneously in murine lupus: implications for pathogenesis. Immunol Rev 164:201–208

[74] Das UN (1991) Interaction(s) between essential fatty acids, eicosanoids, cytokines, growth factors and free radicals: relevance to new therapeutic strategies in rheumatoid arthritis and other collagen vascular diseases. Prostaglandins Leukot Essent Fatty Acids 44:201–210

[75] Mageed RA, Isenberg DA (2002) Tumour necrosis factor alpha in systemic lupus erythematosus and anti-DNA autoantibody production. Lupus 11:850–855

[76] Mok MY (2010) The immunological basis of B-cell therapy in systemic lupus erythematosus. Int J Rheum Dis 13:3–11

[77] Shah D, Kiran R, Wanchu A, Bhatnagar A (2010) Oxidative stress in systemic lupus erythematosus: relationship to Th1 cytokine and disease activity. Immunol Lett 129:7–12

[78] Pan HF, Tao JH, Ye DQ (2010) Therapeutic potential of IL-27 in systemic lupus erythematosus. Expert Opin Ther Targets 14:479–484

[79] Wozniacka A, Lesiak A, Boncela J, Smolarczyk K, McCauliffe DP, Sysa-Jedrzejowska A (2008) The influence of antimalarial treatment on IL-1beta, IL-6 and TNF-alpha mRNA expression on UVB-irradiated skin in systemic lupus erythematosus. Br J Dermatol 159:1124–1130

[80] Hoi AY, Iskander MN, Morand EF (2007) Macrophage migration inhibitory factor: a therapeutic target across inflammatory diseases. Inflamm Allergy Drug Targets 6:183–190

[81] Graham KL, Lee LY, Higgins JP, Steinman L, Utz PJ, Ho PP (2010) Treatment with a toll-like receptor inhibitory GpG oligonucleotide delays and attenuates lupus nephritis in NZB/W mice. Autoimmunity 43:140–155

[82] Visentini M, Conti V, Cagliuso M, Tinti F, Siciliano G, Trombetta AC, Mitterhofer AP, Fiorilli M, Quinti I (2009) Regression of systemic lupus erythematosus after development of an acquired toll-like receptor signaling defect and antibody deficiency. Arthritis Rheum 60:2767–2771

[83] Wong CK, Wong PT, Tam LS, Li EK, Chen DP, Lam CW (2010) Elevated production of B cell chemokine CXCL13 is correlated with systemic lupus erythematosus disease activity. J Clin Immunol 30:45–52

[84] Das UN (2006) Clinical laboratory tools to diagnose inflammation. Adv Clin Chem 41:189–229

[85] Padma M, Das UN (1996) Effect of cis-unsaturated fatty acids on cellular oxidant stress in macrophage tumor (AK-5) cells in vitro. Cancer Lett 109:63–75
[86] Das UN (1991) Arachidonic acid as a mediator of some of the actions of phorbolmyristate acetate, a tumor promotor and inducer of differentiation. Prostaglandins Leukot Essent Fatty Acids 42:241–244
[87] Das UN, Padma M, Sangeetha P et al (1990) Stimulation of free radical generation in human leukocytes by various stimulants including tumor necrosis factor is a calmodulin dependent process. Biochem Biophys Res Commun 167:1030–1036
[88] Serhan CN (2005) Lipoxins and aspirin-triggered 15-epi-lipoxins are the first lipid mediators of endogenous anti-inflammation and resolution. Prostaglandins Leukot Essent Fatty Acids 73:141–162
[89] Claria J, Serhan CN (1995) Aspirin triggers previously undescribed bioactive eicosanoids by human endothelial cell-leukocyte interactions. Proc Natl Acad Sci U S A 92:9475–9479
[90] Das UN (2005) COX-2 inhibitors and metabolism of essential fatty acids. Med Sci Monit 11:RA233–RA237
[91] Leng RX, Pan HF, Chen GM, Wang C, Qin WZ, Chen LL, Tao JH, Ye DQ (2010) IL-23: a promising therapeutic target for systemic lupus erythematosus. Arch Med Res 41:221–225
[92] Mantovani A, Sozzani S, Introna M (1997) Endothelial activation by cytokines. Ann N Y Acad Sci 832:93–116
[93] Robak E, Sysa-Jeorzejewska A, Dziankowska B, Torzecka D, Chojnowski K, Robak T (1998)Association of interferon gamma, tumor necrosis factor alpha and interleukin-6 serum levels with systemic lupus erythematosus activity. Arch Immunol Ther Exp (Warsz) 46:375–380
[94] Tucci M, Calvani N, Richards HB, Aro CQ, Silvestris F (2005) The interplay of chemokines and dendritic cells in the pathogenesis of lupus nephritis. Ann N Y Acad Sci 1051:421–432
[95] Falk RJ, Jennette JC (1988) Anti-neutrophil cytoplasmic autoantibodies with specificity for myeloperoxidase in patients with systemic vasculitis and idiopathic necrotizing and crescentic glomerulonephritis. N Engl J Med 318:1651–1657
[96] Shen JY, Chen SL, Wu YX, Tao RQ, Gu YY, Bao CD, Wang Q (1999) Pulmonary hypertension in systemic lupus erythematosus. Rheumatol Int 18:147–151
[97] Willeke P, Schluter B, Schotte H, Erren M, Mickholz E, Domschke W, Gaubitz M (2004) Increased frequency of GM-CSF secreting PBMC in patients with active systemic lupus erythematosus can be reduced by immunoadsorption. Lupus 13:57–62
[98] Williams RC Jr, Harmon ME, Burlingame R, Du Clos TW (2005) Studies of serum C-reactive protein in systemic lupus erythematosus. J Rheumatol 32:454–461
[99] Lindsey NJ, Henderson FI, Malia R, Milford-Ward MA, Graves M, Hughes P (1994) Inhibition of prostacyclin release by endothelial binding anticardiolipin antibodies in thrombosis-prone patients with systemic lupus erythematosus and the antiphospholipid syndrome. Br J Rheumatol 33:20–26
[100] Das UN (1995) Beneficial action(s) of eicosapentaenoic acid/docosahexaenoic acid and nitric oxide in systemic lupus erythematosus. Med Sci Res 23:723–726
[101] Mohan IK, Das UN (1997) Oxidant stress, anti-oxidants and essential fatty acids in systemic lupus erythematosus. Prostaglandins Leukot Essent Fatty Acids 56:193–198
[102] Zamora MR, O'Brien RF, Rutherford RF, Weil JV (1990) Serum endothelin-1 concentrations and cold provocation in primary Raynaud's phenomenon. Lancet 336:1144–1147
[103] Reimann J, Diamantstein T (1981) Interleukin-2 allows in vivo induction of anti-erythrocyte autoantibody production in nude mice associated with the injection of rat erythrocytes. Clin Exp Immunol 43:641–644
[104] Schleusner HJ, Lassmann H (1986) Recombinant interleukin 2 (IL-2) promotes T cell line-mediated neuroautoimmune disease. J Neuroimmunol 11:87–91
[105] Kolb H, Zielasek J, Treichel U, Freytag G, Wrann M, Kiesel U (1986) Recombinant interleukin 2 enhances spontaneous insulin-dependent diabetes in BB rats. Eur J Immunol 16:209–212
[106] Folkman J, Klagsbrun M (1987) Angiogenic factors. Science 235:442–449

[107] Redini R, Galera A, Mauriel A, Layan G, Pujol J-P (1988) Transforming growth factor-beta stimulates collagen and glycosaminoglycan biosynthesis in cultured rabbit articular chondrocytes. FEBS Lett 234:172–176
[108] Falanga V, Julien JM (1990) In: Piez KA, Sporn MB (eds) Observations in the potential role of transforming growth factor-beta in cutaneous fibrosis. Ann N Y Acad Sci 593:161–171
[109] Letterio JJ, Roberts AB (1998) Regulation of immune responses by TGF-beta. Annu Rev Immunol 16:137–161
[110] Geiser AG, Letterio JJ, Kulkarni AB, Karlsson S, Roberts AB, Sporn MB (1993) Transforming growth factor-beta$_1$ (TGF-beta$_1$) controls expression of major histocompatibility genes in the postnatal mouse-aberrant histocompatibility antigen expression in the pathogenesis of the TGF-beta$_1$ null mouse phenotype. Proc Natl Acad Sci U S A 90:9944–9948
[111] Dang H, Geiser AG, Letterio JJ, Nakabayashi T, Kong L, Fernandes G, Talal N (1995) SLE-like autoantibodies and Sjogren's syndrome-like lymphoproliferation in TGF-beta knockout mice. J Immunol 155:3205–3212
[112] Ohtsuka K, Gray JD, Stimmler MM, Toro B, Horwitz DA (1998) Decreased production of TGF-beta by lymphocytes from patients with systemic lupus erythematosus. J Immunol 160:2539–2545
[113] Ohtsuka K, Gray JD, Quismorio FP Jr, Lee W, Horwitz DA (1999) Cytokine-mediated down-regulation of B cell activity in SLE: effects of interleukin-2 and transforming growth factor-beta. Lupus 8:95–102
[114] Jackson M, Ahmad Y, Bruce IN, Coupes B, Brenchley PEC (2006) Activation of transforming growth factor-β_1 and early atherosclerosis in systemic lupus erythematosus. Arthritis Res Ther 8:R81
[115] Foster MH, Kelley VR (1999) Lupus nephritis: update on pathogenesis and disease mechanisms. Semin Nephrol 19:173–181
[116] Wong CK, Ho CY, Li EK, Lam CW (2000) Elevation of proinflammatory cytokine (IL-18, IL-17, IL-12) and Th2 cytokine (IL-4) concentrations in patients with systemic lupus erythematosus. Lupus 9:589–593
[117] Horwitz DA, Gray JD, Behrendsen SC et al (1998) Decreased production of interleukin-12 and other Th1 type cytokines in patients with recent onset systemic lupus erythematosus. Arthritis Rheum 41:838–844
[118] Takahashi S, Fossati L, Iwamoto M et al (1996) Imbalance towards Th1 predominance is associated with acceleration of lupus-like autoimmune syndrome in MRL mice. J Clin Invest 97:1597–1604
[119] Peng SL, Szabo SJ, Glimcher LH (2002) T-bet regulates IgG class switching and pathogenic autoantibody production. Proc Natl Acad Sci U S A 99:5545–5550
[120] Akahoshi M, Nakashima H, Tanaka Y et al (1999) Th1/Th2 balance of peripheral T helper cells in systemic lupus erythematosus. Arthritis Rheum 42:1644–1648
[121] Chan RW, Tam LS, Li EK et al (2003) Inflammatory cytokine gene expression in the urinary sediment of lupus nephritis patients. Arthritis Rheum 48:1326–1331
[122] Lighvani AA, Frucht DM, Jankovic D et al (2001) T-bet rapidly induced by interferon-gamma in lymphoid and myeloid cells. Proc Natl Acad Sci U S A 98:15137–15142
[123] Lantelme E, Mantovani S, Palermo B, Campanelli R, Sallusto F, Giachino C (2001) Kinetics of GATA-3 gene expression in early polarizing and committed human T cells. Immunology 102:123–130
[124] Lee HJ, Takemoto N, Kurata H et al (2000) GATA-3 induces T helper cell type 2 (Th2) cytokine expression and chromatin remodeling in committed Th1 cells. J Exp Med 192:105–115
[125] Arbuckle MR, McClain MT, Rubertone MV, Scofield RH, Dennis GJ, James JA, Harley JB (2003) Development of autoantibodies before the clinical onset of systemic lupus erythematosus. N Engl J Med 349:1526–1533
[126] Thompson C, Powrie F (2004) Regulatory T cells. Curr Opin Pharmacol 4:408–414
[127] Kammer GM, Perl A, Richardson BC, Tsokos GC (2002) Abnormal T cell signal transduction in systemic lupus erythematosus. Arthritis Rheum 46:1139–1154

[128] Jiang H, Chess L (2004) An integrated view of suppressor T cell subsets in immunoregulation. J Clin Invest 114:1198–1208
[129] Nelson BH (2004) IL-2, regulatory T cells, and tolerance. J Immunol 172:3983–3988
[130] Malek TR, Bayer AL (2004) Tolerance, not immunity, crucially depends on IL-2. Nat Rev Immunol 4:665–674
[131] Lee JH, Wang LC, Lin YT, Yang YH, Lin DT, Chiang BL (2006) Inverse correlation between CD4+ regulatory T-cell population and autoantibody levels in paediatric patients with systemic lupus erythematosus. Immunology 117:280–286
[132] Valencia X, Yarboro C, Illei G, Lipsky PE (2007) Deficient CD4+CD25high T regulatory cell function in patients with active systemic lupus erythematosus. J Immunol 178:2579–2588
[133] Lyssuk EY, Torgashina AV, Soloviev SK, Nassaonov EL, Bykovskaia SN (2007) Reduced number and function of CD4+CD25highFoxP3+ regulatory T cells in patients with systemic lupus erythematosus. Adv Exp Med Biol 601:113–119
[134] Lin SC, Chen KH, Lin CH, Kuo CC, Ling OD, Chan CH (2007) The quantitative analysis of peripheral blood FOXP3-expressing T cells in systemic lupus erythematosus and rheumatoid arthritis patients. Eur J Clin Invest 37:987–996
[135] Shi YX, Zhang XS, Liu DG, Li YQ, Guan ZZ, Jiang WQ (2004) CD4+CD25+T regulatory cells in peripheral blood of B-NHL patients with or without chemotherapy. Ai Zheng 23:597–601
[136] Zhang Q, Qian FH, Liu H, Huang M, Zhang XL, Yin KS (2008) Expression of surface markers on peripheral CD4+CD25high T cells in patients with atopic asthma: role of inhaled corticosteroid. Chin Med J (Engl) 121:205–212
[137] Kawai M, Kitade H, Mathieu C, Waer M, Pirenne J (2005) Inhibitory and stimulatory effects of cyclosporine A on the development of regulatory T cells in vivo. Transplantation 79:1073–1077
[138] Schwarz T (2008) 25 years of UV-induced immunosuppression mediated by T cells-from disregarded T suppressor cells to highly respected regulatory T cells. Photochem Photobiol 84:10–18
[139] Kuhn A, Krammer PH, Kolb-Bachofen V (2006) Pathophysiology of cutaneous lupus erythematosus-novel aspects. Rheumatology 45:iii14–iii16
[140] Kuhn A, Beissert S (2005) Photosensitivity in lupus erythematosus. Autoimmun 38:519–529
[141] Miyara M, Amoura Z, Parizot C et al (2005) Global natural regulatory T cell depletion in active systemic lupus erythematosus. J Immunol 175:8392–8400
[142] Hart PH, Townley SL, Grimbaldeston MA, Khalil Z, Finlay-Jones JJ (2002) Mast cells, neuropeptides, histamine, and prostaglandins in UV-induced systemic immunosuppression. Methods 28:79–89
[143] Zhang Y, Ramos BF, Jakschik BA (1991) Augmentation of reverse arthus reaction by mast cells in mice. J Clin Invest 88:841–846
[144] Danilewicz M, Wagrrowska-Danilewicz M (2001) Quantitative analysis of interstitial mast cells in lupus and non-lupus membranous glomerulopathy. Pol J Pathol 52:211–217
[145] Eklund KK (2007) Mast cells in the pathogenesis of rheumatic diseases and as potential targets for anti-rheumatic therapy. Immunol Rev 217:38–52
[146] Woolley DE, Tetlow LC (2000) Mast cell activation and its relation to proinflammatory cytokine production in the rheumatoid lesion. Arthritis Res 2:65–74
[147] Shin K, Nigrovic PA, Crish J, Boilard E, McNeil HP, Larabee KS, Adachi R, Gurish MF, Gobezie R, Stevens RL, Lee DM (2009) Mast cells contribute to autoimmune inflammatory arthritis via their tryptase/heparin complexes. J Immunol 182:647–656
[148] Bonventre JV (1992) Phospholipase A2 and signal transduction. J Am Soc Nephrol 3:128–150
[149] Levy BD, Clish CB, Schmidt B, Gronert K, Serhan CN (2001) Lipid mediator class switching during acute inflammation: signals in resolution. Nat Immunol 2:612–619
[150] Godson C, Mitchell S, Harvey K, Petasis NA, Hogg N, Brady HR (2000) Cutting edge: lipoxins rapidly stimulate nonphlogistic phagocytosis of apoptotic neutrophils by monocyte-derived macrophages. J Immunol 164:1663–1667

[151] Gilroy DW Colville-Nash PR, Willis D, Chivers J, Paul-Clark MJ, Willoughby DA (1999) Inducible cyclooxygenase may have anti-inflammatory properties. Nat Med 5:698–701

[152] Gilroy DW, Newson J, Sawmynaden P, Willoughby DA, Croxtall JD (2004) A novel role for phospholipase A_2 isoforms in the checkpoint control of acute inflammation. FASEB J 18:489–498

[153] Serhan CN, Hong S, Gronert K, Colgan SP, Devchand PR, Mirick G, Moussignac R-L (2002) Resolvins: a family of bioactive products of omega-3 fatty acid transformation circuits initiated by aspirin treatment that counter proinflammation signals. J Exp Med 196:1025–1037

[154] Croxtall JD, Choudhury Q, Tokumoto H, Flower RJ (1995) Lipocortin-1 and the control of arachidonic acid release in cell signalling. Glucocorticoids inhibit G protein-dependent activation of cPLA2 activity. Biochem Pharmacol 50:465–474

[155] Nakano T, Ohara O, Teraoka H, Arira H (1990) Glucocorticoids suppress group II phospholipase A2 production by blocking mRNA synthesis and post-transcriptional expression. J Biol Chem 205:12745–12748

[156] Cominelli F, Nast CC, Llerena R, Dinarello CA, Zipser RD (1990) Interleukin 1 suppresses inflammation in rabbit colitis. Mediation by endogenous prostaglandins. J Clin Invest 85:582–586

[157] Schwab JH, Anderle SK, Brown RR, Dalldorf FG, Thompson RC (1991) Pro- and anti-inflammatory roles of interleukin-1 in recurrence of bacterial cell wall-induced arthritis in rats. Infect Immun 59:4436–4442

[158] Ichinose M, Hara N, Sawada M, Maeno T (1991) A flow cytometric assay reveals an enhancement of phagocytosis by platelet activating factor in murine peritoneal macrophages. Cell Immunol 156:508–518

[159] Takano T, Panesar M, Papillon J, Cybulsky AV (2000) Cyclooxygenases-1 and 2 couple to cytosolic but not group IIA phospholipase A2 in COS-1 cells. Prostaglandins Other Lipid Mediat 60:15–26

[160] Reddy ST, Herschman HR (1997) Prostaglandin synthase-1 and prostaglandin synthase-2 are coupled to distinct phospholipases for the generation of prostaglandin D2 in activated mast cells. J Biol Chem 272:3231–3237

[161] Munck A, Guyre PM, Holbrook NJ (1984) Physiological function of glucocorticoids in stress and their relation to pharmacological actions. Endocr Rev 5:25–44

[162] Walev I, Klein J, Husmann M, Valeva A, Strauch S, Wirtz H, Weichel O, Bhakdi S (2000) Potassium regulates IL-1 beta processing via calcium-independent phospholipase A2. J Immunol 164:5120–5124

[163] Serhan CN, Clish CB, Brannon J, Colgan SP, Chiang N, Gronert K (2000) Novel functional sets of lipid-derived mediators with antiinflammatory actions generated from omega-3 fatty acids via cyclooxygenase 2-nonsteroidal antiinflammatory drugs and transcellular processing. J Exp Med 192:1197–1204

[164] Wu C-C, Croxtall JD, Perretti M, Bryant CE, Thiemermann C, Flower RJ, Vane JR (1995) Lipocortin 1 mediates the inhibition by dexamethasone of the induction by endotoxin of nitric oxide synthase in the rat. Proc Natl Acad Sci U S A 92:3473–3477

[165] Lasa M, Abraham SM, Boucheron C, Saklatvala J, Clark AR (2002) Dexamethasone causes sustained expression of mitogen-activated protein kinase (MAPK) phosphatase 1 and phosphatase-mediated inhibition of MAPK p38. Mol Cell Biol 22:7802–7811

[166] Crafford LJ, Wilder RL, Ristimaki AP, Sano M, Remmers EF, Epps HR, Hla T (1994) Cyclooxygenase-1 and –2 expression in rheumatoid arthritis synovial tissues. J Clin Invest 93:109–1101

[167] Radomski MW, Palmer RMJ, Moncada S (1990) Glucocorticoids inhibit the expression of an inducible, but not the constitutive, nitric oxide synthase in vascular endothelial cells. Proc Natl Acad Sci U S A 87:10043–10047

[168] Hoeck WG, Ramesha CS, Chang DJ, Fan N, Heller RA (1993) Cytoplasmic phospholipase A_2 activity and gene expression are stimulated by tumor necrosis factor: dexamethasone blocks the inducible synthesis. Proc Natl Acad Sci U S A 90:4475–4479

[169] Kunicka JE, Talle MA, Denhardt GH, Brown M, Prince LA, Goldstein G (1993) Immunosuppression by glucocorticoids: inhibition of production of multiple lymphokines by in vivo administration of dexamethasone. Cell Immunol 149:39–49
[170] Calandra T, Bernhagen J, Metz CN, Spiegel LA, Bacher M, Donnelly T, Cerami A, Bucala R (1995) MIF as a glucocorticoid-induced modulator of cytokine production. Nature 377:68–71
[171] Salvemini D, Seibert K, Masferrer JL, Misko TP, Currie MG, Needleman P (1994) Endogenous nitric oxide enhances prostaglandin production in a model of renal inflammation. J Clin Invest 93:1940–1947
[172] Zaitsu M, Hamasaki Y, Tsuji K, Matsuo M, Fujita I, Aoki Y, Ishii E, Kohashi O (2003) Dexamethasone accelerates catabolism of leukotriene C_4 in bronchial epithelial cells. Eur Respir J 22:35–42
[173] Dworski R, Fitzgerald GA, Oates JA, Sheller JR (1994) Effect of oral prednisone on airway inflammatory mediators in atopic asthma. Am J Respir Crit Care Med 149(4 Pt 1):953–959
[174] Ferrante JV, Ferrante A (2005) Novel role of lipoxygenases in the inflammatory response: promotion of TNF mRNA decay by 15-hydroperoxyeicosatetraenoic acid in a monocytic cell line. J Immunol 174:3169–3172
[175] Ariel A, Chiang N, Arita M, Petasis NA, Serhan CN (2003) Aspirin-triggered lipoxin A4 and B4 analogs block extracellular signal-regulated kinase-dependent TNF-alpha secretion from human T cells. J Immunol 170:6266–6272
[176] Wu SH, Lu C, Dong L, Zhou GP, He ZG, Chen ZQ (2005) Lipoxin A4 inhibits TNF-alpha-induced production of interleukins and proliferation of rat mesangial cells. Kidney Int 68:35–46
[177] Hayakawa M, Ishida N, Takeuchi K, Shibamoto S, Hori T, Oku N, Ito F, Tsujimoto M (1993) Arachidonic acid-selective cytosolic phospholipase A_2 is crucial in the cytotoxic action of tumor necrosis factor. J Biol Chem 268:11290–11295
[178] Das UN (2002) A perinatal strategy for preventing adult disease: the role of long-chain polyunsaturated fatty acids. Kluwer Academic, Boston, MA
[179] Huang YS, Drummond R, Horrobin DF (1987) Protective effect of gamma-linolenic acid on aspirin induced gastric hemorrhage in rats. Digestion 36:36–41
[180] Manjari V, Das UN (2000) Effect of polyunsaturated fatty acids on dexamethasone-induced gastric mucosal damage. Prostaglandins Leukot Essent Fatty Acids 62:85–96
[181] Das UN (2008) Essential fatty acids and their metabolites could function as endogenous HMG-CoA reductase and ACE enzyme inhibitors, anti-arrhythmic, anti hypertensive, anti-atherosclerotic, anti-inflammatory, cytoprotective, and cardioprotective molecules. Lipids Health Dis 7:37
[182] Das UN (2008) Can essential fatty acids reduce the burden of disease(s)? Lipids Health Dis 7:9
[183] Das UN (2006) Essential fatty acids – a review. Curr Pharm Biotechnol 7:467–482
[184] Mohan IK, Das UN (2001) Prevention of chemically induced diabetes mellitus in experimental animals by polyunsaturated fatty acids. Nutrition 17:126–151
[185] Suresh Y, Das UN (2003) Long-chain polyunsaturated fatty acids and chemically-induced diabetes mellitus: effect of ω-6 fatty acids. Nutrition 19:93–114
[186] Suresh Y, Das UN (2003) Long-chain polyunsaturated fatty acids and chemically-induced diabetes mellitus: effect of ω-3 fatty acids. Nutrition 19:213–228
[187] Suresh Y, Das UN (2006) Differential effect of saturated, monounsaturated, and polyunsaturated fatty acids on alloxan-induced diabetes mellitus. Prostaglandins Leukot Essent Fatty Acids 74:199–213
[188] Das UN, Mohan IK, Raju TR (2001) Effect of corticosteroids and eicosapentaenoic acid/docosahexaenoic acid on pro-oxidant and anti-oxidant status and metabolism of essential fatty acids in patients with glomerular disorders. Prostaglandins Leukot Essent Fatty Acids 65:197–203
[189] Das UN (2006) Essential fatty acids: biochemistry, physiology, and pathology. Biotechnology J 1:420–439

[190] Kambe T, Murakami M, Kudo I (1999) Polyunsaturated fatty acids potentiate interleukin-1-stimulated arachidonic acid release by cells overexpressing type IIA secretory phospholipase A2. FEBS Lett 453:81–84

[191] Das UN (2006) Can perinatal supplementation of long-chain polyunsaturated fatty acids prevent atopy, bronchial asthma and other inflammatory conditions? Med Sci Monit 12:RA99–RA111

[192] Belmont HM, Levartovsky D, Goel A, Amin A, Giorno R, Rediske J, Skovron ML, Abramson SB (1997) Increased nitric oxide production accompanied by the up-regulation of inducible nitric oxide synthase in vascular endothelium from patients with systemic lupus erythematosus. Arthritis Rheum 40:1810–1816

[193] Wanchu A, Khuller M, Deodhar SD, Bambery P, Sud A (1998) Nitric oxide synthesis is increased in patients with systemic lupus erythematosus. Rheumatol Int 18:41–43

[194] Oates JC, Christensen EF, Reilly CM, Self SE, Gilkenson GS (1999) Prospective measure of serum 3-nitrotyrosine levels in systemic lupus erythematosus: correlation with disease activity. Proc Assoc Am Physicians 111:611–621

[195] Svollenhoven R, Khademi M, Tarkowski A, Greitz D, Dahlstrom M, Lundberg I, Klareskog L, Olsson T (2001) Increased levels of proinflammatory cytokines and nitric oxide metabolites in neuropsychiatric lupus erythematosus. Ann Rheum Dis 60:372–379

[196] Gilkeson G, Cannon C, Oates J, Reilly C, Goldman D, Petri M (1999) Correlation of serum measures of nitric oxide production with lupus disease activity. J Rheumatol 26:318–324

[197] Yu CC, Yang CW, Wu MS, Ko YC, Huang CT, Hong JJ, Huang CC (2001) Mycophenolate mofetil reduces renal cortical inducible nitric oxide synthase mRNA expression and diminishes glomerulosclerosis in MRL/lpr mice. J Lab Clin Med 138:69–77

[198] Reilly CM, Oates JC, Sudian J, Crosby MB, Halushka PV, Gilkeson GS (2001) Prostaglandin J(2) inhibition of mesangial cell iNOS expression. Clin Immunol 98:337–345

[199] Habib S, Moinuddin, Ali R (2006) Peroxynitrite-modified DNA: a better antigen for systemic lupus erythematosus anti-DNA autoantibodies. Biotechnol Appl Biochem 43(Pt 2):65–70

[200] Robak E, Sysa-Jeorzejewska A, Dziankowska B, Torzecka D, Chojnowski K, Robak T (1998) Association of interferon gamma, tumor necrosis factor alpha and interleukin-6 serum levels with systemic lupus erythematosus activity. Arch Immunol Ther Exp (Warsz) 46:375–380

[201] Yoshizumi M, Kunihara H, Morita T et al (1990) Interleukin-1 increases the production of endothelin-1 by cultured endothelial cells. Biochem Biophys Res Commun 166:324–329

[202] Das UN (1993) Beneficial effect of L-arginine in collagen vascular diseases: a role for nitric oxide. Nutrition 9:277–278

[203] Fries R, Shariat K, von Wilmowsky H, Böhm M (2005) Sildenafil in the treatment of Raynaud's phenomenon resistant to vasodilatory therapy. Circulation 112:2980–2985

[204] Mahler F, Baumgartner I (2005) More potential for sildenafil than potency. Circulation 112:2894–2895

[205] Tsoukas CD, Watry D, Escobar SS, Provvedini DM, Dinarello CA, Hustmyer FG, Manolagas SC (1989) Inhibition of interleukin-1 production by 1,25-dihydroxyvitamin D_3. J Clin Endocrinol Metab 127–133

[206] Lemire JM, Adams JS, Kermani-Arab V, Bakke AC, Sakai R, Jordan SC (1985) 1,25-Dihydroxyvitamin D3 suppresses human T helper/inducer lymphocyte activity in vitro. J Immunol 134:3032–3035

[207] Haq AU (1986) 1,25-Dihydroxyvitamin D3 (calcitriol) suppresses IL-2 induced murine thymocyte proliferation. Thymus 8:295–306

[208] Merino F, Alvarez-Mon M, de la Hera A, Alés JE, Bonilla F, Durantez A (1989) Regulation of natural killer cytotoxicity by 1,25-dihydroxyvitamin D3. Cell Immunol 118:328–336

[209] Pichler J, Gerstmayr M, Szépfalusi Z, Urbanek R, Peterlik M, Willheim M (2002) 1 alpha,25(OH)2D3 inhibits not only Th1 but also Th2 differentiation in human cord blood T cells. Pediatr Res 52:12–18

[210] Tang J, Zhou R, Luger D, Zhu W, Silver PB, Grajewski RS, Su SB, Chan CC, Adorini L, Caspi RR (2009) Calcitriol suppresses antiretinal autoimmunity through inhibitory effects on the Th17 effector response. J Immunol 182:4624–4632

[211] Huang LW, Chang KL, Chen CJ, Liu HW (2001) Arginase levels are increased in patients with rheumatoid arthritis. Kaohsiung J Med Sci 17:358–363
[212] Bultink IE, Teerlink T, Heijst JA, Dijkmans BA, Voskuyl AE (2005) Raised plasma levels of asymmetric dimethylarginine are associated with cardiovascular events, disease activity, and organ damage in patients with systemic lupus erythematosus. Ann Rheum Dis 64:1362–1365
[213] Kiani AN, Mahoney JA, Petri M (2007) Asymmetric dimethylarginine is a marker of poor prognosis and coronary calcium in systemic lupus erythematosus. J Rheumatol 34:1502–1505
[214] Maas R, Dentz L, Schwedhelm E, Thoms W, Kuss O, Hiltmeyer N, Haddad M, Klöss T, Standl T, Böger RH (2007) Elevated plasma concentrations of the endogenous nitric oxide synthase inhibitor asymmetric dimethylarginine predict adverse events in patients undergoing noncardiac surgery. Crit Care Med 35:1876–1881
[215] Svenungsson E, Cederholm A, Jensen-Urstad K, Fei GZ, de Faire U, Frostegard J (2008) Endothelial function and markers of endothelial activation in relation to cardiovascular disease in systemic lupus erythematosus. Scand J Rheumatol 37:352–359
[216] Gustafsson J, Gunnarsson I, Börjesson O, Pettersson S, Möller S, Fei GZ, Elvin K, Simard JF, Hansson LO, Lundberg IE, Larsson A, Svenungsson E (2009) Predictors of the first cardiovascular event in patients with systemic lupus erythematosus-a prospective cohort study. Arthritis Res Ther 11:R186
[217] Das UN (2008) Albumin infusion therapy in stroke, sepsis and the critically ill. Curr Nutr Food Sci 4:217–226
[218] Das UN (2006) Pyruvate is an endogenous anti-inflammatory and anti-oxidant molecule. Med Sci Monit 12:RA79–RA84
[219] Das UN (2006) Is pyruvate an endogenous anti-inflammatory molecule? Nutrition 22:965–972
[220] Das UN (2007) Ethyl pyruvate in sepsis. Adv Sepsis 6:10–15
[221] Das UN (2006) Glucose, insulin, and coronary heart disease. Eur Heart J 27:2141–2142
[222] Das UN (2000) Possible beneficial action(s) of glucose-insulin-potassium regimen in acute myocardial infarction and inflammatory conditions: a hypothesis. Diabetologia 43:1081–1082
[223] Das UN (2001) Can glucose-insulin-potassium regimen suppress inflammatory bowel disease? Med Hypotheses 57:183–185
[224] Das UN (2001) Hypothesis: can glucose-insulin-potassium regimen in combination with polyunsaturated fatty acids suppress lupus and other inflammatory diseases? Prostaglandins Leukot Essent Fatty Acids 65:109–113
[225] Das UN (2010) Metabolic syndrome is a low-grade systemic inflammatory condition. Expert Rev Endocrinol Metab 4:577–592

Chapter 14
Cancer

The World Cancer Report suggested that action on smoking, diet and infections can prevent one third of cancers and another third can be cured [1].

Cancer rates could increase by 50% to 15 million new cases in the year 2020, according to the World Cancer Report, the most comprehensive global examination of the disease to date. However, the report also provides clear evidence that healthy lifestyles and public health action by governments and health practitioners could stem this trend, and prevent as many as one third of cancers worldwide.

In the year 2000, malignant tumours were responsible for 12% of the nearly 56 million deaths worldwide from all causes. In many countries, more than a quarter of deaths are attributable to cancer. In 2000, 5.3 million men and 4.7 million women developed a malignant tumour and altogether 6.2 million died from the disease. The report also revealed that cancer is a major public health problem in developing countries, matching its effect in industrialized nations.

The World Cancer Report is a concise manual describing the global burden, the causes of cancer, major types of malignancies, early detection and treatment. The 351-page global report is issued by IARC, which is part of the World Health Organization (WHO).

It has been suggested that examples of areas where action can make a difference to stemming the increase of cancer rates and preventing a third of cases are:

- Reduction of tobacco consumption. It remains the most important avoidable cancer risk. In the twentieth century, approximately 100 million people died world-wide from tobacco-associated diseases
- A healthy lifestyle and diet can help. Frequent consumption of fruit and vegetables and physical activity can make a difference.
- Early detection through screening, particularly for cervical and breast cancers, allow for prevention and successful cure.

The predicted sharp increase in new cases—from 10 million new cases globally in 2000, to 15 million in 2020—will mainly be due to steadily ageing populations in both developed and developing countries and also to current trends in smoking prevalence and the growing adoption of unhealthy lifestyles.

U. N. Das, *Molecular Basis of Health and Disease*,

DOI 10.1007/978-94-007-0495-4_14,

Tobacco and Cancer

Tobacco consumption remains the most important avoidable cancer risk. In the twentieth century, approximately 100 million people died world-wide from tobacco-associated diseases (cancer, chronic lung disease, cardiovascular disease and stroke).

The lung cancer risk for regular smokers as compared to non-smokers (relative risk, RR) is between 20 and 30-fold. In countries with a high smoking prevalence and where many women have smoked cigarettes throughout adult life, roughly 90% of lung cancers in both men and women are attributable to cigarette smoking. For bladder and renal pelvis, the RR is five-six but this means that more than 50% of cases are caused by smoking.

The RR for cancers of the oral cavity, pharynx, larynx and squamous cell carcinoma of the esophagus is greater than six, and three-four for carcinomas of the pancreas. These risk estimates are higher than previously estimated and unfortunately, additional cancer sites with a RR of two-three have been identified as being associated with tobacco smoking, including cancers of the stomach, liver, uterine cervix, kidney (renal cell carcinoma) nasal cavities and sinuses, esophagus (adenocarcinoma) and myeloid leukemia.

Involuntary (passive) tobacco smoke is carcinogenic and may increase the lung cancer risk by 20%.

While it is best never to start smoking, epidemiological evidence supports the enormous benefits of cessation. The greatest reduction in the number of cancer deaths within the next several decades will be due to those who stop the habit. The greatest effect results from stopping smoking in the early 30 s, but a very impressive risk reduction of more than 60% is obtained even when the habit is quit after the age of 50 years.

Infection and Cancer

In developing countries, up to 23% of malignancies are caused by infectious agents, including hepatitis B and C virus (liver cancer), human papillomaviruses (cervical and ano-genital cancers), and Helicobacter pylori (stomach cancer). In developed countries, cancers caused by chronic infections only amount to approximately 8% of all malignancies. This discrepancy is particularly evident for cervical cancer. In developed countries with an excellent public health infrastructure and a high compliance of women, early cytological detection of cervical cancer (PAP smear) has led to an impressive reduction of mortality while in other world regions, including Central America, South East Africa and India, incidence and mortality rates are still very high. Today, more than 80% of all cervical cancer deaths occur in developing countries.

In the gastro-intestinal tract (GIT), any chronic tissue damage with necrosis and regeneration carries an increased cancer risk, e.g., consumption of very hot beverages (squamous cell carcinoma of the esophagus), gastro-oesophageal reflux (adenocarcinoma of the esophagus), chronic gastritis induced by H. pylori infection (stomach cancer), Crohn's disease (cancer of the small intestines) and ulcerative colitis (colon cancer).

Tobacco and Inflammation

Tobacco smoke is carcinogenic possibly, by producing chronic inflammation. Tobacco smoke exposure of mice produces interstitial granulomatous inflammation similar to Langerhans cell granulomatosis in humans [2]. After ceasing exposure to tobacco smoke, the density of pulmonary Langerhans cells returned to that of the control level; interstitial granulomatous lesions disappeared, but the bronchial epithelial metaplasia did not reverse, suggesting the chronic nature of tobacco-induced changes in the target tissues. The concentration of carcinoembryonic antigen (CEA), known marker of malignant transformation and chronic inflammation, is increased in bronchoalveolar lavage fluid obtained from smokers compared with fluid from non-smokers. Consistent with this proposal, it was noted that mRNA and protein expression of CEA were increased in the normal lung tissue from smokers compared with non-smokers or ex-smokers. In vitro studies showed that cigarette smoke could induce CEA mRNA expression in fetal lung derived cells, indicating that CEA might play a part in recruitment of neutrophils into the lower respiratory tract [3] and thus, neutrophil-derived free radical-induced tissue damage could play a role in tobacco-induced lung cancer development.

It was reported that the bronchitis index scores were significantly higher in tobacco smokers (TS) than in nonsmokers. Mucosal biopsies were positive for the presence of vascular hyperplasia, submucosal edema, inflammatory cell infiltrates, and goblet cell hyperplasia compared to non-smokers. In addition, neutrophil counts and interleukin-8 (IL-8) concentrations were significantly higher in bronchial lavage fluid in smokers compared with non-smokers suggesting that smoking causes significant airway inflammation [4]. In this context, it is interesting to note that cigarette smoking promoted inflammation-associated adenoma/adenocarcinoma formation in the mouse colon in a dose-dependent manner that was found to be associated with the inhibition of cellular apoptosis and supported by increased angiogenesis [5]. Cigarette smoking-induced inflammation-associated adenoma formation in the mouse colon was found to enhance the 5-LOX protein expression in the inflammation-associated colonic adenomas accompanied with an up-regulation of matrix metalloproteinase-2 (MMP-2) and vascular endothelial growth factor (VEGF). These results suggest that cigarette smoke induces 5-LOX expression and thus, promotes angiogenic process and inflammation-associated adenoma formation in mice [6]. In support of this, it was noted that 5-LOX inhibitor is more effective than COX-2 inhibitor, and blocker of both COX-2 and 5-LOX showed a superior anticancer profile in cigarette smokers [7]. These results suggest that inflammation induced by COX and LOX enzyme dependent products play a significant role in cigarette smoke-induced cancer. This is supported by the observation that smoking triggered inflammation could be responsible for pancreatic inflammation and cancer [8, 9]. Furthermore, cigarette smoke exposed experimental animals had impaired acetylcholine-induced relaxations of carotid arteries that were improved by the NADPH oxidase inhibitor; showed significantly increased vascular $O_2^{-\cdot}$ production both in endothelial and smooth muscle cells, enhanced vascular H_2O_2 production and vascular mRNA expression of the proinflammatory cytokines IL-1β, IL-6, and TNF-α and that of inducible nitric oxide synthase was observed. In cultured endothelial cells, cigarette smoke exposure elicited NF-kB activation and increased monocyte adhesiveness that was prevented by NADPH oxidase inhibitor and catalase [10]. Thus cigarette smoke induced significant proinflammatory actions that are likely to promote development of cancer and atherosclerosis.

These evidences suggest a complex interplay occurs between cigarette smoke, inflammation and host immune cells during neoplastic development. Both environmental and genetic factors enhance the risk of cigarette smoking, Helicobacter pylori, hepatitis B/C, human papilloma virus, solar irradiation, asbestos, pancreatitis, or other causes of chronic inflammation and their associated cancer [11]. These results imply that it is not only the induction of inflammatory events that lead to the development of cancer but it could equally be the failure of anti-inflammatory events that protect against these pro-inflammatory insults that lead to the onset of cancer. In other words, it can be said that there is inappropriate inflammation that leads to the development of cancer. When inflammation is severe it leads to the death of target cells/tissue due to the release of excess free radicals, whereas subacute or chronic and persistent inflammation as is seen with chronic smoking, hepatitis B and C viral infections cancer develops. In the latter instance, generation of free radicals due to chronic persistent inflammation are sufficient to turn a normal cell to cancer cell by inducing DNA damage but not sufficient to induce apoptosis of the cell. This implies that if adequate amount of free radicals are generated in tumor cells, it could lead to their death and if free radical damage to normal cells is prevented the conversion of normal to tumor cell can be halted [12, 13].

Thus, controlled generation of free radicals specifically in the tumor cells but not in the normal cells could be a strategy to eliminate cancer cells. This is evident from the fact that currently available drugs and radiation produce free radicals [12–16] and thus, bring about their tumoricidal action. But, these free radicals are also responsible for the adverse effects seen with these agents.

Inflammation of Chronic Infections and Cancer are due to TNF-α and IL-1

This dual role of inflammation both in infection and cancer is evident from the involvement of TNF-α in both these processes. For example, cachexia characterized by anorexia, weight loss, and protein wasting that frequently complicates both chronic inflammation such as tuberculosis and cancer. TNF-α, a humoral mediator of cachexia, stimulates the breakdown of energy stores from adipocytes and myocytes in vitro, and when sublethal doses of recombinant human TNF-α administered twice daily for 7–10 days caused cachexia in rats, as evidenced by reduced food intake, weight loss, depletion of whole-body lipid and protein stores, and anemia as a result of decreased red blood cell mass were seen. Leukocytosis and histopathological evidence of tissue injury and inflammation were observed in several organs, including omentum, liver, spleen, and heart, suggesting that the exposure of the normal host to TNF-α is capable of inducing a pathophysiological syndrome of cachexia, anemia, and inflammation similar to that observed during inflammatory states or malignancy [17].

Similar to TNF-α, IL-1 also mediates several components of both acute and chronic pathological processes observed in patients with cancer and chronic infection such as cachexia. A single injection of recombinant IL-1β or IL-1α induced a 40% reduction in food intake in experimental animals, whereas daily injections slowed normal weight gain. The anorexic response to IL-1 could be prevented by cyclooxygenase inhibitors, suggesting a role for eicosanoids. On the other hand, reduced production of cyclooxygenase products such as PGE_2 induced by feeding experimental animals with ω-3 (also referred to as N-3) PUFAs was associated with a decreased anorexic response to IL-1. Thus, one mechanism by which IL-1 induces anorexia appears to require cyclooxygenase metabolites, such as PGE_2. N-3 fatty acids also reduced the severity of host responses to inflammation and infection that is due to decreased cyclooxygenase products and also due to reduced synthesis of IL-1. These data are supported by the fact that leukocytes from human subjects taking oral N-3 PUFAs produced 60% less IL-1 and the ability of N-3 fatty acids to reduce IL-1 synthesis appears to be via the lipoxygenase pathway [18, 19]. I and my colleagues observed that in tumor bearing rats, n-PUFAs improved food intake; restored normal eating pattern, delayed onset of anorexia, tumor appearance, and growth; and prevented body weight loss [20]. Furthermore, tumor resection and n-PUFAs were found to modify hypothalamic food intake activity by up-regulating NPY and down-regulating α-MSH and 5-HT(1B)-receptors. Tumor resection in anorexic rats on chow diet restored hypothalamic NPY, α-MSH, and food intake quantitatively more than in rats fed n-3 PUFA enriched diet. These results suggest that products produced by the tumor cells and n-PUFAs are able to act on the brain to restore hypothalamic peptides and neurotransmitters to normal. Since, both COX and LOX products seem to have a role in the anorectic actions induced by TNF-α and IL-1 and as inhibition of formation of PGE_2 restored food intake and body weight only partially, it is likely that certain other products of n-3 PUFAs may also be involved in the anorectic actions of cytokines. I propose that both TNF-α and IL-1 are able to inhibit the formation of lipoxins, resolvins and protectins and thus bring about some of their adverse actions. Thus, it is likely that IL-1 and TNF-α enhance PGE_2 formation and inhibit the synthesis of PGD_2, lipoxins, resolvins and protectins (see Fig. 14.1).

It is rather interesting to note that both IL-1 and TNF-α augment free radical generation [21–26], while lipoxins, resolvins and protectins inhibit cytokine-induced free radical generation [27]. Hence, it is likely that the ability of n-3 PUFAs to prevent/reverse anorexia and cachexia, inhibition of tumor growth, up-regulation of

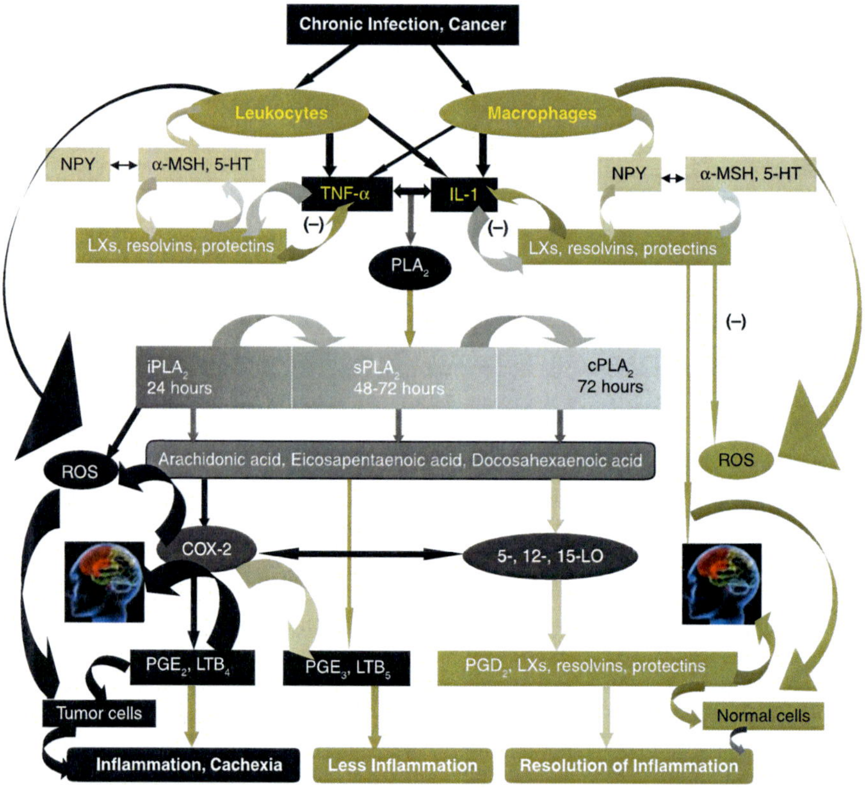

Fig. 14.1 Scheme showing the role of cytokines, hypothalamic neurotransmitters, eicosanoids and lipoxins in inflammation and cachexia of chronic infections such as tuberculosis and cancer

NPY and down-regulation of α-MSH and 5-HT(1B)-receptors, inhibition of IL-1 and TNF-α could be attributed to enhanced formation of lipoxins, resolvins and protectins and suppression of IL-1 and TNF-α-induced free radical generation (see Fig. 14.1).

Glucose Sensing by Neuronal and Tumor Cells and Its Relationship to ATP-Sensitive K^+ Channels and ROS

In this context, it is interesting to note that regulation of ATP-sensitive K^+ channels is a common pathway by which nutrients such as glucose and other factors modulate hypothalamic neuronal sensing of fuels. This so since, a primary increase in hypothalamic glucose levels lowers blood glucose through inhibition of glucose production and this effect of glucose requires its conversion to lactate followed by stimulation of pyruvate metabolism, which activates ATP-sensitive K^+ channels [28]. Incidentally, pyruvate has anti-oxidant and anti-inflammatory actions (pyruvate inhibits NF-κB activation, TNF-α, IL-6, MIF, and HMGB1 production) and is an insulin secretagogue [29, 30]. This suggests that glucose and pyruvate influence glucose sensing by neurons possibly through a free radical dependent process.

There is also evidence to indicate that hypothalamic arcuate neurons may function independent of ATP level [31, 32] that could involve free radicals. For example, transient increase in glucose metabolism generates NADH and $FADH_2$ from the mitochondria and their use increases superoxide anion production (also called as mitochondrial reactive oxygen species, mROS). Hypothalamic slices *ex vivo* exposed to 5–20 mmol/l glucose generated ROS and glucose-induced increased neuronal activity in arcuate nucleus and insulin release are suppressed by antioxidants, implying that the brain glucose-sensing mechanism involves ROS signaling [33]. This is supported by the observation that ATP-sensitive K^+ channels control transmitter release in dorsal striatum through an H_2O_2-dependent mechanism [34]. Furthermore, insulin (10^{-10}–10^{-7} mol/l) caused a dose- and time-dependent (5–90 min) stimulation of H_2O_2 release by human polymorphonuclear leukocytes [35]. Thus, glucose not only induces glucose uptake by cells but also stimulates free radical generation in the same cells.

ATP-sensitive K^+ channels that regulate insulin release, vascular tone, and neuronal excitability are also influenced by NO in an indirect fashion that requires Ras and mitogen-activated protein kinase (MEK) kinase activities. Inhibition of mitogen-activated protein kinase (MEK) kinase abolished the NO activation of ATP-sensitive K^+ channels. The NO precursor L-arginine also stimulated ATP-sensitive K^+ channels via endogenous NO synthase and the Ras signaling pathway. In addition, in rat hippocampal neurons, the protective effect of ischemic preconditioning induced by oxygen-glucose deprivation required ATP-sensitive K^+ channels and NO synthase activity during preconditioning. Thus, neuroprotection caused by NO released during the short episode of sublethal ischemia may be mediated partly by stimulation of ATP-sensitive K^+ channels [36].

In a similar fashion, glucose-sensing mechanisms could be similar, if not identical, in glucose responsive cells such as pancreatic β cells and hypothalamic neurons [35, 36] and cancer cells. Extrapolating the data obtained from studies performed with pancreatic β cells and neuronal cells to cancer cells, it is suggested that glucose sensing by tumor cells is regulated by ATP-sensitive K^+ channels, ROS and the metabolites of glucose namely pyruvate and lactate. For example, multidrug-resistant H69AR cells showed increases in both K^+ channel and volume-regulated Cl^- channel current [37], and K^+ channel activity modulated Ca^{2+} influx into colon cancer cells that, in turn, modulated the proliferation of these (colon cancer) cells [38].

Glibenclamide, a blocker of ATP-sensitive potassium channels, suppressed the progression of many cancers including human gastric cancer cells (MGC-803) in vitro. Glibenclamide induced cellular viability decline, coupled with cell apoptosis and reactive oxygen species (ROS) generation in human gastric cancer cell line MGC-803. Glibenclamide increased NADPH oxidase catalytic subunit gp91(phox) expression and superoxide anion generation, and caused mitochondrial respiration dysfunction in MGC-803 cells. Glibenclamide also led to loss of mitochondrial membrane potential, release of cytochrome c and apoptosis-inducing factor (AIF), and activation of c-jun NH_2-terminal kinase (JNK) in MGC-803 cells. Thus, glibenclamide exerted its antitumor activity in MGC-803 cells by activating ROS-dependent, JNK-driven cell apoptosis suggesting that a close relationship exists between free radical generation and ATP-sensitive potassium channels [39]. Since ion channels are found in a variety of cancer cells and necessary for cell cycle and cell proliferation, it is understandable that ATP-sensitive potassium channel activity plays a critical role in the proliferation of various tumor cells. This is supported by the report that the expression of ATP-sensitive potassium channels in glioma tissues was greatly increased than that in normal tissues. Treatment of glioma cells with tolbutamide, another ATP-sensitive potassium channel inhibitor, suppressed the proliferation of glioma cells and blocked glioma cell cycle in G(0)/G(1) phase. Similarly, down-regulation of ATP-sensitive potassium channel by small interfering RNA (siRNA) inhibited glioma cell proliferation, whereas ATP-sensitive potassium channel agonist diazoxide and overexpression of ATP-sensitive potassium channels promoted the proliferation of glioma cells. In addition, inhibiting ATP-sensitive potassium channels slowed the formation of tumor in nude mice generated by injection of glioma cells, whereas activating ATP-sensitive potassium channels promoted development of tumor in vivo. The effect of ATP-sensitive potassium channel activity on glioma cells proliferation was found to be mediated by extracellular signal-regulated kinase (ERK) activation whereas the mitogen-activated protein kinase kinase (MAPK kinase) inhibitors blocked ERK activation and cell proliferation induced by diazoxide that were reversed by the inhibitory effects of tolbutamide on glioma proliferation. These results suggest that ATP-sensitive potassium channels control glioma cell proliferation via regulating ERK pathway [40].

Thus, the metabolism of oxygen, although central to life, may have harmful actions by producing reactive oxygen species (ROS) that have been implicated in processes as diverse as cancer, cardiovascular disease and ageing. It is important to note that central nervous system stem cells and haematopoietic stem cells and early progenitors contain lower levels of ROS than their more mature progeny, and that these differences are critical for maintaining stem cell function. It was reported that normal mammary epithelial stem cells contained lower concentrations of ROS than their more mature progeny cells. Furthermore, subsets of cancer stem cells in some human and murine breast tumours contained lower ROS levels than corresponding non-tumorigenic cells. Cancer stem cells of some tumours developed less DNA damage and were preferentially spared after irradiation compared to non-tumorigenic cells. Lower ROS levels in cancer stem cells were found to be associated with increased expression of free radical scavenging systems, while depletion of ROS scavengers in cancer stem cells markedly decreased their clonogenicity. These results indicate that, similar to normal tissue stem cells, subsets of cancer stem cells in some tumors contain lower ROS levels and possessed enhanced ROS defenses compared to their non-tumorigenic progeny, which may contribute to tumor radioresistance [41].

In addition, tumor cells have enhanced rates of glucose transport and glycolysis and this enhanced glycolytic flux is due to increased activities of hexokinase, phosphofructokinase and pyruvate kinase and not directly to the increased glucose transport [42]. Tumor cells generate enhanced levels of lactate and pyruvate, a potent antioxidant [43, 44]. Furthermore, tumor cells are more sensitive to oxidative stress and free radical induced apoptosis suggesting that the lower ROS levels and enhanced ROS defenses [45, 46] seen in them are an adoptive and a defensive response to ward off free radical induces apoptosis. This implies that methods designed to enhance free radical generation in tumor cells could form an effective strategy to selectively eliminate them. Such an effort is made by the immune cells of the body by inducing low-grade systemic inflammation but is futile. It is rather interesting that attempts made by the immune cells to recognize tumor cell antigens and generate pro-inflammatory cytokines such as IL-6 and TNF-α by the tumor-infiltrating T cells and macrophages instead of eliminating tumor cells, in fact, promote tumor growth and metastasis [47–51].

Eicosanoids, Free Radicals and Inflammation in Cancer

Eicosanoids have an important regulatory role in inflammation. Prostaglandins (PGs) such as PGE_2, $PGF_{2\alpha}$, and leukotrienes (LTs) such as LTA_4, LTB_4 have pro-inflammatory actions and are produced in large amounts by tumor cells. It was reported that forced expression of cyclooxygenase-2 (COX-2) leads to inhibition of programmed cell death in intestinal epithelial cells and growth of human colonic cancer xenografts could be inhibited by treatment with selective COX-2 inhibitor in tumors that express COX-2 but not in those that lack COX-2 expression. PGE_2 treatment of human colon cancer cells leads to increased clonogenicity of these cells, whereas treatment with selective COX-2 inhibitor decreased colony formation in monolayer culture and this growth inhibition was reversed by treatment with PGE_2. PGE_2 inhibited programmed cell death caused by COX-2 inhibitor and induced Bcl-2 expression, but did not affect Bcl-x or Bax expression in human colon cancer cells. These results suggested that PGE_2 augment tumor cell growth by decreasing cell death [52]. Furthermore, PGE_2 enhanced invasive potential of colon cancer cells [53], and seem to stimulate colon cancer cell growth through its heterotrimeric guanine nucleotide-binding protein (G protein)-coupled receptor, EP2, by a signaling route that involves the activation of phosphoinositide 3-kinase and the protein kinase Akt by free G protein beta-gamma subunits and the direct association of the G protein alphas subunit with the regulator of G protein signaling (RGS) domain of axin. This leads to the inactivation and release of glycogen synthase kinase 3-β from its complex with axin, thereby relieving the inhibitory phosphorylation of β-catenin and activating its signaling pathway [54]. PGE_2 increased the phosphorylation of glycogen synthase kinase-3 and consequently accumulated β-catenin and induced the expression of T cell factor-4 transcription factor, which formed transcriptionally active complex with β-catenin. In animal experiments, administration of 16,16-dimethyl PGE_2 increased the expression of cyclin D1 and VEGF (vascular endothelial growth factor) in APC (min/+) mouse polyps. These results clearly suggested that COX-2/PGE_2 exert pro-oncogenic actions through stimulating the β-catenin/T cell factor-mediated transcription and thus, plays a critical role in colorectal carcinogenesis [55].

In addition, PGE_2 enhanced tumor cell proliferation by transactivation of the epidermal growth factor (EGF) receptor (EGFR) as well as EGFR-independent pathways. Treatment of the hepatocytes with PGE_2 amplified the stimulatory effect on DNA synthesis of EGF and advanced and augmented the cyclin D1 expression in response to EGF in hepatocytes. The pretreatment of the hepatocytes with PGE_2 resulted in an increase in the magnitude of EGF-stimulated Akt phosphorylation and ERK1/2 phosphorylation and kinase activity, including an extended duration of the responses, particularly of ERK, to EGF in PGE_2-treated cells [56].

PGE_2 induced the expression of IL-1α in colon cancer cells, which plays critical roles in tumor metastasis and neoangiogenesis in a variety of cancers. PGE_2 increased the levels of both IL-1α mRNA and protein, suggesting a positive feedback loop between the IL-1 pathway and PGE_2 signaling. Moreover, IL-1α enhanced colorectal neoplasia, stimulating cell migration and neoangiogenesis, whereas knockdown of the expression of IL-1α by small-interfering RNA resulted in a reduction of VEGF secretion in colon cancer cells. Thus, PGE_2 is not only a pro-inflammatory molecule by itself but also has the ability to induce the expression of proinflammatory cytokine IL-1α [57].

These and other results led to the suggestion that COX-2 inhibitors are of significant use in the prevention of colon and other cancers [58]. But, it should be noted that COX-2 inhibitors are not always useful in inhibiting the growth of various tumors and there could occur a pathway that is independent of PGE_2 formation in carcinogenesis. It is likely that the beneficial actions of COX-2 inhibitors could be due to the shunting of the precursor PUFA pool to yet another pathway such as lipoxins, resolvins and protectins with or without the inhibition of PGE_2 formation. It is also possible that availability of significant amounts of the eicosanoid precursor pool (PUFAs) due to the inhibition of the COX-2 pathway, the PUFAs may give rise to higher amounts of lipoxins, resolvins, and protectins, PUFAs may undergo lipid peroxidation or influence glucose metabolism by altering the cell membrane fluidity.

PUFAs, Pro- and Anti-inflammatory Metabolites of PUFAs and Lipid Peroxidation and Cancer

The role of various eicosanoids (such as PGE_2 and LTB_4) and lipoxins, resolvins and protectins; and pro-inflammatory cytokines (IL-6 and TNF-α) in the proliferation and apoptotic pathways seen in normal and tumor cells are rather puzzling. For example, excess production of PGE_2 and LTB_4 and IL-6 and TNF-α are involved in the pathobiology of anorexia and cachexia seen in cancer and chronic infections such as tuberculosis whereas lipoxins, resolvins and protectins formed from n-3 PUFAs (since n-3 PUFAs prevent cachexia) may have ameliorating action on anorexia and cachexia. In a similar fashion, there is reasonable evidence to suggest that tumor cells undergo apoptosis on exposure to PUFAs due to the enhanced formation of ROS and lipid peroxides, whereas under the identical circumstances normal cells (exposed to PUFAs) will form higher amounts of lipoxins, resolvins and protectins and far less amounts of ROS and lipid peroxides that protect them from the cytotoxic actions of ROS and lipid peroxides. This differential formation of various eicosanoids and lipoxins, resolvins and protectins, and ROS and lipid peroxides by normal and tumor cells on exposure to the same amount and types of PUFAs may underlie the differential toxicity shown by PUFAs against normal and tumor cells (see discussion below). In fact, it was observed that apoptosis of tumor cells could occur despite the generation of excess of PGE_2 and LTB_4 (though incubation with n-3 PUFAs inhibit the formation of PGE_2 and LTB_4 but could still enhance the formation of lipid peroxides and ROS generation; whereas AA enhances the formation of PGE_2 and LTB_4 but still causes apoptosis of tumor cells) due to the generation of excess of ROS and lipid peroxides, suggesting that it is the ROS and lipid peroxides that are crucial in determining the fate of the cell (whether it is the normal or tumor cell). The failure of IL-6 and TNF-α to induce tumor cell apoptosis could be as a result of the failure of these cytokines to generate adequate amounts of ROS and lipid peroxides possibly, due to decreased tumor cell PLA_2 activity or PUFA deficiency. In fact, pre-treatment of tumor cells with n-3 and n-6 PUFAs may render them highly susceptible to the cytotoxic action of TNF-α [59–69].

To bypass or neutralize the cytotoxic actions of ROS and lipid peroxides, tumor cells have evolved aerobic glycolysis pathway that dampens ROS generation and produces excess pyruvate, a potent antioxidant. Thus, tumor cells are relatively deficient in pro-oxidant pathways and have evolved an efficient anti-oxidant defenses (see Fig. 14.1). Hence, strategies developed to specifically augment free radical generation and formation of lipid peroxides selectively in the tumor cells could form a new therapeutic approach in the prevention and management of cancer.

Free Radicals Have both Beneficial and Harmful Actions

Free radicals that are involved in inflammation also have direct effects on cell growth and development, cell survival and have a significant role in various diseases including cancer [70–72] and have both beneficial and harmful actions. They are needed for signal-transduction pathways that regulate cell growth, reduction-oxidation (redox) status, and for defense against infections by polymorphonuclear leukocytes, macrophages and other immunocytes [70–78]. Excessive amounts of free radicals start lethal chain reactions that inactivate vital enzymes, proteins and other important subcellular elements needed for cell survival and lead to cell death [70, 76–78]. On the other hand, glucose-stimulated free radical generation seems to have a role in β-cell glucose signaling and insulin secretion [79, 80]. Thus, free radicals are like a double-edged sword.

Lipid Peroxidation in Tumor Cells

The role of free radicals in cell growth and function is evident from the fact that lipid peroxides formed as a result of free radical generation have a regulatory role in cell proliferation. For example, an inverse relationship exists between the concentrations of lipid peroxides and the rate of cell proliferation, i.e., the higher the rate of lipid peroxidation in the cells the lower the rate of cell division and tumor cells are resistant to lipid peroxidation compared to normal cells [81–83]. It was reported that lipid peroxidation decreases with increasing growth rate [84]. In hepatomas: the higher the growth rate of the tumor, the lower the microsomal phospholipid content and the degree of fatty acid unsaturation [85, 86]. The reduced rates of lipid peroxidation observed in Yoshida hepatoma cells and their microsomes when compared with appropriate control tissue (normal liver tissue) under the same pro-oxidant conditions was found to be due to the much reduced levels of polyunsaturated fatty acids, NADPH-cytochrome c reductase and the NADPH-cytochrome P-450 electron transport chain in the Yoshida hepatoma cells [87, 88]. Thus, the low rate of lipid peroxidation in tumor cells could be due to a combination of low levels of polyunsaturated fatty acids and cytochrome P-450 and elevated levels of lipid-soluble anti-oxidant α-tocopherol. Tumor plasma membranes had extremely low rates of malondialdehyde accumulation and LOOH was practically undetectable in the hepatoma cell membranes compared to the normal rat liver membranes [88, 89]. Such a high degree of resistance to peroxidation in the tumor cells has been attributed to a marked decrease in lipid content [88, 89].

These results [81–89] coupled with our observation that incubation of cells with PUFAs augmented free radical generation and formation of lipid peroxidation products selectively in the tumor cells compared to normal cells despite the fact that the uptake of fatty acids was at least two to three times higher in the normal cells compared to tumor cells [90–93], suggests that a close correlation exists between the rate of lipid peroxidation and degree of malignant deviation of the tumor cell, and the susceptibility of the tumor cell to free radical induced cytotoxicity. In other words, the higher the degree of malignant nature of the tumor cell, the lower the rate of lipid peroxidation and higher the degree of susceptibility to free radical-induced toxicity. In fact, resistance to lipid peroxidation appears to start at the premalignant stage of the carcinogenesis process itself, as administration of diethylnitrosamine and 2-acetylaminofluorene leads to inhibition of peroxidation in normal liver and in preneoplastic nodules as well as in the neoplasms that result from this treatment [94].

PUFA Deficiency Exists in Tumor Cells

The low content of PUFAs reported in the tumor cells can be attributed to the loss of or decreased activity of Δ^6 and Δ^5 desaturases [95–97] that results in decreased conversion of dietary linoleic acid (LA, 18:2 n-6) and α-linolenic acid (ALA, 18:3 n-3) to their long chain metabolites γ-linolenic (GLA, 18:3 n-6), dihomo-γ-linolenic

(DGLA, 20:3 n-6), arachidonic (AA, 20:4 n-6) and eicosapentaenoic (EPA, 20:5 n-3) and docosahexaenoic acids (DHA, 22:6 n-3) respectively. In addition, tumor cells have elevated levels of lipid-soluble anti-oxidant α-tocopherol [88]. The higher vitamin E to PUFA ratio in rapidly growing tumors is due to markedly decreased content of PUFA, while the vitamin E quantitated on a per mg protein basis is virtually unchanged. On the other hand, tumor cells have low or almost no superoxide dismutase (SOD), glutathione peroxidase and catalase enzymes [98–100]. Thus, it is the relatively high content of vitamin E that contributes to the low rate of lipid peroxidation seen in the tumor cells. Thus both substrate (i.e., PUFAs) deficiency and a relatively high content of vitamin E are responsible for the low rate of lipid peroxidation seen in tumor cells that, in turn, contributes to their high mitotic index.

Superoxide Dismutase and Free Radicals in Tumor Cells

The two forms of human superoxide dismutase (SOD)-the mitochondrial isoform (manganese-containing SOD, MnSOD) and the copper-zinc containing SOD (CuZn-SOD) that is primarily localized in the cytosol in mammalian cells, although some may be present in the nucleus, mitochondrial intermembrane space, lysosomes and peroxisomes. But both types of SOD are essential for healthy aerobic life.

Superoxide and other free radicals cause cell death by apoptosis. Excess of radicals or severe oxidative stress produces necrosis [101, 102]. Most other death stimuli also increase formation of free radicals and cause death of the cells by apoptosis. Free radical scavengers often, but not always, delay apoptosis [102, 103]. For example, TNF-α induced tumor cell death and host toxicity can be related to its ability to induce the production of free radicals [104]. High intracellular glutathione levels induce tumor cell resistance to recombinant human TNF (rhTNF), whereas low glutathione levels enhanced sensitivity to rhTNF. Pretreatment of the tumor-bearing hosts with DL-buthionine-(S, R)-sulfoximine, an inhibitor of GSH biosynthesis, resulted in an increased sensitivity of tumor cells to rhTNF [104]. Induction of apoptosis by arachidonic acid (AA) in human retinoblastoma cells and other polyunsaturated fatty acids (PUFAs) was found to be accompanied by increased formation of the lipid peroxidation end products, suggesting a role for free radicals [90–93, 105].

Both total SOD and MnSOD specific activities were lower in the tumor cell homogenates compared to normal liver: the lowest activity was associated with the fastest growing tumor [106]. MnSOD activity was decreased in the fast- and medium-growth rate hepatomas but was slightly increased in the tumor with the slowest growth rate compared to normal liver. This suggests that decreased MnSOD specific activity is not a characteristic of all tumors. In addition, it was observed that antimycin A (an inhibitor of the electron transport chain) stimulated the production of $O_2^{-\cdot}$ in normal rat liver and slow-growth-rate tumor cells but not in the submitochondrial particles of fast-growth-rate tumor cells [106], lending support to the concept that the rate of lipid peroxidation is generally low in tumor cells.

A significant decrease in the PUFA content of Novikoff cells or Novikoff microsomes, especially arachidonic acid (AA, 20:4 n-6) and eicosapentaenoic acid (EPA, 20:5 n-3) with a concomitant reduction in NADPH-cytochrome c reductase was reported [107]. A substantial increase in α-tocopherol relative both to total lipid and to methylene-interrupted double bonds in fatty acids was noted. Thus both tumor cells seem to possess relatively high vitamin E content, less PUFAs and as a result decreased rates of lipid peroxidation [87, 107]. In contrast, tumor cells showed varying levels of SOD [108–113] and paradoxically in some human gliomas the malignant phenotype could be suppressed by overexpressing MnSOD [114].

The SH groups of the caspases are essential for their catalytic activity. On exposure to free radicals these SH groups may be inactivated. Thus, $O_2^{-\cdot}$, H_2O_2 and NO may inactivate to promote tumor cell survival by inactivating caspases [115]. But, in general, it appears that majority of the tumors have low SOD concentrations, relatively high amounts of vitamin E, low PUFA content and lipid peroxides, which may render them more susceptible and sensitive to free radical attack. Based on these evidences, the best strategy to make use of free radical toxicity to tumor cells would be to devise strategies that lower SOD, catalase, glutathione and vitamin E concentrations, and at the same time enhance free radical generation and lipid peroxidation process in the tumor cells.

Free Radicals Induce Translocation of p53

P53 protein is an example of a gene product that affects both cell cycle progression and apoptosis [116]. It is known that p53 overexpression induces apoptosis and tumor cells, more often than not, disable p53 during the transformation process.

The p53 tumor suppressor functions to maintain the integrity of the genome. Its nuclear localization is critical to this regulation. Free radicals have the ability to signal p53 translocation to the nucleus. H_2O_2 induced apoptosis coincides with p53 nuclear translocation and is cell cycle related. Thus, free radicals could be considered as powerful inducers of p53 activity that leads to the onset of p53-dependent apoptosis [117, 118].

In addition, tumor cells that lack p53 are significantly more resistant to the cytotoxic effect of a number of pro-oxidants [119]. Furthermore, MnSOD activity was increased in liver tissue from p53-deficient mice in comparison with wild type, while transient transfection of cells with p53 led to a significant reduction in MnSOD levels suggesting that expression of MnSOD is negatively regulated by p53. Increased expression of MnSOD rendered cells resistant to p53-dependent cytotoxic treatments and, in co-transfection experiments, counteracted the growth inhibitory effect of p53 [119]. These results imply that normally a balance maintained between p53 and SOD levels and that overexpression of p53 can lead to a decrease in the levels of SOD. This suggests that p53 has a pro-oxidant type of activity. Thus, when tumor cells are exposed to free radicals such as $O_2^{-\cdot}$, H_2O_2 or NO, SOD present in the cells/tissues is not only utilized to quench these free radicals but, will also lead to an increase in the expression of p53 due to free radical induced stress [117, 118]. This, in turn, suppresses the expression of SOD, tilting the balance more towards a pro-oxidant state. The increase in the pro-oxidant state can induce apoptosis.

Free radicals, especially H_2O_2, can cause ATP depletion in the cells by activating PARP (poly-ADP-ribose-polymerase), the substrate of caspase-3 [120], though this is ontroversial [121]. But, it should be noted that H_2O_2 induces apoptosis at low concentrations, whereas at higher concentrations causes necrosis. Necrosis is known to occur if the oxidative stress is severe. Hence, the final effect of H_2O_2 on tumor cell death, whether it is by apoptosis or by necrosis, is dependent on the concentration of H_2O_2 present.

Oxidant Stress and Telomere

Telomeres of human somatic cells shorten with each cell division but are stabilized at constant length in tumors by the enzyme telomerase [122]. Oxidative stress can shorten telomere [123].

Bcl-2 Opposes the Action of p53

Bcl-2 opposes the pro-oxidant action of p53 by its anti-oxidant action [124] and ability to suppress SOD activity. This suggests that a balance exists between the levels of p53, which induces a pro-oxidant status in cells, and the expression and anti-oxidant activity of Bcl-2. It is possible that increased expression of p53 that occurs during exposure to radiation and free radicals, may lead to inhibition of Bcl-2 expression that, in turn, augments increase in free radical generation and apoptosis as evidenced by the observation that Bcl-2 is down-regulated by p53 [125, 126] and blocks lipid peroxidation that induces apoptosis [127]. Following an apoptotic signal, progressive lipid peroxidation occurs in the cells, whereas over-expression of Bcl-2 suppresses lipid peroxidation. Thus, a close association seems to exist between lipid peroxidation, Bcl-2 and apoptosis [128]. We observed that when tumor cells are treated with PUFAs, there is not only an increase in the formation of lipid peroxides, there was also an increase in protein phosphorylation [129]. Thus, it is possible that enhanced lipid peroxidation in the tumor cells could lead to phosphorylation of Bcl-2 and a reduction in its anti-apoptotic potential that induces apoptosis. Such an interaction between lipid peroxidation and Bcl-2 is understandable since Bcl-2 is localized at intracellular sites of oxygen free radical generation including mitochondria, endoplasmic reticulum and nuclear membranes.

Thus, optimum expression of Bcl-2 and presence of adequate amounts of anti-oxidants: SOD, catalase, glutathione, and vitamin E prevent apoptosis, whereas activation of p53, excess production of free radicals, and an increase in the levels of lipid peroxides in the cells would trigger apoptosis.

It is evident from the preceding discussion that methods designed to augment free radical generation specifically in tumor cells can induce their apoptosis and to protect themselves from the toxic action of free radicals, tumor cells seem to have evolved potent anti-oxidant defenses.

Polyunsaturated Fatty Acids Inhibit Cell Proliferation by Augmenting Free Radical Generation and Lipid Peroxidation

PUFAs (especially GLA, AA, EPA and DHA) when used at appropriate doses inhibited tumor cell proliferation in vitro, whereas anti-oxidants blocked this inhibitory action [130–132], indicating that free radicals and lipid peroxides are the mediators of their (PUFAs) cytotoxic action. Prostaglandins (PGs) derived from PUFAs also inhibited the proliferation of human and animal tumor cells in vitro [133–135] but their effects were variable. When used at appropriate concentrations, GLA, AA, EPA and DHA, were toxic to tumor cells with little or no effect on normal cells in vitro [136–143]. This selective tumoricidal action of fatty acids was not blocked by cyclo-oxygenase and lipoxygenase inhibitors, though in some cells they did block the tumoricidal action of PUFAs, suggesting that prostaglandins and leukotrienes may not always participate in this process. Furthermore, GLA, AA and EPA treated tumor cells but not normal cells produced—a two to threefold increase in free radicals and lipid peroxidation products, suggesting that the low rates of lipid peroxidation observed in tumor cells are, at least in part, due to deficiency of PUFAs and to a relative increase in their anti-oxidant content. Since there is a direct correlation between the rate of lipid peroxidation and the degree of deviation in hepatomas (as discussed previously, see above) and as the rate of lipid peroxidation is low in several tumors, it is likely that lipid peroxidation might act as a physiological inhibitor of mitosis and regulate cell multiplication [144–147].

Tumor cell death caused by TNF-α is associated with release of endogenous AA [148, 149], whereas TNF-α-induced apoptosis can be prevented by the removal of unesterified AA [148]. Thus, the cellular level of unesterified AA may be a general mechanism by which apoptosis is induced in tumor cells. Since other PUFAs, such as GLA, DGLA, EPA and DHA, also induce apoptosis of tumor cells, it is likely that methods designed to enhance the cellular content of unesterified PUFAs may trigger apoptosis in tumor cells. This may explain the beneficial action of EPA- and DHA-rich fish oils in the prevention of colon cancer [150, 151]. Further, tumor cells exposed to PUFAs show low levels of various anti-oxidants [142, 143], which may cause an increase in oxidant stress and an enhancement in cytotoxicity. PUFAs, especially n-3 fatty acids, suppress carcinogen induced *ras* activation [152], Bcl-2 expression, and inhibit the activity of cyclo-oxygenase enzyme. Thus PUFAs have several activities that contribute to their anti-cancer actions.

Normal and Tumor Cells May Process PUFAs Differentially

In nude mouse model LA rich diet enhanced breast cancer progression, where as n-3 fatty acids (rich in EPA and DHA) exerted suppressive effects [144, 145]. In cell culture studies, LA-stimulated growth of tumor cells [146]. These studies led to the suggestion that n-6 fatty acids augment tumor growth. It may be mentioned here that in these studies oils rich in LA were used that did not contain any GLA or AA, which are also n-6 fatty acids, but are metabolites of LA. On the other hand, studies performed with oils, which contain both LA and GLA (such as primrose oil) showed suppression of tumor growth [147–149]. This suggests that LA may promote whereas GLA, DGLA and AA suppress tumor growth [149, 150]. So a cautious approach is needed while extrapolating results obtained with LA rich oils to all n-6 fatty acids. This is supported by the observation that GLA, salts of GLA and a chemical formulation containing both GLA and EPA inhibit tumor growth in vitro and in vivo [151–155].

In addition, there is evidence to suggest that the way PUFAs are handled by normal and tumor cells could be entirely different. For example, DHA that is toxic to tumor cells protects normal neural cells from stress-induced apoptosis. DHA induces apoptosis in neuroblastoma cells. Neuroblastoma cells metabolized DHA to 17-hydroxydocosahexaenoic acid (17-HDHA) via 17-hydroperoxydocosahexaenoic acid (17-HpDHA) through 15-lipoxygenase and autoxidation and did not produce

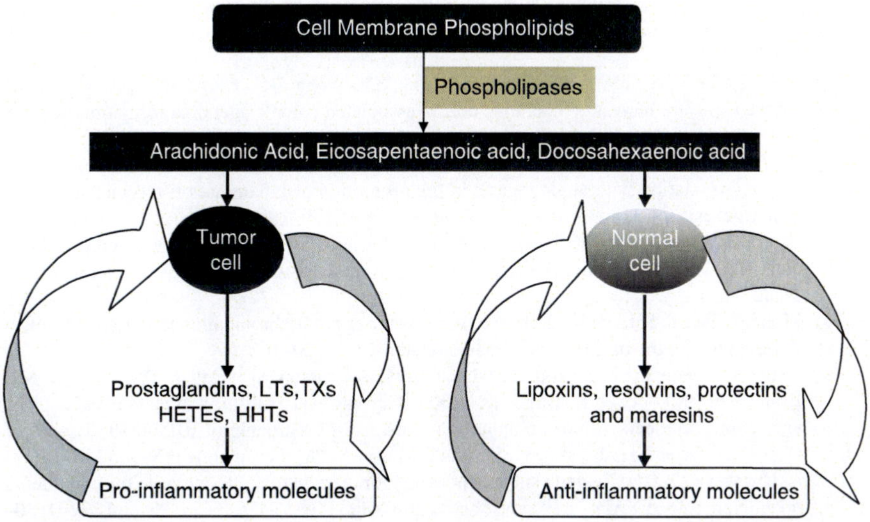

Fig. 14.2 Scheme showing the metabolism of AA/EPA/DHA in normal and tumor cells. Tumor cells produce more prostaglandins, leukotrienes, thromboxanes, HETEs and HHTs, while normal cells generate lipoxins, resolvins, and protectins that are cytoprotective. Normal cells produce lipoxins from AA; lipoxins and resolvins from EPA and protectins from DHA. Thus, cancer is an inflammatory condition

the anti-inflammatory and protective lipid mediators, resolvins and protectins. 17-HpDHA had significant cytotoxic potency on tumor cells. DHA inhibited secretion of PGE_2. These results suggest that the cytotoxic effect of DHA in neuroblastoma is mediated through production of hydroperoxy fatty acids that accumulate to toxic intracellular levels with restricted production of its products, resolvins and protectins that are cytoprotective in nature [156]. In a similar fashion, it is possible that when normal cells are exposed to AA and EPA significant amounts of lipoxins and resolvin are formed, whereas tumor cells would accumulate respective, prostanoids, leukotrienes, thromboxanes and cyclopentanone prostaglandins. Thus, normal cells when exposed to PUFAs produce cytoprotective lipids such as lipoxins, resolvins and protectins while tumor cells generate toxic hydroperoxy fatty acids [155–157]. This differential metabolism of PUFAs by normal and tumor cells may explain why PUFAs are toxic to tumor but not to normal cells (see Fig. 14.2).

References

[1] WHO Report on the Global Tobacco Epidemic (2009)
[2] Zeid NA, Muller HK (1995) Tobacco smoke induced lung granulomas and tumors: association with pulmonary Langerhans cells. Pathology 27:247–254
[3] Ohwada A, Takahashi H, Nagaoka I, Iwabuchi K, Mikami O, Kira S (1995) Effect of cigarette smoke on the mRNA and protein expression of carcinoembryonic antigen (CEA), a possible chemoattractant for neutrophils in human bronchioloalveolar tissues. Thorax 50:651–657
[4] Roth MD, Arora A, Barsky SH, Kleerup EC, Simmons M, Tashkin DP (1998) Airway inflammation in young marijuana and tobacco smokers. Am J Respir Crit Care Med 157(3 Pt 1):928–937
[5] Liu ES, Ye YN, Shin VY, Yuen ST, Leung SY, Wong BC, Cho CH (2003) Cigarette smoke exposure increases ulcerative colitis-associated colonic adenoma formation in mice. Carcinogenesis 24:1407–1413
[6] Ye YN, Liu ES, Shin VY, Wu WK, Cho CH (2004) Contributory role of 5-lipoxygenase and its association with angiogenesis in the promotion of inflammation-associated colonic tumorigenesis by cigarette smoking. Toxicology 203:179–188
[7] Ye YN, Wu WK, Shin VY, Bruce IC, Wong BC, Cho CH (2005) Dual inhibition of 5-LOX and COX-2 suppresses colon cancer formation promoted by cigarette smoke. Carcinogenesis 26:827–834
[8] Malfertheiner P, Schütte K (2006) Smoking—a trigger for chronic inflammation and cancer development in the pancreas. Am J Gastroenterol 101:160–162
[9] Wittel UA, Pandey KK, Andrianifahanana M, Johansson SL, Cullen DM, Akhter MP, Brand RE, Prokopczyk B, Batra SK (2006) Chronic pancreatic inflammation induced by environmental tobacco smoke inhalation in rats. Am J Gastroenterol 101:148–159
[10] Orosz Z, Csiszar A, Labinskyy N, Smith K, Kaminski PM, Ferdinandy P, Wolin MS, Rivera A, Ungvari Z (2007) Cigarette smoke-induced proinflammatory alterations in the endothelial phenotype: role of NAD(P)H oxidase activation. Am J Physiol Heart Circ Physiol 292:H130–H139
[11] Demaria S, Pikarsky E, Karin M, Coussens LM, Chen YC, El-Omar EM, Trinchieri G, Dubinett SM, Mao JT, Szabo E, Krieg A, Weiner GJ, Fox BA, Coukos G, Wang E, Abraham RT, Carbone M, Lotze MT (2010) Cancer and inflammation: promise for biologic therapy. J Immunother 33:335–351
[12] Das UN (1993) Oxy radicals and their clinical implications. Curr Sci 65:964–968

[13] Das UN (1990) Free radicals: biology and relevance to disease. J Assoc Physicians India 38:495–498

[14] Caspary WJ, Niziak C, Lanzo DA, Friedman R, Bachur NR (1979) Bleomycin A2: a ferrous oxidase. Mol Pharmacol 16:256–260

[15] Young RC, Ozols RF, Myers CE (1981) The anthracycline antineoplastic drugs. N Engl J Med 305:139–153

[16] Gajewski E, Rao G, Nackerdien Z, Dizdaroglu M (1990) Modification of DNA bases in mammalian chromatin by radiation-generated free radicals. Biochemistry 29:7876–7882

[17] Tracey KJ, Wei H, Manogue KR, Fong Y, Hesse DG, Nguyen HT, Kuo GC, Beutler B, Cotran RS, Cerami A et al (1988) Cachectin/tumor necrosis factor induces cachexia, anemia, and inflammation. J Exp Med 167:1211–1227

[18] Dinarello CA, Endres S, Meydani SN, Meydani M, Hellerstein MK (1990) Interleukin-1, anorexia, and dietary fatty acids. Ann N Y Acad Sci 587:332–338

[19] Hellerstein MK, Meydani SN, Meydani M, Wu K, Dinarello CA (1989) Interleukin-1-induced anorexia in the rat. Influence of prostaglandins. J Clin Invest 84:228–235

[20] Ramos EJ, Middleton FA, Laviano A, Sato T, Romanova I, Das UN, Chen C, Qi Y, Meguid MM (2004) Effects of omega-3 fatty acid supplementation on tumor-bearing rats. J Am Coll Surg 199:716–723

[21] Tsujimoto M, Yokota S, Vilcek J, Weissmann G (1986) Tumor necrosis factor provokes superoxide anion generation from neutrophils. Biochem Biophys Res Commun 137:1094–1100

[22] Das UN, Padma M, Sagar PS, Ramesh G, Koratkar R (1990) Stimulation of free radical generation in human leukocytes by various agents including tumor necrosis factor is a calmodulin dependent process. Biochem Biophys Res Commun 167:1030–1036

[23] Bautista AP, Schuler A, Spolarics Z, Spitzer JJ (1991) Tumor necrosis factor-alpha stimulates superoxide anion generation by perfused rat liver and Kupffer cells. Am J Physiol 261(6 Pt 1):G891–G895

[24] Test ST (1991) Effect of tumor necrosis factor on the generation of chlorinated oxidants by adherent human neutrophils. J Leukoc Biol 50:131–139

[25] Kilpatrick LE, Sun S, Li H, Vary TC, Korchak HM (2010) Regulation of TNF-induced oxygen radical production in human neutrophils: role of delta-PKC. J Leukoc Biol 87:153–164

[26] Sample AK, Czuprynski CJ (1991) Priming and stimulation of bovine neutrophils by recombinant human interleukin-1 alpha and tumor necrosis factor alpha. J Leukoc Biol 49:107–115

[27] Hachicha M, Pouliot M, Petasis NA, Serhan CN (1999) Lipoxin (LX)A4 and aspirin-triggered 15-epi-LXA4 inhibit tumor necrosis factor 1alpha-initiated neutrophil responses and trafficking: regulators of a cytokine-chemokine axis. J Exp Med 189:1923–1930

[28] Lam TKT, Gutierrez-Juarez R, Pocai A, Rossetti L (2005) Regulation of blood glucose by hypothalamic pyruvate metabolism. Science 309:943–947

[29] Das UN (2006) Pyruvate is an endogenous anti-inflammatory and anti-oxidant molecule. Med Sci Monit 12:RA79–RA84

[30] Das UN (2006) Is pyruvate an endogenous anti-inflammatory molecule? Nutrition 22:965–972

[31] Ainscow EK, Mirshamsi S, Tang T, Ashford ML, Rutter GA (2002) Dynamic imaging of free cytosolic ATP concentration during fuel sensing by rat hypothalamic neurones: evidence for ATP-independent control of ATP-sensitive K^+ channels. J Physiol 544:429–445

[32] Flormonti X, Lorsignol A, Taupignon A, Penicaud L (2004) A new ATP-sensitive K^+ channel-independent mechanism is involved in glucose-excited neurons of mouse arcuate nucleus. Diabetes 53:2767–2775

[33] Leloup C, Magnan C, Benani A, Bonnet E, Alquier T, Offer G, Carriere A, Periquet A, Fernandez Y, Ktorza A, Casteilla L, Penicaud L (2006) Mitochondrial reactive oxygen species are required for hypothalamic glucose sensing. Diabetes 55:2084–2090

[34] Avshalumov MV, Rice ME (2003) Activation of ATP-sensitive K^+ (KATP) channels by H_2O_2 underlies glutamate-dependent inhibition of striatal dopamine release. Proc Natl Acad Sci U S A 100:11729–11734

[35] Spagnoli A, Spadoni GL, Sesti G, Del Principe D, Germani D, Boscherini B (1995) Effect of insulin on hydrogen peroxide production by human polymorphonuclear leukocytes. Studies with monoclonal anti-insulin receptor antibodies, and an agonist and an inhibitor of protein kinase C. Horm Res 43:286–293
[36] Lin YF, Raab-Graham K, Jan YN, Jan LY (2004) NO stimulation of ATP-sensitive potassium channels: involvement of Ras/mitogen-activated protein kinase pathway and contribution to neuroprotection. Proc Natl Acad Sci U S A 101:7799–7804
[37] Jirsch J, Deeley RG, Cole SP, Stewart AJ, Fedida D (1993) Inwardly rectifying K^+ channels and volume-regulated anion channels in multidrug-resistant small cell lung cancer cells. Cancer Res 53:4156–4160
[38] Yao X, Kwan HY (1999) Activity of voltage-gated K^+ channels is associated with cell proliferation and Ca2+ influx in carcinoma cells of colon cancer. Life Sci 65:55–62
[39] Qian X, Li J, Ding J, Wang Z, Duan L, Hu G (2008) Glibenclamide exerts an antitumor activity through reactive oxygen species-c-jun NH_2-terminal kinase pathway in human gastric cancer cell line MGC-803. Biochem Pharmacol 76:1705–1715
[40] Huang L, Li B, Li W, Guo H, Zou F (2009) ATP-sensitive potassium channels control glioma cells proliferation by regulating ERK activity. Carcinogenesis 30:737–744
[41] Diehn M, Cho RW, Lobo NA, Kalisky T, Jo Dorie M, Kulp AN, Qian D, Lam JS, Ailles LE, Wong M, Joshua B, Kaplan MJ, Wapnir I, Dirbas FM, Somlo G, Garberoglio C, Paz B, Shen J, Lau SK, Quake SR, Brown JM, Weissman IL, Clarke MF (2009) Association of reactive oxygen species levels and radioresistance in cancer stem cells. Nature 458:780–785
[42] Singh VN, Singh M, August JT, Horecker BL (1974) Alterations in glucose metabolism in chick-embryo cells transformed by Rous sarcoma virus: intracellular levels of glycolytic intermediates. Proc Natl Acad Sci U S A 71:4129–4132
[43] Das UN (2007) Ethyl pyruvate and sepsis. Adv Sepsis 6:10–15
[44] Das UN (2006) Is pyruvate an endogenous antiinflammatory molecule? Nutrition 22:965–972
[45] Aykin-Burns N, Ahmad IM, Zhu Y, Oberley LW, Spitz DR (2009) Increased levels of superoxide and H_2O_2 mediate the differential susceptibility of cancer cells versus normal cells to glucose deprivation. Biochem J 418:29–37
[46] Lopez-Lazaro M (2010) A new view of carcinogenesis and an alternative approach to cancer therapy. Mol Med 16:144–153
[47] Lawrence T, Hagemann T, Balkwill F (2007) Sex, cytokines, and cancer. Science 317:51–52
[48] Mantovani A (2009) Inflaming metastasis. Nature 457:36–37
[49] Balkwill F (2009) Tumor necrosis factor and cancer. Nat Rev 9:361–371
[50] Dougan M, Dranoff G (2008) Inciting inflammation: the RAGE about tumor promotion. J Exp Med 205:267–270
[51] Gebhardt C, Riehl A, Durchdewald M, Nemeth J, Furstenberger G, Muller-Decker K, Enk A, Arnold B, Bierhaus A, Naworth PP, Hess J, Angel P (2008) RAGE signaling sustains inflammation and promotes tumor development. J Exp Med 205:275–285
[52] Sheng H, Shao J, Morrow JD, Beauchamp RD, DuBois RN (1998) Modulation of apoptosis and Bcl-2 expression by prostaglandin E2 in human colon cancer cells. Cancer Res 58:362–366
[53] Sheng H, Shao J, Washington MK, DuBois RN (2001) Prostaglandin E2 increases growth and motility of colorectal carcinoma cells. J Biol Chem 276:18075–18081
[54] Castellone MD, Teramoto H, Williams BO, Druey KM, Gutkind JS (2005) Prostaglandin E2 promotes colon cancer cell growth through a Gs-axin-beta-catenin signaling axis. Science 310:1504–1510
[55] Shao J, Jung C, Liu C, Sheng H (2005) Prostaglandin E2 Stimulates the beta-catenin/T cell factor-dependent transcription in colon cancer. J Biol Chem 280:26565–26572
[56] Dajani OF, Meisdalen K, Guren TK, Aasrum M, Tveteraas IH, Lilleby P, Thoresen GH, Sandnes D, Christoffersen T (2008) Prostaglandin E2 upregulates EGF-stimulated signaling in mitogenic pathways involving Akt and ERK in hepatocytes. J Cell Physiol 214:371–380

[57] Shao J, Sheng H (2007) Prostaglandin E2 induces the expression of IL-1alpha in colon cancer cells. J Immunol 178:4097–4103
[58] Csiki I, Morrow JD, Sandler A, Shyr Y, Oates J, Williams MK, Dang T, Carbone DP, Johnson DH (2005) Targeting cyclooxygenase-2 in recurrent non-small cell lung cancer: a phase II trial of celecoxib and docetaxel. Clin Cancer Res 11:6634–6640
[59] Neale ML, Fiera RA, Matthews N (1988) Involvement of phospholipase A2 activation in tumour cell killing by tumour necrosis factor. Immunology 64:81–85
[60] Reid T, Ramesha CS, Ringold GM (1991) Resistance to killing by tumor necrosis factor in an adipocyte cell line caused by a defect in arachidonic acid biosynthesis. J Biol Chem 266:16580–16586
[61] Vadas P, Pruzanski W, Stefanski E, Ellies LG, Aubin JE, Sos A, Melcher A (1991) Extracellular phospholipase A2 secretion is a common effector pathway of interleukin-1 and tumour necrosis factor action. Immunol Lett 28:187–193
[62] Suffys P, Beyaert R, De Valck D, Vanhaesebroeck B, Van Roy F, Fiers W (1991) Tumour-necrosis-factor-mediated cytotoxicity is correlated with phospholipase-A2 activity, but not with arachidonic acid release per se. Eur J Biochem 195:465–475
[63] Hayakawa M, Ishida N, Takeuchi K, Shibamoto S, Hori T, Oku N, Ito F, Tsujimoto M (1993) Arachidonic acid-selective cytosolic phospholipase A2 is crucial in the cytotoxic action of tumor necrosis factor. J Biol Chem 268:11290–11295
[64] Levrat C, Louisot P (1996) Increase of mitochondrial PLA2-released fatty acids is an early event in tumor necrosis factor alpha-treated WEHI-164 cells. Biochem Biophys Res Commun 221:531–538
[65] Wu YL, Jiang XR, Newland AC, Kelsey SM (1998) Failure to activate cytosolic phospholipase A2 causes TNF resistance in human leukemic cells. J Immunol 160:5929–5935
[66] El Mahdani NE, Ameyar M, Cai Z, Colard O, Masliah J, Chouaib S (2000) Resistance to TNF-induced cytotoxicity correlates with an abnormal cleavage of cytosolic phospholipase A2. J Immunol 165:6756–6761
[67] Wolf LA, Laster SM (1999) Characterization of arachidonic acid-induced apoptosis. Cell Biochem Biophys 30:353–368
[68] Brekke OL, Espevik T, Bardal T, Bjerve KS (1992) Effects of n-3 and n-6 fatty acids on tumor necrosis factor cytotoxicity in WEHI fibrosarcoma cells. Lipids 27:161–168
[69] Brekke OL, Sagen E, Bjerve KS (1999) Specificity of endogenous fatty acid release during tumor necrosis factor-induced apoptosis in WEHI 164 fibrosarcoma cells. J Lipid Res 40:2223–2233
[70] Das UN (1993) Oxy radicals and their clinical implications. Curr Sci 65:964–968
[71] Das UN (1990) Free radicals: biology and relevance to disease. J Assoc Physicians India 38:495–498
[72] Halliwell BA (2000) Superway to kill cancer cells? Nat Med 6:1105–1106
[73] Cleveland JL, Kastan MB (2000) A radical approach to treatment. Nature 407:309–311
[74] Sunderesan M, Yu ZX, Ferrans VJ, Irani K, Finkel T (1995) Requirement for generation of H_2O_2 for platelet-derived growth factor signal transduction. Science 270:296–299
[75] Suderesan M, Yu ZX, Ferrans VJ, Sulciner DJ, Gutkind JS, Irani K, Goldschmidt-Clermont PJ, Finkel T (1996) Regulation of reactive-oxygen-species generation in fibroblasts by Rac1. Biochem J 318:379–382
[76] Jayanthi S, Ordonez S, McCoy MT, Cadet JL (1999) Dual mechanism of Fas-induced cell death in neuroglioma cells: a role for reactive oxygen species. Brain Res Mol Brain Res 72:158–165
[77] Wang S, Leonard SS, Ye J, Ding M, Shi X (2000) The role of hydroxyl radical as a messenger in Cr(VI)-induced p53 activation. Am J Physiol Cell Physiol 279:C868–C875
[78] Das UN (1990) Tuning free radical metabolism to kill tumor cells selectively with emphasis on the interaction(s) between essential fatty acids, free radicals, lymphokines and prostaglandins. Indian J Pathol Microbiol 33:94–103
[79] Dukes ID, McIntyre MS, Mertz RJ, Philipson LH, Roe MW, Spencer B, Worley JF 3rd (1994) Dependence on NADH produced during glycolysis for beta-cell glucose signaling. J Biol Chem 269:10979–10982

[80] Eto K, Tsubamoto Y, Terauchi Y, Sugiyama T, Kshimoto T, Takahashi N, Yamauchi N, Kubota N, Murayama S, Aizawa T, Akanuma Y, Aizawa S, Kasai H, Yazaki Y, Kadowaki T (1999) Role of NADH shuttle system in glucose-induced activation of mitochondrial metabolism and insulin secretion. Science 283:981–985
[81] Das UN (1992) Cis-unsaturated fatty acids as potential anti-mutagenic, tumoricidal and anti-metastatic agents. Asia Pacific J Pharmacol 7:305–327
[82] Galeotti T, Borrello S, Masoti L (1990) Oxy radical sources, scavenger systems and membrane damage in cancer cells. In: Das DK, Essman R (eds) Oxygen radicals: systemic events and disease processes. S. Karger, Basel, pp 129–148
[83] Dianzani MU, Rossi MA (1981) Lipid peroxidation in tumors. In: Pani P, Feo F, Columbano A, Cagliari ESA (eds) Recent trends in chemical carcinogenesis, vol 1. Cagliari, Italy, pp 243–257
[84] Bartoli GM, Galeotti T (1979) Growth-related lipid peroxidation in tumor microsomal membranes and mitochondria. Biochim Biophys Acta 574:537–541
[85] Hostetler KY, Zenner BD, Morris HP (1979) Phospholipid content of mitochondrial and microsomal membranes from Morris hepatomas of varying growth rates. Cancer Res 39:2978–2983
[86] Hartz JW, Morton RE, Waite MM, Morris HP (1982) Correlation of fatty acyl composition of mitochondrial and microsomal phospholipid with growth rate of rat hepatomas. Lab Invest 46:73–78
[87] Cheeseman KH, Emery S, Maddix SP, Slater TF, Burton GW, Ingold K (1988) Studies on lipid peroxidation in normal and tumour tissue. The Yoshida rat liver tumour. Biochem J 250:247–252
[88] Cheeseman KH, Burton GW, Ingold KU, Slater TF (1984) Lipid peroxidation and lipid antioxidants in normal and tumor cells. Toxicol Pathol 12:235–239
[89] Borrello S, Minotti G, Galeotti T (1988) Factors influencing O_2 and t-Bu OOH-dependent lipid peroxidation of tumor microsomes. In: Rotilio G (ed) Superoxide and superoxide dismutase in chemistry, biology and medicine. Elsevier, Amsterdam, pp 323–324
[90] Das UN, Begin ME, Ells G, Huang YS, Horrobin DF (1987) Polyunsaturated fatty acids augment free radical generation in tumor cells in vitro. Biochem Biophys Res Commun 145:15–24
[91] Das UN, Huang YS, Begin ME, Ells G, Horrobin DF (1987) Uptake and distribution of cis-unsaturated fatty acids and their effect on free radical generation in normal and tumor cells in vitro. Free Radic Biol Med 3:9–14
[92] Das UN (1990) Selective enhancement of free radicals in tumor cells as a strategy to kill tumor cells both in vitro and in vivo. In: Reddy CC et al (eds) Biological oxidation systems. Academic, New York, pp 607–624
[93] Bendetti A (1984) Loss of lipid peroxidation as an histochemical marker for preneoplastic hepatocellular foci of rats. Cancer Res 44:5712–5717
[94] Dunbar LM, Bailey JM (1975) Enzyme deletions and essential fatty acid metabolism in cultured cells. J Biol Chem 250:1152–1153
[95] Morton RE, Hartz JW, Reitz RC, Waite BM, Morris H (1979) The acyl-CoA desaturases of microsomes from rat liver and the Morris 7777 hepatoma. Biochim Biophys Acta 573:321–331
[96] Nassar BA, Das UN, Huang YS, Ells G, Horrobin DF (1992) The effect of chemical hepatocarcinogenesis on liver phospholipid composition in rats fed n-6 and n-3 fatty acid-supplemented diets. Proc Soc Exp Biol Med 199:365–368
[97] Oberley LW, Buettner G (1979) Role of SOD in cancer. A review. Cancer Res 39:1141–1149
[98] Tisdale MJ, Mahmoud MD (1983) Activities of free radical metabolizing enzymes in tumours. Br J Cancer 47:809–812
[99] Bize IB, Oberley LW, Morris HP (1980) Superoxide dismutase and superoxide radical in Morris hepatomas. Cancer Res 40:3686–3693
[100] Halliwell B, Gutteridge JMC (1999) Free radicals in biology and medicine, 3rd edn. Oxford University Press, Oxford

[101] Hampton MB, Fadeel B, Orrenius S (1998) Redox regulation of the caspases during apoptosis. Ann N Y Acad Sci 854:328–335

[102] Jacobson MD, Raff MC (1995) Programmed cell death and Bcl-2 protection in very low oxygen. Nature 374:814–816

[103] Zimmerman RJ, Marafino BJ Jr, Chan A, Landre P, Winkelhake JL (1989) The role of oxidant injury in tumor cell sensitivity to recombinant human tumor necrosis factor in vivo. Implications for mechanisms of action. J Immunol 142:1405–1409

[104] Vento R, D'Alessandro N, Giuliano M, Lauricella M, Carabillo M, Tesoriere G (2000) Induction of apoptosis by arachidonic acid in human retinoblastoma Y79 cells: involvement of oxidative stress. Exp Eye Res 70:503–517

[105] Bize IB, Oberley LW, Morris HP (1980) Superoxide dismutase and superoxide radical in Morris hepatomas. Cancer Res 40:3686–3693

[106] Das UN (2011) Selective enhancement of free radicals in tumor cells as a strategy to eliminate cancer. In: K Pantopoulos (ed) Principles of free radical biomedicine (in press)

[107] Cheeseman KH, Collins M, Proudfoot K, Slater TF, Burton GW, Webb AC, Ingold KU (1986) Studies on lipid peroxidation in normal and tumour tissues. The Novikoff rat liver tumour. Biochem J 235:507–514

[108] Malafa M, Margenthaler J, Webb B, Neitzel L, Christophersen M (2000) MnSOD expression is increased in metastatic gastric cancer. J Surg Res 88:130–134

[109] Westman NG, Marklund SL (1981) Copper- and zinc-containing superoxide dismutase and manganese-containing superoxide dismutase in human tissues and human malignant tumors. Cancer Res 41:2962–2966

[110] Toh Y, Kuninaka S, Oshiro T, Ikeda Y, Nakashima H, Baba H, Kohnoe S, Okamura T, Mori M, Sugimachi K (2000) Overexpression of manganese superoxide dismutase mRNA may correlate with aggressiveness in gastric and colorectal adenocarcinomas. Int J Oncol 17:107–112

[111] Izutani R, Asano S, Imano M, Kuroda D, Kato M, Ohyanagi H (1998) Expression of manganese superoxide dismutase in esophageal and gastric cancers. J Gastroenterol 33:816–822

[112] Zhong W, Yan T, Lim R, Oberley LW (1999) Expression of superoxide dismutases, catalase, and glutathione peroxidase in glioma cells. Free Radic Biol Med 27:1334–1345

[113] Huang Y, He T, Domann FE (1999) Decreased expression of manganese superoxide dismutase in transformed cells is associated with increased cytosine methylation of the SOD2 gene. DNA Cell Biol 18:643–652

[114] Zhong W, Oberley LW, Oberley TD, St Clair DK (1997) Suppression of the malignant phenotype of human glioma cells by overexpression of manganese superoxide dismutase. Oncogene 14:481–490

[115] Clement MV, Pervaiz S (1999) Reactive oxygen intermediates regulate cellular apoptosis response to apoptotic stimuli: an hypothesis. Free Radic Biol Med 30:247–252

[116] Kastan MB, Canman CE, Leonard CJ (1995) P53, cell cycle control and apoptosis: implications for cancer. Cancer Metastasis Rev 14:3–15

[117] Uberti D, Yavin E, Gil S, Ayasola KR, Goldfinger N, Rotter V (1999) Hydrogen peroxide induces nuclear translocation of p53 and apoptosis in cells of oligodendroglia origin. Brain Res Mol Brain Res 65:167–175

[118] Kitamura Y, Ota T, Matsuoka Y, Tooyama I, Kimura H, Shimohama S, Nomura Y, Gebicke-Haerter PJ, Taniguchi T (1999) Hydrogen peroxide-induced apoptosis mediated by p53 protein in glial cells. Glia 25:154–164

[119] Pani G, Bedogni B, Anzevino R, Colavitti R, Palazzotti B, Borrello S, Galeotti T (2000) Deregulated manganese superoxide dismutase expression an resistance to oxidative injury in p53-deficient cells. Cancer Res 60:4654–4660

[120] Hilf R, Murant RS, Narayana U, Gibson SL (1986) Relationship of mitochondrial function and cellular adensine triphosphate levels to hematoporphyrin derivative-induced photosensitization in R 3230 AC mammary tumors. Cancer Res 46:211–217

[121] Lee Y, Shacter E (2000) Hydrogen peroxide inhibits activation, not activity, of cellular caspase-3 in vivo. Free Radic Biol Med 29:684–692
[122] Saretzki G, von Zglinicki T (1999) Replicative senescence as a model of aging: the role of oxidative stress and telomere shortening-an overview. Z Gerontol Geriatr 32:69–75
[123] Oikawa S, Kawanishi S (1999) Site-specific DNA damage at GGG sequence by oxidative stress may accelerate telomere shortening. FEBS Lett 453:365–368
[124] Tyurina YY, Tyurina VA, Certa G, Quinn PJ, Schor NF, Kagan VE (1997) Direct evidence for antioxidant effect of BCL-2 in PC 12 rat pheochromocytoma cells. Arch Biochem Biophys 344:413–423
[125] Haldar S, Negrini M, Monne M, Sabbioni S, Croce CM (1994) Down regulation of bcl-2 by p53 in breast cancer cells. Cancer Res 54:2095–2097
[126] Haldar S, Jena N, Croce CM (1995) Inactivation of Bcl-2 by phosphorylation. Proc Natl Acad Sci U S A 92:4507–4511
[127] Hockenbery DM, Oltvai ZN, Yin XM, Milliman CL, Korsmeyer SJ (1993) Bcl-2 functions in an antioxidant pathway to prevent apoptosis. Cell 75:241–251
[128] Das UN (1999) Essential fatty acids, lipid peroxidation and apoptosis. Prostaglandins Leukot Essent Fatty Acids 61:157–163
[129] Padma M, Das UN (1999) Effect of cis-unsaturated fatty acids on the activity of protein kinases and protein phosphorylation in macrophage tumor (AK-5) cells in vitro. Prostaglandins Leukot Essent Fatty Acids 60:55–63
[130] Morisaki N, Lindsey JA, Stitts JM, Zhang H, Cornwell DG (1984) Fatty acid metabolism and cell proliferation. V. Evaluation of pathways for the generation of lipid peroxides. Lipids 19:381–394
[131] Morisaki N, Sprecher H, Milo GE, Cornwell DG (1982) Fatty acid specificity in the inhibition of cell proliferation and its relationship to lipid peroxidation and prostaglandin biosynthesis. Lipids 17:893–899
[132] Liepkalns VA, Icard-Liepkalns C, Cornwell DG (1982) Regulation of cell division in a human glioma cell clone by arachidonic acid and alpha-tocopherol quinone. Cancer Lett 15:173–178
[133] Tolnai S, Morgan JF (1962) Studies on the in vitro anti-tumor activity of fatty acids. V. Unsaturated fatty acids. Can J Biochem Physiol 40:869–875
[134] Rossi MA, Cecchini G (1983) Lipid peroxidation in hepatomas of different degrees of deviation. Cell Biochem Funct 1:49–54
[135] Burlakova EB, Palmina NP (1967) On the possible role of free radical mechanism on the regulation of cell replication. Biofizika 12:82–88
[136] Gonzalez M, Schemmel R, Dugan L, Gray J, Welsch C (1993) Dietary fish oil inhibits human breast carcinoma growth: a function of increased lipid peroxidation. Lipids 28:827–832
[137] Das UN (2008) Can essential fatty acids reduce the burden of disease(s)? Lipids Health Dis 7:9
[138] Das UN (2006) Tumoricidal and anti-angiogenic actions of gamma-linolenic acid and its derivatives. Curr Pharm Biotechnol 7:457–466
[139] Cao Y, Pearman AT, Zimmerman GA, McIntyre TM, Prescott SM (2000) Intracellular unesterified arachidonic acid signals apoptosis. Proc Natl Acad Sci U S A 97:11280–11285
[140] Brekke OL, Sagen E, Bjerve KS (1999) Specificity of endogenous fatty acid release during tumor necrosis factor-induced apoptosis in WEHI 164 fibrosarcoma cells. J Lipid Res 40:2223–2233
[141] Ramesh G, Das UN (1996) Effect of free fatty acids on two stage skin carcinogenesis in mice. Cancer Lett 100:199–209
[142] Calviello G, Palozza O, Piccioni E, Maggiano N, Frattucci A, Franceschelli P, Bartoli GM (1998) Supplementation with eicosapentaenoic and docosahexaenoic acid inhibits growth of Morris hepatocarcinoma 3924A in rats: effects on proliferation and apoptosis. Int J Cancer 75:699–705
[143] Chapkin RS, Jiang YH, Davidson LA, Lupton JR (1997) Modulation of intracellular second messengers by dietary fat during colonic tumor development. Adv Exp Med Biol 422:85–96

[144] Rose DP, Hatala MA, Connolly JM, Rayburn J (1993) Effect of diets containing different levels of linoleic acid on human breast cancer growth and lung metastasis in nude mice. Cancer Res 53:4686–4690

[145] Rose DP, Connolly JM, Liu X-H (1994) Dietary fatty acids and human breast cancer cell growth, invasion, and metastasis. Adv Exp Med Biol 364:83–91

[146] Rose DP, Connolly JM, Liu X-H (1994) Effects of linoleic acid on the growth and metastasis of two human breast cancer cell lines in nude mice and the invasive capacity of these cell lines in vitro. Cancer Res 54:6557–6562

[147] El-Ela SHA, Prasse KW, Carroll R (1987) Effects of dietary primrose oil on mammary tumorigenesis induced by 7, 12-dimethyl bez(a)anthracene. Lipids 22:1041–1044

[148] El-Ela SHA, Prasse KW, Carroll R, Wade AE, Dharwadkar S, Bunce OR (1988) Eicosanoids synthesis in 7, 12-dimethylbenz(a)anthracene-induced mammary carcionomas in Sprague-Dawley rats fed primrose oil, menhaden oil or corn oil diet. Lipids 23:948–954

[149] Cameron E, Bland J, Marcuson R (1989) Divergent effects of omega-6 and omega-3 fatty acids on mammary tumor development in C3H/Heston mice treated with DMBA. Nutr Res 9:383–393

[150] Ramesh G, Das UN, Koratkar R, Padma M, Sagar PS (1992) Effect of essential fatty acids on tumor cells. Nutrition 8:343–347

[151] Menendez JA, del Mar Barbacid M, Montero S, Sevilla E, Escrich E, Solanas M, Cortes-Funes H, Colomer R (2001) Effects of gamma-linolenic acid and oleic acid on paclitaxel cytotoxicity in human breast cancer cells. Eur J Cancer 37:402–413

[152] Ravichandran D, Cooper A, Johnson CD (1998) Growth inhibitory effect of lithium gammalinolenate on pancreatic cancer cell lines: influence of albumin and iron. Eur J Cancer 34:188–192

[153] Ravichandran D, Cooper A, Johnson CD (2000) Effect of 1-γlinolenyl-3-eicosapentaenoyl propane diol on the growth of human pancreatic carcinoma in vitro and in vivo. Eur J Cancer 36:423–427

[154] Kenny FS, Gee JM, Nicholson RI, Morris T, Watson S, Bryce RP, Hartley J, Robertson JFR (1998) Effect of dietary GLA +/- tamoxifen on growth and ER in a human breast cancer xenograft model. Eur J Cancer 34:S20

[155] Kenny FS, Pinder S, Ellis IO, Bryce RP, Hartley J, Robertson JFR (1998) Gamma linolenic acid with tamoxifen as primary therapy in breast cancer. Eur J Cancer 34:S18–S19

[156] Gleissman H, Yang R, Martinod K, Lindskog M, Serhan CN, Johnsen JI, Kogner P (2010) Docosahexaenoic acid metabolome in neural tumors: identification of cytotoxic intermediates. FASEB J 24:906–915

[157] Madhavi N, Das UN (1994) Effect of n-6 and n-3 fatty acids on the survival of vincristine sensitive and resistant human cervical carcinoma cells in vitro. Cancer Lett 84:31–41

Chapter 15
Aging

Introduction

Ageing or aging is the accumulation of changes in an organism over time. Aging in humans refers to a multidimensional process of physical, psychological, and social change. Some dimensions of ageing grow and expand over time, while others decline. Reaction time, for example, may slow with age, while knowledge of world events and wisdom may expand. Evidence suggests that even late in life potential exists for physical, mental, and social growth and development. Aging is an important part of all human societies reflecting the biological changes that occur, but also reflecting cultural and societal conventions. It is estimated that approximately 100,000 people worldwide die each day of age-related causes.

The term "aging" is somewhat ambiguous. Distinctions may be made between "universal ageing" (age changes that all people share) and "probabilistic ageing" (age changes that may happen to some, but not all people as they grow older, such as the onset of type 2 diabetes mellitus). Chronological ageing, referring to how old a person is, is arguably the most straightforward definition of aging and may be distinguished from "social ageing" (society's expectations of how people should act as they grow older) and "biological ageing" (an organism's physical state as it ages).

Differences are sometimes made between populations of elderly people. Divisions are sometimes made between the young old (65–74), the middle old (75–84) and the oldest old (85+). However, chronological age does not correlate perfectly with functional age, but differ in their mental and physical capacities.

Population aging is the increase in the number and proportion of older people in society. Aging has a significant impact on society. Young people tend to commit most crimes; they are more likely to push for political and social change, to develop and adopt new technologies, and to need education. Older people have different requirements from society and government as opposed to young people, and frequently differing values as well. Older people are also far more likely to vote, and hence, the aged have comparatively more political influence.

In biology, *senescence* is the state or process of aging. *Cellular senescence* is a phenomenon where isolated cells demonstrate a limited ability to divide in culture,

U. N. Das, *Molecular Basis of Health and Disease*,
DOI 10.1007/978-94-007-0495-4_15, © Springer Science+Business Media B.V. 2011

while *organismal senescence* is the ageing of organisms. After a period of near perfect renewal (in humans, between 20 and 35 years of age), organismal senescence is characterized by the declining ability to respond to stress, increasing homeostatic imbalance and increased risk of disease. This irreversible series of changes inevitably ends in death. As genes and environmental factors influence aging are being discovered, ageing is increasingly being regarded as a disease that is potentially treatable like other diseases.

Telomere and Aging

Indeed, aging is not an unavoidable property of life. Instead, it is the result of a genetic program. In humans and other animals, cellular senescence has been attributed to the shortening of telomeres with each cell cycle; when telomeres become too short, the cells die. The length of telomeres is therefore can be considered as the "molecular clock," of aging process.

Telomere length is maintained in immortal cells (e.g., germ cells and keratinocyte stem cells, but not other skin cell types) by the telomerase enzyme. It is possible to immortalize mortal cells by the activation of their telomerase gene. Cancerous cells are almost immortal due to the reactivation of their telomerase gene by mutation. Since this mutation is rare, the telomere "clock" can be seen as a protective mechanism against cancer [1]. Other genes are known to affect the aging process include the sirtuin family of genes that have been shown to have a significant effect on the lifespan of yeast and nematodes. Overexpression of the RAS2 gene increases lifespan in yeast substantially.

In addition, diet has been shown to substantially affect lifespan in many animals. Specifically, caloric restriction (that is, restricting calories to 30–50% less than an *ad libitum* animal would consume, while still maintaining proper nutrient intake), has been shown to increase lifespan in mice up to 50%. Caloric restriction increases lifespan in primates although the increase in lifespan is only notable if the caloric restriction is started early in life. Since, at the molecular level, age is counted not as time but as the number of cell doublings, this effect of calorie reduction could be mediated by the slowing of cellular growth and, therefore, the lengthening of the time between cell divisions.

Theories of Aging

At present, the biological basis of ageing is unknown. It is common knowledge that substantial variability exists in the rates of ageing across different species, and that this to a large extent is genetically based. In model organisms and laboratory settings, researchers have been able to demonstrate that selected alterations in specific genes can extend lifespan (quite substantially in nematodes, less so in fruit flies, and even less in mice). Nevertheless, even in the relatively simple organisms, the mechanisms of aging remain to be elucidated. Because the lifespan of even the simple lab mouse is around 3 years, very few experiments directly test specific ageing theories.

Telomere and Aging

Telomeres (structures at the ends of chromosomes) have experimentally been shown to shorten with each successive cell division. Shortened telomeres activate a mechanism that prevents further cell multiplication. This may be an important mechanism of ageing in tissues like bone marrow and the arterial lining where active cell division is necessary. Importantly though, mice lacking telomerase enzyme do not show a dramatically reduced lifespan, as the simplest version of this theory would predict.

In a study wherein telomere length, telomerase activity and chromosome rearrangements in human cells before and after transformation with SV40 or Ad5 was measured, it was noted that in all mortal populations, telomeres shortened by approximately 65 bp/generation during the lifespan of the cultures. In immortal cells, telomere length and frequency of dicentric chromosomes stabilized. Telomerase activity was not detectable in control or extended lifespan populations but was present in immortal populations. These results suggest that critically shortened telomeres may be incompatible with cell proliferation and stabilization of telomere length by telomerase may be required for immortalization especially, in cancer cells [2]. These results are supported by the observation that telomerase activation may be a common step in immortalization [3].

In addition, telomeres, the G/C-rich DNA sequences capping the ends of all eukaryotic chromosomes, have been shown to shorten during replicative aging of normal cells both in vitro and in vivo (Fig. 15.1). Moreover, variation in the initial length of terminal restriction fragments (TRF) accounts for much of the variation in replicative capacity of fibroblast cultures from different donors. There appears to be a critical or threshold length that acts as a signal for cell senescence. This is supported by the observation that replicative capacity was found to be directly proportional to mean TRF length ($m = 7.2$ population doublings/kbp, $r = 0.65$, $P = 0.0004$) and total signal intensity ($m = 25.0$ population doublings/unit, $r = 0.63$, $P < 0.003$) at early passage in fibroblasts in culture. More importantly, the variability in both mean TRF length and signal intensity ($F = 2.0$ and 2.9; $P = 0.02$ and 0.03, respectively) at senescence was markedly less than that at early passage. Thus, there exists a critical telomere length in senescing cells and a causal role of telomere shortening in cell senescence [4]. Furthermore, telomeres in somatic cells are progressively shortened with aging since, shortening of telomeric repeats correlated with aging ($p < 0.0001$), but not with white blood cell count, neutrophil count, and smoking habit [5].

It is important to note that in most multicellular eukaryotic organisms, telomerase (an enzyme that is involved in synthesis of telomere) is active only in germ cells, stem cells and leukocytes. Telomeres act as a sort of time-delay "fuse", eventually

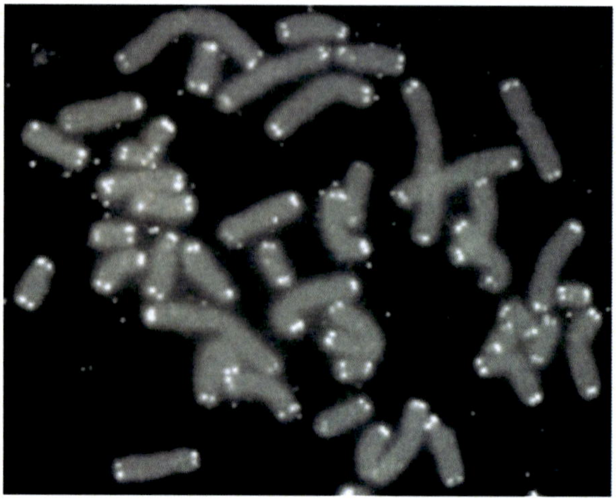

Fig. 15.1 Human chromosomes (*grey*) capped by telomeres (*white*)

running out after a certain number of cell divisions and resulting in the eventual loss of vital genetic information from the cell's chromosome with future divisions.

Eukaryotic telomeres normally terminate with 3′ single-stranded-DNA overhang which is essential for telomere maintenance and capping. Telomeres form large loop structures called telomere loops, or T-loops. Here, the single-stranded DNA curls around in a long circle stabilized by telomere-binding proteins. At the very end of the T-loop, the single-stranded telomere DNA is held onto a region of double-stranded DNA by the telomere strand disrupting the double-helical DNA and base pairing to one of the two strands. This triple-stranded structure is called a displacement loop or D-loop.

Telomere shortening in humans induces replicative senescence which blocks cell division, a mechanism that is meant to prevent genomic instability and development of cancer in human aged cells by limiting the number of cell divisions. On the other hand, malignant cells which bypass this arrest become immortalized by telomere extension mostly due to the activation of telomerase, the reverse transcriptase enzyme responsible for synthesis of telomeres. However, 5–10% of human cancers activate the Alternative Lengthening of Telomeres (ALT) pathway which relies on recombination-mediated elongation.

Thus, in theory it is possible to extend human life by lengthening the telomeres in cells that could be achieved by the activation of telomerase (by drugs), or possibly permanently by gene therapy. But, so far these ideas have not been proven in humans.

However, it has been hypothesized that there is a trade-off between cancerous tumor suppression and tissue repair capacity, in that lengthening telomeres might slow aging and in exchange increase vulnerability to cancer [6].

It is important to realize that we still do not know the exact relationship between telomeres and aging. Changing telomere lengths are usually associated with changing speed of senescence. This telomere shortening, however, might be a consequence of, and not a reason for, aging. That the role of telomeres is far from being understood is demonstrated by two recent studies on long-lived seabirds. The telomeres of Leach's Storm-petrel (*Oceanodroma leucorhoa*) seem to lengthen with chronological age, the first observed instance of such behaviour of telomeres [7]. Juola et al. [8] reported that in the long-lived seabird species, the Great Frigatebird (*Fregata minor*), telomere length did decrease until at least c.40 years of age (i.e., probably over the entire lifespan), but the speed of decrease slowed down massively with increasing ages, and that rates of telomere length decrease varied strongly between individual birds. This suggests that, at least, in this species (and probably in frigatebirds and their relatives in general), telomere length could not be used to determine a bird's age sufficiently well. Thus, it seems that there is much more variation in the behavior of telomere length than initially believed.

In a study of 30 men with low-risk prostate cancer a study was made on the possible effects of lifestyle changes on telomeres. The men were asked to make several lifestyle changes, including attending a 3-day retreat; eating a diet low in refined sugars and rich in whole foods, fruits, and vegetables, with only 10% of calories derived from fat; and engaging in several other activities, such as moderate aerobic exercise, relaxation techniques and breathing exercises. Telomerase levels were measured at baseline, and again after 3 months, when it was found that telomerase in the blood had increased by 29%, suggesting that comprehensive lifestyle changes may cause improvements in telomerase and telomeres that may be beneficial. These results need to be confirmed in a large scale study. But, the results of this interesting study do indicate that specific lifestyle changes can increase telomerase activity and thus, prolong lifespan or decrease the rate of aging [9].

Telomere in Type 2 Diabetes Mellitus

The lengths of the terminal restriction fragments (TRFs) of DNA of leukocytes from 234 white men comprising 54 patients with type 1 diabetes mellitus, 74 patients with type 2 diabetes mellitus and 106 control subjects when examined, it was noted that the TRF length in type 1 diabetes was significantly shorter than that of nondiabetic control subjects (mean $\pm$ SE: 8.6 ± 0.1 vs. 9.2 ± 0.1, $P = 0.002$). No significant difference was observed between the TRF length from leukocytes of patients with type 2 diabetes versus nondiabetic subjects. Neither the duration nor the complications of type 1 diabetes mellitus (i.e., nephropathy and hypertension) had an effect on the TRF length of leukocytes from patients with type 1 diabetes. The shortened TRF length of leukocytes of patients with type 1 diabetes led to the suggestion that a marked reduction in the TRF length of subsets of leukocytes may play a role in the pathogenesis of type 1 diabetes [10]. But, subsequent studies showed that telomere shortening occurs in Asian Indian Type 2 diabetic patients [11]. When peripheral venous monocyte and T-cell telomere length was examined, mean monocyte telomere length in the type 2 diabetic group was highly significantly lower than in control subjects without significant differences in lymphocyte telomere length. A trend toward increased oxidative DNA damage in all diabetes cell types examined and a significant inverse relationship between oxidative DNA damage and telomere length in the diabetic group was observed. The telomere length was found to be unrelated to plasma CRP concentration or insulin resistance. Thus, monocyte telomere shortening in type 2 diabetes could be due to increased oxidative DNA damage that suggests that monocytes adhering to vascular endothelium and entering the vessel wall in type 2 diabetes are from a population with shorter telomeres and at increased risk of replicative senescence within vascular plaque [12]. It was also reported that telomere shortening is seen even at the stage of impaired glucose tolerance and among subjects with Type 2 diabetes, those with atherosclerotic plaques had greater shortening of telomere length compared to those without plaques [13]. What is more interesting is the report that subjects with type 2 diabetes and microalbuminuria have shorter TRF length and increased arterial stiffness than those without microalbuminuria. Additionally, TRF length was found to be associated with age, albumin excretion rate, and nitrosative stress, implying that shorter TRF length could indicate older biological age and the increased arterial stiffness in patients with type 2 diabetes who have microalbuminuria may be due to the more pronounced "aging" of these subjects [14].

It is important to note that despite the fact that telomeres were significantly shorter in the arteries of diabetic versus non-diabetic patients ($p = 0.049$) and in mononuclear cells of both type I and type II diabetes, in diabetics good glycemic control attenuated shortening of the telomeres. But surprisingly, in arterial cells good glycemic control attenuated, but not abolished, the telomere shortening, whereas in type II diabetics the mononuclear telomere attrition was completely prevented by adequate glycemic control. Telomere shortening in mononuclear cells of type I diabetic patients was attenuated but not prevented by good glycemic control. These results suggest that diabetes is associated with premature cellular senescence which can be prevented by good glycemic control in type II DM and reduced in type I DM [15].

In addition, telomere shortening was described in diabetics with complications such as nephropathy and myocardial infarction [16, 17]. Moreover, glucose, HbA1C and waist-to-hip ratio, variables related to glycemic control, showed a significant inverse correlation with leukocyte telomeres length, indicating that it is the complications due to diabetes are responsible for the telomere shortening rather than diabetes itself. Thus, it can be said that hyperglycemia might be driving the oxidative-induced telomere loss in diabetes and the contribution of inflammation cannot be excluded. The leukocyte telomere length probably reflects the lifelong accumulating burden of increased oxidative stress and inflammation. The rate of telomere shortening seem to depend on the baseline telomere length indicating that in longer telomeres, greater loss per cell division is more likely to occur. This, coupled with the high heritability in telomere length shown by twin studies [18, 19], supports the hypothesis that telomere length is, to a large extent, genetically determined. It also supports the theory, that predisposition to CVD and/or diabetes might be expressed through inherited short telomeres. The shorter leukocyte telomere length associated with the presence of type 2 diabetes mellitus could be partially attributed to the high oxidative stress in these patients. Since an association of the UCP2 functional promoter variant with the leukocyte telomere length has been found, it suggests a link between mitochondrial production of reactive oxygen species and shorter telomere length in type 2 diabetes mellitus [20, 21].

Telomere and Hypertension

Similar to the changes found in telomere length seen in diabetes mellitus, even hypertension is associated with changes in telomere length. Pulse pressure rises with age, and it might serve as a phenotype of biological aging of the vasculature. In a study conducted in twins as to the relationship between telomere length in white blood cells and pulse pressure, it was reported that terminal restriction fragment (TRF) length showed an inverse relation with pulse pressure. Both TRF length and pulse pressure were found to be highly familial indicating that telomere length, which is under genetic control, plays a role in the regulation of pulse pressure, including vascular aging [22]. These results are supported by the observation that the rate of age-dependent telomere attrition was higher in both the intima and media of the distal versus proximal abdominal aorta and telomere length was negatively correlated with atherosclerotic grade. However, after adjustment for age, this relationship was not statistically significant. The high rate of age-dependent telomere attrition in the distal abdominal aorta probably reflects enhanced cellular turnover rate due to local factors such as an increase in shear wall stress in this vascular segment. Thus, the rate of telomere attrition can be considered as a function of age and atherosclerosis in cells of the human abdominal aorta [23]. In addition, it was reported that telomerase activity was higher in patients under 45 year-old with uncontrolled hypertension as compared with healthy individuals and patients under 45 year-old with well controlled hypertension ($p < 0.05$). The white blood cell count was higher in the hypertensive than the control group ($p < 0.05$) and in the latter group telomerase activity was significantly lower than in the other two groups ($p < 0.05$). Since white blood cell count was higher in the hypertensive than the control group ($p < 0.05$) these results indicate that a relationship exists between telomerase activity in peripheral leukocytes, the proliferation of these white blood cells and the presence of essential arterial hypertension [24]. Since hypertension is a major risk factor for atherosclerotic lesions, shorter telomere length in white blood cells can be associated with an increased predilection to carotid artery atherosclerosis [25] and could be used as a marker of the latter. Since telomere attrition in white blood cells is more closely associated with endothelial damage and atherosclerosis than is chronological aging, lends support to the hypothesis that mean telomere restriction fragment length (mTRFL) in white blood cells could be used as a marker for biological aging of the cardiovascular system [26]. In addition, in the Bogalusa Heart Study, over 10.1–12.8 years, the relative changes in telomere length correlated with the homeostasis model assessment of insulin resistance ($r = -0.531$, $P < 0.001$) and changes in the body mass index ($r = -0.423$, $P < 0.001$), suggesting a close association among telomere biology with insulin resistance and adiposity in humans [27]. These results are further supported by the observation that insulin resistance, leptin, and CRP were inversely correlated with leukocyte TRFL in premenopausal but not postmenopausal women, while insulin resistance, CRP, but not leptin independently accounted for variation in white blood cell TRFL in premenopausal women indicating that menopausal status impacts leukocyte telomere length and its relation with insulin resistance and inflammation in women [28]. Thus, there appears to be a close association between inflammation and telomere length: the higher the inflammation the shorter the telomere length.

In a study that explored the relations of leukocyte telomere length, expressed by terminal restriction fragment (TRF) length, with insulin resistance, oxidative stress and hypertension, TRF length was inversely correlated with age ($r = -0.41$, $P < 0.0001$) and age-adjusted TRF length was inversely correlated with the Homeostatic Model Assessment of Insulin Resistance (HOMA-IR) ($r = -0.16$, $P = 0.007$) and urinary 8-epi-PGF(2alpha) ($r = -0.16$, $P = 0.005$)—an index of systemic oxidative stress. Hypertensive subjects exhibited shorter age-adjusted TRF length (hypertensives = 5.93 ± 0.042 kb, normotensives = 6.07 ± 0.040 kb, $P = 0.025$). These observations further strengthened the association noted previously that hypertension, increased insulin resistance and oxidative stress are associated with shorter leukocyte telomere length and that shorter leukocyte telomere length in hypertensives is largely due to insulin resistance [29]. Leukocyte telomere length was shorter in individuals with a higher renin-to-aldosterone ratio, especially those with hypertension [30], and activation of the renin-angiotensin-aldosterone system is associated with increased oxidative stress and inflammation, yet another piece of evidence that supports the contention that oxidative stress induced by inflammation induces telomere shortening.

It is interesting to note that leukocyte telomere length (LTL) was positively associated with serum IGF-I concentration in elderly men [31, 32]. It was reported that an increase of 0.08 kb in LTL for each standard deviation increase of IGF-1 ($p = 0.04$), while IGFBP-1 and IGFBP-3 were not significantly associated with LTL. Thus, high IGF-1 may be an independent predictor of longer LTL, suggesting a role for IGF-1 in mechanisms relating to telomere maintenance.

Endothelial Dysfunction, Insulin Resistance, Obesity, Hypertension, Type 2 Diabetes, Inflammation and Telomere

Despite these evidences that suggest that insulin resistance, hypertension, type 2 diabetes mellitus and aging are associated with shorter telomere, still the biological meaning of the associations between LTL and aging-related diseases is not clear. But, what is known is that LTL is highly variable at birth and throughout life [33], age-dependent LTL shortening is much faster in early life than during adulthood [34] probably because of the rapid proliferation of hematopoietic stem cells (HSCs) during growth and development and LTL shortening throughout life largely mirrors telomere shortening in HSCs. In HSCs, like in other somatic cells with rudimentary activity of telomerase, telomere shortening records replication [35]. Since oxidative stress exerts major influence on telomere shortening, because the GGG triplets on the telomeres are highly sensitive to the hydroxyl radical, telomere dynamics (telomere length and its shortening), is a record of not only the replicative history but also the accruing burden of oxidative stress of cell populations that undergo replication. Because inflammation and oxidative stress are at the center of the aging process and low-grade systemic inflammation is present in obesity, insulin resistance, hypertension and type 2 diabetes mellitus, and all these diseases are associated with endothelial dysfunction; it is reasonable to suggest that shorter LTL is associated with aging-related diseases and particularly atherosclerosis. From this point of view, shortened LTL, and, by implication, telomere length in HSCs, are biomarkers of the atherosclerotic process. However, it also suggests that LTL might somehow relate to the function of endothelial progenitor cells. Thus, shortened LTL might be an index of reduced HSC reserves (and so to endothelial dysfunction) expressed in a limited ability of the bone marrow to supply adequately functioning endothelial progenitor cells. The development of most diseases is as a result of the outcome of an imbalance between injurious factors and elements that serve to counter their effects and/or the healing process. Endothelial progenitor cells originate from the HSC pool and possess the unique ability of homing to sites of injured endothelium, where they integrate themselves into the vascular wall and engage in endothelial repair. Atherosclerosis, obesity, insulin resistance, hypertension, and type 2 diabetes mellitus that apparently start with endothelial injury, is marked by diminished numbers of endothelial progenitor cells in the circulation and reduced function of these cells, a flaw that might arise from shortened telomere length [36, 37]. If this argument is true, it implies that factors that enhance the repair process or suppress inflammation and oxidative stress could prevent telomere shortening.

PUFAs and Their Anti-inflammatory Products and Telomere

As already discussed in the previous chapters, polyunsaturated fatty acids and their anti-inflammatory products such as lipoxins, resolvins, protectins and maresins might be one such endogenous factor(s) that protect telomere from oxidative stress. This argument is supported by the report that breast cancer risk may be affected by telomere length among premenopausal women or women with low dietary intake of antioxidants or antioxidant supplements [38]. Furthermore, in a recent study performed in a cohort of patients with CHD it was reported that individuals in the lowest quartile of DHA + EPA experienced the fastest rate of telomere shortening (0.13 telomere-to-single-copy gene ratio[T/S] units over 5 years; 95% confidence interval[CI], 0.09–0.17), whereas those in the highest quartile experienced the slowest rate of telomere shortening (0.05 T/S units over 5 years; 95% CI, 0.02–0.08; $P < 0.001$ for linear trend across quartiles). Levels of DHA + EPA were associated with less telomere shortening before and after sequential adjustment for established risk factors and potential confounders. Each 1-SD increase in DHA + EPA levels was associated with a 32% reduction in the odds of telomere shortening, suggesting in patients with coronary artery disease an inverse relationship exists between baseline blood levels of marine omega-3 fatty acids and the rate of telomere shortening over 5 years [39]. Though the exact mechanism by which EPA/DHA prevented reduction in the telomere length, I propose that this could be attributed to the increased formation of their anti-inflammatory compounds such as lipoxins, resolvins, protectins and maresins.

Several studies have reported cross sectional associations between longer telomeres and nutritional supplements, including multivitamins, vitamin C, vitamin D, vitamin E, and folic acid that possibly have anti-oxidant actions [40–44].

In contrast, in cultured colorectal adenocarcinoma cells, EPA and DHA suppressed telomerase activity and reduced telomerase levels [45]. But, there are no studies to date that explored the biological effect of omega-3 fatty acids on telomerase in noncancerous tissues. But, based on the above studies, it is clear that omega-3 fatty acids might exert bidirectional effects on telomerase depending on cellular context: in healthy tissues, they may enhance telomerase activity while suppressing it in cancerous cells. Such properties would be highly desirable in the development of treatments targeting telomeric aging. In part, these bidirectional effects of EPA/DHA on telomere length in normal vs tumor cells could be attributed to the products that are formed in normal and tumor cells. As already discussed in Chap. 14 on cancer, it is likely that in normal cells there is a preferential formation of anti-inflammatory products such as lipoxins, resolvins, protectins and maresins that prevent telomere shortening (or enhance the activity of telomerase) whereas in the tumor cells EPA/DHA might give rise to pro-inflammatory compounds such as PGs, LTs, TXs that enhance oxidative stress and shorten telomere (see Fig. 15.2). This interesting proposal needs to be verified and confirmed.

P53, Telomere, Aging, Type 2 Diabetes Mellitus, Cancer

In this context, the relationship among p53 gene, oxidative stress, aging and cancer is rather puzzling. Matheu et al. [46] showed that genetically manipulated mice with increased, but otherwise normally regulated, levels of Arf (positive regulator of p53) and p53 present strong cancer resistance and have decreased levels of aging-associated damage. These observations suggested a protective role of Arf/p53 to aging and indicated a rationale for the co-evolution of cancer resistance and longevity. In contrast, other investigators have found that permanent activation of p53 results in premature ageing [47, 48], and that the absence of p53 may alleviate the premature ageing of mice with high levels of constitutive endogenous damage [49]. It is of critical importance to note that, in these mouse models of premature ageing, p53 is permanently activated, owing either to truncation of p53 domains or to constitutive endogenous damage. When p53 activity is enhanced but normal regulation is retained there is no acceleration of ageing, as observed by Matheu et al. [46] with increased p53 and increased Arf or decreased Mdm2 [50–53]. Moreover, a combined increase in Arf and p53 resulted in detectable anti-ageing activity. These results suggest that p53 under normal physiological conditions may have a global anti-oxidant effect, thus

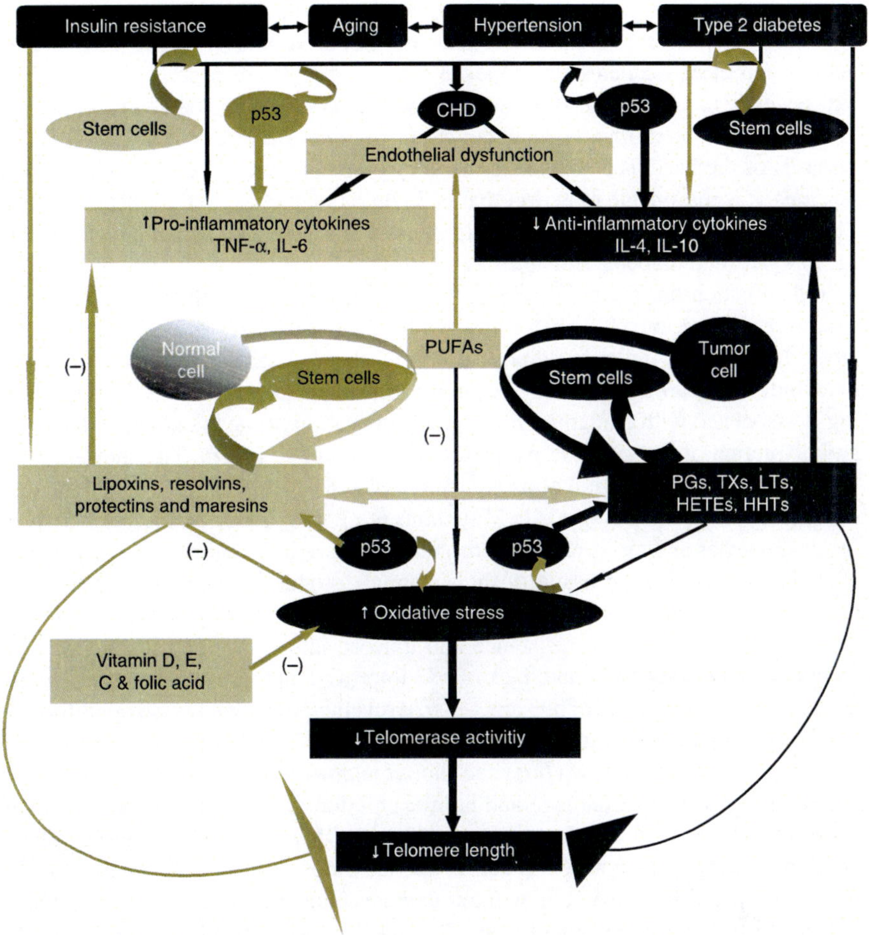

Fig. 15.2 Scheme showing relationship among aging and insulin resistance, hypertension, type 2 diabetes mellitus and CHD and oxidative stress, PUFAs, lipoxins, resolvins, protectins, maresins, eicosanoids and telomere length. Obesity resulting from overnutrition stimulates the generation of ROS that would overwhelm the antioxidant protection in adipose and other tissues, enhance the production of pro-inflammatory cytokines, decrease the release of anti-inflammatory cytokines and thus, induce low-grade systemic inflammation, accelerate DNA damage and aging. The obesity-mediated aging of adipose tissue and other tissues especially endothelial cells is also associated with telomere shortening, which leads to alteration in the expression of p53. These events trigger endothelial dysfunction and insulin resistance that ultimately lead to the development of hypertension, type 2 diabetes mellitus, atherosclerosis and aging. Overnutrition and insulin resistance suppress the activities of Δ^6 and Δ^5 desaturases leading to reduced formation of PUFAS, the precursors of lipoxins, resolvins and protectins. Decreased levels of lipoxins, resolvins and protectins results in impaired resolution of inflammation, DNA damage, telomere shortening, p53 dysfunction, impaired stem cell function that ultimately initiate the development and progression of aging and age-associated diseases. Response to treatment or progression of diseases such as hypertension, type 2 diabetes, atherosclerosis and aging process will be slow if adequate numbers of stem cells are present in various tissues and target organs. PUFAs and their various metabolites influence the stem cell biology and thus, affect aging process and aging-associated diseases

decreasing ageing-associated oxidative damage and intimately linked to longevity and cancer resistance. It is possible that p53 may have both pro- and anti-oxidant actions and these diametrically opposite actions of p53 may depend on other local and systemic factors, circumstances under which p53 is being coerced to act, etc.

Since age-related illnesses include: cardiovascular disease, cancer, degenerative diseases of the brain such as Alzheimer's disease and metabolic disorders such as diabetes, it is reasonable to expect that p53 could be involved in their pathobiology. Since, type 2 diabetes is also a recognized cause of accelerated aging, understanding the link between diabetes and aging is important.

Inflammation has a crucial role in the development of endothelial dysfunction, insulin resistance, type 2 diabetes and cardiovascular diseases associated with obesity. Macrophages infiltrate adipose tissue in obese states, and pro-inflammatory cytokines levels are elevated and cause insulin resistance and type 2 diabetes. Aging is associated with oxidative stress, genetic instability (due to oxidative damage), and disruption of homeostatic pathways and to telomere length. The shortening of telomeres leads to activation of tumor suppressors, in particular p53, which induces cell cycle arrest and aging [54]. Minamino et al. [55] reported that p53 derived from adipocytes and macrophages contributed to the aging of adipose tissue in obese animals and the adipose tissue of these animals produced more free radicals, secreted higher amounts of pro-inflammatory cytokines such as TNF-α, had genetic instability, showed insulin resistance and glucose intolerance and lower levels of adiponectin. On the other hand, lack of p53 improved all these abnormalities, while transgenic overexpression of p53 and *Cdkn1a* (cyclin-dependent kinase inhibitor 1A) in adipose tissue induced inflammation and insulin resistance.

Mice that lack telomerase (*Tert*) had shorter telomeres and G4 *Tert*-deficient mice showed increased DNA damage and high expression of senesence markers such as senescence-associated β-galactosidase, p53 and *Cdkn1a* in adipose tissue. These abnormalities were also noted in adipose tissue biopsies from individuals with diabetes compared to individuals without diabetes, suggesting that adipose tissue is associated with an aging phenotype. The G4 *Tert*-deficient mice developed glucose intolerance and insulin resistance on a high-fat-high-sugar diet compared to age-matched wild-type mice on a similar diet. Macrophages infiltrated the adipose tissue of G4 *Tert*-deficient mice, and they developed insulin resistance in the liver and muscle independent of body weight. Surgical removal of adipose tissue improved glucose metabolism in G4 *Tert*-deficient mice, whereas transplantation of G4 *Tert*-deficient adipose tissue into normal mice induced insulin resistance that was attenuated when adipose tissue deficient in both telomerase and p53 was transplanted into normal mice, thus establishing a functional connection between telomere dysfunction and p53 activation. Primary human adipocytes when treated with H_2O_2 expressed high levels of p53 and TNF-α, suggesting that oxidative stress could underlie the metabolic abnormalities [54, 55]. As proposed previously [56–62], obesity resulting from overnutrition stimulates the generation of ROS that would overwhelm the antioxidant protection in adipose and other tissues, thus accelerating DNA damage and aging. The obesity-mediated aging of adipose tissue is also associated with telomere shortening, which leads to activation of p53. These alterations trigger inflammatory responses in adipose and other tissues and stimulate cytokine production, which then lead to insulin resistance locally and systemically (Fig. 15.2).

This sequence of events suggested above imply that if adequate endogenous resolution mechanisms are in place to repair the damage induced by free radicals and inflammation, it is possible that aging and its associated conditions including cancer can be halted, prevented or postponed. In other words, it is the failure of events that lead to successful resolution of damage induced by free radicals and inflammation that are responsible for the development of hypertension, type 2 diabetes mellitus, CHD and aging. The support to this proposal comes from the observation that atherosclerosis results from a failure in the resolution of local inflammation. It was noted that in apolipoprotein E-deficient mice 12/15-lipoxygenase expression protects mice against atherosclerosis *via* its role in the local biosynthesis of pro-resolving anti-inflammatory lipid molecules such as lipoxin A_4, resolvin D_1, and protectin D_1. These anti-inflammatory lipid molecules acted on macrophages and vascular endothelial cells and exerted specific actions to control the magnitude of the local inflammatory response. Adequate expression of 12/15-lipoxygenase resulted in increased production of lipoxins, resolvins and protectins from AA, EPA and DHA that resulted in a delay in the development of atherosclerosis, decrease in the production of pro-inflammatory cytokines and adhesion molecules and increased phagocytosis of macrophages towards apoptotic thymocytes and reduced activation of vascular endothelial cells [63].

Similar results were obtained in Alzheimer's disease. It looks like 12/15-lipoxygenase deficiency leads to impairment in the generation of lipoxins, resolvins and protectins that result in impaired wound healing as a result of defective resolution phase of inflammation. Lipoxins, resolvins and protectins are not only anti-inflammatory molecules by their ability to suppress the production of pro-inflammatory cytokines but are also capable of enhancing the phagocytic activity of macrophages toward apoptotic cells, an anti-inflammatory and pro-resolution function that is important in both acute inflammation and in atherosclerosis. In addition, lipoxins, resolvins and protectins derived from the action of 12/15-lipoxygenase also suppress the expression of adhesion molecules and chemokines by vascular endothelial cells that will allow the resolution phase to set in and give the vascular wall and other sites of inflammation to return to normal.

Our own studies previously showed that alloxan-induced diabetes mellitus could be prevented completely by pre-treatment or simultaneous treatment with AA, EPA and DHA (AA > EPA = DHA). This beneficial action of PUFAs was not abrogated by both cyclo-oxygenase and lipoxygenase inhibitors suggesting that PGs, LTs and TXs are not involved in this protective action of PUFAs [64–68]. Though we did not measure the plasma/tissue levels of lipoxins, resolvins and protectins, it is now obvious that in these experimental animals alloxan-induced diabetes was prevented by the formation of these anti-inflammatory compounds. On the basis of these studies [63–68], it can be said that a deficiency of various PUFAs, 12/15 lipoxygenase enzymes and phospholipases that are necessary for the release of adequate amounts of necessary PUFAs for the formation of anti-inflammatory lipoxins, resolvins and protectins could lead to continued inflammation once it is incited, non-resolution of inflammation and tissue/organ/organ damage. Hence, it is possible that as age advances deficiency of various PUFAs, 12/15 lipoxygenase enzymes and phospholipases could occur resulting in decreased formation of anti-inflammatory lipid mediators and increase in the incidence of diseases of aging such as hypertension, type 2 diabetes mellitus, atherosclerosis, cancer, Alzheimer's disease and aging itself (see Fig. 15.2).

Since lipoxins, resolvins and protectins enhance the formation of NO; suppress the generation of MPO (myeloperoxidase) and free radicals they may also serve as genome protectors. For example, previously we showed that radiation and chemical-induced chromosomal damage could be prevented by PUFAs [69–73] that could be attributed to the formation of lipoxins, resolvins and protectins. This implies that PUFAs and their anti-inflammatory products lipoxins, resolvins and protectins could prevent shortening of telomere. Since cancer is also an inflammatory condition (see Chap. 14), it is reasoned that production of adequate amounts of lipoxins, resolvins and protectins will also prevent the development of cancer.

Other Theories of Aging

Several other theories of aging such as wear-and-tear theory, accumulative-waste theory, autoimmune theory, cross-linkage theory, free-radical theory, mitohormesis and misrepair-accumulation theory can all be explained in terms of the arguments presented above. In all these theories of aging, the basic premises is that there is failure of adequate repair or resolution of inflammation to either to remove the accumulated waste material, enhanced generation of free radicals that trigger damage to various cells and tissues.

Restricting calories while maintaining adequate amounts of other nutrients can extend lifespan in laboratory animals. Calorie restriction is known to enhance the activities of enzymes Δ^6 and Δ^5 desaturases that are necessary for the formation of PUFAs form dietary essential fatty acids [74]. Thus, calorie restriction could modulate the formation of PUFAs [65, 75] that, in turn, may ultimately result in the formation of anti-inflammatory lipoxins, resolvins and protectins that prevent free radical accumulation and enhance repair process and resolve inflammation ultimately leading to the prevention of aging and aging-associated diseases.

Stems cells are necessary to replace the worn cells and tissues. There is reasonable evidence to suggest that PUFAs and their products have the ability to modulate stem cell biology [76] implying that lipoxins, resolvins and protectins may have a role in the proliferation and differentiation of embryonic stem cells in addition to their capacity to suppress inflammation.

Aging is a Low-grade Systemic Inflammatory Condition

It is evident from the preceding discussion that advancing age is associated with low-grade systemic inflammation. A direct relationship seems to occur between aging and increasing incidences of chronic diseases such as insulin resistance, obesity, hypertension, type diabetes mellitus and the development of cancer. In fact, with advancing age associated diseases individuals manifest an underlying chronic inflammatory state as evidenced by local infiltration of inflammatory cells, such as macrophages, and higher circulatory levels of pro-inflammatory cytokines IL-6, TNF-α, complement components and adhesion molecules. This implies that treatment with anti-inflammatory agents provide symptomatic relief to several aging-associated diseases, including Alzheimer's or Parkinson's disease, indicating that chronic inflammation plays a role in the pathogenesis of these diseases. The molecular mechanism(s) underlying this low-grade systemic inflammatory state during cellular senescence is unclear. In part, this could be attributed to a gradual and steady increase in the plasma and cellular concentrations of pro-inflammatory cytokines and lipid molecules and a decrease in anti-inflammatory cytokines and lipid mediators. These alterations in the pro- and anti-inflammatory molecules could lead to an inappropriate increase in the production of oxygen free radicals and decrease in anti-oxidant defenses that trigger cellular damage. Thus, oxygen free radicals is a primary driving force for aging and increased activation of redox-regulated transcription factors, such as NF-kB that regulate the expression of pro-inflammatory molecules that has been documented in aged animals/individuals versus their young counterparts. Human polynucleotide phosphorylase (hPNPase(old-35)), a RNA degradation enzyme shown to be upregulated during differentiation and cellular senescence, may represent a molecular link between aging and its associated inflammation. hPNPase(old-35) promotes reactive oxygen species (ROS) production, activates the NF-kB pathway and initiates the production of pro-inflammatory cytokines IL-6 and IL-8 [77–79].

Aging and its associated low-grade systemic inflammation may favor tumorigenesis. With advancing age, dysregulation of oxygen-, heme-, and proteolysis-dependent metabolic pathways may occur that may promote inflammation that creates a pro-cancer microenvironment that may facilitate the survival and growth of cancer cells. There are certain features that are common between low-grade systemic inflammation and pro-cancer microenvironment. These include: enhanced oxidative cell resistance against apoptosis, increased production of matrix-degrading enzymes, switching to glycolytic metabolism, angiogenesis and vasorelaxation thus providing nutrient delivery, but restriction of the immune cell mobilization and/or its activation. The pro-cancer microenvironment is somewhat similar to the non-healing wound state that often occurs around carcinomas [80] and non-healing wounds seen in diabetics with diabetic foot ulcer. The non-healing ulcer in the diabetics could be due to neuropathy, vascular insufficiency, local deficiency of growth factors, and defective macrophage function in removing the debris, local inflammation in the form of infection, enhanced free radical generation and failure of resolution of inflammation, possibly, due to local deficiency and/or insufficiency of anti-inflammatory lipid mediators such as lipoxins, resolvins and protectins.

The aging process is often paralleled by decreases in muscle and increases in fat mass that can lead to the development what is called as "sarcopenic obesity" [81, 82]. This is due to the inflammatory cytokines produced by adipose tissue, especially visceral fat, which accelerate muscle catabolism and thus contribute to the initiation and sustenance of sarcopenic obesity. This is supported by the observation that sarcopenic obesity was associated with elevated levels of IL-6, C-reactive protein, IL-1 receptor antagonist, and soluble IL-6 receptor ($P < 0.05$). These findings confirm the proposal that global obesity and, to a greater extent, central obesity directly affect inflammation, which in turn negatively affects muscle strength, contributing to the development and progression of sarcopenic obesity and indicate that proinflammatory cytokines are critical in both the development and progression of sarcopenic obesity [83].

In fact, it was reported that insulin sensitivity deteriorates with age leading to many metabolic complications. In a study that compared the differences in fasting glucose, glucose tolerance, and inflammatory markers between two generations it was noted that the first generation subjects were substantially insulin resistant, compared with their young descendents, evidenced by exaggerated glucose and insulin responses (> 100% greater area under curves above baseline) under oral glucose challenged condition. Their waist circumference, diastolic blood pressure, and cholesterol levels were significantly greater than controls. Furthermore, CRP of the first generation was approximately 2.3 folds of the control value suggesting a low grade systemic inflammation, yet the levels of physical activity and dietary intake were not different between groups. Based on this study, it was determined that OGTT (oral glucose tolerance test) and CRP reflect the age-dependent metabolic deterioration than fasting glucose value and suggest that with age glucose tolerance deteriorates and inflammatory markers increase [84].

In a study performed in centenarians, it was found that they had low IGF-1-mediated responses and high levels of anti-inflammatory cytokines such as IL-10 and TGF-β and well-preserved p53-mediated responses that seem to contribute to their protection from cancer [85]. Thus, centenarians are unique in that, despite high levels of pro-inflammatory markers, they also exhibit anti-inflammatory markers that delay disease onset, suggesting that the balance between pro- and anti-inflammatory cytokines is crucial to successful aging and longevity [86] (see Fig. 15.3).

Exercise is Anti-inflammatory in Nature

One possible physiological stimulus to successful aging and longevity seems to be regular exercise that has anti-inflammatory property [87, 88]. In a recent study, we observed that obese rats that exercised regularly showed decreased plasma uric acid, IL-6 and TNF-α levels and a marked decrease in the expression of TNF-α and IL-6 in the pancreatic islet cells, confirming that aerobic exercise is anti-inflammatory in nature [89].

In a study in which we evaluated the association of physical activity and diet with plasma CRP levels among subjects with abdominal obesity, compared with those with low CRP levels, subjects with high CRP levels (i.e., > 3.0 mg/l) were physically inactive ($P = 0.01$), were less likely to adopt the Mediterranean diet ($P = 0.008$), had higher glucose levels, had a higher prevalence of hypertension, had a lower

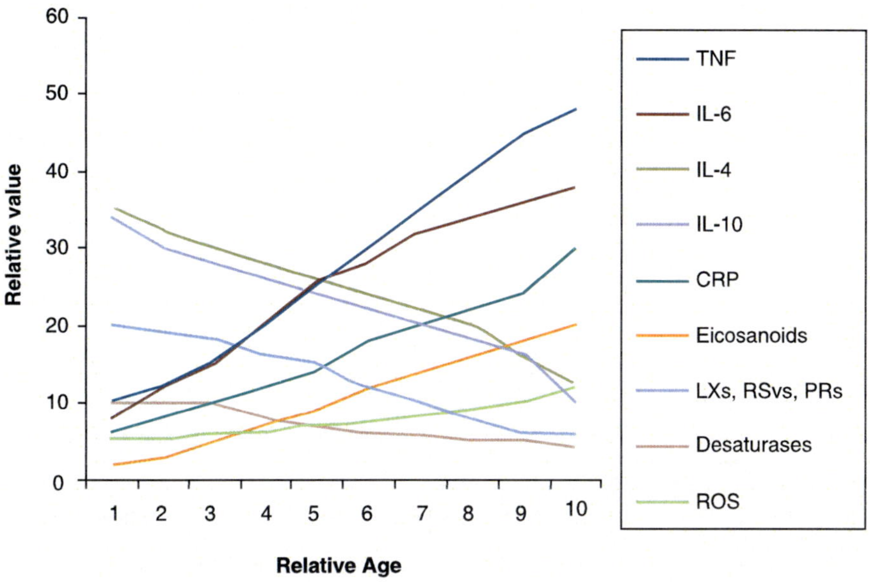

Fig. 15.3 Scheme showing relative changes in the plasma/tissue concentrations of pro- and anti-inflammatory cytokines, CRP, pro-inflammatory eicosanoids, and ROS and anti-inflammatory lipoxins (LXs), resolvins (RSVs) and protectins (PRs). It may be noted that these are only relative values. Though the values have been shown as changing with advancing age, it may not happen always. For example, interventions in the form of diet control, exercise and pharmaceuticals may halt the increase or even decrease the concentrations of pro-inflammatory cytokines, ROS and eicosanoids and/or enhance the concentrations of anti-inflammatory cytokines and anti-inflammatory lipid mediators. It may be noted that under physiological conditions, a balance is maintained between pro- and anti-inflammatory molecules (such as cytokines) and pro- and anti-inflammatory lipid molecules. In addition, an interaction(s) among pro- and anti-inflammatory cytokines and pro- and anti-inflammatory lipid mediators occurs. It is predicted that with advancing age, the levels of pro-inflammatory cytokines and lipid molecules increase and those of anti-inflammatory cytokines and lipid mediators decrease. With advancing age, a gradual decline in the activities of Δ^6 and Δ^5 desaturases is expected leading to a decrease in the plasma and cellular content of PUFAs. Each value on the horizontal axis refers to 10 years

high-density lipoprotein cholesterol, and had increased smoking habits and higher anthropometric indices (all $P < 0.05$). Moreover, adoption of the Mediterranean diet in combination with medium physical activity reduced the likelihood of having high CRP levels by 72% ($P = 0.018$), irrespective of smoking and various clinical and biological characteristics. Among subjects with abdominal obesity, low-grade systemic inflammation was found to be associated with the adoption of an unfavorable lifestyle, including physical inactivity and unhealthy dietary habits, as well as increased blood pressure levels and low high-density lipoprotein cholesterol [90]. This study once again emphasized the importance of careful diet [91] and exercise to limit low-grade systemic inflammation and high risk of adult diseases such as hypertension, type 2 diabetes mellitus, and CHD and their associated atherosclerosis.

In summary, aging is a low-grade systemic inflammation; and as we age the plasma and tissue levels of pro-inflammatory cytokines increase and anti-inflammatory cytokines and lipid molecules likely to decrease that parallels with the increasing incidence of obesity, hypertension, type 2 diabetes mellitus, atherosclerosis, CHD and cancer. Hence, anti-inflammatory measures in the form of Mediterranean diet, exercise and perhaps, anti-inflammatory drugs may retard the aging process and its associated diseases.

References

[1] Hanahan D, Weinberg RA (2000) The hallmarks of cancer. Cell 100:57–70
[2] Counter CM, Avilion AA, LeFeuvre CE, Stewart NG, Greider CW, Harley CB, Bacchetti S (1992) Telomere shortening associated with chromosome instability is arrested in immortal cells which express telomerase activity. EMBO J 11:1921–1929
[3] Counter CM, Botelho FM, Wang P, Harley CB, Bacchetti S (1994) Stabilization of short telomeres and telomerase activity accompany immortalization of Epstein-Barr virus-transformed human B lymphocytes. J Virol 68:3410–3414
[4] Allsopp RC, Harley CB (1995) Evidence for a critical telomere length in senescent human fibroblasts. Exp Cell Res 219:130–136
[5] Satoh H, Hiyama K, Takeda M, Awaya Y, Watanabe K, Ihara Y, Maeda H, Ishioka S, Yamakido M (1996) Telomere shortening in peripheral blood cells was related with aging but not with white blood cell count. Jpn J Hum Genet 41:413–417
[6] Weinstein BS, Ciszek D (2002) The reserve-capacity hypothesis: evolutionary origins and modern implications of the trade-off between tumor-suppression and tissue-repair. Exp Gerontol 37:615–627
[7] Nakagawa S, Gemmell NJ, Burke T (2004) Measuring vertebrate telomeres: applications and limitations. Mol Ecol 13:2523–2533
[8] Juola FA, Haussmann MF, Dearborn DC, Vleck CM (2006) Telomere shortening in a long-lived marine bird: cross-sectional analysis and test of an aging tool. Auk 123:775
[9] Ornish D, Lin J, Daubenmier J, Weidner G, Epel E, Kemp C, Magbanua MJM, Marlin R, Yglecias L, Carroll PR, Blackburn EH (2008) Increased telomerase activity and comprehensive lifestyle changes: a pilot study. Lancet Oncol 9:1048–1057
[10] Jeanclos E, Krolewski A, Skurnick J, Kimura M, Aviv H, Warram JH, Aviv A (1998) Shortened telomere length in white blood cells of patients with IDDM. Diabetes 47:482–486
[11] Adaikalakoteswari A, Balasubramanyam M, Mohan V (2005) Telomere shortening occurs in Asian Indian type 2 diabetic patients. Diabet Med 22:1151–1156
[12] Sampson MJ, Winterbone MS, Hughes JC, Dozio N, Hughes DA (2006) Monocyte telomere shortening and oxidative DNA damage in type 2 diabetes. Diabetes Care 29:283–289
[13] Adaikalakoteswari A, Balasubramanyam M, Ravikumar R, Deepa R, Mohan V (2007) Association of telomere shortening with impaired glucose tolerance and diabetic macroangiopathy. Atherosclerosis 195:83–89
[14] Tentolouris N, Nzietchueng R, Cattan V, Poitevin G, Lacolley P, Papazafiropoulou A, Perrea D, Katsilambros N, Benetos A (2007) White blood cells telomere length is shorter in males with type 2 diabetes and microalbuminuria. Diabetes Care 30:2909–2915
[15] Uziel O, Singer JA, Danicek V, Sahar G, Berkov E, Luchansky M, Fraser A, Ram R, Lahav M (2007) Telomere dynamics in arteries and mononuclear cells of diabetic patients: effect of diabetes and of glycemic control. Exp Gerontol 42:971–978
[16] Verzola D, Gandolfo MT, Gaetani G, Ferraris A, Mangerini R, Ferrario F, Villaggio B, Gianiorio F, Tosetti F, Weiss U, Traverso P, Mji M, Deferrari G, Garibotto G (2008) Accelerated

senescence in the kidneys of patients with type 2 diabetic nephropathy. Am J Physiol Renal Physiol 295:F1563–F1573

[17] Olivieri F, Lorenzi M, Antonicelli R, Testa R, Sirolla C, Cardelli M, Mariotti S, Marchegiani F, Marra M, Spazzafumo L, Bonfigli AR, Procopio A (2009) Leukocyte telomere shortening in elderly Type2DM patients with previous myocardial infarction. Atherosclerosis 206:588–593

[18] Slagboom PE, Droog S, Boomsma DI (1994) Genetic determination of telomere size in humans: a twin study of three age groups. Am J Hum Genet 55:876–882

[19] Graakjaer J, Pascoe L (2004) Der-Sarkissian H. The relative lengths of individual telomeres are defined in the zygote and strictly maintained during life. Aging Cell 3:97–102

[20] Salpea KD, Talmud PJ, Cooper JA, Maubaret CG, Stephens JW, Abelak K, Humphries SE (2010) Association of telomere length with type 2 diabetes, oxidative stress and UCP2 gene variation. Atherosclerosis 209:42–50

[21] Salpea KD, Humphries SE (2010) Telomere length in atherosclerosis and diabetes. Atherosclerosis 209:35–38

[22] Jeanclos E, Schork NJ, Kyvik KO, Kimura M, Skurnick JH, Aviv A (2000) Telomere length inversely correlates with pulse pressure and is highly familial. Hypertension 36:195–200

[23] Okuda K, Khan MY, Skurnick J, Kimura M, Aviv H, Aviv A (2000) Telomere attrition of the human abdominal aorta: relationships with age and atherosclerosis. Atherosclerosis 152:391–398

[24] Tristano A, Eugenia Chollet M, Willson ML, Adjounian H, Fernanda Correa M, Borges A (2003) Telomerase activity in peripheral blood leukocytes from patients with essential hypertension. Med Clin (Barc) 120:365–369

[25] Benetos A, Gardner JP, Zureik M, Labat C, Xiaobin L, Adamopoulos C, Temmar M, Bean KE, Thomas F, Aviv A (2004) Short telomeres are associated with increased carotid atherosclerosis in hypertensive subjects. Hypertension 43:182–185

[26] Nakashima H, Ozono R, Suyama C, Sueda T, Kambe M, Oshima T (2004) Telomere attrition in white blood cell correlating with cardiovascular damage. Hypertens Res 27:319–325

[27] Gardner JP, Li S, Srinivasan SR, Chen W, Kimura M, Lu X, Berenson GS, Aviv A (2005) Rise in insulin resistance is associated with escalated telomere attrition. Circulation 111:2171–2177

[28] Aviv A, Valdes A, Gardner JP, Swaminathan R, Kimura M, Spector TD (2006) Menopause modifies the association of leukocyte telomere length with insulin resistance and inflammation. J Clin Endocrinol Metab 91:635–640

[29] Demissie S, Levy D, Benjamin EJ, Cupples LA, Gardner JP, Herbert A, Kimura M, Larson MG, Meigs JB, Keaney JF, Aviv A (2006) Insulin resistance, oxidative stress, hypertension, and leukocyte telomere length in men from the Framingham heart study. Aging Cell 5:325–330

[30] Vasan RS, Demissie S, Kimura M, Cupples LA, Rifai N, White C, Wang TJ, Gardner JP, Cao X, Benjamin EJ, Levy D, Aviv A (2008) Association of leukocyte telomere length with circulating biomarkers of the renin-angiotensin-aldosterone system: the Framingham heart study. Circulation 117:1138–1144

[31] Movérare-Skrtic S, Svensson J, Karlsson MK, Orwoll E, Ljunggren O, Mellström D, Ohlsson C (2009) Serum insulin-like growth factor-I concentration is associated with leukocyte telomere length in a population-based cohort of elderly men. J Clin Endocrinol Metab 94:5078–5084

[32] Kaplan RC, Fitzpatrick AL, Pollak MN, Gardner JP, Jenny NS, McGinn AP, Kuller LH, Strickler HD, Kimura M, Psaty BM, Aviv A (2009) Insulin-like growth factors and leukocyte telomere length: the cardiovascular health study. J Gerontol A Biol Sci Med Sci 64:1103–1106

[33] Okuda K, Bardeguez A, Gardner JP, Rodriguez P, Ganesh V, Kimura M, Skurnick J, Awad G, Aviv A (2002) Telomere length in the newborn. Pediatr Res 52:377–381

[34] Rufer N, Brümmendorf TH, Kolvraa S, Bischoff C, Christensen K, Wadsworth L, Schulzer M, Lansdorp PM (1999) Telomere fluorescence measurements in granulocytes and T lymphocyte subsets point to a high turnover of hematopoietic stem cells and memory T cells in early childhood. J Exp Med 190:157–167

[35] Allsopp RC, Vaziri H, Patterson C, Goldstein S, Younglai EV, Futcher AB, Greider CW, Harley CB (1992) Telomere length predicts replicative capacity of human fibroblasts. Proc Natl Acad Sci U S A 89:10114–10118

[36] Aviv A (2009) Leukocyte telomere length, hypertension, and atherosclerosis: are there potential mechanistic explanations? Hypertension 53:590–591
[37] Satoh M, Ishikawa Y, Takahashi Y, Itoh T, Minami Y, Nakamura M (2008) Association between oxidative DNA damage and telomere shortening in circulating endothelial progenitor cells obtained from metabolic syndrome patients with coronary artery disease. Atherosclerosis 198:347–353
[38] Shen J, Gammon MD, Terry MB, Wang Q, Bradshaw P, Teitelbaum SL, Neugut AI, Santella RM (2009) Telomere length, oxidative damage, antioxidants and breast cancer risk. Int J Cancer 124:1637–1643
[39] Farzaneh-Far R, Lin J, Epel ES, Harris WS, Blackburn EH, Whooley MA (2010) Association of marine omega-3 fatty acid levels with telomeric aging in patients with coronary heart disease. JAMA 303:250–257
[40] Richards JB, Valdes AM, Gardner JP et al (2007) Higher serum vitamin D concentrations are associated with longer leukocyte telomere length in women. Am J Clin Nutr 86:1420–1425
[41] Xu Q, Parks CG, DeRoo LA, Cawthon RM, Sandler DP, Chen H (2009) Multivitamin use and telomere length in women. Am J Clin Nutr 89:1857–1863
[42] Furumoto K, Inoue E, Nagao N, Hiyama E, Miwa N (1998) Age-dependent telomere shortening is slowed down by enrichment of intracellular vitamin C via suppression of oxidative stress. Life Sci 63:935–948
[43] Tanaka Y, Moritoh Y, Miwa N (2007) Age-dependent telomere-shortening is repressed by phosphorylated alpha-tocopherol together with cellular longevity and intracellular oxidative-stress reduction in human brain microvascular endotheliocytes. J Cell Biochem 102:689–703
[44] Paul L, Cattaneo M, D'Angelo A et al (2009) Telomere length in peripheral blood mononuclear cells is associated with folate status in men. J Nutr 139:1273–1278
[45] Eitsuka T, Nakagawa K, Suzuki T, Miyazawa T (2005) Polyunsaturated fatty acids inhibit telomerase activity in DLD-1 human colorectal adenocarcinoma cells: a dual mechanism approach. Biochim Biophys Acta 1737:1–10
[46] Matheu A, Maraver A, Klatt P, Flores I, Garcia-Cao I, Borras C, Flores JM, Vin J, Blasco MA, Serrano M (2007) Delayed ageing through damage protection by the Arf/p53 pathway. Nature 448:375–380
[47] Tyner SD et al (2002) p53 mutant mice that display early ageing-associated phenotypes. Nature 415:45–53
[48] Maier B et al (2004) Modulation of mammalian life span by the short isoform of p53. Genes Dev 18:306–319
[49] Varela I et al (2005) Accelerated ageing in mice deficient in Zmpste24 protease is linked to p53 signalling activation. Nature 437:564–568
[50] Garcia-Cao I et al (2002) 'Super p53' mice exhibit enhanced DNA damage response, are tumor resistant and age normally. EMBO J 21:6225–6235
[51] Matheu A et al (2004) Increased gene dosage of Ink4a/Arf results in cancer resistance and normal aging. Genes Dev 18:2736–2746
[52] Garcia-Cao I et al (2006) Increased p53 activity does not accelerate telomere-driven ageing. EMBO Rep 7:546–552
[53] Mendrysa SM et al (2006) Tumor suppression and normal aging in mice with constitutively high p53 activity. Genes Dev 20:16–21
[54] Ahima RS (2009) Connecting obesity, aging and diabetes. Nat Med 15:996–997
[55] Minamino T, Orimo M, Shimizu I, Kunieda T, Yokoyama M, Ito T, Nojima A, Nabetani A, Oike Y, Matsubara H, Isikawa F, Komuro I (2009) A crucial role for adipose tissue p53 in the regulation of insulin resistance. Nat Med 15:1082–1088
[56] Das UN (2007) Metabolic syndrome X is a low-grade systemic inflammatory condition with its origins in the perinatal period. Curr Nutr Food Sci 3:277–295
[57] Das UN (2008) Is metabolic syndrome X a disorder of the brain? Curr Nutr Food Sci 4:73–108
[58] Das UN (2010) Metabolic syndrome is a low-grade systemic inflammatory condition. Expert Rev Endocrinol Metab 4:577–592

[59] Das UN (2010) Metabolic syndrome pathophysiology: the role of essential fatty acids fatty acids and their metabolites. Wiley-Blackwell, Ames

[60] Das UN (2002) Obesity, metabolic syndrome X, and inflammation. Nutrition 18:430–432

[61] Das UN (2002) Is metabolic syndrome X an inflammatory condition? Exp Biol Med 227:989–997

[62] Das UN (2004) Metabolic syndrome X: an inflammatory condition? Curr Hypertens Rep 6:66–73

[63] Merched AJ, Ko K, Gotlinger KH, Serhan CN, Chan L (2008) Atherosclerosis: evidence for impairment of resolution of vascular inflammation governed by specific lipid mediators. FASEB J 22:3595–3606

[64] Krishna Mohan I, Das UN (2001) Prevention of chemically-induced diabetes mellitus in experimental animals by polyunsaturated fatty acids. Nutrition 17:126–151

[65] Suresh Y, Das UN (2001) Protective action of arachidonic acid against alloxan-induced cytotoxicity and diabetes mellitus. Prostaglandins Leukot Essent Fatty Acids 64:37–52

[66] Suresh Y, Das UN (2003) Long-chain polyunsaturated fatty acids and chemically-induced diabetes mellitus: effect of ω-6 fatty acids. Nutrition 19:93–114

[67] Suresh Y, Das UN (2003) Long-chain polyunsaturated fatty acids and chemically-induced diabetes mellitus: effect of ω-3 fatty acids. Nutrition 19:213–228

[68] Suresh Y, Das UN (2006) Differential effect of saturated, monounsaturated, and polyunsaturated fatty acids on alloxan-induced diabetes mellitus. Prostaglandins Leukot Essent Fatty Acids 74:199–213

[69] Das UN et al (1985) Benzo (a) pyrene and gamma-radiation-induced genetic damage in mice can be prevented by gamma-linolenic acid but not by arachidonic acid. Nutr Res 5:101

[70] Das UN, Devi GR, Rao KP, Rao MS (1985) Prostaglandins and their precursors can modify genetic damage induced by benzo (a,) pyrene and gamma-radiation. Prostaglandins 29:911–916

[71] Das UN et al (1985) Benzo (a) pyrene-induced genetic damage in mice can be prevented by evening primrose oil. IRCS Med Sci 13:316

[72] Das UN, Rao KP (2006) Effect of γ-linolenic acid and prostaglandins E1 on gamma-radiation and chemical-induced genetic damage to the bone marrow cells of mice. Prostaglandins Leukot Essent Fatty Acids 74:165–173

[73] Shivani P, Rao KP, Chaudhury JR, Ahmed J, Rao BR, Kanjilal S, Hasan Q, Das UN (2009) Effect of polyunsaturated fatty acids on diphenyl hydantoin-induced genetic damage in vitro and in vivo. Prostaglandins Leukot Essent Fatty Acids 80:43–50

[74] Laganiere S, Yu BP (1993) Modulation of membrane phospholipid fatty acid composition by age and food restriction. Gerontology 39:7–18

[75] Ford JH (2010) Saturated fatty acid metabolism is key link between cell division, cancer, and senescence in cellular and whole organism aging. Age (Dordr) 32:231–237

[76] Das UN (2011) Influence of polyunsaturated fatty acids and their metabolites on stem cell biology: hypothesis. Nutrition 27:21–25

[77] Sarkar D, Fisher PB (2006) Molecular mechanisms of aging-associated inflammation. Cancer Lett 236:13–23

[78] Sarkar D, Lebedeva IV, Emdad L, Kang DC, Baldwin AS Jr, Fisher PB (2004) Human polynucleotide phosphorylase (hPNPaseold-35): a potential link between aging and inflammation. Cancer Res 64:7473–7478

[79] Sarkar D, Park ES, Emdad L, Randolph A, Valerie K, Fisher PB (2005) Defining the domains of human polynucleotide phosphorylase (hPNPaseOLD-35) mediating cellular senescence. Mol Cell Biol 25:7333–7343

[80] Schwartsburd PM (2004) Age-promoted creation of a pro-cancer microenvironment by inflammation: pathogenesis of dyscoordinated feedback control. Mech Ageing Dev 125:581–590

[81] Roubenoff R (2000) Sarcopenic obesity: does muscle loss cause fat gain? Lessons from rheumatoid arthritis and osteoarthritis. Ann N Y Acad Sci 904:553–557

[82] Roubenoff R, Hughes VA (2000) Sarcopenia: current concepts. J Gerontol A Biol Sci Med Sci 55:M716–M724

[83] Schrager MA, Metter EJ, Simonsick E, Ble A, Bandinelli S, Lauretani F, Ferrucci L (2007) Sarcopenic obesity and inflammation in the InCHIANTI study. J Appl Physiol 102:919–925
[84] Ho CT, Su CL, Chen MT, Liou YF, Lee SD, Chien KY, Kuo CH (2008) Aging effects on glycemic control and inflammation for politicians in Taiwan. Chin J Physiol 51:402–407
[85] Salvioli S, Capri M, Bucci L, Lanni C, Racchi M, Uberti D, Memo M, Mari D, Govoni S, Franceschi C (2009) Why do centenarians escape or postpone cancer? The role of IGF-1, inflammation and p53. Cancer Immunol Immunother 58:1909–1917
[86] Franceschi C (2007) Inflammaging as a major characteristic of old people: can it be prevented or cured? Nutr Rev 65(12 Pt 2):S173–S176
[87] Das UN (2004) Anti-inflammatory nature of exercise. Nutrition 20:323–326
[88] Das UN (2006) Exercise and inflammation. Eur Heart J 27:1385–1386
[89] Teixeira de Lemos E, Reis F, Baptista S, Pinto R, Sepodes B, Vala H, Rocha-Pereira P, Correia de Silva G, Teixeira N, Santos Silva A, Carvalho L, Teixeira F, Das UN (2009) Exercise training decreases proinflammatory profile in Zucker diabetic (type 2) fatty rats. Nutrition 25:330–339
[90] Pitsavos C, Panagiotakos DB, Chrysohoou C, Tzima N, Das UN, Stefanadis C (2007) Diet, exercise and C-reactive protein levels in people with abdominal obesity: the ATTICA study. Angiology 58:225–233
[91] Merry BJ (2002) Molecular mechanisms linking calorie restriction and longevity. Int J Biochem Cell Biol 34:1340–1354

Chapter 16
Adult Diseases and Low-Grade Systemic Inflammation Have Their Origins in the Perinatal Period

Introduction

It is evident from the discussion in the preceding chapters on various diseases/disorders that low-grade systemic inflammation plays a significant role in the pathobiology of obesity, hypertension, type 2 diabetes mellitus, dyslipidemia, atherosclerosis, coronary heart disease, cancer, aging, Alzheimer's disease, schizophrenia, depression, dementia and even stroke (though this disease was not discussed in details, in general, it occurs as a result of underlying hypertension, diabetes mellitus, hyperlipidemia and hence, could be considered as a consequence of these diseases rather than as a separate disease entity). The presence of low-grade systemic inflammation as evidenced by increased plasma levels of CRP, IL-6, TNF-α, HMGB-1, MIF, ROS, iNO, and a concomitant decrease in anti-inflammatory cytokines such as IL-4, IL-10, IL-12, TGF-β, and anti-oxidants, and decreased plasma and tissue levels of various PUFAs such as AA, EPA, DHA, GLA, DGLA and their anti-inflammatory products such as lipoxins, resolvins, protectins and maresins may underlie all these diseases. Thus, an imbalance between pro- and anti-inflammatory molecules seems to be a common feature in these diseases. Thus, the molecular events in all these diseases are similar but the target tissues are different. This implies that methods designed to suppress the production of pro-inflammatory molecules and/or increase the synthesis and secretion of anti-inflammatory molecules could be of benefit in their prevention and management (Fig. 16.1).

Based on the evidence that low-grade systemic inflammation is a common event in several adult diseases, it is reasonable to suggest that different features seen in specific conditions can be attributed to damage or dysfunction of specific tissues relevant to a given condition/disease. In other words, the underlying pathophysiology is similar but the clinical features of the diseases are different simply because the tissues/organs involved in the said disease processes are different. For example, in hypertension endothelial cells are the target of the adverse action of the pro-inflammatory molecules that causes endothelial dysfunction; in schizophrenia it is the neuronal cells and various neurotransmitters; in atherosclerosis it is the endothelial and smooth muscle cells; and in type 2 diabetes mellitus it is the adipose tissue to start with but

U. N. Das, *Molecular Basis of Health and Disease*,
DOI 10.1007/978-94-007-0495-4_16,

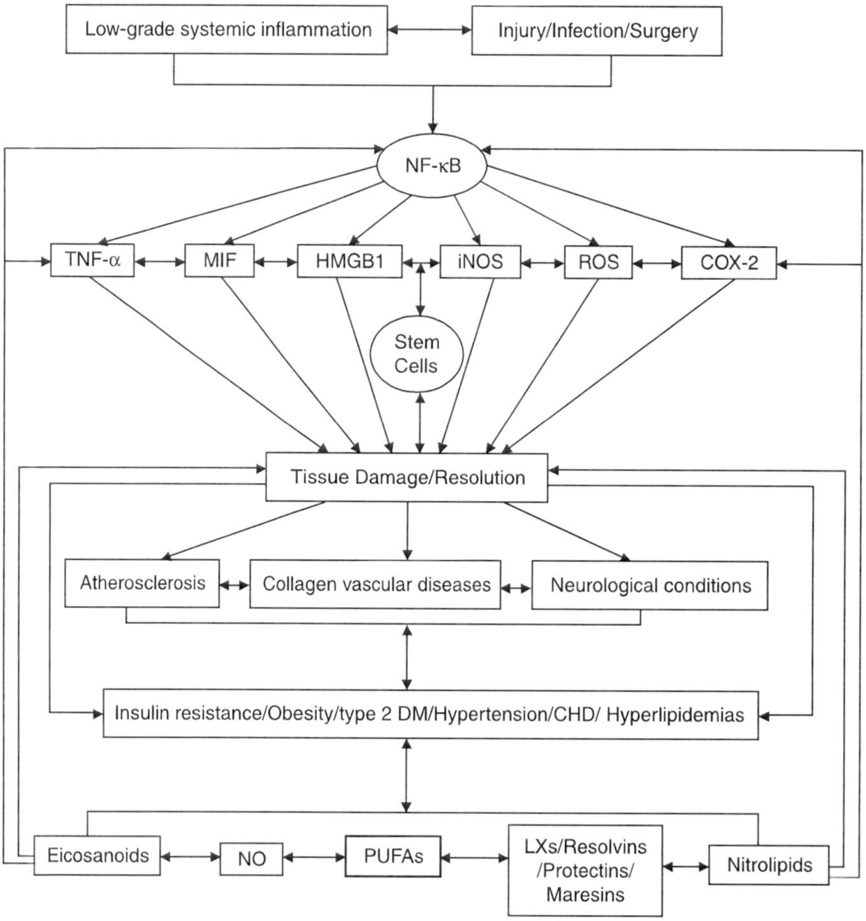

Fig. 16.1 Scheme showing relationship between various mediators of tissue damage/resolution and clinical conditions and the role of PUFAs and their metabolites in these processes

later muscle, endothelial cells, liver and hypothalamic neurons are also affected. Thus, the local imbalance between the pro- and anti-inflammatory molecules that is tilted more in favor of pro-inflammatory molecules in the specific cells/tissues/organs leads to damage or dysfunction of those specific tissues/organs that ultimately leads to the development of specific disease and the varied clinical features seen in those diseases. It is important to note that in some diseases the target tissues/organs could be more than one and yet times it may be difficult to determine which tissue/organ is the first to be affected by the inflammatory molecules. For instance, the target tissues in type 2 diabetes mellitus could be pancreatic β cells, endothelial cells, ventromedial hypothalamic neurons or adipose cells or a combination there of. In addition, in these diseases the initial trigger could be dietary or other environmental factors with or without underlying genetic predisposition or factors that initiate low-grade systemic inflammatory process. Once the initial trigger is cleared or abrogated by the innate and adaptive immune systems and/or anti-inflammatory homeostatic mechanisms that are set in motion to help in the resolution of the inflammation, healing of the tissue injury occurs that leads to restoration of the function of the involved tissues and organs and normalcy is restored and health is regained. But, when the target tissues/organ dysfunction persists for prolonged periods of time, collateral damage to others tissues/organs would occur that leads to major complications and so it is extremely difficult, if not impossible, to restore normalcy. For the repair process and restoration of normal physiological function to occur in several of the adult diseases, it is necessary for the stem cells to proliferate and differentiate to replace the damaged tissues and restore normal physiological function. In this context, it is important to note that cytokines, PUFAs, lipoxins, hormones and several vitamins and minerals have a significant role in the growth, differentiation and survival of stem cells [1].

When and How the Inflammatory Process is Initiated?

Despite the evidence that several adult diseases are low-grade systemic inflammatory conditions, it is still not clear whether inflammation is the cause or affect of the disease. It is not known when and how these diseases are initiated. In this context, the role of breast-feeding and perinatal feeding on fetal and childhood growth, the development of brain growth and development and programming of the hypothalamic centers that regulate blood pressure, insulin secretion, programming of body weight/appetite/satiety set point, and autonomic nervous system deserve special attention. Breast-fed children have decreased incidence of a number of low-grade systemic inflammatory conditions such as insulin resistance, obesity, hypertension, type 1 and type 2 diabetes, schizophrenia, metabolic syndrome and some types of cancer. But, the exact mechanism(s) for this beneficial action is not clear. Since hormonal signals and/or nutritional factors and infections to which the subject is exposed during the fetal and early childhood period may serve as programming stimuli that can have lifetime consequences, it is reasonable to propose that majority, if not all, of the adult diseases have their origin in the perinatal period. If this is true, it implies that even the low-grade systemic inflammation that participates in the pathophysiology of these diseases have their origins in the perinatal period.

Perinatal Programming of Adult Diseases

Stimuli or insults induced during the perinatal period can have lifetime consequences and is called as "programming". Hormonal signals or nutritional factors may serve as programming stimuli. Smallness and thinness at birth, continued slow growth in early childhood, followed by acceleration of growth so that height and weight approach the population means is considered as the most unfavorable growth pattern that can result in fetal adaptations that may programme the development of insulin resistance, obesity, hypertension, diabetes mellitus, and ischemic heart disease (IHD) in later life [2–5]. This suggests that perinatal nutrition is an important determinant of adult diseases. One endogenous factor that has a negative feed-back control on TNF-α production that also plays an important role in the growth and development of brain is long chain polyunsaturated fatty acids (PUFAs). Since the development of brain occurs during the period between 2nd trimester to 5 years of age and again during the adolescence, it is reasonable to assume that perinatal nutrition and childhood nutrition plays a significant role in this process. Several studies showed that PUFAs and their long-chain metabolites, PUFAs such as arachidonic acid (AA), eicosapentaenoic acid (EPA), and docosahexaenoic acid (DHA) are essential not only for brain growth and development but also regulate the synthesis of various cytokines; modulate insulin action, and concentrations of various neuropeptides. This suggests that various factors that influence the metabolism of PUFAs; action and levels of cytokines; synthesis, release and action of various neurotransmitters and the activity of autonomic nervous system; and the growth and development and function of various tissues and organs such as liver, adipose tissue, muscle, and the humoral and neural factors that influence the interaction and cross-talk among various organs and brain play a significant role in the pathobiology of adult diseases. In this context, the metabolism of PUFAs (see Chap. 4) and their products that modulate autonomic nervous system, neurotransmitters, and their ability to participate in the inflammation and resolution of inflammation is of particular interest.

Factors Influencing the Metabolism of EFAs

In Chap. 4, a detailed discussion of the metabolism of EFAs and the various factors that influence their metabolism and actions has been given. Here only a brief mention of factors that participate in the metabolism of EFAs is given.

Saturated fats, cholesterol, trans-fatty acids formed by vegetable oil processing, alcohol, adrenaline, and glucocorticoids inhibit Δ^6 and Δ^5 desaturases. Pyridoxine, zinc, and magnesium are necessary co-factors for normal Δ^6 desaturase activity. Insulin activates Δ^6 desaturase whereas diabetics have reduced Δ^6 desaturase activity. The activity of Δ^6 desaturase falls with age. Oncogenic viruses and radiation inhibit Δ^6 desaturase activity. Total fasting and protein deficiency reduce the activity of Δ^6 desaturase. A fat-free diet and partial caloric restriction enhances Δ^6 desaturase activity. A glucose-rich diet inhibits Δ^6 desaturase activity.

Peroxisome proliferator-activated receptor-α (PPAR-α) activates the transcription of hepatic Δ^6 desaturase by more than 500%. Hepatic expression of Δ^5 desaturase as well as Δ^6 desaturase was highly activated in transgenic mice overexpressing nuclear SREBP-1a, -1c, and -2. Disruption of the SREBP-1 gene significantly reduced the expression of both desaturases in the livers of SREBP-1-deficient mice refed after fasting. The hepatic expression of both desaturases was downregulated by dietary PUFAs, which suppressed SREBP-1c gene expression. In contrast, sustained expression of hepatic nuclear SREBP-1c protein in the transgenic mice abolished the PUFA suppression of both desaturases. Fasting induced both the desaturases. These data suggest that both Δ^6 and Δ^5 desaturases are regulated by SREBP-1c and PPAR-α, two reciprocal transcription factors for fatty acid metabolism, and that some of their lipogenic actions are brought about by their ability to regulate the producing PUFAs [6].

Activities of Δ^6 and Δ^5 desaturases are decreased in diabetes mellitus, hypertension, hyperlipidemia and the metabolic syndrome. Trans-fats interfere with the metabolism of EFAs and promote inflammation, atherosclerosis and coronary heart disease. The pro-inflammatory action of trans-fats can be attributed to their ability to interfere with the metabolism of EFAs. Several PUFAs, especially EPA and DHA are known to inhibit the production of pro-inflammatory cytokines: interleukin-6 (IL-6), tumor necrosis factor-α (TNF-α), IL-1, and IL-2 (reviewed in [4, 5, 7, 8]). Saturated fatty acids and cholesterol also interfere with the metabolism of EFAs and thus, promote the production of pro-inflammatory cytokines, which explains their ability to cause atherosclerosis and coronary heart disease (CHD). This suggests that trans-fats, saturated fats, and cholesterol have pro-inflammatory actions whereas PUFAs such as GLA, DGLA, EPA and DHA and their products namely lipoxins, resolvins, protectins, maresins and nitrolipids possess anti-inflammatory properties. By interfering with the metabolism of EFAs, saturated fats, cholesterol and trans-fats could reduce the formation of their long-chain metabolites GLA, DGLA, AA, EPA, and DHA (PUFAs) that are essential for the formation of biologically active and beneficial prostacyclin (PGI_2), PGI_3, lipoxins, resolvins, and NPD_1.

PUFAs Modulate Glucose and Glutamine Uptake and Their Metabolism

PUFAs modulate the fluidity of the cell membrane and thus, determine and influence the behaviour of membrane-bound enzymes and receptors. Such an action of PUFAs on neurons is particularly significant since, this suggests that PUFAs will be able to modulate the synthesis, release and binding of various neurotransmitters to their respective receptors and thus modulate their action. In this context, it is noteworthy that infants preferentially accumulate AA, EPA and DHA in the brain during the last trimester of pregnancy and the first months of life. Adequate amounts of AA and DHA are essential for optimal development and function of central nervous system (reviewed in [9–15]). Infants are capable of forming AA and DHA by elongation and desaturation of EFAs, LA and ALA, respectively. But, vegetable oil based infant feed formulas lead to sub-optimal neural development and performance due to decrease in brain PUFA content [16, 17].

It is generally believed that omega-3 and omega-6 PUFA are not only critical for infant and childhood brain development and somatic growth, but that their levels especially of EPA and DHA are often low in the Western diet. Both epidemiological and intervention studies, indicated that DHA and AA supplementation, during pregnancy, lactation, or childhood plays an important role in childhood neurodevelopment and for infant growth and development [18–21]. A positive association between blood DHA levels and improvement on tests of cognitive and visual function in healthy children was reported. Controlled trials showed that supplementation with DHA and EPA may help in the management of childhood psychiatric disorders, and improve visual and motor functions in children with phenylketonuria. In all studies, DHA and EPA supplementation was found to be well tolerated [18–21].

Human infants accumulate AA, EPA and DHA from maternal and/or placental transfer, consumption of human milk, and synthesis from LA and ALA. AA regulates energy metabolism in the cerebral cortex by stimulating glucose uptake in cerebral cortical astrocytes [22]. AA metabolites LTB4 and 5-HETE also stimulated the uptake of glucose in human leukocytes [23]. Exposure of adipocytes to AA rapidly enhanced basal 2-deoxyglucose uptake, reaching maximal effect at approximately 8 h, while insulin-stimulated 2-deoxyglucose uptake was not altered. AA increased the apparent Vmax of basal 2-deoxyglucose uptake was more than doubled, while the apparent Km for 2-deoxyglucose remained unchanged and enhanced the content of the ubiquitous glucose transporter (GLUT-1) in both total cellular and plasma membranes by a PKC-independent mechanism [24].

It is interesting to note that growth of the murine B-lympho-cyte cell line CC9C10 and the myeloma SP2/0 was enhanced significantly by the presence of the unsaturated fatty acids, oleic and linoleic acids in serum-free culture. The cellular content of linoleic and oleic acids gradually increased during continuous culture passage, with no evidence of regulatory control. Over 10 culture passages in the presence of these fatty acids, the unsaturated/saturated fatty acid ratio of all cellular lipid fractions increased substantially. Most of the fatty acid accumulated in the polar lipid fraction (more than 74%) and only a small proportion was oxidized to CO_2 (0.5%). LA caused a decrease to one-eighth in the rate of metabolism of glutamine and a 1.4-fold increase in the rate of metabolism of glucose with no change in the relative flux of glucose through the pathways of glycolysis, pentose phosphate or the tricarboxylic acid cycle. The changes in energy metabolism were reversed when the cells were removed from fatty acid-supplemented medium. It appeared that LA decreased the rate of uptake of glutamine into cells as evidence by the observation that growth of the CC9C10 cells in the presence of LA caused the Km of glutamine uptake to increase from 2.7 to 23 mM, whereas glucose uptake was unaffected [25]. This change in glutamine uptake in the presence of LA and the resultant increase in the growth of the cells is understandable since, tumor cells use large amounts of glutamine to form glutamate and then to α-ketoglutarate that is fed into the Krebs cycle [26]. In addition, it was also reported that $PGF_{2\alpha}$ may act with cAMP (cyclic AMP) in a synergistic way to increase glucose transport by enhanced GLUT1 expression by a PKC-dependent mechanism in adipose cells [27]. This evidence suggests that both PUFAs and their products such as PGs and LTs modulate glucose uptake and its metabolism by neuronal, adipose and tumor cells.

On the other hand, glucose enhances ACh release in the brain [28]. Since AA enhances glucose uptake and, in turn, glucose augments ACh release, it is likely that AA augments ACh release [29]. DHA, another PUFA, enhances cerebral ACh levels and improves learning ability in rats [30]. ACh modulates long-term potentiation and synaptic plasticity in neuronal circuits and interacts with dopamine receptor in the hippocampus [31]. In obesity, a decrease in the number of dopamine receptors or dopamine concentrations occurs [32] and obesity is common in type 2 diabetes. These results imply that PUFAs, glucose and glutamine uptake and their metabolism, Ach release and dopamine function and the synaptic plasticity are interrelated and function in a cohesive manner that may have relevance to several neurological conditions including schizophrenia, Alzheimer's disease, depression, and the role of hypothalamic neurons in obesity, satiety and appetite control.

PUFAs, Insulin, and Acetylcholine Function as Endogenous Cyto- and Neuroprotectors

Insulin receptor tyrosine kinase substrate p58/53 and the insulin receptor are components of synapses in the CNS [33]. Insulin and calorie restriction augment the activities of desaturases (reviewed in [4, 5, 7, 8]) and this increases the formation of PUFAs from their precursors. Insulin-like growth factor-1 (IGF-1) and insulin antagonize neuronal death induced by TNF-α [34, 35]. AA, DHA, and EPA and other PUFAs have neuroprotective and cytoprotective actions [36–41] and are also potent inhibitors of IL-1, IL-2 and TNF-α production [42–44]. Insulin and PUFAs regulate superoxide anion generation and enhance the production of eNO [45–51]. NO is anti-inflammatory in nature [49] and quenches superoxide anion. IGF-I and, possibly, insulin enhance ACh release from rat cortical slices [52, 53]. ACh inhibits the synthesis and release of TNF-α both *in vitro* and *in vivo* and thus, has anti-inflammatory actions [54] and is also a potent stimulator of eNO synthesis [55]. These data suggest that insulin and IGF-I enhance the formation of PUFAs in the brain by their action on desaturases, and PUFAs, in turn, enhance ACh levels in the brain (this is in addition to the ability of insulin and IGF-I to directly enhance ACh levels in the brain) and inhibit the production of TNF-α. Thus, insulin, ACh, and PUFAs suppress TNF-α production and augment the synthesis of eNO. ACh and eNO are not only neuroprotective in nature but also interact with other neurotransmitters and regulate their secretion, release and action. Thus, insulin, IGF-I, ACh, and PUFAs protect brain from insults induced by TNF-α and other molecules.

In addition, there is evidence to suggest that PUFAs and insulin have cytoprotective actions as well. For example, we showed that PUFAs prevent radiation and chemical-induced cytotoxicity and genotoxic actions both in vitro and in vivo [41, 56, 57]. Alloxan-induced cytotoxicity to pancreatic β cells was prevented by AA, EPA and DHA both in vitro and in vivo [58–61], suggesting that PUFAs may have anti-diabetic actions. PUFAs, especially n-3 PUFAs, were found to prevent status epilepticus-associated neuropathological changes in the hippocampal formation of rats with epilepsy [62]. In our studies, we noted that both cyclo-oxygenase and lipoxygenase inhibitors did not prevent the cytoprotective actions of PUFAs against alloxan-induced damage to pancreatic β cells, suggesting that either the fatty acids themselves are active or other products are formed that have cytoprotective actions. Recent studies suggested that AA, EPA and DHA form precursors to anti-inflammatory and potent cytoprotective products such as lipoxins, resolvins, protectins and maresins [63–70] that could be responsible for the cytoprotective actions of various PUFAs. It is likely that normal cells when exposed to PUFAs produce significant amounts of lipoxins, resolvins and protectins that protect them from the cytotoxic actions of various chemicals and radiation, while tumor cells when supplemented with the same fatty acids do not produce these cytoprotective lipid molecules but produce cytotoxic molecules such as 17-hydroxydocosahexaenoic acid (17-HDHA) via 17-hydroperoxydocosahexaenoic acid (17-HpDHA) through 15-lipoxygenase and autoxidation. In contrast to normal neural cells, neuroblastoma cells did not produce the anti-inflammatory and protective lipid mediators, resolvins and protectins. The cytotoxic effect of DHA in neuroblastoma seems to be mediated through production of hydroperoxy fatty acids that accumulate to toxic intracellular levels. These evidences suggest that normal and tumor cells metabolize PUFAs in a differential fashion that seems to underlie the cytoprotective action in normal cells and at the same time PUFAs are directed to form toxic hydroperoxy fatty acids to kill tumor cells.

In addition, PUFAs and insulin interact with each other to bring about some of their actions. Incorporation of significant amounts of PUFAs into the cell membranes increase their fluidity that, in turn, enhances the number of insulin receptors on the membranes and the affinity of insulin to its receptors. Thus PUFAs attenuate insulin resistance [71–78]. Hereditary hypertriglyceridemic (hHTg) rats have reduced activity of the Δ^6 desaturase in liver without any changes in gene expression for this enzyme; and the concentration of AA was significantly decreased in hHTg rat liver suggesting that impaired insulin action in hHTg rat is due to a deficiency of PUFAs. Feeding these animals with fish oil, a rich source of EPA and DHA, not only reduced plasma levels of triglycerides but also restored insulin sensitivity [79, 80]. These results were supported by the observation that supplementation of fish oil to high fat diet fed experimental animals improved *in vivo* insulin action; and this insulin sensitizing effect of fish oil was accompanied by a decrease of circulating triglycerides, free fatty acids and glycerol levels in the postprandial state and by a lower lipid content in liver and skeletal muscle [80]. These results are interesting since it is known that increase in IMCL is associated with insulin resistance and increased expression of perilipins, whereas EPA/DHA reduce IMCL and possibly that of perilipins. Thus, one mechanism by which EPA/DHA are beneficial in the metabolic syndrome could be by reducing IMCL and the expression of perilipins.

Since brain is rich in PUFAs, especially AA, EPA, and DHA, one important function of PUFAs in the brain could be to ensure the presence of adequate number of insulin receptors. Thus a defect in the metabolism of PUFAs or when adequate amounts of PUFAs are not incorporated into the neuronal cell membranes during the fetal development and infancy, it may cause a defect in the expression or function of insulin receptors in the brain. This may lead to the development of type 2 diabetes as seen in NIRKO mice [81]. Furthermore, systemic injections of either glucose or insulin in *ad libitum* fed rats resulted in an increase in extracellular acetylcholine in the amygdala [82]. Acetylcholine (ACh) modulates dopamine release that, in turn, regulates appetite [83]. ACh inhibits the production of pro-inflammatory cytokines (IL-1, IL-2 and TNF-α) in the brain and thus, may also protect the neurons.

The cytoprotective actions of insulin, which is similar to that of PUFAs and acetylcholine, is further evident from our previous study wherein it was noted that insulin infusion protected cardiac tissue from ischemia-reperfusion induced injury by inhibiting ischemia/reperfusion-induced TNF-α production through the Akt-activated and eNOS-NO-dependent pathway in cardiomyocytes. The antiinflammatory property elicited by insulin may contribute to its cardioprotective and prosurvival effects both in vitro and in vivo [84]. These results are in support of the previous proposal that insulin has anti-inflammatory actions, shows cytoprotective and cardioprotective actions [85–90].

In addition to their ability to possess cytoprotective actions, PUFAs also have a significant role in the growth and development of brain.

PUFAs in Brain Growth and Development

Brain is rich in AA, EPA and DHA which constitute as much as 30–50% of the total fatty acids in the brain, where they are predominantly associated with membrane phospholipids. Hence, when the concentrations of these fatty acids are inadequate, especially, during the critical period of brain growth, which is from third trimester to 2 years post-term and adolescence; the development, maturation, synaptic connections of hypothalamic neurons (especially in the VMH), the synthesis, release and functionality of various neurotransmitters is expected to be inappropriate or inadequate. Such a developmental aberration of the hypothalamic neurons will lead to a defect in the expression or function of insulin receptors in the brain, various neurotransmitters and their receptors that, in turn, predisposes to defective blood glucose sensing both in the brain and periphery resulting in failure of pancreatic β cells to produce adequate amounts of insulin. These events could eventually result in the development of the metabolic syndrome. In this context, it is noteworthy that PUFAs to have a critical and direct regulatory role in the growth and development of brain. PUFAs also have a regulatory role in the synthesis, release and function of various neurotransmitters and hypothalamic peptides.

Syntaxin, SNARE Complex and PUFAs

Increase in cell membrane surface area and growth of neurite processes from the cell body are critical for proper neuronal development and synapse formation [91]. Nerve growth cones are highly enriched with AA-releasing phospholipases, which have been implicated in neurite outgrowth [92, 93]. The fusion of transport organelles with plasma membrane leads to cell membrane expansion [94]. Syntaxin 3, a plasma membrane protein that has an important role in the growth of neurites has been shown to be a direct target for AA, DHA and other PUFAs [95]. It was reported that AA, DHA, and other PUFAs but not saturated and monounsaturated fatty acids activated syntaxin 3. Of all the fatty acids tested, AA and DHA were found to be the most potent compared to LA and ALA. Even syntaxin-1 that is specifically involved in fast calcium-triggered exocytosis of neurotransmitters is sensitive to AA [96]. These results suggest that AA is involved both in exocytosis of neurotransmitters and neurite outgrowth. SNAP-25 (synaptosomal-associated protein of 25 kDa), a syntaxin partner, implicated in neurite outgrowth interacted with syntaxin 3 only in the presence of AA that allowed the formation of the binary syntaxin 3-SNAP 25 complex. AA stimulated syntaxin 3 to form the ternary SNARE complex (soluble N-ethylmaleimide-sensitive factor attachment protein receptor), which is needed for the fusion of plasmalemmal precursor vesicles into the cell surface membrane that leads to membrane fusion. The intrinsic tyrosine fluorescence of syntaxin 3 showed marked changes upon addition of AA, DHA, LA, and ALA, whereas saturated and monounsaturated (oleic acid) fatty acids were ineffective. These results indicate that AA and DHA change the α-helical syntaxin structure to expose SNARE motif for immediate SNAP 25 engagement and thus, facilitate neurite outgrowth.

PUFAs Modulate RAR-RXR and Other Nuclear Receptors and are Essential for Brain Growth and Development

Retinoic acid (RA) has profound effects on the development of vertebrate limb and nervous system, and in epithelial cell differentiation that are transduced by its binding to a nuclear retinoic acid receptor (RAR) which, in the presence of ligand, is transformed into a transcription factor. The differential expression of RAR gene family receptors: RAR-α, RAR-β, and RAR-γ, is important for correct transduction of the RA signal in various tissues. The other subtype of retinoid receptor is the retinoid X receptor (RXR), which also could be α, β, and γ. RXRs are also transcription factors that can act as ligand-dependent and -independent partners for RARs and other nuclear receptors. There is evidence to suggest that RAR-RXR dimmers act on the β-catenin signaling pathway to produce some of their actions. RAR-RXR nuclear receptors are essential for the development brain and other neural structures [97]. AA, DHA, and possibly, EPA are known to serve as endogenous ligands of RAR-RXR and activate them [98–100]. Several RXR heterodimerization partners such as peroxisome proliferator-activated receptors (PPARs), the liver X receptors (LXR) and farnesoid X receptor (FXR) are essential for regulating energy and nutritional homeostasis and in the development of brain and other neural structures. This suggests that AA, DHA, and EPA participate in the growth and development of brain and other neuronal structures by their ability to bind to RAR-RXR, LXR, FXR and other nuclear receptor heterodimers. This is supported by the observation that EPA/DHA and other fatty acids alter gene expression in the developing brain [101].

Myelin-specific mRNA levels were found to be developmentally regulated and influenced by dietary fat. Neonatal brain stearoyl CoA desaturase and LDL receptor mRNA levels were altered by neonatal fat intake. The neonatal response to dietary fat is tissue-specific at the mRNA level. [101].

Since PUFAs are structural components of all tissues and are indispensable for cell membrane synthesis, especially of the brain, retina and other neural tissues; and serve as precursors for eicosanoids, which regulate numerous cell and organ functions, it is reasonable to expect that n-3 and n-6 fatty acids are essential for the growth and development of human brain, particularly in early life. It is known that light sensitivity of retinal rod photoreceptors is significantly reduced in newborns with n-3 fatty acid deficiency, and that DHA significantly enhanced visual acuity maturation and cognitive functions. Clinical studies revealed that dietary supplementation with EPA/DHA-rich oils resulted in increased blood levels of DHA and AA, as well as an associated improvement in visual function in formula-fed infants matching that of human breast-fed infants. These beneficial effects are not only due to the known effects of these fatty acids on membrane biophysical properties, neurotransmitter content, and the corresponding electrophysiological correlates but also because of their ability to alter gene expression of the developing retina and brain. Intracellular fatty acids or their metabolites regulate transcriptional activation of gene expression during adipocyte differentiation and retinal and nervous system development. Regulation of gene expression by PUFAs occurs at the transcriptional level and is probably mediated by nuclear transcription factors activated by fatty acids and by modulating micro-RNAs. These nuclear receptors are part of the family of steroid hormone receptors. AA/EPA/DHA have significant effects on photoreceptor membranes and neurotransmitters involved in the signal transduction process; rhodopsin activation, rod and cone development, neuronal dendritic connectivity, and functional maturation of the central nervous system [102].

For example, PGD_2-synthesizing enzyme that is expressed in antigen-presenting cells, mast cells, and other immunocompetent cells is also present in microglia and the migration pathway of microglia in the developing mouse brain. The expression of PGD_2 synthase enzyme mRNA peaked at postnatal day 10, decreased gradually thereafter, and reached a plateau at postnatal day 20. Most of the PGD_2 synthase positive cells at postnatal day 10 had morphological characteristics of ameboid microglia and gave positive immunostaining with microglia-specific markers, which became less detectable later on, but PGD_2 synthase was still expressed even in resting microglia. These evidences suggest that PGD_2 synthase enzyme is a useful marker for microglial development. In addition, spaciotemporal evaluation of microglial development and migration with PGD_2 synthase immunostaining revealed that the migration pathways of microglia in the postnatal brain could be: from the lateral ventricle via subventricular zones to brain parenchyma; from the leptomeninges around the cerebellopontine angle to the cerebellar white matter; and from the overlying leptomeninges to the hippocampus, basal forebrain, and brainstem [103]. This suggests that one can use PGD_2 synthase marker to trace the growth and migration pathways of microglia.

In this context, it is noteworthy that albumin, a serum protein present in the developing brain, could serve as a stimulator of fatty acid synthesis and thus, may aid in brain growth and development. It is known that albumin binds tightly to PUFAs and could carry various fatty acids from place to place. For instance, when albumin is infused it could mobilize DHA and, possibly, other PUFAs from liver to the target tissues [104]. It was shown that albumin stimulates the synthesis of oleic acid by cultured astrocytes by inducing stearoyl-CoA 9-desaturase, the rate-limiting enzyme in oleic acid synthesis, through activation of the sterol regulatory element-binding protein-1. In experimental animals, albumin reaches maximal brain level by day 1 after birth, coinciding with activation of the sterol response element binding protein-1, which is responsible for the transcription of the enzymes required for oleic acid synthesis. The developmental profile of stearoyl-CoA 9-desaturase-1 mRNA expression follows that of sterol regulatory element-binding protein-1 activation, indicating that these phenomena are tightly linked. Since, oleic acid induces neuronal differentiation, as indicated by the expression of growth associated protein-43 and the expression of growth associated protein-43 mRNA peaks at about day 7 after birth, following the maximal expression of stearoyl-CoA 9-desaturase-1 mRNA that occurs between days 3 and 5 postnatally, it is reasonable to conclude that the synthesis of oleic acid is linked to neuronal differentiation during rat brain development [105]. It is possible that similar function could be attributed to other fatty acids such as AA/EPA/DHA. Furthermore, DHA/EPA/AA appears to be versatile molecules with a wide range of actions spanning from participation in cellular oxidative processes and intracellular signaling to modulatory roles in gene expression and growth regulation [106].

The essentiality of PUFAs in brain and growth development is further evident from the fact that maternal α-linolenic acid (ALA, 18:3 n-3) dietary deficiency in postnatal rat brain showed a marked decrease of the dopamine-synthesizing enzyme tyrosine hydroxylase accompanied by a down-regulation of the vesicular monoamine transporter (VMAT-2) and a depletion of VMAT-associated vesicles in the hippocampus compared with adequately fed controls. The dopamine transporter (DAT) was not affected by the ALA deficiency indicative of a DAT/VMAT-2 ratio increase that may enhance the risk of damage of the dopaminergic terminal. A robust increase in dopamine receptor (DAR1 and DAR2) levels was noticed in the cortex and striatum structures possibly to compensate for the low levels of DA in synaptic clefts. Microglia activation was noticed following ALA deficiency. Since ALA deficiency could lead to decreased DHA synthesis, it has been proposed that reduced levels of anti-oxidants in the developing brain might be responsible for microglial activation and enhanced oxidative stress that increased the risk of dopamine-associated neurological disorders [107, 108] that include obesity, schizophrenia, depression and anxiety.

It is important to note that DHA/EPA supplementation significantly reduced DNA fragmentation and caspase-3 activation in developing cerebellum of hypothyroid pups. This anti-apoptotic actions of EPA/DHA is due to their ability to decrease the levels of pro-apoptotic basal cell lymphoma protein-2 (Bcl-2)-associated X protein (Bax) and increase the levels of anti-apoptotic proteins like Bcl-2 and Bcl-extra large

(Bcl-x(L)). Furthermore, EPA/DHA restored levels of cerebellar phospho (p)-AKT, phospho-extracellular regulated kinase (p-ERK) and phospho-c-Jun N-terminal kinase (p-JNK). These results suggest that some of the beneficial actions of EPA/DHA in brain growth and development include their ability to regulate apoptotic signaling pathways under stress [109].

Furthermore, the vital role of PUFAs particularly during embryonic development became evident from the observation that the expression of genes encoding enzymes involved in PUFA biosynthesis, namely fatty acyl desaturase (Fad) and Elovl5- and Elovl2-like elongases, showed temporal expression of all three genes from the beginning of embryogenesis (zygote), suggesting maternal mRNA transfer to the embryo. When spatial expression was studied by whole-mount in situ hybridisation in 24 embryos, both fad and elovl2 were found to be highly expressed in the head area where neuronal tissues are developing. Of all, elovl5 showed specific expression in the pronephric ducts, suggesting an as yet unknown role in fatty acid metabolism during zebrafish early embryonic development. The yolk syncytial layer also expressed all three genes, suggesting an important role in remodelling of yolk fatty acids during zebrafish early embryogenesis. Tissue distribution in zebrafish adults demonstrates that the target genes are expressed in all tissues but more particularly in liver, intestine and brain showing the highest expression [110]. These results suggest that PUFAs are essential during embryo development and more so for brain growth and development and zebrafish could be used as a model organism to delve more deeply into the role of PUFAs in the development of various organs.

PUFAs Modulate Gene Expression and Interact with Cytokine TNF-α and Insulin to Influence Neuronal Growth and Synapse Formation

It was reported that mRNA level of genes involved in myelination were affected by a diet lacking essential fatty acids [101]. The expression level of 102 cDNAs, representing 3.4% of the total 3200 DNA elements on the microarray, were significantly altered (either upregulated or downregulated) in brains of rats fed with ω-3 DHA/ALA diets [111–114]. Of all the genes examined, 55 genes were upregulated and 47 were downregulated relative to controls. The altered genes included those involved in synaptic plasticity, cytoskeleton, signal transduction, ion channel formation, energy metabolism, and regulatory proteins. Of all, the 15 genes that responded more intensely to the ALA/DHA diet include those that encode a clathrin-associated adaptor protein, farnesyl pyrophosphatase synthetase, Sec24 protein, NADH dehydrogenase/cytochrome *c* oxidase, cytochrome b, cytochrome c oxidase subunit II, ubiquitin-protein ligase Nedd42, and transcription factor-like protein. In addition, several genes that participate in signal transduction, like RAB6B, small GTPase and calmodulins were also upregulated. α- and γ-synuclein and D-cadherin genes were upregulated in response to ALA/DHA-rich diet, which are specifically enriched at synaptic contacts and are known to play a significant role in neural plasticity, development and maturation of neurons [115]. The overexpression of mitochondrial enzymes observed in ALA/DHA diet supplemented rats suggests that the brain was in an elevated metabolic state. Perinatal supply of ω-3 fatty acids influences brain gene expression later in life and is critical to the development and maturation of several brain centers that are specifically involved in the regulation of appetite and satiety. It is possible that the quality and quantity of PUFAs available during the perinatal period may determine the expression level of various genes, their response to the environmental agents, and determine the quality and levels of expression of various pro-oxidant and anti-oxidant enzymes, cytokines, pro-resolution and wound healing molecules, etc., timing of their expression, duration of expression and their interaction(s) with other concerned genes. Thus, in essence PUFAs might be the master regulators of gene expression and they may be able to regulate and determine gene expression at various stages of growth and development and at different periods of age and the response of genes to different environmental and endogenous stimuli and molecules.

For example, it was reported that perinatal supplementation of ω-3 fatty acids (especially DHA) induces overexpression of genes coding for cytochrome c and TNF receptor (TNFRSF1A), while omega-3 lipids decreased TNF-α and PGE_2 production in LPS-stimulated macrophages [116], probably, through decreasing NF-kappaB activation. It should also be noted that PUFAs may have a direct effect on the expression of genes and/or may bring about their actions through their products such as eicosanoids, lipoxins, resolvins, protectins and nitrolipids. It was reported that aspirin-treated enterocytes generated 15R-HETE, a precursor of 15-epi-LXA_4 biosynthesis, which sharply inhibited TNF-α-induced IL-8 and thus, downregulated mucosal inflammatory events [117]. Similarly, eicosanoids derived from PUFAs may inhibit proinflammatory gene expression by acting on the PPAR-γ expression to bring about their biological actions [118]. Hence, the actions of PUFAs may occur at several stages and brought about by several of their products. It is also important to note that sometimes the actions of products of various PUFAs may have antagonistic actions (for example: some eicosanoids have pro-inflammatory actions whereas lipoxins have anti-inflammatory actions). Hence, the final outcome of the actions of various PUFAs and their products on a physiological function in a given tissue or organ will depend on the balance between these mutually antagonistic molecules. Similarly, even in a given pathological process the balance between mutually antagonistic actions of various PUFAs and their very many metabolites will determine the continuation of the diseases process or recovery from the same.

The actions of PUFAs on the expression of neurotransmitter genes is particularly relevant while considering the role of PUFAs in brain growth and development and their involvement in various neurological diseases. For instance, it was reported that supplementation of AA and EPA/DHA increased the expression of serotonin receptor in hypothalamus [119]. 5-HT_4 receptor increases in expression have been shown to augment hippocampal acetylcholine outflow. It was also reported that AA and EPA/DHA feeding enhanced the expression of POMC in hippocampus suggesting that AA/EPA/DHA can influence appetite and satiety and thus, control energy metabolism. Changes in the expression of acetylcholine is of particular interest since, it has a regulatory role in the release and action of various other neurotransmitters such as serotonin, dopamine and catecholamines and is also a potent anti-inflammatory molecule [120–124], while catecholamines have pro-inflammatory actions [125]. Thus, yet another mechanism by which PUFAs regulate inflammation and immune response is by altering the levels of acetylcholine in the brain. Thus, PUFAs, its anti-inflammatory products such as lipoxins, resolvins, protectins, nitrolipids and acetylcholine are essential to prevent inflammation in the brain to ensure its proper growth and development in perinatal period.

These results are interesting since; there is now evidence to suggest that TNF-α produced by glial cells enhances synaptic efficacy by increasing surface expression of AMPA receptors. Continued presence of TNF-α is required for preservation of synaptic strength at excitatory synapses [126, 127]. TNF-α production is suppressed by EPA/DHA and acetylcholine, whereas excess TNF-α induces apoptosis of neurons. On the other hand, hepatic vagus nerve stimulation attenuated Fas-induced hepatocyte apoptosis through alpha7 nicotinic AChR by causing the Kupffer cells to reduce their generation of an excessive amount of reactive oxygen species [128] and there it is likely that similar anti-apoptotic function could be served by acetylcholine in the brain also. Thus, there appears to be a balance maintained between pro-apoptotic action of TNF-α and anti-apoptotic function of acetylcholine (and incidentally acetylcholine suppresses the synthesis and release of TNF-α) that are essential to regulate the neuronal generation, their synaptic formation and the strength of the synapses that are formed. PUFAs and their metabolites by regulating the acetylcholine outflow and TNF-α synthesis may play a pivotal role in the neurogenesis. Since, during the development of brain in the perinatal period there is a constant cycles and waves of neuronal death and generation and regeneration from the neuronal stem cells, it is plausible that the levels of PUFAs and their metabolites, acetylcholine and TNF-α and possibly, other cytokines change (increase and decrease) in a wave form to modulate the brain growth during its various phases of development. It is likely that there is a sort of a celestial dance among the levels of various PUFAs, their metabolites, acetylcholine, TNF-α and other cytokines and a close interaction(s) that ultimately moulds the growth and development of the brain.

Neurogenesis and Neuronal Movement During the Growth and Development of Brain and PUFAs

The most important event(s) that happen in the growth and development of brain is a massive rearrangement of the neuronal cells that transforms a relatively uniform ball of cells into a multilayered organ with innumerable connections (synapses) at the right time and at the right place. This has to occur in a precise choreography that is strikingly similar among organisms from flies to fish to people. Although neuronal cell movements are crucial for the development of brain and sometimes involve longer journeys [129], if the intricate dance of neuronal cells goes awry, the resulting defects are usually catastrophic.

Just what causes the neuronal cells to move and guides them to their designated places is fascinating. There could be a link between genetic signaling cascades to molecules that actually affect the movements of neuronal cells, including those that cause cells to stick together and those that promote movement. It is likely that rearrangements of neuronal cells arose from cell division—that certain cells divide faster than others and change the architecture of the brain. It is also likely that cells constantly shift places in a specified pattern. Before cells can move they must first loosen the adhesives holding them together called cadherins, which protrude from the cell surface allowing cells to stick to each other. Reelin and mDab1 are two proteins that are involved in neuronal migration [130] though; there could be many more such proteins. These proteins might interact with the intracellular signaling enzymes tyrosine kinases and have the ability to bind to Src and thus, can link a tyrosine kinase like Src to another protein in a signaling pathway.

Fetal stem cells can multiply and differentiate to neurons and glia. The adult nervous system contains multipotential precursors for neurons, astrocytes, and oligodendrocytes. Cultured cells from both the fetal and adult CNS that have proliferated in vitro can differentiate to show morphological and electrophysiological features characteristic of neurons: regenerative action potentials and synaptic structures, suggesting the multipotential nature of cells derived from the CNS. Sonic hedgehog and members of the transforming growth factor-β (TGF-β), basal-fibroblast growth factor (b-FGF), platelet-derived growth factor (PDGF), ciliary neurotrophic factor (CNTF), neurotrophins, epidermal growth factor (EGF), BMPs (bone morphogenetic factors), angiopoietin are some of the factors that seem to be involved in the growth, differentiation and proliferation of neural stem cells [131]. PTEN is also involved in the control of neural cell size, and in the proliferation and self-renewal of neural stem cells [132]. Wnt signaling has recently emerged as a key factor in controlling stem cell expansion. There is now evidence to suggest that many of these factors involved in neuronal stem cell proliferation and differentiation interact with PUFAs as discussed below.

Insulin, PUFAs and Neuronal Proliferation

Insulin is needed for neuronal growth and differentiation and synaptic plasticity in the CNS [133, 134] but also stimulates the formation of AA/EPA/DHA by activating of Δ^6 and Δ^5 desaturases, and suppresses TNF-α production. Insulin has been shown to determine final size of the cells and body possibly, by regulating metabolism [134]. Calorie restriction activates Δ^6 and Δ^5 desaturases; partly, by enhancing insulin action, and promotes the formation of AA/EPA/DHA. Calorie restriction also promotes mitochondrial biogenesis by inducing the expression of eNOS [134] and the enhanced formation of NO that occurs as a result, is a neurotransmitter and vasodilator that may aid the rapidly growing brain during perinatal period. Furthermore, as already described above, both insulin and AA/EPA/DHA stimulate eNO formation. This close interaction and feed-back regulation between TNF-α, EPA/DHA, insulin, Δ^6 and Δ^5 desaturases, and neuronal growth and synapse formation, and the fact that TNF-α is needed for synaptic strength whereas AA/EPA/DHA is needed for the activation of syntaxin 3 and neurite outgrowth suggests that growth of neurons and synaptic formation will be optimum only when all these factors are present in physiological concentrations. In contrast, when AA/EPA/DHA concentrations are sub-optimal, TNF-α levels tend to be high. High TNF-α concentrations have neurotoxic actions and hence, could cause damage to VMH neurons. This will lead to hyperphagia, hyperglycemia, hyperinsulinemia, hypertriglyceridemia and IGT. Thus, TNF-α may participate in the pathogenesis of metabolic syndrome by two mechanisms: (a) inducing peripheral and central insulin resistance, and (b) damage or interfere with the action of VMH neurons.

Catenin, *wnt* and Hedgehog Signaling Pathway in Brain Growth and Development and PUFAs

It is well recognized that during brain development, proliferation of neural progenitor cells is tightly controlled to produce the organ of predetermined size. To achieve this objective, cell-cell communication is essential so that information concerning the density of their cellular neighborhood is provided. Adherens junctions, which contain cadherins, β-catenins, and α-catenins, mediate intercellular adhesion in neural progenitors [135]. In a recent study, it was observed that mice with a conditional α*E-catenin* allele (α*E-catenin*$^{loxP/loxP}$) crossed with mice carrying nestin-promoter driven Cre recombinase (*Nestin-Cre*$^{+/-}$), which is expressed in CNS stem/neural progenitors starting at embryonic day 10.5 (E10.5) resulted in α*E-catenin*$^{loxP/loxP}$*/Nestin-Cre*$^{+/-}$ animals that displayed loss of αE-catenin in neural progenitor cells [136]. The knockout α*E-catenin*$^{loxP/loxP}$*/Nestin-Cre*$^{+/-}$ mice were born with bodies similar to their littermates, but with enlarged heads due to shortening of the cell cycle, decreased apoptosis, and cortical hyperplasia as a result of abnormal activation of hedgehog pathway. Hedgehog pathway plays a critical role in mammalian CNS development and brain cancer. Hedgehog pathway promotes survival and blocks apoptosis of neuroepithelial cells and hence, its activation may produce cortical hyperplasia in α*E-catenin*$^{loxP/loxP}$*/Nestin-Cre*$^{+/-}$ mice. These results suggest that the increase in cell density is sensed by an increase in the per cell area occupied by adherens junctions that is transmitted to the hedgehog pathway. This, in turn, provides a negative feedback loop resulting in a decrease in cell proliferation that ultimately controls the size of developing brain [136]. It is interesting to note that β-catenin is required for the mitogenic activity of PGE_2 in colon cancer cells [137], whereas GLA, the precursor of AA (from which PGE_2 is formed) inhibits the expression of catenin both in vascular endothelial cells and human cancer cells [138, 139]. This suggests that PUFAs have a negative feedback control on catenin expression and thus, they may regulate brain size, development and growth. Thus, one of the functions of AA, EPA and DHA in the brain could be not only to regulate synapse formation and neurite growth but also to control brain growth and size. It was reported that TNF-α also induced a significant decrease of E-cadherin and β-catenin expression [140] suggesting that cytokines play a role in brain growth and development. This is especially interesting in the light of the known fact that at high concentrations TNF-α induces apoptosis of neuronal cells [34, 35]. Thus, there seems to be a close interaction(s) between the expression of catenins, and their modulation by TNF-α and possibly, other cytokines, and PUFAs are crucial to neurite growth, synapse formation, and brain growth and development. Proper development of neurons and synaptic connections between different neurons ultimately determines the response of various neurons, especially those of hypothalamic neurons, to various neurotransmitters and plasma glucose that, in turn, regulates insulin secretion by pancreatic β cells and glucose production by liver. This is supported by the observation that an increase in circulating glucose and a primary increase in hypothalamic glucose levels inhibits glucose production in the liver and thus, lowers blood glucose [141]. Activation of neuronal pyruvate flux is required for hypothalamic (especially the arcuate nucleus) glucose sensing and for control of blood glucose and liver glucose metabolism through the activation of ATP-sensitive potassium channels in the glucose sensing hypothalamic neurons [141]. These results suggest that specific hypothalamic neurons play a significant role in the control of blood glucose levels, glucose production by liver and insulin secretion by pancreatic β cells. The ability of these specific hypothalamic neurons to control glucose homeostasis may, in turn, depend on the health of these neurons and their synaptic connections with other neurons and their ability to respond to various neurotransmitters in an appropriate manner. Impairment in the biochemical sensing of carbohydrates (especially glucose) by the hypothalamic neurons may represent a basic underpinning for defects in the regulation of food intake [142, 143], β-cell function [144], and liver glucose homeostasis [145]. Both type 2 diabetes mellitus and metabolic syndrome are typical examples of diseases the prevalence of which is dependent on environmental, nutritional factors operating on genetic susceptibility.

One important regulatory factor that controls β-catenin-dependent transcription of target genes is *Wnt* proteins that signal through seven-pass transmembrane receptors of the frizzled family to activate β-catenin. The *Wnt* family of secreted glycoproteins regulates a large number of developmental processes including cell growth, cell polarity, cell-fate determination, tissue patterning, tissue specification, and tumorigenesis. *Wnt*s are crucial cell signaling molecules during development and in adult life. In the absence of *Wnt* receptor activation, the modular protein Axin provides a scaffold for the binding of glycogen synthase kinase-3β (GSK3β), Adenomatous polyposis coli protein (APC) and β-catenin. This, in turn, facilitates β-catenin phosphorylation by GSK3β [146, 147] and leads to the degradation of β-catenin via the ubiquitin pathway [148]. Upon *Wnt*'s binding of the frizzled receptor, the Axin-GSK3β-APC-β-catenin complex is disrupted. As a result, β-catenin is no longer targeted for ubiquitin degradation and so accumulates in the nuclei [149], where it interacts with the members of the lymphoid enhancer factor/T-cell factor classes of transcription factors to regulate the expression of target genes. Overexpression of GSK3β and Axin or depletion of maternal β-catenin RNA causes deficiencies in dorsal structures [150–152]. β-catenin induces growth of cardiomyocytes *in vitro* and is necessary for hypertrophic stimulus-induced growth of cardiomyocytes *in vivo* [153]. β-catenin is stabilized in cardiomyocytes on exposure to hypertrophic stimuli. But, in this instance, the stabilization of β-catenin was independent of *Wnt* signaling though inhibition of GSK3β remained central to hypertrophic stimulus-induced stabilization of β-catenin.

Wnt signaling leads to stabilization of β-catenin [154] and inappropriate activation of *Wnt* signaling has been described in many tumors. Transcriptional targets directly activated by β-catenin include: cyclin D1, *c-myc*, matrilysin, PPAR-δ, and upregulation of COX-2 [155–159]. *Wnt* expressing mammary epithelial cells and under conditions of nuclear β-catenin accumulation showed transcriptional upregulation of COX-2 [160, 161], and there is evidence to suggest that β-catenin causes upregulation of COX-2, whereas EPA suppresses COX-2 and catenin expression [138, 139, 214, 215] and also functions as an endogenous ligand of PPARs [162]. But it is not known whether PUFAs can directly influence the expression of *Wnt*. In this context, it is noteworthy that *Wnt* pathway plays a major role in cardiac myogenesis, myocardial hypertrophy, and heart failure, possibly, by inhibiting GSK-3β activity [163, 164], which leads to stabilization of β-catenin complex. This leads to β-catenin translocation to the nucleus where it participates in the transcription processes. In obesity, there is an overexpression of SRP4, an endogenous antagonist of *Wnt* protein and a repressor of *Wnt* receptors, FDZ6 and FDZ4, and also of Dsh3, a direct inhibitor of GSK-3β activity. These changes favor β-catenin ubiquitination and degradation in proteasomes and direct repression of several factors that favor a role in cardiac hypertrophy such as c-myc, GATA4, and MEF2B [165]. This regulation of the *Wnt*/β-catenin pathway noted in the obese heart has the potential to prevent the development of cardiac hypertrophy since the volume overload observed in obesity-related hypertension decreases the expression of β-catenin and connexin 43, whereas hearts from hypertensive patients showed decreased GSK-3β activity, nuclear accumulation of β-catenin that could lead to myocardial hypertrophy [166]. Thus, *Wnt*/β-catenin/GSK3β and hedgehog signaling pathway is not only involved in the growth and development of brain but also in cardiac hypertrophy. Since PUFAs have a negative feedback control on catenin expression and TNF-α synthesis [41–43, 138, 139] and TNF-α also induced a significant decrease of E-cadherin and β-catenin expression [140], it implies that in an indirect fashion PUFAs play a regulatory role in the expression and action of *Wnt*/β-catenin/GSK3β and hedgehog signaling pathway and thus, in brain growth and development (Fig. 16.2).

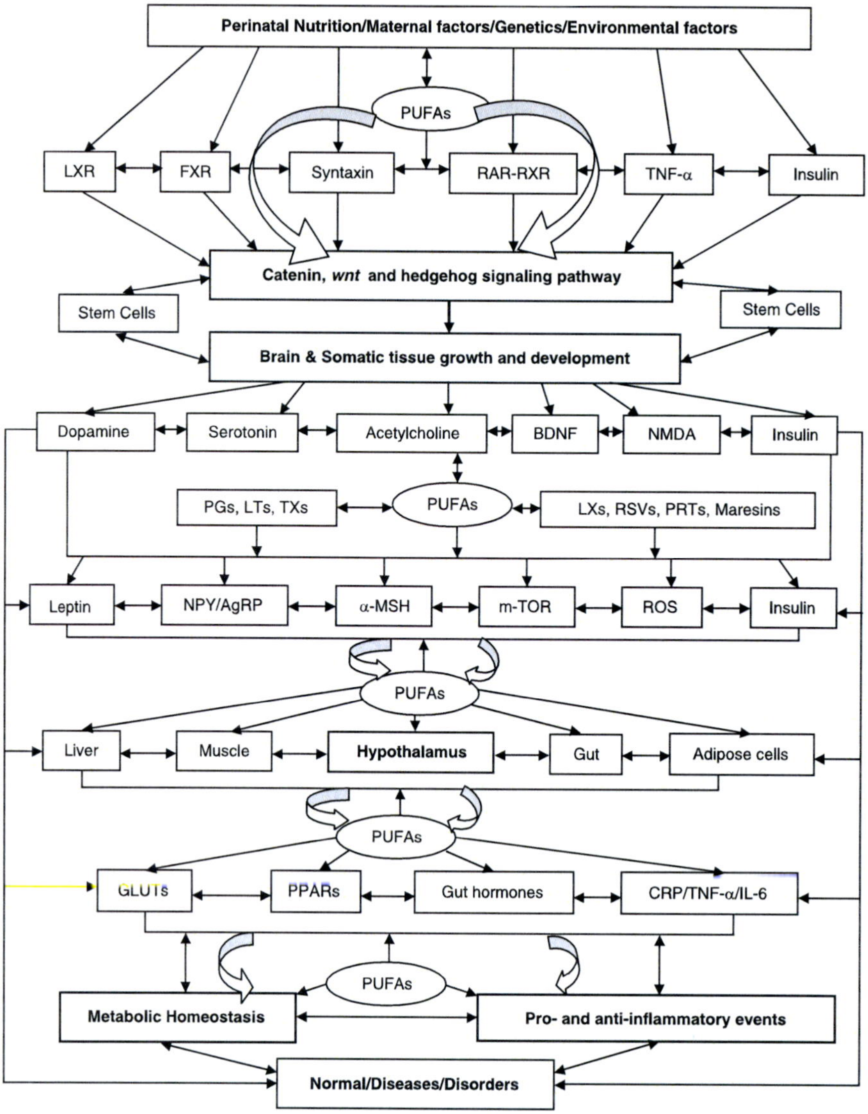

Fig. 16.2 Scheme showing possible relationship among various molecules/molecular events and PUFAs, brain growth and development, neuropeptides, inflammation and diseases/disorders. For further details see text

PUFAs Modulate NMDA, γ-Aminobutyric Acid (GABA), Serotonin and Dopamine in the Brain

Since PUFAs play a significant role in the growth and development of brain, it is possible that they (PUFAs) also regulate the fetal brain nerve growth cone membranes and monoaminergic neurotransmitters. This is especially so since, it is known that AA, DHA and other PUFAs but not saturated and monounsaturated fatty acids activate syntaxin 3, a plasma membrane protein that has an important role in the growth of neurites [95]. Further, syntaxin1 that is involved in fast calcium-triggered exocytosis of neurotransmitters is modulated by AA [96], implying that AA is involved both in exocytosis of neurotransmitters and neurite outgrowth. SNAP25 (synaptosomal-associated protein of 25 kDa), a syntaxin partner implicated in neurite outgrowth, interacted with syntaxin 3 only in the presence of AA, DHA, LA, and ALA, whereas saturated and monounsaturated fatty acids were ineffective, to form the ternary SNARE complex (soluble N-ethylmaleimide-sensitive factor attachment protein receptor), which is needed for the fusion of plasmalemmal precursor vesicles into the cell surface membrane that leads to membrane fusion, an event that facilitates neurite outgrowth.

Rats fed purified diets containing safflower oil, a rich source of LA, soybean oil as a source of LA and ALA, and high fish oil, rich in DHA, through gestation showed that offspring of rats fed fish oil had significantly higher DHA in their brain nerve growth cone membrane phosphatidylserine (PS), phosphatidylethanolamine (PE), and phosphatidylinositol (PI) than the soybean oil group. The growth cone membrane phosphatidylcholine (PC), PE and PS AA was significantly lower in the fish oil than in the soybean or safflower oil groups. Serotonin concentration was significantly higher in brain of offspring in the safflower oil compared with the soybean oil group. The newborn brain dopamine was inversely related to PE DHA and PS DHA, but positively related to PC AA. These results suggest that maternal dietary fatty acids alter fetal brain growth cone fatty acid content and neurotransmitters involved in neurite extension, target finding and synaptogenesis [167].

In a study that investigated the effect of feeding formula from birth to 18 days with different PUFAs on the concentrations of monoaminergic neurotransmitters in various regions of the brain, it was observed that animals that received LA + ALA in formula had a significant effect on frontal cortex dopamine, 3,4-dihydroxyphenylacetic acid, homovanillic acid, serotonin, and 5-hydroxyindolacetic acid; striatum serotonin and inferior colliculus serotonin, resulting in lower concentrations in piglets fed the low compared with adequate LA + ALA formula. Inclusion of AA and DHA in the low, but not in the adequate LA + ALA formula, resulted in increased concentrations of all monoamines in the frontal cortex, and in striatum and inferior colliculus serotonin, increased dopamine and 5-hydroxyindolacetic acid in superior and inferior colliculus, areas related to processing and integration of visual and auditory information. Higher dopamine and 5-hydroxyindolacetic acid were found in superior and inferior colliculus regions even when AA and DHA were added to the LA + ALA adequate formula [168]. Thus, it can be said that functional changes among animals and infants fed diets varying in ω-6 and ω-3 fatty acids could involve altered neurotransmitter metabolism that may explain the improvements in visual, auditory, and learning tasks reported for infants and animals given diets rich in ω-3 fatty acids [169–173]. In addition, piglets fed diets deficient in LA and ALA from birth to 18 days not only had lower amounts of AA in frontal cortex PC and PI and lower DHA in PC and PE but also had significantly lower frontal cortex dopamine,

3,4-dihydroxyphenylacetic (DOPAC), homovanillic acid (HVA), serotonin and 5-hydroxyindoleacetic acid (5-HIAA) concentrations. These indices were restored to normal or were even higher in piglets that received AA and DHA suggesting that dietary PUFAs fed for 18 days from birth affects frontal cortex neurotransmitters in rapidly growing piglets and that these changes are specifically due to AA and/or DHA [174]. These results coupled with the observation that both AA and DHA influence the expression of dopamine receptor genes and their products [175], modify monoaminergic neurotransmitters in frontal cortex and hippocampus [176, 177], and facilitate release and actions of GABA [178–181] and acetylcholine [182–185] lends support to the concept that PUFAs have a modulatory influence on the release, action and properties of various neurotransmitters in the brain. Exogenously added AA (20–160 μM) stimulated dopamine uptake when pre-incubated for short times (15–30 min); whereas at 160 μM AA inhibited following longer pre-exposures (45–60 min) in glioma cells [186]; markedly stimulated, in a dose-dependent manner, the spontaneous release of dopamine, inhibited in a dose-dependent manner dopamine uptake into synaptosomes, but still stimulated dopamine spontaneous release in the presence of dopamine uptake inhibitors in purified synaptosomes from the rat striatum indicating that AA both inhibits dopamine reuptake and facilitates its release process [187].

In Chinese hamster ovary (CHO) cells transfected with the D2 receptor complementary DNA, D2 agonists potently enhanced AA release that has been initiated by stimulating constitutive purinergic receptors or by increasing intracellular Ca2+. In contrast, CHO cells expressing D1 receptors, D1 agonists exerted no such effect. When D1 and D2 receptors are coexpressed, however, activation of both subtypes results in a marked synergistic potentiation of AA release. In view of the numerous actions of AA and its metabolites in neuronal signal transduction, these results suggest that facilitation of its release may be implicated in dopaminergic responses, such as feedback inhibition mediated by D2 autoreceptors, and may constitute a molecular basis for D1/D2 receptor synergism [188]. In this context, it is interesting to note that in obesity, a decrease in the number of dopamine receptors or dopamine concentrations occurs [32] and obesity is common in type 2 diabetes. Both in obesity and type 2 diabetes mellitus, plasma concentrations of PUFAs especially AA, EPA, and DHA are decreased [189–193]. Numerous studies showed an association between poor fetal growth and adult insulin resistance and increased incidence of type 2 diabetes mellitus and metabolic syndrome. Early growth retardation, as a result of maternal protein restriction, could lead to alterations in desaturase activities similar to those observed in human insulin resistance. This is supported by the observation that in both muscle and liver the ratio of DHA to docosapentaenoic acid (DPA) was reduced in low protein offspring. Δ^5 desaturase activity in hepatic microsomes was reduced in the low protein offspring that was negatively correlated ($r = -0.855$) with fasting plasma insulin. No such correlation was observed in controls. These results suggest that it is possible to programme the activity of key enzymes involved in the desaturation of PUFAs by perinatal factors such as maternal protein intake [194]. Since, the PUFA composition of skeletal muscle membranes and insulin sensitivity are closely related [189–193] it is suggested that maternal protein restriction decreases Δ^5 desaturase activity such that fetal tissue content of PUFAs are decreased (including muscle) that, in turn, programmes the development of insulin resistance and metabolic syndrome during their adult life, a mechanism linking fetal growth retardation to insulin resistance. Maternal factors (such as maternal protein restriction) could also influence PUFA content in the brain. Since PUFAs such as AA and DHA have profound influence on the secretion and actions of various neurotransmitters, it is reasonable to propose that alterations in the concentrations of various LCPUFAs in the brain (especially in the hypothalamus) during the perinatal period could lead to changes in the levels and actions of dopamine, serotonin, acetylcholine and other neurotransmitters that, in turn, lead to the development of insulin resistance and metabolic syndrome in adult life. This is so, since VMH-lesioned rats that develop all features of type 2 DM showed selectively decreased concentrations of norepinephrine and dopamine in the hypothalamus, long-term infusion of norepinephrine plus serotonin into the VMH impairs pancreatic islet function in as much as VMH norepinephrine and serotonin levels are elevated in hyperinsulinemic and insulin-resistant animals [195–197], suggesting that dysfunction of VMH, impaired pancreatic β cell function, insulin resistance, tissue concentrations of PUFAs, alterations in the actions and levels of various neurotransmitters, and the development of metabolic syndrome are closely related to each other (see Fig. 16.2). It is not only that perturbations in the concentrations of PUFAs in the brain as a result of maternal protein restriction induce changes in the concentrations and actions of various neurotransmitters serotonin, dopamine, acetylcholine, and food intake regulating peptides such as NPY, AgRP (agouti related peptide), POMC (pro-opiomelanocortin) and the number of their receptors and insulin action in the brain (as discussed above), neurotransmitters are also known to influence the metabolism and actions of PUFAs. For instance, it was reported that in the intact rat brain, D2 but not D1 receptors are coupled to the activation of PLA_2 and the release of AA [198]. This suggests that there is both positive and negative feedback control between PUFAs and various neurotransmitters and their actions. In a similar fashion, various perinatal and maternal factors including PUFAs may regulate the expression, release and function of various other neurotransmitters and hypothalamic peptides such as leptin, NPY, AgRP and melanocortins. Such an interaction between PUFAs and hypothalamic peptides and neurotransmitters may program the hypothalamic bodyweight/appetite/satiety set point that could influence the development of obesity, metabolic syndrome, type 2 diabetes mellitus and hypertension in adult life. Such an influence of PUFAs in brain growth and development may also set the tone for the development of various neurological conditions such as schizophrenia, depression and Alzheimer's disease.

Such a concept may explain the relationship between perinatal and *in utero* nutrition and its long-term effects into adulthood. The excitatory and inhibitory inputs/outputs onto the NPY/AgRP and POMC/CART neurons reported [199–205] also suggests that leptin affects not only the transcription and release of neuropeptides but also the functional activity of neurotransmitters such as GABA (inhibitory) and glutamine (excitatory) that are ultimately the mediators of the metabolic signals of leptin, ghrelin, and other neuropeptides. If this concept is true, it suggests that maternal diet could influence EFA metabolism and leptin expression and action in the fetus and the newborn [205].

Maternal Diet Influences EFA Metabolism and Leptin Levels

Low birth weight is associated with high prevalence of metabolic syndrome in later life [206, 207]. Babies with low birth weights have 10 times greater chance of developing metabolic syndrome compared to those whose birth weight were normal. In addition, postnatal nutrition and growth also play a role in the development of metabolic syndrome in later life [208]. Though, the exact cause for this is not known, at least, in part, this could be attributed to the maternal and perinatal factors especially their diet. Maternal protein restriction or increased consumption of saturated and/or trans-fatty acids and energy rich diets (maternal over-nutrition) during pregnancy decrease the activity of Δ^6 and Δ^5 desaturase enzymes that are essential for the metabolism of dietary essential fatty acids LA and ALA and the formation of their long-chain metabolites such as AA, EPA and DHA. Perinatal protein depletion leads to almost complete absence of activities of Δ^6 and Δ^5 desaturases in fetal liver and placenta [209–212]. Thus, both protein deficiency and high-energy diet decreases the activities of Δ^6 and Δ^5 desaturases that, in turn, leads to maternal and fetal deficiency of EPA, DHA and AA.

Dietary quantity and quality has been shown to affect serum leptin levels [213–215]. A diet rich in PUFAs increases leptin levels in diet-induced obese adult rats [213], suggesting that variation in the type of diet during pregnancy and lactation significantly modulate fetal and neonatal growth and development by leptin-associated mechanisms since leptin influences NPY/AgRP and POMC/CART neurons and their connections [199–204]. Plasma leptin levels were found to be low in the lactating dams fed the EFA-deficient diet and their suckling pups compared with controls [216]. The suckling pups showed decreased concentrations of leptin even in their adipose tissue [217], suggesting that maternal EFA deficiency can produce a decrease in leptin levels in several tissues, possibly, even in the hypothalamus. These low leptin levels during the perinatal period alters NPY/AgRP and POMC/CART homeostasis [199–204] that may lead to the hypothalamic "body weight/ appetite/ satiety set point" set at a higher level that is long-lasting and potentially irreversible onto adulthood. Thus, maternal malnutrition, low perinatal PUFAs and consequent low leptin concentrations could lead to the development of metabolic syndrome in adulthood.

EPA, DHA, and AA inhibit TNF-α and IL-6 synthesis. Hence, PUFAs deficiency due to maternal malnutrition increases the generation of TNF-α and IL-6 both in the maternal and fetal tissues that, in turn, induces insulin resistance. Prenatal exposure to TNF-α produces obesity [218], and obese children and adults have high levels of TNF-α and IL-6 [219, 220]. Low plasma and tissue concentrations of EPA, DHA, and AA also decrease adiponectin levels that further aggravate insulin resistance. TNF-α and IL-6 increase the activity of 11β-HSD-1 that causes abdominal obesity, a characteristic feature of metabolic syndrome [221]. Since a close positive and negative feedback regulation between perilipins, TNF-α, adipocyte size, PPAR-γ, exercise and insulin resistance exists, low plasma and tissue concentrations of PUFAs and leptin due to maternal malnutrition will also explain abnormalities of perilipins and IMCL seen in obese subjects who are prone to develop metabolic syndrome.

In addition, AA and DHA enhance cerebral ACh levels and improve learning ability in rats [29, 30] and ACh modulates long-term potentiation and synaptic plasticity in neuronal circuits and interacts with dopamine receptor in the hippocampus [31]. ACh also inhibits synthesis and release of TNF-α and thus, has anti-inflammatory actions [54] and is a potent stimulator of eNO synthesis [55].

Perinatal PUFA Deficiency May Initiate Low-Grade Systemic Inflammation and Adult Diseases

It is evident from the preceding discussion that a deficiency of PUFAs during the critical period of brain growth and development and somatic growth leads to a deficiency of leptin, ACh, and an imbalance in the NPY/AgRP and POMC/CART homeostasis, changes in the concentrations of dopamine, serotonin, GABA and other neuropeptides, and an increase in the levels of TNF-α, an inflammatory cytokine that has neurotoxic actions. All these adverse events as a result of perinatal PUFA inadequacy, could lead to the initiation of low-grade systemic inflammation (due to enhanced TNF-α production) and neuronal damage may predispose to the development of various neurological conditions such as Alzheimer's disease, schizophrenia and depression and obesity, hypertension, osteoporosis and type 2 diabetes mellitus later in life. Thus, various adult diseases may have their origins in perinatal period. These evidences imply that metabolic syndrome, Alzheimer's disease, depression schizophrenia, hypertension, type 2 diabetes mellitus and obesity could be due to perinatal deficiency of EPA, DHA and AA and their metabolites such as lipoxins, resolvins, protectins and nitrolipids (Figs. 16.1 and 16.2). Thus, it is proposed that adult diseases enumerated above have their origins in the perinatal period [4, 5]. This also implies that the low-grade systemic inflammation starts in the perinatal period of life itself and that these diseases/disorders are disorders of the brain as discussed in previous chapters. In view of this, PUFAs and their metabolites play a significant role in all these diseases as already discussed in previous chapters.

In this context, the significance of breast-feeding lies in the fact that human breast milk is rich in AA, EPA, DHA, GLA, DGLA, LA and AA. It is likely that when the child is adequately breast fed, the tissue and plasma concentrations of various PUFAs will be optimal that leads to formation of optimal amounts of lipoxins, resolvins, protectins, maresins and nitrolipids so that (a) inflammatory processes are under control; (b) brain growth and development is adequate; (c) neuronal synaptic connections are perfect; (d) neurotransmitters are produced in adequate amounts and at the right time and right place; and (e) various tissues and organs are able to meet the endogenous and external challenges in a favorable fashion so that tissue damage is minimal and the repair process and wound healing is normal and restoration of target organs to normal is easily reestablished. This implies that supplementation of various PUFAs and their anti-inflammatory products and other endogenous molecules involved in the restoration of homeostasis are provided in optimal amounts when homeostatic mechanisms are disturbed, so that health can be restored.

References

[1] Das UN (2011) Influence of polyunsaturated fatty acids and their metabolites on stem cell biology. Nutrition 27:21–25
[2] Barker DJP (ed) (1992) Fetal and infant origins of adult disease. BMJ Publishing, London
[3] Robinson R (2001) The fetal origins of adult disease. BMJ 322:375–376
[4] Das UN (2002) A perinatal strategy for preventing adult disease. The role of polyunsaturated fatty acids. Kluwer Academic, Norwell
[5] Das UN (2010) Metabolic syndrome pathophysiology. The role of essential fatty acids. Wiley-Blackwell, Ames
[6] Matsuzaka T, Shimano H, Yahagi N, Amemiya-Kudo M, Yoshikawa T, Hasty AH, Tamura Y, Osuga J, Okazaki H, Iizuka Y, Takahashi A, Sone H, Gotoda T, Ishibashi S, Yamada N (2002) Dual regulation of mouse Delta(5)- and Delta(6)-desaturase gene expression by SREBP-1 and PPARalpha. J Lipid Res 43:107–114
[7] Das UN (2006) Essential fatty acids: biochemistry, physiology, and pathology. Biotechnol J 1:420–439
[8] Das UN (2006) Biological significance of essential fatty acids. J Assoc Physicians India 54:309–319
[9] Auestad N, Scott DT, Janowsky JS, Jacobson C, Carroll RE, Montalto MB, Halter R, Qiu W, Jacobs JR, Connor WE, Conner SL, Taylor JA, Neuringer M, Fitzgerald KM, Hall RT (2003) Visual, cognitive, and language assessments at 39 months: a follow-up study of children fed formulas containing long-chain polyunsaturated fatty acids to 1 year of age. Pediatrics 112(3 Pt 1):e177–e183
[10] Fewtrell MS, Morley R, Abbott RA, Singhal A, Isaacs EB, Stephenson T, MacFadyen U, Lucas A (2002) Double-blind, randomized trial of long-chain polyunsaturated fatty acid supplementation in formula fed to preterm infants. Pediatrics 110(1 Pt 1):73–82
[11] SanGiovanni JP, Parra-Cabrera S, Colditz GA, Berkey CS, Dwyer JT (2000) Meta-analysis of dietary essential fatty acids and long-chain polyunsaturated fatty acids as they relate to visual resolution acuity in healthy preterm infants. Pediatrics 105:1292–1298
[12] O'Connor DL, Hall R, Adamkin D, Auestad N, Castillo M, Connor WE, Connor SL, Fitzgerald K, Groh-Wargo S, Hartmann EE, Jacobs J, Janowsky J, Lucas A, Margeson D, Mena P, Neuringer M, Nesin M, Singer L, Stephenson T, Szabo J, Zemon V, Ross Preterm Lipid Study (2001) Growth and development in preterm infants fed long-chain polyunsaturated fatty acids: a prospective, randomized controlled trial. Pediatrics 108:359–371
[13] Boehm G, Borte M, Bohles HJ, Muller H, Kohn G, Moro G (1996) Docosahexaenoic and arachidonic acid content of serum and red blood cell membrane phospholipids of preterm infants fed breast milk, standard formula or formula supplemented with n-3 and n-6 long-chain polyunsaturated fatty acids. Eur J Pediatr 155:410–416
[14] Fewtrell MS, Abbott RA, Kennedy K, Singhal A, Morley R, Caine E, Jamieson C, Cockburn F, Lucas A (2004) Randomized, double-blind trial of long-chain polyunsaturated fatty acid supplementation with fish oil and borage oil in preterm infants. J Pediatr 144:471–479
[15] Willatts P, Forsyth JS, DiModugno MK, Varma S, Colvin M (1998) Effect of long-chain polyunsaturated fatty acids in infant formula on problem solving at 10 months of age. Lancet 352:688–691

[16] Salem N Jr, Wegher B, Mena P, Uauy R (1996) Arachidonic and docosahexaenoic acids are biosynthesized from their 18-carbon precursors in human infants. Proc Natl Acad Sci U S A 93:49–54
[17] Farquharson J, Jamieson EC, Abbasi KA, Patrick WJA, Logan RW, Cockburn F (1995) Effect of diet on the fatty acid composition of the major phospholipids of infant cerebral cortex. Arch Dis Child 72:198–203
[18] Ryan AS, Astwood JD, Gautier S, Kuratko CN, Nelson EB, Salem N Jr (2010) Effects of long-chain polyunsaturated fatty acid supplementation on neurodevelopment in childhood: a review of human studies. Prostaglandins Leukot Essent Fatty Acids 82:305–314
[19] Auestad N, Scott DT, Janowsky JS, Jacobsen C, Carroll RE, Montalto MB, Halter R, Qiu W, Jacobs JR, Connor WE, Connor SL, Taylor JA, Neuringer M, Fitzgerald KM, Hall RT (2003) Visual, cognitive, and language assessments at 39 months: a follow-up study of children fed formulas containing long-chain polyunsaturated fatty acids to 1 year of age. Pediatrics 112(3 Pt 1):e177–e183
[20] Chang CY, Ke DS, Chen JY (2009) Essential fatty acids and human brain. Acta Neurol Taiwan 18:231–241
[21] Makrides M, Smithers LG, Gibson RA (2010) Role of long-chain polyunsaturated fatty acids in neurodevelopment and growth. Nestle Nutr Workshop Ser Pediatr Program 65:123–133
[22] Yu N, Martin J-L, Stella N, Magistretti PJ (1993) Arachidonic acid stimulates glucose uptake in cerebral cortical astrocytes. Proc Natl Acad Sci U S A 90:4042–4046
[23] Goetzl EJ, Goldman DW, Naccache PH, Sha'afi RI, Pickett WC (1982) Mediation of leukocyte components of inflammatory reactions by lipoxygenase products of arachidonic acid. Adv Prostaglandin Thromboxane Leukot Res 9:273–282
[24] Fong JC, Chen CC, Liu D, Chai SP, Tu MS, Chu KY (1996) Arachidonic acid stimulates the intrinsic activity of ubiquitous glucose transporter (GLUT1) in 3T3-L1 adipocytes by a protein kinase C-independent mechanism. Cell Signal 8:179–183
[25] Butler M, Huzel N, Barnabé N (1997) Unsaturated fatty acids enhance cell yields and perturb the energy metabolism of an antibody-secreting hybridoma. Biochem J 322(Pt 2):615–623
[26] Garber K (2010) Oncology's energetic pipeline. Nat Biotechnol 18:888–891
[27] Chiou GY, Fong JC (2005) Synergistic effect of prostaglandin F2alpha and cyclic AMP on glucose transport in 3T3-L1 adipocytes. J Cell Biochem 94:627–634
[28] Ragozzino ME, Unick KE, Gold PE (1996) Hippocampal acetylcholine release during memory testing in rats: augmentation by glucose. Proc Natl Acad Sci U S A 93:4693–4698
[29] Das UN (2002) Alcohol consumption and risk of dementia. Lancet 360:490
[30] Minami M, Kimura S, Endo T, Hamaue N, Horafuji M, Togashi H, Matsumoto M, Yaoshioka M, Saito H, Watanabe S, Kobayashi T, Okuyama H (1997) Dietary docosahexaenoic acid increases cerebral acetylcholine levels and improves passive avoidance performance in stroke-prone spontaneously hypertensive rats. Pharmacol Biochem Behav 58:1123–1129
[31] Hersi AI, Kitaichi K, Srivastava LK, Gaudreau P, Quirion R (2000) Dopamine D-5 receptor modulates hippocampal acetylcholine release. Brain Res Mol Brain Res 76:336–340
[32] Wang G-J, Volkow ND, Logan J, Pappas NR, Wong CT, Zhu W, Netusil N, Fowler JS (2001) Brain dopamine and obesity. Lancet 357:354–357
[33] Abott MA, Wells DG, Fallon JR (1999) The insulin receptor tyrosine kinase substrate p58/53 and the insulin receptor are components of CNS synapses. J Neurosci 19:7300–7308
[34] Venters HD, Dantzer R, Kelley KW et al (2000) A new concept in neurodegeneration: TNFalpha is a silencer of survival signals. Trends Neurosci 23:175–180
[35] Venters HD, Tang Q, Liu Q et al (1999) A new mechanism of neurodegeneration: proinflammatory cytokine inhibits receptor signaling by a survival peptide. Proc Natl Acad Sci U S A 96:9879–9884
[36] Lauritzen I, Blondeau N, Heurteaux C, Widmann C, Romey G, Lazdunski M (2000) Polyunsaturated fatty acids are potent neuroprotectors. EMBO J 19:1784–1793
[37] Wang X, Zhao X, Mao ZY, Wang XM, Liu ZL (2003) Neuroprotective effect of docosahexaenoic acid on glutamate-induced cytotoxicity in rat hippocampal cultures. Neuroreport 14:2457–2461

[38] Cao D, Xue R, Xu J, Liu Z (2005) Effects of docosahexaenoic acid on the survival and neurite outgrowth of rat cortical neurons in primary cultures. J Nutr Biochem 16:538–546
[39] Hogyes E, Nyakas C, Kiliaan A, Farkas T, Penke B, Luiten PG (2003) Neuroprotective effect of developmental docosahexaenoic acid supplement against excitotoxic brain damage in infant rats. Neuroscience 119:999–1012
[40] Lonergan PE, Martin DS, Horrobin DF, Lynch MA (2002) Neuroprotective effect of eicosapentaenoic acid in hippocampus of rats exposed to gamma-irradiation. J Biol Chem 277:20804–20811
[41] Das UN, Rao KP (2006) Effect of γ-linolenic acid and prostaglandins E1 on gamma-radiation and chemical-induced genetic damage to the bone marrow cells of mice. Prostaglandins Leukot Essent Fatty Acids 74:165–173
[42] Kumar SG, Das UN (1994) Effect of prostaglandins and their precursors on the proliferation of human lymphocytes and their secretion of tumor necrosis factor and various interleukins. Prostaglandins Leukot Essent Fatty Acids 50:331–334
[43] Kumar SG, Das UN, Kumar KV, Tan BKH, Das NP (1992) Effect of n-6 and n-3 fatty acids on the proliferation and secretion of TNF and IL-2 by human lymphocytes in vitro. Nutr Res 12:815–823
[44] Endres S, Ghorbani R, Kelley VE et al (1989) The effect of dietary supplementation with n-3 polyunsaturated fatty acids on the synthesis of interleukin-1 and tumor necrosis factor by mononuclear cells. N Engl J Med 320:265–271
[45] Kuboki K, Jiang ZY, Takahara N, Ha SW, Igarashi M, Yamauchi T, Feener EP, Herbert TP, Rhodes CJ, King GL (2000) Regulation of endothelial constitutive nitric oxide synthase gene expression in endothelial cells and in vivo: a specific vascular action of insulin. Circulation 101:676–681
[46] Das UN (1994) Beneficial effect of eicosapentaenoic and docosahexaenoic acids in the management of systemic lupus erythematosus and its relationship to the cytokine network. Prostaglandins Leukot Essent Fatty Acids 51:207–213
[47] Satomi N, Sakurai A, Haranaka K (1985) Relationship of hypoglycemia to tumor necrosis factor production and antitumor activity: role of glucose, insulin and macrophages. J Natl Cancer Inst 74:1255–1260
[48] Boichot E, Sannomiya P, Escofier N et al (1999) Endotoxin-induced acute lung injury in rats: role of insulin. Pulm Pharmacol Ther 12:285–290
[49] Guidot DM, Hybertson BM, Kitlowski RP, Repine JE (1996) Inhaled nitric oxide prevents IL 1 induced neutrophil accumulation and associated acute edema in isolated rats lungs. Am J Physiol 271:L225–L229
[50] Suresh Y, Das UN (2003) Long-chain polyunsaturated fatty acids and chemically induced diabetes mellitus: effect of ω-3 fatty acids. Nutrition 19:213–228
[51] Vecchione C, Aretini A, Maffei A, Marino G, Selvetella G, Poulet R, Trimarco V, Frati G, Lembo G (2003) Cooperation between insulin and leptin in the modulation of vascular tone. Hypertension 42:166–170
[52] Nilsson L, Sara VR, Norberg A (1988) Insulin-like growth factor 1 stimulates the release of acetylcholine from rat cortical slices. Neurosci Lett 88:221–226
[53] Napoli I, Blusztajn JK, Mellott TJ (2008) Prenatal choline supplementation in rats increases the expression of IGF2 and its receptor IGF2R and enhances IGF2-induced acetylcholine release in hippocampus and frontal cortex. Brain Res 1237:124–135
[54] Borovikova LV, Ivanova S, Zhang M, Yang H, Botchkina GI, Watkins LR, Wang H, Abumrad N, Eaton JW, Tracey KJ (2000) Vagus nerve stimulation attenuates the systemic inflammatory response to endotoxin. Nature 405:458–462
[55] Xu Z, Tong C, Eisenach JC (1996) Acetylcholine stimulates the release of nitric oxide from rat spinal cord. Anesthesiology 85:107–111
[56] Ponnala S, Rao KP, Chaudhury JR, Ahmed J, Rama Rao B, Kanjilal S, Hasan Q, Das UN (2009) Effect of polyunsaturated fatty acids on diphenyl hydantoin-induced genetic damage in vitro and in vivo. Prostaglandins Leukot Essent Fatty Acids 80:43–50

[57] Das UN, Ramadevi G, Rao KP, Rao MS (1985) Prostaglandins and their precursors can modify genetic damage-induced by gamma-radiation and benzo(a)pyrene. Prostaglandins 29:911–920
[58] Krishna Mohan I, Das UN (2001) Prevention of chemically induced diabetes mellitus in experimental animals by polyunsaturated fatty acids. Nutrition 17:126–151
[59] Suresh Y, Das UN (2003) Long-chain polyunsaturated fatty acids and chemically induced diabetes mellitus: effect of omega-6 fatty acids. Nutrition 19:93–114
[60] Suresh Y, Das UN (2003) Long-chain polyunsaturated fatty acids and chemically induced diabetes mellitus. Effect of omega-3 fatty acids. Nutrition 19:213–228
[61] Suresh Y, Das UN (2001) Protective action of arachidonic acid against alloxan-induced cytotoxicity and diabetes mellitus. Prostaglandins Leukot Essent Fatty Acids 64:37–52
[62] Ferrari D, Cysneiros RM, Scorza CA, Arida RM, Cavalheiro EA, de Almeida AC, Scorza FA (2008) Neuroprotective activity of omega-3 fatty acids against epilepsy-induced hippocampal damage: quantification with immunohistochemical for calcium-binding proteins. Epilepsy Behav 13:36–42
[63] Das UN (2008) Essential fatty acids and their metabolites could function as endogenous HMG-CoA reductase and ACE enzyme inhibitors, anti-arrhythmic, anti-hypertensive, anti-atherosclerotic, anti-inflammatory, cytoprotective, and cardioprotective molecules. Lipids Health Dis 7:37
[64] Belayev L, Khoutorova L, Atkins KD, Bazan NG (2009) Robust docosahexaenoic acid-mediated neuroprotection in a rat model of transient, focal cerebral ischemia. Stroke 40:3121–3126
[65] Bazan NG (2009) Neuroprotectin D1-mediated anti-inflammatory and survival signaling in stroke, retinal degenerations, and Alzheimer's disease. J Lipid Res 50(Suppl):S400–S405
[66] Pan HC, Kao TK, Ou YC, Yang DY, Yen YJ, Wang CC, Chuang YH, Liao SL, Raung SL, Wu CW, Chiang AN, Chen CJ (2009) Protective effect of docosahexaenoic acid against brain injury in ischemic rats. J Nutr Biochem 20:715–725
[67] Cao DH, Xu JF, Xue RH, Zheng WF, Liu ZL (2004) Protective effect of chronic ethyl docosahexaenoate administration on brain injury in ischemic gerbils. Pharmacol Biochem Behav 79:651–659
[68] Bazan NG (2005) Neuroprotectin D1 (NPD1): a DHA-derived mediator that protects brain and retina against cell injury-induced oxidative stress. Brain Pathol 15:159–166
[69] Mukherjee PK, Marcheselli VL, Serhan CN, Bazan NG (2004) Neuroprotectin D1: a docosahexaenoic acid-derived docosatriene protects human retinal pigment epithelial cells from oxidative stress. Proc Natl Acad Sci U S A 101:8491–8496
[70] Gleissman H, Yang R, Martinod K, Lindskog M, Serhan CN, Johnsen JI, Kogner P (2010) Docosahexaenoic acid metabolome in neural tumors: identification of cytotoxic intermediates. FASEB J 24:906–915
[71] Borkman M, Stolien LH, Pan DA, Jenkins AB, Chisholm DJ, Campbell LV (1993) The relation between insulin sensitivity and the fatty acid composition of skeletal muscle phospholipids. N Engl J Med 328:238–244
[72] Das UN (2005) A defect in the activity of Δ^6 and Δ^5 desaturases may be a factor predisposing to the development of insulin resistance syndrome. Prostaglandins Leukot Essent Fatty Acids 72:343–350
[73] Ginsberg BH, Jabour J, Spector AA (1982) Effect of alterations in membrane lipid unsaturation on the properties of the insulin receptor of Ehrlich ascites cells. Biochim Biophys Acta 690:157–164
[74] Somova L, Moodley K, Channa ML, Nadar A (1999) Dose-dependent effect of dietary fish-oil (n-3) polyunsaturated fatty acids on in vivo insulin sensitivity in rat. Methods Find Exp Clin Pharmacol 21:275–278
[75] Huang Y-J, Fang VS, Chou Y-C, Kwok C-F, Ho L-T (1997) Amelioration of insulin resistance and hypertension in a fructose-fed rat model with fish oil supplementation. Metabolism 46:1252–1258

[76] Mori Y, Murakawa Y, Katoh S, Hata S, Yokoyama J, Tajima N, Ikeda Y, Nobukata H, Ishikawa T, Shibutani Y (1997) Influence of highly purified eicosapentaenoic acid ethyl ester on insulin resistance in the Otsuka Long-Evans Tokushima fatty rat, a model of spontaneous non-insulin dependent diabetes mellitus. Metabolism 46:1458–1464
[77] Nobukata H, Ishikawa T, Obata M, Shibutani Y (2000) Long-term administration of highly purified eicosapenatenoic acid ethyl ester prevents diabetes and abnormalities of blood coagulation in male WBN/Kob rats. Metabolism 49:912–919
[78] Demcakova E, Sebokova E, Ukropec J, Gasperikova D, Klimes I (2001) Delta-6 desaturase activity and gene expression, tissue fatty acid profile and glucose turnover rate in hereditary hypertriglyceridemic rats. Endocrinol Regul 35:179–186
[79] Gasperikova D, Demcakova E, Ukropec J, Klimes I, Sebokova E (2002) Insulin resistance in the hereditary hypertriglyceridemic rat is associated with an impairment of delta-6 desaturase expression in liver. Ann N Y Acad Sci 967:446–453
[80] Simoncikova P, Wein S, Gasperikova D, Ukropec J, Certik M, Klimes I, Sebokova E (2002) Comparison of the extrapancreatic action of gamma-linolenic acid and n-3 PUFAs in the high fat diet-induced insulin resistance. Endocr Regul 36:143–149
[81] Bruning JC, Gautam D, Burks DJ, Gillette J, Schubert M, Orban PC, Klein R, Krone W, Muller-Wieland D, Kahn CR (2000) Role of brain insulin receptor in control of body weight and reproduction. Science 289:2122–2125
[82] Hajnal A, Pothos EN, Lenard L, Hoebel BG (1998) Effects of feeding and insulin on extracellular acetylcholine in the amygdala of freely moving rats. Brain Res 785:41–48
[83] Wang G-J, Volkow ND, Logan J, Pappas NR, Wong CT, Zhu W, Netusil N, Fowler JS (2001) Brain dopamine and obesity. Lancet 357:354–357
[84] Li J, Zhang H, Wu F, Nan Y, Ma H, Guo W, Wang H, Ren J, Das UN, Gao F (2008) Insulin inhibits tumor necrosis factor-alpha induction in myocardial ischemia/reperfusion: role of Akt and endothelial nitric oxide synthase phosphorylation. Crit Care Med 36:1551–1558
[85] Das UN (2000) Possible beneficial action(s) of glucose-insulin-potassium regimen in acute myocardial infarction and inflammatory conditions: a hypothesis. Diabetologia 43:1081–1082
[86] Das UN (2001) Is insulin an anti-inflammatory molecule? Nutrition 17:409–413
[87] Das UN (2001) Can glucose-insulin-potassium regimen suppress inflammatory bowel disease? Med Hypotheses 57:183–185
[88] Das UN (2001) Hypothesis: can glucose-insulin-potassium regimen in combination with polyunsaturated fatty acids suppress lupus and other inflammatory diseases? Prostaglandins Leukot Essent Fatty Acids 65:109–113
[89] Das UN (2002) Insulin and the critically ill. Crit Care 6:262–263
[90] Das UN (2002) Is insulin an endogenous cardioprotector? Crit Care 6:389–393
[91] Futerman AH, Banker GA (1996) The economics of neurite outgrowth: the addition of new membrane to growing axons. Trends Neurosci 19:144–149
[92] Negre-Aminou P, Nemenoff RA, Wood MR, de la Houssaye BA, Pfenninger KH (1996) Characterization of phospholipase A2 activity enriched in the nerve growth cone. J Neurochem 67:2599–2608
[93] Hornfelt M, Ekstrom PA, Edstrom A (1999) Involvement of axonal phospholipase A2 activity in the outgrowth of adult mouse sensory axons in vitro. Neuroscience 91:1539–1547
[94] Kelly RB (1999) Deconstructing membrane traffic. Trends Cell Biol 9:M29–M33
[95] Darios F, Davletov B (2006) Omega-3 and omega-6 fatty acids stimulate cell membrane expansion by acting on syntaxin 3. Nature 440:813–817
[96] Rickman C, Davletov B (2005) Arachidonic acid allows SNARE complex formation in the presence of Munc18. Chem Biol 12:545–553
[97] van der Wees J, Schilthuis JG, Koster CH, Diesveld-Schipper H, Folkers GE, van der Saag PT, Dawson MI, Shudo K, van der Burg B, Durston AJ (1998) Inhibition of retinoic acid receptor-mediated signalling alters positional identity in the developing hindbrain. Development 125:545–556

[98] Mata de Urquiza A, Liu S, Sjoberg M, Zetterstrom RH, Griffiths W, Sjovall J, Perlmann T (2000) Docosahexaenoic acid, a ligand for the retinoid X receptor in mouse brain. Science 290:2140–2144
[99] Fan Y-Y, Spencer TE, Wang N, Moyer MP, Chapkin RS (2003) Chemopreventive n-3 fatty acids activate RXRα in colonocytes. Carcinogenesis 24:1541–1548
[100] Lengqvist J, Mata de Urquiza, Bergman A-C, Willson TM, Sjovall J, Perlmann T, Griffiths W (2004) Polyunsaturated fatty acids including docosahexaenoic and arachidonic acid bind to the retinoid X receptor α ligand-binding domain. Mol Cell Proteomics 3:692–703
[101] DeWille JW, Farmer SJ (1992) Postnatal dietary fat influences mRNAS involved in myelination. Dev Neurosci 14:61–68
[102] Uauy R, Hoffman DR, Peirano P, Birch DG, Birch EE (2001) Essential fatty acids in visual and brain development. Lipids 36:885–895
[103] Mohri I, Eguchi N, Suzuki K, Urade Y, Taniike M (2003) Hematopoietic prostaglandin D synthase is expressed in microglia in the developing postnatal mouse brain. Glia 42:263–274
[104] Das UN (2009) Albumin and lipid enriched albumin for the critically ill. J Assoc Physicians India 57:53–58
[105] Velasco A, Tabernero A, Medina JM (2003) Role of oleic acid as a neurotrophic factor is supported in vivo by the expression of GAP-43 subsequent to the activation of SREBP-1 and the up-regulation of stearoyl-CoA desaturase during postnatal development of the brain. Brain Res 977:103–111
[106] Yavin E (2006) Versatile roles of docosahexaenoic acid in the prenatal brain: from pro- and anti-oxidant features to regulation of gene expression. Prostaglandins Leukot Essent Fatty Acids 75:203–211
[107] Kuperstein F, Eilam R, Yavin E (2008) Altered expression of key dopaminergic regulatory proteins in the postnatal brain following perinatal n-3 fatty acid dietary deficiency. J Neurochem 106:662–671
[108] Kuperstein F, Yakubov E, Dinerman P, Gil S, Eylam R, Salem N Jr, Yavin E (2005) Overexpression of dopamine receptor genes and their products in the postnatal rat brain following maternal n-3 fatty acid dietary deficiency. J Neurochem 95:1550–1562
[109] Sinha RA, Khare P, Rai A, Maurya SK, Pathak A, Mohan V, Nagar GK, Mudiam MK, Godbole MM, Bandyopadhyay S (2009) Anti-apoptotic role of omega-3-fatty acids in developing brain: perinatal hypothyroid rat cerebellum as apoptotic model. Int J Dev Neurosci 27:377–383
[110] Monroig O, Rotllant J, Sànchez E, Cerdà-Reverter JM, Tocher DR (2009) Expression of long-chain polyunsaturated fatty acid (LC-PUFA) biosynthesis genes during zebrafish Danio rerio early embryogenesis. Biochim Biophys Acta 1791:1093–1101
[111] Kitajka K, Sinclair AJ, Weisinger RS, Weisinger HS, Mathai M, Jayasooriya AP, Halver JE, Puskas LG (2004) Effects of dietary omega-3 polyunsaturated fatty acids on brain gene expression. Proc Natl Acad Sci U S A 101:10931–10936
[112] Kitajka K, Puskas LG, Zvara A, Hackler L, Barcelo-Coblijn G, Yeo YK, Farkas T (2002) The role of n-3 polyunsaturated fatty acids in brain: modulation of rat brain gene expression by dietary n-3 fatty acids. Proc Natl Acad Sci U S A 99:2619–2624
[113] Barcelo-Coblijn G, Hogyes E, Kitajka LG, Zvara A, Hackler L Jr, Nyakas C, Penke Z, Farkas T (2003) Modification by docosahexaenoic acid of age-induced alterations in gene expression and molecular composition of rat brain phospholipids. Proc Natl Acad Sci U S A 100:11321–11326
[114] Puskas LG, Kitajka K, Nyakas C, Barcelo-Coblijn G, Farkas T (2003) Short-term administration of omega 3 fatty acids from fish oil results in increased transthyretin transcription in old rat hippocampus. Proc Natl Acad Sci U S A 100:1580–1585
[115] Galvin JE, Schuck TM, Lee VM, Trojanowski JQ (2001) Differential expression and distribution of alpha-, beta-, and gamma-synuclein in the developing human substantia nigra. Exp Neurol 168:347–355

[116] Babcock TA, Helton WS, Anwar KN, Zhao YY, Espat NJ (2004) Synergistic anti-inflammatory activity of omega-3 lipid and rofecoxib pretreatment on macrophage proinflammatory cytokine production occurs via divergent NF-kappaB activation. JPEN J Parenter Enteral Nutr 28:232–239
[117] Gronert K, Gewirtz A, Madara JL, Serhan CN (1998) Identification of a human enterocyte lipoxin A4 receptor that is regulated by interleukin (IL)-13 and interferon gamma and inhibits tumor necrosis factor alpha-induced IL-8 release. J Exp Med 187:1285–1294
[118] Hinz B, Brune K, Pahl A (2003) 15-Deoxy-Delta(12,14)-prostaglandin J2 inhibits the expression of proinflammatory genes in human blood monocytes via a PPAR-gamma-independent mechanism. Biochem Biophys Res Commun 302:415–420
[119] Berger A, Mutch DM, German JB, Roberts MA (2002) Dietary effects of arachidonate-rich fungal oil and fish oil on murine hepatic and hippocampal gene expression. Lipids Health Dis 1:2
[120] Parrish WR, Rosas-Ballina M, Gallowitsch-Puerta M, Ochani M, Ochani K, Yang LH, Hudson L, Lin X, Patel N, Johnson SM, Chavan S, Goldstein RS, Czura CJ, Miller EJ, Al-Abed Y, Tracey KJ, Pavlov VA (2008) Modulation of TNF release by choline requires alpha7 subunit nicotinic acetylcholine receptor-mediated signaling. Mol Med 14:567–574
[121] Hamano R, Takahashi HK, Iwagaki H, Yoshino T, Nishibori M, Tanaka N (2006) Stimulation of alpha7 nicotinic acetylcholine receptor inhibits CD14 and the toll-like receptor 4 expression in human monocytes. Shock 26:358–364
[122] Yoshikawa H, Kurokawa M, Ozaki N, Nara K, Atou K, Takada E, Kamochi H, Suzuki N (2006) Nicotine inhibits the production of proinflammatory mediators in human monocytes by suppression of I-kappaB phosphorylation and nuclear factor-kappaB transcriptional activity through nicotinic acetylcholine receptor alpha7. Clin Exp Immunol 146:116–123
[123] Borovikova LV, Ivanova S, Nardi D, Zhang M, Yang H, Ombrellino M, Tracey KJ (2000) Role of vagus nerve signaling in CNI-1493-mediated suppression of acute inflammation. Auton Neurosci 85:141–147
[124] Borovikova LV, Ivanova S, Zhang M, Yang H, Botchkina GI, Watkins LR, Wang H, Abumrad N, Eaton JW, Tracey KJ (2000) Vagus nerve stimulation attenuates the systemic inflammatory response to endotoxin. Nature 405:458–462
[125] Flier MA, Rittirsch D, Nadeau BA et al (2007) Phagocyte-derived catecholamines enhance acute inflammatory injury. Nature 449:721–726
[126] Stellwagen D, Malenka RC (2006) Synaptic scaling mediated by glial TNF-α. Nature 440:1054–1059
[127] Beattie EC, Stellwagen D, Morishita W, Bresnahan JC, Ha BK, Von Zastrow M, Beattie MS, Malenka RC (2002) Control of synaptic strength by glial TNFα. Science 295:2282–2285
[128] Hiramoto T, Chida Y, Sonoda J, Yoshihara K, Sudo N, Kubo C (2008) The hepatic vagus nerve attenuates Fas-induced apoptosis in the mouse liver via alpha7 nicotinic acetylcholine receptor. Gastroenterology 134:2122–2131
[129] Lois C, Alvarez-Buylla A (1994) Long-distance neuronal migration in the adult mammalian brain. Science 264:1145–1148
[130] Barinaga M (1997) Researchers find signals that guide young brain neurons. Science 278:385–386
[131] Gage FH (1997) Mammalian neural stem cells. Science 287:1433–1438
[132] Penninger JM, Woodgett J (2001) PTEN—coupling tumor suppression to stem cells? Science 294:2116–2118
[133] Colombani J, Bianchini L, Layalle S, Pondeville E, Dauphin-Villemant C, Antoniewski C, Carre C, Noselli S, Leopold P (2005) Antagonistic actions of ecdysone and insulins determine final size in Drosophila. Science 310:667–670
[134] Nisoli E, Tonello C, Cardile A, Cozzi V, Bracale R, Tedesco L, Falcone S, Valerio A, Cantoni O, Clementi E, Moncada S, Carruba MO (2005) Calorie restriction promotes mitochondrial biogenesis by inducing the expression of eNOS. Science 310:314–317
[135] Chenn A, Zhang YA, Chang BT, McConnell SK (1998) Intrinsic polarity of mammalian neuroepithelial cells. Mol Cell Neurosci 11:183–193

[136] Lien W-H, Klezovitch O, Fernandez TE, Delrow J, Vasioukhin V (2006) αE-catenin controls cerebral cortical size by regulating the hedgehog signaling pathway. Science 311:1609–1611
[137] Castellone MD, Teramoto H, Williams BO, Druey KM, Gutkind JS (2005) Prostaglandin E2 promotes colon cancer growth through a Gs-axin-β-catenin signaling axis. Science 310:1504–1510
[138] Cai J, Jiang WG, Mansel RE (1999) Inhibition of the expression of VE-cadherin/catenin complex by gamma linolenic acid in human vascular endothelial cells, and its impact on angiogenesis. Biochem Biophys Res Commun 258:113–118
[139] Jiang WG, Hiscox S, Horrobin DF, Hallett MB, Mansel RE, Puntis MC (1995) Expression of catenins in human cancer cells and its regulation by n-6 polyunsaturated fatty acids. Anticancer Res 15:2569–2573
[140] Carayol N, Vachier I, Campbell A, Crampette L, Bousquet J, Godard P, Chanez P (2002) Regulation of E-cadherin expression by dexamethasone and tumour necrosis factor-alpha in nasal epithelium. Eur Respir J 20:1430–1436
[141] Lam TKT, Gutierrez-Juarez R, Pocai A, Rossetti L (2005) Regulation of blood glucose by hypothalamic pyruvate metabolism. Science 309:943–947
[142] Woods SC, McKay LD (1978) Intraventricular alloxan eliminates feeding elicited by 2-deoxyglucose. Science 202:1209–1211
[143] Woods SC, Lotter EC, McKay LD, Porte D Jr (1979) Chronic intracerebroventricular infusion of insulin reduces food intake and body weight of baboons. Nature 282:503–505
[144] Rossetti L, Shulman GI, Zawalich W, Defronzo RA (1987) Effect of chronic hyperglycemia on in vivo insulin secretion in partially pancreatectomized rats. J Clin Invest 80:1037–1044
[145] Mevorach M, Giacca A, Aharon Y, Hawkins M, Shamoon H, Rossetti L (1998) Regulation of endogenous glucose production by glucose per se is impaired in type 2 diabetes mellitus. J Clin Invest 102:744–753
[146] Ikeda S, Kishida S, Yamamoto H, Murai H, Koyama S, Kikuchi A (1998) Axin, a negative regulator of the Wnt signaling pathway, forms a complex with GSK-3beta and beta-catenin and promotes GSK-3beta-dependent phosphorylation of beta-catenin. EMBO J 17:1371–1384
[147] Itoh K, Krupnik VE, Sokol SY (1998) Axis determination in Xenopus involves biochemical interactions of axin, glycogen synthase kinase 3 and beta-catenin. Curr Biol 8:591–594
[148] Aberle H, Bauer A, Stappert J, Kispert A, Kemler R (1997) Beta-catenin is a target for the ubiquitin-proteasome pathway. EMBO J 16:3797–3804
[149] Fagotto F, Gluck U, Gumbiner BM (1998) Nuclear localization signal-independent and importin/karyopherin-independent nuclear import of beta-catenin. Curr Biol 8:181–190
[150] Yost C, Farr GH III, Pierce SB, Ferke DM, Chen MM, Kimerlman D (1998) GBP, an inhibitor of GSK-3, is implicated in Xenopus development and oncogenesis. Cell 93:1031–1041
[151] Brannon M, Gomperts M, Sumoy L, Moon RT, Kimelman D (1997) A beta-catenin/XTcf-3 complex binds to the siamois promoter to regulate dorsal axis specification in Xenopus. Genes Dev 11:2359–2370
[152] Carron C, Pascal A, Djiane A, Boucaut J-C, Shi D-L, Umbhauer M (2003) Frizzled receptor dimerization is sufficient to activate the Wnt/β-catenin pathway. J Cell Sci 116:2541–2550
[153] Haq S, Michael A, Andreucci M, Bhattacharya K, Dotto P, Walters B, Woodgett J, Kilter H, Force T (2003) Stabilization of β-catenin by a Wnt-independent mechanism regulates cardiomyocyte growth. Proc Natl Acad Sci U S A 100:4610–4615
[154] Mikels AJ, Nusse R (2006) Purified Wnt5a protein activates or inhibits β-catenin-TCF signaling depending on receptor context. PLoS Biol 4:e115
[155] He TC, Sparks AB, Rago C, Hermeking H, Zawel L, de Costa LT, Morin PJ, Vogelstein B, Kinzler KW (1998) Identification of c-MYC as a target of the APC pathway. Science 281:1509–1512
[156] He TC, Chan TA, Vogelstein B, Kinzler KW (1999) PPARdelta is an APC-regulated target of nonsteroidal anti-inflammatory drugs. Cell 99:335–345
[157] Shtutman M, Zhurinsky J, Simcha I, Albanese C, D'Amico M, Pestell R, Bez-Ze'ev A (1999) The cyclin D1 gene is a target of the beta-catenin/LEF-1 pathway. Proc Natl Acad Sci U S A 96:5522–5527

[158] Tetsu O, McCormick F (1999) Beta-catenin regulates expression of cyclin D1 in colon carcinoma cells. Nature 398:422–426
[159] Howe LR, Subbaramaiah K, Chung WJ, Dannenberg AJ, Brown AMC (1999) Transcriptional activation of cyclooxygenase-2 in Wnt-1-transformed mouse mammary epithelial cells. Cancer Res 59:1572–1577
[160] Howe LR, Crawford HC, Subbaramaiah K, Hassell JA, Dannenberg AJ, Brown AMC (2001) PEA3 is up-regulated in response to Wnt1 and activates the expression of cyclooxygenase-2. J Biol Chem 276:20108–20115
[161] Haertel-Wiesmann M, Liang Y, Fantl WJ, Williams LT (2000) Regulation of cyclooxygenase-2 and periostin by Wnt-3 in mouse mammary epithelial cells. J Biol Chem 275:32046–32051
[162] Xu HE, Lambert MH, Montana VG, Parks DJ, Blanchard SG, Brown PJ, Sternbach DD, Lehmann JM, Wisely GB, Willson TM, Kliewer SA, Milburn MV (1999) Molecular recognition of fatty acids by peroxisome proliferator-activated receptors. Mol Cell 3:397–403
[163] Nakamura T, Sano M, Songyang Z, Schneider MD (2003) A Wnt- and beta-catenin-dependent pathway for mammalian cardiac myogenesis. Proc Natl Acad Sci U S A 100:5834–5839
[164] Salic A, Lee E, Mayer L, Kirschner MW (2000) Control of beta-catenin stability: reconsitution of the cytoplasmic steps of the wnt pathway in Xenopus egg extracts. Mol Cell 5:523–532
[165] Hardt SE, Sadoshima J (2002) Glycogen synthase kinase-3veta: a novel regulator of cardiac hypertrophy and development. Circ Res 90:1055–1063
[166] Phili-Couderc P, Pathak A, Smith F, Dambrin C, Harmancey R, Buys S, Galinier M, Massabuau P, Roncalli J, Senard J-M, Rouet P (2004) Uncomplicated human obesity is associated with a specific cardiac transcriptome: involvement of the Wnt pathway. FASEB J 18:1539–1540
[167] Innis SM, de La Presa Owens S (2001) Dietary fatty acid composition in pregnancy alters neurite membrane fatty acids and dopamine in newborn rat brain. J Nutr 131:118–122
[168] de La Presa Owens S, Innis SM (2000) Diverse, region-specific effects of addition of arachidonic and docosahexaenoic acids to formula with low or adequate linoleic and alpha-linolenic acids on piglet brain monoaminergic neurotransmitters. Pediatr Res 48:125–130
[169] Helland IB, Smith L, Saarem K, Saugstad OD, Drevon CA (2003) Maternal supplementation with very-long-chain n-3 fatty acids during pregnancy and lactation augments children's IQ at 4 years of age. Pediatrics 111:39–44
[170] Whalley LJ, Fox HC, Wahle KW, Starr JM, Deary IJ (2004) Cognitive aging, childhood intelligence, and the use of food supplements: possible involvement of n-3 fatty acids. Am J Clin Nutr 80:1650–1657
[171] Gustafsson PA, Duchen K, Birberg U, Karlsson T (2004) Breastfeeding, very long polyunsaturated fatty acids (PUFA) and IQ at 6 1/2 years of age. Acta Paediatr 93:1280–1287
[172] Auestad N, Scott DT, Janowsky JS, Jacobsen C, Carroll RE, Montalto MB, Halter R, Qiu W, Jacobs JR, Connor WE, Connor SL, Taylor JA, Neuringer M, Fitzgerald KM, Hall RT (2003) Visual, cognitive, and language assessments at 39 months: a follow-up study of children fed formulas containing long-chain polyunsaturated fatty acids to 1 year of age. Pediatrics 112(3 Pt 1):e177–e183
[173] Willatts P, Forsyth JS, DiModugno MK, Varma S, Colvin M (1998) Effect of long-chain polyunsaturated fatty acids in infant formula on problem solving at 10 months of age. Lancet 352:688–691
[174] de La Presa Owens S, Innis SM (1999) Docosahexaenoic and arachidonic acid prevent a decrease in dopaminergic and serotoninergic neurotransmitters in frontal cortex caused by a linoleic and alpha-linolenic acid deficient diet in formula-fed piglets. J Nutr 129:2088–2093
[175] Kuperstein F, Yakubov E, Dinerman P, Gil S, Eylam R, Salem N Jr, Yavin E (2005) Overexpression of dopamine receptor genes and their products in the postnatal rat brain following maternal n-3 fatty acid dietary deficiency. J Neurochem 95:1550–1562
[176] Acar N, Chardigny JM, Berdeaux O, Alamnza S, Sebedio JL (2002) Modification of the monoaminergic neurotransmitters in frontal cortex and hippocampus by dietary trans alpha-linolenic acid in piglets. Neurosci Lett 331:198–202

[177] Delion S, Chalon S, Guilloteau D, Besnard JC, Durand G (1996) Alpha-Linolenic acid dietary deficiency alters age-related changes of dopaminergic and serotoninergic neurotransmission in the rat frontal cortex. J Neurochem 66:1582–1591
[178] Cunha RA, Ribeiro JA (1999) Facilitation of GABA release by arachidonic acid in rat hippocampal synaptosomes. Eur J Neurosci 11:2171–2174
[179] Cheramy A, Artaud F, Godeheu G, L'hirondel M, Glowinski J (1996) Stimulatory effect of arachidonic acid on the release of GABA in matrix-enriched areas from the rat striatum. Brain Res 742:185–194
[180] Nabekura J, Noguchi K, Witt MR, Neilsen M, Akaike N (1998) Functional modulation of human recombinant gamma-aminobutyric acid type A receptor by docosahexaenoic acid. J Biol Chem 273:11056–11061
[181] Hamano H, Nabekura J, Nishikawa M, Ogawa T (1996) Docosahexaenoic acid reduces GABA response in substantia nigra neuron of rat. J Neurophysiol 75:1264–1270
[182] Almeida T, Cunha RA, Ribeiro JA (1999) Facilitation by arachidonic acid of acetylcholine release from the rat hippocampus. Brain Res 826:104–111
[183] Blanchet F, Gauchy C, Perez S, Glowinski J, Kemel ML (1999) Role of arachidonic acid in the regulation of the NMDA-evoked release of acetylcholine in striatal compartments. Synapse 31:140–150
[184] Aid S, Vancassel S, Linard A, Lavialle M, Guesnet P (2005) Dietary docosahexaenoic acid[22:6(n-3)] as a phospholipid or a triglyceride enhances the potassium chloride-evoked release of acetylcholine in rat hippocampus. J Nutr 135:1008–1013
[185] Minami M, Kimura S, Endo T, Hamaue N, Hirafuji M, Togashi H, Matsumoto M, Yoshioka M, Saito H, Watanabe S, Kobayashi T, Okuyama H (1997) Dietary docosahexaenoic acid increases cerebral acetylcholine levels and improves passive avoidance performance in stroke-prone spontaneously hypertensive rats. Pharmacol Biochem Behav 58:1123–1129
[186] Zhang L, Reith ME (1996) Regulation of the functional activity of the human dopamine transporter by the arachidonic acid pathway. Eur J Pharmacol 315:345–354
[187] L'hirondel M, Cheramy A, Godeheu G, Glowinski J (1995) Effects of arachidonic acid on dopamine synthesis, spontaneous release, and uptake in striatal synaptosomes from the rat. J Neurochem 64:1406–1409
[188] Piomelli D, Pilon C, Giros B, Sokoloff P, Martres MP, Schwartz JC (1991) Dopamine activation of the arachidonic acid cascade as a basis for D1/D2 receptor synergism. Nature 353:164–167
[189] Rodriguez Y, Christophe AB (2005) Long-chain omega6 polyunsaturated fatty acids in erythrocyte phospholipids are associated with insulin resistance in non-obese type 2 diabetics. Clin Chim Acta 354:195–199
[190] Enriquez YR, Giri M, Rottiers R, Christophe A (2004) Fatty acid composition of erythrocyte phospholipids is related to insulin levels, secretion and resistance in obese type 2 diabetics on Metformin. Clin Chim Acta 346:145–152
[191] Borkman M, Storlien LH, Pan DA, Jenkins AB, Chisholm DJ, Campbell LV (1993) The relation between insulin sensitivity and the fatty-acid composition of skeletal-muscle phospholipids. N Engl J Med 328:238–244
[192] Das UN, Vijay Kumar K, Krishna Mohan I (1994) Lipid peroxides and essential fatty acids in patients with diabetes mellitus and diabetic nephropathy. J Nutr Med 4:149–155
[193] Warensjo E, Ohrvall M, Vessby B (2006) Fatty acid composition and estimated desaturase activities are associated with obesity and lifestyle variables in men and women. Nutr Metab Cardiovasc Dis 16:128–136
[194] Ozanne SE, Martensz ND, Petry CJ, Loizou CL, Hales CN (1998) Maternal low protein diet in rats programmes fatty acid desaturase activities in the offspring. Diabetologia 41:1337–1342
[195] Ohtani N, Ohta M, Sugano T (1997) Microdialysis study of modification of hypothalamic neurotransmitters in streptozotocin-diabetic rats. J Neurochem 69:1622–1628
[196] Takahashi A, Ishimaru H, Ikarashi Y, Maruyama Y (1994) Aspects of hypothalamic neuronal systems in VMH lesion-induced obese rats. J Auton Nerv Syst 48:213–219

[197] Barber M, Kasturi BS, Austin ME, Patel KP, Mohan Kumar SM, Mohan Kumar PS (2003) Diabetes-induced neuroendocrine changes in rats: role of brain monoamines, insulin and leptin. Brain Res 964:128–135
[198] Bhattacharjee AK, Chang L, Lee HJ, Bazinet RP, Seemann R, Rapoport SI (2005) D2 but not D1 dopamine receptor stimulation augments brain signaling involving arachidonic acid in unanesthetized rats. Psychopharmacology (Berl) 180:735–742
[199] Bouret SG, Draper SJ, Simerly RB (2004) Trophic action of leptin on hypothalamic neurons that regulate feeding. Science 304:108–110
[200] Pinto S, Roseberry AG, Liu H, Diano S, Shanabrough M, Cai X, Friedman JM, Horvath TL (2004) Rapid rewiring of arcuate nucleus feeding circuits by leptin. Science 304:110–115
[201] Cowley M, Smart JL, Rubenstein M, Cerdan MG, Diano S, Horvath TL, Cone RD, Low RD (2001) Leptin activates anorexigenic POMC neurons through a neural network in the arcuate nucleus. Nature 411:480–484
[202] Schwartz MW, Seeley RJ, Woods SC, Weigle DS, Campfield LA, Burn P, Baskin DG (1997) Leptin increases hypothalamic pro-opiomelanocortin mRNA expression in the rostral arcuate nucleus. Diabetes 46:2119–2123
[203] Stephens TW, Basinski M, Bristow PK, Bue-Valleskey JM, Burgett SG, Craft L, Hale J, Hoffmann J, Hsiung HM, Kriauciunas A et al (1995) The role of neuropeptide Y in the antiobesity action of the obese gene product. Nature 377:530–532
[204] Pinto S, Roseberry AG, Liu H, Diano S, Shanabrough M, Cai X, Friedman JM, Horvath TL (2004) Rapid rewiring of arcuate nucleus feeding circuits by leptin. Science 304:110–115
[205] Elmquist JK, Flier JS (2004) The fat-brain axis enters a new dimension. Science 304:63–64
[206] Phipps K, Barker DJ, Hales CN et al (1993) Fetal growth and impaired glucose tolerance in men and women. Diabetologia 36:225–228
[207] Barker DJ, Hales CN, Fall CH et al (1993) Type 2 (non-insulin dependent) diabetes mellitus, hypertension, and hyperlipidemia (syndrome X): relation to reduced fetal growth. Diabetologia 36:62–67
[208] Lucas A, Fewtrell MS, Cole TJ (1999) Fetal origins of adult disease-the hypothesis revisited. BMJ 319:245–249
[209] Das UN (1991) Essential fatty acids: biology and their clinical implications. Asia Pacific J Pharmacol 16:317–330
[210] Das UN (1999) Essential fatty acids in health and disease. J Assoc Physicians India 47:906–911
[211] Das UN (2006) Essential fatty acids-a review. Curr Pharm Biotechnol 7:467–482
[212] Mercuri O, de Tomas E, Itarte H (1979) Prenatal protein depletion and $\delta 9$, $\delta 6$, and $\delta 5$ desaturases in the rat. Lipids 14:822–825
[213] Trottier G, Koski KG, Brun T, Toufexis DJ, Richard D, Walker CD (1998) Increased fat intake during lactation modifies hypothalamic-pituitary-adrenal responsiveness in developing rat pups: a possible role for leptin. Endocrinology 139:3704–3711
[214] Frederich RC, Hamann A, Anderson S, Lollmann B, Lowell BB, Flier JS (1995) Leptin levels reflect body lipid content in mice: evidence for diet-induced resistance to leptin action. Nat Med 1:1311–1314
[215] Cha MC, Jones PJH (1998) Dietary fat type and energy restriction interactively influence plasma leptin concentrations in rats. J Lipid Res 39:1655–1660
[216] Korotkova M, Gabrielsson B, Hanson LA, Strandvik B (2001) Maternal essential fatty acid deficiency depresses serum leptin levels in suckling rat pups. J Lipid Res 42:359–365
[217] Korotkova M, Gabrielsson B, Hanson LA, Strandvik B (2002) Maternal dietary intake of essential fatty acids affects adipose tissue growth and leptin mRNA expression in suckling rat pups. Pediatr Res 52:78–84
[218] Dhalgren J, Nilsson C, Jennische E et al (2001) Prenatal cytokine exposure results in obesity and gender-specific programming. Am J Physiol 281:E326–E334
[219] Zahorska-Markiewicz B, Janowska J, Olszanecka-Glinianowicz M et al (2000) Serum concentrations of TNF-α and soluble TNF-α receptors in obesity. Int J Obes 24:1392–1395

[220] Fried SK, Bunkin DA, Greenberg AS (1998) Omental and subcutaneous adipose tissues of obese subjects release interleukins-6: depot differences and regulation by glucocorticoid. J Clin Endocrinol Metab 83:847–850

[221] Rask E, Walker BR, Soderberg S, Livingstone DEW, Eliasson M, Johnson O, Andrew R, Olsson T (2002) Tissue-specific changes in peripheral cortisol metabolism in obese women: increased 11β-hydroxysteroid dehydrogenase type 1 activity. J Clin Endocrinol Metab 87:3330–3336

Chapter 17
Clinical Implications

Introduction

It is evident from the preceding chapters that several adult diseases: obesity, insulin resistance, type 2 diabetes mellitus, hypertension, dyslipidemia, coronary heart disease, metabolic syndrome, some cancers, schizophrenia, depression, Alzheimer's disease, atherosclerosis, aging, osteoporosis, stroke, lupus, rheumatoid arthritis and other autoimmune diseases are all low-grade systemic inflammatory conditions. The enhanced production of pro-inflammatory cytokines, ROS, reactive nitrogen species, pro-inflammatory eicosanoids, a decrease in the cellular anti-oxidants and a simultaneous decrease in the levels of anti-inflammatory cytokines and certain PUFAs and their products such as lipoxins, resolvins, protectins, maresins and nitrolipids seem to occur in all these conditions. As already discussed in the previous chapter, the target tissues/organs are different depending on the underlying condition though low-grade systemic inflammation is common in all these diseases. In certain conditions, such as lupus and rheumatoid arthritis, local inflammatory events seem to be more evident. On the other hand, in obesity, type 2 diabetes mellitus, insulin resistance, hypertension, metabolic syndrome, aging, dyslipidemia, coronary heart disease, some cancers, schizophrenia, depression, Alzheimer's disease, atherosclerosis, osteoporosis and stroke, low-grade systemic inflammation is more common unlike the localized inflammation seen in lupus and rheumatoid arthritis. Even in rheumatoid arthritis and lupus, especially, when these patients have attained relative remission or under immunosuppressive therapy one can see little localized inflammation and more of low-grade systemic inflammation.

This imbalance between the pro- and anti-inflammatory molecules seen in all these diseases and the specific tissue/organ involvement depending on the disease, it calls for therapeutic strategies that suppress the production of pro-inflammatory molecules and enhancement or boosting the action of anti-inflammatory molecules. Such an approach may need both local and systemic strategies.

But, what is heartening to note is the fact that there are many anti-inflammatory molecules produced by the body that themselves could be exploited in the management of these diseases. Some of these endogenous anti-inflammatory molecules are

U. N. Das, *Molecular Basis of Health and Disease*,
DOI 10.1007/978-94-007-0495-4_17,

highly unstable and produced in relatively small amounts by various tissues that calls for their synthesis in the lab so that more stable and long-acting versions could be developed in order to enhance their duration of action and have specific therapeutic properties depending on the necessity.

Since obesity, metabolic syndrome, cancer, schizophrenia and other adult diseases are complex disorders, they may need multiple therapeutic strategies and one magic bullet may not work for them. At times, in order to suppress the underlying low-grade systemic inflammation and restore health one may need to resort to the use more than one molecule or strategy. It is possible that a combination of molecules need to be administered to obtain optimal therapeutic effect in some, if not all, of these diseases.

Based on the discussions in the previous chapters and the current knowledge, I propose certain therapeutic strategies for some of these adult diseases. It is likely that the suggested therapeutic strategies may look either too simple or too complex but are certainly novel and may even be surprising. Even if one or two of the strategies proposed turn out to be correct, even for the management of one or two diseases, that itself can be considered as an advance in the management of these adult diseases. The proposed therapeutic strategies are simple, feasible with the currently available technology and are easily implementable.

In the first instance, I will present the therapeutic strategies that are based on the principles of enhancing natural process(s) of resolution of inflammation, wound healing and repair so that relatively few side-effects are likely to occur. Such a natural process of resolution and repair could depend on the proper use of certain endogenous anti-inflammatory molecules that may resolve inflammation, enhance wound healing and repair process and restore normal physiology. Subsequently, I will discuss how these therapeutic molecules could be used in the management of specific diseases.

Glucose-Insulin-Potassium Regimen

Glucose-insulin-potassium (GIK) regimen is one regimen that is familiar to almost all physicians. It is used to treat patients with a moderate degree of hyperglycemia, even in the absence of ketoacidosis and those with diabetic ketoacidosis. There is now ample evidence to suggest that insulin has potent anti-inflammatory actions [1]. Satomi et al. [2] showed that both glucose administration and insulin injections to pre-sensitized mice inhibited TNF production. Fraker et al. [3] reported that recombinant human TNF-cachectin-induced decrease in food intake, decrease in nitrogen balance and decrease in body weight gain compared to saline controls in rats can be reversed by concurrent administration of insulin without causing any treatment deaths. Further, 5 days of TNF-cachectin treatment induced severe interstitial pneumonitis, periportal inflammation in the liver, and an increase in wet organ weight in the heart, lungs, kidney, and spleen in the rats could be reverted to normal when insulin treatment was given concurrently. These results suggest that administration of insulin has the ability to reverse the nutritional and histopathologic toxicity of sublethal doses of TNF. Ottlecz et al. [4–6] showed that insulin has anti-inflammatory action against carrageenan-induced paw oedema. Biochot et al. [7] demonstrated that luminol-dependent chemiluminescence by the mononuclear cells in the bronchoalveolar lavage (BAL) fluid and the levels of TNF-α in the BAL supernatant reverted to normal levels after treatment of Wistar diabetic rats with insulin, indicating that insulin regulates superoxide anion generation and TNF-α synthesis and release, and thus produces its antiinflammatory action [1–7].

Insulin decreases the mortality and prevents the incidence of infection and sepsis in critically ill patients. When endotoxemic rats were administered insulin without producing any change in glucose or electrolyte levels, a significant decrease in the proinflammatory signal transcription factors[CCAAT/enhancer-binding protein-β, signal transducer and activator of transcription -3 and -5, RANTES (regulated on activation, normal T cell expressed and secreted)] and cytokine expression in the liver and serum levels of IL-1β, IL-6, macrophage inflammatory factor, and TNF-α was observed [8]. Insulin administration further decreased serum HMGB1 levels compared with controls. In addition, insulin increased antiinflammatory cytokine expression in the liver; serum levels of IL-2, IL-4, and IL-10; and hepatic suppressor of cytokine signaling-3 mRNA expression. Thus, insulin suppressed inflammation in this animal model by decreasing the proinflammatory and increasing the antiinflammatory molecules. Since, in this study, plasma glucose and electrolyte levels did not differ between insulin-treated and controls, it is reasonable to assume that the effects are direct antiinflammatory mechanisms of insulin as proposed previously [1].

Hyperglycemia in critical illness is common and is considered as an independent risk factor for morbidity and death. It is known that intensive insulin therapy decreases this risk by up to 50% [9]. But, it is not clear to what extent this benefit is due to reversal of glucotoxicity or to a direct effect of insulin; in view of its antiinflammatory effects and what are the underlying mechanisms. The insulin receptor is expressed on resting neutrophils, monocytes, and B cells, but is not detectable on T cells. However, significant up-regulation of insulin receptor expression is observed on activated T cells, which suggests an important role during T cell activation. Exogenous insulin in vitro induced a shift in T cell differentiation toward a TH_2-type response, decreasing the T helper type 1 to TH_2 ratio by 36%. These changes correlated with a corresponding change in cytokine secretion, with the IFN-γ to IL-4 ratio being decreased by 33%. These changes were associated with increased TH_2-promoting ERK phosphorylation in the presence of insulin. Thus, insulin has the ability to influences T cell differentiation promoting a shift toward a TH_2-type response [10]. This ability of insulin in changing T cell polarization may contribute to its antiinflammatory action that may have relevance to its role not only in sepsis, but also in chronic inflammation associated with obesity and type 2 diabetes mellitus.

In severe burns, the liver plays a pivotal role by modulating inflammatory processes, metabolic pathways, immune functions, and the acute phase response. Hence, liver integrity and function are important for recovery. On the other hand, thermal injury causes hepatic damage by inducing hepatic edema, fatty infiltration, hepatocyte apoptosis, and metabolic derangements associated with insulin resistance and impaired insulin signaling. It was reported that insulin administration improved survival and decreased the rate of infections in severely burned and critically ill patients.

Insulin administration decreased the synthesis of proinflammatory cytokines and signal transcription factors and improved hepatic structure and function after a severe burn injury; insulin also restored hepatic homeostasis and improved hepatic dysfunction postburn via alterations in the signaling cascade [11]. These results attest to the fact that insulin has antiinflammatory actions [1]. These results are supported by the observation that severely burned pediatric patients, who received insulin to maintain blood glucose at a range from 120 to 180 mg/dl, insulin administration decreased pro-inflammatory cytokines and proteins, while increasing constitutive-hepatic proteins. Burned children receiving insulin required significantly less albumin substitution to maintain normal levels compared with control. Insulin decreased free fatty acids and serum triglycerides when compared with controls, while serum IGF-I and IGFBP-3 significantly increased with insulin administration [12]. These results suggested that insulin attenuated the inflammatory response by decreasing the pro-inflammatory and increasing the anti-inflammatory cascade, thus restoring systemic homeostasis in critically ill patients.

Type 2 diabetics with microalbuminuria when treated with multiple insulin injections daily for 2 weeks showed significantly decreased urinary monocyte chemoattractant protein-1 (MCP-1) and intercellular adhesion molecule-1 (ICAM-1), the two important inflammatory chemokines, compared to the control [13]. These results suggested that intensive insulin treatment may protect against renal injury in early diabetic nephropathy.

Insulin enhances the expression of eNOS gene both in vitro and in vivo [14]. NO quenches the superoxide anion [15] and thus, NO is of benefit in inflammatory conditions. The expression of MIF (macrophage migration inhibitory factor) in adipocytes can be modulated by insulin and glucose [16]. MIF is secreted together with insulin from pancreatic β cells and acts as an autocrine factor to stimulate insulin release [17]. During systemic inflammatory process, MIF is secreted from the pituitary gland accompanied by an increase in glucocorticoid secretion. The increase in plasma glucose levels that occurs as a result of this glucocorticoid secretion is, in turn, controlled by MIF by its positive effect on insulin secretion. TNF-α up-regulates MIF secretion [18]. On the other hand, TNF-α induces insulin resistance [19]. Thus, glucose homeostasis during systemic inflammatory process is regulated by glucocorticoids, TNF-α, MIF and insulin. This feed-back loop among MIF, glucose and insulin, implies that infusion of glucose and insulin inhibits MIF production similar to their (insulin + glucose) inhibitory action on TNF-α [20–22]. In support of this postulation, we recently showed that insulin inhibited ischemia/reperfusion-induced TNF-α and MPO (myeloperoxidase) production through the Akt-activated and eNOS-NO-dependent pathway in cardiomyocytes both in vitro and in vivo [23]. Thus, insulin has potent antiinflammatory property that may contribute to its cardio-protective and cytoprotective actions that explains its ability to improve survival in the critically ill.

In view of these evidences [1–23], it is suggested that glucose+insulin regimen is likely to be useful in the management of several inflammatory conditions including lupus, RA and other rheumatological conditions, ulcerative colitis, Crohn's disease, burns, and other critically ill patients both in the medical and surgical wards and those with coronary heart disease [24].

It is also suggested that plasma levels of TNF-a, MIF, HMGB1, IL-6, IL-4, IL-10, pyruvate, and various free radicals (including NO) need to be measured in addition to plasma glucose concentrations to ensure that insulin regimen adopted is adequate. The plasma glucose levels need to be maintained ~80–100 mg%, such that the production of pro-inflammatory cytokines is suppressed and synthesis and release of antiinflammatory cytokines is enhanced, and failure to do so would give negative results. Since there could be individual variations in response to the anti-inflammatory actions of insulin, it is important that the plasma levels of pro- and anti-inflammatory cytokines and free radicals should be measured in order to know the adequacy of the dose of insulin instituted as suggested previously [24].

Ethyl Pyruvate

Pyruvic acid, present in the cells and extracellular fluids as its conjugate anion, is the final product of glycolysis and the starting substrate for tricarboxylic acid (TCA) cycle. It plays a crucial role in intermediary metabolism. Pyruvate is unstable in solutions, and spontaneously undergoes condensation and cyclization. Ethyl pyruvate (EP), a derivative of pyruvic acid, in a calcium- and potassium-containing balanced salt solution (called as Ringer ethyl pyruvate solution) is not only stable and non-toxic but is an effective anti-inflammatory molecule compared to pyruvate [25]. Ethyl pyruvate is approved by Food and Drug Administration (FDA) as a food additive. EP is used in a calcium-containing solution because it is a hydrophilic compound and the calcium prevents emulsion and increases the solubility. Since EP is chemically related to lactate, substituting lactate for EP can provide a therapeutic anti-inflammatory property to the Ringer's solution. EP in a Ringer-type Ca^{2+}- and K^{+}-containing balanced salt solution is not only stable but also more effective than sodium pyruvate [26].

Ringer ethyl pyruvate solution prolonged survival of rats that were in hemorrhagic shock [25, 26] by effectively scavenging the free radicals [27–29]. Ethyl pyruvate inhibited the release of TNF-α and HMGB1 from endotoxin-stimulated murine macrophages and attenuated activation of NF-κB signaling pathways. In LPS-induced endotoxic shock animal model, ethyl pyruvate improved survival by lowering circulating concentrations of nitrite/nitrate (metabolites of nitric oxide, NO) and IL-6 and enhanced plasma levels of IL-10, an anti-inflammatory cytokine [30], suggesting that ethyl pyruvate has significant anti-inflammatory actions. In view of the significant anti-inflammatory actions [31–33], it is suggested that administration of ethyl pyruvate may be of significant benefit in suppressing inflammatory events in lupus, RA and other rheumatological conditions.

Despite the fact that ethyl pyruvate has potent anti-inflammatory actions both in vitro and in vivo [31–33] and protected animals in several models of critical illness including myocardial or mesenteric ischemia/reperfusion injury, sepsis, and hemorrhagic shock, when ethyl pyruvate (7,500 mg administered intravenously starting after the induction of general anesthesia followed by 5 more doses of 7,500 mg administered every 6 h, corresponding to a dose of 90 mg/kg at each of the 6 dosing intervals, exceeding the dose of 40 mg/kg shown to be effective in many animal models) was administered to patients undergoing higher-risk cardiac surgery in a double-blind, randomized, placebo-controlled study no statistically significant differences were observed between groups with regard to clinical parameters or markers of systemic inflammation. These results suggested that despite positive results in numerous animal models, the administration of ethyl pyruvate did not appear to confer any benefit to cardiac surgical patients undergoing coronary artery bypass graft and/or cardiac valvular surgery with cardiopulmonary bypass [34]. In spite of these negative results, perhaps, it may be worth to study the effectiveness of ethyl pyruvate in combination with insulin or other anti-inflammatory molecules.

Lipid-enriched Albumin

Albumin, the major protein produced by hepatocytes in the liver, maintains oncostatic pressure. Albumin traps oxygen radical and quenches free radicals; inhibits copper ion-dependent lipid peroxidation and retards the formation of hydroxyl radicals and thus, has both neuroprotective and cytoprotective action. This beneficial action of albumin has been attributed to the ability of albumin to mobilize docosahexaenoic acid (DHA) and, possibly, other polyunsaturated fatty acids (PUFAs) from liver and other tissues which, in turn, are converted to anti-inflammatory molecules such as protectins, lipoxins and resolvins. These results are interesting in the light of the fact that PUFAs when incorporated into the cell membranes could alter their proliferation, especially that of tumor cells.

When the fatty acid composition of HT-29 human colon cancer cells was altered by supplementing the cells with stearic acid (18:0; SA), γ-linolenic acid (GLA), ALA, EPA and DHA as a fatty acid/bovine serum albumin complex, the cells incorporated and modified the supplemented fatty acids by desaturation, elongation and retroconversion. The unsaturation index (UI) of membranes of cells supplemented with EPA and DHA was higher than other groups. A negative correlation between the activity of phospholipase C in the presence of G protein activation and phosphatidylethanolamine GLA (since GLA is incorporated into phosphatidylethanolamine) content without affecting unsaturation index was noted, suggesting that G protein may be sensitive to the level of GLA content and not to the general fluidity of the membranes [35]. These results are interesting since, G proteins, which belong to the larger group of enzymes GTPases, communicate signals from hormones and neurotransmitters to regulate metabolic enzymes, ion channels, and transporters, and control transcription, motility, contractility, and secretion that, in turn, regulate systemic functions such as embryonic development, learning and memory, and homeostasis [36]. These results suggest that GLA and possibly other

PUFAs have a regulatory role in several cellular functions by modulating G proteins that may explain the anti-cancer actions of some of the fatty acids [37–41].

Administration of albumin-DHA complex containing 2.1 ± 0.1 μmol DHA per milliliter of albumin to the 2-h middle cerebral artery suture-occlusion animal model, a high degree neurobehavioral and histological neuroprotection was noted [42]. DHA pretreatment was also reported to have improved functional outcome and reduced volume loss after hypoxia-ischemia in neonatal rats [43].

Albumin-DHA complex facilitates DHA delivery to the brain so that significant amounts of NPD_1 (neuroprotectin D_1) is formed that prevented ischemia-reperfusion injury. DHA confers neuroprotection by opening background K^+ channels and inhibiting apoptosis. Administration of albumin-DHA complex increased the formation of NPD_1 and infusion of NPD_1 reduced infarct size, diminished polymorphonuclear leukocyte infiltration, NF-κB activation, and pro-inflammatory cyclo-oxygenase-2 expression [44, 45].

In view of these anti-inflammatory actions, it is possible that albumin-PUFA complexes could be of significant benefit in the management of rheumatological conditions RA and lupus; sepsis and the critically ill due to surgical and medical conditions. It is possible that such albumin-PUFA complex could be developed as nanoparticles that could be given orally, parentarally and other systemic routes depending on the necessity. Albumin-PUFA complex could be developed into a slow-release, sub-cutaneous delivery system for the prevention and treatment of obesity, type 1 and type 2 diabetes mellitus, hypertension and metabolic syndrome. This is so since, previously, we showed that pre-treatment with PUFAs can prevent chemical-induced diabetes mellitus [46–50].

In a recent study [51], it was reported that mfat-1 transgenic mouse (in which endogenous production of n-3 PUFAs was achieved through overexpressing a C. elegans n-3 fatty acid desaturase gene) islets contained much higher levels of n-3 PUFAs and lower levels of n-6 PUFAs than the wild type and showed significantly elevated insulin secretion stimulated by glucose, amino acids, and glucagon-like peptide-1 (GLP-1). These mice when challenged with TNF-α, Il-1β, and IFN-γ, completely resisted cytokine-induced cell death. In addition, the expression of mfat-1 decreased PGE_2 production, which contributed to the elevation of insulin secretion. Furthermore, cytokine-induced activation of NF-kB and extracellular signal-related kinase 1/2 (ERK(1/2)) was significantly attenuated and that the expression of pancreatic duodenal hemeobox-1 (PDX-1), glucokinase, and insulin-1 was increased as a result of n-3 PUFA production. These results are in support of our previous studies wherein we showed that PUFAs, especially EPA and DHA prevent chemical-induced damage to pancreatic β cells [46–50].

In a similar study [52] but performed in the transgenic fat-1 mouse (in which ω-3 fatty acid desaturase from C. elegans was transgenically expressed to enhance n-3 fatty acid levels), high fat diet inhibited the production of protectin D_1, a potent anti-inflammatory compound, in muscle and adipose tissue that resulted in impaired capacity to resolve an acute inflammatory response and display elevated adipose macrophage accrual and chemokine/cytokine expression. This was found to be associated with insulin resistance and higher activation of iNOS and JNK in muscle and liver. These defects could be reversed when protectin D_1 synthesis was restored to normal without altering food intake, weight gain or adiposity.

In fact, recently, it was shown that n-3 fatty acids bind to G protein-coupled receptor GPR120 on macrophages and fat cells and thus, inhibit inflammation induced by macrophages and reverses insulin resistance in obese mice [53, 54]. PUFAs are known to regulate gut incretin GLP-1 secretion through GPR-120 [55]. GPR-40, which is abundantly expressed in the pancreas, functions as a receptor for PUFAs and thus, amplifies glucose-stimulated insulin secretion from pancreatic β cells by activating GPR-40 [56].

These results [35–56] results suggest that PUFAs by themselves and/or by giving raise to their anti-inflammatory products such as lipoxins, resolvins, protectins and maresins are able to prevent inappropriate inflammation, suppress chemical-induced damage to pancreatic β cells, neurons and possibly, other cells and thus bring about their cytoprotective actions. These beneficial actions of PUFAs are responsible for the prevention of both type 1 and type 2 diabetes mellitus and hypoxia, ischemia-reperfusion-induced damage to cells and tissues. In view of this, it is reasonable to suggest that PUFAs may be supplemented to infants, children and adults and pregnant women and lactating mothers to prevent various diseases. In order to know whether the supplemented PUFAs are sufficient enough to bring about their beneficial actions, one may measure plasma and tissue concentrations of not only PUFAs but also of lipoxins, resolvins, protectins, maresins and nitrolipids to make sure that adequate amounts of these beneficial products are being generated. It may be necessary to do more studies to know the exact amounts of and types of PUFAs to be provided for various diseases and the combination of PUFAs to be given. It is also necessary to determine what co-factors need to be provided so that PUFAs are optimally utilized in the body. Nevertheless, it is clear that PUFAs and their products could form a useful tool to prevent and manage many adult diseases.

Vagal Nerve Stimulation (VNS) Suppresses Inflammation

Vagus nerve has the ability to regulate the production of pro-inflammatory cytokines: TNF, IL-1, HMGB1, IL-6, and MIF [57–59]. Acetylcholine, the principal vagus neurotransmitter, inhibits the production of pro-inflammatory cytokines through a mechanism dependent on the $\alpha 7$ nicotinic acetylcholine receptor subunit. Strong expression of a7nAChR in the synovium of RA and psoriatic arthritis patients was detected [60]. Both peripheral macrophages and synovial fibroblasts respond in vitro to specific a7nAChR cholinergic stimulation with potent inhibition of proinflammatory cytokines [57–59]. It is likely that fibroblasts, especially in the lining layer, may have the ability to balance inflammatory mechanisms and arthritis development through feedback cholinergic stimulation by nearby immune cells. It is possible that specific cholinergic mechanisms may be involved in regulation of antibody production also locally in the joint. This implies that new therapies directed at regulation of the cholinergic and a7nAChR mediated mechanisms in the management of lupus and

RA could be developed. This is supported by the observation that measurement of RR interval variability (heart rate variability, HRV) as a marker of vagus nerve tone (that reflects parasympathetic activity), in RA patients revealed that vagus nerve activity was significantly depressed in patients compared to control [61]. This suggests that targeting the alpha7nAChR dependent control of cytokine release in RA patients could form a novel approach to suppress inflammation. As alpha7nAChR agonists ameliorate the clinical course of collagen induced arthritis in animals, it is tempting to suggest that alpha7nAChR agonists may be able of help to RA patients. It was reported that collagen induced arthritis in alpha7nAChR$^{(-/-)}$ mice was significantly severe and showed increased synovial inflammation and joint destruction compared to the wild-type mice. This exacerbation of arthritis was due to elevation in the levels of proinflammatory cytokines and enhanced T-helper cell 1 (Th1)-cytokine and TNF-α production by spleen cells. These results indicate that immune cell function in a model of rheumatoid arthritis is regulated by the cholinergic system and is mediated by the alpha7nAChR [62]. The clinical implications of these findings are that vagus nerve stimulation could be employed in the treatment of lupus, RA and other autoimmune diseases.

VNS for Obesity, Hypertension, Type 2 Diabetes Mellitus and Metabolic Syndrome

Since, obesity, hypertension, type 2 diabetes mellitus and metabolic syndrome are low-grade systemic inflammatory conditions, it is reasonable to propose that VNS will also be of significant benefit in these conditions.

It is known that pancreatic islets are extensively innervated- fine unmyelinated nerve fibres spreading over the blood vessels of the islets and ending on the endocrine cells. Subdiaphragmatic stimulation of the cut right vagal trunk (at 10 impulses/s, duration 1–5 ms, for 10 min) produced a mean increase of 50% over resting levels in inferior vena cava (IVC) insulin concentration and an increase in splenic vein insulin of 30% over resting levels in fasting baboons [63]. Subsequent studies showed that stimulation of either the right or the left cervical vagus released the same amount of insulin, whereas bilateral stimulation released twice as much. It was also reported that following a stimulation that depleted the "vagally-releasable pool", a recovery period of 15–20 min was needed before the same maximal output could be obtained again in anesthetized and eviscerated cats [64]. It was also shown that direct enhancement of insulin secretion occurs by vagal stimulation of the isolated pancreas [65]. These evidences clearly suggest that vagus nerve is a significant regulator of insulin secretion from the pancreatic β cells. Furthermore, vagus serves as the neuronal pathway in the cross-talk between the liver and adipose tissue [66], modulates pancreatic β cell mass [67–69] and facilitates the communication between liver and pancreatic β cells (Fig. 17.1).

The insulin secretory response to the glucose load was greater in obese than in lean rats. Atropine that blocks vagal action significantly reduced basal and stimulated

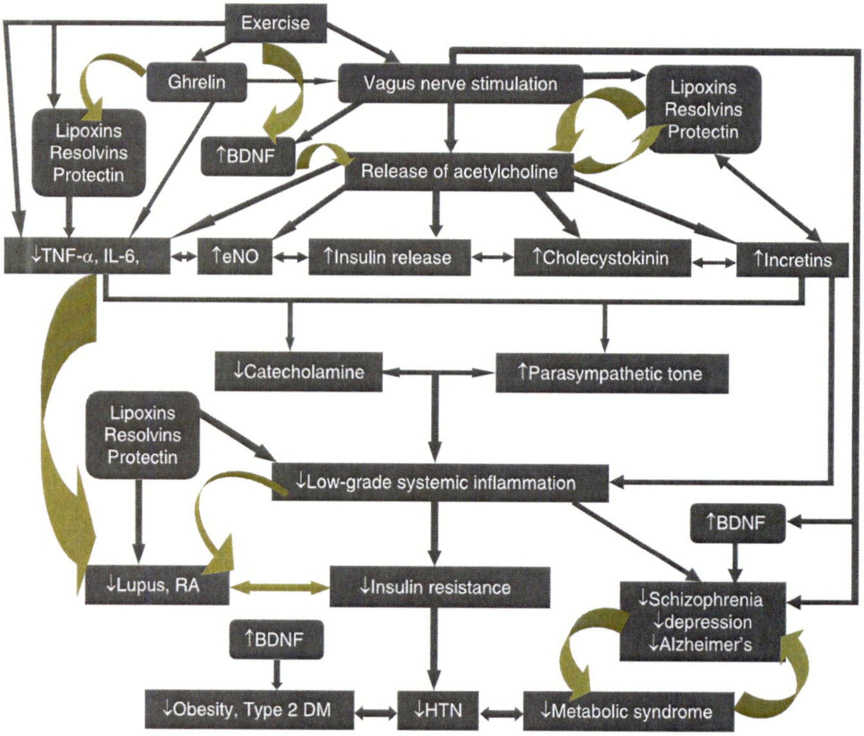

Fig. 17.1 Scheme showing the relationship among vagus nerve stimulation, inflammation, low-grade systemic inflammatory conditions and autoimmune diseases. Vagus nerve is a significant regulator of insulin secretion from the pancreatic β cells. Furthermore, vagus serves as the neuronal pathway in the cross-talk between the liver and adipose tissue [66], modulates pancreatic β cell mass [67–69] and facilitates the communication between liver and pancreatic β cells. Insulin resistance is associated with a reduction in vagal activity. Acetylcholine, the principal neurotransmitter of vagus nerve, is a potent anti-inflammatory molecule that suppresses the production of IL-6 and TNF-α. Vagus nerve stimulation also increases the production of incretins that enhance insulin secretion. Ghrelin, another intestinal peptide, also has anti-inflammatory actions and it increases acetylcholine levels in the brain and is a stimulator of the vagus nerve. Thus, ghrelin and acetylcholine interact with each other to ultimately suppress inflammation and enhance insulin secretion and reduce insulin resistance. Vagus nerve stimulation increases BDNF levels in the brain. BDNF infusion/injection reduces obesity, decreases insulin resistance, and ameliorates type 2 diabetes mellitus. Incretins and BDNF also modulate inflammation. It is possible that acetylcholine and vagus nerve stimulation enhances the formation of lipoxins, resolvins, protectins and maresins. But, this needs to be established. Preliminary evidence suggests that ghrelin (a gut peptide that increases appetite) levels are low in patients with lupus, and possibly, in other inflammatory conditions. Since ghrelin has anti-inflammatory actions, it can be deduced that low ghrelin leads to decrease in appetite and increase in inflammation. It is possible, but yet to be confirmed, that ghrelin may increase the formation of lipoxins, resolvins and protectins. Ghrelin is known to enhance endothelial nitric oxide generation and hence, when ghrelin levels are low endothelial dysfunction is likely to occur as seen in lupus, RA and other inflammatory conditions. Exercise is of benefit in the prevention and management of insulin resistance, obesity, type 2 diabetes mellitus, metabolic syndrome, and is known to prevent Alzheimer's disease and of significant help in lupus and RA.

levels of insulin in obese but not in lean rats. Adrenalectomy reduced basal insulin levels and the secretory response in obese but not lean rats and also abolished the atropine-blockable component of the response. Peripheral corticosterone replacement of adrenalectomized fa/fa (fat) rats restored the hyperinsulinemia, whereas chronic infusion of dexamethasone intracerebroventricularly to adrenalectomized fa/fa rats increased basal insulin and the secretory response to glucose an effect that was blocked by atropine. In contrast, intracerebroventricular infusion of obese rats with corticotropin releasing factor reduced basal and stimulated insulin levels. These results suggest that the hypersecretion of insulin in obese fa/fa rats results, at least in part, from a central glucocorticoid-mediated stimulation of vagal drive to the pancreatic B-cells [70]. The increased insulin secretion seen in preobese Zucker fa/fa rats is an early abnormality that is mediated by the vagus nerve, and increased secretion of insulin in adult obese fa/fa rats is also, partly, vagus-nerve mediated [71, 72], suggesting that in the early stages of obesity a compensatory enhanced vagal activity occurs in response to insulin resistance. In fact, it was reported that insulin resistance is associated with a reduction in vagal activity with no affect in baroreflex sensitivity [73].

Insulin resistance that is produced by bilateral cervical vagotomy can be partially reversed by acetylcholine. The parasympathetic nerves that regulate hormonal control of insulin resistance pass through the cervical vagus and the hepatic branch, and finally, through the anterior hepatic plexus along the common hepatic artery and denervation at any of these sites leads to functional elimination of all hepatic parasympathetic input regulating insulin sensitivity [74]. In addition, vagus nerve stimulation also increases the production of incretins that are also capable of enhancing insulin secretion [75].

These evidences suggest that insulin resistance and low-grade systemic inflammation seen in obesity, hypertension, type 2 diabetes mellitus and metabolic syndrome could be due to decreased vagal activity and consequent reduced release and action of acetylcholine that leads to the loss of the "cholinergic anti-inflammatory pathway" [57–59]. In view of the anti-inflammatory and insulin stimulating actions of acetylcholine, the principal vagal neurotransmitter, it is reasonable to propose that vagus nerve stimulation could be tried in the management of insulin resistance and its associated diseases such as obesity, type 2 diabetes mellitus, hypertension and

---

Exercise enhances vagal tone, increases BDNF levels in the brain and reduces the production of IL-6 and TNF-α and thus, is anti-inflammatory in nature. Hence, exercise need to form an integral part of any management strategy of these diseases.

Vagus nerve stimulation could be of significant benefit in the prevention and treatment of obesity, insulin resistance, hypertension, type 2 diabetes mellitus, metabolic syndrome and neurological conditions such as Alzheimer's disease, schizophrenia and depression, and autoimmune diseases such as lupus and RA in view of its anti-inflammatory properties. It is recommended that a combination of vagus nerve stimulation (or acetylcholine agonists or synthetic analogues), ghrelin, BDNF, lipoxins, resolvins, protectins and maresins in various permutations and combinations may be tried to find the most suitable combination for the prevention and management of these diseases metabolic syndrome. Vagus nerve stimulation enhances the production of BDNF in the brain [76] that is known to be beneficial in the prevention and treatment of obesity, type 2 diabetes mellitus and metabolic syndrome [77–80], while BDNF enhances acetylcholine levels in the brain [81].

It is suggested that vagal nerve stimulation may also be useful in the management of schizophrenia, Alzheimer's disease and other neurological conditions in which low-grade systemic inflammation is present. On the other hand, PUFAs need to be supplemented from childhood to adulthood, probably throughout life, to prevent schizophrenia, Alzheimer's disease, depression and other neurological conditions as there is evidence that some, if not, of these diseases have their origins in the perinatal period as discussed in previous chapters. One method of knowing whether the supplemented PUFAs and vagus nerve stimulation will be useful is to measure plasma lipoxins, resolvins, protectins and maresins and acetylcholine, serotonin, dopamine and catecholamines in the peripheral leukocytes of these subjects.

Vagus nerve stimulation (VNS) is currently in use as an adjunctive treatment for certain types of intractable epilepsy and major depression [82]. VNS uses an implanted stimulator that sends electric impulses to the left vagus nerve in the neck via a lead wire implanted under the skin. VNS implantation devices consist of a titanium-encased generator about the size of a pocket watch with a lithium battery to fuel the generator, a lead wire system with electrodes, and an anchor tether to secure leads to the vagus nerve. The battery life for the pulse generator is between 1 to 16 years, depending on the settings i.e. how strong the signal is being sent, the length of time the device stimulates the nerve each time, and how frequently the device stimulates the nerve. Implantation of the VNS device is usually done as an out-patient procedure. Since VNS device is already in use and is relatively safe, this could be tried for the management of insulin resistance and its associated conditions and lupus and other autoimmune diseases. Vagus nerve stimulation could be tried as a standalone procedure or could be clubbed with other conventional therapeutic drugs that are currently in use for all these diseases.

Lipoxins, Resolvins, Protectins or Their Synthetic Analogues

As discussed previously, lipoxins, resolvins and protectins are potent endogenous anti-inflammatory compounds whose deficiency could lead to unabated inflammation. Hence, methods designed to enhance their formation such as co-administration of aspirin with PUFAs and/or development and use of their stable synthetic analogues may prove to be useful in various rheumatological conditions. Such an approach is urgently needed. Yet another possibility that should be seriously considered is the preparation and administration of PUFAs+BDNF stable complexes in the prevention of management of obesity, type 2 diabetes mellitus and metabolic syndrome. Alternatively, lipoxins/resolvins/protectins+BDNF stable complexes could be prepared and in obesity, type 2 diabetes mellitus and metabolic syndrome.

Ghrelin

Ghrelin is a growth hormone secretagogue produced by the gut, and is expressed in the hypothalamus and other tissues as well. Ghrelin not only plays an important role in the regulation of appetite, energy balance and glucose homeostasis but also shows anti-bacterial activity, suppresses pro-inflammatory cytokine production and restores gut barrier function. Ghrelin inhibited proinflammatory cytokine production, mononuclear cell binding, and nuclear factor-kappaB activation in human endothelial cells in vitro and endotoxin-induced cytokine production in vivo [83]. Ghrelin stimulates the vagus nerve and thus, could produce "cholinergic anti-inflammatory pathway" into action [57–59]. Studies showed that vagotomy prevented ghrelin's down-regulatory effect on TNF-α and IL-6 production confirming that ghrelin down-regulates proinflammatory cytokines in sepsis through activation of the vagus nerve [84]. Ghrelin has sympathoinhibitory properties that are mediated by central ghrelin receptors involving a NPY/Y1 receptor-dependent pathway [85]. Ghrelin inhibited the production of HMGB1 by activated macrophages [86] that may explain its beneficial action in sepsis and other inflammatory conditions [87, 88]. In view of these evidences, it is proposed that ghrelin infusion could be of significant benefit in lupus, RA and other autoimmune diseases not only by its direct action of the production of pro-inflammatory cytokines but also its ability to stimulate vagus that has anti-inflammatory actions.

Similar to the combination of PUFAs+BDNF and lipoxins/resolvins/protectins + BDNF stable complexes suggested above, ghrelin+PUFAs/lipoxins/resolvins/protectins complexes also need to be considered for the management of obesity, type 2 diabetes mellitus, metabolic syndrome, sepsis, lupus and RA. It is recommended that various permutations and combinations of BDNF, ghrelin, PUFAs, lipoxins, resolvins and protectins need to prepared and tried to find the best combination that is most suitable in the prevention and management of low-grade systemic inflammatory conditions and blatantly inflammatory conditions such as sepsis, lupus and RA.

PUFAs as Potential Anti-cancer Drugs

A detailed discussion of the role of PUFAs, eicosanoids, lipoxins, resolvins, protectins, lipid peroxides, free radicals, anti-oxidants, cytokines, macrophages and leukocytes in cancer has been discussed in detail in Chap. 14. It is evident from this discussion that PUFAs are handled differently by normal and tumor cells. The cross-talk between macrophages and tumor and normal cells is important in this context. It is evident from this discussion that tumor cells are deficient in PUFAs (especially AA, EP and DHA, some tumor cells are also deficient in GLA), have relatively higher content of anti-oxidants and as a result are more susceptible to the cytotoxic action of free radicals and lipid peroxides induced by supplementation of PUFAs. Even

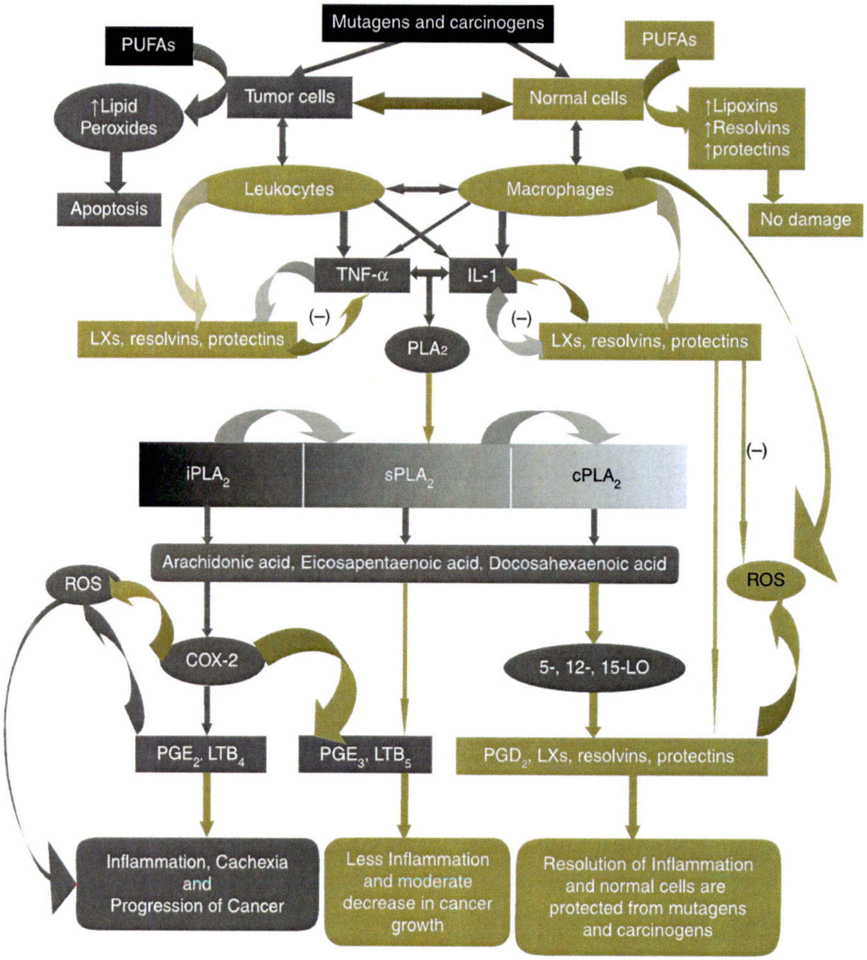

Fig. 17.2 Scheme showing the relationship among normal and cancer cells, cytokines, ROS, PUFAs, eicosanoids and lipoxins, resolvins and protectins and their relationship to inflammation and cancer growth.
When normal cells are exposed to mutagens and carcinogens, there will be increased production of ROS by normal cells, local leukocytes and macrophages. In response to these external stimuli, normal cells produce enhanced amounts of lipoxins, resolvins and protectins from the cell membrane lipids that are released by the activation of phospholipase A_2. Infiltrating or local leukocytes and macrophages produce enhanced amounts of pro-inflammatory cytokines such as IL-6 and TNF-α that, in turn, enhance ROS generation and cause local inflammation. If the cell stores of PUFAs are adequate, activation of phospholipase A_2 (the type of phospholipase activated in normal and cells may also be distinct) will lead to the release of adequate amounts of various PUFAs that will get converted to lipoxins, resolvins and protectins that, in turn, suppress leukocyte and macrophage activation, decrease ROS generation and inhibit inflammation and finally protect the normal cells from apoptosis and prevent from becoming cancer cells. In the case of tumor cells, infiltrating leukocytes and macrophages produce enhanced amounts of IL-6 and TNF-α that produce exaggerated amounts of ROS leading to augmented inflammation. ROS produce further DNA damage and progression of cancer. Tumor cells have decreased amounts of PUFAs in their

the tumoricidal action of TNF-α seems to be dependent on the activation of phospholipase A_2 and release of AA. It appears that normal cells synthesize and release significant amounts of lipoxins, resolvins, maresins, and other anti-inflammatory eicosanoids (such as PGD_2) and fewer amounts of pro-inflammatory PGE_2, $PGF_{2\alpha}$, and other hydroperoxy fatty acids, while tumor cells seem to do the opposite (produce more amounts of PGE_2, PGE_3, $PGF_{2\alpha}$ and other hydroperoxy fatty acids and fewer amounts of lipoxins, resolvins, protectins, maresins and PGD_2). The products of significant amounts of PGE_2, PGE_3, $PGF_{2\alpha}$ and other hydroperoxy fatty acids that are pro-inflammatory molecules produced by tumor cells could be responsible for the low-grade systemic inflammation seen in cancer (Fig. 17.2).

PUFAs, Especially GLA, for Glioma

Based on these observations, we hypothesized that supplementation of adequate amounts of PUFAs may induce apoptosis of tumor cells. Several studies that have been discussed in Chap. 14 did suggest that certain PUFAs, especially GLA, AA, EPA and DHA do induce apoptosis of tumor cells without any cytotoxic action on normal cells.

Based on these and other studies (see Chap. 14), it was reasoned that selective delivery of PUFAs may be useful in the treatment of cancer.

In a preliminary clinical study, it was observed that intra-tumoral injection of GLA can induce regression of glioma without any significant side-effects [89–91]. In a similar study performed in animal glioma models, it was shown that intra-tumoral

---

cell membranes that are converted to pro-inflammatory eicosanoids due to the activation of COX-2. Thus, inflammation perpetuates cancer growth. When tumor and normal cells are supplemented with PUFAs, normal cells produce adequate amounts of lipoxins, resolvins and protectins that protect them from ROS, lipid peroxides and suppress inflammation and so they do not undergo apoptosis. On the other hand, tumor cells when incorporate significant amounts of supplemented PUFAs generate more free radicals, show enhanced lipid peroxidation that cause further DNA damage leading to their apoptosis. Thus, PUFAs bring about their differential cytotoxicity against tumor cells without any damage to normal cells. In this context, the cross-talk between normal cells and leukocytes and macrophages on one hand and tumor cells and leukocytes and macrophages on the other hand is important that may determine the production of anti-inflammatory lipoxins, resolvins and protectins that are anti-inflammatory by normal cells and pro-inflammatory eicosanoids and lipid peroxides by tumor cells. When tumor cells are supplemented with EPA, the production of pro-inflammatory eicosanoids is decreased without resulting in significant increase in the production of lipid peroxides and hence, there will be only a moderate decrease in tumor growth. Thus, normal cells when exposed to PUFAs produce cytoprotective lipids such as lipoxins, resolvins and protectins while tumor cells generate toxic hydroperoxy fatty acids. This differential metabolism of PUFAs by normal and tumor cells may explain why PUFAs are toxic to tumor but not normal cells. There may also be some cross-talk between normal and tumor cells. Drug-resistant tumor cells may produce fewer amounts of lipoxins, resolvins and protectins that are sufficient to prevent their apoptosis but not suppress inflammation completely. It is likely that normal cells release preformed lipoxins, resolvins and protectins that may be taken up by tumor cells and used to protect them from apoptosis. But this remains to be established

GLA is indeed non-toxic to normal neuronal cells but induces apoptosis of glioma cells [92–95]. These studies revealed that intra-tumoral injection of GLA is safe and could be exploited as a potential drug for glioma.

Modified GLA (and Other PUFAs) for Cancer

In an extension of these studies [89–91], GLA has been converted into lithium-GLA to make it more water soluble and then conjugated to iodized salt solution (called as LGIOC = lithium GLA conjugated to iodized oily lymphographic agent) and injected intra-arterially into tumor-feeding blood vessels hepatoma, giant cell tumor and renal cell tumors with the idea that intra-arterial infusion would reach the tumor bed easily and so GLA will be able to induce apoptosis of tumor cells. But, to our surprise, LGIOC induced rapid and irreversible occlusion of only tumor-feeding vessels without any action on normal blood vessels [96, 97]. These results suggest that under certain circumstances (especially when the PUFA molecule is modified), it could behave as a potent anti-vascular and anti-angiogenic molecule, in addition to its ability to induce apoptosis of tumor cells.

Studies have shown that EGF (epidermal growth factor) stimulates cells to divide by activating members of the EGFR (EGF receptor). EGFR activation plays an important role in cancerous tumor survival. Several types of human cancer exhibit sustained activation of EGFRs by secreted growth factors. Amplification and rearrangement of the gene encoding EGFR occur in a significant fraction of glioblastomas and squamous-cell carcinomas and correlate with reduced patient survival. Consistent with their pivotal role in stimulating cell proliferation, blocking EGFR function is seen to result in retarded tumor growth. Erbitux® is a chimeric monoclonal antibody which is specific for the EGFR. Over expression of EGFR is common in many solid tumors, such as colorectal and lung carcinomas as well as cancers of the head and neck, and glioblastoma multiforme. It correlates with increased metastasis, decreased survival and a poor prognosis. EGFR protects malignant tumor cells from the cytotoxic effects of chemotherapy and radiotherapy, making these treatments less effective. Erbitux® binds to the extracellular domain of EGFR on the tumor cell, thereby inhibiting receptor-associated tyrosine kinase. This inhibition blocks the intracellular pathways associated with tumor cell proliferation, so preventing tumor growth and dissemination as well as inducing tumor cell death or apoptosis. There is evidence to suggest that EGF and EGFR have angiogenic actions and that Erbitux® prevents angiogenesis [98–100].

VEGF is secreted by hypoxic cells, including those that are cancerous. VEGF stimulates new blood vessel formation or angiogenesis by binding to specific receptors on nearby blood vessels to stimulate extensions of existing blood vessels. Angiogenesis plays an important role in both tumor growth and metastasis. Monoclonal antibodies are designed to bind to VEGF preventing it from binding to its receptors and therefore potentially inhibiting tumor growth. Bevacizumab is a humanized monoclonal antibody to VEGF developed by Genentech and is called as

Avastin®. By inhibiting VEGF, Avastin interferes with the blood supply to tumors, a process that is critical to tumor growth and metastasis.

Several clinical studies showed that both Erbitux® and Avastin®, which are humanized monoclonal antibodies to EGFR and VEGF respectively, are useful in the treatment of colon cancer. Erbitux® shrank tumors in 22.9% of advanced colon cancer patients when combined with chemotherapy. Avastin® in combination with chemotherapy extended colon cancer patients' lives by 5 months in a trial of 900 patients.

EGF and VEGF are mentioned here only as examples on the role of various growth factors and their receptors in cancer. Several studies have indicated that blocking or neutralizing the actions of various growth factors and their receptors suppresses cancer [101–104].

PUFAs+Growth Factors for Cancer

Tumor cells depend on various growth factors for their growth and proliferation and survival. To attract these growth factors for their own survival, several tumors express growth factor receptors on their cell surface. In order to gain survival advantage over the surrounding normal cells, tumor cells express more number of such growth factor receptors on their cell surface. Thus, there is a differential and overexpression of growth factor receptors by tumor cells compared to normal cells. This property of amplification or overexpression of growth factor receptors on the surface of tumor cells can be to deliver PUFAs selectively to tumor cells to induce their apoptosis.

I propose that monoclonal and polyclonal antibodies developed against various growth factors, receptors, and other cell surface and intracellular markers and proteins; and growth factors can be coupled to PUFAs such that the actions of these antibodies and growth factors are potentiated and PUFAs are selectively delivered to the tumor cells. It is likely that the beneficial actions of compounds formed as a result of such coupling of monoclonal and polyclonal antibodies and growth factors/antibodies to growth factors with PUFAs will be more than the sum effect of these antibodies or growth factors and PUFAs when are administered separately. In fact, it is likely that when growth factor and PUFAs are coupled and administered to the tumors, the growth factors instead of enhancing the growth of the tumor cells produce the opposite actions namely death of the tumor cells by selectively delivering the tumoricidal PUFAs. Thus, in this instance, growth factors instead of enhancing the growth of cancer cells promote the death of tumor cells by forming a vehicle to deliver PUFAs to tumor cells. Similarly, when monoclonal antibodies or polyclonal antibodies against various growth factors (such as EGF and VEGF) are coupled to PUFAs and injected will be able to reach the tumor selectively and induce the apoptosis of tumor cells and also potentiate the anti-angiogenic actions of these monoclonal antibodies. Such studies may prove to be interesting and may form a new therapeutic approach to cancer.

PUFAs for Rheumatological Conditions

Lupus, RA and other autoimmune diseases are chronic inflammatory conditions that often have acute and intermittent inflammatory episodes. More often than not, these are life-long diseases and are associated with considerable renal, pulmonary and cardiac complications that could lead to death. Current therapeutic approaches depend on the use of synthetic and potent anti-inflammatory and immunosuppressive drugs that are often associated with significant side-effects. Some of the current drugs in use include: non-steroidal anti-inflammatory compounds, chloroquine (hydroxychloroquine), corticosteroids (oral or parenteral), D-penicillamine, sulfasalazine, methotrexate, anti-TNF antibodies {treatment with anti-TNF monoclonal antibodies (infliximab, adalimumab, and certolizumab pegol) has been shown to provide substantial benefit to patients of RA, Crohn's disease, and psoriasis through reductions in both localized and systemic expression of markers associated with inflammation} and immunosuppressive drugs (such as cyclosporine, cyclophosphamide and azathioprine). Though there have been significant advances in the management of autoimmune diseases by the use of these drugs, especially biologics, their actions are often unpredictable, not all patients respond adequately to these therapeutic approaches and could cause significant side-effects. Hence, it is suggested that PUFAs/lipoxins/resolvins/protectins may be coupled or complexed with anti-TNF antibodies (such as infliximab, adalimumab, and certolizumab pegol) for possible use in various rheumatological conditions.

References

[1] Das UN (2001) Is insulin an anti-inflammatory molecule? Nutrition 17:409–413

[2] Satomi N, Sakurai A, Haranaka K (1985) Relationship of hypoglycemia to tumor necrosis factor production and antitumor activity: role of glucose, insulin, and macrophages. J Natl Cancer Inst 74:1255–1260

[3] Fraker DL, Merino MJ, Norton JA (1989) Reversal of the toxic effects of cachectin by concurrent insulin administration. Am J Physiol 256:E725–E731

[4] Ottlecz A, Koltai M, Blazso G, Minker E (1978) Contribution to the regulatory role of insulin in inflammation and anaphylaxis. Int Arch Allergy Appl Immunol 56:284–286

[5] Ottlecz A, Koltai M, Gecse A (1977) Further studies on the antiinflammatory effect of insulin. Agents Actions 7:487–491

[6] Ottlecz A, Gecse A, Koltai M, West GB (1977) Involvement of the kinin system in the insulin-induced inhibition of carrageenan oedema in rats. Monogr Allergy 12:131–137

[7] Boichot E, Sannomiya P, Escofier N et al (1999) Endotoxin-induced acute lung injury in rats. Role of insulin. Pulm Pharmacol Ther 12:285–290

[8] Jeschke MG, Klein D, Bolder U, Einspanier R (2004) Insulin attenuates the systemic inflammatory response in endotoxemic rats. Endocrinology 145:4084–4093

[9] Van den Berghe G, Wouters P, Weekers F et al (2001) Intensive insulin therapy in critically ill patients. N Engl J Med 345:1359–1367

[10] Viardot A, Grey ST, Mackay F, Chisholm D (2007) Potential antiinflammatory role of insulin via the preferential polarization of effector T cells toward a T helper 2 phenotype. Endocrinology 148:346–353

[11] Jeschke MG, Boehning DF, Finnerty CC, Herndon DN (2007) Effect of insulin on the inflammatory and acute phase response after burn injury. Crit Care Med 35(9 Suppl):S519–S523
[12] Jeschke MG, Klein D, Herndon DN (2004) Insulin treatment improves the systemic inflammatory reaction to severe trauma. Ann Surg 239:553–560
[13] Ye SD, Zheng M, Zhao LL, Qian Y, Yao XM, Ren A, Li SM, Jing CY (2009) Intensive insulin therapy decreases urinary MCP-1 and ICAM-1 excretions in incipient diabetic nephropathy. Eur J Clin Invest 39:980–985
[14] Kuboki K, Jiang ZY, Takahara N et al (2000) Regulation of endothelial constitutive nitric oxide gene expression in endothelial cells and in vivo: a specific vascular action of insulin. Circulation 101:676–681
[15] Kojda G, Harrison D (1999) Interaction between NO and reactive oxygen species: pathophysiological importance in artherosclerosis, hypertension, diabetes and heart failure. Cardiovasc Res 43:562–571
[16] Sakaue S, Nishihira J, Hirokawa J et al (1999) Regulation of macrophage migration inhibitory factor (MIF) expression by glucose and insulin in adipocytes in vitro. Mol Med 5:361–371
[17] Waeber G, Calandra T, Roduit R et al (1997) Insulin secretion is regulated by the glucose-dependent production of islet beta cell macrophage inhibitory factor. Proc Natl Acad Sci U S A 94:4782–4787
[18] Hirokawa J, Sakaue S, Furuya Y et al (1998) Tumor necrosis factor alpha regulates the gene expression of macrophage migration inhibitory factor through tyrosine kinase-dependent pathway in 3T3-L1 adipocytes. J Biochem (Tokyo) 123:733–739
[19] Hotamisligil GS (1999) The role of TNFalpha and TNF receptors in obesity and insulin resistance. J Intern Med 245:621–625
[20] Das UN (2000) Possible beneficial action(s) of glucose-insulin-potassium regimen in acute myocardial infarction and inflammatory conditions: a hypothesis. Diabetologia 43:1081–1082
[21] Das UN (2001) Can glucose-insulin-potassium regimen suppress inflammatory bowel disease? Med Hypotheses 57:183–185
[22] Das UN (2001) Hypothesis: can glucose-insulin-potassium regimen in combination with polyunsaturated fatty acids suppress lupus and other inflammatory diseases? Prostaglandins Leukot Essent Fatty Acids 65:109–113
[23] Li J, Zhang H, Wu F, Nan Y, Guo W, Wang H, Ren J, Das UN, Gao F (2008) Insulin inhibits tumor necrosis factor-α- induction in myocardial ischemia/reperfusion: role of Akt and endothelial nitric oxide synthase phosphorylation. Crit Care Med 36:1551–1558
[24] Das UN (2008) Glucose, insulin, and coronary heart disease. Eur Heart J 29:1075–1076
[25] Reade MC, Fink MP (2005) Bench-to-bedside review: amelioration of acute renal impairment using ethyl pyruvate. Crit Care 9:556–560
[26] Sims CA, Wattanasirichaigoon S, Menconi MJ, Ajami AM, Fink MP (2001) Ringer's ethyl pyruvate solution ameliorates ischemia/reperfusion-induced intestinal mucosal injury in rats. Crit Care Med 29:1513–1518
[27] Tawadrous ZS, Delude RL, Fink MP (2002) Resuscitation from hemorrhagic shock with Ringer's ethyl pyruvate solution improves survival and ameliorates intestinal mucosal hyperpermeability in rats. Shock 17:473–477
[28] Dobsak P, Courdertot-Masuyer C, Zeller M, Vergely C, Laubriet A, Assem M, Eicher JC, Teyssier JR, Wolf JE, Rochetter L (1999) Antioxidative properties of pyruvate and protection of the ischemic rat heart during cardioplegia. J Cardiovasc Pharmacol 34:651–659
[29] Cicalese L, Lee K, Schraut W, Watkins S, Borle A, Stanko R (1999) Pyruvate prevents ischemia-reperfusion mucosal injury of rat small intestine. Am J Surg 171:97–100
[30] Venkataraman R, Kellum JA, Song M, Fink MP (2002) Resuscitation with Ringer's ethyl pyruvate solution prolongs survival and modulates plasma cytokine and nitrite/nitrate concentrations in a rat model of lipopolysaccharide-induced shock. Shock 18:507–512
[31] Das UN (2006) Pyruvate is an endogenous anti-inflammatory and anti-oxidant molecule. Med Sci Monit 12:RA79–RA84

[32] Das UN (2006) Is pyruvate an endogenous anti-inflammatory molecule? Nutrition 22:965–972
[33] Das UN (2007) Ethyl pyruvate in sepsis. Adv Sepsis 6:10–15
[34] Bennett-Guerrero E, Swaminathan M, Grigore AM, Roach GW, Aberle LG, Johnston JM, Fink MP (2009) A phase II multicenter double-blind placebo-controlled study of ethyl pyruvate in high-risk patients undergoing cardiac surgery with cardiopulmonary bypass. J Cardiothorac Vasc Anesth 23:324–329
[35] Awad AB, Young AL, Fink CS (1996) The effect of unsaturated fatty acids on membrane composition and signal transduction in HT-29 human colon cancer cells. Cancer Lett 108:25–33
[36] Neves SR, Ram PT, Iyengar R (2002) G protein pathways. Science 296:1636–1639
[37] Begin ME, Ells G, Das UN, Horrobin DF (1986) Differential killing of human carcinoma cells supplemented with n-3 and n-6 polyunsaturated fatty acids. J Natl Cancer Inst 77:1053–1062
[38] Das UN (1990) Gamma-linolenic acid, arachidonic acid and eicosapentaenoic acid as potential anti-cancer drugs. Nutrition 6:429–434
[39] Das UN (1991) Tumoricidal action of cis-unsaturated fatty acids and its relationship to free radicals and lipid peroxidation. Cancer Lett 56:235–243
[40] Madhavi N, Das UN (1994) Effect of n-6 and n-3 fatty acids on the survival of vincristine sensitive and resistant human cervical carcinoma cells in vitro. Cancer Lett 84:31–41
[41] Das UN, Prasad VSSV, Reddy DR (1995) Local application of gamma-linolenic acid in the treatment of human gliomas. Cancer Lett 94:147–155
[42] Belayev L, Marcheselli VL, Khoutorova L, Rodriguez de Turco EB, Busto R, Ginsberg MD, Bazan NG (2005) Docosahexaenoic acid complexed to albumin elicits high-grade ischemic neuroprotection. Stroke 36:118–123
[43] Berman DR, Mozurkewich E, Liu Y, Barks J (2009) Docosahexaenoic acid pretreatment confers neuroprotection in a rat model of perinatal cerebral hypoxia-ischemia. Am J Obstet Gynecol 200:305.e1–305.e6
[44] Marcheselli VL, Hong S, Lukiw WJ, Tian XH, Gronert K, Musto A, Hardy M, Gimenez JM, Chiang N, Serhan CN, Bazan NG (2003) Novel docosanoids inhibit brain ischemia-reperfusion-mediated leukocyte infiltration and pro-inflammatory gene expression. J Biol Chem 278:43807–43817
[45] Das UN (2008) Albumin infusion therapy in stroke, sepsis and the critically ill. Curr Nutr Food Sci 4:217–226
[46] Krishna Mohan I, Das UN (2001) Prevention of chemically-induced diabetes mellitus in experimental animals by polyunsaturated fatty acids. Nutrition 17:126–151
[47] Suresh Y, Das UN (2001) Protective action of arachidonic acid against alloxan-induced cytotoxicity and diabetes mellitus. Prostaglandins Leukot Essent Fatty Acids 64:37–52
[48] Suresh Y, Das UN (2003) Long-chain polyunsaturated fatty acids and chemically-induced diabetes mellitus: effect of ω-6 fatty acids. Nutrition 19:93–114
[49] Suresh Y, Das UN (2003) Long-chain polyunsaturated fatty acids and chemically-induced diabetes mellitus: effect of ω-3 fatty acids. Nutrition 19:213–228
[50] Suresh Y, Das UN (2006) Differential effect of saturated, monounsaturated, and polyunsaturated fatty acids on alloxan-induced diabetes mellitus. Prostaglandins Leukot Essent Fatty Acids 74:199–213
[51] Wei D, Li J, Shen M, Jia W, Chen N, Chen T, Su D, Tian H, Zheng S, Dai Y, Zhao A (2010) Cellular production of n-3 PUFAs and reduction of n-6-to-n-3 ratios in the pancreatic beta-cells and islets enhance insulin secretion and confer protection against cytokine-induced cell death. Diabetes 59:471–478
[52] White PJ, Arita M, Taguchi R, Kang JX, Marette A (2010) Transgenic restoration of long chain {omega}-3 fatty acids in insulin target tissues improves resolution capacity and alleviates obesity-linked inflammation and insulin resistance in high fat-fed mice. Diabetes 59:3066–3073
[53] Saltiel AR (2010) Fishing out a sensor for anti-inflammatory oils. Cell 142:672–674

[54] Oh DY, Talukdar S, Bae EJ, Imamura T, Morinaga H, Fan W, Li P, Lu WJ, Watkins SM, Olefsky JW (2010) GPR120 is an omega-3 fatty acid receptor mediating potent anti-inflammatory and insulin-sensitizing effects. Cell 142:687–698
[55] Hirasawa A, Tsumaya K, Awaji T, Katsuma S, Adachi T, Yamada M, Sugimoto Y, Miyazaki S, Tsujimoto G (2005) Free fatty acids regulate gut incretin glucagon-like peptide-1 secretion through GPR120. Nat Med 11:90–94
[56] Itoh Y, Kawamata Y, Harada M, Kobayashi M, Fujii R, Fukusumi S, Ogi K, Hosoya M, Tanaka Y, Uejima et al (2003) Free fatty acids regulate insulin secretion from pancreatic β cells through GPR40. Science 422:173–176
[57] Borovikova LV, Ivanova S, Zhang M, Yang H, Botchkina GI, Watkins LR, Wang H, Abumrad N, Eaton JW, Tracey KJ (2000) Vagus nerve stimulation attenuates the systemic inflammatory response to endotoxin. Nature 405:458–462
[58] Wang H, Yu M, Ochani M, Amella CA, Tanovic M, Susarla S, Li JH, Wang H, Yang H, Ulloa L, Al-Abed Y, Czura CJ, Tracey KJ (2003) Nicotinic acetylcholine receptor alpha7 subunit is an essential regulator of inflammation. Nature 421:384–388
[59] Karimi K, Bienenstock J, Wang L, Forsythe P (2010) The vagus nerve modulates CD4+ T cell activity. Brain Behav Immun 24:316–323
[60] Westman M, Engström M, Catrina AI, Lampa J (2009) Cell specific synovial expression of nicotinic alpha 7 acetylcholine receptor in rheumatoid arthritis and psoriatic arthritis. Scand J Immunol 70:136–140
[61] Bruchfeld A, Goldstein RS, Chavan S, Patel NB, Rosas-Ballina M, Kohn N, Qureshi AR, Tracey KJ (2010) Whole blood cytokine attenuation by cholinergic agonists ex vivo and relationship to vagus nerve activity in rheumatoid arthritis. J Intern Med 268:94–101
[62] van Maanen MA, Stoof SP, Larosa GJ, Vervoordeldonk MJ, Tak PP (2010) Role of the cholinergic nervous system in rheumatoid arthritis: aggravation of arthritis in nicotinic acetylcholine receptor alpha7 subunit gene knockout mice. Ann Rheum Dis 69:1717–1723
[63] Daniel PM, Henderson JR (1967) The effect of vagal stimulation on plasma insulin and glucose levels in the baboon. J Physiol 192:317–327
[64] Uvnäs-Wallensten K, Nilsson G (1978) A quantitative study of the insulin release induced by vagal stimulation in anesthetized cats. Acta Physiol Scand 102:137–142
[65] Bergman RN, Miller RE (1973) Direct enhancement of insulin secretion by vagal stimulation of the isolated pancreas. Am J Physiol 225:481–486
[66] Uno K, Katagiri H, Yamada T, Ishigaki Y, Ogihara T, Imai J, Hasegawa Y, Gao J, Kaneko K, Iwasaki H, Ishihara H, Sasano H, Inukai K, Mizuguchi H, Asano T, Shiota M, Nakazato M, Oka Y (2006) Neuronal pathway from the liver modulates energy expenditure and systemic insulin sensitivity. Science 312:1656–1659
[67] Gautam D, Han SJ, Duttaroy A, Mears D, Hamdan FF, Li JH, Cui Y, Jeon J, Wess J (2007) Role of the M3 muscarinic acetylcholine receptor in beta-cell function and glucose homeostasis. Diabetes Obes Metab 9(Suppl 2):158–169
[68] Edvell A, Lindstrom P (1998) Vagotomy in young obese hyperglycemic mice: effects on syndrome development and islet proliferation. Am J Physiol 274(6 Pt 1):E1034–E1039
[69] Kiba T, Tanaka K, Hoshino M, Misugi K, Inoue S (1996) Ventromedial hypothalamic lesion-induced vagal hyperactivity stimulates rat pancreatic cell proliferation. Gastroenterology 110:885–893
[70] Stubbs M, York DA (1991) Central glucocorticoid regulation of parasympathetic drive to pancreatic B-cells in the obese fa/fa rat. Int J Obes 15:547–553
[71] Rohner-Jeanrenaud F, Hochstrasser AC, Jeanrenaud B (1983) Hyperinsulinemia of preobese and obese fa/fa rats is partly vagus nerve mediated. Am J Physiol 244:E317–E322
[72] Miller AW, Sims JJ, Canavan A, Hsu T, Ujhelyi MR (1999) Impaired vagal reflex activity in insulin-resistant rats. J Cardiovasc Pharmacol 33:698–702
[73] Latour MG, Lautt WW (2002) The hepatic vagus nerve in the control of insulin sensitivity in the rat. Auton Neurosci 95:125–130
[74] Rocca AS, Brubaker PL (1999) Role of the vagus nerve in mediating proximal nutrient-induced glucagon-like peptide-1 secretion. Endocrinology 140:1687–1694

[75] Follesa P, Biggio F, Gorini G, Caria S, Talani G, Dazzi L, Puligheddu M, Marrosu F, Biggio G (2007) Vagus nerve stimulation increases norepinephrine concentration and the gene expression of BDNF and bFGF in the rat brain. Brain Res 1179:28–34
[76] Tonra JR, Ono M, Liu X, Garcia K, Jackson C, Yancopoulos GD, Wiegand SJ, Wong V (1999) Brain-derived neurotrophic factor improves blood glucose control and alleviates fasting hyperglycemia in C57BLKS-Lepr(db)/lepr(db) mice. Diabetes 48:588–594
[77] Yamanaka M, Itakura Y, Inoue T, Tsuchida A, Nakagawa T, Noguchi H, Taiji M (2006) Protective effect of brain-derived neurotrophic factor on pancreatic islets in obese diabetic mice. Metabolism 55:1286–1292
[78] Das UN (2010) Obesity: genes, brain, gut, and environment. Nutrition 26:459–473
[79] Ono M, Itakura Y, Nonomura T, Nakagawa T, Nakayama C, Taiji M, Noguchi H (2000) Intermittent administration of brain-derived neurotrophic factor ameliorates glucose metabolism in obese diabetic mice. Metabolism 49:129–133
[80] Sala R, Viegi A, Rossi FM, Pizzorusso T, Bonanno G, Raiteri M, Maffei L (1998) Nerve growth factor and brain-derived neurotrophic factor increase neurotransmitter release in the rat visual cortex. Eur J Neurosci 10:2185–2191
[81] Jobst BC (2010) Electrical stimulation in epilepsy: vagus nerve and brain stimulation. Curr Treat Options Neurol 12:443–453
[82] Li WG, Gavrila D, Liu X, Wang L, Gunnlaugsson S, Stoll LL, McCormick ML, Sigmund CD, Tang C, Weintraub NL (2004) Ghrelin inhibits proinflammatory responses and nuclear factor-kappaB activation in human endothelial cells. Circulation 109:2221–2226
[83] Wu R, Dong W, Cui X, Zhou M, Simms HH, Ravikumar TS, Wang P (2007) Ghrelin down-regulates proinflammatory cytokines in sepsis through activation of the vagus nerve. Ann Surg 245:480–486
[84] Wu R, Zhou M, Das P, Dong W, Ji Y, Yang D, Miksa M, Zhang F, Ravikumar TS, Wang P (2007) Ghrelin inhibits sympathetic nervous activity in sepsis. Am J Physiol Endocrinol 293:E1697–E1702
[85] Chorny A, Anderson P, Gonzalez-Rey E, Delgado M (2008) Ghrelin protects against experimental sepsis by inhibiting high-mobility group box 1 release and by killing bacteria. J Immunol 180:8369–8377
[86] Wu R, Dong W, Zhou M, Zhang F, Marini CP, Ravikumar TS, Wang P (2007) Ghrelin attenuates sepsis-induced acute lung injury and mortality in rats. Am J Respir Crit Care Med 176:805–813
[87] Chang L, Zhao J, Yang J, Zhang Z, Tang C (2003) Therapeutic effects of ghrelin on endotoxic shock in rats. Eur J Pharmacol 473:171–176
[88] Naidu MRC, Das UN, Kishan A (1992) Intratumoral gamma-linolenic acid therapy of human gliomas. Prostaglandins Leukot Essent Fatty Acids 45:181–184
[89] Das UN, Prasad VSSV, Reddy DR (1995) Local application of gamma-linolenic acid in the treatment of human gliomas. Cancer Lett 94:147–155
[90] Bakshi A, Mukherjee D, Bakshi A, Banerji AK, Das UN (2003) Gamma-linolenic acid therapy of human gliomas. Nutrition 19:305–309
[91] Miyake JA, Benadiba M, Colquhoun A (2009) Gamma-linolenic acid inhibits both tumour cell cycle progression and angiogenesis in the orthotopic C6 glioma model through changes in VEGF, Flt1, ERK1/2, MMP2, cyclin D1, pRb, p53 and p27 protein expression. Lipids Health Dis 8:8
[92] Das UN (2007) Gamma-linolenic acid therapy of human glioma-a review of in vitro, in vivo, and clinical studies. Med Sci Monit 13:RA119–RA131
[93] Leaver HA, Wharton SB, Bell HS, Leaver-Yap IM, Whittle IR (2002) Highly unsaturated fatty acid induced tumour regression in glioma pharmacodynamics and bioavailability of gamma linolenic acid in an implantation glioma model: effects on tumour biomass, apoptosis and neuronal tissue histology. Prostaglandins Leukot Essent Fatty Acids 67:283–292
[94] Leaver HA, Bell HS, Rizzo MT, Ironside JW, Gregor A, Wharton SB, Whittle IR (2002) Antitumour and pro-apoptotic actions of highly unsaturated fatty acids in glioma. Prostaglandins Leukot Essent Fatty Acids 66:19–29

[95] Das UN (2002) Abrupt and complete occlusion of tumor-feeding vessels by gamma-linolenic acid. Nutrition 18:662–664

[96] Das UN (2004) Occlusion of infusion vessels on gamma-linolenic acid infusion. Prostaglandins Leukot Essent Fatty Acids 70:23–32

[97] Barry JB, Giguere V (2005) Epidermal growth factor-induced signaling in breast cancer cells results in selective target gene activation by orphan nuclear receptor estrogen-related receptor alpha. Cancer Res 65:6120–6129

[98] Vallbohmer D, Lenz HJ (2005) Epidermal growth factor receptor as a target for chemotherapy. Clin Colorectal Cancer 5(Suppl 1):S19–S27

[99] Stewart R, Nelson J, Wilson DJ (1989) Epidermal growth factor promotes chick embryonic angiogenesis. Cell Biol Int Rep 13:957–965

[100] De S, Razorenova O, McCabe NP, O'Toole T, Qin J, Byzova TV (2005) VEGF-integrin interplay controls tumor growth and vascularization. Proc Natl Acad Sci U S A 102:7589–7594

[101] Grose R, Dickson C (2005) Fibroblast growth factor signaling in tumorigenesis. Cytokine Growth Factor Rev 16:179–186

[102] Hicklin DJ, Ellis LM (2005) Role of the vascular endothelial growth factor pathway in tumor growth and angiogenesis. J Clin Oncol 23:1011–1027

[103] Nakayama T, Yao L, Tosato G (2004) Mast cell-derived angiopoietin-1 plays a critical role in the growth of plasma cell tumors. J Clin Invest 114:1317–1325

[104] Li Y, Lu Y, Xing G, Zhu Y, He F (2004) Macrophage migration inhibitory factor directly interacts with hepatopoietin and regulates the proliferation of hepatoma cell. Exp Cell Res 300:379–387

Index

DOI 10.1007/978-94-007-0495-4, © Springer Science+Business Media B.V. 2011